XVII Workshop on High Energy Spin Physics (D-SPIN 2017)

Journal of Physics: Conference Series Volume 938

Dubna, Russia
11 – 15 September 2017

ISBN: 978-1-5108-5518-2
ISSN: 1742-6588

Printed from e-media with permission by:

Curran Associates, Inc.
57 Morehouse Lane
Red Hook, NY 12571

Some format issues inherent in the e-media version may also appear in this print version.

Printed by Curran Associates, Inc. (2018)

For permission requests, please contact the Institute of Physics
at the address below.

Institute of Physics
Dirac House, Temple Back
Bristol BS1 6BE UK

Phone: 44 1 17 929 7481
Fax: 44 1 17 920 0979

techtracking@iop.org

Additional copies of this publication are available from:

Curran Associates, Inc.
57 Morehouse Lane
Red Hook, NY 12571 USA
Phone: 845-758-0400
Fax: 845-758-2633
Email: curran@proceedings.com
Web: www.proceedings.com

Table of contents

Volume 938

XVII Workshop on High Energy Spin Physics "DSPIN-2017"

11–15 September 2017, Dubna, Russian Federation

Accepted papers received: 13 December 2017
Published online: 5 January 2018

Preface

Papers

SECTION 1: Experimental results

SECTION 4: Related problems

SECTION 6: Spintronics and physics of nanostructures

IOP Conf. Series: Journal of Physics: Conf. Series **938** (2017) 011001 doi:10.1088/1742-6596/938/1/011001

Preface

17th Workshop on High Energy Spin Physics was a continuation of the series of workshops, started in Dubna in 1981 due to the initiative of the prominent theorist Lev Iosifovich Lapidus. Since than, such workshops are taking place in Dubna or Protvino each odd year, between Spin Physics . These workshops provide an opportunity to discuss the news of the respective year. From the other side, their important specifics was the possibility of participation of the numerous participants from Former Soviet Union and Eastern Europe countries, for which the long travel s may be difficult due to financial (and earlier also bureaucratic) reasons.

There were more than 100 participants from Russia, Germany, Czech Republic, Italy, Poland, Slovakia, USA, Armenia, Bulgaria, Estonia, Netherlands, Great Britain.
The last years brought many new experimental results, starting from the Higgs boson discovery at LHC. In the talk of I. Ginzburg (Novosibirsk) the two-doublet Higgs model was considered focusing on the C and P symmetries important for spin physics. The theoretical possibilities of new particles production and searches at LHC and future ILC were considered in the talks of A. Pankov and A. Tsytrinov from Gomel.

The search of CP-violation and EDM appearance were considered by A. Saleev (Germany) and A. Silenko (JINR). In the talk of I. Gorbunov (JINR) the review of latest results of CMS was given, and A. Artamonov (IHEP) presented the data on quarkonia polarization at LHCb and their interpretation.

The classical experiments on high energy nucleons spin structure are using both the scattering of leptons on polarized nucleons (COMPASS, JLab, HERMES) and polarized hadrons collisions (RHIC, IHEP,JINR, COMPASS). The review talks were presented from practically all the centers where spin experiments are performed by A. Biselli, V Punjabi, K. Barish (USA), G. Marukyan (Armenia), V. Rykov (MEPhI).

The special attention was given to the COMPASS experiment at CERN, where JINR is actively participating: talks of F. Bradamante (Trieste) and J. Matousek (Trieste/Prague), A. Sandacz and B. Marianski (Warsaw), M.Pesek (Prague), J. Bedfer (France). The data in SIDIS were shown to allow one to test the sign change of Sivers function (being the fine test of QCD) in the COMPASS DY experiment. The exclusive hard processes at COMPASS were also discussed.

The special session was devoted to the plans of the SPD at NICA (JINR) development, aimed for the experiments with polarized protons and deuterons: talks of A. Kovalenko, A. Nagajtsev, A. Guskov, I. Anikin, S. Goloskokov, Yu. Filatov, S. Sharov (JINR), M. Kondratenko (Novosibirsk). The spin effects in the heavy-ion collisions at NICA MPD were also considered by O. Teryaev (JINR).

Several talks were dedicated to the development of the data processing and extracting both unpolarized and polarized parton distributions. In particular, talk of D. Strozik-Kotlorz (JINR,Poland) was dedicated to the development of the method of truncated Mellin moments and its application, while that of B. Ermolaev to the new method of large and short distances factorization in QCD, One may specially mark the talk of A. Kerbizi (Trieste) who realized in the computer code very interesting ideas presented by X. Artru on previous DSPIN workshops.

There were several talks describing the applications of ideas and methods of spin physics to astrophysical problems.: Yu. Obukhov (NSI RAS) P. Zavada (Prague), M. Dvornikov (IZMIRAN). In

IOP Conf. Series: Journal of Physics: Conf. Series **938** (2017) 011001 doi:10.1088/1742-6596/938/1/011001

the talk of V. Neznamov (VNIIEF) the interesting analogs of confinement in electrodynamics of heavy nuclei were considered. The talk of F. Delcarro (Pavia), dedicated to the problem of momentum and spin localization, was also close to these problems.

The Workshop was supported by Russian Foundation for Basic Research, International Spin Physics Committee, European Physical Society, Heisenberg-Landau (Germany-JINR), Bogolyubov-Infeld (Poland-JINR) and Votruba-Blokhintsev (Czech Republik-JINR) collaboration programs.

All information about Workshop is placed on The Workshop cite (http://theor.jinr.ru/~spin/2017/).

IOP Conf. Series: Journal of Physics: Conf. Series **938** (2017) 011002 doi:10.1088/1742-6596/938/1/011002

Peer review statement

All papers published in this volume of *Journal of Physics: Conference Series* have been peer reviewed through processes administered by the proceedings Editors. Reviews were conducted by expert referees to the professional and scienti c standards expected of a proceedings journal published by IOP Publishing.

$\Upsilon(\mathrm{nS})$ polarizations in pp collisions at $\sqrt{s} = 7$ and 8 TeV by the LHCb collaboration

Alexander Artamonov on behalf of the LHCb Collaboration

Institute for High Energy Physics, National Research Center Kurchatov Institute, Protvino, Moscow region, 142281 Russia

E-mail: Alexander.Artamonov@ihep.ru

Abstract. A polarization measurement carried out for the $\Upsilon(1S)$, $\Upsilon(2S)$ and $\Upsilon(3S)$ mesons produced in pp collisions at $\sqrt{s} = 7$ and 8 TeV is presented. Data samples used for the polarization measurement were collected by the LHCb experiment during the 2011 and 2012 data taking runs with integrated luminosities of 1 and 2 fb^{-1}, respectively. The measurement has been performed in three polarization frames, using an angular distribution analysis of the $\Upsilon \to \mu^+\mu^-$ decays in the kinematic region of the Υ transverse momentum $p_T^{\Upsilon} < 30$ GeV/c and rapidity $2.2 < y^{\Upsilon} < 4.5$. No large polarization is observed.

1. Introduction

It is already forty years since the discovery of the first bottomonium state, the $\Upsilon(1S)$ meson [1], but studies of heavy quarkonium production continue to play an important role in the development of quantum chromodynamics (QCD) [2]. According to the current theoretical framework, nonrelativistic QCD (NRQCD) [3, 4], inclusive production of a heavy quarkonium is viewed as a two-step process. In the first step, a heavy quark-antiquark pair, $Q\overline{Q}$, is perturbatively created in a color singlet or color octet state, and then, in the second step, the hadronization process non-perturbatively transforms the $Q\overline{Q}$ pair into an observable colorless bound state. The non-perturbative transitions are described by long-distance matrix elements which are conjectured to be independent of production processes in the first step, and need to be extracted experimentally. The most distinct signature of the first NRQCD calculations [5, 6, 7, 8] was the prediction of transverse polarization for S-wave quarkonium states (such as the J/ψ, $\psi(2S)$ and $\Upsilon(\mathrm{nS})$ mesons) directly produced (i.e. not coming from decays) at large transverse momentum in high-energy hadron collisions. NRQCD calculations for promptly produced heavy quarkonia (if charmonia then those not coming from b-hadron decays) are complicated by feed-down (electromagnetic or hadronic transitions) from higher level states.

Full NLO calculations [9] performed for the $\Upsilon(1S)$, $\Upsilon(2S)$ and $\Upsilon(3S)$ mesons (shortly denoted by Υ), including effects of feed-down contributions for the $\Upsilon(1S)$ and $\Upsilon(2S)$ but not for the $\Upsilon(3S)$, give very small transverse polarizations for the first two bottomonium states and a large transverse polarization for the third state. These calculations were done before LHCb obtained that the fractions of Υ mesons originating from χ_b decays are around $30-40\,\%$ for high transverse momenta, $p_T^{\Upsilon} \gtrsim 20\,\mathrm{GeV}/c$ [10]. The $\Upsilon(3S)$ was considered as almost feed-down free state before. Although taking into account feed-down contributions improves the description of Υ polarization, there are still problems in describing the J/ψ polarization [11] (even with considering feed-down) and the $\psi(2S)$ polarization [12] (which includes negligible feed-down contributions).

The angular distribution of muons from the $\Upsilon \to \mu^+\mu^-$ decay can be written as [13, 14, 15]

$$\frac{1}{\sigma}\frac{d\sigma}{d\Omega} = \frac{3}{4\pi}\,\frac{1}{3+\lambda_\theta}\left(1+\lambda_\theta\cos^2\theta+\lambda_{\theta\phi}\sin 2\theta\cos\phi+\lambda_\phi\sin^2\theta\cos 2\phi\right), \tag{1}$$

where the angular quantities $\Omega = (\cos\theta, \phi)$ describe a direction of μ^+ in the Υ rest frame with respect to some specified axes, $\vec{\lambda} \equiv (\lambda_\theta, \lambda_{\theta\phi}, \lambda_\phi)$ are the angular distribution parameters directly related to the spin-1 density-matrix elements [13, 16, 17]. The parameter λ_θ is a measure of spin-alignment, and can be expressed as $\lambda_\theta = (\sigma_T - 2\sigma_L)\,/\,(\sigma_T + 2\sigma_L)$, where σ_T (σ_L) is the transverse (longitudinal) component of the cross section. If the spin-alignment parameter $\lambda_\theta > 0$ ($\lambda_\theta < 0$), the Υ meson is called to be transversely (longitudinally) polarized in a specified frame, while the case $\lambda_\theta = \lambda_{\theta\phi} = \lambda_\phi = 0$ means that the Υ meson is unpolarized. The parameters $\vec{\lambda}$ depend on a definition of coordinate axes specified in the Υ rest frame. The following three coordinate systems are widely used in polarization analyses: helicity (HX) [18], Collins-Soper (CS) [19] and Gottfried-Jackson (GJ) [19]. The frames are specified by different directions of the spin-quantization axis, z axis, defined in the production plane of Υ meson [15]. In all these frames, the y axis is normal to the production plane [15, 21], and the remaining x axis completes a right-handed coordinate system.

Until recently, quarkonium polarization measurements have been reduced only to studies of the parameter λ_θ, some times in different polarization frames. As pointed out in [15], measuring all the three polarization parameters is important from the theoretical and experimental points of view. Since from having the $\vec{\lambda}$ in one frame, it is possible to transform them into another [15, 22], and perform some cross checks of results. In particular, an important cross check is provided by a polarization parameter $\tilde{\lambda} = (\lambda_\theta + 3\lambda_\phi)\,/\,(1-\lambda_\phi)$ [23, 24], which is invariant for all rotations around the y axis, that is invariant in the HX, CS and GJ frames. The physical meaning of the parameter $\tilde{\lambda}$ was first recognized in [25] (see also [26, 27]).

The first full angular distribution analysis of muons from the $\Upsilon \to \mu^+\mu^-$ decays was performed by the CDF collaboration [28] using data of $p\bar{p}$ collisions at $\sqrt{s} = 1.96$ TeV. CDF found that the angular distributions of muons from all the three Υ states are nearly isotropic in the central rapidity region $|y^\Upsilon| < 0.6$ and $p_T^\Upsilon < 40$ GeV/c. This result is consistent with the previous CDF measurement [29], and inconsistent with the measurement performed by the D0 collaboration [30]. D0 observed the significant p_T^Υ dependent longitudinal polarization for the $\Upsilon(1S)$ mesons produced in $p\bar{p}$ collisions at $\sqrt{s} = 1.96$ TeV, for $|y^\Upsilon| < 1.8$ and $p_T^\Upsilon < 20$ GeV/c. The next full angular distribution analysis for the $\Upsilon \to \mu^+\mu^-$ was performed by the CMS collaboration [31] using pp collisions data at $\sqrt{s} = 7$ TeV, for the rapidity ranges $|y^\Upsilon| < 0.6$ and $0.6 < |y^\Upsilon| < 1.2$, and for $10 < p_T^\Upsilon < 50$ GeV/c [31]. CMS found no evidence of large transverse or longitudinal polarization for any of the three Υ mesons in the explored kinematic region. The experimental situation is complicated by the result of the fixed-target experiment E866 [32], which performed the polarization measurement of the Υ mesons produced in p-Cu collisions at $\sqrt{s} = 38.8$ GeV in $0.0 < x_F^\Upsilon < 0.6$ and $p_T^\Upsilon < 4$ GeV/c. The E866 collaboration found that the $\Upsilon(1S)$ meson is produced weakly polarized, while the $\Upsilon(2S)$ and $\Upsilon(3S)$ mesons are produced with a maximal transverse polarization. Although different production energies may determine different dominant contributions in the production processes, all these results underscore the need for further experimental study of the Υ polarization.

The LHCb collaboration performed the full angular distribution analysis [21] for the Υ mesons produced in pp collisions at $\sqrt{s} = 7$ and $8\,$TeV in the LHCb setup during the 2011 and 2012 data taking runs with integrated luminosities of 1 and $2\,\mathrm{fb}^{-1}$, respectively. The polarization measurement was done in the HX, CS and GJ frames in the Υ kinematic range defined by $p_T^\Upsilon < 30$ GeV/c and $2.2 < y^\Upsilon < 4.5$.

IOP Conf. Series: Journal of Physics: Conf. Series **938** (2017) 012001 doi:10.1088/1742-6596/938/1/012001

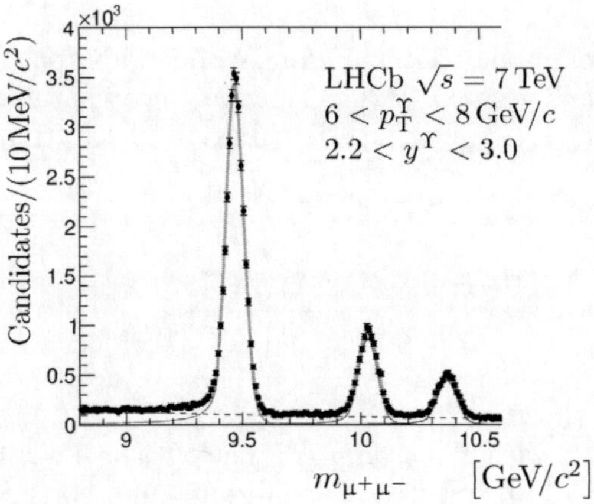

Figure 1: Dimuon mass distribution in the region $6 < p_T^\Upsilon < 8\,\text{GeV}/c$, $2.2 < y^\Upsilon < 3.0$ for data obtained at $\sqrt{s} = 7\,\text{TeV}$. The thick dark yellow solid curve shows the result of the fit, as described in the text. The three peaks, shown with thin red solid lines, correspond to the $\Upsilon(1S)$, $\Upsilon(2S)$ and $\Upsilon(3S)$ signals (left to right). The background component is indicated with a dashed blue line.

2. LHCb detector and selection of $\Upsilon \to \mu^+\mu^-$ decays

The LHCb detector is a single-arm forward spectrometer primarily designed to look for indirect evidence of new physics in *CP* violation and rare decays of charm and beauty hadrons. The detector covers a unique pseudorapidity range $2 < \eta < 5$ corresponding to $\sim 4\%$ of the solid angle, but $\sim 25\%$ of all produced $b\bar{b}$ pairs fall into this geometrical acceptance. A detailed description of the LHCb detector and its performance are given in [33, 34].

The selection of Υ candidates is similar to those used in the previous LHCb analyses [35, 36, 37, 38]. The Υ candidates are formed from pairs of oppositely charged tracks reconstructed in the tracking system. Each track is required to have a good reconstruction quality [39] and to be identified as a muon [40]. Each muon is then required to have momentum satisfying $1 < p_T < 25\,\text{GeV}/c$, $10 < p < 400\,\text{GeV}/c$ and pseudorapidity within the region $2.2 < \eta < 4.5$. The two muons are required to originate from a common vertex with a good χ^2 probability of the vertex fit. In addition, the consistency of the dimuon vertex with a primary vertex is ensured via the quality requirement of a global fit, performed for each dimuon candidate using the primary vertex position as a constraint [41]. This global fit requirement also reduces the background caused by genuine muons coming from decays of long-lived charm and beauty hadrons. A large fraction of the combinatorial background is populated at large values of $|\cos\theta_{GJ}|$, where θ_{GJ} is the polar angle of μ^+ in the GJ frame. To reduce this background, a requirement $|\cos\theta_{GJ}| < 0.8$ is applied. Finally, the mass of the muon pair is required to be in the range $8.8 < m_{\mu^+\mu^-} < 11.0\,\text{GeV}/c^2$.

As an example, a dimuon mass distribution of the $\Upsilon \to \mu^+\mu^-$ candidates finally selected in $6 < p_T^\Upsilon < 8\,\text{GeV}/c$ and $2.2 < y^\Upsilon < 3.0$ for $\sqrt{s} = 7\,\text{TeV}$ data set is shown in Fig. 1. The dimuon mass distribution is parametrized by the sum of three double-sided Crystal Ball functions [42, 43] for describing the Υ mesons and an exponential function for the combinatorial background. The parametrization is done by an unbinned extended maximum likelihood fit. The dimuon mass fit is performed in each $(p_T^\Upsilon, y^\Upsilon)$ bin, and results of the fit are then used by the *sPlot* analysis [44] for obtaining so called signal *sWeights*: $w_i^{\Upsilon(1S)}$, $w_i^{\Upsilon(2S)}$ and $w_i^{\Upsilon(3S)}$. The *sWeights* are assigned to each, i^{th}, dimuon candidate for extracting the appropriate Υ signal component. The total signal yields obtained from the dimuon mass fit in the full explored range of p_T^Υ and y^Υ are 0.8×10^6 (1.8×10^6) $\Upsilon(1S)$ candidates, 0.2×10^6 (0.5×10^6) $\Upsilon(2S)$ candidates and 0.1×10^6 (0.2×10^6) $\Upsilon(3S)$ candidates in $\sqrt{s} = 7\,\text{TeV}$ ($\sqrt{s} = 8\,\text{TeV}$) data set. The average mass resolution of the $\Upsilon(1S)$ peak is $42\,\text{MeV}/c^2$.

3. Polarization analysis

The polarization measurement is performed using an unbinned maximum likelihood approach [45] already applied in the J/ψ and ψ(2S) polarization analyses [11],[12]. The polarization parameters are determined from fits to the two-dimensional $(\cos\theta, \phi)$ angular distribution of μ^+ from the $\Upsilon \to \mu^+\mu^-$ decay, described by Eq. 1. In each $\left(p_T^\Upsilon, y^\Upsilon\right)$ bin, the following logarithm of the likelihood function is constructed for each Υ state:

$$\log \mathcal{L}(\lambda_\theta, \lambda_{\theta\phi}, \lambda_\phi)_\Upsilon = s_w \sum_{i=1}^{N_{\text{tot}}} w_i^\Upsilon \times \log\left[\frac{\mathcal{P}(\cos\theta_i, \phi_i|\lambda_\theta, \lambda_{\theta\phi}, \lambda_\phi)}{\mathcal{N}(\lambda_\theta, \lambda_{\theta\phi}, \lambda_\phi)}\right], \tag{2}$$

where $\mathcal{P}(\cos\theta_i, \phi_i|\lambda_\theta, \lambda_{\theta\phi}, \lambda_\phi) \equiv 1 + \lambda_\theta \cos^2\theta_i + \lambda_{\theta\phi} \sin 2\theta_i \cos\phi_i + \lambda_\phi \sin^2\theta_i \cos 2\phi_i$, $\mathcal{N}(\lambda_\theta, \lambda_{\theta\phi}, \lambda_\phi)$ is normalization factor determined by Monte Carlo for each Υ meson, w_i^Υ is one of the *sWeights* for the ith Υ candidate, and N_{tot} is the total number of all selected Υ candidates in a considered $\left(p_T^\Upsilon, y^\Upsilon\right)$ bin. The constant scale factor $s_w = \sum_{i=1}^{N_{\text{tot}}} w_i^\Upsilon / \sum_{i=1}^{N_{\text{tot}}} (w_i^\Upsilon)^2$ is introduced to take into account correctly the effect of the *sWeights* on statistical uncertainties of the $\vec{\lambda}$ parameters obtained after the polarization fit. The influence of the s_w was validated by pseudoexperiments. The normalization factor $\mathcal{N}(\lambda_\theta, \lambda_{\theta\phi}, \lambda_\phi)$ is defined as

$$\mathcal{N}(\lambda_\theta, \lambda_{\theta\phi}, \lambda_\phi) \equiv \int \mathrm{d}\Omega\ \mathcal{P}(\cos\theta, \phi|\lambda_\theta, \lambda_{\theta\phi}, \lambda_\phi) \times \varepsilon(\cos\theta, \phi) \tag{3}$$

and is calculated using simulated events. In the simulation, where the Υ mesons are generated unpolarized, the two-dimensional $(\cos\theta, \phi)$ angular distribution of μ^+ from decays of selected Υ candidates is proportional to the total efficiency $\epsilon(\cos\theta, \phi)$, so $\mathcal{N}(\lambda_\theta, \lambda_{\theta\phi}, \lambda_\phi)$ is evaluated by summing $P(\cos\theta_j, \phi_j|\lambda_\theta, \lambda_{\theta\phi}, \lambda_\phi)$ over the selected Υ candidates in the simulated sample

$$\mathcal{N}(\lambda_\theta, \lambda_{\theta\phi}, \lambda_\phi) \propto \sum_j \epsilon^{\mu^+\mu^-} \kappa^\Upsilon \mathcal{P}(\cos\theta_j, \phi_j|\lambda_\theta, \lambda_{\theta\phi}, \lambda_\phi), \tag{4}$$

where $\epsilon^{\mu^+\mu^-}$ is a muon identification efficiency measured directly from data using a large sample of low-background $J/\psi \to \mu^+\mu^-$ events (no muon identification requirement was applied when selecting the Υ candidates in the simulated samples); κ^Υ is a correction factor for MC, obtained using data-driven techniques to account for small differences between data and simulation in a tracking efficiency of muons [39, 40] and in the p_T^Υ and y^Υ spectra [46, 47].

4. Results and conclusions

Different sources of systematic uncertainty have been considered when determining the polarization parameters, namely systematic uncertainty related to: a) the Υ signal extraction procedure; b) the muon identification efficiency; c) the track reconstruction efficiency; d) a possible small difference in the trigger efficiency between data and simulation; e) correction factors for the muon identification efficiency; f) the finite size of the simulated samples. All these sources have been studied for the polarization parameters λ_θ, $\lambda_{\theta\phi}$, λ_ϕ and $\tilde{\lambda}$ in the HX, CS and GJ frames for each $\left(p_T^\Upsilon, y^\Upsilon\right)$ bin. It was found that the most dominant systematic uncertainty is related to the finite size of MC samples, varying between 30% and 70% of the statistical uncertainty. The total systematic uncertainty for each polarization parameter is calculated as the quadratic sum of systematic uncertainties coming from all the considered sources, assuming no correlation between them. For some high-p_T^Υ bins the systematic and statistical uncertainties are comparable.

All the Υ polarization results obtained by the LHCb collaboration for data collected at $\sqrt{s} = 7$ TeV and 8 TeV are given in [21]. Here we outline the main features of the results. The

IOP Conf. Series: Journal of Physics: Conf. Series **938** (2017) 012001 doi:10.1088/1742-6596/938/1/012001

values of the parameter λ_θ obtained for the Υ mesons do not show any significant transverse or longitudinal polarization in all frames over the considered kinematic region. The values of the parameters $\lambda_{\theta\phi}$ and λ_ϕ are small in all frames over all $(p_{\mathrm{T}}^{\Upsilon},\, y^{\Upsilon})$ bins. All the three polarization parameters do not manifest a distinct dependence on the y^{Υ}. The values of the frame invariant parameter $\tilde{\lambda}$ measured in the HX, CS and GJ frames are consistent with each other. Moreover, all values of the $\tilde{\lambda}$ are close to zero in all phase-space bins. In the considered phase space domain, the Υ polarization results corresponding to $\sqrt{s} = 7$ and 8 TeV are in good agreement with each other.

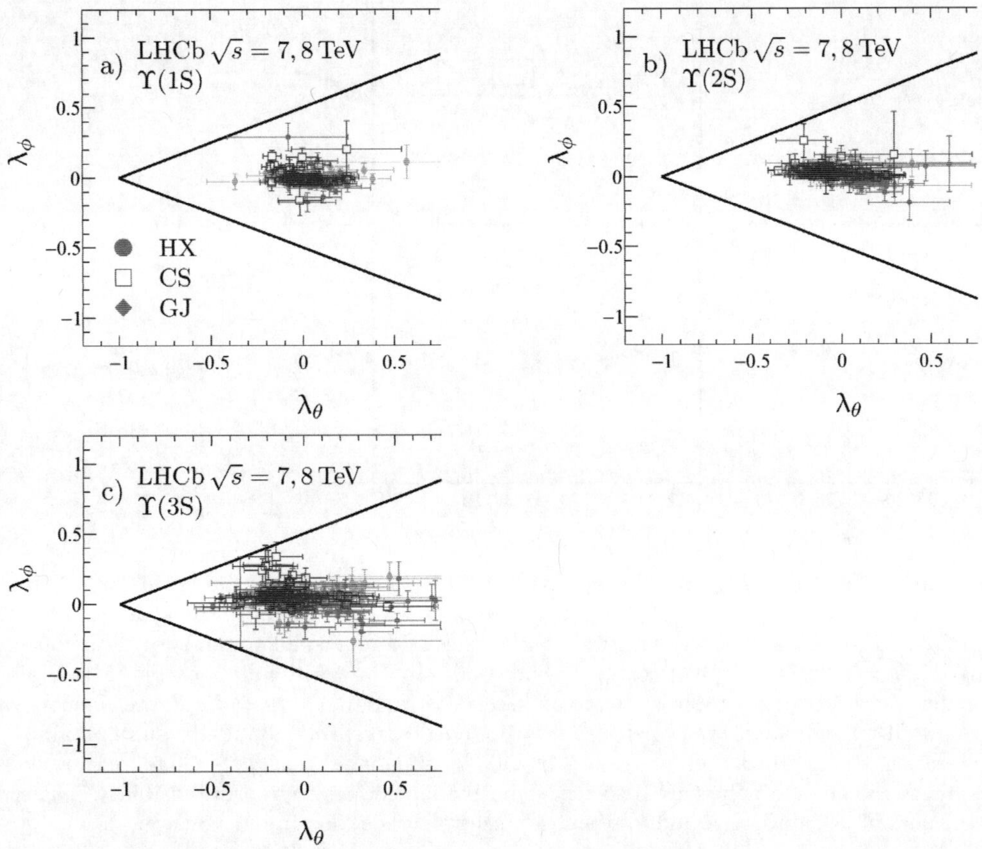

Figure 2: The measured values of $\lambda_\theta, \lambda_\phi$ for a) Υ(1S), b) Υ(2S) and c) Υ(3S) mesons. The red solid circles, blue open squares and green solid diamonds correspond to the helicity (HX), Collins-Soper (CS) and Gottfried-Jackson (GJ) frames, respectively. The thick black lines show the regions allowed by the positivity constraints.

The $\vec{\lambda}$ parameters have been checked for positivity constraints imposed on the spin-1 density-matrix [15, 21, 48, 26, 27]. The values of the $\vec{\lambda}$ satisfy all the six positivity constraints in all frames over all phase-space bins. In particular, Fig. 2 shows regions allowed by the positivity constraints together with the parameters λ_θ and λ_ϕ measured in all $(p_{\mathrm{T}}^{\Upsilon},\, y^{\Upsilon})$ bins, for data collected at $\sqrt{s} = 7$ and 8 TeV. Further, since the z axes of the HX, CS and GJ frames coincide in the limit $p_{\mathrm{T}}^{\Upsilon} \to 0$ [15, 22], we checked this constraint and found that all values of the $\vec{\lambda}$ are very similar for low-$p_{\mathrm{T}}^{\Upsilon}$ bins in all frames. It was also found that the parameters $\lambda_{\theta\phi}$ and λ_ϕ are very close to zero in the limit $p_{\mathrm{T}}^{\Upsilon} \to 0$, in accordance with kinematic constraints pointed out in [14].

Figs. 3 and 4 show a comparison of the LHCb results [21] with results obtained by the

IOP Conf. Series: Journal of Physics: Conf. Series **938** (2017) 012001 doi:10.1088/1742-6596/938/1/012001

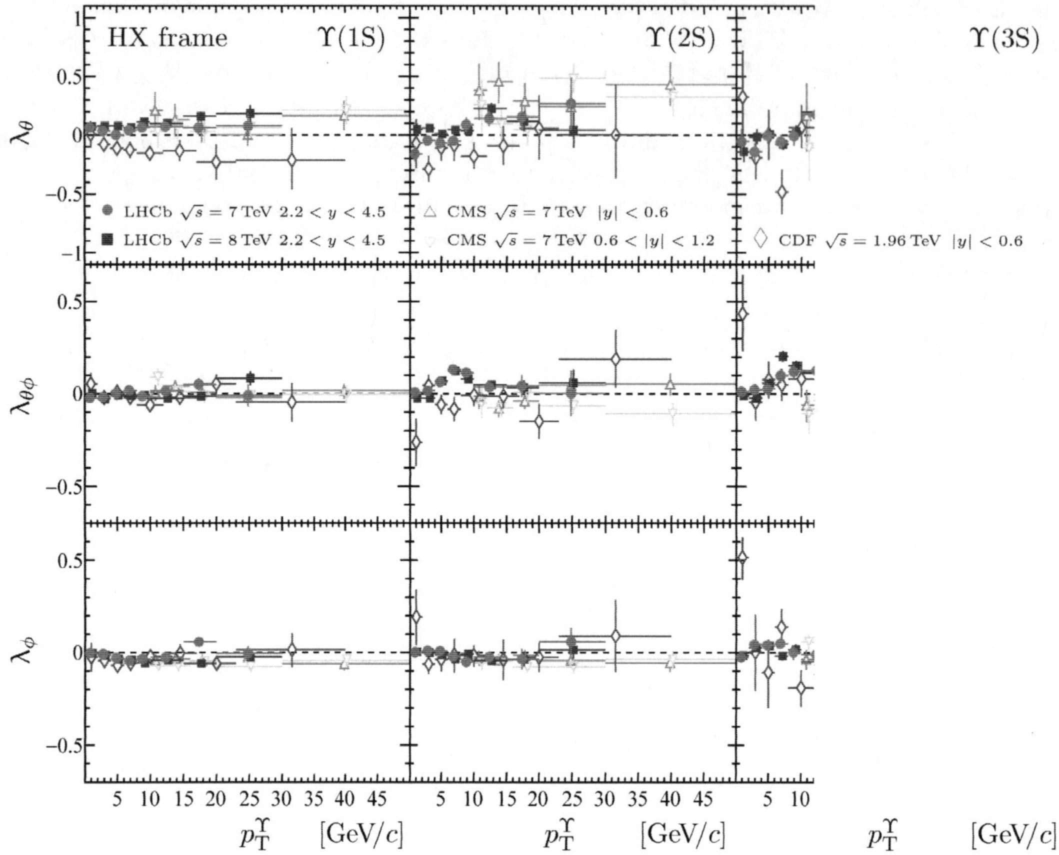

Figure 3: The values of λ_θ (top), $\lambda_{\theta\phi}$ (middle) and λ_ϕ (bottom) parameters, measured in the HX frame for Υ(1S) (left), Υ(2S) (center) and Υ(3S) (right) mesons. Results of the LHCb analysis for the rapidity region $2.2 < y^{\Upsilon} < 4.5$ are shown with red solid circles and blue solid squares for data collected in pp collisions at $\sqrt{s} = 7$ and 8 TeV, respectively. The results by the CMS collaboration [31] obtained in pp collision at $\sqrt{s} = 7$ TeV for rapidity regions $|y^{\Upsilon}| < 0.6$ and $0.6 < |y^{\Upsilon}| < 1.2$ are shown with magenta open upward triangles and cyan open downward triangles, respectively. The results obtained by the CDF collaboration [28] in p$\bar{\text{p}}$ collision at $\sqrt{s} = 1.96$ TeV for rapidity region $|y^{\Upsilon}| < 0.6$ are shown with green open diamonds. Some data points are displaced from the bin centers to improve visibility. The error bars indicate the sum of the statistical and systematic uncertainties added in quadrature.

CDF [28] and CMS [31] collaborations for the HX and CS frames, respectively. There is good agreement with CMS results for both frames, and with CDF for the CS frame.

The LHCb collaboration continues to perform measurements devoted to b-hadron and quarkonium physics [49, 50, 51, 52]. In particular, the polarization studies performed by LHCb [11, 12, 21] allowed one to shed some new facts on production mechanism of heavy quarkonium states, and to aggravate old ones.

Acknowledgments
I wish to thank A.V. Efremov and O.V. Teryaev for their invitation to the DSPIN-17 workshop and for their warm hospitality, and also S.R. Slabospitsky and O.V. Teryaev for useful and interesting discussions.

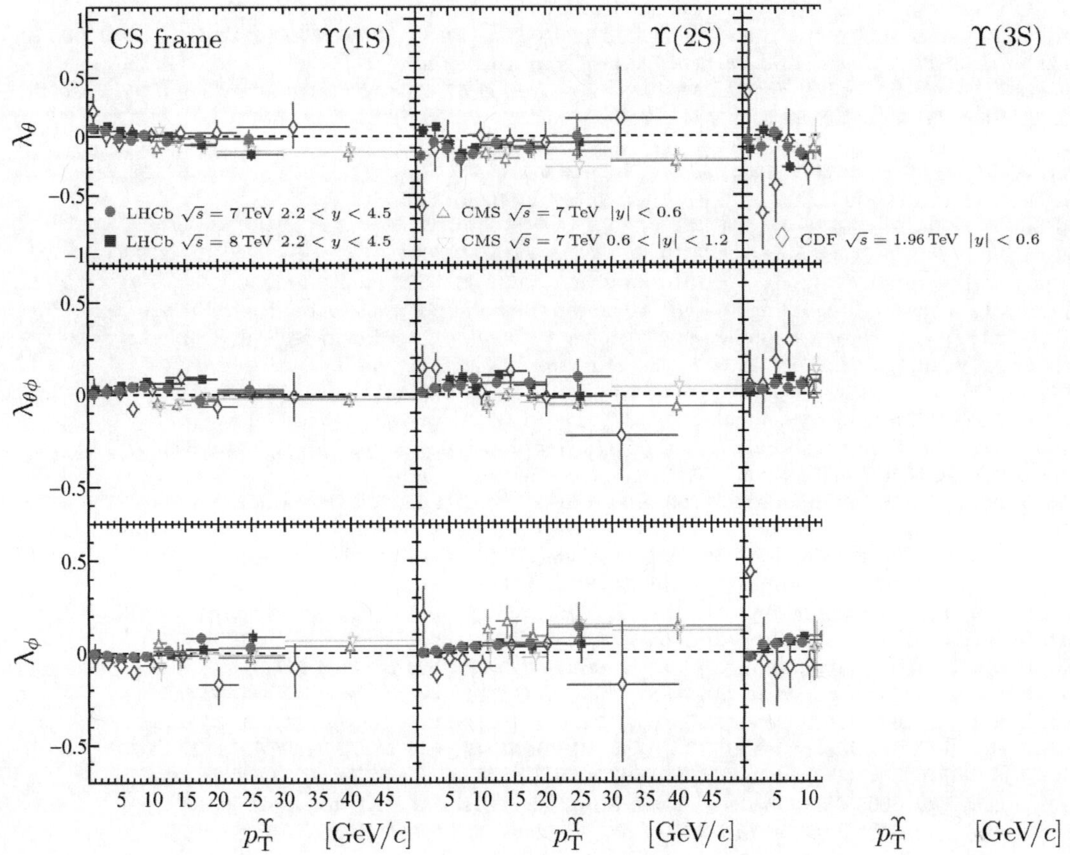

Figure 4: The values of λ_θ (top), $\lambda_{\theta\phi}$ (middle) and λ_ϕ (bottom) parameters, measured in the CS frame for $\Upsilon(1S)$ (left), $\Upsilon(2S)$ (center) and $\Upsilon(3S)$ (right) mesons. Results of the LHCb analysis for the rapidity region $2.2 < y^{\Upsilon} < 4.5$ are shown with red solid circles and blue solid squares for data collected in pp collisions at $\sqrt{s} = 7$ and 8 TeV, respectively. The results by the CMS collaboration [31] obtained in pp collision at $\sqrt{s} = 7$ TeV for rapidity regions $|y^{\Upsilon}| < 0.6$ and $0.6 < |y^{\Upsilon}| < 1.2$ are shown with magenta open upward triangles and cyan open downward triangles, respectively. The results obtained by the CDF collaboration [28] in $p\bar{p}$ collision at $\sqrt{s} = 1.96$ TeV for rapidity region $|y^{\Upsilon}| < 0.6$ are shown with green open diamonds. Some data points are displaced from the bin centers to improve visibility. The error bars indicate the sum of the statistical and systematic uncertainties added in quadrature.

References

[1] Herb S W *et al*, 1977 *Phys. Rev. Lett.* **39** 252
[2] Braaten E and Russ J, 2014 *Annu. Rev. Nucl. Part. Sci.* **64** 221 (*Preprint* arXiv:1401.7352)
[3] Caswell W E and Lepage G P, 1986 *Phys. Lett.* B **167** 437
[4] Bodwin G T, Braaten E and Lepage G P, 1995 *Phys. Rev.* D **51** 1125 (*Preprint* arXiv:hep-ph/9407339)
Bodwin G T, Braaten E and Lepage G P, 1997 *Phys. Rev.* D **55** 5853
[5] Cho P and Wise M B, 1994 *Phys. Lett.* B **346** 129 (*Preprint* arXiv:hep-ph/9411303)
[6] Beneke M and Rothstein I Z, 1995 *Phys. Lett.* B **372** 157 (*Preprint* arXiv:hep-ph/9509375)
[7] Beneke M and Krämer M, 1996 *Phys. Rev.* D **55** R5269 (*Preprint* arXiv:hep-ph/9611218)
[8] Leibovich A K, 1997 *Phys. Rev.* D **56** 4412 (*Preprint* arXiv:hep-ph/9610381)
[9] Gong B, Wan L-P, Wang J-X and Zhang H-F, 2014, *Phys. Rev. Lett.* **112** 032001 (*Preprint* arXiv:1305.0748)
[10] Aaij R *et al* [LHCb collaboration], 2014 *Eur. Phys. J.* C **74** 3092 (*Preprint* arXiv:1407.7734)
[11] Aaij R *et al* [LHCb collaboration], 2013 *Eur. Phys. J.* C **73** 2631 (*Preprint* arXiv:1307.6379)
[12] Aaij R *et al* [LHCb collaboration], 2014 *Eur. Phys. J.* C **74** 2872 (*Preprint* arXiv:1403.1339)
[13] Oakes R J, 1966 *Nuovo Cim.* A **44** 440

IOP Conf. Series: Journal of Physics: Conf. Series **938** (2017) 012001 doi:10.1088/1742-6596/938/1/012001

[14] Lam C S and Tung W K, 1978 *Phys. Rev.* D **18** 2447
[15] Faccioli P, Lourenço C, Seixas J and Wöhri H K, 2010 *Eur. Phys. J.* C **69** 657 (*Preprint* arXiv:1006.2738)
[16] Pilkuhn H M, 1979 *"Relativistic Particle Physics"* Springer-Verlag
[17] Beneke M, Krämer M and Vänttinen M, 1998 *Phys. Rev.* D **57** 4258 (*Preprint* arXiv:hep-ph/9709376)
[18] Jacob M and Wick G C, 1959 *Ann. Phys.* **7** 404
[19] Collins J C and Soper D E, 1977 *Phys. Rev.* D **16** 2219
[20] Gottfried K and Jackson J D, 1964 *Nuovo Cim.* **33** 309
[21] Aaij R *et al* [LHCb collaboration], 2017 *Preprint* arXiv:1709.01301
[22] Falciano S *et al*, 1986 *Z. Phys.* C **31** 513
[23] Faccioli P, Lourenço C and Seixas J, 2010 *Phys. Rev. Lett.* **105** 061601 (*Preprint* arXiv:1005.2601)
[24] Faccioli P, Lourenço C and Seixas J, 2010 *Phys. Rev.* D **81** 111502 (*Preprint* arXiv:1005.2855)
[25] Teryaev O V, 2006 Proceedings of XI Advanced Research Workshop on High Energy Spin Physics, *DUBNA-SPIN-05*, Dubna, September 27–October 1, 2005 (ed. by Efremov A V and Goloskokov S V)
[26] Teryaev O V, 2011 *Nucl. Phys. Proc. Suppl.* **214** 118
[27] Teryaev O V, 2011 Proceedings of XIV Advanced Research Workshop on High Energy Spin Physics, *DUBNA-SPIN-11*, Dubna, September 20–24, 2011 (ed. by Efremov A V and Goloskokov S V)
[28] Aaltonen T *et al* [CDF collaboration], 2012 *Phys. Rev. Lett.* **108** 151802 (*Preprint* arXiv:1112.1591)
[29] Acosta D *et al* [CDF collaboration], 2002 *Phys. Rev. Lett.* **88** 161802
[30] Abazov V M *et al* [D0 collaboration], 2008 *Phys. Rev. Lett.* **101** 182004 (*Preprint* arXiv:0804.2799)
[31] Chatrchyan S *et al* [CMS collaboration], 2013 *Phys. Rev. Lett.* **110** 081802 (*Preprint* arXiv:1209.2922)
[32] Brown C N *et al* [E866 collaboration], 2001 *Phys. Rev. Lett.* **86** 2529
[33] Alves A A *et al* [LHCb collaboration], 2008 *JINST* **3** S08005
[34] Aaij R *et al* [LHCb collaboration], 2014 *Int. J. Mod. Phys.* A **30** 1530022 (*Preprint* arXiv:1412.6352)
[35] Aaij R *et al* [LHCb collaboration], 2012 *Eur. Phys. J.* C **72** 2025 (*Preprint* arXiv:1202.6579)
[36] Aaij R *et al* [LHCb collaboration], 2013 *JHEP* **06** 064 (*Preprint* arXiv:1304.6977)
[37] Aaij R *et al* [LHCb collaboration], 2014 *Eur. Phys. J.* C **74** 2835 (*Preprint* arXiv:1402.2539)
[38] Aaij R *et al* [LHCb collaboration], 2015 *JHEP* **11** 103 (*Preprint* arXiv:1509.02372)
[39] Aaij R *et al* [LHCb collaboration], 2014 *JINST* **10** P02007 (*Preprint* arXiv:1408.1251)
[40] Aaij R *et al* [LHCb collaboration], 2013 *JINST* **8** P10020 (*Preprint* arXiv:1306.0249)
[41] Hulsbergen W D, 2005 *Nucl. Instrum. Meth.* A **552** 566 (*Preprint* arXiv:physics/0503191)
[42] Skwarnicki T, 1986 *PhD thesis, Institute of Nuclear Physics, Krakow* (DESY-F31-86-02)
[43] Aaij R *et al* [LHCb collaboration], 2011 *Phys. Lett.* B **707** 52 (*Preprint* arXiv:1109.0963)
[44] Pivk M and Le Diberder F R, 2004 *Nucl. Instrum. Meth.* A **555** 356 (*Preprint* arXiv:physics/0402083)
[45] Xie Y, 2009 *Preprint* arXiv:0905.0724
[46] Sjöstrand T, Mrenna S and Skands P, 2006 *JHEP* **05** 026 (*Preprint* arXiv:hep-ph/0603175)
[47] Belyaev I *et al*, 2011 *J. Phys. Conf. Ser.* **331** 032047
[48] Palestini S, 2010 *Phys. Rev.* D **83** 031503 (*Preprint* arXiv:1012.2485)
[49] Altarelli M P, 2013 *EPJ Conf* **60** 15005 (*Preprint* arXiv:1307.1110)
[50] Manca G, 2014 *Int. J. Mod. Phys.* A **29** 1430014
[51] Yang Z, 2014 *Nucl. Phys.* A **931** 643
[52] Belyaev I M and Egorychev V Yu, 2015 *Phys. Atom. Nucl.* **78** 977

IOP Conf. Series: Journal of Physics: Conf. Series **938** (2017) 012002 doi:10.1088/1742-6596/938/1/012002

Longitudinal spin structure of the nucleon: COMPASS legacy

Yann Bedfer, on behalf of the COMPASS collaboration
Irfu/DPhN, CEA-Saclay, France

E-mail: Yann.Bedfer@cern.ch

Abstract. The COMPASS collaboration has dedicated a large fraction of its already twenty year long existence to the study of the polarised structure of the nucleon. This presentation addresses two of the investigated aspects: inclusive measurements and direct measurements of the gluon polarised PDF, Δg. In the first case, a summary of what constitutes COMPASS legacy is presented. This covers the actual measurements of structure function g_1, at both high and low Q^2, and the information extracted from them in terms of polarised PDFs, axial charge and Bjorken sum rule. In particular, the quark spin contribution to the nucleon spin is determined to be $0.26 < \Delta\Sigma < 0.36$ and from COMPASS data alone the Bjorken sum rule is verified to 9% accuracy. In the second case, the latest developments in the analysis of our data on semi-inclusive single-hadron production at high-p_T are reported.

1. Introduction

Polarised aspects of the nucleon structure are still less known than unpolarised ones. COMPASS is one of the several experiments set up to gain further insight into this polarised structure over the last twenty years. It fulfils its specifications by measuring double spin cross-section asymmetries in lepton-nucleon scattering using muons of energy between 160 and 200 GeV.

2. Experimental apparatus and analysis methods

The experiment is a fixed target magnetic spectrometer. It is described in details in [1] and [2]. As far as polarised nucleon structure is concerned, it operates in conjunction with two other equally important pieces of equipment: a polarised muon beam and a polarised target, which fulfil the requisites of double spin measurements. The muon beam delivered leptons with a natural, and hence stable, longitudinal polarisation of $\sim$ 80% and an intensity of about 5×10^7 per second. The polarised target was designed with an impressive 1.2 m of length, compensating the weakness of the beam intensity. It provided protons and deuterons polarised to $\sim$90% and $\sim$50%, respectively. It was subdivided into cells arranged along the beam axis and polarised in opposite direction, allowing a simultaneous recording of the two nucleon spin states. The polarisation configuration was regularly inverted.

The cross-section asymmetry A_{LL} is extracted from the difference in number of interactions for parallel and anti-parallel spins of muon and nucleon. For inclusive measurements, this asymmetry when divided by the depolarisation factor is approximately equal to A_1. The latter, together with the knowledge of the unpolarised structure function F_2 and the ratio R, give the

IOP Conf. Series: Journal of Physics: Conf. Series **938** (2017) 012002 doi:10.1088/1742-6596/938/1/012002

spin-dependent structure function g_1. The parametrisation of F_2 and R were taken from Refs [3] and [4], respectively.

3. Results for A_1^p and g_1^p in the DIS region ($Q^2 > 1$ (GeV/c)^2)

Results presented here, Fig. 1, are based on data collected in 2007 [5] and 2011 [6]. In 2011 the beam energy was increase… Results on A_1^p and g_1^p for the … data, Fig. 1, thus illustrating … that it reaches down to low … $g_1^p(x)$ stays finite and positive …

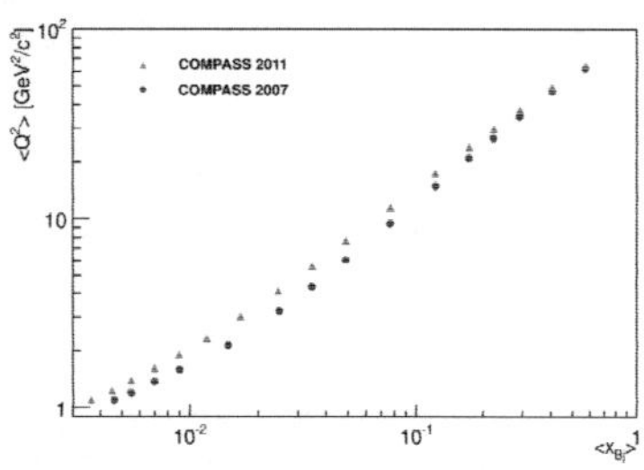

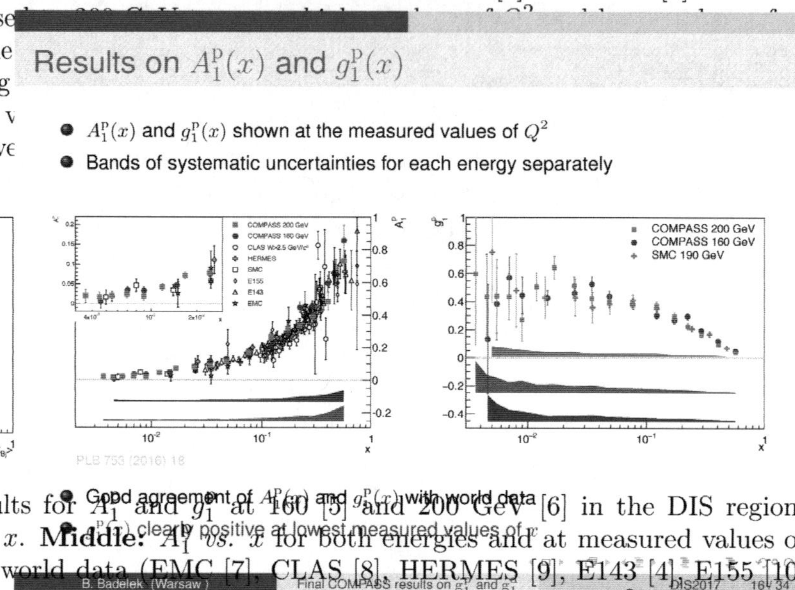

Figure 1. COMPASS results for A_1^p and g_1^p at 160 [5] and 200 GeV [6] in the DIS region. **Left:** mean value of Q^2 *vs.* x. **Middle:** A_1^p *vs.* x for both energies and at measured values of Q^2 , compared to the other world data (EMC [7], CLAS [8], HERMES [9], E143 [4], E155 [10] and SMC [3]). **Right:** g_1^p *vs.* x for both energies and at measured values of Q^2, compared to the SMC measurements [3]. Bands at the bottom indicate the systematic uncertainties of the COMPASS data at 160 GeV (blue), 200 GeV (red) and SMC at 190 GeV (green).

4. Results for A_1^d and g_1^d in the DIS region ($Q^2 > 1$ (GeV/c)^2)

Results presented here, Fig. 2, are based on data collected in 2002-2004 [11] and 2006 [12]. Results on A_1^d and g_1^d from both samples agree very well with each other and with the world data, Fig. 2, thus illustrating their weak dependence on Q^2. Contrary to the behaviour of g_1^p and to the hints from SMC [3], g_1^d is compatible with zero at lowest measured x.

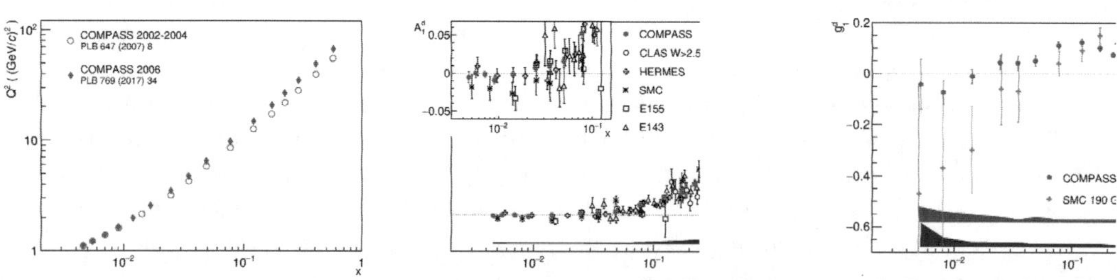

Figure 2. COMPASS results for A_1^d and g_1^d for the data collected in 2002-2004 [11] and 2006 [12] in the DIS region. **Left:** mean values of Q^2 *vs.* x. **Middle:** A_1^d *vs.* x for both samples and at measured values of Q^2, compared to the other world data (CLAS [8], HERMES [9], SMC [3], E143 [4] and E155 [13]). **Right:** g_1^d *vs.* x for both samples combined and at measured values of Q^2 compared to the SMC measurements [3]. Bands at the bottom indicate the systematic uncertainties of the COMPASS data (blue) and SMC (green).

IOP Conf. Series: Journal of Physics: Conf. Series **938** (2017) 012002 doi:10.1088/1742-6596/938/1/012002

5. A_1 and g_1 in the non-perturbative $Q^2 < 1$ (GeV/c)^2 region

The behaviour of g_1 in the region of lower Q^2, as opposed to that of DIS, is largely unknown. It is governed by soft physics processes and the transition to DIS is still not understood. For fixed-target experiments, low values of Q^2 are strongly correlated with low values of x, where also physics of large parton densities sets up.

Measurements at low x and low Q^2 are scarce as they put high demands on event triggering and reconstruction. They were performed only by the Spin Muon Collaboration on proton and deuteron [3] and by COMPASS on the deuteron [14]. The latter, very precise results do not reveal any spin effects over the whole measured interval of x. This very fact was understood as an early indication that ΔG, the total contribution of the gluon spin to the nucleon spin, is small [15].

Here we present new results obtained on the proton in the region 0.0062 (GeV/c)$^2 < Q^2 <$ 1 (GeV/c)^2, $4\,10^5 < x < 4\,10^2$, $0.1 < y < 0.9$ and for $W > 5$ GeV. Four different 2-dimensional grids of kinematic variables are used: (x, Q^2), (ν, Q^2), (x, ν) and (Q^2, x); an example is shown in Fig. 3 (see Ref. [16] for the details). Resulting asymmetry $A_1^{\rm p}$ and structure function $g_1^{\rm p}$ are presented in Fig. 3. Very clear spin effects are seen at all x.

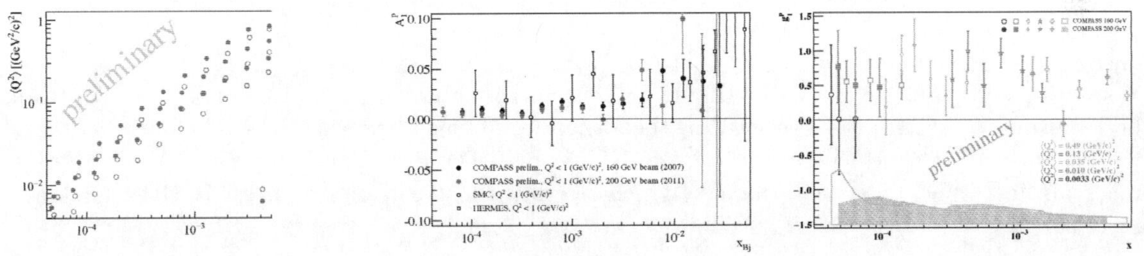

Figure 3. COMPASS results for $A_1^{\rm p}$ and $g_1^{\rm p}$ at low Q^2 for two muon energies. **Left:** kinematic range in (Q^2, x) variables. **Middle:** $A_1^{\rm p}$ *vs.* x at measured values of Q^2, compared to data of HERMES [9] and SMC [3, 17]. **Right:** $g_1^{\rm p}$ *vs.* x at measured values of Q^2.

6. NLO QCD analysis of the g_1 world DIS data

The COMPASS results on the spin-dependent structure functions are used in an NLO QCD fit, together with the world data on $g_1^{\rm p}$, $g_1^{\rm d}$ and $g_1^{^3{\rm He}}$ with $Q^2 > 1$ (GeV/c)^2 and $W^2 >$ 10 (GeV/c^2)^2 [6, 12]. Renormalisation/factorisation scheme is $\overline{\rm MS}$. Fitted are the gluon helicity Δg, the singlet $\Delta q^S = \Delta(u+\bar{u}) + \Delta(d+\bar{d}) + \Delta(s+\bar{s})$ and two non-singlet distributions $\Delta q_3 = \Delta(u+\bar{u}) - \Delta(d+\bar{d})$ and $\Delta q_8 = \Delta(u+\bar{u}) + \Delta(d+\bar{d}) - 2\Delta(s+\bar{s})$. The functional shape at input scale Q_0^2 is $\Delta f_k(x) = \eta_k x^{\alpha_k}(1-x)^{\beta_k}(1+\gamma_k x)/\int_0^1 x^{\alpha_k}(1-x)^{\beta_k}(1+\gamma_k x){\rm d}x$ where $k = S, 3, 8, g$ and η_k is the first moment of $\Delta f_k(x)$ at Q_0^2. But several simplifications are applied, leaving a maximum of 11 free parameters. In particular, the non-singlet moments are fixed by the baryon decay constants: $\eta_3 = {\rm F} + {\rm D}$, $\eta_8 = 3{\rm F} - {\rm D}$, assuming SU(2) and SU(3) symmetries. In addition, for the non-singlet and gluon distributions, $\gamma = 0$. For the gluons, β_g is fixed at the value of the unpolarised distribution in the MSTW PDF set [18]. The positivity constraint on $q + \bar{q}$ and g is enforced by a χ^2 penalty. Only statistical errors are considered in the fit; normalisations of data sets are allowed to vary, constrained by systematical uncertainties. Varying the input scale allows to generate sets of solutions. Doing this while fixing parameter γ_S to zero or not, generates two such sets, all elements of which describe the data equally well. Their envelope represents the systematical uncertainty.

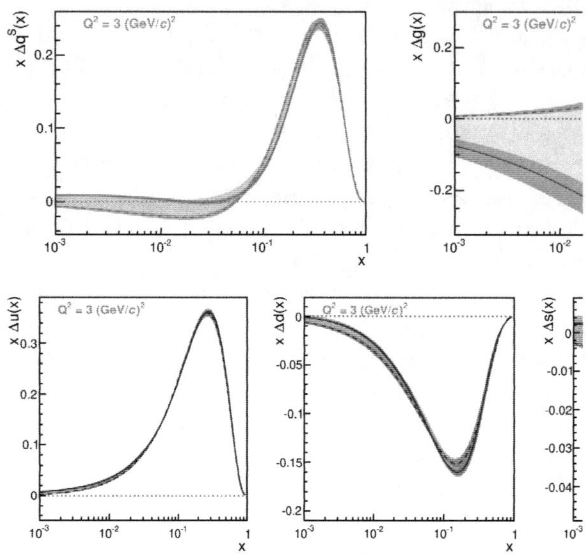

Figure 4. Results of the QCD fit to g_1 world data, at $Q^2 = 3$ (GeV/c)^2 for two sets of solutions [6], see text. Top: singlet $x\Delta q^S(x)$ and gluon distribution $x\Delta g(x)$. Bottom: distributions of $x[\Delta q(x) + \Delta\bar{q}(x)]$ for the different flavours (u, d and s). Continuous and dashed lines correspond the fit with $Q_0^2 = 1$ (GeV/c)^2. The dark bands represent the statistical uncertainties, only. The light bands, which overlay the dark ones, represent the systematic uncertainties.

Results of the fit are shown in Fig. 4. The contribution of quark spin to the nucleon spin $\Delta\Sigma$ is found to be 0.31 with an uncertainty of ± 0.05 that can be traced to the large uncertainty on Δg.

7. First moments of structure functions $g_1^{\rm d}$ and $g_1^{\rm NS}$

The first moment $\Gamma_1^{\rm d}$ of $g_1^{\rm d}$ allows for a determination of the flavour-singlet axial charge a_0, with the axial charge a_8 as an additional input. COMPASS data alone, when integrated over x and extrapolated into the high x region, give ([12]):

$$a_0(Q^2 = 3\,({\rm GeV/c})^2) = 0.32 \pm 0.02_{stat.} \pm 0.04_{syst.} \pm 0.05_{evol.}$$

which is consistent with the value obtained from the COMPASS NLO QCD fit. The largest contribution to the uncertainty comes from the evolution to common Q^2 that data points need to undergo before integration, and hence, ultimately, from the uncertainty of Δg.

The non-singlet structure function, $g_1^{\rm NS}$ is defined as:

$$g_1^{\rm NS} = g_1^{\rm p} - g_1^{\rm n} = 2[g_1^{\rm p} - g_1^{\rm d}/(1 - 1.5\omega_D)]$$

where $\omega_D = 0.05 \pm 0.02$ is the contribution of the D-state in the deuteron. Its first moment is the right hand side of the fundamental Bjorken sum rule. The extraction of the $\Gamma_1^{\rm NS}$ from COMPASS data alone, following a procedure similar to the above gives:

$$\Gamma_1^{\rm NS} = 0.192 \pm 0.007_{stat.} \pm 0.0015_{syst.}$$

which leads to a validation of the sum rule at the level of 9%, see Ref. [12] for the details.

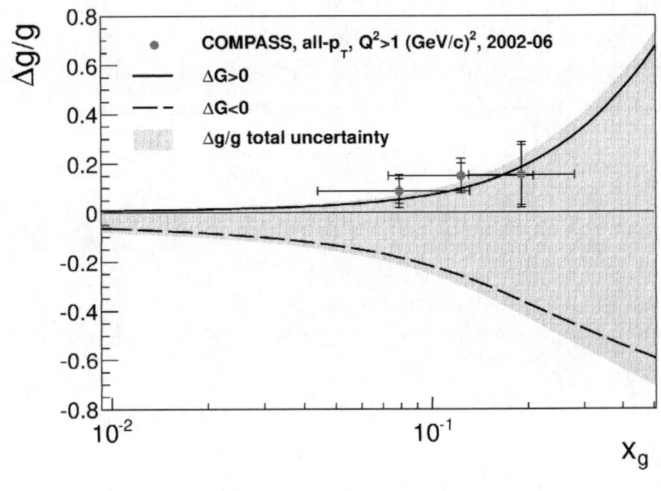

Figure 5. $\Delta g/g$ *vs.* x: LO analysis of COMPASS DIS data [19] in three ranges of x_g compared to predictions from COMPASS NLO QCD fit [6] at $Q^2 = 3$ (GeV/c)^2. The three data points are not fully independent. Horizontal bars represent 1σ confidence intervals. Inner error bars represent statistical uncertainties and outer ones statistical and systematic uncertainties combined in quadrature.

8. Direct measurement of Δg

COMPASS has explored several avenues to better constrain the gluon polarised PDF, Δg, left largely undetermined by inclusive measurements. One of these is the semi-inclusive single-hadron production as a function of p_T, p_T being the hadron transverse momentum with respect to the virtual photon.

A re-analysis [19] was performed on the data with $Q^2 > 1$ (GeV/c)^2 taken on polarised deuterons. It uses a novel method, putting the data to better use. It yields:

$$\langle \Delta g/g \rangle = 0.113 \pm 0.038_{(stat.)} \pm 0.036_{(syst.)}$$

for a weighted average of gluon momentum fraction $\langle x_g \rangle \approx 0.10$ and an average Q^2 of 3 (GeV/c)^2. This result is compatible with and supersedes our previous result [20] obtained from the same $Q^2 > 1$ (GeV/c)^2 data. It favours a positive gluon polarisation in the measured x_g range. The gluon polarisation is also extracted in three bins which correspond to three partially overlapping x_g ranges. These are shown in Fig. 5. Within experimental uncertainties, the values do not show any significant dependence on x_g.

The above analysis relies on a Monte Carlo generator, *viz.* LEPTO [21], to go from the hadron to the parton level. And at parton level, it performs calculations at order $\mathcal{O}(\alpha_S)$, *i.e* LO for the gluon-induced subprocesses. For this reason, we consider our approach to be a LO analysis, although LEPTO is not strictly speaking a fixed-order pQCD program. This is a strong limitation, given that state-of-the-art QCD fits of polarised PDFs are now based on NLO pQCD calculations. Remains that the results demonstrate the potential of COMPASS single-hadron production data to constrain Δg. These are made available in Ref. [20], in the shape of raw double spin asymmetries in bins of p_T.

Single-hadron production at high p_T was also measured by COMPASS in the low Q^2 quasi-real photoproduction region [22]. This time, a pQCD framework that was developed up to NLO is available to interpret the data [23]. Provided a technique known as "threshold resummation" at next-to-leading logarithm (NLL) is applied, the calculations reproduce COMPASS unpolarised cross-section measurements [24] within theoretical uncertainties. This hence demonstrates the applicability of the framework to our kinematical domain. The resummation technique has recently been extended to the polarised case [25] and applied to the COMPASS case. Results show improved agreement with polarised PDFs from DSSV2014 [26], but the interpretation of the data is still affected by uncertainties arising from fragmentation functions.

9. Conclusions

Results presented in this paper constitute the COMPASS legacy on $g_1^{\rm p}$ and $g_1^{\rm p}$ both for the DIS ($Q^2 > 1$ (GeV/c)^2) and nonperturbative ($Q^2 < 1$ (GeV/c)^2) regions.

Direct measurements of the gluon distribution *via* semi-inclusive single-hadron production show promising results, although they are not yet in a shape that can be included in NLO QCD fits of polarised PDFs.

10. References

[1] COMPASS Collaboration, P. Abbon *et al.*, Nucl. Instrum. Meth. A **577** (2007) 455.
[2] COMPASS Collaboration, P. Abbon *et al.*, Nucl. Instrum. Meth. A **779** (2015) 69.
[3] SMC, B. Adeva *et al.*, Phys. Rev. D **58** (1998) 112001.
[4] E143 Collaboration, K. Abe *et al.*, Phys. Rev. D 58 (1998) 112003.
[5] COMPASS Collaboration, M.G. Alekseev, *et al.*, Phys. Lett. B 690 (2010) 466.
[6] COMPASS Collaboration, C. Adolf, *et al.*, Phys. Lett. B 753 (2016) 18.
[7] EM Collaboration, J. Ashman *et al.*, Phys. Lett. B 206 (1988) 364; Nucl. Phys. B 328 (1989) 1.
[8] CLAS Collaboration, K. V. Dharmawardane *et al.*, Phys. Lett. B 641 (2006) 11.
[9] HERMES Collaboration, A. Airapetian *et al.*, Phys. Rev. D 75 (2007) 012007.
[10] E155 Collaboration, P.L. Anthony *et al.*, Phys. Lett. B 493 (2000) 19.
[11] COMPASS Collaboration, V.Yu. Alexakhin *et al.*, Phys. Lett. B 647 (2007) 8.
[12] COMPASS Collaboration, C. Adolph *et al.*, Phys. Lett. B 769 (2017) 34.
[13] E155 Collaboration, P. L. Anthony, et al., Phys. Lett. B 463 (1999) 339.
[14] COMPASS Collaboration, V.Yu. Alexakhin *et al.*, Phys. Lett. B 647 (2007) 330.
[15] A. W. Thomas, Int. J. Mod. Phys. E **18** (2009) 1116
[16] COMPASS Collaboration, C. Adolph *et al.*, submitted to Phys. Lett. B.
[17] SMC, B. Adeva *et al.*, Phys. Rev. D **60** (1999) 072004; erratum *ibid.*, D **62** (2000) 079902.
[18] MSTW, A.D. Martin *et al.*, Eur. Phys. J. C **63** (2009) 189.
[19] COMPASS Collaboration, C. Adolph *et al.*, Eur. Phys. J. C **77** (2017) no.4, 209
[20] COMPASS Collaboration, C. Adolph *et al.*, Phys. Lett. B **718** (2013) 922
[21] G. Ingelman, A. Edin and J. Rathsman, Comput. Phys. Commun. **101** (1997) 108.
[22] COMPASS Collaboration, C. Adolph *et al.*, Phys. Lett. B **753** (2016) 573
[23] B. Jäger, M. Stratmann and W. Vogelsang, Eur. Phys. J. C **44** (2005) 533
[24] COMPASS Collaboration, C. Adolph *et al.*, Phys. Rev. D **88** (2013) no.9, 091101
[25] C. Uebler, A. Schäfer and W. Vogelsang,
[26] D. de Florian, R. Sassot, M. Stratmann and W. Vogelsang, Phys. Rev. Lett. **113** (2014) no.1, 012001

Studies of GPDs at Jefferson Lab: results and future experiments

Angela Biselli
for the CLAS collaboration
Professor, Fairfield University, Fairfield, 06824 CT, USA
E-mail: biselli@jlab.org

Abstract. The Generalized Parton Distributions (GPDs) have emerged as a universal tool to describe hadrons in terms of their elementary constituents, the quarks and the gluons. Deeply Virtual Compton Scattering (DVCS) on a proton or neutron (N), $eN \to e'N'\gamma$, is one of the simplest processes that can be described in terms of GPDs. The amplitudes of DVCS and Bethe-Heitler, process where a photon is emitted by the incident or scattered electron, can be accessed via cross section measurements or exploiting their interference which give rise to spin asymmetries. Spin asymmetries, cross sections and cross-section differences can be connected to different combinations of the four leading order GPDs (H, E, $\tilde{H}$, $\tilde{E}$) for the two quark flavors depending on the observable and the type of target.

1. Introduction

Elastic scattering and deep inelastic scattering have been for years the methods of choice to study the structure of the nucleon. Elastic scattering gives access to the form factors which are related to the transverse spatial distribution of quarks, whereas deep inelastic scattering gives access to parton distributions which are longitudinal momentum and spin distribution of quarks. While both these quantities are important, it is clear that they are a subset of more fundamental quantities which encompass all the dimensions in space and momentum. The generalized parton distributions (GPDs) give fully correlated quark distributions in both coordinate and momentum space. These distributions allow access to crucial information such as the angular momentum distribution of quarks in the nucleon [?, ?, ?].

The cleanest way to access GPDs is via deeply virtual Compton scattering (DVCS), where the virtual photon interacts with a single quark of the nucleon radiating a real photon. As shown in Fig. ??, this exclusive process can be factorized, at high photon virtualities, into a hard scattering part, that can be treated perturbatively, and a nucleon-structure part, parameterized by the GPDs.

At leading twist the soft process is described by four chiral-even GPDs: H, $\tilde{H}$, E, and $\tilde{E}$, which depend on the longitudinal momentum fraction transferred to the proton, $\xi \simeq x_B/(2-x_B)$, the momentum transfer, t, between the virtual and the real photons and the momentum fraction of the struck quark, $x+\xi$, which is not experimentally accessible. All these four GPDs all involve processes that conserve the quark helicity, but while H and $\tilde{H}$ preserve the nucleon helicity, E and $\tilde{E}$ do not. The GPD H (E) is an average over the two possible possible quark-helicity

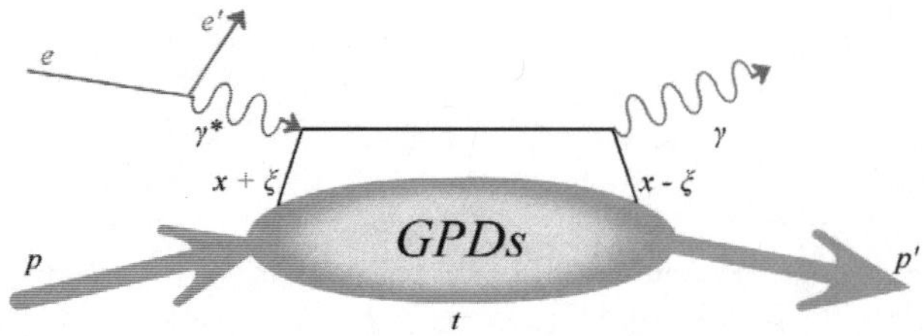

Figure 1. Handbag diagram for the DVCS process.

conserving processes, whereas $\tilde{H}$ ($\tilde{E}$) is a difference. GPDs are defined at the quark level for each flavor.

From the experimental point of view, the DVCS cannot be disentangled from the Bethe Heitler process, where the final-state photon is emitted by either the incoming or the outgoing electron. To extract the DVCS amplitude $T_{\rm DVCS}$, which contains the GPDs, one can measure both cross-section

$$\frac{d^4\sigma}{dQ^2 dx_B dt d\phi} \propto |T_{\rm DVCS} + T_{\rm BH}|^2 = |T_{\rm DVCS}|^2 + |T_{\rm BH}|^2 + I, \tag{1}$$

or asymmetries, which at leading twist can be written as:

$$A = \frac{\sigma^+ - \sigma^-}{\sigma^+ + \sigma^-} \propto \frac{I}{|T_{\rm DVCS}|^2 + |T_{\rm BH}|^2 + I}. \tag{2}$$

Here $I = T_{\rm DVCS} T^*_{\rm BH} + T^*_{\rm DVCS} T_{\rm BH}$ is the interference between the two processes. The DVCS amplitude $T_{\rm DVCS}$ depends on linear combinations of Compton form factors $\mathcal{F}$, whose real and imaginary parts are connected to the GPDs by

$$\Re{\rm e}\mathcal{F} = \mathcal{P} \int_{-1}^{1} dx \left[\frac{1}{x-\xi} \mp \frac{1}{x+\xi}\right] F(x,\xi,t) \tag{3}$$

$$\Im{\rm m}\mathcal{F} = \pi \left[F(\xi,\xi,t) \mp F(-\xi,\xi,t)\right] \tag{4}$$

Here the "$\mp$" sign apply, respectively, to the quark-helicity independent, or unpolarized, GPDs (H, E) and to the quark-helicity dependent, or polarized, GPDs $(\tilde{H}, \tilde{E})$. This means that the experimental observables depend on eight GPD-related quantities. Moreover the CFFs accessed experimentally are not directly the quark ones but CFFs for the type of target, proton, neutron or nuclei, which are linear combination of CFFs of different quark flavors. Luckily different observables (e.g. beam and target asymmetries) and different targets have different sensitivities to the various CFFs and therefore by performing several measurements one can separate the different contributions for a certain target and ultimately, by combining different target measurements one can perform flavor separation. For instance, the beam-spin asymmetry $A_{\rm LU}$ can be expressed as [?]

$$A_{\rm LU}(\phi) \propto \Im{\rm m}\left\{F_1\mathcal{H} + \frac{x_{\rm B}}{2-x_{\rm B}}(F_1+F_2)(\tilde{\mathcal{H}} - \frac{t^2}{4M^2}F_2\mathcal{E}) + ...\right\}\sin\phi\,, \tag{5}$$

where F_1 and F_2 are the form factors, and is sensitive to $\mathcal{H}$, $\tilde{\mathcal{H}}$, and $\mathcal{E}$, particularly to $\mathcal{H}_p$ for the proton and to $\mathcal{E}_n$ for the neutron. The longitudinal target-spin asymmetry $A_{\rm UL}$

$$A_{\rm UL}(\phi) \propto \Im{\rm m}\left\{F_1\tilde{\mathcal{H}} + \frac{x_{\rm B}}{2-x_{\rm B}}(F_1+F_2)(\mathcal{H} + \frac{x_{\rm B}}{2}\mathcal{E}) + ...\right\}\sin\phi\,, \tag{6}$$

is equally sensitive to $\mathcal{H}_p$ and $\tilde{\mathcal{H}}_p$ for the proton and $\mathcal{H}_n$ for the neutron. Furthermore the cross section and the double spin asymmetry are sensitive to the real part CFFs.

2. Results

This paper focuses on the last decade DVCS results at Jefferson Lab from Hall A and Hall B and the future measurements planned at this facility (see Section **??**). Jefferson lab operated at a maximum energy of 6 GeV, a much lower energy as compared to other facilities, which allows to access the large x_{B} region and therefore have an insight into the quarks valence region. Hall A used a high resolution spectrometer to detect the electron and an electromagnetic calorimeter to detect the photon, while the presence of the proton or neutron was ensured via missing mass cuts. Hall B used the CEBAF Large Acceptance Spectrometer (CLAS) [?] in its standard configuration with an additional calorimeter (inner calorimeter) for the detection of forward photons. In this configuration the complete final state $ep\gamma$ was simultaneously detected and its exclusivity was ensured by applying a set tight kinematic cuts. CLAS is a large acceptance detector which allowed to measure the DVCS process over a large kinematic range in Q^2, x_{B} and t, providing rich data sets to explore the GPD dependence on such variables. Experiments conducted at Jefferson Laboratory includes measurements of asymmetries and cross sections on protons, neutrons and nuclei

2.1. Asymmetries

The most accessible quantities that can be measured to study DVCS are the beam and target spin asymmetries, BSA and TSA respectively. These quantities depend on the interference between BH and DVCS and therefore show a clear DVCS signal. Also, by virtue of the fact that asymmetries are a ratio, they are less sensitive to certain systematic effects, albeit they have much larger statistical uncertainties than cross sections. Two exploratory measurements in Hall B of the BSA [?] and TSA [?] showed the handbag dominance and a clear twist-2 $\sin\phi$ dependence and triggered a series of dedicated experiments with both unpolarized and polarized targets focused on the extraction these asymmetries over a large kinematic range [?, ?, ?]. For example, Figure **??** shows the $-t$-dependence of the $\sin\phi$ term of the BSA [?] and TSA [?]. The proton BSA, sensitive to $\Im\mathrm{m}\mathcal{H}_p$, shows a steeper drop than $\Im\mathrm{m}\tilde{\mathcal{H}}_p$. Since the Compton form factors $\Im\mathrm{m}\mathcal{H}_p$ and $\Im\mathrm{m}\tilde{\mathcal{H}}_p$ are related to the Fourier transforms of the electric charge and axial charge respectively, this behavior indicates that the axial charge is more concentrated in the proton center than the electric one.

Using the polarized proton data, the double spin asymmetry (DSA) was also measured [?]. The DSA is important since it provides information on the real part of the Compton form factors but conversely to the single spin asymmetries it has a large BH component. The measurement of the three asymmetries at the same kinematic points allowed a simultaneous fit to extract the Compton form factors for the proton. This was done using a quasi model-independent technique [?] in which the bounds of the domains of variation of the CFFs is limited to ± 5 times the value predicted by the VGG model [?], and $\tilde{\mathcal{E}}_p$=0. Figure **??** shows the results of the fit for $\Im\mathrm{m}\mathcal{H}_p$ and $\Im\mathrm{m}\tilde{\mathcal{H}}_p$. In addition to confirming the fact that the axial charge is more concentrated than the electrical charge, one can see that the slope of $\Im\mathrm{m}\mathcal{H}_p$ decreases as x_{B} becomes bigger, indicating that the electric charge is more concentrated for valence quarks than sea quarks.

In Hall B Tthe BSA on nuclei was also measured [?]. The experiment used a ^{4}He target, which being a spin-0 nucleus at twist-2 is sensitive only to the GPD H_A. The analysis measured the BSA for coherent ($e^4\mathrm{He} \to e^4\mathrm{He}\gamma$) scattering which was used to extract, in a model-independent way, the real and the imaginary parts of the ^{4}He CFF, H_A providing a first insight on the partonic structure of nuclei.

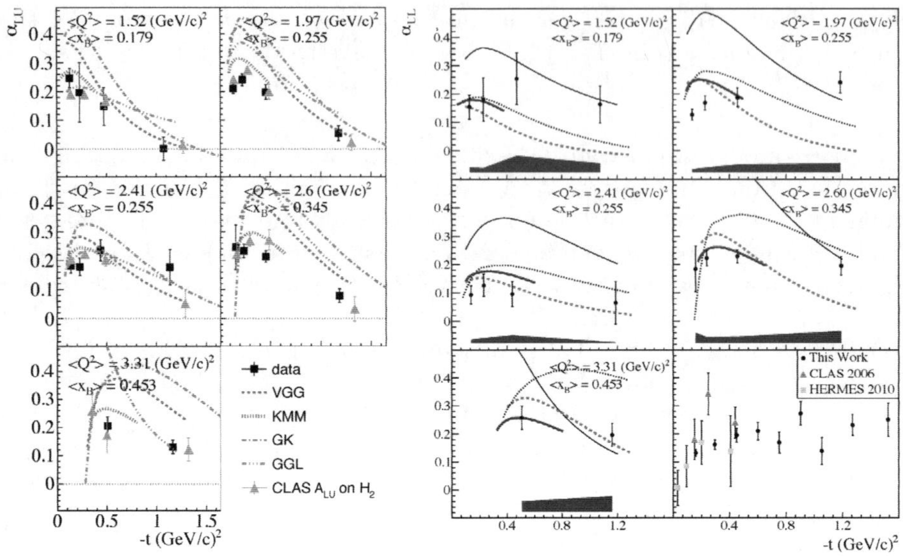

Figure 2. BSA (left panel) and TSA (right panel) $-t$ dependence of the $\sin\phi$ term. Left panel: the curves show the VGG prediction [?] (red dashed), KMM12 [?] (green dotted), GK [?] (blue dash-dotted), and GGL [?] (orange dashed-three- dotted). The triangular green data come from the previous CLAS experiment with unpolarized proton target [?]. Right panel: the curves show the predictions of 4 GPDs models: VGG (red dashed line), GK (black dotted lube), KMM12 (blue thick solid line), GGL (black solid line). The data agree qualitatively with the model predictions but clearly provide new constraints to the GPDs. More details on the models comparison can be found in [?, ?]

2.2. Cross-sections, and cross section differences

Cross sections and cross section differences are essential to measure with high precision the real part of Compton form factors. Both Hall A and Hall B have extracted cross sections with a proton target. In all these experiments the beam was polarized allowing the extraction of the cross section differences as well. Hall A results from 2006 [?] and the recent re-analysis of the same dataset [?] extracted the cross section and the cross-section differences over a limited Q^2 and $-t$ with high precision. A fit of the data allowed to extract both the DVCS and interference Compton form factor, as well as twist-2 and twist-3 contributions. The same Hall B experiment that lead to the first dedicated measurement of BSA [?] was analyzed to extract the cross sections and cross-section differences over a large kinematic range [?]. Results of this work show a clear non-zero contribution from the DVCS process. Moreover, using [?] the real and imaginary part of $\mathcal{H}$ was found with $\mathcal{E}_p$ and $\widetilde{\mathcal{E}}_p$ set to zero. The x_B trend of the fit indicate that the transverse size and partonic content are bigger at smaller momentum fractions. Additional data with polarized beam on unpolarized protons was taken in 2008-2009. This data was not included in either [?] and [?] and they were analyzed independently [?]. Preliminary results of cross-sections and cross-sections difference are shown in Figure ??. The publication of these results is currently in preparation and it will provide new constraints to the GPDs fits.

Hall A has recently published new results on cross-section and cross-section differences [?]. In this work data were taken at different beam energies and, leveraging the beam energy dependence on the various factors that multiply cross section contributions, a Rosenbluth separation was performed to separate the $|T_{DVCS}|^2$ from the $DVCS$-BH interference I amplitudes. Figure ?? shows a sample of the cross section and cross section differences results (right two panels) and

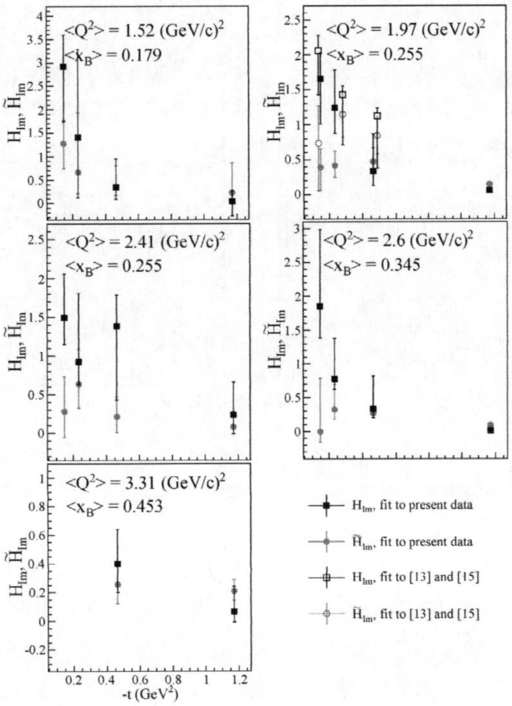

Figure 3. Compton form factor extraction from the simultaneous fit of BSA, TSA and DSA. The solid black squares and the solid red circles show the results for the imaginary parts of $\mathcal{H}_p$ and $\widetilde{\mathcal{H}}_p$ respectively. The results are compared to fits to previous CLAS data [?, ?]

the results for the separation of the amplitudes (left two panels). One can see the high sensitivity of the data to the higher twist (HT) or next-to-leading order (NLO) and therefore to the gluons contributions.

Cross-section and cross-section difference were also extracted for the neutron [?]. Neutron measurements are essential since they are needed, once combined with the proton measurements, to do flavor separation. The $\sin\phi$ moment for the neutron, which is sensitive to $\mathcal{E}_n$ was found very small, nevertheless comparison with models where different values of the angular momentum of the u and d quarks, shows that the data are sensitive to this quantity.

3. Future experiment

The experimental program at Jefferson lab on DVCS has been producing numerous compelling results. Compton form factors for proton, neutron and helium were extracted in several kinematic bins offering a first insight on the distribution of the electric and axial charge for valence and sea quark and hopefully to extract the angular momentum of the d and u quarks. To have a full picture of the GPDs, more data are needed. The Continuous Electron Beam Accelerator Facility (CEBAF) facility has been upgraded to a maximum of beam energy of 12 GeV. In addition to the new experimental Hall, Hall D, focusing mostly on hadron spectroscopy, the detectors in the three existing halls have been upgraded as well to perform at the new maximum energy. Hall A and C still host high resolution spectrometers while Hall B upgraded its detector to the so called CLAS12. Similarly to its predecessor, CLAS12 features six superconducting coils which define six sectors, each equipped with three regions of drift chambers, a high and a low threshold Cherenkov detector, a pre-shower and a electromagnetic

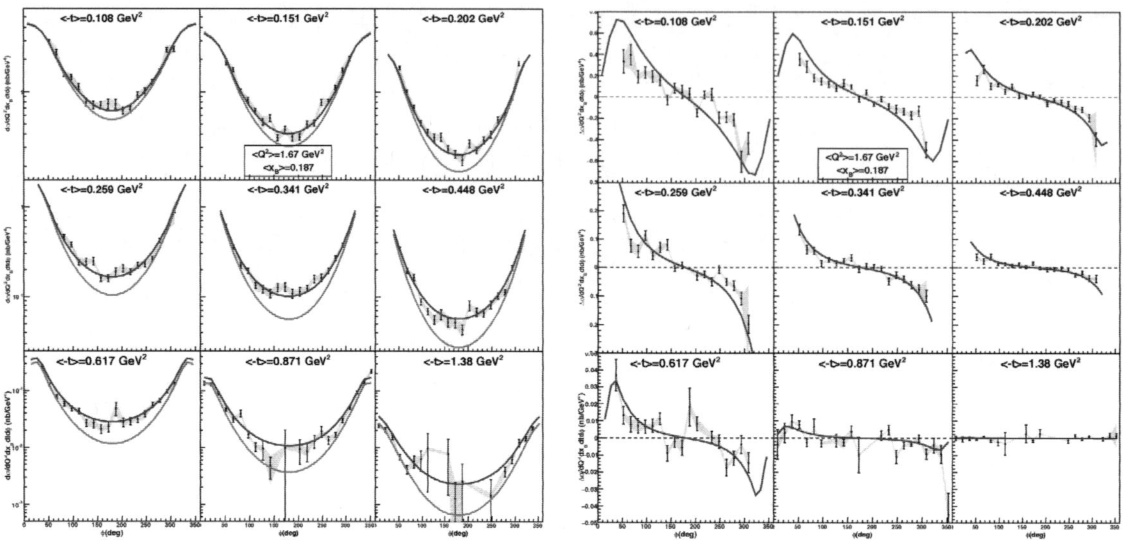

Figure 4. The unpolarized cross section (left) and cross-section differences (right) as a function of ϕ for the Bx=0.187 GeV^2 and Q^2=1.67 GeV^2. The blue curve is the result of the the VGG model and the the red (only left panel) is BH only. The grey band represents the systematic uncertainty.

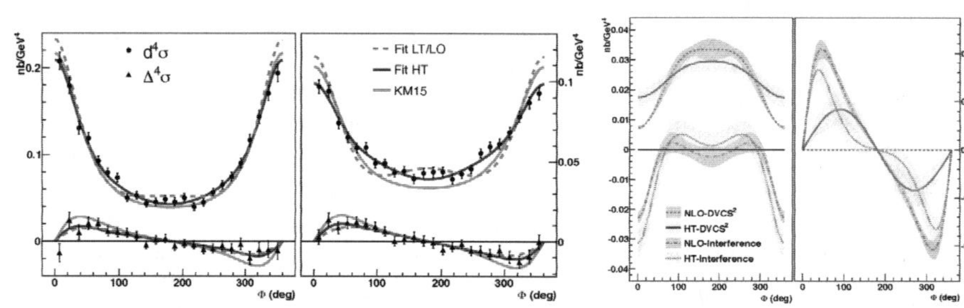

Figure 5. Cross sections and cross-section differences at Q^2=1.75 GeV^2, Bx=0.36, and t = 0.30 GeV^2 for two different beam energies (left). Dashed lines represent the LT/LO fit, solid blu lines is the HT fit. $|T_{DVCS}|^2$ and DVCS-BH interference I contributions for cross section and cross-section differences (right).

calorimeter and a forward time-of-flight scintillators. The new spectrometer has also a central detector around the target which allows the detection of particles at large angles. The central detector consist of a solenoid magnet, a silicon vertex tracker and the central time-of-flight. CLAS12 is shown in left panel of Figure ??. The right panel of Figure ?? shows the kinematic coverage (Q^2 vs x_B) of Jefferson lab as compared to other facilities. As one can see Jefferson lab is uniquely suited to study the valence quark region (high x_B) and the large acceptance of CLAS12 will allow mapping of the GPDs up to a $x_{\rm B} = 0.65$. There are several experiments dedicated to the DVCS program in all three experimental halls. The first experiment has taken already data in Hall A and focused on extracting the beam polarized and unpolarized cross sections on the proton. The experiment will run at three different beam energies and will extract the $|T_{DVCS}|^2$ term via Rosenbluth separation similarly to [?]. This experiment will be followed in Fall 2017 by the first CLAS12 experiment, which will focus on the extraction of the DVCS beam spin

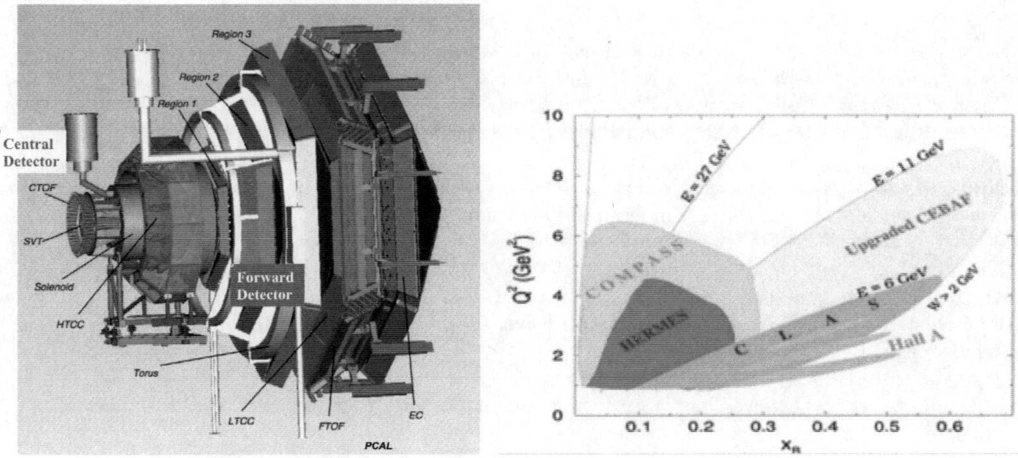

Figure 6. CLAS12 (left) and Jefferson Laboratory kinematic range (right).

asymmetry and cross section on the proton. Other experiments are planned in the following years to measure target and double spin asymmetries on the proton and to conduct the same measurements on the neutron. Figure ?? shows a sample of the expected results for the target spin asymmetry of the neutron. Moreover proposals to measure DVCS at different energies and using the transverse target with CLAS12 have been approved. This set of experiments represent a comprehensive study of the DVCS on both proton and neutron and the completion of such program will give a first insight on quarks' GPDs and, possibly, will allow the testing of Ji sum's rule.

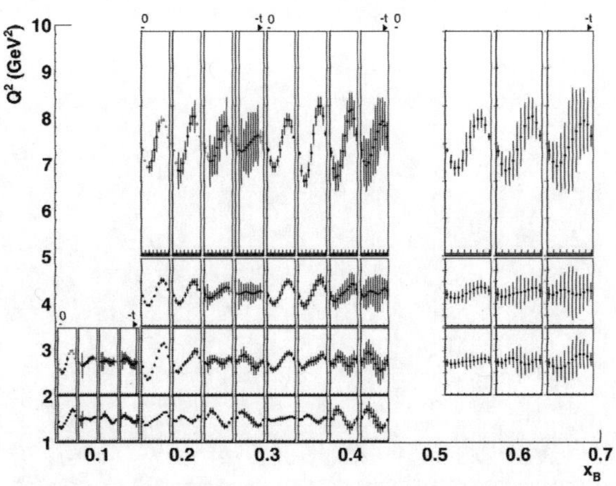

Figure 7. Projected target-spin asymmetry. The y-scale range, common to all bins, is -0.6-0.6. The black and red points are obtained, respectively, without and with the Forward Tagger.

References

[1] D. Mueller, D. Robaschik, B. Geyer, F. M. Dittes and J. Horejsi, Fortsch. Phys. **42**, 101 (1994).
[2] X. D. Ji, Phys. Rev. Lett. **78**, 610 (1997).
[3] A. V. Radyushkin, Phys. Rev. D **56**, 5524 (1997).

[4] A. V. Belitsky, D. Mueller and A. Kirchner, Nucl. Phys. B **629**, 323 (2002).
[5] B. A. Mecking *et al.* [CLAS Collaboration], Nucl. Instrum. Meth. A **503**, 513 (2003).
[6] S. Stepanyan *et al.* [CLAS Collaboration], Phys. Rev. Lett. **87**, 182002 (2001)
[7] S. Chen *et al.* [CLAS Collaboration], Phys. Rev. Lett. **97**, 072002 (2006).
[8] F. X. Girod *et al.* [CLAS Collaboration], Phys. Rev. Lett. **100**, 162002 (2008).
[9] E. Seder *et al.* [CLAS Collaboration], Phys. Rev. Lett. **114**, no. 3, 032001 (2015)
[10] S. Pisano *et al.* [CLAS Collaboration], Phys. Rev. D **91**, no. 5, 052014 (2015).
[11] M. Vanderhaeghen, P. A. M. Guichon and M. Guidal, Phys. Rev. D **60**, 094017 (1999); M. Guidal, M. V. Polyakov, A. V. Radyushkin and M. Vanderhaeghen, Phys. Rev. D **72**, 054013 (2005).
[12] K. Kumericki, D. Muller and M. Murray, Phys. Part. Nucl. **45**, no. 4, 723 (2014).
[13] P. Kroll, H. Moutarde and F. Sabatie, Eur. Phys. J. C **73**, no. 1, 2278 (2013).
[14] G. R. Goldstein, J. O. Hernandez and S. Liuti, Phys. Rev. D **84**, 034007 (2011).
[15] M. Guidal, Eur. Phys. J. A **37**, 319 (2008).
[16] M. Hattawy *et al.*, arXiv:1707.03361 [nucl-ex].
[17] C. M. Camacho *et al.* [Jefferson Lab Hall A and Hall A DVCS Collaborations], Phys. Rev. Lett. **97**, 262002 (2006).
[18] M. Defurne *et al.* [Hall A Collaboration], arXiv:1504.05453 [nucl-ex].
[19] H. S. Jo *et al.* [CLAS Collaboration], arXiv:1504.02009 [hep-ex].
[20] N. Hirlinger Saylor. *in preparation.*
[21] M. Defurne *et al.*, arXiv:1703.09442 [hep-ex].
[22] M. Mazouz *et al.* [Jefferson Lab Hall A Collaboration], Phys. Rev. Lett. **99**, 242501 (2007) doi:10.1103/PhysRevLett.99.242501

IOP Conf. Series: Journal of Physics: Conf. Series **938** (2017) 012004 doi:10.1088/1742-6596/938/1/012004

Recent COMPASS results on Transverse Spin Asymmetries in SIDIS

Franco Bradamante
Trieste Section of INFN, Trieste, Italy,
on behalf of the COMPASS Collaboration

Abstract. After reviewing the most important COMPASS results on transverse spin effects in SIDIS, I will present the recent work we have done on weighted Sivers asymmetries and "transversity induced" Λ polarisation. Using the high statistics data collected in 2010 on a transversely polarised proton target COMPASS has evaluated the transverse momentum weighted Sivers asymmetries both in x- and in z-bins. The results are also compared with the standard unweighted asymmetries. Using the whole data sample collected over the years on transversely polarised deuteron and proton targets COMPASS has measured the transversity induced Λ polarisation in the reaction $\mu N \rightarrow \mu' \Lambda X$. Possible future SIDIS COMPASS measurements will also be briefly mentioned.

1. Introduction

In 1996, the COMPASS spectrometer was proposed as a "COmmon Muon and Proton Apparatus for Structure and Spectroscopy", capable of addressing a large variety of open problems in both hadron structure and spectroscopy. By now, we have published a large number of results, and in this workshop several COMPASS contributions will cover the various facets of the hadron structure, namely the longitudinal spin structure of the nucleon in semi-inclusive deeply inelastic scattering (SIDIS), the transverse spin structure of the nucleon in SIDIS, in exclusive reactions, and in Drell-Yan scattering, and the hadron multiplicities in unpolarised SIDIS [1]. In this review I will shortly remind the most important results we have obtained on the transverse spin structure of the nucleon in SIDIS, then I will present our most recent work on the weighted Sivers asymmetries and on the Λ polarisation, and I will conclude mentioning the future measurements we are planning.

2. The TMD PDFs

The description of the nucleon structure in terms of collinear parton distributions functions (PDFs), i. e. distributions depending only on Bjorken x and the negative squared four-momentum transfer Q^2, has recently been generalized to take into account the parton transverse momentum k_T. A complete picture of the nucleon at leading order needs a total of eight transverse momentum dependent (TMD) PDFs. After integration over the transverse momentum three of them reduce to the "collinear" PDFs, namely the number density, the helicity and the transversity distributions. The other five TMD PDFs are sensitive to the direction of the quark transverse momentum, they vanish when integrating over k_T, and their measurement provides important information on the dynamics of the partons in the transverse plane in momentum space.

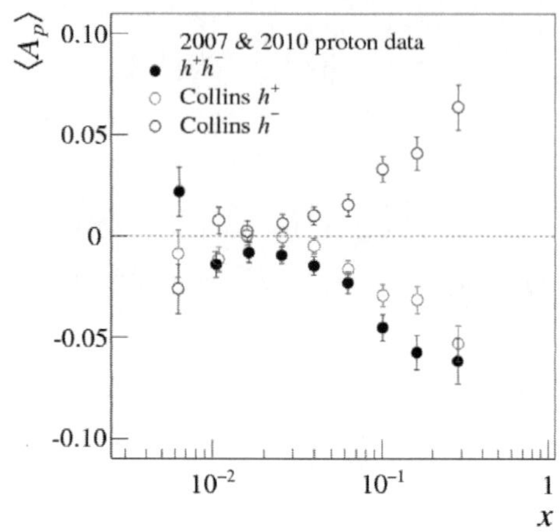

Figure 1. Comparison of the di-hadron asymmetry with the Collins asymmetry from the combined 2007 and 2010 proton data [6].

Particularly interesting is the measurement of the SIDIS cross-section when the target nucleon is transversely polarised. In this case 8 (5 in case of unpolarised lepton beam) different spin-dependent azimuthal modulations of the produced hadrons are expected, from which invaluable information on the TMD PDFs can be extracted. All these target spin dependent azimuthal modulations have been investigated by the HERMES and COMPASS Collaborations in pioneering experiments. In particular, HERMES [2, 3] and COMPASS [4, 5, 6] are up to now the only SIDIS experiments that have shown that the Sivers function [7], the transversity function and the Collins function [8] are different from zero. It is also of interest to remark that out of the 8 possible transverse spin azimuthal modulations in the SIDIS cross-section only the Collins and the Sivers asymmetries have been found to be different from zero, while the others have all been measured to be compatible with zero. And that there is no clear experimental evidence that the second T-odd TMD PDF, the so called Boer-Mulders function, which should show up in the azimuthal modulations of the unpolarised SIDIS cross-section, is different from zero. It has also to be added that the non-zero results for the asymmetries have been obtained scattering leptons on a proton target, while the corresponding asymmetries on a deuteron target (as measured by COMPASS) are compatible with zero [9, 10], hinting at a cancellation between the u- and d- quarks PDFs.

3. Transversity induced asymmetries

Some of the most beautiful results of COMPASS are shown in Fig. 1, where the Collins asymmetries for positive and negative hadrons are compared to the dihadron asymmetry, namely to the azimuthal modulation of the plane containing two oppositely charged hadrons. This dihadron asymmetry can be expressed as the product of the quark transversity distribution and a chiral-odd dihadron fragmentation function, which survives after integration over the two hadron momenta, and thus can be analyzed in the framework of collinear factorization. The mirror symmetry of the Collins asymmetry of positive and negative hadrons and the similarity between the positive hadron Collins asymmetry and the dihadron asymmetry suggested us an in-depth comparison between the two asymmetries. The conclusion of this investigation [11] was that both the single hadron and the dihadron transverse-spin dependent fragmentation functions are driven by the same elementary mechanism, which is very well described in the $^3P_0$ recursive string fragmentation model [12].

4. The Sivers asymmetries

The Sivers effect was experimentally observed in SIDIS on transversely polarised proton targets, first by the HERMES Collaboration [2, 3] and then by the COMPASS Collaboration [13, 5], while the COMPASS measurements on a transversely polarised deuteron target [9, 10] gave asymmetries compatible with zero. Combined analysis of the proton and deuteron data soon allowed for first extractions of the Sivers function for u- and d-quarks [14, 15, 16, 17], which turned out to be different from zero, with similar strength and opposite sign, a most important result in TMD physics.

In the standard Amsterdam notation the Sivers asymmetry can be written as

$$A_{Siv}(x,z) = \frac{\sum_q e_q^2 x f_{1T}^{\perp q}(x) \otimes D_1^q(z)}{\sum_q e_q^2 x f_1^q(x) \cdot D_1^q(z)} \tag{1}$$

where x is the Bjorken variable, z is the fraction of the available energy carried by the hadron, and $\otimes$ indicates a convolution over the transverse momenta of the Sivers function $f_{1T}^{\perp}$ and of the fragmentation function D_1. In all those analyses, due to the presence of the convolution, some functional form had to be assumed for the transverse momentum dependence of the quark distribution functions and of the fragmentation functions. In most of the analyses these functions were assumed to be Gaussian. In this case the Sivers asymmetry becomes

$$A_{Siv,G}(x,z) = a_G \frac{\sum_q e_q^2 x f_{1T}^{\perp\,(1)\,q}(x) \cdot D_1^q(z)}{\sum_q e_q^2 x f_1^q(x) \cdot D_1^q(z)}\,. \tag{2}$$

where

$$f_{1T}^{\perp(1)}(x) = \int d^2\vec{k}_T \frac{k_T^2}{2M^2} f_{1T}^{\perp}(x, k_T^2) \tag{3}$$

and $a_G = \sqrt{\pi}M/\sqrt{\langle k_T^2\rangle_S + \langle p_\perp^2\rangle/z^2}$ and $\vec{p}_\perp$ the transverse momentum of the hadron with respect to the quark direction. The procedure allows to evaluate from the measured Sivers asymmetries the first moment of the Sivers function $f_{1T}^{\perp(1)}$, and not the Sivers function itself.

Already twenty years ago an alternative method was proposed [18, 19, 20] to determine $f_{1T}^{\perp(1)}$ without making any assumption on the functional form neither of the distribution functions nor of the fragmentation functions. The method consists in measuring asymmetries weighted by the measured transverse momentum of the hadron P_T^h as described in the next Section. For some reasons the method was not pursued: the only results (still preliminary) came from HERMES [21]. Recently, much interest has been dedicated again to the weighted asymmetries (see, e.g. [22]). In this contribution the first COMPASS results [23, 24] on the weighted Sivers asymmetries from the data collected in 2010 with the transversely polarised proton target are presented.

5. $P_T^h/(zM)$ weighted Sivers asymmetries

To extract the weighted asymmetries, only the spin-dependent part of the cross-section has to be weighted, leaving unweighted the unpolarised cross-section. In this section the weighting is done with $P_T/(zM)$, where M is the nucleon mass. After some algebra one gets the simple result that in the numerator of eq. (1) the convolution becomes the product of the first transverse moment of the Sivers function and the fragmentation function D_1, so that the weighted Sivers asymmetry is

$$A_{Siv}^w(x,z) = 2\frac{\sum_q e_q^2 x f_{1T}^{\perp\,(1)\,q}(x)\, D_1^q(z)}{\sum_q e_q^2 x f_1^q(x)\, D_1^q(z)}\,. \tag{4}$$

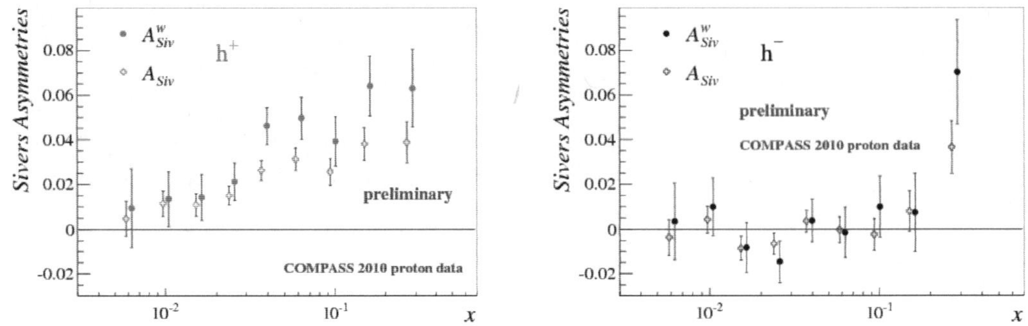

Figure 2. Full points: A^w_{Siv} in the nine x bins for positive (left panel) and negative (right panel) hadrons. The open crosses are the standard Sivers asymmetries A_{Siv} from Ref. [5].

Assuming u-dominance for positive hadrons produced on a proton target, the fragmentation function cancels out and the asymmetry simply becomes

$$A^w_{Siv}(x,z) \simeq 2\frac{f_{1T}^{\perp(1)\,u}(x)}{f_1^u(x)}\,. \tag{5}$$

Since the comparison with the standard Sivers asymmetry is also important, the data production and all the cuts to select the muons and the hadrons are the same as for the published data [5]. In particular, the selected phase space is defined by $0.004 < x < 0.7$, $Q^2 > 1$ GeV^2/c^2, $0.1 < y < 0.9$, $W > 5$ GeV/c^2, $P_T > 0.1$ GeV/c, and $z > 0.2$.

As shown in [24], the distributions of the weight factor $P_T/(zM)$ in the different x bins are very similar. Also, the P_T/z acceptance of the spectrometer is about 60% and rather flat. The results in the case of positive hadrons are given in Fig. 2 left, while Fig. 2 right gives the results for negative hadrons. In both cases the standard Sivers asymmetries published in [5] are also plotted for comparison. As expected, the trend of the asymmetries is similar both for positive and negative hadrons. Assuming u-dominance, the results for positive hadrons which are clearly different from zero in particular at large x, where $\langle Q^2\rangle$ reaches ~ 20 GeV^2/c^2, constitute the first direct measurement of $f_{1T}^{\perp(1)\,u}(x)/f_1^u(x)$.

Given the similar trend of the weighted asymmetries A^w_{Siv} and of the standard asymmetries A_{Siv}, we have evaluated in each x bin their ratios $R^w = A^w_{Siv}/A_{Siv}$. The values of R^w are given in Fig. 3 for positive hadrons, which exhibit a large Sivers asymmetry. In spite of the large statistical uncertainties the ratios are compatible with a constant value, which is rather well determined (1.6 ± 0.1). Also, the values in the different x bins agree rather well with those of $\langle P_T/(zM)\rangle$, as expected.

In Fig. 4 the weighted Sivers asymmetries measured in our "standard" z-range ($z > 0.2$) are compared with the corresponding ones in the range $0.1 < z < 0.2$. It is interesting to see that the positive hadron asymmetries are basically unchanged, hinting again at u-dominance, and supporting the idea that factorisation works also at small z in the COMPASS kinematical range. The negative hadron asymmetries receive contributions from both the u- and the d- quark, and become larger and similar to the positive hadron asymmetries. It is of interest to look at A^w_{Siv} as a function of z, after integration over x. The results in the range $0.1 < z < 1$ are shown in Fig. 5. For positive hadrons the values look constant, within the statistical uncertainties, as expected in the case of u dominance.

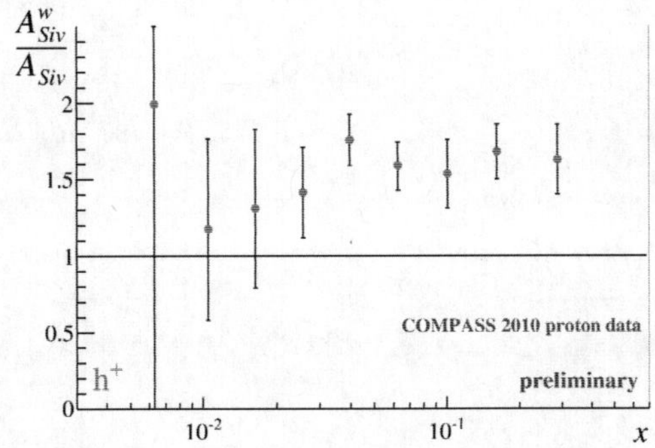

Figure 3. The ratios R^w between A^w_{Siv} and A_{Siv} in the nine x-bins for positive hadrons.

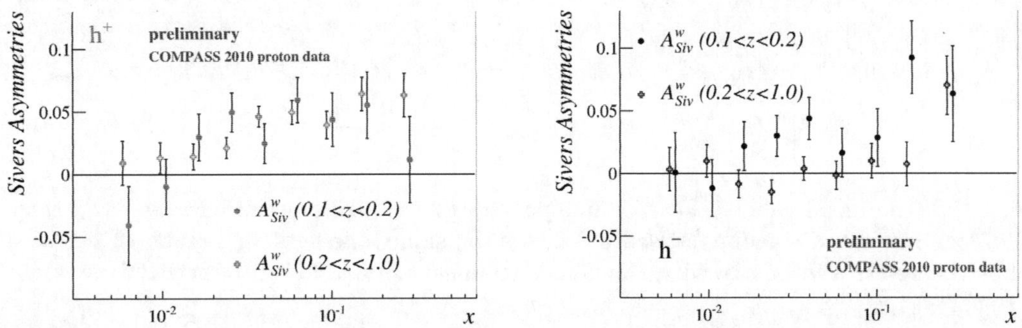

Figure 4. Comparison of the weighted asymmetries vs. x measured in our "standard" z range ($z > 0.2$) and the corresponding ones in the range ($0.1 < z < 0.2$) for positive (left) and negative (right) hadrons.

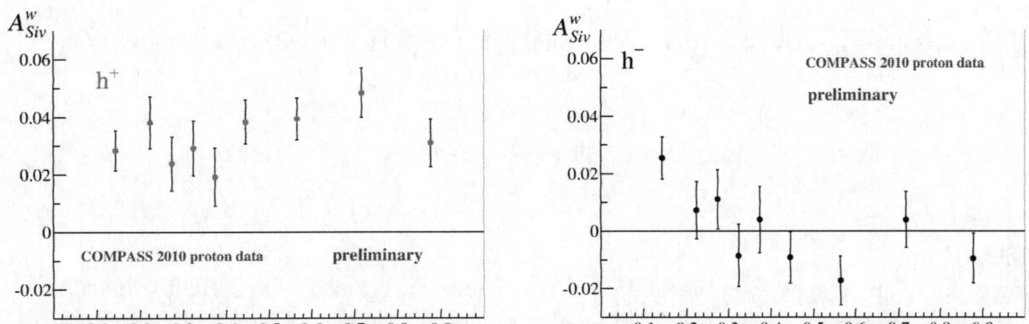

Figure 5. Weighted Sivers asymmetries vs z for positive (left) and negative (right) hadrons.

6. P_T^h/M weighted Sivers asymmetries

These weighted asymmetries are of interest because their values should exhibit a z dependence close to that of the Sivers asymmetries in the Gaussian model. Using as weight $w' = P_T^h/M$, eq.

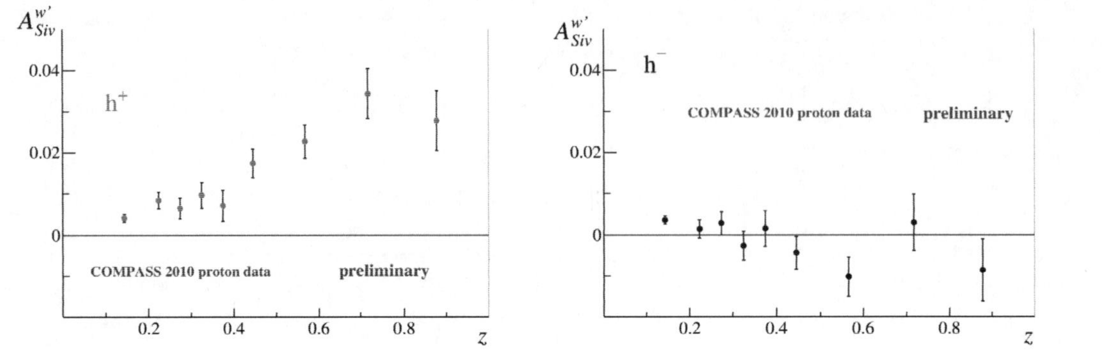

Figure 6. P_T^h/M weighted Sivers asymmetries vs z for positive (left) and negative (right) hadrons.

(4) becomes

$$A^{w'}_{Siv}(x,z) \quad = \quad 2\frac{\sum_q e_q^2 x f_{1T}^{\perp(1)\,q}(x)\, z D_1^q(z)}{\sum_q e_q^2 x f_1^q(x)\, D_1^q(z)} \tag{6}$$

By assuming that the mean value $< P_T^h >$ is independent on z in the measured range and that the slopes of the Sivers and the unpolarised PDFs are the same, one gets $a_G \simeq \pi M z/2 < P_T^h >$ [25] and thus the Sivers asymmetry with the Gaussian ansatz in eq.(2) can be written as

$$A_{Siv,G}(x,z) \quad = \quad \frac{\pi M}{2\langle P_T^h\rangle}\frac{\sum_q e_q^2 x f_{1T}^{\perp(1)\,q}(x)\, z D_1^q(z)}{\sum_q e_q^2 x f_1^q(x)\, D_1^q(z)}\,, \tag{7}$$

which is similar to the expression $A^{w'}_{Siv}(x,z)$ given in eq. (6). In particular it is $R_G^{w'} = A^{w'}_{Siv}/A_{Siv,G} = 4\langle P_T^h\rangle/\pi M \simeq 0.7$. The results for the x-integrated asymmetry $A^{w'}_{Siv}$ obtained using $w = P_T^h/M$ are shown in fig. 6 as a function of z for positive (left) and negative (right) hadrons. The values for positive hadrons are in qualitative agreement with the expectation:

$$A^{w'}_{Siv}(z) \quad = \quad 2\,z\,\frac{\int_{\Omega_x} dx C(x) x f_{1T}^{\perp(1)\,u}(x)}{\int_{\Omega_x} dx C(x) x f_1^u(x)} \tag{8}$$

7. Λ polarisation

The measurement of the transverse polarisation of Λ hyperons produced in SIDIS off transversely polarised nucleons has always been indicated as a promising channel to access transversity [26]. The basic idea is that, if transversity is different from zero, the polarisation of the fragmenting quark is transferred to the Λ according to the transversely polarised fragmentation function $H_1^{\Lambda/q}$, so that the Λ is polarised. Now we know that both the u- and the d- quark transversity are different from zero, so it is possible to check whether there is any transfer of polarisation from them to the lambda, and obtain information on $H_1^{\Lambda/q}$. Summing over all quark species, we get:

$$\vec{P}_\Lambda = \frac{\sum_q e_q^2 h_1^q(x) H_1^{\Lambda/q}(z)}{\sum_q e_q^2 f_1^q(x) D_1^{\Lambda/q}(z)} D_{NN} R \vec{P}_N \tag{9}$$

where R is the rotation about the normal to the scattering plane which brings the initial proton momentum along the Λ momentum [27]. Λ hyperons are a natural choice to investigate such transversity transmitted polarisation: their weak decay channel $\Lambda \to p\pi^-$ is in fact self-analysing, that is, the polarisation is revealed through the angular asymmetry of the decay proton:

$$\frac{dN}{dcos\theta} \propto 1 + \alpha P_\Lambda \cos\theta \qquad (10)$$

where α is the Λ weak decay parameter and θ is the angle between the proton momentum and the polarisation vector, assumed to be coincident with that of the outgoing quark.

Studies of Λ transversity transmitted polarisation were already carried out in COMPASS using 2002-2004 deuteron data and 2007 proton data, but the results have never been published. Now the analysis has been resumed and extended to the proton data collected in 2010. Since some aspects of the analysis are new, the 2007 data have also been re-analysed and results are presented here for the whole set of proton data. The $\bar{\Lambda}$ polarisation has also been extracted from the complete proton data set.

In Fig. 7 both the Armenteros-Podolanski plots and the mass distribution for Λ candidates from 2010 data are shown after all cuts. The results for the polarisation are given in Fig 8 as a function of x. All the values are compatible with zero.

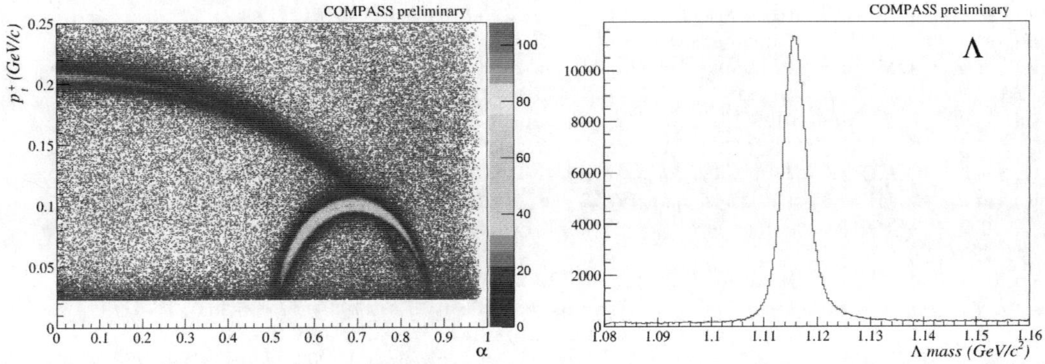

Figure 7. Armenteros plot after all but mass cuts and invariant mass spectrum for Λ candidates.

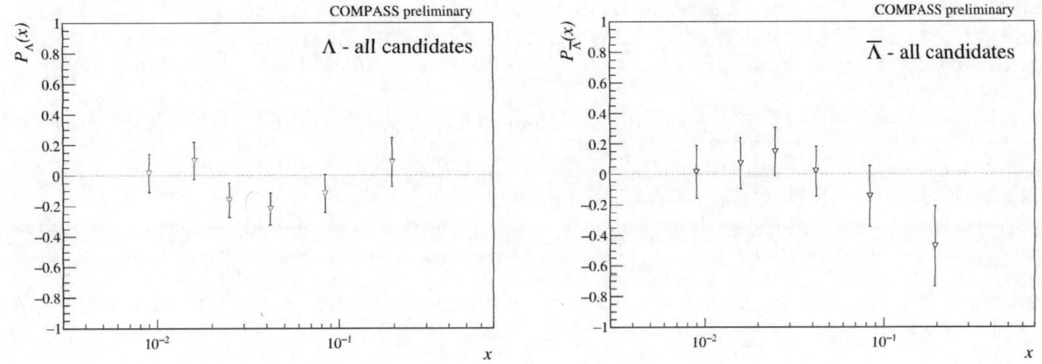

Figure 8. Polarisations for all Λ (left) and $\bar{\Lambda}$ (right) candidates as function of x.

8. Future SIDIS measurements

The Collaboration has submitted to the CERN SPS Committee a proposal for an extension of the present COMPASS-II experimental programme to perform in the years 2021 and 2022 two new measurements, muon SIDIS on a transversely polarised deuteron target and elastic muon proton scattering [28]. The first program is the missing piece in the COMPASS data sets on transverse target spin orientations. While in 2010 a good statistics sample of SIDIS data on transversely polarised proton target was collected, in the years 2002, 2003 and 2004, due to the late delivery of the new polarised target magnet and to the shortness of the periods of data taking, we provided only a marginal (albeit unique) data set for the isoscalar deuteron target. With one additional year of deuteron target data taking accurate flavor separation for the new PDFs will be possible, improving in particular our knowledge of the d-quark distribution functions. The second measurement hopefully will provide a decisive contribution to the solution of the so-called proton radius puzzle.

References

[1] Contributions of Y. Bedfer, of A. Ivanov, of B. Marianski, of A. Sandacz, of M. Pesek, of J. Matousek, and of N. Mitrofanov to this workshop
[2] Airapetian A *et al.* (HERMES) 2005 *Phys. Rev. Lett.* **94** 012002 (*Preprint* `hep-ex/0408013`)
[3] Airapetian A *et al.* (HERMES) 2009 *Phys. Rev. Lett.* **103** 152002 (*Preprint* `0906.3918`)
[4] Adolph C *et al.* (COMPASS) 2012 *Phys. Lett.* **B717** 376–382 (*Preprint* `1205.5121`)
[5] Adolph C *et al.* (COMPASS) 2012 *Phys. Lett.* **B717** 383–389 (*Preprint* `1205.5122`)
[6] Adolph C *et al.* (COMPASS) 2014 *Phys. Lett.* **B736** 124–131 (*Preprint* `1401.7873`)
[7] Sivers D W 1990 *Phys. Rev.* **D41** 83
[8] Collins J C 1993 *Nucl. Phys.* **B396** 161–182 (*Preprint* `hep-ph/9208213`)
[9] Alexakhin V Yu *et al.* (COMPASS) 2005 *Phys. Rev. Lett.* **94** 202002 (*Preprint* `hep-ex/0503002`)
[10] Ageev E S *et al.* (COMPASS) 2007 *Nucl. Phys.* **B765** 31–70 (*Preprint* `hep-ex/0610068`)
[11] Adolph C *et al.* (COMPASS) 2016 *Phys. Lett.* **B753** 406–411 (*Preprint* `1507.07593`)
[12] Artru X 2002 *10th Rhodanien Seminar: The Spin in Physics Turin, Italy, March 3-8, 2002* (*Preprint* `hep-ph/0207309`)
[13] Alekseev M G *et al.* (COMPASS) 2010 *Phys. Lett.* **B692** 240–246 (*Preprint* `1005.5609`)
[14] Efremov A V, Goeke K and Schweitzer P 2003 *Phys. Lett.* **B568** 63–72 (*Preprint* `hep-ph/0303062`)
[15] Vogelsang W and Yuan F 2005 *Phys. Rev.* **D72** 054028 (*Preprint* `hep-ph/0507266`)
[16] Collins J C, Efremov A V, Goeke K, Menzel S, Metz A and Schweitzer P 2006 *Phys. Rev.* **D73** 014021 (*Preprint* `hep-ph/0509076`)
[17] Anselmino M, Boglione M and Melis S 2012 *Phys. Rev.* **D86** 014028 (*Preprint* `1204.1239`)
[18] Kotzinian A M and Mulders P J 1996 *Phys. Rev.* **D54** 1229–1232 (*Preprint* `hep-ph/9511420`)
[19] Kotzinian A M and Mulders P J 1997 *Phys. Lett.* **B406** 373–380 (*Preprint* `hep-ph/9701330`)
[20] Boer D and Mulders P J 1998 *Phys. Rev.* **D57** 5780–5786 (*Preprint* `hep-ph/9711485`)
[21] Gregor I M (HERMES) 2005 *Acta Phys. Polon.* **B36** 209–215
[22] Kang Z B, Vitev I and Xing H 2013 *Phys. Rev.* **D87** 034024 (*Preprint* `1212.1221`)
[23] Sbrizzai G (COMPASS) *QCD-N 4rd Workshop on the QCD Structure of the Nucleon, Getxo, Bilbao, Spain, 11-15 July, 2016.*
[24] Bradamante F (COMPASS) 2017 *22nd International Symposium on Spin Physics (SPIN 2016) Urbana, IL, USA, September 25-30, 2016* (*Preprint* `1702.00621`)
[25] Martin A, Bradamante F and Barone V 2017 *Phys. Rev.* **D95** 094024 (*Preprint* `1701.08283`)
[26] Artru X and Mekhfi M 1991 *Nucl. Phys.* **A532** 351–358
[27] Artru X 1993 *QCD and high-energy hadronic interactions. Proceedings, Hadronic Session of the 28th Rencontres de Moriond, Moriond Particle Physics Meeting, Les Arcs, France, March 20-27, 1993* pp 47–52
[28] The COMPASS Collaboration, Addendum to the COMPASS-II Proposal,CERN-SPSC-2017-034 ; SPSC-P-340-ADD-1. October 2, 2017.

IOP Conf. Series: Journal of Physics: Conf. Series **938** (2017) 012005 doi:10.1088/1742-6596/938/1/012005

Dp breakup reaction investigation using polarized and unpolarized deuteron beam

M Janek[1], V P Ladygin[2], Yu V Gurchin[2], A Yu Isupov[2], J-T Karachuk[3], A N Khrenov[2], A K Kurilkin[2], P K Kurilkin[2], A N Livanov[2], G Martinska[4], S M Piyadin[2], S G Reznikov[2], G Tarjanyiova[1], A A Terekhin[2] and I E Vnukov[5]

[1]Physics Department, University of Zilina, Univerzitna 1, 01001 Zilina, Slovakia
[2]Veksler and Baldin Laboratory of High Energies, Joint Institute for Nuclear Research, Joliot-Curie 6, 141980 Dubna, Moscow region, Russia
[3]Advanced Research Institute for Electrical Engineering, Splaiul Unirii 313, Bucharest, Romania
[4]Institute of Physics, University of P. J. Shafarik, Park Angelinum 9, 04001 Kosice, Slovakia
[5]Belgorod State University, Pobedy 85, 308015 Belgorod, Russia

E-mail: `janek@fyzika.uniza.sk`

Abstract. Deuteron Spin Structure (DSS) collaboration program is aimed on few nucleon correlations investigation using unpolarized and polarized deuteron beam at intermediate energy range. Data of the *dp* breakup reaction have been obtained at energy range from 300 - 500 MeV of deuteron energy for various detector configurations in region where few nucleon correlations and relativistic effects can play significant role. Results are obtained in the form of S-curve and T_{20} and iT_{11} tensor analyzing powers in case of unpolarized and polarized data, respectively. Future plans on extracted deuteron beam of Nuclotron are concerned to tensor analyzing power T_{20} and spin correlation C_{yy} of $^3He(d,p)^4He$ reaction in the kinetic energy range between 1.0 and 1.75 GeV.

1. Introduction

The most used contribution in theoretical models which include three nucleon forces (3NFs) are two pion exchange, e.g. Urbana IX [1] and Tucson-Melbourne [2]. Models can describe binding energies of three and four nucleon systems. Generally, cross section of reactions with few nucleons in which is nucleon unpolarized is better described than for the polarized case. Much worst situation is when the beam or target is polarized, especially in case of polarization observables measurement. Spin structure of the *np* SRCs has been investigated via the measurements of the tensor analyzing power A_{yy} in deuteron inclusive breakup at various energies in the wide regions of the longitudinal x_F and transverse proton momentum p_T [3]. The A_{yy} data demonstrate the dependence on two internal variables, x_F and p_T or their combinations. However none of used approach [4],[5] describe the data. In the vicinity of the Sagara discrepancy the currently known 3NFs contribute by up to 30% to the *dp* elastic scattering cross section at intermediate energies [6]. The investigation of *dp* breakup reaction at deuteron energy of 270 MeV at RIKEN [7] and IUCF [8] reveals that vector analyzing power A_y can be described using NN forces only. But other polarization observables need 3NFs to describe the data. *Dp* breakup reaction investigation measured at KVI at energy of 130 MeV [9] shows similar behavior. Inclusion of 3NFs improve the description of a part of the data but breaks other. Relativistic effects for the *nd* elastic scattering cross section at 70 MeV and 250 MeV were investigated in [10]. It was found that relativistic effects contribution is located mainly at backward angles, but their contribution is not large enough to fill discrepancy between experimental data and theory, even in the case when standard three nucleon forces [11] are used. For the d(n,np)n breakup reaction at 200 MeV large relativistic effects were observed in configuration where one arm is fixed and second arm scans the angular range. The contribution which comes from relativistic effects can reach up to 60%.

2. Deuteron breakup reaction investigation using unpolarized and polarized beam

The goal of the Deuteron Spin Structure (DSS) experimental program is to obtain the information about two and three nucleon forces, including their spin dependent parts, from *dp* elastic scattering at the energies between 300 – 2000 MeV and *dp* breakup reactions with registration of two protons at deuteron energies of 300 500 MeV [12].

Deuterons are accelerated by Nuclotron up to required energy. *Dp* breakup reaction has been investigated at 300 MeV, 400 MeV and 500 MeV of deuteron energy. Polyethylene and Carbon targets are enclosed in a spherical hull of Internal Target Station (ITS) [13]. Up to six various targets can be placed inside of ITS. The *dp* breakup reaction is investigated by the simultaneous registration of two protons by two detectors – the two detectors work in coincidence. Up to eight $\Delta E - E$ detectors can be used in experiment. Effect on protons is obtained by subtracting Carbon spectra from Polyethylene one. Various detector setups were used for *dp* breakup reaction investigation. Details of the $\Delta E - E$ detector construction can be found in [14]. Reasonable agreement was obtained in comparison of energy spectra and missing mass distributions of *dp* breakup reaction between experiment and GEANT4 simulation which give us opportunity to handle with the signal and background in more flexible way [15]. The stability of amplitude of all photomultiplier tubes (PMTs) were monitored during all data acquisition. Mean values of LED amplitudes have been obtained at 200 MeV/n data taking. All PMT amplitudes from calibration and physical data obtained on Polyethylene and Carbon targets were corrected by multiplied factor defined as a ratio between mean LED amplitude value at 200 MeV/n to mean LED amplitude one between two closest LED runs in time to calibration or physical ones. By this procedure possible temperature dependence of ΔE detectors was suppressed. Detailed description of LED system of PMTs can be found in [14]. Detectors used for *dp* breakup investigation at intermediate energy range from 300 to 500 MeV have been calibrated using following procedure. Amplitude and time information from all PMTs are used to reconstruct amplitude spectra of ΔE detectors. Assuming linear dependence of amplitudes on particle energy, calibration coefficients for particular ΔE and E detector is obtained by solving of the set of linear equation. For this purpose detectors have been placed under *pp* quasi elastic kinematics at 300 MeV, 400 MeV and 500 MeV where energies can be calculated from kinematics. Good agreement between experimental and simulated ΔE vs. E plot is observed. Obtained missing mass and energy spectra were compared with results obtained form GEANT4. Satisfactorily agreement is obtained, see Fig.1. Missing mass spectra obtained by subtracting of Carbon content from Polyethylene one at angles of 31° and 43° for the particles which are stopped in detector at deuteron energy of 200 MeV/n is shown in Fig.2. Energy resolution of the detector can be found with gaussian fit. Obtained mean value of the missing spectra (941 MeV) is in agreement with neutron missing mass, with energy resolution of 11 MeV. Detailed information about calibration procedure can be found in [16].

The largest part of *dp* breakup data obtained at 200 MeV/n in 2014 follows inverse kinematics of the reaction in which relativistic effects can play a role [10]. The *dp* breakup data have been obtained under condition when one arm is fixed and and second scans the angular interval allowed

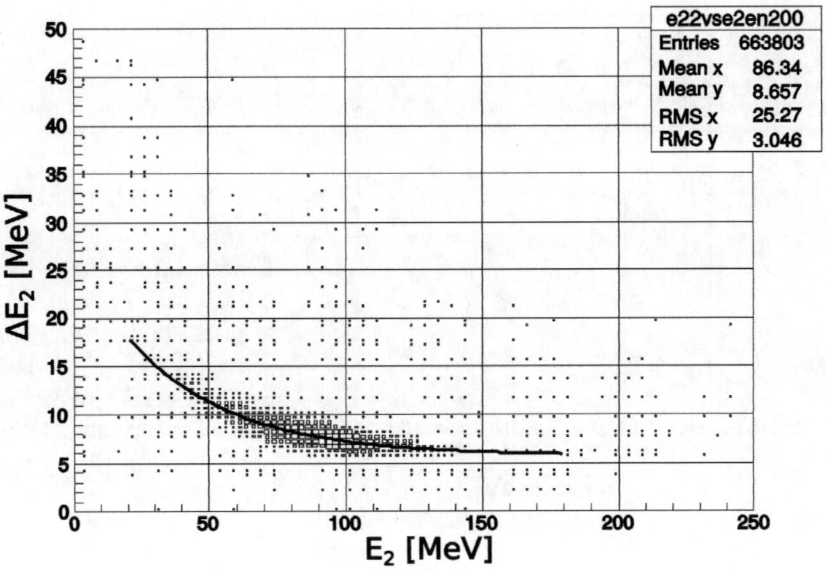

Figure 1. $\Delta E - E$ correlation obtained for one of arm in the *pp* quasi elastic kinematics at 90° cm and deuteron energy of 200 MeV/n. Curve represents the $\Delta E - E$ correlation obtained from GEANT4 simulation [17] at the same kinematic configuration.

by mechanics. Relativistic effects can play more significant role in the cross section under this condition as it was investigated in [10]. In our case one arm was fixed at polar angle (θ_2) of 43.0° and second was placed at angle (θ_1) of 27.0°, 31.0°, 35.0°, 39.0° and 43.0°. The kinematic variable S corresponds to the arc-length along the kinematic curve with zero point chosen closest to axis origin. Points of kinematic S-curve are calculated with energy step of 5 MeV for the detector arms placed at polar angles of $\theta_1 = 39.0°$, $\theta_2 = 43.0°$ and azimuthal angle between arms of 179.2° at deuteron energy of 400 MeV. The S is increasing in anti-clockwise direction. The number of breakup events in an interval $S - \Delta S/2$, and $S + \Delta S/2$ was obtained by projecting the events on a line perpendicular to the S-curve. Preliminary energy distribution of deposited energy and S distribution (with the step of 5 MeV) in the first (E_1) and second arm (E_2) placed

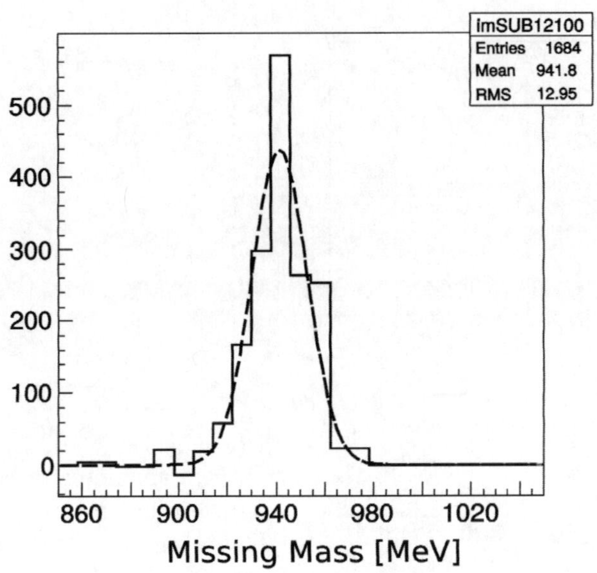

Figure 2. Missing mass spectra with gauss fit (941 ± 11 MeV) obtained by subtracting of Carbon content from Polyethylene one at angles of 31° and 43° for the particles which are stopped in detector at deuteron energy of 200 MeV/n.

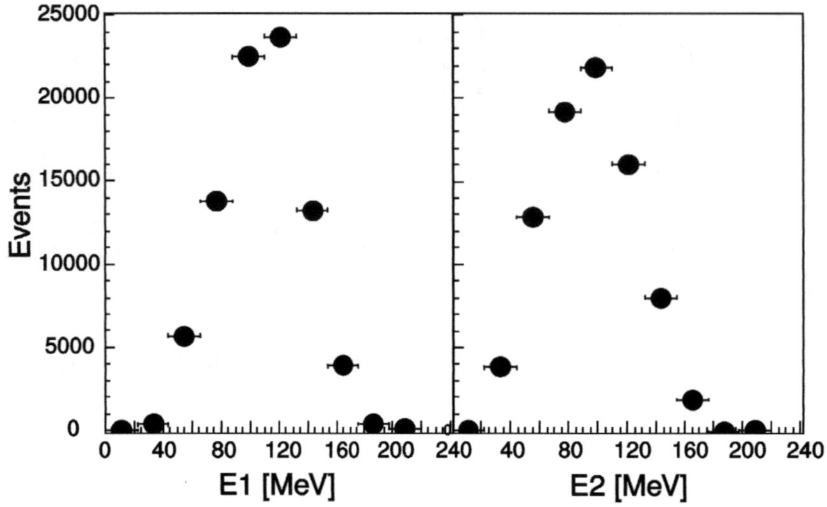

Figure 3. Energy distribution of deposited energy in the first (E_1) and second arm (E_2) placed at polar angles of 39° and 43°, respectively. Difference in azimuthal angles is 179.2°.

at polar angles of 39° and 43° are shown in Fig.3 and Fig.4, respectively. Difference in azimuthal angles is 179.2°. The energy resolution is taken from the detector calibration.

Polarization of deuteron beam is provided by new polarization ion source (PIS). Ideal values of polarizations were $p_z, p_{zz} = (-1/3, \pm 1)$. PIS can provide unique opportunity for the studies of the spin effects and polarization phenomena in few body systems at ITS and also on extracted beam with the BM@N setup. Commissioning of the upgraded DSS setup has been performed at 270 MeV using unpolarized and polarized deuteron beam in June-July 2016. Good values of vector and tensor polarizations [18] were obtained at low energy polarimeter. The polarization was flipped each spill between "spin-up" and "spin-down" (perpendicular to the plane of the accelerator) and no polarization. Polarization of deuteron beam was monitored continuously during whole experiment. Vector and tensor polarizations are obtained from the asymmetries and known values of analyzing powers of dp elastic reaction at energy of 270 MeV. Measured values of polarization for "+ mode" when tensor component is positive are: $p_z = -0.190 \pm 0.009$, $p_{zz} = 0.533 \pm 0.017$ and for "- mode": $p_z = -0.230 \pm 0.007$, $p_{zz} = -0.705 \pm 0.013$. Statistical

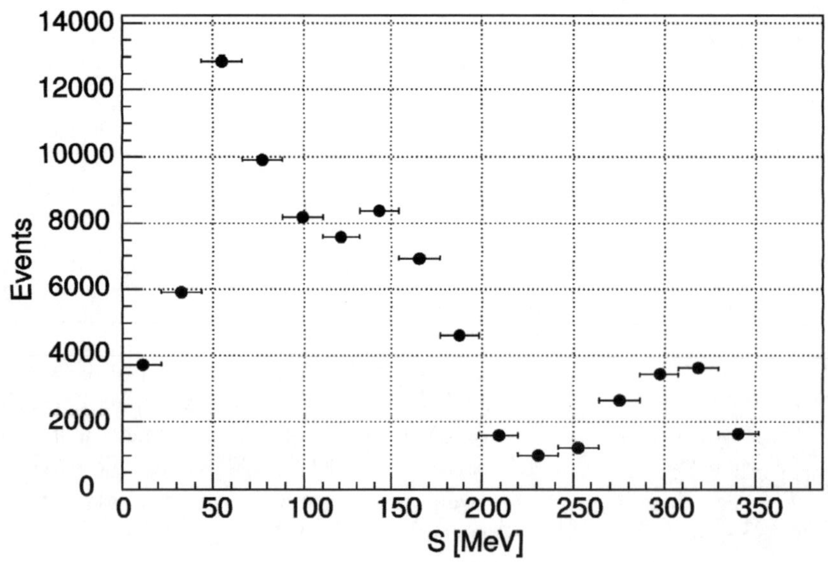

Figure 4. S distribution for the detector configuration of 39° and 43°. Difference in azimuthal angles is 179.2°.

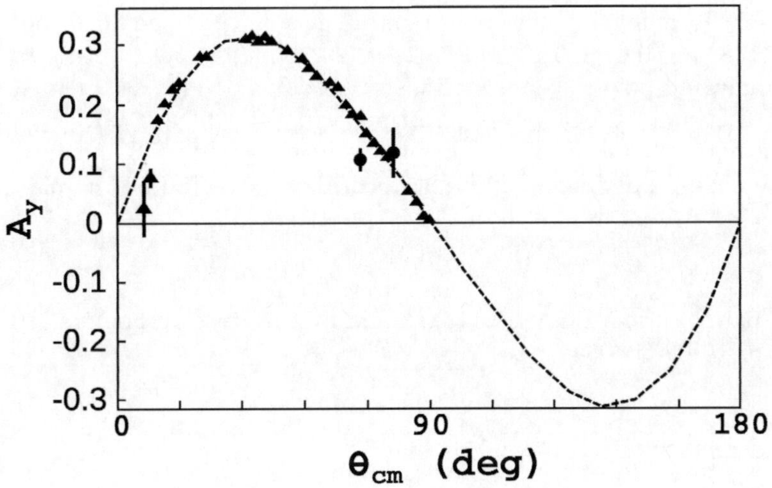

Figure 5. Angular dependence of the vector analyzing power A_y at energy of 200 MeV/n. Data obtained at Nuclotron at 72.3° and 76.5° in cm are represented by full circle symbols, triangle symbols represent data from [19].

and systematical errors are taken into account. Statistical error comes from ratio of *dp* elastic to *pp* quasi elastic events.

Analyzing power iT_{11} at 72.3° and 76.5° in cm was measured under pp quasi conditions. Obtained values at 72.3° and 76.5° are 0.10 ± 0.02 and 0.11 ± 0.06, respectively. Results are in agreement with world *pp*- elastic scattering data within experimental errors , see Fig.5. Values of the vector iT_{11} and tensor T_{20} analyzing powers at polar angles of 34.8° and 36.8° and difference in azimuthal angles of 135° are 0.47 ± 0.10 and 0.02 ± 0.20 [20].

3. $d^3He \to p^4He$ reaction investigation at Nuclotron

Polarization correlation coefficient $C_{//}$ for the $d^3He \to p^4He$ reaction can be a unique probe to the D-state admixture in deuteron in one nucleon exchange approximation [21]. Measurement of tensor analyzing power T_{20} can give the key for understanding its behaviour in *dp*- backward elastic scattering. The possibility to perform the $d^3He \to p^4He$ reaction at extracted beam of Nuclotron at BM@N area was investigated using GEANT4 and ROOT simulation package at energy of 1.5 GeV. Proposed experiment is assumed to be performed in energy range from 1.0 – 1.75 GeV. The simulation [22] has been performed for the initial deuteron kinetic energy $T_d = 1.5$ GeV and 30 cm of the ^3He target cell. In simplified model deuteron beam hits the ^3He target which is placed in the one of the focuses of the beam transportation line of Nuclotron. Quartz radiator which can serve as a TOF start counter spreads the size of beam and the number of scattered particles in the secondary beam due to additional scattering on it in comparison with the option without TOF, however, the number of scattered events is quite small for both cases. The further simulation used quartz radiator. The momentum reconstruction of particles is based on the information derived from 12 GEM stations placed in the inhomogeneous magnetic field defined by a field map. From obtained result follows that protons passed through 12 GEM tracker stations can be tracked from 0.4 T to maximal magnetic field of 0.9 T. Simplified BM@N setup with 12 stations of GEM tracker situated in the magnet can provide the momentum resolution of few percents. The time difference between mRPC and start detector is about 20 ns which is good enough to separate protons from the background deuterons.

4. Conclusion

Energy and S distributions of *dp* breakup reaction obtained in coplanar kinematics at 200 MeV/n were discussed. Obtained values of vector analyzing power at 72.3° and 76.5° in cm measured under *pp* quasi conditions are in agreement with world data within experimental accuracy.

Values of the vector iT_{11} and tensor T_{20} analyzing powers at polar angles of 34.8° and 36.8° and difference in azimuthal angles of 135° are 0.47 ± 0.10 and 0.02 ± 0.20, respectively. In order to substantially decrease errors of analyzing powers beam luminosity needs to be increase at least 10 times. Study of $^3He(d,p)^4He$ reaction at 1.5 GeV was performed using ROOT and GEANT4 package. Suitability of experiment performed on extracted beam was investigated using maximal and one half of maximal magnetic field. Experiment can be performed at extracted beam under investigated conditions.

Acknowledgments
The work has been supported in part by the VEGA 1/0113/18 and by the Russian Foundation for Basic Research under grant no. 16-02-00203a.

References

[1] Rauprich G et al. 1991 *Nucl. Phys.* A **535** 313
[2] Patberg H et al. 1996 *Phys. Rev.* C **53** 1497
[3] Azhgirey L S et al. 1996 *Phys.Lett.* B **387** 37
[4] Machleidt R 2001 *Phys. Rev.* C **63** 024001
[5] Karmanov V A and Smirnov A V 1992 *Nucl.Phys.* A **546** 691
[6] Sakamoto N et al. 1996 *Phys. Lett.* B **367** 60
[7] Sekiguchi K et al. 2009 *Phys. Rev.* C **79** 054008
[8] Meyer H O et al. 2004 *Phys. Rev. Lett.* **93** 112502
[9] Kistyn S et al. 2005 *Phys. Rev.* C **72** 044006
[10] Witala H et al. 2011 *Few-Body Syst* **49** 61
[11] Coon S A, Han H K 2001 *Few-Body Syst* **30** 131
[12] Ladygin V P et al. 2011 *J. Phys. Conf. Ser.* 012131 295
[13] Malakhov A I et al. 2000 *Nucl. Instr. Meth. in Phys. Res.* A **440** 320
[14] Piyadin S M et al. 2011 *Phys. Part. Nucl. Lett.* V. 8 no. 2 107
[15] Janek M et al. 2014 *Physics of Particles and Nuclei Letters* no. 4 552
[16] Janek M et al. 2018 *Particles and Nuclei, Letters* Vol.15 No.1
[17] Janek M, Trpisova B, Ladygin V P, Piyadin S M 2014 *Physics of Particles and Nuclei Letters* V. 11 no. 4. 552
[18] Janek M et al. 2017 *Few-Body Systems* Vol. 58 Issue 2 40
[19] Rahm S et al. 1998 *Phys. Rev.* C **57** 1077
[20] Janek M et al. 2017 *Communications: Scientific letters of the University of Zilina* 3 62
[21] Uesaka T et al. 1999 *Phys.Lett.* B **467** 199
[22] Janek M et al. 2016 *Journal of physics: conf. series* 678 012032

Simultaneous measurements of spin observables A_N and A_{NN} in elastic *pp* scattering (extension of the SPASCHARM program at U-70).

V V Abramov[1], A A Bogdanov[2], M A Chetvertkov[3], V A Chetvertkova[4], V V Mochalov[1,2], V V Moiseev[1], K D Novikov[1,2], S B Nurushev[1,2], M B Nurusheva[2], V A Okorokov[2], P A Semenov[1,2], M N Strikhanov[2], A N Vasiliev[1,2]

[1]NRC «Kurchatov Institute» - IHEP, Protvino, Moscow region, 142281, Russia
[2]National Research Nuclear University MEPhI (Moscow Engineering Physics Institute), Moscow, 115409, Russia
[3]MD Anderson Cancer Center, Department of Radiation Physics, 1515 Holcombe Blvd, Unit 1420, Houston, TX 77030, USA
[4]GSI, Plankstrasse 1, 64291, Darmstadt, Germany

E-mail: mbnurusheva@mail.ru

Abstract. We propose to measure spin observables A_N and A_{NN} in elastic pp scattering by using the transversely polarized proton beam and target at momenta p = 12-45 GeV/*c*. Existence of both polarized target and beam gives us unique possibility to measure A_N simultaneously and independently for polarized beam (A_B) and target (A_T) to carry out and verify experimental measurements of single-spin and double-spin (A_{NN}) measurements in diffractive region.

1. Introduction

Spin correlation parameter A_{NN} was measured earlier in diffractive region at the momentum less than 12 GeV/*c* [1,2,3]. We propose to measure analyzing power A_N for both polarized beam and target and the initial state spin-spin correlation parameter A_{NN} in proton-proton elastic scattering at -t from 0.075 to 0.6 (GeV/*c*)^2 at the new IHEP beam line 24A in the momenta range12-45 GeV/*c* [4].

The results of this experiment can be described by a set of spin parameters by the formula [5]:

$$\left[\frac{d\sigma}{d\Omega}\right]_{ij} = \left[\frac{d\sigma}{d\Omega}\right]_0 [1 + P_B(ij)A_B + P_T(ij)A_T + P_B(ij)P_T(ij)A_{nn}] \quad (1),$$

where P_B and P_T are the beam and target polarizations, A_B and A_T are the beam and target analyzing powers[1]. The A_{NN}, A_B, and A_T can be expressed by equations (2-4), respectively:

$$A_{NN} = \frac{1}{P_B P_T} \frac{N(\Uparrow\uparrow) - N(\Uparrow\downarrow) - N(\Downarrow\uparrow) + N(\Downarrow\downarrow)}{N(\Uparrow\uparrow) + N(\Uparrow\downarrow) + N(\Downarrow\uparrow) + N(\Downarrow\downarrow)} \quad (2),$$

$$A_B = \frac{1}{P_B} \frac{N(\Uparrow\uparrow) + N(\Uparrow\downarrow) - N(\Downarrow\uparrow) - N(\Downarrow\downarrow)}{N(\Uparrow\uparrow) + N(\Uparrow\downarrow) + N(\Downarrow\uparrow) + N(\Downarrow\downarrow)} \quad (3),$$

$$A_T = \frac{1}{P_T} \frac{N(\Uparrow\uparrow) - N(\Uparrow\downarrow) + N(\Downarrow\uparrow) - N(\Downarrow\downarrow)}{N(\Uparrow\uparrow) + N(\Uparrow\downarrow) + N(\Downarrow\uparrow) + N(\Downarrow\downarrow)} \quad (4),$$

[1] A_B and A_T are the same values in the physical sense and corresponds to different polarized objects (beam or target)

IOP Conf. Series: Journal of Physics: Conf. Series **938** (2017) 012006 doi:10.1088/1742-6596/938/1/012006

where N(ij) is the number of events normalized to the incident proton flux with the beam's (i) and target's (j) spin direction up (↑) or down (↓). Our main goal is to determine the momentum dependence of the spin-correlation parameters in the range of 12-45 GeV/*c* in the diffractive kinematic region.

2. Existing experimental data of A_N and A_{NN} in proton-proton elastic scattering

A_N is fairly well measured at various energies (see figure 1 which was presented originally in Ref. [6]). The solid line is the fitting function with poles, suggested by the Regge model, namely, $A_N = a1 + a2/\sqrt{p_{beam}} + a3/p_{beam}$ [7].

The measured value of A_N at 45 GeV/*c* is equal to (2.3±0.2) % [8], and this value is used here for estimating the required beam time. The experimental data on the spin correlation parameter A_{NN} in the diffractive region are very scarce. This parameter has been measured only at the momenta of 3.3; 6; 11.75 GeV/*c* [1,2,3] in laboratory system (see figure 2). The solid line presents the fit to experimental data with the function $A_{NN} = a(p/p_0)^b{}_{beam}$, where a=(57±23)% and b= – 0.86±0.33, and p_0=1 GeV/*c*. The expected value for A_{NN} at 45 GeV/*c* is (2.2±1.1)%.

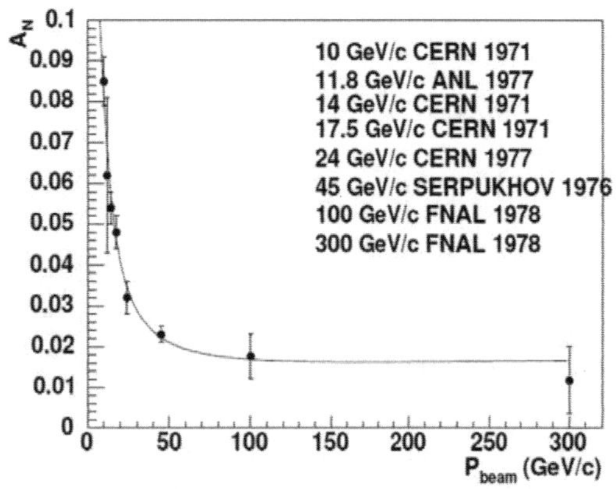

Figure 1. A_N for pp elastic scattering at $-t$=0.15 (GeV/*c*)^2 as a function of beam momentum.

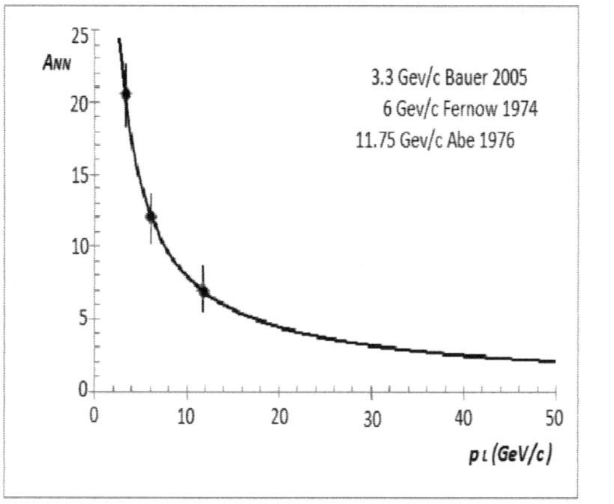

Figure 2. A_{NN} (in %) for pp elastic scattering at $-t$=0.6 (GeV/*c*)^2 as a function of beam momentum.

3. Beam characteristics and beam polarization measurement

The polarized proton and antiproton beams are described in detail in Ref. [9]. We plan to carry out these measurements at the polarized-proton beam at p_b= 12-45 Gev/*c* with the narrowest possible momentum band Δp/p=1.2% [9], striking the polarized target of a diameter 18 mm [10]. The beam properties are as follows [9]: an average beam polarization P_B=40%, transverse dimensions $\sigma(x)\times\sigma(y)$ = 11×8.7 mm^2, angular divergences σ(x')×σ(y)=1.4×1.7 mrad^2, intensity I=5×10^6 protons per cycle hitting the target.

The transverse polarization dependence on the beam-ray position along the vertical axis in the intermediate focus is presented in figure 3 [9]. It can be described by the function $\langle P_N \rangle$=ay+by3, where a=(0.025 ± 0.027) mm^{-1}, b= -(0. 68 ±0.19) × 10^{-5} mm^{-3}. The beam-tagging system will be used for measuring the momentum of each particle at the accuracy of about 1% and the polarization accuracy of few percent [11].

4. Experimental setup

The design of the experimental setup is presented in figure 4. The main detectors to be used are the trigger scintillator counters (S1-S8) and hodoscopes (H1-H8).

The parameters for all hodoscopes are presented in table 1. The hodoscopes are constructed in such a manner that each scintillator strip overlaps its two adjacent neighbors by one-third of the width. This kind of overlapping arrangement provides twice as many of the spatial segments for the same number of channels of encoding electronics as for the edge-butted array. It also removed the loss of tracking efficiency by eliminating the inter-segment gaps.

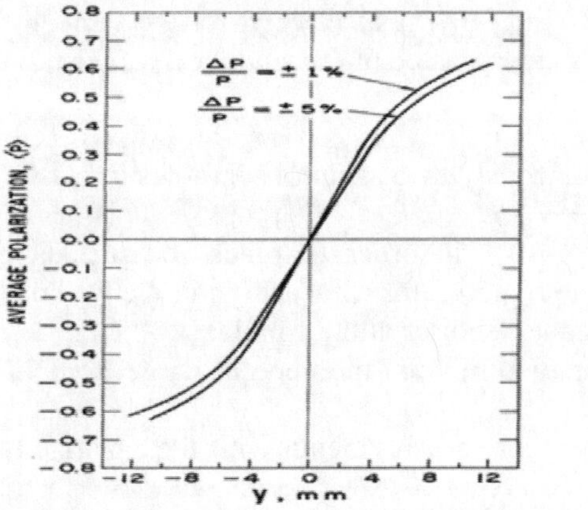

Figure 3. Correlation between the average vertical transverse polarization of protons and the vertical ray coordinate y in the plane of intermediate focus.

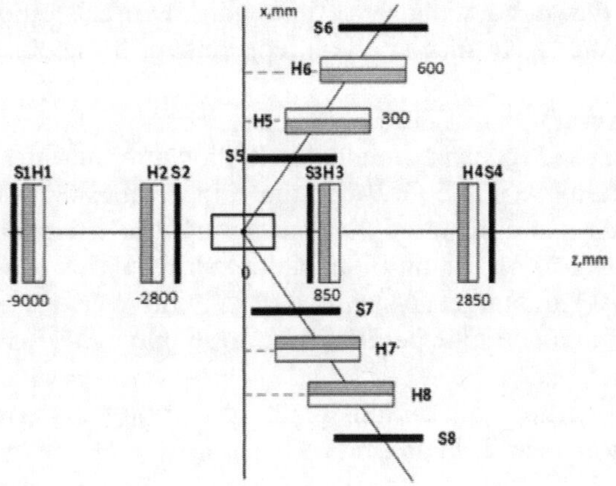

Figure 4. Design of the experimental setup.

In order to suppress inelastic events, a special fast trigger will be organized on the base of coincidences between scattered and recoil protons in forward (H3-H4) and backward (H5-H8) detectors. It worth mentioning, that elastic scattering measurements can be carried out simultaneously with the inclusive double-spin studies, using the main detectors of SPASCHARM experiment.

Table 1. Design of the hodoscopes

Hodoscope	Distance from target (mm)	Total dimensions (mm)	Dimension of scintillator strips: Width × Thickness × Length (mm^3)		Numbers of channels	
	Z	X × Y	X	Y	X	Y
		Beam	Hodoscopes			
H1	-9000	40 × 40	6 × 3 × 40	6 × 3 × 40	9	9
H2	-2800	40 × 40	6 × 3 × 40	6 × 3 × 40	9	9
		Forward	Hodoscopes			
H3	850	62 × 45	6 × 3 × 62	6 × 3 × 45	15	11
H4	2850	182 × 86	6 × 3 × 182	6 × 3 × 86	45	21
		Recoil	Hodoscopes			
	X	Z × Y	Z	Y	Z	Y
H5 - H7	300	322 × 228	9 × 5 × 322	9 × 5 × 228	51	37
H6 - H8	600	432 × 430	9 × 5 × 432	9 × 5 × 420	71	69

5. Beam time estimation

Numerical calculations for the number of elastic scattering events has been done at 45 Gev/c and for the worst case scenario, i.e. at the maximum value of $|t|= 0.6$ (GeV/c)^2 and $\Delta t=\pm 0.1$ (GeV/c)^2, where the cross-section is the smallest. For measuring the t-dependence of A_{NN}, the following input parameters have been assumes in simulations: the main beam characteristics as they are described in Sec. 3, the beam duty factor $\nu=0.1$, the polarized target proton density $\rho = 0.12$ g/sm^3, the vertical component of magnetic field in the frozen polarized target zone 0.45 T, the target polarization $P_T = (75\pm 5)$ %. These correspond to the luminosity $L=10^{28}$ sm^{-2}s^{-1} [9].

IOP Conf. Series: Journal of Physics: Conf. Series **938** (2017) 012006 doi:10.1088/1742-6596/938/1/012006

An event rate is defined as $n=L\ \Delta t\ d\sigma/dt\ \Delta\phi/\phi$, where $d\sigma/dt$ =0.23 mb/(GeV/c)^2 [12] and $\Delta\phi/\phi$=0.1. Finally, at the assumptions above, the counting rate n=0.04644 s^{-1}. In order to measure A_{NN} with the statistical error ΔA_{NN}=A_{NN} = 2.2%, and taking into account both the beam and target polarizations and data taking inefficiency[2], the required beam time would be about 10 days. The parameters A_B and A_T will be measured at the same time with the precision of ~50% and 30%, respectively, for the same |t|= 0.6 (GeV/c)^2. The much better precisions for these observables are expected at the smaller |t| values.

6. Summary

We proposed the experiment which allows to measure simultaneously three spin observables in elastic pp scattering at 12-45 GeV/c in the $|t|$ interval of 0.07÷0.60 (GeV/c)^2.

10 days are required for measuring the t-dependence of A_{NN} in order to reach the precision ΔA_{NN}=A_{NN} at maximum momentum value 45 Gev/c. The time required for measuring the A_N for both beam (A_B) and target (A_T) with ~10 % relative error is significantly shorter at the smaller $|t|$.

The momentum dependence of these parameters can be studied in wide momentum range from 12 to 45 GeV/c.

The work on estimations of the possibility to carry out similar measurements with the polarized-antiproton beam is in progress.

This work has been supported in part by the RFBR Grant No. 16-02-00667 and by the Competitiveness Programme of National Research Nuclear University MEPhI.

References

[1] Auer P et al 1985 *Phys. Rev.* **D32** 1609
[2] Abe K et al 1976 *Phys. Lett.* **B63** 239
[3] Bauer F et al 2005 *Phys. Rev.* **C71** 054002
[4] Abramov V V et al 2011 *J.Phys.Conf.Ser. 295* 012018
[5] Khiari F Z et al 1989 *Phys. Rev.* **D39** 45
[6] Okada H 2006 *Measurement of the Analyzing Power AN in pp Elastic scattering in the CNI Region with the Polarized Atomic Hydrogen Gas Target* PhD Thesis (Osaka)
[7] Trueman T L 1996 *Proc. 12th International Symposium in High Energy Spin Physics* (Amsterdam)
[8] Gaidot A et al 1976 *Phys.Lett.* **61B** 103
[9] Azhgirey I I et al 2017 *J.Phys.Conf.Ser. 798 (2017) no 1* 012177
[10] Apokin V D et al 1985 *Nucl. Phys.* **B255** 253
[11] Semenov P A et al 2016 *Int.J.Mod.Phys.Conf.Ser. 40 (2016) no.01* 1660086
[12] Bruneton C et al 1977 *Nucl. Phys* **B124** 391

[2] Target polarization has to be changed at least each two days, which takes about 10 hours for each change of polarization sign

First results on the energy scan of the vector A_y and tensor A_{yy} and A_{xx} analyzing powers in deuteron-proton elastic scattering at Nuclotron[1]

V P Ladygin[1†], A V Averyanov[1], E V Chernykh[1], D Enache[2] Yu V Gurchin[1], A Yu Isupov[1], M Janek[3], J-T Karachuk[1,2], A N Khrenov[1], D O Krivenkov[1], P K Kurilkin[1], N B Ladygina[1], A N Livanov[1], S M Piyadin[1], S G Reznikov[1], Ya T Skhomenko[1,4], A A Terekhin[1], A V Tishevsky[1], T Uesaka[5]

[1] Joint Institute for Nuclear Research, Dubna, Russia
[2] National Institute for R&D in Electrical Engineering ICPE-CA, Bukharest, Romania
[3] Physics Department, University of Žilina, Žilina, Slovakia
[4] Belgorod State National Research University, Belgorod, Russia
[5] Nishina Center for Accelerator-Based Science, RIKEN, Wako, Japan

E-mail: †`vladygin@jinr.ru`

Abstract. New results on the vector A_y and tensor A_{yy} and A_{xx} analyzing powers in deuteron-proton elastic scattering obtained at Nuclotron in the energy range 400-1800 MeV are presented. These data have been obtained in 2016-2017 at DSS setup at internal target station using polarized deuteron beam from new source of polarized ions. The preliminary data on the deuteron analyzing powers in in the wide energy range demonstrate the sensitivity to the short-range spin structure of the nucleon-nucleon correlations.

1. Introduction

One of the tools to investigate the equation-of-state (EOS) of dense nuclear matter is the study of the short range correlations (SRC) of nucleons in nuclei which is the subject of intensive theoretical and experimental works during last years. Since SRC have densities comparable to the density in the center of a nucleon which is about $\rho \sim 5\rho_0$ ($\rho_0 \approx 0.17$ fm^{-3}), they can be considered as the drops of cold dense nuclear matter [1]. These studies explore a new part of the phase diagram and very essential to understand the evolution of neutron stars.

The results obtained at BNL [2], SLAC [3] and JLAB [4, 5] clearly demonstrate that more than 90% all nucleons with momenta $k \geq 300$ MeV/c belong to 2N SRC; the probability for a given proton with momenta $300 \leq k \leq 600$ MeV/c to belong to pn correlation is ~18 times larger than for pp correlations; the probability for a nucleon to have momentum ≥ 300 MeV/c in medium nuclei is ~25%; 3N SRC are present in nuclei with a significant probability [6]. However, still many open questions persist and further investigations are required both from the experimental and theoretical sides. For instance, the experimental data on the spin structure of 2N (I=1) and 3N SRC are almost absent.

[1] Dedicated to the memory of Prof. L.S.Zolin

The main tools to study SRCs at hadronic facilities can be deuteron structure investigations at large internal momenta allowing to explore 2N SRC with $I = 0$; ^{3}He structure to understand the role of 2N SRC with $I = 1$ and 3N SRC; nuclei breakup $A(p,pp)X$, $A(p,pn)X$, $A(p,ppp)X$ etc. with the detection of few nucleons in the final state. The greate importance is the study of the spin effects in these reactions because the data on the SRCs spin structure are scarce. Nuclotron and NICA will allow to investigate the spin effects for multi-nucleon correlations in a wide energy range.

The main goal of the Deuteron Spin Structure (DSS) experimental program is to obtain the information on the spin - dependent parts of two-nucleon ($2N$) and three-nucleon ($3N$) forces from two processes: *dp*- elastic scattering in a wide energy range and *dp*- nonmesonic breakup with two protons detection at energies 300 – 500 MeV [7, 8, 9] using the Nuclotron internal target station (ITS) [10]. The motivation of this program at low and intermediate energies is based on theoretical analysis of the experimental results obtained for the deuteron induced reactions (see recent reviews [11, 12] and references therein). Importance of the *dp*- elastic scattering studies at high energy is discussed in [13].

Such experimental program at Nuclotron was started by the measurements of the vector A_y and tensor A_{yy} and A_{xx} analyzing powers in *dp*- elastic scattering at T_d of 880 MeV [14] and 2000 MeV [15]. The systematic measurements of the differential cross section have been performed also in recent years [16, 17, 18].

In this paper we report new results of the energy scan of the vector A_y and tensor A_{yy} and A_{xx} analyzing powers in *dp*- elastic scattering obtained at the Nuclotron ITS [10] in the energy range of 400-1800 MeV.

2. Experimental setup at ITS

The internal target station (ITS) setup is well suited for study of the energy dependence of polarization observables for the deuteron-proton elastic scattering and deuteron breakup reaction with the detection of two protons at large scattering angles. For these purposes the CH_2-target of 10 μm thick is used for the measurements. The yield from carbon content of the CH_2-target is estimated in separate measurements using several twisted 8μm carbon wires. The monitoring of the intensity is done from the detection of *pp*- quasielastic scattering at 90° in cms by the scintillation counters placed in the horizontal plane. The detection of the *dp*- elastic events is done by the coincidence measurements of the proton and deuteron. The detectors are placed in the both horizontal and vertical planes for the analyzing powers measurements. The selection of the *dp*- elastic events is done by the correlation of the energy losses in plastic scintillators for deuteron and proton and their time-of-flight difference. The use of large amount of the scintillation counters allowed to cover wide angular range [19]. Such a metod has been used to obtain the polarization data in *dp*- elastic scattering at T_d of 880 MeV [14] and 2000 MeV [15].

The upgraded setup at ITS has been used to measure the vector A_y and tensor A_{yy} and A_{xx} analyzing powers in *dp*- elastic scattering between 400 MeV and 1800 MeV using polarized deuteron beam from new source of polarized ions (SPI) developed at LHEP-JINR [20]. These measurements were performed using internal target station at Nuclotron [10] with new control and data acquisition system [21]. The existing setup [19] has been upgraded by new VME based DAQ [22], new MPod based high voltage system [23], new system of monitors etc.

The same setup has been used as a polarimeter based on the use of *dp*- elastic scattering at large angles ($\theta_{\rm cm} \geq 60°$) at 270 MeV[19], where precise data on analyzing powers [24, 25, 26] exist, has been developed at internal target station (ITS) at Nuclotron [10]. The accuracy of the determination of the deuteron beam polarization achieved with this method is better than 2% because of the values of the analyzing powers were obtained for the polarized deuteron beam, which absolute polarization had been calibrated via the $^{12}C(d,\alpha)^{10}B^*[2^+]$ reaction[26].

3. Measurements of the analyzing power in *dp*- elastic scattering

New SPI [20] has been used to provide polarized deuteron beam. In the current experiment the spin modes with the maximal ideal values of (P_z,P_{zz})= (0,0), (-1/3,+1) and (-1/3, +1) were used. The deuteron beam polarization has been measured at 270 MeV [19]. The *dp*- elastic scattering events at 270 MeV were selected using correlation of the energy losses and time-of-flight difference for deuteron and proton detectors. The values of the beam polarization for different spin have been obtained as weighted averages for 8 scattering angles for *dp*- elastic scattering in the horizontal plane only. The typical values of the beam polarization were ~65-75% from the ideal values.

After deuteron beam polarization measurements at 270 MeV, the beam has been accelerated up to the required energy T_d between 400 MeV and 1800 MeV. The scintillation detectors were positioned in the horizontal and vertical plane in accordance with the kinematic of *dp*- elastic scattering for the investigated energy The main part of the measurements were performed using CH_2 target. Carbon target was used to estimate the background. The selection of the *dp*-elastic events is done by the correlation of the energy losses in plastic scintillators for deuteron and proton and their time-of-flight difference. The normalized numbers of *dp*-elastic scattering events for each spin mode were used to calculate the values of the analyzing powers A_y, A_{yy} and A_{xx}.

4. Discussion of the *dp*- elastic scattering results

A prior of the experiment with polarized deuteron beam DSS collaboration measured the differential cross section of *dp*- elastic scattering in a wide angular range at the energies T_d between 400 and 2000 MeV. The results obtained at ITS at Nuclotron at 1400 MeV [18] given by the full squares are compared with the data obtained earlier at 1283 MeV, 1595 MeV [27] and 1600 MeV [28] shown by the open diamonds, circles and triangles in Fig.1.

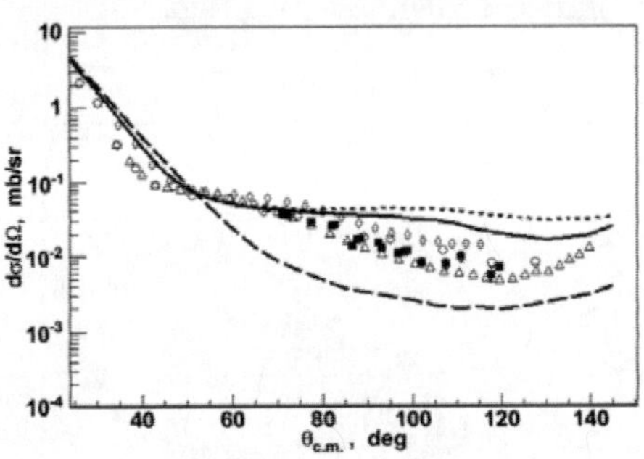

Figure 1. The differential cross section as a function of the cms scattering angle at the deuteron kinetic energy T_d of 1400 MeV [18], 1283 MeV, 1595 MeV [27] and 1600 MeV [28] are shown by the full squares, open diamonds, circles and triangles, respectively. The curves obtained in framework of the relativistic multiple scattering model [29, 30, 31] are explained in the text.

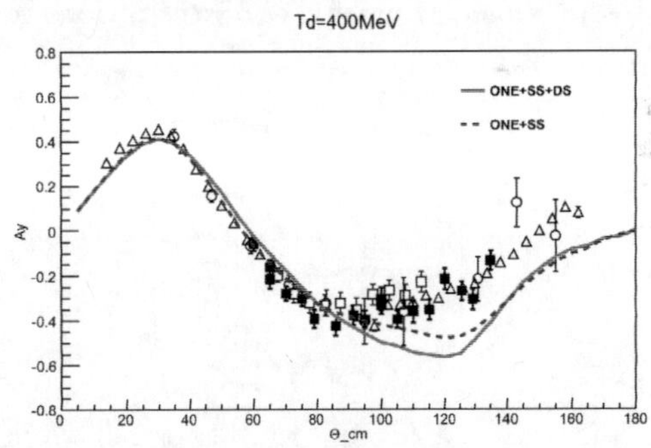

Figure 2. The angular dependence of the deuteron vector analyzing power A_y at the deuteron kinetic energy T_d of 400 MeV. The full squares are the preliminary results of the present experiment at ITS at Nuclotron. Open symbols are the world data [32, 33, 34]. Lines are the calculation within relativistic multiple scattering model [29, 30].

The theoretical calculations were performed in the relativistic multiple scattering expansion formalism [29, 30, 31]. The four contributions are taken into account: one-nucleon-exchange,

IOP Conf. Series: Journal of Physics: Conf. Series **938** (2017) 012007 doi:10.1088/1742-6596/938/1/012007

single- and double- scattering, and Δ- isobar excitation. The presented approach was applied earlier to describe the differential cross sections at deuteron energies between 500 and 1300 MeV in a whole angular range [31]. The dashed line in Fig 1 is the result of the calculations taking into account one-nucleon exchange (ONE) and single-scattering (SS) diagrams only, while the dotted line is obtained when double-scattering (DS) term is added into the consideration [29, 30]. The full calculation with Δ- isobar excitation included is given by the solid line [31]. One can see that the differential cross section is described quite well up to the scattering angles of $\sim 60^\circ$ taking into account only the ONE and SS terms. If we consider the dp -elastic scattering at the angles larger than 60° , it is necessary to include the DS term into consideration. It should be noted the double-scattering contribution into the reaction amplitude increases with the deuteron energy growing and may change the value of the differential cross section on a few orders in comparison with the result obtained without inclusion of the DS-term. The Δ-excitation begins to manifest itself at the angle equal to about 120° and describes the behaviour of the experimental data at the angle above 140° where the differential cross section sharply increases. It has been demonstrated also that the contribution of the Δ- isobar mechanism grows with the initial deuteron energy. It is negligible at $T_d = 500$ MeV and very significant at higher energies [31].

The angular dependence of the deuteron vector analyzing power A_y at the deuteron kinetic energy T_d of 400 MeV is presented in Fig. 2. The full squares are the results of the present experiment at ITS at Nuclotron. Open circles, squares and triangles are the data obtained at Saclay [32] and IUCF [33], [34], respectively. Dashed and solid lines are the calculations performed within relativistic multiple scattering model [29, 30] considering ONE+SS terms only and with the DS contribution added, respectively. Note that the contribution of the Δ- isobar mechanism is negligible at this energy [31]. One can see good agreement of new data obtained at Nuclotron with the data from earlier experiments [32, 33, 34]. The relativistic multiple scattering model [29, 30] describes the data up to $\sim 90^\circ$ only, while it fails to reproduce the data at larger angles. The considering of the DS term does not improve the agreement.

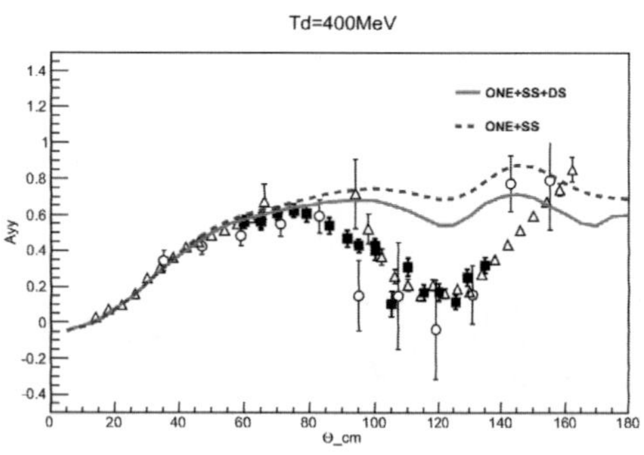

Figure 3. The angular dependence of the tensor analyzing power A_{yy} at the deuteron kinetic energy T_d of 400 MeV. The full squares are the preliminary results of the present experiment at ITS at Nuclotron. Open symbols are the world data [32, 34]. Lines are the same as in Fig. 2.

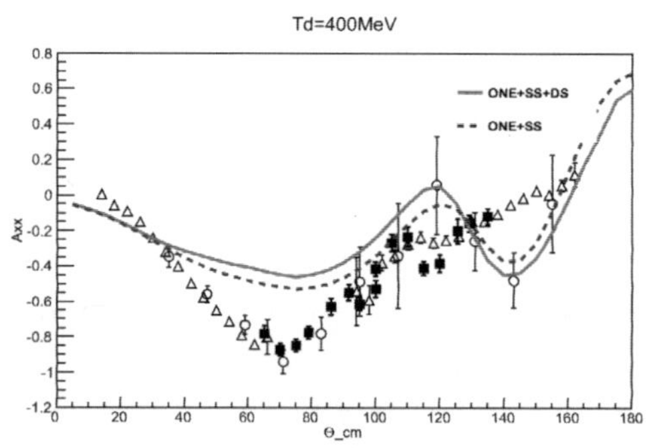

Figure 4. The angular dependence of the tensor analyzing power A_{xx} at the deuteron kinetic energy T_d of 400 MeV. The full squares are the preliminary results of the present experiment at ITS at Nuclotron. Open symbols are the world data [32, 34]. Lines are the same as in Fig. 2.

The angular dependencies of the tensor analyzing powers A_{yy} and A_{xx} at the deuteron kinetic energy T_d of 400 MeV are presented in Fig. 3 and Fig. 4, respectively. The full squares are the results of the present experiment at ITS at Nuclotron. Open circles and triangles are the data obtained at Saclay [32] and IUCF [34], respectively. The lines are the same as in Fig.2. The consideration of the ONE+SS terms allows to describe the behaviour of the A_{yy} analyzing power up to ~80° only. The DS term gives a significant contribution at larger angles, however, its taking into account does not remove the discrepancy of the calculation with the data. The A_{xx} behaviour is not described by the model [29, 30] over the whole angular range. The considering of the contribution of the three-nucleon forces or N^{4}LO calculations performed within chiral effective field theory (χEFT) [35] do not allow to get an agreement with the data on the tensor analyzing powers. The reason of the deviation can be the neglecting by the 3N SRCs.

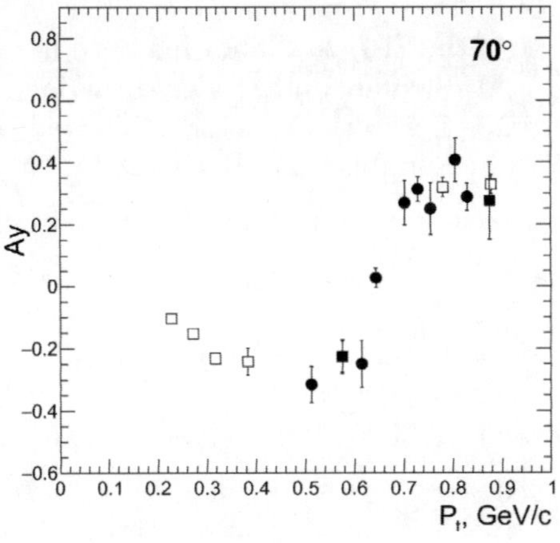

Figure 5. The energy dependence of the vector analyzing power A_y at 70° in the cms. The full circles are the preliminary results of the present experiment. The full squares are the data obtained at ITS at Nuclotron in 2005 [14, 15]. Open symbols are the world data [24, 25, 32, 36, 37].

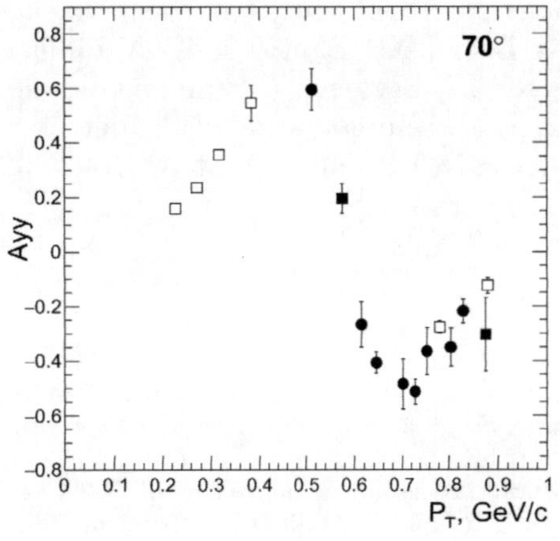

Figure 6. The energy dependence of the tensor analyzing power A_{yy} at 70° in the cms. The full circles are the preliminary results of the present experiment. The full squares are the data obtained at ITS at Nuclotron in 2005 [14, 15]. Open symbols are the world data [24, 25, 32, 36, 37].

The energy dependencies of the vector A_y and tensor A_{yy} analyzing powers at 70° in the cms are presented as a function of the transverse momentum P_T in Figs 5 and 6, respectively. The full circles are the preliminary results of the present experiment. The full squares are the data obtained at ITS at Nuclotron in 2005 [14, 15]. Open symbols are the world data [24, 25, 32, 36, 37]. Both A_y and A_{yy} analyzing powers change the sign at $P_T \sim 600$ MeV/c and have the tendencies at larger P_T to reach the positive and negative constant values, respectively. These features of the data indicate the serious deviation of the spin structure of the 2N SRCs on the standard description of the nucleon-nucleon interaction. Further theoretical investigations are required to understand the behaviour of the data at large P_T.

5. Conclusions

Upgraded Nuclotron with new SPI [20] provides quite unique opportunity for the studies of the spin effects and polarization phenomena in few body systems.

The realization of the DSS program at ITS allows to obtain the crucial data on the spin structure of 2-nucleon and 3- nucleon short range correlations. The first natural step in these studies, namely, the energy scan of the deuteron analyzing powers in dp- elastic scattering has been performed in 2016-2017.

Next stages of the DSS experiment using polarized deuterons and protons at ITS are the systematic studies of the analyzing powers A_y, A_{yy} and A_{xx} in *dp*- elastic scattering at 270-700 MeV and in *dp*- nonmesonic breakup at 300-500 MeV using new SPI at Nuclotron to study the relativistic effects in 2NFs and manifestation of the short-range 3NFs.

The availability of the polarized proton beam allows to extend the DSS physics program at ITS [13], namely, to perform the experiments on the measurements of the nucleon analyzing power A_y^p in *pd*- elastic scattering at 135-1000 MeV, in *pd*- nonmesonic breakup at the energies between 135-250 MeV for different kinematic configurations, in the $pA \to ppX$ reactions etc.

Acknowledgments

The authors thank the Nuclotron staff for providing good conditions of the experiment. They thank A.S. Belov, V.B. Shutov and V.V Fimushkin for the tune of the SPI [20]. They express the gratitude to S.N. Bazylev, V.I. Maximenkova, I.V. Slepnev, V.M. Slepnev and A.V. Shutov for the help during the preparation of the detector and DAQ system. The work has been supported in part by the RFBR under grant $N^0$16-02-00203a, JINR- Slovak Republic and JINR-Romania scientific cooperation programs in 2016-2017.

References

[1] Frankfurt L, Sargsian M and Strikman M 2008 *Int.J.Mod.Phys. A* **23** 2991.
[2] Piasetzky E, Sargsian M, Frankfurt L, Strikman M and Watson J W 2006 *Phys.Rev.Lett.* **97** 162504.
[3] Frankfurt L L, Strikman M I, Day D B and Sargsian M M 1993 *Phys.Rev. C* **48** 2451.
[4] Egiyan K Sh *et al.* 2003 *Phys.Rev. C* **68** 014313.
[5] Egiyan K Sh *et al.* 2006 *Phys.Rev.Lett.* **96** 082501.
[6] Frankfurt L, Sargsian M and Strikman M 2008 *AIP Conf.Proc.* **1056** p.322.
[7] Ladygin V P *et al.* 2014 *Phys.Part.Nucl.* **45** 327.
[8] Ladygin V P *et al.* 2014 *Few Body Syst.* **55** 709.
[9] Janek M *et al.* 2017 *Few Body Syst.* **58** 40.
[10] Malakhov A I *et al.* 2000 *Nucl.Instrum.Meth. in Phys.Res. A* **440** 320.
[11] Glöckle W, Witala H, Hüber D, Kamada H and Golak J 1996 *Phys.Rept.* **274** 107.
[12] Kalantar-Nayestanaki N, Epelbaum E, Messchendorp J G and Nogga A 2012 *Rept.Prog.Phys.* **75** 016301.
[13] Ladygin V P *et al.* 2016 *Int. J. Mod. Phys. Conf. Ser.* **40** 1660074.
[14] Kurilkin P K *et al.* 2012 *Phys.Lett. B* **715** 61.
[15] Kurilkin P K *et al.* 2011 *Phys.Part.Nucl.Lett.* **8** 1081.
[16] Gurchin Yu V *et al.* 2013 *Phys.Part.Nucl.Lett.* **10** 243.
[17] Terekhin A A *et al.* 2015 *Phys.Part.Nucl.Lett.* **12** 695.
[18] Terekhin A A *et al.* 2017 *JINR preprint R1-2017-33 (in Russian)*, to be published in *Phys.Atom.Nucl.*.
[19] Kurilkin P K *et al.* 2011 *Nucl.Instr.Meth. in Phys.Res. A* **642** 45.
[20] Fimushkin V V *et al.* 2016 *J.Phys.Conf.Ser.* **678** 012058.
[21] Isupov A Yu *et al.* 2013 *Nucl.Instrum.Meth. in Phys.Res. A* **698** 127.
[22] Isupov A Yu *talk at this Conference.*
[23] Skhomenko Ya T *et al.* *talk at this Conference.*
[24] Sekiguchi K *et al.* 2002 *Phys.Rev.* C **65** 034003.
[25] Sekiguchi K *et al.* 2004 *Phys.Rev.* C **70** 014001
[26] Suda K *et al.* 2007 *Nucl.Instr.Meth. in Phys.Res.* A **572** 745.
[27] Gülmez E *et al.* 1991 *Phys.Rev.* C **43** 2067.
[28] Winkelmann E *et al.* 1980 *Phys.Rev.* C **421** 2535.
[29] Ladygina N B 2008 *Phys.Atom.Nucl.* **71** 2039.
[30] Ladygina N B 2009 *Eur.Phys.J.* A **42** 91.
[31] Ladygina N B 2016 *Eur.Phys.J.* A **52** 199.
[32] Garçon M *et al.* 1986 *Nucl.Phys.* A **458** 287.
[33] Cadman R V *et al.* 2001 *Phys.Rev.Lett.* **86** 967.
[34] Przewoski B v *et al.* 2006 *Phys.Rev.* C **74** 064003.
[35] Binder S *et al.* 2016 Phys.Rev. C **93** 044002.
[36] Haji-Saied M *et al.* 1987 *Phys.Rev.* C **36** 2010.
[37] Ghazikhanian V *et al.* 1991 *Phys.Rev.* C **43** 1532.

IOP Conf. Series: Journal of Physics: Conf. Series **938** (2017) 012008 doi:10.1088/1742-6596/938/1/012008

First results on the measurements of the proton beam polarization at internal target at Nuclotron[1]

V P Ladygin[1†], Yu V Gurchin[1], A Yu Isupov[1], M Janek[2], A N Khrenov[1], P K Kurilkin[1], A N Livanov[1], S M Piyadin[1], S G Reznikov[1], Ya T Skhomenko[1,3], A A Terekhin[1], A V Tishevsky[1], A V Averyanov[1], S N Bazylev[1], A S Belov[4], A V Butenko[1], E V Chernykh[1], Yu N Filatov[5], V V Fimushkin[1], D O Krivenkov[1], A M Kondratenko[6], M A Kondratenko[6], A D Kovalenko[1], I V Slepnev[1], V M Slepnev[1], A V Shutov[1], A O Sidorin[1], I E Vnukov[3], V S Volkov[3]

[1] Joint Institute for Nuclear Research, Dubna, Russia
[2] Physics Department, University of Žilina, Žilina, Slovakia
[3] Belgorod State National Research University, Belgorod, Russia
[4] Institute of Nuclear Physics, Moscow, Russia
[5] Moscow Institute of Physics and Technology, Moscow, Russia
[6] Science and Technique Laboratory "Zaryad", Novosibirsk, Russia

E-mail: †vladygin@jinr.ru

Abstract. The spin program at NICA using SPD and MPD requires high intensity polarized proton beam with high value of the beam polarization. First results on the measurements of the proton beam polarization performed at internal target at Nuclotron are reported. The polarization of the proton beam provided by new source of polarized ions has been measured at 500 MeV using quasielastic proton-proton scattering and DSS setup at internal target. The obtained value of the vertical polarization of ~35% is consistent with the calculations taking into account the current magnetic optics of the Nuclotron injection line.

1. Introduction

The availability of the polarized proton beam is required by the spin program at NICA [1]. The main part of the physics program with SPD detector consists of the measurements of asymmetries in the lepton pair (Drell-Yan) production in collisions of non-polarized, longitudinally and transversally polarized protons and deuterons beams. These measurements can provide an access to all leading twist collinear and Transverse-Momentum Dependent ditsribution functions of quarks and anti-quarks in nucleons. The measurements of asymmetries in production of $J\Psi$ and direct photons, which supply complimentary information on the nucleon structure, will be performed simultaneously with Drell-Yan data using dedicated triggers. The set of these measurements permits to tests the quark-parton model of nucleons at the QCD twist-2 level with minimal systematic errors [2, 3]. This program can be extended by the measurements of the single and double spin asymmetries in the pion, kaon, proton inclusive production, vector

[1] Dedicated to the memory of Prof. L.S.Zolin

mesons and hyperon production in polarized proton-proton collisions etc. Moreover, polarized proton beam is needed for the spin studies with fixed target, namely, measurements of spin observables proton-proton, proton-neutron and proton-deuteron elastic scattering., investigation of the spin structure of the short-range nucleon correlations and three nucleon forces [4]. The realization of this program requires good knowledge of the proton beam polarization.

In this paper we report first results of the proton beam polarization measurements performed using upgraded polarimeter [5] at the Nuclotron internal target station.

2. Experiment at ITS

The polarimeter based on the use of *dp*- elastic scattering at large angles ($\theta_{\rm cm} \geq 60^\circ$) at 270 MeV[5], where precise data on analyzing powers [6, 7, 8] exist, has been developed at internal target station (ITS) at Nuclotron[9]. The accuracy of the determination of the deuteron beam polarization achieved with this method is better than 2% because of the values of the analyzing powers were obtained for the polarized deuteron beam, which absolute polarization had been calibrated via the $^{12}\mathrm{C}(d,\alpha)^{10}\mathrm{B}^*[2^+]$ reaction[8].

The use of large amount of the scintillation counters allowed to cover wide angular range. The measurement of the beam polarization has been performed at 270 MeV where the precise data on the tensor and vector analyzing powers based on the absolute calibration of the beam polarization exist [8]. These measurements were performed using internal target station at Nuclotron [9] with new control and data acquisition system [10]. The polarimeter has been upgraded [5] by new VME based DAQ [11], new MPod based high voltage system [12, 13], new system of monitors etc.

The same setup has been used to measure the vector A_y and tensor A_{yy} and A_{xx} analyzing powers in *dp*- elastic scattering between 400 MeV and 1800 MeV [14] using polarized deuteron beam from new source of polarized ions (SPI) developed at LHEP-JINR [15]. The setup has been also adopted for the measurements of the proton beam polarization at 500 MeV.

3. Measurements of the analyzing power for *pp*- elastic scattering

The classical method to measure the proton beam polarization at intermediate and high energies is the use of the left-right *pp*- elastic or quasi-elastic scattering (see, for instance, [16] and references therein). The maximal value of the analyzing power at the energies below 1000 MeV is close to $\sim 40^\circ$ in cms [17], that corresponds roughly 14-15° in the laboratory. Unfortunately, this angle is inaccessible due to design of ITS and detector support.

This method has been modified to increase the polarimeter figure of merit. For this purpose the measurements of the proton beam polarization using several pairs of the detectors placed in the kinematic coincidences corresponding to *pp*- elastic scattering in the horizontal plane (orbit plane of Nuclotron) have been proposed. The similar method has been used at COSY using EDDA setup [18, 19]. The feasibility of the proposed method at ITS has been checked in November 2016 run at Nuclotron using polarized deuteron beam with the energy of 500 MeV/nucleon.

New SPI [15] has been used to provide polarized deuteron beam. In the current experiment the spin modes with the maximal ideal values of (P_z,P_{zz})= (0,0), (-1/3,+1) and (-1/3, +1) were used. The deuteron beam polarization has been measured at 270 MeV [5]. The *dp*- elastic scattering events at 270 MeV were selected using correlation of the energy losses and time-of-flight difference for deuteron and proton detectors. The values of the beam polarization for different spin have been obtained as weighted averages for 8 scattering angles for *dp*- elastic scattering in the horizontal plane only. They were measured as (P_z,P_{zz})= (-0.232±0.018,+0.595±0.013) and (-0.243±0.013,-0.736±0.011) for spin modes "2-6" and "3-5", respectively.

IOP Conf. Series: Journal of Physics: Conf. Series **938** (2017) 012008 doi:10.1088/1742-6596/938/1/012008

After deuteron beam polarization measurements at 270 MeV, the beam has been accelerated up to 1000 MeV (or 500 MeV/nucleon). Eight pairs of the scintillation detectors were positioned in the horizontal plane covering angular range 55°-125° in the cms for *pp*- elastic scattering at 500 MeV on the left. Since analyzing power is antisymmetric with respect to θ^*=90°, the scattering in the backward hemisphere on the left at the angle $\pi - \theta$ can be considered as the scattering in the forward hemisphere on the right at the angle θ [20].

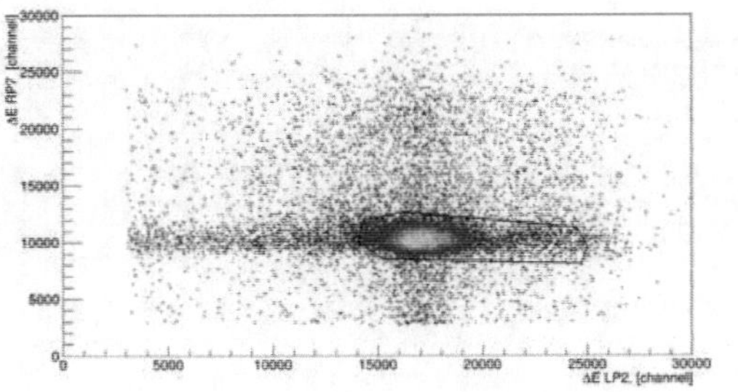

Figure 1. Correlation of the signal amplitudes in scintillation counters for proton-proton coincidences at 65° in cms. The solid line is a graphical cut to select *pp*- quasielastic scattering events.

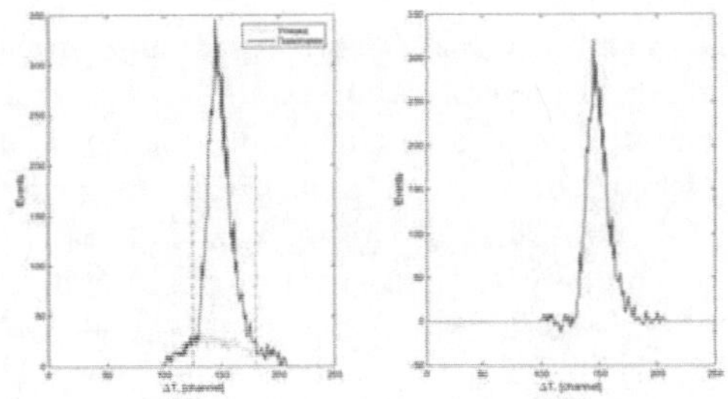

Figure 2. The time-of-flight difference between the signals from two proton detectors placed at 65° in cms. Left and right panels are the spectra obtained on CH_2 target and after carbon subtraction, respectively.

The main part of the measurements were performed using CH_2 target. Carbon target was used to estimate the background. The *pp*- quasielastic scattering events were selected using the energy losses and time-of-flight difference for two proton detectors placed in the kinematic coincidences. The correlation of the signal amplitudes in scintillation counters for proton-proton coincidences at 65° in cms is shown in Fig.1. The time-of-flight difference between the signals from two proton detectors placed at 65° in cms is presented in Fig.2. Left and right panels are the spectra obtained on CH_2 target and after CH_2-C subtraction, respectively.

The normalized numbers of *pp*-elastic scattering events for each spin mode were used to calculate the values of the analyzing power of *pp*- elastic scattering at 500 MeV. The obtained results were found to be in good agreement with the SP07 solution of SAID PWA [17], that proved the feasibility of the proposed method.

4. Results of the proton beam polarization measurement

The unpolarized and polarized proton beam provided by SPI [15] has been accelerated in March 2017 run up to 500 MeV. The typical intensity of the beam was $\sim$1.5$\cdot 10^8$ ppp and $\sim$2-3$\cdot 10^7$ ppp for unpolarized and polarized cases, respectively. SPI provided proton beam polarization using WFT 1$\rightarrow$3 with ideal value of the polarization P=-1. The polarization of the proton beam has been obtained using the data from eight pairs of the detectors placed in the kinematic coincidences. The values of the analyzing power for *pp* elastic scattering were taken from SAID PWA [17].

The weighted average values of the proton beam polarization were found as 0.017$\pm$0.021 and -0.354$\pm$0.022 for unpolarized and polarized cases, respectively. According to spin transport calculations taking into account the current magnetic option of the Nuclotron injection line [21] the polarization at the exit of SPI equals -0.90$\pm$0.06. The vertical component of the proton beam polarization at the ITS point can be increased up to this value by the installation of two solenoids at the exit of SPI and in to the Nuclotron injection line [21].

5. Conclusions

The ITS deuteron beam polarimeter [5] has been used to measure the proton beam polarization at 500 MeV using *pp*- quasi-elastic scattering.

The obtained value of the vertical proton polarization is -0.354±0.022. This value corresponds to polarization at the exit of SPI of -0.90±0.06 taking into account the current magnetic option of the Nuclotron injection line. Installation of two solenoids at the exit of SPI and in to the Nuclotron injection line will increase the vertical component of the proton beam polarization at the ITS point.

The current version of the ITS deuteron beam polarimeter can be applied for the proton beam polarization measurement at the energy range of 200-1000 MeV. The extension of the proton polarimetry at ITS to the higher energies (up to 3500 MeV) is possible by the enlargement of the angular span of the polarimeter using new detector support, new scintillation counters etc.

The availability of the polarized proton beam allows to extend the DSS physics program at ITS [4], namely, to perform the experiments on the measurements of the nucleon analyzing power A_y^p in *pd*- elastic scattering at 135-1000 MeV and in *pd*- nonmesonic breakup at the energies between 135-250 MeV for different kinematic configurations etc.

Acknowledgments

The authors thank the Nuclotron staff for providing good conditions of the experiment. They are grateful to Dr.V.A. Mikhaylov for the advise during acceleration of polarized protons. They thank V.B. Shutov and V.I. Maximenkova for the help during the preparation of the detector. The work has been supported in part by the RFBR under grant $N^0$16-02-00203a and JINR-Slovak Republic scientific cooperation program in 2016-2017.

References

[1] *http://nica.jinr.ru*
[2] Nagaitsev A P *et al. talk at this Conference.*
[3] Guskov A V *et al. talk at this Conference.*
[4] Ladygin V P *et al.* 2016 *Int. J. Mod. Phys. Conf. Ser.* **40** 1660074.
[5] Kurilkin P K *et al.* 2011 *Nucl.Instr.Meth. in Phys.Res. A* **642** 45.
[6] Sekiguchi K *et al.* 2002 *Phys.Rev.* C **65** 034003.
[7] Sekiguchi K *et al.* 2004 *Phys.Rev.* C **70** 014001.
[8] Suda K *et al.* 2007 *Nucl.Instr.Meth. in Phys.Res.* A **572** 745.
[9] Malakhov A I *et al.* 2000 *Nucl.Instrum.Meth. in Phys.Res. A* **440** 320.
[10] Isupov A Yu *et al.* 2013 *Nucl.Instrum.Meth. in Phys.Res. A* **698** 127.
[11] Isupov A Yu *talk at this Conference.*
[12] Skhomenko Ya T *et al.* 2016 *Scientific Statements of Belgorod State University. Series: Mathematics and Physics* **43** 115.
[13] Skhomenko Ya T *et al. talk at this Conference.*
[14] Ladygin V P *et al. talk at this Conference.*
[15] Fimushkin V V *et al.* 2016 *J.Phys.Conf.Ser.* **678** 012058.
[16] Azhgirey L S *et al.* 2003 *Nucl.Instrum.Meth. in Phys.Res. A* **497** 340.
[17] *http://gwdac.phys.gwu.edu*
[18] Altmeier M *et al.* 2005 *Eur.Phys.J.* A **23** 351.
[19] Bauer F *et al.* 2005 *Phys.Rev.* C **71** 054002.
[20] Lechanoine-LeLuc C and Lehar F 1993 *Rev.Mod.Phys.* **65** 47.
[21] Filatov Yu N *et al. talk at this Conference.*

IOP Conf. Series: Journal of Physics: Conf. Series **938** (2017) 012009 doi:10.1088/1742-6596/938/1/012009

Extraction of helicity amplitude ratios from exclusive ρ^0-meson electroproduction on transversely polarized protons

S I Manaenkov[1] on behalf of the HERMES collaboration

[1] NRC "Kurchatov Institute"– PNPI, Gatchina 188300, Russia

E-mail: manaenkov_si@pnpi.nrcki.ru

Abstract. Exclusive ρ^0-meson electroproduction is studied by the HERMES experiment, using the 27.6 GeV longitudinally polarized electron/positron beam of HERA and a transversely polarized hydrogen target, in the kinematic region 1.0 GeV^2 < Q^2 < 7.0 GeV^2, 3.0 GeV < W < 6.3 GeV, and $-t'$ < 0.4 GeV^2. Using an unbinned maximum-likelihood method, 25 parameters are extracted. They determine the real and imaginary parts of the ratios of certain helicity amplitudes (describing ρ^0-meson production by a virtual photon) and the dominant amplitude $F_{0\frac{1}{2}0\frac{1}{2}}$ without the nucleon-helicity flip. The latter amplitude describes the production of a longitudinal ρ^0 meson by a longitudinal virtual photon. The transverse target polarization allows for the first time the extraction of ratios of a number of nucleon-helicity-flip amplitudes to $F_{0\frac{1}{2}0\frac{1}{2}}$. The ratios of nucleon-helicity-non-flip amplitudes to $F_{0\frac{1}{2}0\frac{1}{2}}$ are found to be in good agreement with those from the previous HERMES analysis. A comparison of the extracted amplitude ratios with the Goloskokov-Kroll model shows the necessity to add pion exchange amplitudes with positive $\pi\rho$ form factor to the amplitudes based on generalized parton distributions to improve the HERMES data description.

1. Introduction

Exclusive electroproduction of vector mesons (V) on nucleons (N) gives information both on the reaction mechanisms and the nucleon structure [1]. Electroproduction at high energies can be considered to consist of three subprocesses: i) the incident lepton emits a virtual photon γ^*, which dissociates into a quark-antiquark pair; ii) this $q\bar{q}$ pair interacts strongly with the nucleon; iii) the observed vector meson is formed from the scattered $q\bar{q}$ pair. In Regge phenomenology, the interaction of the $q\bar{q}$ pair with the nucleon proceeds through the exchange of a pomeron or/and exchange of secondary reggeons. If the quantum numbers of the particle lying on the Regge trajectory are $J^P = 0^+$, 1^-, etc. (pomeron, ρ, f_2, ...), the process is denoted Natural Parity Exchange (NPE). Alternatively, the case of $J^P = 0^-$, 1^+, etc. (π, a_1, ...) corresponds to Unnatural Parity Exchange (UPE). In the framework of perturbative quantum chromodynamics valid at large photon virtuality Q^2 and high photon-nucleon center-of-mass (CM) energy W, the nucleon structure can also be studied through hard exclusive meson production, as the process amplitude contains Generalized Parton Distributions (GPDs) (see review [1]). However, the factorization property that permits to extract GPDs is rigorously proved only for the amplitudes $F_{0\frac{1}{2}0\pm\frac{1}{2}}$ of longitudinal vector meson production by longitudinal virtual photons [2]. In the Goloskokov-Kroll (GK) model (see [3] and references therein), the validity of factorization is

IOP Conf. Series: Journal of Physics: Conf. Series **938** (2017) 012009 doi:10.1088/1742-6596/938/1/012009

assumed for some other amplitudes in addition to $F_{0\frac{1}{2}0\pm\frac{1}{2}}$ and this assumption is justified with a good description of the existing data. The presented HERMES data can be well described in the GK model if the pion exchange is taken into account (see [4] and references therein). This means that the GPD-based approach should be modified at intermediate energies and Q^2.

All observables in vector-meson electroproduction can be expressed in terms of the helicity amplitudes in the CM system describing ρ^0-meson production by a virtual photon. In particular, Spin-Density-Matrix Elements (SDMEs) are functions of the helicity amplitude ratios (HARs). Therefore HARs can be extracted from the data as was shown in Ref. [5]. For the first time, HARs with nucleon-helicity flip are obtained in the present analysis since the data on ρ^0-meson production by the longitudinally polarized electrons/positrons on a transversely polarized hydrogen target are accumulated. The most recent HERMES results on the ρ^0-meson production presented at this workshop are published in Ref. [6] where more details can be found.

2. Spin-density-matrix elements and helicity amplitudes

The SDMEs are the Fourier coefficients in the angular distribution of the final-state particles. In the present paper, the formalism proposed in Ref. [7] for SDMEs is used. Any SDME can be expressed as a ratio of two sums of bilinear products of the helicity amplitudes $F_{\lambda_V \lambda'_N \lambda_\gamma \lambda_N}$ and their complex conjugate quantities $F^*_{\lambda_V \lambda'_N \lambda_\gamma \lambda_N}$. Here, $F_{\lambda_V \lambda'_N \lambda_\gamma \lambda_N}$ is the helicity amplitude of the process $\gamma^*(\lambda_\gamma)+N(\lambda_N) \to V(\lambda_V)+N(\lambda'_N)$, where the particle helicities are given in parentheses. The helicity amplitudes depend on W, Q^2, and $t' = t - t_{min}$, where t is the Mandelstam variable and $-t_{min}$ represents the smallest kinematically allowed value of $-t$ at fixed W and Q^2.

Any helicity amplitude can be decomposed into a sum of a NPE ($T_{\lambda_V \lambda'_N \lambda_\gamma \lambda_N}$) and an UPE ($U_{\lambda_V \lambda'_N \lambda_\gamma \lambda_N}$) amplitude: $F_{\lambda_V \lambda'_N \lambda_\gamma \lambda_N} = T_{\lambda_V \lambda'_N \lambda_\gamma \lambda_N} + U_{\lambda_V \lambda'_N \lambda_\gamma \lambda_N}$, for details see Refs. [7, 8]. The amplitudes obey the symmetry relations that hold because of parity conservation

$$T_{\lambda_V \lambda'_N \lambda_\gamma \lambda_N} = (-1)^{-\lambda_V+\lambda_\gamma} T_{-\lambda_V \lambda'_N -\lambda_\gamma \lambda_N} = (-1)^{-\lambda_N+\lambda'_N} T_{\lambda_V -\lambda'_N \lambda_\gamma -\lambda_N}, \quad (1)$$

$$U_{\lambda_V \lambda'_N \lambda_\gamma \lambda_N} = -(-1)^{-\lambda_V+\lambda_\gamma} U_{-\lambda_V \lambda'_N -\lambda_\gamma \lambda_N} = -(-1)^{-\lambda_N+\lambda'_N} U_{\lambda_V -\lambda'_N \lambda_\gamma -\lambda_N}. \quad (2)$$

These symmetry properties of the helicity amplitudes permit to introduce the notations:

$$T^{(1)}_{\lambda_V \lambda_\gamma} \equiv T_{\lambda_V \frac{1}{2} \lambda_\gamma \frac{1}{2}}, \; U^{(1)}_{\lambda_V \lambda_\gamma} \equiv U_{\lambda_V \frac{1}{2} \lambda_\gamma \frac{1}{2}}, \; T^{(2)}_{\lambda_V \lambda_\gamma} \equiv T_{\lambda_V \frac{1}{2} \lambda_\gamma -\frac{1}{2}}, \; U^{(2)}_{\lambda_V \lambda_\gamma} \equiv U_{\lambda_V \frac{1}{2} \lambda_\gamma -\frac{1}{2}}. \quad (3)$$

All other amplitudes can be obtained from Eqs. (1 − 2). The HARs extracted in the present analysis are defined for $n = 1,\ 2$ as $t^{(n)}_{\lambda_V \lambda_\gamma} = T^{(n)}_{\lambda_V \lambda_\gamma}/T^{(1)}_{00}$, $u^{(n)}_{\lambda_V \lambda_\gamma} = U^{(n)}_{\lambda_V \lambda_\gamma}/T^{(1)}_{00}$.

3. Experiment and data analysis

3.1. Experiment

The data were accumulated with the HERMES spectrometer using the 27.6 GeV longitudinally polarized electron or positron beam of HERA and a gaseous hydrogen target. The HERMES setup included a forward spectrometer [9], in which the scattered lepton and the produced hadrons were detected within an angular acceptance of ±170 mrad horizontally and ±(40 − 140) mrad vertically. The tracking system had a momentum resolution of about 1.5% and an angular resolution of about 1 mrad. Lepton identification was accomplished using a transition-radiation detector, a preshower scintillator counter, and an electromagnetic calorimeter. The particle-identification system included also a dual-radiator ring-imaging Cherenkov detector [10] to identify hadrons. Combining the responses of the detectors in a likelihood method leads to an average lepton-identification efficiency of 98%, with a hadron contamination of less than 1%. The helicity of the beam was typically reversed every two months. The beam polarization was continuously measured by two Compton polarimeters [11, 12]. The average value of the beam polarization for the events used in the analysis is ± 0.30 with a 0.03 uncertainty. A small part of the data was collected with an unpolarized target and the main part with a transversely polarized target [13], for which the polarization direction was reversed every 60 s to 180 s. The measured mean value of the target polarization is $\langle |P_T| \rangle = 0.72 \pm 0.06$ [14].

3.2. Data analysis

The ρ^0 mesons are produced and decay in the following exclusive reactions: $e + p \to e + p + \rho^0$, and $\rho^0 \to \pi^+ + \pi^-$. The event sample used in this analysis is practically the same as that used in Ref. [14]. The most important improvement is the application of a new tracking algorithm, which is based on a Kalman filter [15]. For the present analysis, the data are required to fulfil the following criteria: i) The scattered lepton has to have an energy larger than 3.5 GeV in order to not introduce effects from varying trigger thresholds. ii) The longitudinal beam polarization is restricted to the interval $15\% < |P_B| < 80\%$. iii) Events with exactly two oppositely charged hadrons and one lepton with the same charge as the beam lepton are selected. All tracks are required to originate from the same vertex. iv) No calorimeter clusters were observed (no π^0 is detected). v) The two-pion invariant mass is required to obey 0.6 GeV$\leq$ M$(\pi^+\pi^-)\leq$ 1.0 GeV. vi) The recoiling proton was not detected, but instead reconstructed through the missing energy. Taking into account the spectrometer resolution, the missing energy ΔE has to lie in the interval -1.0 GeV $< \Delta E <$ 0.8 GeV. Here, $\Delta E = \frac{M_X^2 - M_p^2}{2M_p}$, with M_p being the proton mass and $M_X^2 = (p + q - p_{\pi^+} - p_{\pi^-})^2$ the missing mass squared, where p, q, p_{π^+}, and p_{π^-} are the four-momenta of target nucleon, virtual photon, and each of the two pions respectively. The distribution of missing energy ΔE, shown in Ref. [14], exhibits a clearly visible exclusive peak. vii) The kinematic constraints $Q^2 > 1.0$ GeV^2, 6.3 GeV $> W >$ 3 GeV, $-t' < 0.4$ GeV^2 are applied. After application of all these requirements, the data sample contains 8741 events. These data are referred to in the following as data in the "entire kinematic region". The applied constraints do not fully suppress the background. The exclusive sample contains contributions from double-diffractive processes, which should be negligible in the low ΔE region, from non-resonant $\pi^+\pi^-$ pair production, which is of the order of $1 - 2\%$ [14], from Δ-isobar excitation, which in the HERMES acceptance is less than 7% [16] and from semi-inclusive deep-inelastic scattering (SIDIS) events. The presented results are not corrected for the former three processes, while a correction is applied for SIDIS background. The shaded histogram of missing energy ΔE (presented in Ref. [14]) shows SIDIS background obtained from a PYTHIA Monte Carlo (MC) simulation. It is normalized to the data in the region of 2 GeV $< \Delta E <$ 20 GeV. The simulation is used to determine the fraction of SIDIS background under the exclusive peak. This fraction f_{bg} increases from 7% to 23% for increasing $-t'$ and amounts on average to 11%. A fit of the angular distribution of the SIDIS MC events under the exclusive peak is performed. The obtained function W_{bg} having no free parameters and the angular distribution W_ρ of the exclusive ρ^0 production are combined to give the total angular distribution $W_{tot} = (1 - f_{bg})W_\rho + f_{bg}W_{bg}$. The function W_{tot} is used in an unbinned maximum-likelihood method, the detector properties are taken into account by a MC simulation. The contribution of background from SIDIS events to the obtained amplitude ratios is considered as one of the main sources of systematic uncertainty. This uncertainty is calculated as the difference between HARs obtained when f_{bg} is put equal to zero and HARs extracted when the background corrections are taken into account.

4. Results

Using an unbinned maximum-likelihood method the HARs are obtained from the 25-parameter fit in the entire kinematic region ($\langle W \rangle = 4.73$ GeV, $\langle Q^2 \rangle = 1.93$ GeV^2, $\langle -t' \rangle = 0.132$ GeV^2). The result is shown in Fig. 1 with red points. While the phase of $u_{11}^{(1)}$ is fixed according to the results of Refs. [17, 18], its modulus is fit so that Re[$u_{11}^{(1)}$] and Im[$u_{11}^{(1)}$] represent the results for

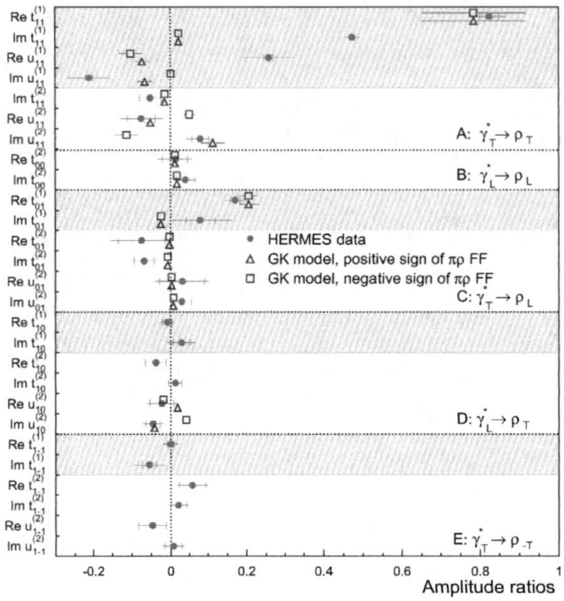

Figure 1. Comparison of amplitude ratios determined in Ref. [6] to those calculated in the GK model. The red, filled circles correspond to the extracted amplitude ratios and the blue, open triangles (squares) represent the result of the GK model calculation using the positive (negative) sign of the $\pi\rho$ transition form factor. The inner (outer) error bars of red points represent the statistical (total) uncertainty. The amplitude ratios that are set to zero in the GK model are not shown. The amplitude ratios are ordered according to the classes proposed in Ref. [8, 14].

one free parameter. The systematic uncertainty of $\mathrm{Re}[\mathrm{u}_{11}^{(1)}]$ and $\mathrm{Im}[\mathrm{u}_{11}^{(1)}]$ is much larger than the statistical one due to the large total experimental uncertainty of the $u_{11}^{(1)}$ phase. The value of $\mathrm{Im}[t_{11}^{(1)}]$ represents the result of Ref. [5]; its error bar shows the total uncertainty. The shadowed area corresponds to ratios of amplitudes without the nucleon-helicity flip obtained here and earlier in Ref. [5] for the unpolarized proton. All other ratios of nucleon-helicity-flip amplitudes to $T_{00}^{(1)}$ are calculated for the first time. As can be seen from Fig. 1 all HARs except $t_{11}^{(1)}$, $t_{01}^{(1)}$, and $u_{11}^{(1)}$ are compatible with zero within two standard deviations of their total uncertainty. The HAR $t_{01}^{(1)}$ is responsible for the violation of the s-channel helicity-conservation approximation, while $u_{11}^{(1)}$ shows the role of UPE in ρ^0-meson electroproduction.

The results of the theoretical predictions in the Goloskokov-Kroll model [3, 19] are depicted with blue triangles and squares in Fig. 1. In order to describe the presented HERMES data the one pion exchange (OPE) diagrams have to be included in addition to the handbag Feynman graphs containing the GPDs. Note that even the HARs compatible with zero are informative and can be used to fix the sign of the $\pi\rho$ form factor $g_{\pi\rho}(Q^2)$ in the OPE amplitude which is proportional to $g_{\pi\rho}(Q^2)/(t - m_\pi^2)$, where m_π denotes the π^0 mass. As seen from Fig. 1 the positive sign of the $\pi\rho$ form factor $g_{\pi\rho}(Q^2)$ (triangles) corresponds to the extracted HARs $u_{11}^{(2)}$ and $u_{10}^{(2)}$, that are sensitive to the sign, much better ($\chi^2/ndf = 1.8/4$) than the negative sign

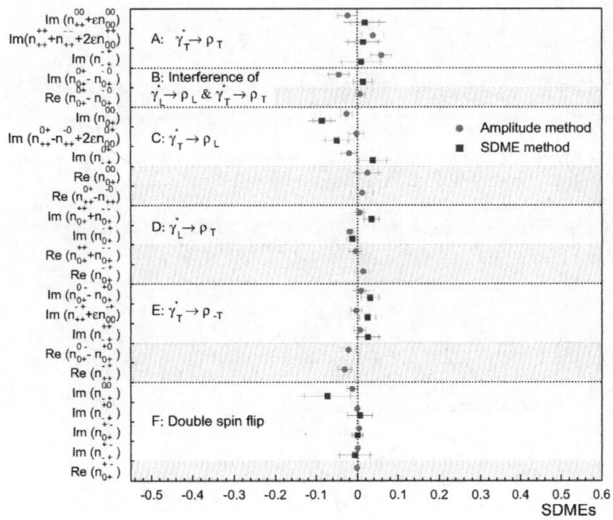

Figure 2. Comparison of the Diehl SDMEs [7] $n^{\lambda_V \lambda'_V}_{\lambda_\gamma \lambda'_\gamma}$ calculated from the helicity-amplitude ratios (red circles) and the SDMEs (blue squares) directly extracted in Ref. [14] in the entire kinematic region. For the first case, a 25-parameter fit is used. The points in the shaded area show SDMEs that can be obtained only if the beam is longitudinally polarized. The inner (outer) error bars represent the statistical (total) uncertainty.

(squares) for which $\chi^2/ndf = 30.3/4$.

The most serious disagreement between the extracted HAR $\mathrm{Im}[t^{(1)}_{11}]$ and that calculated in the GK model was already discussed in the previous HERMES work [5] and can be explained probably with the unsatisfactory description of the $q\bar{q}$ rescattering from the proton in the GK model. As can be seen from Fig. 1 neither the phase nor the modulus of the amplitude ratio $u^{(1)}_{11}$ are described well in the GK model and this problem should be solved in the future.

A comparison of the SDMEs directly obtained from the HERMES data in Refs. [8] and [14] to those calculated with the HARs extracted in the present analysis is performed in Ref. [6] and good agreement is found. The correlation matrix for the 25 parameters is taken into account for the calculation of the statistical uncertainty of the SDMEs in the Diehl representation [7] $u^{\lambda_V \lambda'_V}_{\lambda_\gamma \lambda'_\gamma}$, $n^{\lambda_V \lambda'_V}_{\lambda_\gamma \lambda'_\gamma}$, and $s^{\lambda_V \lambda'_V}_{\lambda_\gamma \lambda'_\gamma}$ obtained from the HARs. The SDMEs $n^{\lambda_V \lambda'_V}_{\lambda_\gamma \lambda'_\gamma}$ and $s^{\lambda_V \lambda'_V}_{\lambda_\gamma \lambda'_\gamma}$, presented in Figs. 2 and 3, have linear contributions of the small HARs $t^{(2)}_{\lambda_V \lambda_\gamma}$ and $u^{(2)}_{\lambda_V \lambda_\gamma}$, respectively. These SDMEs can only be extracted from measurements with a transversely polarized target so that the nucleon-helicity-flip amplitude ratios $t^{(2)}_{\lambda_V \lambda_\gamma}$ and $u^{(2)}_{\lambda_V \lambda_\gamma}$ are extracted in Ref. [6] for the first time. The total uncertainty is the sum in quadrature of the statistical and the total systematic uncertainties. The SDMEs in Figs. 2 and 3 are reordered according to the SDME classes proposed in Refs. [8, 14]. Those SDMEs that can be extracted only from data taken with a longitudinally polarized lepton beam are shown in shaded areas. Figures 2 and 3 show that for some of the calculated SDMEs $n^{\lambda_V \lambda'_V}_{\lambda_\gamma \lambda'_\gamma}$ and $s^{\lambda_V \lambda'_V}_{\lambda_\gamma \lambda'_\gamma}$ no published results from Ref. [14] exist, because the beam polarization was not exploited in that analysis. While in Refs. [8] and [14] a total of 53 SDMEs could be extracted, the amplitude method presented here allows

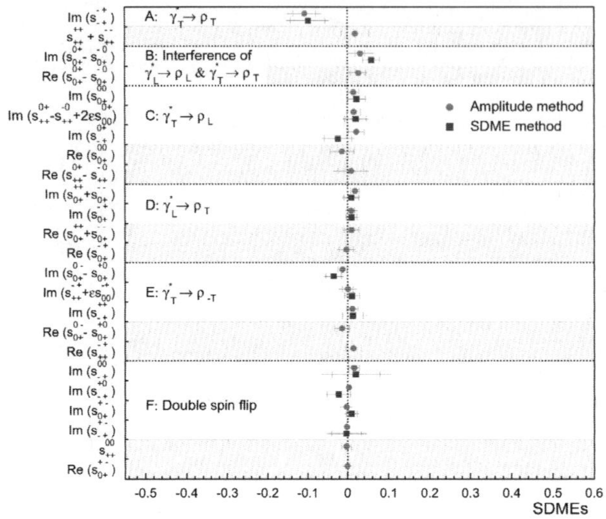

Figure 3. Comparison of the Diehl SDMEs [7] $s^{\lambda_V \lambda'_V}_{\lambda_\gamma \lambda'_\gamma}$ calculated from the helicity-amplitude ratios (red circles) and the SDMEs (blue squares) directly extracted in Ref. [14]. The meaning of the error bars and the futher explanations are the same as for Fig. 2.

for the calculation of 71 SDMEs based on the extraction of 25 parameters. As seen from Figs. 2 and 3, there is reasonable agreement between SDMEs directly extracted in Refs. [8, 14] and those calculated from the HARs in Ref. [6]. The SDMEs $u^{\lambda_V \lambda'_V}_{\lambda_\gamma \lambda'_\gamma}$ which can be obtained with unpolarized targets are not shown in the present paper. But a comparison of the calculated SDMEs $u^{\lambda_V \lambda'_V}_{\lambda_\gamma \lambda'_\gamma}$ with those directly extracted from the HERMES data [8] is also performed in Ref. [6] (for the first time in Ref. [5]) and shows reasonable agreement.

References

[1] Diehl M 2003 *Phys. Rep.* **388** 41
[2] Collins J C, Frankfurt L and Strikman M S 1997 *Phys. Rev.* D **56** 2982
[3] Goloskokov S V and Kroll P 2008 *Eur. Phys. J.* C **53** 367
[4] Goloskokov S V and Kroll P 2014 *Eur. Phys. J.* A **50** 146
[5] Airapetian A et al. (HERMES Collaboration) 2011 *Eur. Phys. J.* C **71** 1609
[6] Airapetian A et al. (HERMES Collaboration) 2017 *Eur. Phys. J.* C **77** 378
[7] Diehl M 2007 *JHEP* **0709** 064
[8] Airapetian A et al. (HERMES Collaboration) 2009 *Eur. Phys. J.* C **62** 659
[9] Ackerstaff K et al. (HERMES Collaboration) 1998 *Nucl. Instr. and Meth.* A **417** 230
[10] Akopov N et al. (HERMES Collaboration) 2002 *Nucl. Instr. and Meth.* A **479** 511
[11] Barber D P et al. (HERMES Collaboration) 1993 *Nucl. Instr. and Meth.* A **329** 79
[12] Beckmann M et al. (HERMES Collaboration) 2002 *Nucl. Instr. and Meth.* A **479** 334
[13] Airapetian A et al. (HERMES Collaboration) 2005 *Nucl. Instr. and Meth.* A **540** 68
[14] Airapetian A et al. (HERMES Collaboration) 2009 *Phys. Lett.* B **679** 100
[15] Fruhwirth R 1987 *Nucl. Instr. and Meth.* A **262** 444
[16] Tytgat M 2001 *DESY-THESIS-2001-014*
[17] Airapetian A et al. (HERMES Collaboration) 2001 *Phys. Lett.* B **513** 301
[18] Airapetian A et al. (HERMES Collaboration) 2003 *Eur. Phys. J.* C **29** 171
[19] Goloskokov S V and Kroll P 2014 *Eur. Phys. J.* C **74** 2725

IOP Conf. Series: Journal of Physics: Conf. Series **938** (2017) 012010 doi:10.1088/1742-6596/938/1/012010

Transverse target spin asymmetries in exclusive ω muoproduction at COMPASS

Bohdan Mariański[1]
(on behalf of the COMPASS Collaboration)
National Institute for Nuclear Research, PL 00-681 Warsaw, Poland
E-mail: `b.marianski@ncbj.gov.pl`

Abstract.
Exclusive production of ω mesons was studied at the COMPASS experiment by scattering 160 GeV/c muons off transversely polarised protons. Five single-spin and three double-spin azimuthal asymmetries were measured in the range of photon virtuality 1 (GeV/c)^2 $< Q^2 <$ 10 (GeV/c)^2, Bjorken scaling variable $0.003 < x_{Bj} < 0.3$ and transverse momentum squared of the ω meson 0.05 (GeV/c)^2 $< p_T^2 <$ 0.5 (GeV/c)^2. The results are compared to recent calculations based on Gloskokov-Kroll GPD model with the $\pi\omega$ transition form factor.

1. Introduction

Exclusive leptoproduction of vector meson in the process $\gamma^* + N \rightarrow V + N'$ $(V = \rho^0,\ \phi,\ \omega)$ provides information both on reaction mechanism and nucleon structure. This contribution is concentrated on transverse target spin asymmetries obtained in exclusive ω production. Studying these asymmetries give the possibility to constrain Generalized Parton Distribution (GPD). Access to $GPDs$ relies on factorization property of the process amplitude. The process amplitude is a convolution of the lepton-quark hard-scattering amplitude with soft part which contains $GPDs$ and vector meson distribution amplitude. At leading twist chiral-even $GPDs$ $H^{q(g)}(\text{x},\ \xi,\ \text{t})$, $E^{q(g)}(\text{x},\ \xi,\ \text{t})$, $\tilde{H}^{q(g)}(\text{x},\ \xi,\ \text{t})$, $\tilde{E}^{q(g)}(\text{x},\ \xi,\ \text{t})$ without parton helicity flip are sufficient to describe exclusive vector mason production on spin 1/2 target. When higher twist effects are included in the Distribution Amplitude, chiral odd $GPDs$ $H_T^{q(g)}(\text{x},\ \xi,\ \text{t})$, $E_T^{q(g)}(\text{x},\ \xi,\ \text{t})$, $\tilde{H}_T^{q(g)}(\text{x},\ \xi,\ \text{t})$, $\tilde{E}_T^{q(g)}(\text{x},\ \xi,\ \text{t})$ with parton helicity flip, appear. The $GPDs$ H and E are of special interest as they are related to the total angular momentum of partons in the nucleon by the Ji's relation [1]. $GPDs$ also describe the nucleon as an extended object which correlates longitudinal momenta of parton and transverse spatial coordinates. $GPDs$ H are well constraint by the experimental data. Less is know about $GPDs$ E. Production of exclusive vector meson with unpolarised target is mainly sensitive to $GPDs$ H , while for transversely polarised target it is also sensitive to $GPDs$ E.

The factorisation of process amplitude is proven rigorously when the lepton-quark interaction is mediated by longitudinely polarised virtual photon. However, phenomenological pQCD-inspired models taking into account parton transverse momenta have been proposed by Goloskokov and Kroll [2, 3, 4], which for both longitudinal and transverse photons describes reasonably well the

[1] Work supported by the Polish NCN Grant 2015/18/M/ST2/00550

behaviour of the cross section, spin density matrix elements, azimuthal and spin asymmetries. It will be referred to as "GK" model. It is known that in ω exlusive production Unnatural Parity Exchange (UPE) processes plays a substantial role. In the GK model the UPE processes are described by a predominant contribution of pion-pole exchange.

2. Theoretical formalism

The cross section for exclusive ω muoproduction, $\mu \mathrm{N} \to \mu' \omega \mathrm{N}'$, on a transversely polarised nucleon is given in Ref. [5]:

$$\left[\frac{\alpha_{em}}{8\pi^3}\frac{y^2}{1-\epsilon}\frac{1-x_{Bj}}{x_{Bj}}\frac{1}{Q^2}\right]^{-1}\frac{\mathrm{d}\sigma}{\mathrm{d}x_{Bj}\mathrm{d}Q^2\mathrm{d}t\mathrm{d}\phi\mathrm{d}\phi_s} =$$
$$\begin{aligned}
&\frac{1}{2}\left(\sigma_{++}^{++}+\sigma_{++}^{--}\right)+\epsilon\sigma_{00}^{++}-\epsilon\cos(2\phi)\,\mathrm{Re}\,\sigma_{+-}^{++}\\
&-\sqrt{\epsilon(1+\epsilon)}\cos\phi\,\mathrm{Re}\left(\sigma_{+0}^{++}+\sigma_{+0}^{--}\right)\\
&-P_\ell\sqrt{\epsilon(1-\epsilon)}\sin\phi\,\mathrm{Im}\left(\sigma_{+0}^{++}+\sigma_{+0}^{--}\right)\\
&-S_T\Big[\sin(\phi-\phi_s)\,\mathrm{Im}\left(\sigma_{++}^{+-}+\epsilon\sigma_{00}^{+-}\right)+\frac{\epsilon}{2}\sin(\phi+\phi_s)\,\mathrm{Im}\,\sigma_{+-}^{+-}\\
&\quad+\frac{\epsilon}{2}\sin(3\phi+\phi_s)\,\mathrm{Im}\,\sigma_{+-}^{-+}+\sqrt{\epsilon(1+\epsilon)}\sin\phi_s\,\mathrm{Im}\,\sigma_{+0}^{+-}\\
&\quad+\sqrt{\epsilon(1+\epsilon)}\sin(2\phi-\phi_s)\,\mathrm{Im}\,\sigma_{+0}^{-+}\Big]\\
&+S_TP_\ell\Big[\sqrt{1-\epsilon^2}\cos(\phi-\phi_s)\,\mathrm{Re}\,\sigma_{++}^{+-}-\sqrt{\epsilon(1-\epsilon)}\cos\phi_s\,\mathrm{Re}\,\sigma_{+0}^{+-}\\
&\quad-\sqrt{\epsilon(1-\epsilon)}\cos(2\phi-\phi_s)\,\mathrm{Re}\,\sigma_{+0}^{-+}\Big],
\end{aligned}\tag{1}$$

where only terms relevant for the present analysis are shown. Defintion of angles ϕ and ϕ_s is shown in Fig 1.

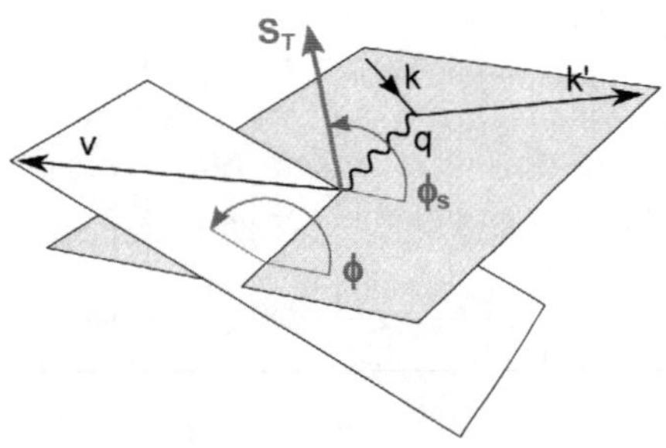

Figure 1. Kinematics of exclusive meson production in the target rest frame. Here $\mathbf{k}$, $\mathbf{k}'$, $\mathbf{q}$ and $\mathbf{v}$ represent the three-momentum vectors of the incident and the scattered muons, the virtual photon and the meson respectively. The component of the target spin vector $\mathbf{S}$ (not shown) perpendicular to the virtual-photon direction is denoted by $\mathbf{S_T}$.

For a transversely polarised target five single (UT) and three double (LT) spin asymmetries can be defined:

IOP Conf. Series: Journal of Physics: Conf. Series **938** (2017) 012010 doi:10.1088/1742-6596/938/1/012010

$$
\begin{aligned}
A_{\mathrm{UT}}^{\sin(\phi-\phi_s)} &= -\frac{\mathrm{Im}\left(\sigma_{++}^{+-}+\epsilon\sigma_{00}^{+-}\right)}{\sigma_0}, & A_{\mathrm{LT}}^{\cos(\phi-\phi_s)} &= \frac{\mathrm{Re}\,\sigma_{++}^{+-}}{\sigma_0},\\
A_{\mathrm{UT}}^{\sin(\phi+\phi_s)} &= -\frac{\mathrm{Im}\,\sigma_{+-}^{+-}}{\sigma_0}, & A_{\mathrm{LT}}^{\cos\phi_s} &= -\frac{\mathrm{Re}\,\sigma_{+0}^{+-}}{\sigma_0},\\
A_{\mathrm{UT}}^{\sin(3\phi-\phi_s)} &= -\frac{\mathrm{Im}\,\sigma_{+-}^{-+}}{\sigma_0}, & A_{\mathrm{LT}}^{\cos(2\phi-\phi_s)} &= -\frac{\mathrm{Re}\,\sigma_{+0}^{-+}}{\sigma_0},\\
A_{\mathrm{UT}}^{\sin\phi_s} &= -\frac{\mathrm{Im}\,\sigma_{+0}^{+-}}{\sigma_0}, & &\\
A_{\mathrm{UT}}^{\sin(2\phi-\phi_s)} &= -\frac{\mathrm{Im}\,\sigma_{+0}^{-+}}{\sigma_0}. & &
\end{aligned}
\tag{2}
$$

Here, σ_0 is the total unpolarised cross section, which is the sum of the cross sections for longitudinally and transversely polarised virtual photons, σ_L and σ_T, respectively:

$$
\sigma_0 = \frac{1}{2}\left(\sigma_{++}^{++}+\sigma_{++}^{--}\right)+\epsilon\sigma_{00}^{++} = \sigma_T+\epsilon\sigma_L. \tag{3}
$$

Each asymmetry is related to a modulation of the cross section as a function of ϕ and/or (see Eq. 1), which is indicated by the superscript.

3. Experimental data

The results presented in this Contribution are based on the data taken with the transversely polarised NH_3 target in 2010. The ω meson is produced in the following process

$$
\begin{aligned}
&\mu N \to \mu' N' \omega\\
&\quad\hookrightarrow \pi^+\pi^-\pi^0 \qquad \mathrm{BR}\approx 89\%\\
&\qquad\quad\hookrightarrow \gamma\gamma\ . \qquad \mathrm{BR}\approx 99\%
\end{aligned}
$$

Therefore, event containes an incident muon track and three outgoing tracks (μ', h^+, h^-), and only two ECAL clusters which are time-correlated with beam. It is checked that clusters are not caused by charged particles.

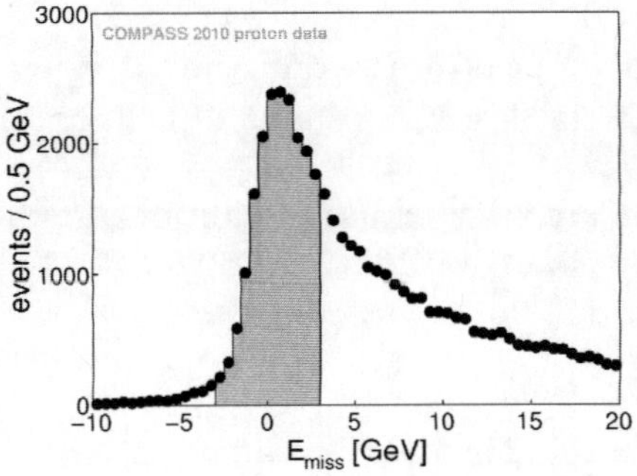

Figure 2. Distribution of E_{miss} . The accepted events are denoted by the shaded area.

Exclusivity is defined by applying a constraint on missing energy. Missing energy is calculated as $E_{miss} = \frac{M_X^2 - M_p^2}{2M_p}$ with $M_X^2 = (p + q - p_{\pi^+} - p_{\pi^-} - p_{\pi^0})^2$, where p, q, p_{π^+}, p_{π^-} and p_{π^0} are 4-momenta of proton, γ^* and each of three pions. The E_{miss} distribution is shown in Fig. 2.

For exclusive production one would expect the missing energy $E_{miss} = 0$. Taking into account experimental resolution the applied cut is -3.0 GeV $< E_{miss} <$ 3.0 GeV.

The constraint on squared transverse momentum of ω with respect to virtual photon γ^* momentum, $0.05 < P_T^2 < 0.5$ GeV^2, is applied, in order to remove coherent production and to suppress non-exclusive background (see Fig. 3). Other constraints and mean kinematic values are the following $Q^2 = -q^2 = -(k-k')^2 = 1.0 \div 10.0$ GeV^2, $\langle Q^2 \rangle = 2.2$ GeV^2, $W = \sqrt{(q+p)^2} >$ 5.0 GeV, $\langle W \rangle = 7.1$ GeV, $x_B = Q^2/2pq = 0.003 \div 0.35$, $\langle x_B \rangle = 0.049$

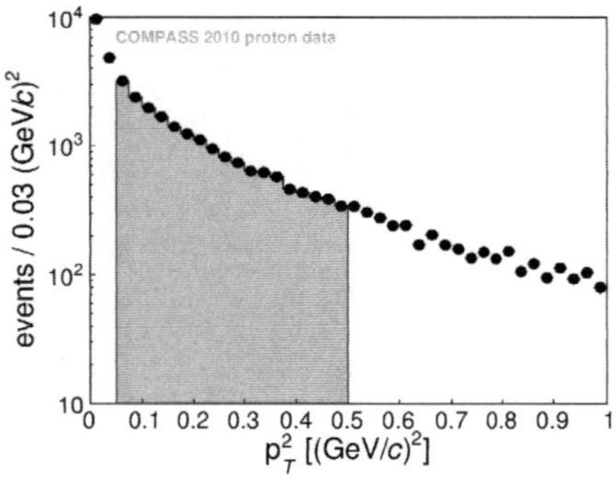

Figure 3. Distribution of p_T^2. The accepted events are denoted by the shaded area.

A π^0 meson is reconstructed from two calorimeter clusters as described in Ref. [6]

The limit on the invariant mass of two photons, $M_{\gamma\gamma}$, depends on the energy $E_{\gamma\gamma}$ of the π^0 candidate:

$$|M_{\gamma\gamma} - M_{\pi^0,\,\mathrm{par}}(E_{\gamma\gamma})| < 3\;\sigma_{\mathrm{par}}(E_{\gamma\gamma}) \qquad (4)$$

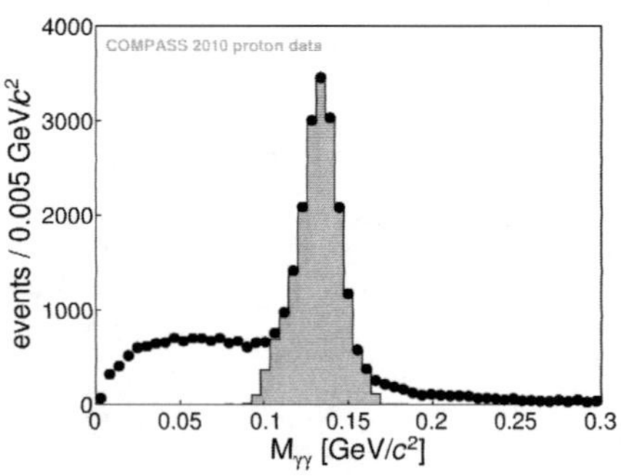

Figure 4. Distribution of the invariant mass of two photons. The accepted events are denoted by the shaded area.

Here, position $M_{\pi^0,\,\mathrm{par}}(E_{\gamma\gamma})$ and width $\sigma_{\mathrm{par}}(E_{\gamma\gamma})$ of the π^0 peak are parameterised using semi-inclusive data for π^0 mesons. In Fig. 4 distribution of the invariant mass of two photons is shown.

Events corresponding to incoherent exclusive ω production are selected using additional cuts on the invariant mass of the $\pi^+\pi^-\pi^0$ system, $M_{\pi^+\pi^-\pi^0}$,

$$\left| M_{\pi^+\pi^-\pi^0} - M_\omega^{\mathrm{PDG}} \right| < 70\ \mathrm{MeV}/c^2, \qquad (5)$$

where $M_\omega^{\mathrm{PDG}} = 782.65$ MeV$/c^2$ is the nominal ω resonance mass. The distribution of $M_{\pi^+\pi^-\pi^0}$ is shown in Fig. 5.

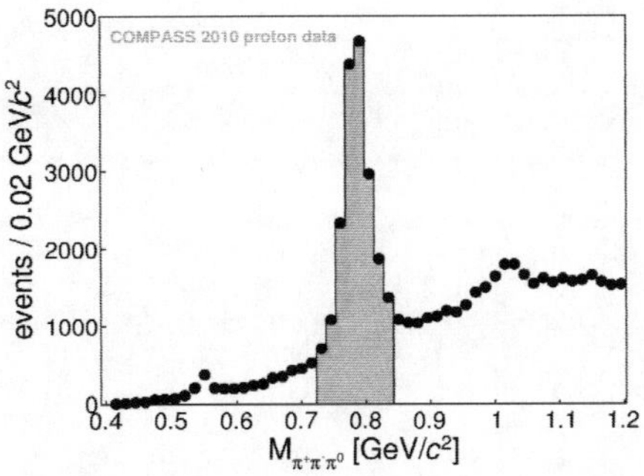

Figure 5. Distribution of $M_{\pi^+\pi^-\pi^0}$. The accepted events are denoted by the shaded area.

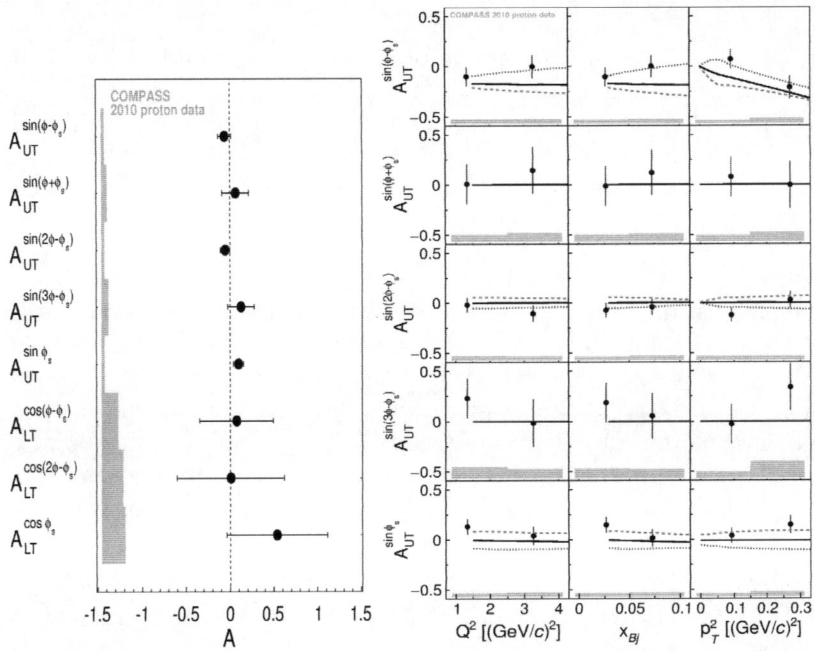

Figure 6. Left: Average azimuthal asymmetries for exclusive ω muoproduction. The error bars (left bands) represent the statistical (systematic) uncertainties. Right: Single spin azimuthal asymmetries as a function of Q^2, x_{Bj} and p_T^2. The curves show the predictions of the *GPD*-based model [7] for the average Q^2, W and p_T^2 values of the COMPASS data. The dashed red and dotted blue curves represent the predictions with the positive and negative $\pi\omega$ form factors, respectively, while the solid black curve represents the predictions without the pion pole.

4. Results

Asymmetries were extracted using unbinned maximum likelihood method with simultanous fit of signal and background asymmetries (16-parameter fit). In this fit it is assumed that asymmetries of semiinclusive deep inelastic scatterinig (SIDIS) background in exclusive region is the same as in region 7.0 GeV $< E_{miss} <$ 20.0 GeV.

The extracted azimuthal asymmetries, for the entire kinematic region, are shown in Fig. 6 (left). In addition, the single-spin asymmetries are measured in bins of Q^2, x_{Bj} or p_T^2 with the results shown in Fig. 6 (right). The double-spin asymmetries are not not shown in separate kinematic bins because of large uncertainties. In Figure 6 (right) the COMPASS results are compared to the calculations of the GK model [7].

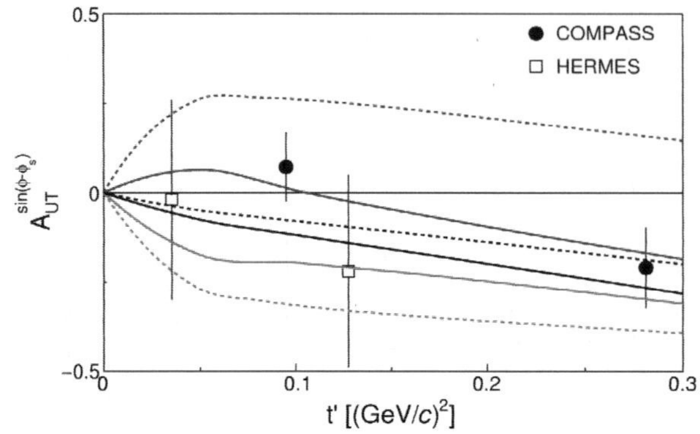

Figure 7. The asymmetry $A_{UT}^{sin(\phi-\phi_S)}$ for exclusive ω muoproduction by the COMPASS (filled circles) and HERMES [8] (open squares) collaborations as a function of t'. The curves show the predictions of the GPD-based model [7] given for the average Q^2 and W values of the COMPASS (solid lines) and HERMES (dashed lines) data. For each set of curves, the upper (blue) and lower (red) ones are for the negative and positive $\pi\omega$ form factors, respectively, while the middle (black) one represents the predictions without the pion pole.

The latter are obtained for the average W, Q^2 and x_{Bj} values of the COMPASS data: $W = 7.1$ GeV$/c^2$ and $p_T^2 = 0.17$ (GeV$/c)^2$ for the x_{Bj} and Q^2 dependences, and $W = 7.1$ GeV$/c^2$ and $Q^2 = 2.2$ (GeV$/c)^2$ for the p_T^2 dependence. The predictions are given for three versions of the model: with the pion-pole contribution using a positive or negative $\pi\omega$ transition form factor, and without the pion-pole contribution. The interpretation of ω results is challenging in terms of $GPDs$, as exclusive ω meson production is significantly influenced by the pion-pole exchange contribution, and at present the sign of $\pi\omega$ transition form factor is unknown. By comparing the COMPASS results with the calculations of the GK model (see Fig. 6 (right)), one finds that the asymmetries $A_{UT}^{sin(\phi-\phi_S)}$ and $A_{UT}^{sin(2\phi-\phi_S)}$ prefer the negative $\pi\omega$ transition form factor, while the asymmetry $A_{UT}^{sin\phi_S}$ prefers the positive one. The other measured asymmetries are not sensitive to the sign of the $\pi\omega$ form factor.

The single-spin azimuthal asymmetries for ω production on transversely polarised protons were measured also by the HERMES collaboration [8]. They conclude that their data seem to favour the positive $\pi\omega$ form factor, although within large experimental uncertainties. A direct comparison of published asymmetry values measured in both experiments is not straightforward, because the HERMES definition of physics asymmetries differs from that given in Eq. 3. Such comparison is only possible for the asymmetry $A_{UT}^{sin(\phi-\phi_S)}$. The results from both experiments are shown as a function of t' in Fig. 7 indicating their compatibility within experimental uncertainties. Note that the COMPASS results cover a wider kinematic range and they have smaller uncertainties, for example for the asymmetry $A_{UT}^{sin(\phi-\phi_S)}$ by a factor larger than two.

References

[1] X.-D. Ji, *Phys. Rev.* **D55** (1997) 7114
[2] S. V. Goloskokov and P. Kroll, *Eur. Phys. J.* **C42** (2005) 281
[3] S. V. Goloskokov and P. Kroll, *Eur. Phys. J.* **C53** (2008) 367
[4] S. V. Goloskokov and P. Kroll, *Eur. Phys. J.* **C59** (2009) 809
[5] M. Diehl and S. Sapeta,*Eur. Phys. J.* **C41** (2005) 515 .
[6] COMPASS Collaboration *Nucl. Phys.* **B 915** (2017) 454
[7] S. V. Goloskokov and P. Kroll, *Eur. Phys. J.* **A50** (2014) 146
[8] A.Airapetian et al [HERMES], *Eur. Phys. J.* **C75** (2015) 600

IOP Conf. Series: Journal of Physics: Conf. Series **938** (2017) 012011 doi:10.1088/1742-6596/938/1/012011

Overview of recent HERMES results

Hrachya Marukyan (On behalf of the HERMES Collaboration)
AANL (Yerevan Physics Institute), Alikhanian Brs. 2, Yerevan 036, Armenia
E-mail: marukyan@mail.desy.de

Abstract. An overview of more recent and important results from the HERMES experiment is presented in this paper. HERMES collected a wealth of data using the 27.6 GeV polarized HERA lepton beam and various polarized and unpolarized gaseous targets. This allows for a series of unique measurements of observables sensitive to the multidimensional structure of the nucleon. Amongst them are the semi-inclusive deep-inelastic scattering measurements of azimuthal modulations sensitive to the leading-twist transverse momentum distributions and the distributions sensitive to convolutions of the twist-2 and twist-3 distribution and fragmentation functions. They all provide information on the three-momentum-dependent quark distributions. Knowledge about the quark distribution as a function of longitudinal momentum and transverse position in impact-parameter space can be accessed, e.g., through exclusive vector−meson leptoproduction. Particularly, the measurement of spin-density-matrix elements on an unpolarized hydrogen target and the measurement of azimuthal modulations on transversely polarized hydrogen target for the $\omega-$meson production are described in this paper. The results of the measurement of ratios of helicity amplitudes for the ρ^0-meson leptoproduction on transversely polarized hydrogen target are also presented.

1. Introduction

The HERMES experiment at DESY collected the data from 1995 until 2007 using the 27.6 GeV HERA electron or positron beams. In the analysis of data presented here, longitudinally polarized leptons were scattered off longitudinally, transversely or an unpolarized hydrogen targets, and unpolarized deuterium target. The scattered lepton and particles produced in the reaction were detected by a forward spectrometer. The lepton-hadron separation was performed using a transition-radiation detector, a scintillator preshower counter, an electromagnetic calorimeter and a threshold gas Cherenkov counter (ring-imaging Cherenkov detector from 1998). Hadron identification was performed by Cherenkov detectors, allowing the discrimination of pions, kaons and protons.

The analysis of the azimuthal distribution of hadrons in semi-inclusive deep-inelastic scattering (DIS) of leptons off a transversely polarized hydrogen target provides access to the Sivers distribution [1] and Collins fragmentation [2] functions. The former describes the distribution of unpolarized quarks in a transversely polarized nucleon, correlating quark transverse momentum with the nucleon's transverse spin, while the latter describes the fragmentation of a transversely polarized quark into an unpolarized hadron. The selected results sensitive to these two quantities are presented in section 2. The results from the recent analysis of amplitudes of the single-spin asymmetry in semi-inclusive DIS leptoproduction of hadrons in case of collision of a longitudinally polarized lepton beam with an unpolarized target are also presented in this section.

Data collected on longitudinally and transversely polarized hydrogen targets are used to extract the Spin-Density-Matrix Elements (SDMEs) and the amplitudes of the single-spin asymmetry of the cross section respectively in hard exclusive leptoproduction of $\omega-$ mesons. They provide information on the sign of $\pi\omega$ transition form factor. The measurement of amplitudes of azimuthal sine modulations is presented in section 3.

The results from the analysis of the ratios of helicity amplitudes in hard exclusive leptoproduction of ρ^0- mesons on transversely polarized hydrogen target are summarized in section 4.

2. Single- and double-spin asymmetries measured in semi-inclusive DIS off an unpolarized and transversely polarized hydrogen target

Semi-inclusive DIS off a transversely polarized target (T) allows to access the single- and double-spin asymmetry amplitudes with characteristic angular modulations. Two angles appear in the description of the cross section of this process: angle ϕ (below also denoted by $\sin\phi_h$), defined as the angle between the lepton-scattering and the hadron-production planes, and ϕ_S, being the azimuthal angle of the transverse component of the target-spin vector about the virtual photon direction with respect to the lepton scattering plane. Each of the azimuthal amplitudes corresponds to convolutions of different distribution and fragmentation functions. Two of them, proportional to the $\sin(\phi - \phi_S)$ and $\sin(\phi + \phi_S)$ modulations respectively, are interpreted as the convolutions of the Sivers distribution function [1] and the spin-independent fragmentation function and transversity and the Collins fragmentation function [2]. The HERMES results on the Sivers and Collins amplitudes for charged and neutral pions and for charged kaons as a function of x, z and $P_{h\perp}$ were published in Ref. [3] and Ref. [4], respectively. Here, x represents the Bjorken scaling variable, z denotes the fractional hadron energy with respect to the virtual photon energy in the target rest frame and $P_{h\perp}$ the magnitude of the transverse momentum of the final state hadron.

Moreover, the fine binned three-dimensional (x, z, $P_{h\perp}$) extraction of various azimuthal amplitudes for charged pions, kaons and protons is possible from HERMES data. This allows to constrain global fits to the experimental measurements in a more profound way. As an example, the HERMES preliminary results of the Collins amplitudes for negative pions vs. x extracted for four $P_{h\perp}$ and z bins are presented in figure 1, showing an increase of the amplitudes with x at large $P_{h\perp}$, and the Sivers amplitudes for positive kaons vs. $P_{h\perp}$ for four z and x bins are presented in figure 2. Note that the values of Sivers amplitudes are larger for positive kaons than for π^+ (not shown here). The three-dimensional extraction of all possible single- and double-spin asymmetry amplitudes sensitive to different combination of distribution and fragmentation functions are also available (not shown here).

In case of longitudinal beam (L) and unpolarized target (U) only target spin-independent parts can contribute to the asymmetry. Here, the structure function of interest is related to the asymmetry amplitude $A_{LU}^{\sin\phi_h}$, which is sensitive to convolutions of twist-2 distribution (fragmentation) functions with twist-3 fragmentation (distribution) functions. HERMES published the results of the single-spin asymmetry A_{LU} for charged and neutral pions as a function of x, z and $P_{h\perp}$ [5]. The HERMES preliminary results of the asymmetry amplitudes $A_{LU}^{\sin\phi_h}$ extracted from data on unpolarized hydrogen and deuterium targets for charged pions and kaons are shown in figure 3 as a function of x, z and $P_{h\perp}$. In order to present a direct relation to the corresponding structure functions the asymmetry amplitudes $A_{LU}^{\sin\phi_h}$ are plotted taking into account the ratio of longitudinal and transverse photon flux ϵ. As an example, the multidimensional extraction of these asymmetry amplitudes for positive pions in projections of x, z and $P_{h\perp}$ in four bins for each of the kinematic variables are shown in figure 4.

In figure 5 the HERMES preliminary results on the x, z and $P_{h\perp}$ dependencies of the asymmetry amplitudes $A_{LU}^{\sin\phi}$ extracted from data on unpolarized protons for π^+ and π^- mesons are compared with those obtained by the CLAS Collaboration at Jefferson Lab [6]. The different behavior of asymmetry amplitudes for the two data sets observed for negative pions in z projection is probably due to the sensitivity of the Collins $e(x)$ term to different $x-$range probed in these experiments. In figure 6 the HERMES preliminary results of the asymmetry amplitudes extracted from unpolarized deuteron data for charged pions are compared with similar results for muon production of charged hadrons in SIDIS from a ^6LiD target obtained by COMPASS Collaboration at CERN [7]. For the isoscaler targets a consistent behavior is observed.

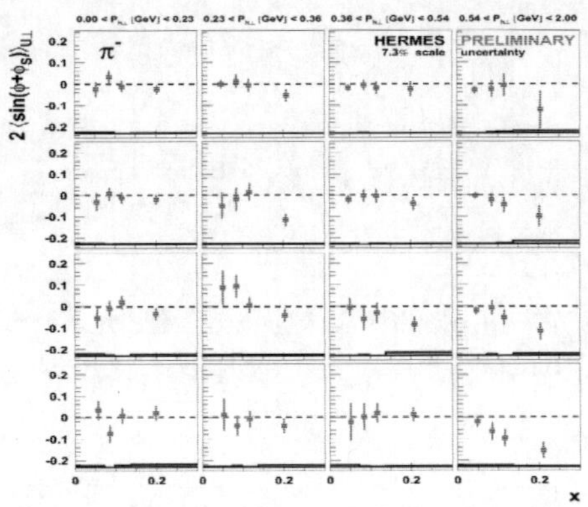

Figure 1. The HERMES preliminary results of Collins amplitudes for π^- vs. x for four bins in $P_{h\perp}$ and z.

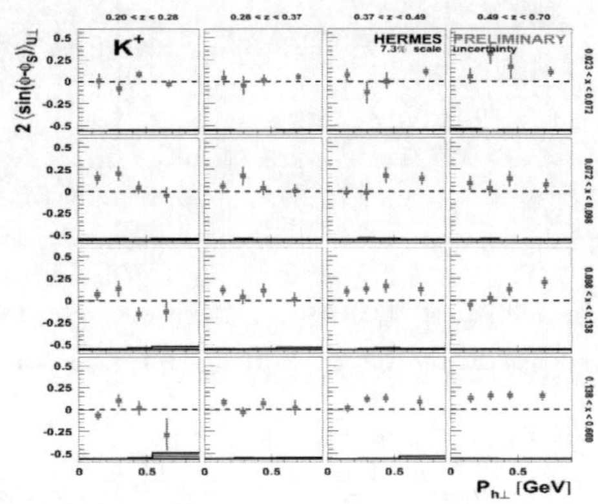

Figure 2. The HERMES preliminary results of Sivers amplitudes for K^+ vs. $P_{h\perp}$ for four bins in z and x.

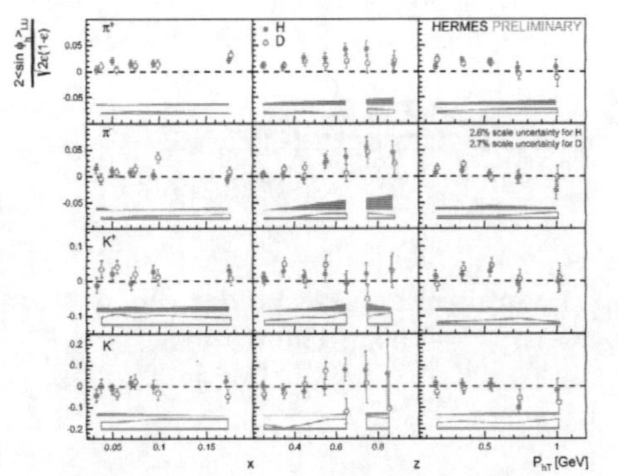

Figure 3. The HERMES preliminary results of $A_{LU}^{\sin\phi_h}$ amplitudes for charged pions and kaons extracted from data on unpolarized hydrogen (H) and deuterium (D) targets.

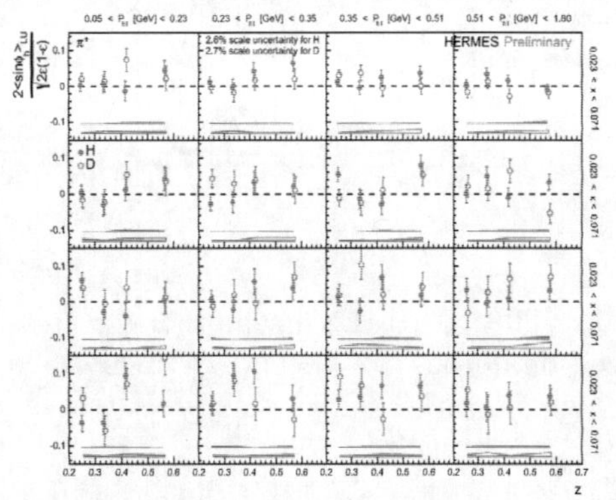

Figure 4. The HERMES preliminary results of the multidimensional extraction of $A_{LU}^{\sin\phi_h}$ amplitudes for π^+ mesons from data on unpolarized hydrogen (H) and deuterium (D) targets.

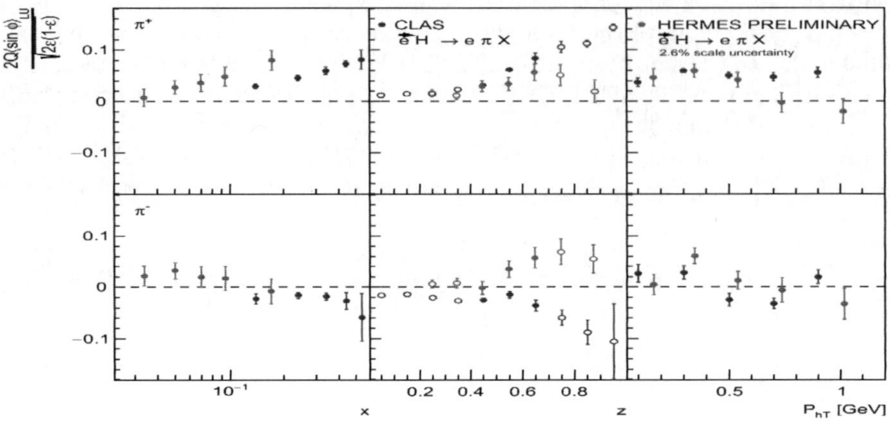

Figure 5. The HERMES preliminary results of $A_{LU}^{\sin\phi}$ amplitudes extracted from data on unpolarized protons for π^+ and π^- mesons compared with those from CLAS [6].

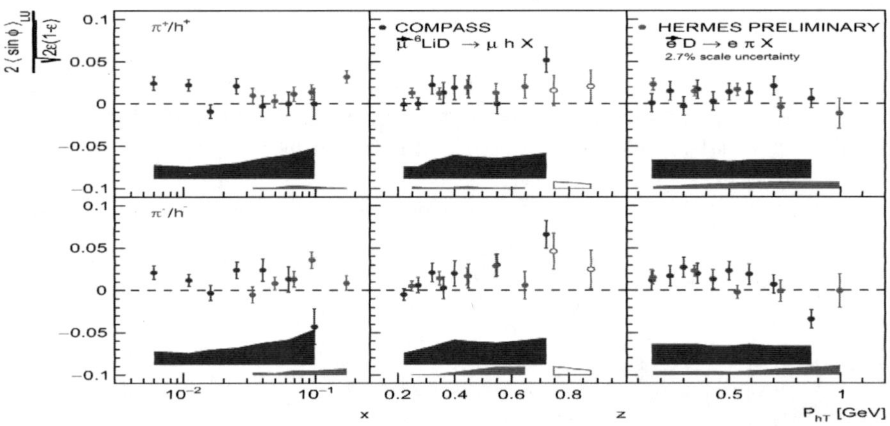

Figure 6. The HERMES preliminary results of $A_{LU}^{\sin\phi}$ amplitudes extracted from data on unpolarized deuterons for π^+ and π^- mesons compared with those from COMPASS [7].

3. Exclusive $\omega-$meson leptoproduction

Hard exclusive leptoproduction of $\omega-$mesons has been studied with the HERMES spectrometer by scattering longitudinally polarized positron and electron beams on unpolarized hydrogen and deuterium targets or on transversely polarized hydrogen target in the kinematic region $Q^2 > 1.0\,\mathrm{GeV}^2, 3.0\,\mathrm{GeV} < \mathrm{W} < 6.3\,\mathrm{GeV}$, and $-\mathrm{t}' < 0.2\,\mathrm{GeV}^2$. Here, Q^2 represents the negative square of the virtual-photon four-momentum, W is the invariant mass of the photon-nucleon system and $-t'$ is the smallest kinematically allowed value of the momentum transfer to the target. Using an unbinned maximum likelihood method, 15 unpolarized and 8 polarized SDMEs are extracted from unpolarized hydrogen and deuterium data [8]. From the analysis of transversely polarized hydrogen data the amplitudes of five sine modulations of the single-spin asymmetry of the cross section with respect to the transverse proton polarization are measured [9], the Q^2 and $-t'$ dependence of which are shown in figure 7. They are determined in the entire kinematic region (open squares) as well as for two bins in photon virtuality and momentum transfer to the nucleon (full circles). The inner error bars represent the statistical uncertainties, while the outer ones indicate the statistical and systematic uncertainties added in quadrature. The results receive an additional 8.2% scale uncertainty corresponding to the target-polarization uncertainty. These results are compared to a phenomenological model that includes the pion pole contribution [10]. As can be seen the data favor, within this model, a positive $\pi\omega$ transition form factor (red curves), the sign of which was not possible to determine from the measurements of SDMEs. Also, a separation of asymmetry amplitudes into longitudinal and transverse components is performed (not shown here).

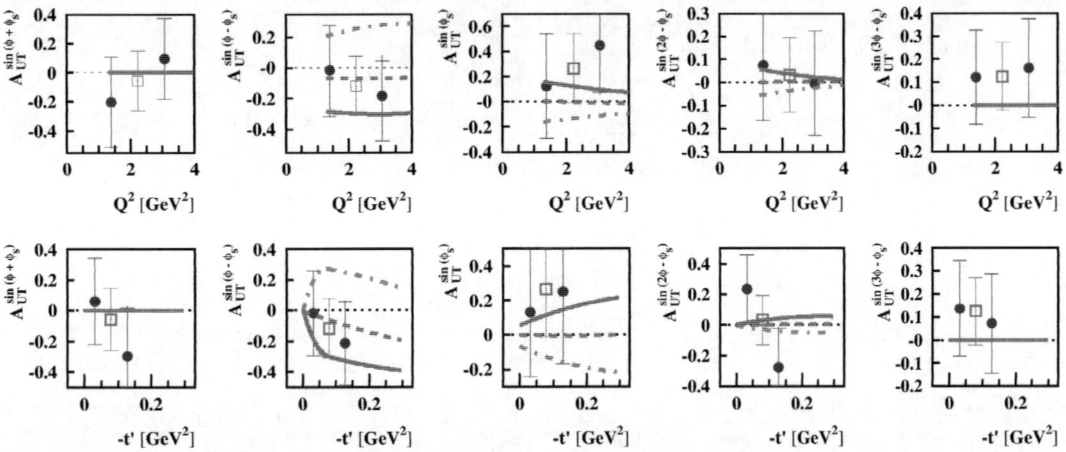

Figure 7. The amplitudes describing the strength of the sine modulations of the cross section for hard exclusive $\omega-$meson leptoproduction.

4. Ratios of the helicity amplitudes in exclusive ρ^0-meson leptoproduction

Hard exclusive leptoproduction of ρ^0-mesons has been studied by the HERMES experiment in scattering longitudinally polarized positron and electron beams on unpolarized, longitudinally or transversely polarized hydrogen targets [11-14]. The results of the analysis of SDMEs on unpolarized and transversely polarized hydrogen target were published in Ref. [11] and Ref. [12], respectively. The results of the ρ^0 helicity amplitudes extracted from transversely polarized hydrogen data were published first in Ref. [13] and updated in Ref. [14]. In the latter case, the ρ^0-mesons were detected by the HERMES spectrometer in the kinematic region of $1.0\,\mathrm{GeV}^2 < \mathrm{Q}^2 < 7.0\,\mathrm{GeV}^2, 3.0\,\mathrm{GeV} < \mathrm{W} < 6.3\,\mathrm{GeV}, \mathrm{and} - \mathrm{t}' < 0.4\,\mathrm{GeV}^2$. Using an unbinned maximum-likelihood fit method 25 parameters are extracted, which determine the real and imaginary parts of the ratios of several helicity amplitudes. (The denominator of these ratios is the dominant nucleon-helicity-non-flip amplitude $F_{0\frac{1}{2}0\frac{1}{2}}$).

The results of the ρ^0 helicity amplitude ratios obtained from the 25-parameter fit is presented in figure 8. Also the results from a previous analysis [12] are plotted in the figure. Furthermore, 71 ρ^0 SDMEs are determined from the 25 parameters (see Figs. 4-6 of Ref [14]), compared to the 53 elements directly determined in earlier analyses [11-13].

All amplitude ratios are compared with those from the phenomenological model that takes into account the contribution from pion exchange [10], and they are found to be in good agreement for most of the helicity amplitude ratios with a few exception (see Fig. 6. of Ref [14]). Within the model, the data favor a positive sign for the $\pi - \rho$ transition form factor.

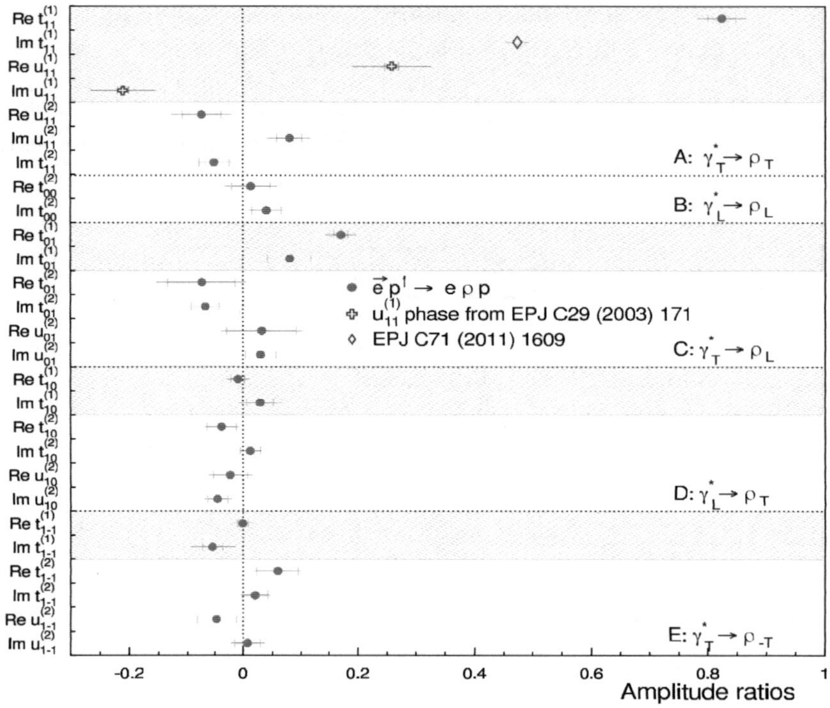

Figure 8. The summary plot of ratios of the helicity amplitudes for hard exclusive ρ^0-meson leptoproduction measured at HERMES [14].

Acknowledgments

I would like to thank the organizers for their support and for a very interesting conference.

References

[1] Sivers D W 1990 *Phys. Rev.* D **41** 83
[2] Collins J C 1993 *Phys. Rev.* B **396** 161
[3] Airapetian A et al. (HERMES Collaboration) 2009 *Phys. Rev. Lett.* **103** 152002
[4] Airapetian A et al. (HERMES Collaboration) 2010 *Phys. Lett.* B **693** 11
[5] Airapetian A et al. (HERMES Collaboration) 2007 *Phys. Lett.* B **648** 164
[6] Gohn W et al. (CLAS Collaboration) 2014 *Phys. Rev.* D **89** 072011
[7] Adolph C et al. (COMPASS Collaboration) 2014 *Nucl. Phys.* B **886** 1046
[8] Airapetian A et al. (HERMES Collaboration) 2014 *Eur. Phys. J.* C **74** 3110
[9] Airapetian A et al. (HERMES Collaboration) 2015 *Eur. Phys. J.* C **75** 600
[10] Goloskokov S V and Kroll P 2014 *Eur. Phys. J.* A **50** 146
[11] Airapetian A et al. (HERMES Collaboration) 2011 *Eur. Phys. J.* C **71** 1609
[12] Airapetian A et al. (HERMES Collaboration) 2009 *Eur. Phys. J.* C **62** 659
[13] Airapetian A et al. (HERMES Collaboration) 2009 *Phys. Lett.* B **679** 100
[14] Airapetian A et al. (HERMES Collaboration) 2017 *Eur. Phys. J.* C **77** 378

IOP Conf. Series: Journal of Physics: Conf. Series **938** (2017) 012012 doi:10.1088/1742-6596/938/1/012012

Measurement of q_{T}-weighted TSAs in 2015 COMPASS Drell–Yan data

Jan Matoušek, on Behalf of the COMPASS Collaboration
Charles University, Faculty of Mathematics and Physics, 180 00 Prague, Czech Republic and Trieste Section of INFN, 341 27 Trieste, Italy
E-mail: `jan.matousek@cern.ch`

Abstract. In the polarised Drell–Yan experiment at the COMPASS facility at CERN the beam of negatively-charged pions with 190 GeV/c momentum and intensity about 10^8 pions/s interacted with transversely polarised NH_3 target. Muon pairs produced in Drell–Yan process (DY) were detected. Recently, the first ever Transverse Spin Asymmetries (TSAs) measurement in DY has been presented by COMPASS. A complementary analysis of the TSAs weighted by powers of the dimuon transverse momentum q_{T} are presented. In the Transverse Momentum Dependent (TMD) PDF formalism, the q_{T}-weighted TSAs can be written in terms of products of the TMD PDFs of two colliding hadrons, unlike the conventional TSAs, which are their convolutions over quarks transverse momenta. The results are compared in a straightforward way with the weighted Sivers asymmetry in the SIDIS process, released by COMPASS in 2016.

1. Introduction

The hadron structure can be described at leading twist by eight Transverse Momentum Dependent (TMD) Parton Distribution Functions (PDFs), which depend on the fraction x of the hadron momentum carried by the parton and the transverse component of the parton momentum $\boldsymbol{k}_{\mathrm{T}}^2$. They have been probed in Semi-Inclusive Deep Inelastic Scattering (SIDIS), where the cross-section contains convolutions of the TMD PDFs and fragmentation functions [1].

Comparison of SIDIS with the Drell–Yan process (DY), giving access to convolutions of TMD PDFs of the two colliding hadrons [2], can provide a test of the PDFs universality. In fact, two PDFs (the Sivers and Boer–Mulders functions) are predicted to bear opposite signs when extracted from SIDIS and DY [3]. Recently, COMPASS has done a pioneering measurement of the Transverse Spin Asymmetries (TSAs) in DY [4]. A complementary analysis of the same data, using the formalism of transverse momentum q_{T} weighted TSAs [5, 6], is presented.

The convolutions of TMDs are usually solved assuming a certain functional form of their dependence on $\boldsymbol{k}_{\mathrm{T}}^2$ (e.g. Gaussian). In SIDIS, it can be avoided using the TSAs weighted with powers of the outgoing hadron transverse momentum P_{T} [7, 8, 9]. Preliminary P_{T}-weighted TSAs from HERMES [10] were used to estimate the q_{T}-weighted Sivers asymmetry expected in DY experiments [5]. Similarly, in Sec. 3 we use the recent COMPASS measurement of the P_{T}/z-weighted Sivers asymmetry [11] to get a projection, which we compare with our DY results.

2. Transverse momentum weighted asymmetries in Drell–Yan process

We study the Drell–Yan reaction with 190 GeV/c pion beam and NH_3 target with the H nuclei transversely polarised $\pi^- p^\uparrow \to \mu^-\mu^+ X$. At Leading Order (LO), the reaction proceeds via annihilation of a quark-antiquark pair into a virtual photon with momentum q, which decays into the dimuon. The LO cross-section contains five orthogonal modulations in ϕ and ϕ_S – the azimuthal angles of the muon momentum in the Collins–Soper frame and of the target spin vector $\boldsymbol{S}_\mathbf{T}$ in the target rest frame with z-axis along the beam momentum and x-axis along dimuon transverse momentum $\boldsymbol{q}_\mathbf{T}$, respectively [12]. The structure functions $F^X_{\mathrm{U,T}}$ can be written as convolutions of TMD PDFs over the intrinsic transverse momenta of the two colliding partons $\boldsymbol{k}_{\pi\mathbf{T}}$ and $\boldsymbol{k}_{N\mathbf{T}}$ [2]. When the structure functions are integrated over q_T with properly chosen weights, the convolutions can be disentangled[1]:

$$\int \mathrm{d}^2\boldsymbol{q}_\mathbf{T}\, F^1_\mathrm{U} = \frac{1}{3}\sum_q e_q^2\left[f^{\bar q}_{1,\pi}(x_\pi)\, f^q_{1,N}(x_N) + f^q_{1,\pi}(x_\pi)\, f^{\bar q}_{1,N}(x_N)\right] \tag{1}$$

$$\int \mathrm{d}^2\boldsymbol{q}_\mathbf{T}\, \frac{q_\mathrm{T}}{M_\mathrm{p}} F^{\sin\phi_S}_\mathrm{T} = -\frac{2}{3}\sum_q e_q^2\left[f^{\bar q}_{1,\pi}(x_\pi)\, f^{\perp(1)q}_{1\mathrm{T,p}}(x_N) + (q\leftrightarrow\bar q)\right], \tag{2}$$

$$\int \mathrm{d}^2\boldsymbol{q}_\mathbf{T}\, \frac{q^3_\mathrm{T}}{2M_\pi M^2_\mathrm{p}} F^{\sin(2\phi+\phi_S)}_\mathrm{T} = -\frac{2}{3}\sum_q e_q^2\left[h^{\perp(1)\bar q}_{1,\pi}(x_\pi)\, h^{\perp(2)q}_{1\mathrm{T,p}}(x_N) + (q\leftrightarrow\bar q)\right], \tag{3}$$

$$\int \mathrm{d}^2\boldsymbol{q}_\mathbf{T}\, \frac{q_\mathrm{T}}{M_\pi} F^{\sin(2\phi-\phi_S)}_\mathrm{T} = -\frac{2}{3}\sum_q e_q^2\left[h^{\perp(1)\bar q}_{1,\pi}(x_\pi)\, h^{q}_{1,\mathrm{p}}(x_N) + (q\leftrightarrow\bar q)\right], \tag{4}$$

where the sums run over quarks and antiquarks q; e_q are fractional electric charges; $M_{\pi,\mathrm{p}}$ are the pion and proton masses; and $f^{(n)}$ or $h^{(n)}$ are the n-th k^2_T-moments of the TMD PDFs, $f^{(n)}(x) = \int \mathrm{d}^2\boldsymbol{k}_\mathbf{T}\, \left[k^2_\mathrm{T}/(2M^2)\right]^n f(x,k^2_\mathrm{T})$. We measure the q_T-weighted TSAs, defined as

$$A^{\sin\Phi\, W_\Phi}_\mathrm{T} = \frac{\int \mathrm{d}^2\boldsymbol{q}_\mathbf{T}\, W_\Phi\, F^{\sin\Phi}_\mathrm{T}}{\int \mathrm{d}^2\boldsymbol{q}_\mathbf{T}\, F^1_\mathrm{U}}, \qquad \Phi = \phi_S, 2\phi+\phi_S, 2\phi-\phi_S \tag{5}$$

where W_Φ denotes the weights. The weighted TSAs are obtained by fit of the so-called modified double ratio $R(\Phi) \propto A^{\sin\Phi W_\Phi}_\mathrm{T} \sin\Phi$. The ratio is constructed from event counts and sums of event weights coming from two oppositely-polarised target cells and from two sub-periods divided by polarisation reversal to cancel the acceptance $a(\Phi)$. It is calculated using eight bins in Φ and one or three bins in four kinematic variables. As in the standard TSA analysis [4] we use the dilution factor to correct the asymmetries for the target composition, so the TSAs refer to proton.

The data have been collected in 2015 in nine data-taking periods (each having two sub-periods). The event sample is almost the same as in the TSA analysis [4]. The same invariant mass range $M \in [4.3, 8.5]\,\mathrm{GeV}/c^2$ is used. The sharp cuts on q_T are replaced by cut on individual muon transverse momenta $l_\mathrm{T} < 7\,\mathrm{GeV}/c$. About 39 000 dimuons pass the event selection. We estimate the background to be at the level of up to 4 % [4]. Several possible systematic effects have been investigated. The major contribution comes from the effect of variation of the data-taking conditions within a given period, estimated by measurement of false asymmetries. They are calculated from events with sub-periods or target cells of origin changed in such a way that the physics asymmetries cancel. The combined systematic uncertainty is about 0.7 times the statistical one. In addition, there are normalisation uncertainties of about 5 % from the polarisation measurement and dilution factor calculation. The results are shown on Fig. 1, the q_T distribution on Fig. 2.

3. Transverse momentum weighted Sivers asymmetry in SIDIS and Drell–Yan

To get a projection for the expected q_T-weighted Sivers asymmetry in DY, we use the corresponding P_T/z-weighted asymmetry in SIDIS $\mu\mathrm{p}^\uparrow \to \mu' hX$, measured by COMPASS for

[1] We use the same Sivers function sign as in Ref. [11, 13] and opposite to Ref. [1, 2] and the Trento convention [14].

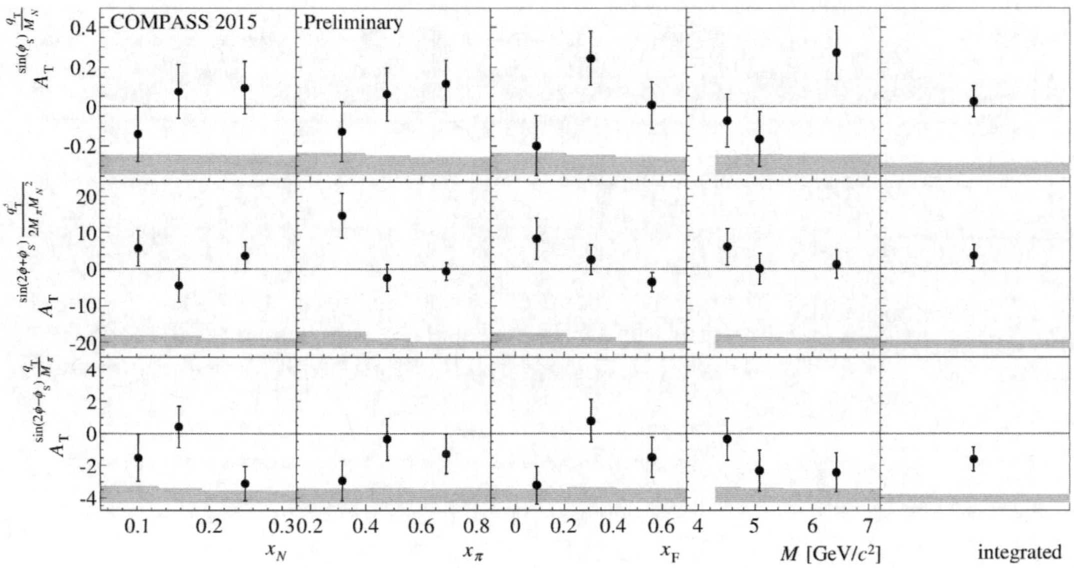

Figure 1. The q_T-weighted TSAs. The systematic uncertainty is denoted by blue bands. Normalisation uncertainties of about 5% (from polarisation and dilution factor) are not shown.

positive and negative hadrons h with $z > 0.2$ [11]. It can be written as [9, 11]:

$$A_{\mathrm{UT,T}}^{\sin(\phi_{\mathrm{h}}-\phi_{\mathrm{S}})\frac{P_{\mathrm{T}}}{zM}}(x,z,Q^2) = \frac{\int \mathrm{d}^2\boldsymbol{P}_{\mathbf{T}}\,\frac{P_{\mathrm{T}}}{zM}\,F_{\mathrm{UT,T}}^{\sin(\phi_{\mathrm{h}}-\phi_{\mathrm{S}})}}{\int \mathrm{d}^2\boldsymbol{P}_{\mathbf{T}}\,F_{\mathrm{UU,T}}^{1}} = 2\frac{\sum_q e_q^2\, f_{1\mathrm{T,p}}^{\perp(1)q}(x,Q^2)\,D_{1,q}^h(z,Q^2)}{\sum_q e_q^2\, f_{1,\mathrm{p}}^q(x,Q^2)\,D_{1,q}^h(z,Q^2)}, \quad (6)$$

where we use the standard SIDIS variables, F indicate the SIDIS structure functions [1], and $D_{1,q}^h(z)$ is the fragmentation function of q into hadron h. In writing explicitly the asymmetry, we only consider u, d, and s quarks and the corresponding antiquarks q within the proton; we assume vanishing Sivers function of sea quarks; and, like in Ref. [13], we integrate the fragmentation functions over the available range in z.

We take the unpolarised PDFs from the CTEQ 5D global fit [15], implemented in the LHAPDF library [16] and the charged hadron fragmentation functions from the DSS 07 LO global fit [17]. As in [13], we use the collinear evolution of the PDFs and FFs, taken at the mean Q^2 at each x (plotted on Fig. 3), as x and Q^2 are correlated. We parametrise the first k_{T}^2-moment of the Sivers function as $x f_{1\mathrm{T}}^{\perp(1)q}(x) = a_q\, x^{b_q}\,(1-x)^{c_q}$. The asymmetries for positive and negative hadrons in bins of x are simultaneously fitted (Fig. 4).

To get the projection for the q_{T}-weighted asymmetry in DY, we use Eq. (2, 5), we assume the change-of-sign prediction [3] and valence quark dominance. The asymmetry simplifies to $A_{\mathrm{T}}^{\sin\phi_{\mathrm{S}} q_{\mathrm{T}}/M_{\mathrm{p}}}(x_N, Q^2) \approx 2 f_{1\mathrm{T,p}}^{\perp(1)\mathrm{u}}(x_N,Q^2)/f_{1,\mathrm{p}}^{\mathrm{u}}(x_N,Q^2)$. We identify the x_N with the Bjorken x from SIDIS and we use the same unpolarised PDF, taken at the mean Q^2 of the DY events used in the analysis (Fig. 3). No evolution of $f_{1\mathrm{T,p}}^{\perp(1)\mathrm{u}}$ between the SIDIS and DY kinematics is considered. The result, compared with the measured asymmetries (Sec. 2), is shown on Fig. 6. A projection for combined analysis of 2015 and 2018 data is shown as well, assuming the statistics in 2018 to be 1.5 times larger than in 2015. The 1σ error-bands account only for the uncertainty of the fit and the statistical errors of the experimental data. Variation of PDF and FF sets has been found to lead to differences of about 0.02.

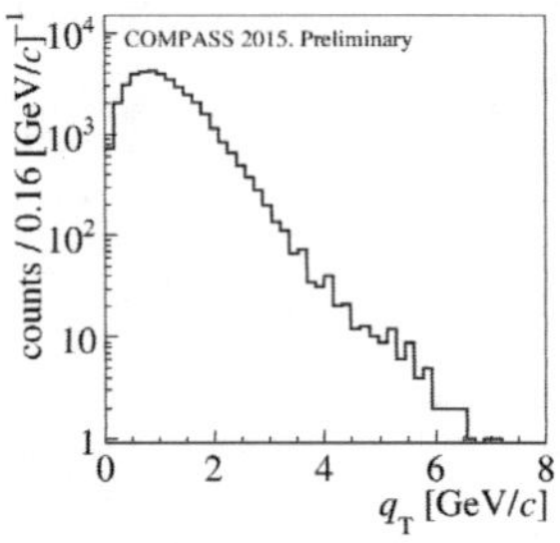

Figure 2. Distribution of q_{T} in the selected Drell–Yan events.

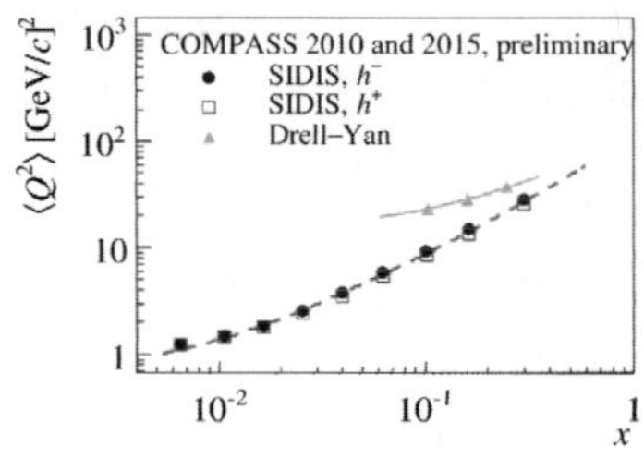

Figure 3. Mean Q^2 of the events in the Drell–Yan analysis and SIDIS analysis [11].

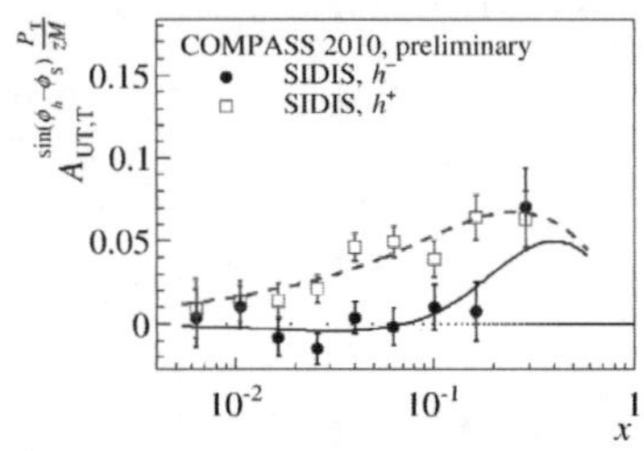

Figure 4. The weighted Sivers asym. in SIDIS [11], fitted. Statistical errors only.

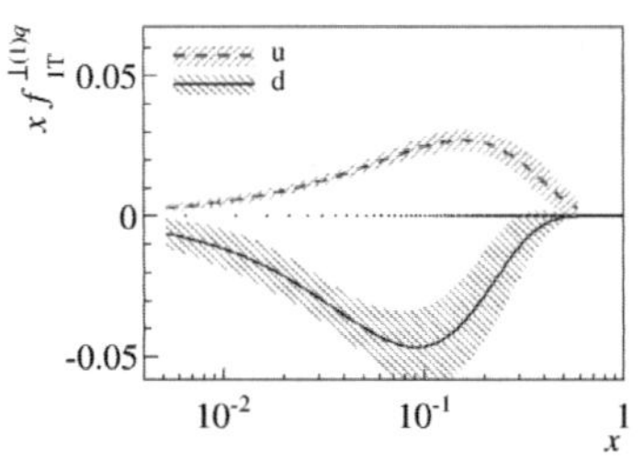

Figure 5. The Sivers PDF first k_{T}^2-moment as a function of x and $Q^2(x)$. The 1σ error-bands account only for the uncertainty of the fit and the statistical errors of the data.

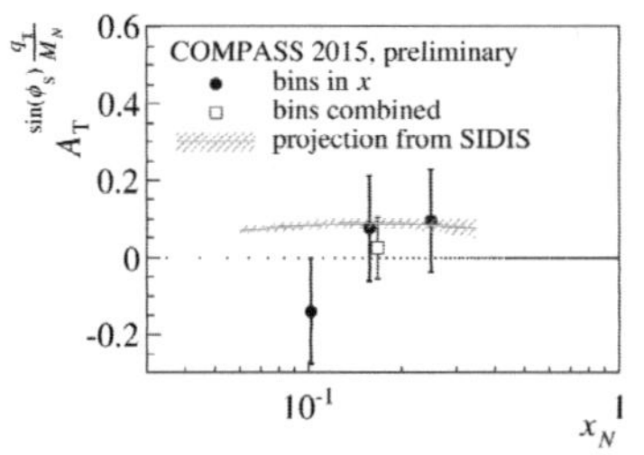

Figure 6. Weighted Sivers asymmetry in Drell–Yan from 2015 data and the projection from corresponding asymmetry in SIDIS. Only statistical errors are shown.

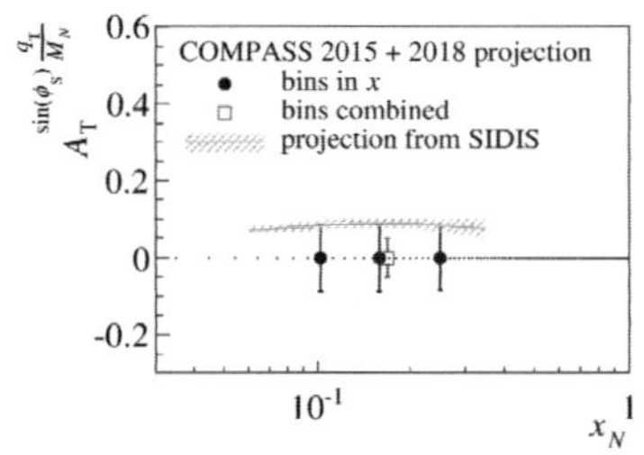

Figure 7. A projection for combined analysis of 2015 and 2018 data. The statistics in 2018 is assumed 1.5 times larger than in 2015. Only statistical errors are shown.

Acknowledgments

The author kindly acknowledges the support from Votruba–Blokhintsev programme and Ministry of Education, Youth and Science of Czech Republic grant LM2015058.

References

[1] Bacchetta A *et al.* 2007 *J. High Energy Phys.* **02** 093 (*Preprint* arXiv:hep-ph/0611265)
[2] Arnold S, Metz A and Schlegel M 2009 *Phys. Rev.* **D79** 034005 (*Preprint* arXiv:0809.2262 [hep-ph])
[3] Collins J C 2002 *Phys. Lett.* **B536** 43–48 (*Preprint* arXiv:hep-ph/0204004)
[4] Aghasyan M *et al.* (COMPASS) 2017 *Phys. Rev. Lett.* **119** 112002 (*Preprint* arXiv:1704.00488 [hep-ex])
[5] Efremov A V *et al.* 2005 *Phys. Lett.* **B612** 233–244 (*Preprint* arXiv:hep-ph/0412353)
[6] Sissakian A *et al.* 2006 *Eur. Phys. J.* **C46** 147–150 (*Preprint* arXiv:hep-ph/0512095)
[7] Kotzinian A M and Mulders P J 1996 *Phys. Rev.* **D54** 1229–1232 (*Preprint* arXiv:hep-ph/9511420)
[8] Kotzinian A M and Mulders P J 1997 *Phys. Lett.* **B406** 373–380 (*Preprint* arXiv:hep-ph/9701330)
[9] Boer D and Mulders P J 1998 *Phys. Rev.* **D57** 5780–5786 (*Preprint* arXiv:hep-ph/9711485)
[10] Gregor I M (HERMES) 2005 *Acta Phys. Polon.* **B36** 209–215
[11] Bradamante F (COMPASS) 2017 *Proceedings of the 22nd International Spin Symposium, Urbana-Champaign, USA, 25–30 September 2016* (*Preprint* arXiv:1702.00621 [hep-ex])
[12] Gautheron F *et al.* (COMPASS) 2010 *COMPASS-II Proposal* CERN-SPSC-2010-014, SPSC-P-340 (Geneva: CERN) URL `http://cds.cern.ch/record/1265628`
[13] Martin A, Bradamante F and Barone V 2017 *Phys. Rev.* **D95** 094024 (*Preprint* arXiv:1701.08283 [hep-ph])
[14] Bacchetta A *et al.* 2004 *Phys. Rev.* **D70** 117504 (*Preprint* arXiv:hep-ph/0410050)
[15] Lai H L *et al.* (CTEQ) 2000 *Eur. Phys. J.* **C12** 375–392 (*Preprint* arXiv:hep-ph/9903282)
[16] Buckley A *et al.* 2015 *Eur. Phys. J.* **C75** 132 (*Preprint* arXiv:1412.7420 [hep-ph])
[17] de Florian D, Sassot R and Stratmann M 2007 *Phys. Rev.* **D75** 114010 (*Preprint* arXiv:hep-ph/0703242)

IOP Conf. Series: Journal of Physics: Conf. Series **938** (2017) 012013 doi:10.1088/1742-6596/938/1/012013

Study of nucleon correlations in nuclei by (p, p') inelastic reaction at 1 GeV

O V Miklukho[1], A Yu Kisselev[2], G M Amalsky, V A Andreev, G E Gavrilov, D S Ilyin, A A Izotov, N G Kozlenko, P V Kravchenko, M P Levchenko, D V Novinskiy, D A Maysuzenko, V I Murzin, A N Prokofiev[3], A V Shvedchikov, S I Trush and A A Zhdanov

Konstantinov Petersburg Nuclear Physics Instinute, National Research Center "Kurchatov Institute", Gatchina, 188300 Russia

E-mail: [1]miklukho_ov@pnpi.nrcki.ru, [2]kisselev@mail.desy.de, [3]prokofiev_an@pnpi.nrcki.ru

Abstract. The secondary proton polarization, differential cross section, and cross section ratios were measured in the (p, p') inelastic reaction with nuclei at 1 GeV and a laboratory scattering angle of $\Theta = 21^0$. The data were obtained over a wide range of the scattered proton momentum covering the pN quasielastic peak and high momentum region up to a momentum corresponding to excited level of the nucleus under investigation. Scattered protons were detected by the magnetic spectrometer equipped with a polarimeter based on multiwire proportional chambers and carbon analyzer. A structure in the polarization and cross section data, possibly related to the in-medium elastic scattering on nucleon correlations arising in nuclei, and a scaling of the scattering cross section ratios of the nuclei were observed in the high momentum range.

1. Introduction

This work is a part of the experimental program in the framework of which the effects from scattering off nucleon associations (nucleon correlations, NCs) in nuclei [1] are studied with a 1 GeV proton beam at the synchrocyclotron of the Petersburg Nuclear Physics Institute (PNPI). These effects also were investigated in other experiments by various nuclear reactions (e.g., [2–9]).

In PNPI exclusive experiments [10], an essential difference in polarization of two secondary protons from the (p, $2p$) reaction with nuclei was found. This effect can be explained as the scattering on the NCs in a low-energy arm of the two-arms magnetic spectrometer.

In the first PNPI inclusive experiment [11], a growth of the polarization with the final proton momentum K in the (p, p') reaction with the ^{40}Ca nucleus at the scattering angle of 21^0 was observed. At a value of K corresponding to kinematics of the quasielastic scattering on a ^{4}He-like nucleon cluster inside of calcium nucleus, the polarization turned out to be close to that in free elastic p^4He scattering.

In new inclusive experiments [12–14], we studied in details the inelastic (p, p') reaction with different nuclei with a better statistic accuracy. The secondary proton polarization P and differential cross section $\sigma^{\rm incl} = \frac{d^2\sigma}{d\Omega dK}$ of the reaction were measured as a function of the scattered proton momentum K in narrow intervals of K (~10 MeV/c).

2. Experimental method and reaction kinematics

A general layout of the experimental setup is presented in Fig. 1 [13].

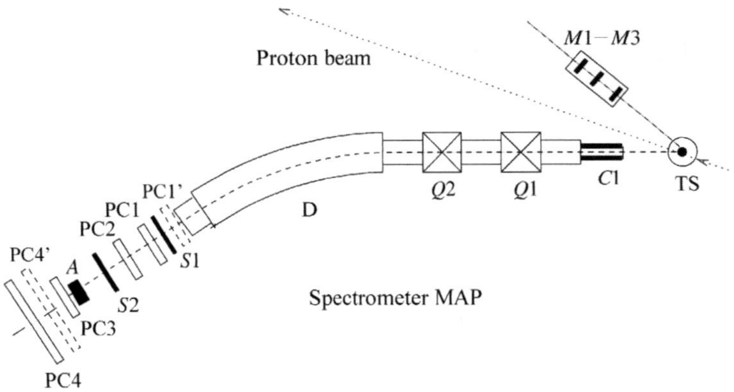

Figure 1. Layout of the experimental setup: (TS) target of the MAP spectrometer, ($Q1$–$Q2$) magnetic quadrupoles, (D) dipole magnet, ($C1$) collimator, ($S1$–$S2$ and $M1$–$M3$) scintillation counters, (PC1–PC4, PC1', and PC4') proportional chambers of the MAP polarimeter, and (A) its carbon analyzer.

The secondary protons from the reaction under investigation were detected by means of the magnetic spectrometer MAP equipped with a polarimeter based on multiwire proportional chambers and carbon analyzer. The spectrometer was installed at an angle of 21^0 with respect to the direction of the proton beam. Its momentum resolution for this scattering angle can be estimated on a width of the clearly separated 2^+ excited level in the reaction with the carbon nucleus (Fig. 2). This resolution was about ±2.5 MeV/c in this experiment.

The scattering differential cross sections of the reaction were found at different momentum settings of the spectrometer using information from the proportional chamber PC2-X, which was located in a focal plane of the spectrometer. The polarization of the secondary protons was defined on an azimuthal angular asymmetry of their scattering off a carbon analyzer. The analyzing power of the MAP polarimeter was calibrated on the basis of polarization data obtained for elastic pp scattering in the present experiment. In order to perform a calibration over a wide range of secondary proton energies, polarization measurements were performed with polyethylene (CH_2) and carbon (C) targets for various angular and proper momentum settings of the spectrometer. The values found for the pp polarization were compared with the predictions of the phase-shift analysis. A correction to the analyzing power of the polarimeter was introduced thereupon [13].

We performed measurements in a wide range of the scattered proton momentum K covering the pN quasielastic peak with its maximum at $K = K_{\rm pN}$ ~1480 MeV/c and a high momentum region ($K >$ 1530 MeV/c) up to the momentum ($K_{\rm L}$) corresponding to excited level of the nucleus under investigation. At $K > K_{\rm pN}$, the quasielastic pNC scattering off the nucleon correlations (NCs) is kinematically preferable since they are more massive than the nucleon. In this region, the four-momentum transfer Q remains almost invariable (~600 MeV/c). Consequently, the Bjorken variable $x_{\rm B} = \frac{Q^2}{2m\nu}$ is actually determined only by the K momentum. The momentum interval between $K_{\rm pN}$ and $K_{\rm L}$ covers a range of $x_{\rm B}$ = 1–4. The momentum $K_{\rm N}^{\rm min}(\Theta)$ is a minimum momentum which a nuclear nucleon must have to scatter a beam proton with the final momentum K. This momentum is a monotonically increasing function of K. The dashed vertical line in Fig. 3 at $K = K_{\rm C}$ ~1575 MeV/c separates a K momentum region where the $K_{\rm N}^{\rm min}(21^0)$ momentum is more than the Fermi momentum $k_{\rm F}$ for the carbon nucleus (~220 MeV/c) [14]. In a range $K > K_{\rm C}$, a contribution from quasielastic scattering off the uncorrelated nuclear nucleons is essentially suppressed [12].

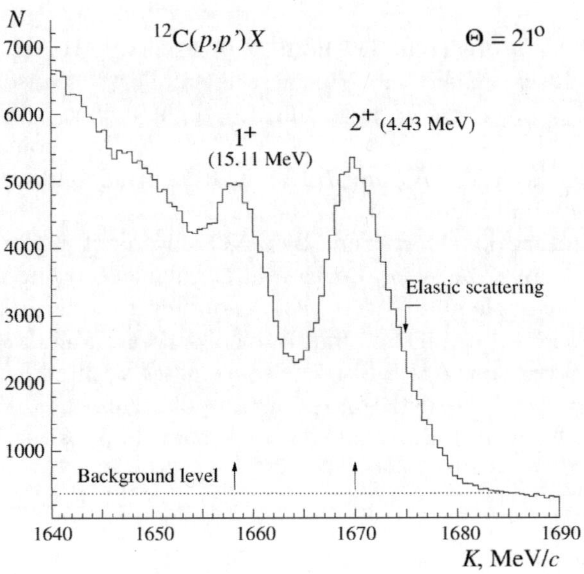

Figure 2. Momentum distribution in the inclusive reaction $^{12}C(p,p')X$ at the scattering angle of $\Theta = 21^0$ [13].

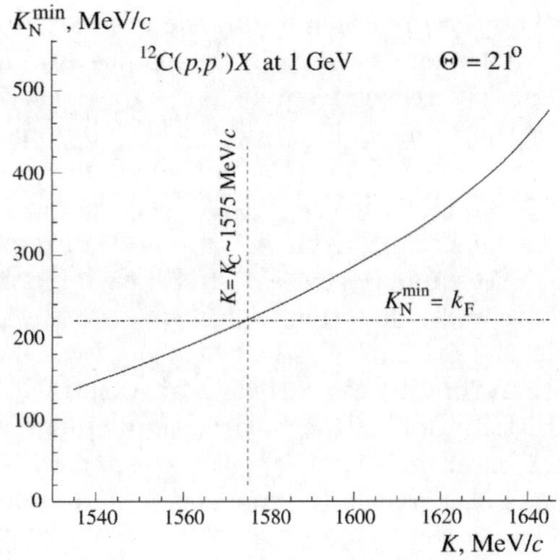

Figure 3. The solid curve is an averaged estimation of the K_N^{min} momentum for the carbon nucleus versus the scattered proton momentum K [14]. The momenta K_N^{min}, k_F, and K_C, the dashed vertical and dash-dotted horizontal lines were defined in the text.

3. Experimental data and observations

In Fig. 4, the secondary proton polarization (black squares) and cross sections (circles) of the (p, p') inelastic reaction with the carbon nucleus are presented as a function of the scattered proton momentum K [13]. The left and right vertical axes refer to the polarization and scattering cross section, respectively. The dashed curve is the result of the polarization calculation in the framework of the spin-dependent Distorted Wave Impulse Approximation (DWIA) taking into account the relativistic distortion of the nucleon spinor in a nuclear medium DWIA* [10]. For the calculations, the known ThreeDee code was used.

At the K momentum greater than the momentum corresponding to a maximum of the quasielastic pN peak (K_{pN} ~1480 MeV/c), it is possible to distinguish the momentum intervals indicated by the dotted-line segments and marked with the Roman numerals II (1535−1570 MeV/c), III (1570−1600 MeV/c), and IV (1600−1635 MeV/c). The characteristics by which these intervals were picked out:

1. The polarization within each interval practically does not change and grows from the interval II to the interval IV.

2. The beginning of each interval is determined by slowing down of the scattering cross section (indicated by an arrow) which may be due to scattering by a heavier nuclear particle.

Suppose that the intervals II, III, and IV correspond to the elastic scattering in nuclear medium on two-, three-, and four-nucleon correlations. Note here that a high-momentum region, just above the IV range, possibly corresponds to the quasielastic scattering off a residual nucleus X from the reaction under investigation.

To verify the last assumption, we calculated the momenta K_2, K_3, and K_4 corresponding to the quasielastic maxima in the scattering on these correlations (Fig. 4). In the kinematical calculations we assumed:

1) A nucleon correlation (NC) is immovable. 2) A NC mass is equal to a mass of the real light nucleus with a simple structure like ^2H, ^3He (^3H), and ^4He. 3) The residual nuclei (X) in

the (p, p') reaction are in the ground state.

One can see that the momenta K_2, K_3, and K_4 are within the momentum intervals II, III, and IV, respectively. This is so in the case inclusive where the NC mass is smaller (because of the in-medium modification [10]) than the mass of the corresponding free light nucleus. A decrease of about 10% in the NC masses reduces the momenta K_2, K_3 (K_3^*), and K_4 by about 12, 8, and 6 MeV/c, respectively. Note, for the ^{12}C data the K_3 and K_3^* momenta corresponding to the scattering off the ^{3}He and ^{3}H nuclei are virtually identical.

We suppose that a width of each momentum interval is determined by a NC movement in the transverse direction with respect to the proton beam. The setup horizontal angular acceptance in the medium elastic scattering by a moving four-nucleon cluster ($\sim 4.5^0$) is many time larger than that in the scattering by a resting correlation ($\sim 1^0$). This enables us to observe an angular distribution of the polarization $P_{\rm IV}$ in the scattering on a ^{4}He-like nucleon cluster within the IV momentum range. We expected to see a homogeneous distribution of the polarization as it was observed in the free elastic p^4He scattering shown in Fig. 5 [15]. Two horizontal oppositely

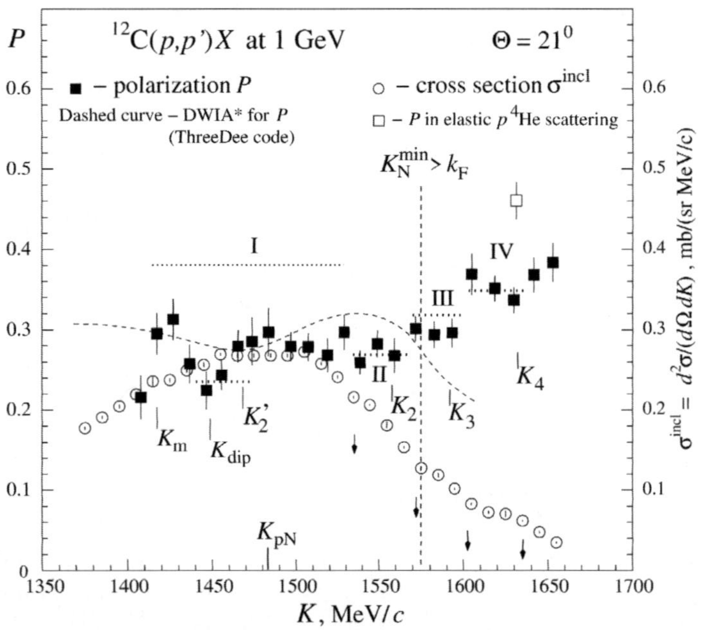

Figure 4. Polarization (black squares) and cross section (circles) of the ^{12}C$(p, p')X$ reaction at the angle of $\Theta = 21^0$ versus the scattered proton momentum K [13]. The empty square is the polarization in free elastic p^4He scattering [15]. The momentum intervals I–IV and corresponding dotted-line segments, the momenta $K_{\rm m}$, $K_{\rm dip}$, K_2', $K_{\rm pN}$, and K_2, K_3, K_4, and $K_{\rm N}^{\rm min}$, $k_{\rm F}$, the dashed curve and dashed vertical line (Fig. 3) were defined in the main body of the text.

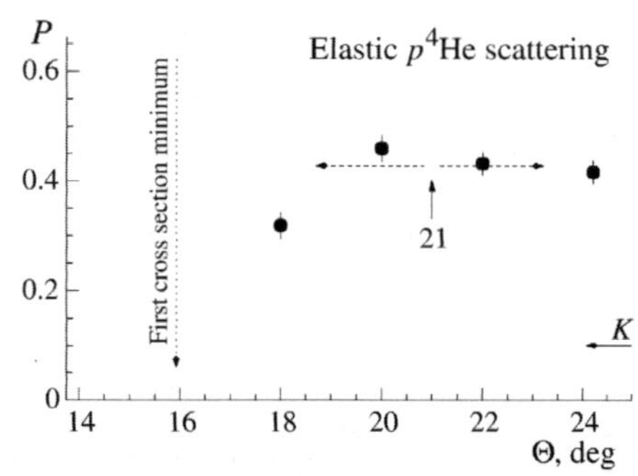

Figure 5. Angular dependence of the polarization in elastic p^4He scattering at 1 GeV [15]. The dashed horizontal arrows were defined in the text.

directed arrows indicate the angular range corresponding to the effective acceptance seen in the inclusive reaction in the momentum interval IV.

Important observation. The polarization ($P_{\rm IV}$) in the interval IV is less than that ($P_{^4{\rm He}}$) in the free elastic p^4He scattering (empty square) [15]. This can be due to the in-medium modification of proton interaction with the four-nucleon cluster. The relative polarization drop in the nuclear medium is about 20%. We found an approximately the same polarization drop in the scattering on the uncorrelated nuclear nucleons from a comparison of the polarizations calculated with (P_{DWIA^*}) and without (P_{DWIA}) taking into account the nuclear medium effect [13].

In the region I (indicated by a dotted-line segment), around the momentum K_{pN}, a contribution from multistep processes of nucleon knockout from nucleus can be sizable [16]. The outgoing proton momentum in these processes is lower than in the one-step (p, p') reaction under investigation. The two-step nucleon knockout process decreases the polarization at K less than K_{m} (Fig. 4) as well as the analyzing power in the similar experiment (LAMPF) at O.8 GeV [16]. We also see a dip in the polarization in a range of K around the momentum K_{dip} (for all nuclei investigated). There is no similar dip in the LAMPF analyzing power A_{y} in both experimental and theoretical data. The latter was obtained taking into account the multistep processes. This dip in the polarization may be due to the inelastic scattering by a two-nucleon correlation leading to its decay into two nucleons. The polarization in the process, as well as in the in-medium elastic scattering off the correlation in the range II, can be essentially smaller than that in the in-medium elastic scattering off the uncorrelated nuclear nucleons. In the short range correlation approach [9], two nucleons belonging to the two-nucleon correlation have oppositely directed momenta of nearly equal magnitude close to the Fermi momentum (~250 MeV/c, it corresponds to nucleon kinetic energy ~35 MeV). The momentum K_2' (Fig. 4) was obtained in the kinematical program at an excitation energy (~70 MeV/c) of the residual nucleus (X = ^{10}B) equal to total kinetic energy of the nucleons into the resting correlation. The K_2' momentum is shifted with respect to the momentum K_2 (for the interval II) toward lower values of K. The momentum range, marked by the dotted-line segment, covering the momentum K_2' is determined by the correlation motion.

In Fig. 6, the data obtained in investigating the inelastic ^{40}Ca$(p, p')X$ reaction is presented [13]. All designations in the figure is the same as in Fig. 4. For the calcium nucleus we determined

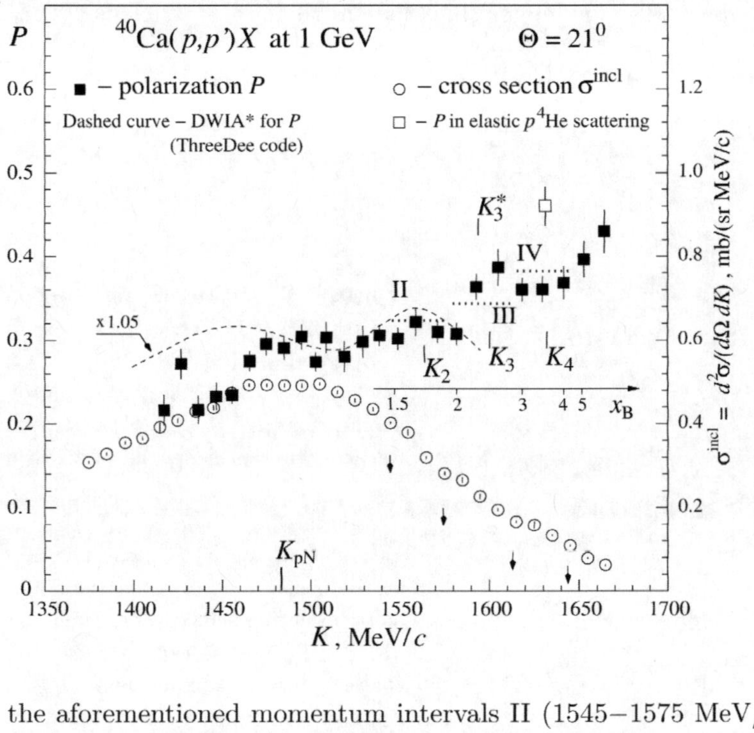

Figure 6. Polarization (black squares) and cross section (circles) of the ^{40}Ca$(p, p')X$ reaction at the angle of $\Theta = 21^0$ versus the scattered proton momentum K [13]. The empty square is the polarization in free elastic p^4He scattering [15]. The momentum intervals II–IV and corresponding dotted-line segments, the momenta K_{pN} and K_2, K_3, K_3^*, K_4, and the dashed curve were defined in the main body of the text. An additional axis for the Bjoken variable indication x_{B} was given.

the aforementioned momentum intervals II (1545–1575 MeV/c), III (1575–1610 MeV/c), and IV (1610–1645 MeV/c). Comparison with the carbon data:

1. The beginning of each interval for the ^{40}Ca nucleus is shifted with respect to that for the carbon nucleus by about 5 to 10 MeV/c toward higher values. The investigation of the ^{56}Fe

IOP Conf. Series: Journal of Physics: Conf. Series **938** (2017) 012013 doi:10.1088/1742-6596/938/1/012013

nucleus showed that this effect is not related to a bigger mass of the calcium nucleus.

2. The polarization in the interval III for the calcium data increases with K unlike the polarization in the scattering off the ^{12}C nucleus. It is possible that this is due to a dispersion effect because of a meaningful difference (~7 MeV/c) of the momenta K_3 and K_3^* corresponding to the in-medium elastic scattering on the ^{3}He and ^{3}H nuclei (the spectrometer momentum resolution was about ~2.5 MeV/c). It is also assumed that the polarization in the elastic scattering on the ^{3}He nucleus is much bigger than that on the tritium nucleus [13]. In this case, when K is more than K_3^*, the scattering on the ^{3}He nucleus dominates and the averaged polarization has a largest value. In the region, where K is less than K_3^* (within the III momentum interval), processes of the scattering off these three-nucleon correlations are mixed and the averaged polarization decreases.

3. The polarizations in the interval IV (P_{IV}) in the scattering of the calcium and carbon nuclei are nearly identical [$P_{\rm IV}$(Ca) $= 0.363 \pm 0.009$, $P_{\rm IV}$(C) $= 0.348 \pm 0.010$].

In Fig. 7, the scattering cross section ratios for the (p, p') reaction with the ^{56}Fe and ^{12}C nuclei η(Fe/C), with the ^{56}Fe and ^{28}Si nuclei η(Fe/Si), and with the ^{56}Fe and ^{40}Ca nuclei η(Fe/Ca) are presented [14]. Main observations:

1. We see a step-function behavior of the scattering cross section ration η(Fe/C) in the region of K from 1560 to 1630 MeV/c covering the momentum intervals III and IV indicated by the dotted-line segments. Within the intervals a value of the η(Fe/C) ration does not depend substantially on K (i.e., scaling). This intervals coincide with analogues intervals observed in the polarization and cross section data for the carbon nucleus (Fig. 4), which are presumably due to the in-medium elastic scattering off three- and four-nucleon correlations [13]. According to the JLAB work [9], observation of such steps in the cross section ratio is a crucial test of the dominance of inclusive scattering off the short range correlations, and the steps are due to that an average nucleon density in the iron nucleus is bigger than that in the carbon nucleus [14].

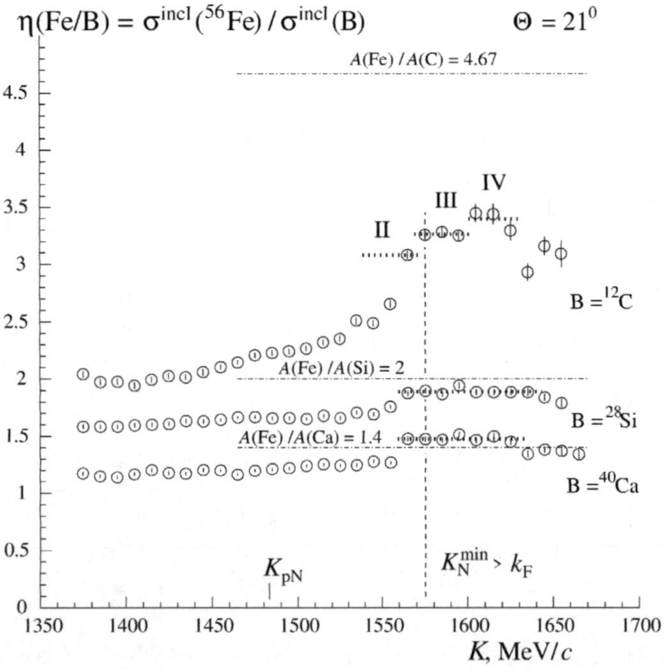

Figure 7. Scattering cross section ratios η(Fe/B) (circles), where B corresponds to the ^{12}C, ^{28}Si, or ^{40}Ca nucleus, versus the scattered proton momentum K [14]. The dashed vertical line represents the same as in Fig. 3 and Fig. 4. The dash-dotted horizontal lines correspond to the ratios of appropriate atomic numbers of the nuclei under investigation. The dotted-line segments cover the intervals II, III, IV, and 1560–1635 MeV/c. These intervals and momenta $K_{\rm pN}^{min}$, $k_{\rm F}$, and $K_{\rm pN}$ were defined in the text.

2. The scaling of the cross section ratios η(Fe/Si) and η(Fe/Ca) are observed in the same K region mentioned above. We do not see any steps in these ratios. It is possible that an average

nucleon density in the nuclei ^{28}Si, ^{40}Ca, and ^{56}Fe is the same. It is interesting that the values of these ratios are close (within the systematic errors of their determination) to those of the atomic number ratios A(Fe/Si) and A(Fe/Ca) (indicated by the dash-dotted line segments) [14].

4. Conclusion

A structure in the polarization and cross section of the (p, p') inelastic reaction with the nuclei ^{12}C and ^{40}Ca at 1 GeV and a laboratory scattering angle of $\Theta = 21^0$ was observed. This structure can be related to the in-medium elastic scattering on nucleon correlations in nuclei.

The scaling of the scattering cross section ratios off the nuclei under investigation (a value of the ratio is independent of the secondary proton momentum K) was observed in the high momentum range of $K = 1560-1635$ MeV/c.

Acknowledgments

We are grateful to the staff of the PNPI 1 GeV proton accelerator for the stable beam operation and to A.A. Vorobyov and S.L. Belostotski for their support and fruitful discussion.

The authors wish to thank the organizers of the DSPIN-17 workshop for the invitation to present this paper.

References

[1] Blokhintsev D I 1958 *Sov. Phys. JETP* **6** 995
[2] Leksin G A 1957 *Sov. Phys. JETP* **5** 371
[3] Azhgirei L S, Vzorov I K, Zrelov V P, Mescheriakov M G, Neganov B S and Shabudin A F 1957 *Sov. Phys. JEPT* **6** 911
[4] Bayukov Y D, Vorobev L S, Kartashov G R, Leksin G A, Fedorov V B and Khovansky V D 1966 *Bull. Acad. Sci. USSR: Phys.* **30** 530
[5] Baldin A M *et al.* 1973 *Sov. J. Nucl. Phys.* **18** 41
[6] Bayukov Y D, Vorobev L S, Leksin G A, Stolin V L, Fedorov V B and Khovanskii V D 1974 *Sov. J. Nucl. Phys.* **18** 639
[7] Efremov A V, Kaidalov A B, Kim V T, Lykasov G L and Slavin N V 1988 *Sov. J. Nucl. Phys.* **47**, 868
[8] Vorobev L S, Leksin G A and Stavinsky A V 1996 *Phys. Atom. Nucl.* **59**, 662
[9] Egiyan K S *et al.* 2006 *Phys. Rev. Lett.* **96**, 082501-6
[10] Miklukho O V *et al.* 2013 *Phys. Atom. Nucl.* **76** 871-80
[11] Miklukho O V *et al.* 2011 *arXiv:* 1103.6113v1 [nucl-ex]
[12] Miklukho O V *et al.* 2015 *JETP Letters* **102** 11-3
[13] Miklukho O V *et al.* 2017 *Phys. Atom. Nucl.* **80** 299-306
[14] Miklukho O V *et al.* 2017 *JETP Letters* **106** 69-72
[15] Miklukho O V *et al.* 2006 *Phys. Atom. Nucl.* **69** 452-9
[16] Smith R D and Wallace S J 1985 *Phys. Rev.* C **32** 1654-66

IOP Conf. Series: Journal of Physics: Conf. Series **938** (2017) 012014 doi:10.1088/1742-6596/938/1/012014

Comparative study of the inclusive asymmetries induced by polarized protons and antiprotons at 16 GeV/c at the U–70 accelerator

V A Okorokov[1], V V Abramov[2], A A Bogdanov[1], M A Chetvertkov[3], V A Chetvertkova[4], V V Mochalov[1,2], S B Nurushev[1,2], M B Nurusheva[1], V L Rykov[1], P A Semenov[1,2], M N Strikhanov[1] and A N Vasiliev[1,2]

[1]National Research Nuclear University MEPhI (Moscow Engineering Physics Institute), Kashirskoe highway 31, 115409 Moscow, Russia
[2]National Research Center "Kurchatov Institute" – Institute for High Energy Physics, Protvino, Moscow region, 142281, Russia
[3]MD Anderson Cancer Center, Department of Radiation Physics, 1515 Holcombe Blvd, Unit 1420, Houston, TX 77030, USA
[4]GSI Helmholtzzentrum für Schwerionenforschung GmbH, Plankstrasse 1, 64291, Darmstadt, Germany

E-mail: VAOkorokov@mephi.ru

Abstract. The only comparative study of the inclusive pion single-spin asymmetries produced in the interactions of the polarized protons and antiprotons in collisions with unpolarized proton was carried out at E–704 experiment. Significant asymmetries were found at large x_F and middle p_T, π^+ and π^0 asymmetries have positive signs while π^- has negative one in the $p^\uparrow + p$ collisions, while in the $\bar{p}^\uparrow + p$ interactions the π^- and π^0 asymmetries have positive signs while π^+ has negative sign. Similar experimental study can be done in the SPASCHARM experiment at U–70 accelerator at IHEP for various secondary particles with the use of 16 GeV polarized proton and antiproton beams.

1. Introduction

Inclusive reactions measured with polarized (anti)proton beam give insight into the spin dependence of the underlying partonic processes and add new input regarding the problem of the spin structure of polarized protons. Significant polarization effects appear already at relatively low values of the transverse momentum p_T ($p_T \sim 1.0$ GeV/c), where perturbative quantum chromodynamics (pQCD) is not expected to be applicable. Therefore the new experimental measurements with high statistics as well as phenomenological studies are essential for future progress in quantitative description of spin observables.

The following inclusive reactions are used for study of single-spin asymmetries for pions

$$p^\uparrow + p \to \pi^{\pm,0} + X, \quad (1)$$

$$\bar{p}^\uparrow + p \to \pi^{\pm,0} + X. \quad (2)$$

The analyzing power A_N is the physics observable under consideration here. The A_N is deduced from the measured yields of pions produced in a well defined azimuthal angular interval around the beam axis using vertically polarized p ($\bar{p}$) of both polarization signs:

$$A_N(z) = \frac{1}{P_B\langle\cos\phi\rangle}\frac{N_\uparrow(z,\phi) - N_\downarrow(z,\phi)}{N_\uparrow(z,\phi) + N_\downarrow(z,\phi)}. \quad (3)$$

Here $z \equiv (p_T, x_F, \sqrt{s})$ – a set of kinematic variables, p_T is the pion transverse momentum, $x_F = 2p_L/\sqrt{s}$ – Feynman variable for pion with longitudinal momentum p_L at the collision energy $\sqrt{s}$, P_B is the beam polarization, and ϕ is the azimuthal angle between the beam polarization axis directed upward and the normal to the production plane. $N_\uparrow$ ($N_\downarrow$) is the number of pions produced for positive (negative) spin orientation of the beam (anti)protons at the target, normalized to the corresponding beam flux. The reactions (1) and (2) were studied at p = 200 GeV/c ($\sqrt{s}$ = 19.4 GeV) with the E–704 setup [1] and here it is suggested to continue such a study in deeply non-perturbative region at p = 16 GeV/c ($\sqrt{s}$ = 5.64 GeV) in the SPASCHARM project [2, 3, 4].

2. The E–704 results for secondary pions

The magnitude of A_N increases for both π^+ and π^- particles with x_F for $p^\uparrow$ beam, but the sign of A_N is negative for the π^- data. The values of A_N are large for high values of x_F, up to 0.29 ± 0.09 for π^+ and down to 0.37 ± 0.07 for π^-. Detailed analysis of data at p = 200 GeV/c shown a threshold effect in which A_N increases dramatically above p_T = 0.7 GeV/c [1]. The E–704 experimental results obtained for $A_N(x_F)$ above and below p_T threshold indicated that the increase of A_N is primarily x_F effect above a p_T threshold. The situation is similar for study of the $A_N(x_F)$ with $\bar{p}^\uparrow$ beam with taken into account the change of signs of electric charge for the beam and secondary particles under consideration. Furthermore the p_T dependence was obtained for A_N for charged pion production in inclusive reaction (2) with $\bar{p}^\uparrow$ beam [5]. As example, figure 1 shows the dependence of A_N on x_F (left) and p_T (right) for secondary charged pions in (2). In this case the E–704 data exhibit an almost mirror symmetric dependence in x_F. The analyzing power for π^- production increases from 0.0 to about +0.25 with increasing x_F above $p_T \sim 0.5$ GeV/c while, for π^+ production, A_N decreases from 0.0 to about -0.35 with increasing x_F above the same p_T. The E–704 results with $\bar{p}^\uparrow$ show a threshold effect about $p_T \sim 0.5$ GeV/c, above which A_N increases in magnitude for both π^+ and π^- and for transverse momenta below this p_T value, A_N is significantly smaller and compatible with zero [5]. The threshold effect was also confirmed by the results of additional analysis for $p_T \geq 0.5$ GeV/c. Therefore the E–704 experimental results for (2) show the magnitude of A_N increases for both types of charged pions with increasing x_F, but the sign of A_N is positive for the π^- data and negative for π^+ data above the same $p_T \sim 0.5$ GeV/c. It appears that A_N depends primarily on x_F namely, and reaches large values above the p_T threshold of about 0.5 GeV/c as well as for inclusive reaction (1) with charged pions at some higher threshold p_T.

Therefore A_N seems similar (in the sense of behavior and magnitude) for $p^\uparrow$ / $\bar{p}^\uparrow$ beams in the case of opposite charges for pions, i.e. the reaction (1) with proton beam and inclusive π^+ the A_N appears to be similar to that in reaction (2) with antiproton beam and inclusive π^- and vise versa. One can note that the models based on non-perturbative approaches, such as a soft pion exchange mechanism [6], resonance-decay interference between real and virtual channels [7] and rotating constituents in the polarized (anti)proton [8] appear to be in good qualitative agreement with the features of the data on the pion production asymmetry measured with the both polarized protons and antiprotons.

The kinematic dependencies of A_N on x_F, p_T were also studied for the π^0 production in the inclusive reactions (1) and (2) in E–704 experiment [1, 9, 10, 11]. For the π^0 data obtained with

IOP Conf. Series: Journal of Physics: Conf. Series **938** (2017) 012014 doi:10.1088/1742-6596/938/1/012014

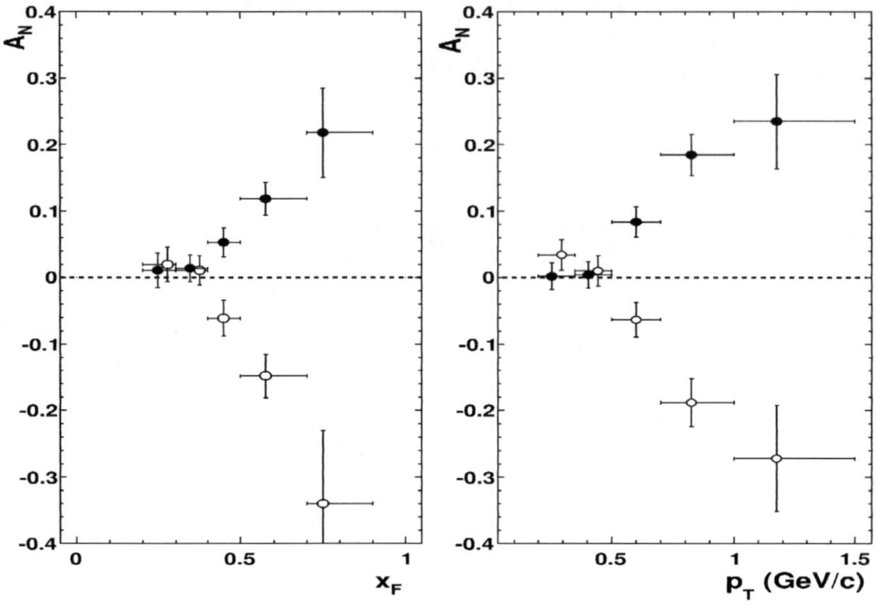

Figure 1. A_N data for $\bar{p}^{\uparrow}$ beam as a function of x_F integrated over p_T in the range $0.2 - 1.5$ GeV/c (left) and depends on p_T in the x_F range of $0.2 - 0.9$ (right) for charged pions. Experimental points for π^- are shown by solid symbols, for π^+ – by open ones. For clarity, first two π^- (π^+) data points are slightly shifted on left (right) in each panel. Data are taken from [5].

$p^{\uparrow}$ beam, $A_N(x_F)$ has the same sign as for π^+ data and is about half as large [1, 10]. The similar situation is observed for $A_N(x_F)$ in the $\bar{p}^{\uparrow} + p$ inclusive reaction for comparison π^0 results with π^- analyzing power [5, 10]. Thus the π^0 productions by polarized $p^{\uparrow}$ and $\bar{p}^{\uparrow}$ are related by charge conjugation of the beam and the produced particle. The measured asymmetries have the same sign and similar x_F dependence [10]. At large x_F, there is an indication that the magnitude of A_N for π^0 production by incident antiprotons is less than for incident protons. This would mean that the interactions involve constituents other than gluons and quark-antiquark pairs in the target proton. The A_N is observed to be zero for single-spin inclusive π^0 production in $p^{\uparrow} + p$ and $\bar{p}^{\uparrow} + p$ inclusive reactions in the $1 < p_T < 3$ GeV/c region within a statistical accuracy [11]. But it should be noted that in this case the amount of data for studying (2) interactions was an order of magnitude less than that for (1) interactions. In general in perturbative QCD single-spin transverse asymmetries are expected to be practically zero. Thus this expectation in the $1 < p_T < 3$ GeV/c region is confirmed by the data from the E–704 experiment, if perturbative QCD is applicable to these p_T values at beam momentum $p = 200$ GeV/c. Moreover the experimental errors are large for A_N at $p_T > 2.5$ GeV/c [11]. Therefore new experimental data with high statistics seem important for verification of some predictions of the QCD.

3. The SPASCHARM project

The SPASCHARM (**SP**in **A**symmetry in **CHARM**onia) is the project for world-class research works in fixed target mode for high energy spin physics [2, 3, 4]. The main goals of the SPASCHARM project are the studies of the (i) spin structure of the nucleon and (ii) possible spin dependence of the strong interaction for matter and antimatter with help of the systematic

IOP Conf. Series: Journal of Physics: Conf. Series **938** (2017) 012014 doi:10.1088/1742-6596/938/1/012014

physical analysis for a wide set of hadronic reactions and secondary particles. The same name experimental setup is the core part of the SPASCHARM project. It is suggested to have two stages of the SPASCHARM project: first of all, studies with unpolarized proton beams using a polarized target and second phase of the project is the using of polarized beams.

The polarized $p^\uparrow$ ($\bar{p}^\uparrow$) beam is obtained by selecting $p^\uparrow$ ($\bar{p}^\uparrow$) from the weak decay of Λ ($\bar{\Lambda}$) produced in a primary target by extracted proton beam. This method is used for both the E–704 experiment and the SPASCHARM setup. The main features are shown in table 1 for U–70 and Tevatron beams. As seen from table 1 there are some advantages of the expectations for U–70 (for instance, intensities for primary and polarized proton beams) with respect to corresponding Tevatron beam parameters. It should be noted the U–70 allows the study of single-spin asymmetry in deeply non-pertubative region for some range of $\sqrt{s}$.

Table 1. Some main characteristics for U–70 and Tevatron beams.

	Beam parameter	U–70	Tevatron
1	primary proton beam, p (Gev/c)	50–60	800
2	primary beam intensity, c^{-1}	$\sim 2 \times 10^{12}$	1.5×10^{11}
3	polarized beam, p (Gev/c)	15–45^a	185 ± 17
4^b	beam intensity at the target, c^{-1}	$(0.9 - 6.8) \times 10^6$	1.5×10^6
		$(0.8 - 4.0) \times 10^4$	1.5×10^5
5^b	beam polarization	$\pm(0.45 \pm 0.05)$	$\pm(0.40 \pm 0.12)$
		–//–	$\pm(0.45 \pm 0.03)$

^{a}The minimum (maximum) relative uncertainty for beam momentum δp is 4.5% (11%) for $p = 15$ GeV/c and 3.0% (9.0%) for $p = 45$ GeV/c.
^{b}The first / second line corresponds to the $p^\uparrow$ / $\bar{p}^\uparrow$ beam.

Dependence of polarized beam intensity on momentum is shown in figure 2 for p (left) and $\bar{p}$ (right). The difference in intensities of $p^\uparrow$ and $\bar{p}^\uparrow$ beams increases dramatically with a growth of beam momentum. As seen for polarized proton beam the contribution of π^+ from neutral kaon decays is small, furthermore this contribution decreases rapidly with beam momentum. Therefore the U–70 allows the polarized proton beam with good quality (purity). Figure 2 (right) shows that the beam of $\bar{p}^\uparrow$ with $p = 16$ GeV/c seems optimal in terms of intensity and background conditions. The number of π^- is approximately 3 times higher than the intensity of the antiproton beam. The separation of antiprotons at this level of background is quite possible with the help of Cherenkov beam counters.

The SPASCHARM detector is an open geometry setup with good particle identification and relatively large acceptance. Schematic view of the SPASCHARM is shown in figure 3. The SPASCHARM setup consists of the following main subsystems [2, 4]:
– various targets (liquid hydrogen, nuclear from Be up to Pb; polarized / unpolarized),
– spectrometer for registration of charged particles (magnet, GEM detector, multiwire proportional chambers – PC and drift ones – DC),
– electromagnetic calorimeter (ECAL) is shown in figure 4 and it is made by "shashlik" technology which is well established and used successfully, for instance, during preparation the PANDA experiment,
– hadronic calorimeter (HCAL),
– set of detectors for particle identification and multiplicity measurement (ring image Cherenkov detector – RICH, muon spectrometer – MuD and time-of-light system – TOF).

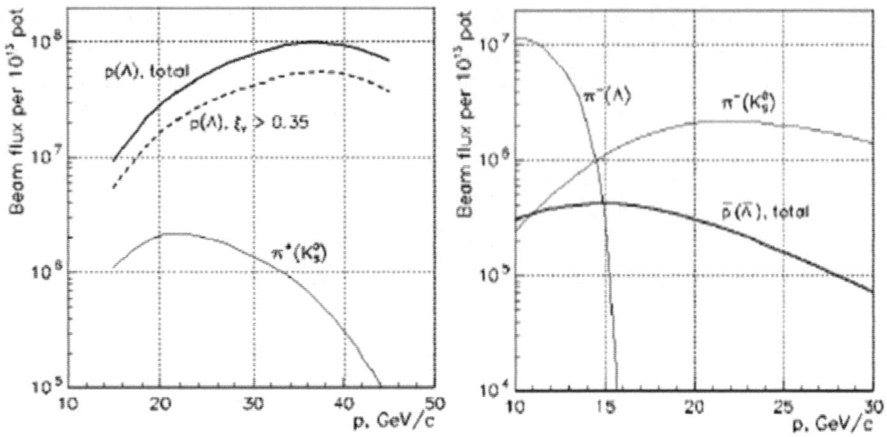

Figure 2. Intensity for beam of $p^{\uparrow}$ (left) and $\bar{p}^{\uparrow}$ (right) along with pion background of appropriate sign of electric charge for maximum δp shown in table 1. The quantity is calculated for 10^{13} primary protons with energy 60 GeV, where ξ_y is the transverse polarization averaged over ensemble. Data are taken from [4].

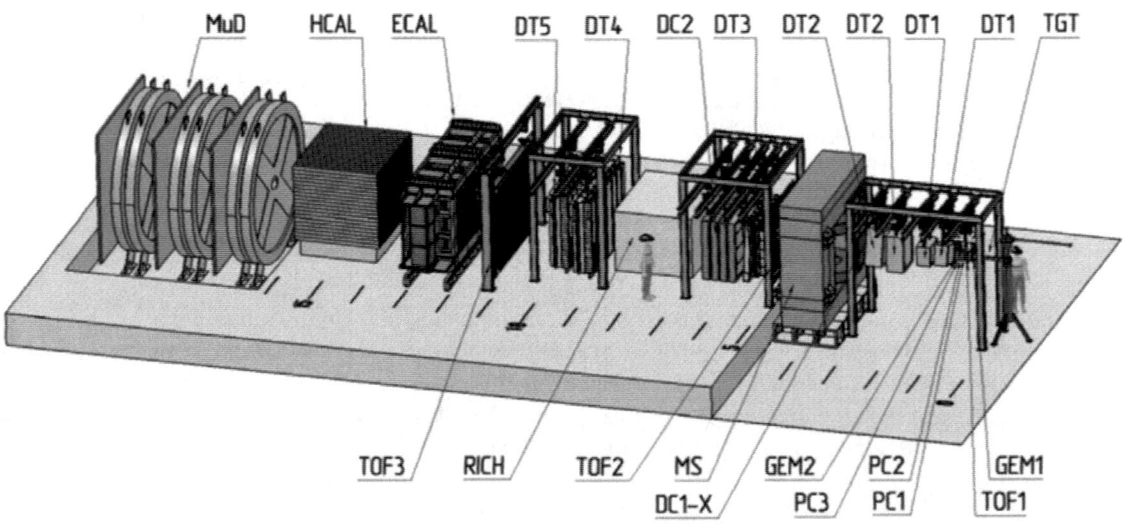

Figure 3. Schematic 3D view of the SPASCHARM setup. Here liquid hydrogen target (TGT) is shown, DT1–5 – a sets of thin-walled drift tubes, other subsystems are described in the text. Distances are shown in meters along the beam direction with respect to the center of the target.

Subsystems of the SPASCHARM detector are designed for high quality particle identification. According to the Conceptual design report (CDR) [4] the SPASCHARM will provide registration of charged particles from 0.5 up to 50 GeV/c in full azimuth 2π, registration of γ from 0.2 up to 50 GeV/c with the same angular range as for charged particles, identification of charged particles for momentum domain 1–20 GeV/c, registration of decay particles from hyperons with future reconstruction of parent particle, high coordinate resolution for beam particles, especially, for elastic scattering study [4]. Furthermore there are following special requests for charmonium

IOP Conf. Series: Journal of Physics: Conf. Series **938** (2017) 012014 doi:10.1088/1742-6596/938/1/012014

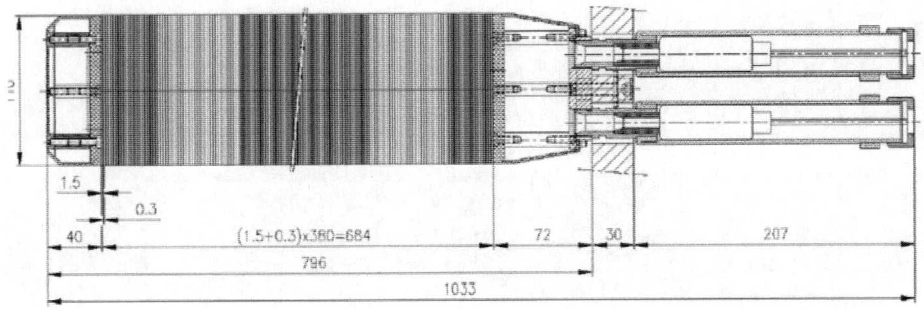

Figure 4. Scheme of supermodul for ECAL ("shashlik" consists from layers of Pb-absorber and scintillator). Picture is taken from [4].

study: momentum resolution is $\sim 2\%$ at 10 GeV/c and energy resolution for electromagnetic calorimeter is $3\%/\sqrt{E}$. Therefore there are some advantages of the SPASCHARM project regarding of earlier experiments, for instance, E–704. In particular, addition of new detectors (GEM, MDC, high quality EMC etc.) allows the increase of statistics significantly compared to the previous experiments. Due to 2π acceptance on azimuthal angle of the SPASCHARM setup one can expect that the systematic errors in A_N will be small. With the use of polarized proton beam at SPASCHARM a precision measurement of A_N for inclusive production in the transverse polarized beam fragmentation region in a wide (x_F, p_T)-region will be worthwhile [2]. One can expect that the kinematic ranges $0.2 < x_F < 1.0$ and $0.5 < p_T < 3.5$ GeV/c will be covered. Also the estimations will be obtained for accuracy of the A_N measurement in the reactions (1), (2) and for corresponding time of data collection for various accelerator parameters (beam intensity, run duration etc.).

Within SPASCHARM project the software is constructed for on-line and off-line data analyses. The software is developed as object-oriented environment and it is based on the ROOT package. Also SPASCHARM software includes GEANT 3, 4 and some event generators (PYTHIA, PLUTO etc.) for Monte-Carlo simulation of hadronic interactions. At present the software environment SpascharmRoot has been developed and partially implemented. The SpascharmRoot allows the simulation, reconstruction, on- and off-line analysis of experimental data. The environment is under active development. The SPASCHARM software is permanently improving, for instance, the possibility is considered for using of the GRID technology for distributed data analysis.

4. Summary

Analyzing power shows the significant magnitude for inclusive pions above $p_T \sim 0.5$ GeV/c but there is only one measurement of the A_N for inclusive pions with polarized antiproton beam. At present phenomenological models describe the experimental data at qualitative level only and precision of experimental results do not allow the clear discrimination between various models. Consequently the new high-precision measurements seem important for better understanding of single-spin asymmetry and more definite physics conclusions. Hopefully, the SPASCHARM experiment will provide the high-statistics data which will shed new light for single-spin pion asymmetry as well as in general for spin structure of the proton.

Acknowledgments

The work has been supported in part by the RFBR Grant No 16–02–00667 and by the NRNU MEPhI Academic Excellence Project (contract No 02.a03.21.0005, 27.08.2013).

References

[1] Adams D L *et al.* 1991 *Phys. Lett.* B **264** 464
[2] Vasiliev A N *et al.* 2007 *arXiv*: 0712.2691 [hep-ex]
[3] Mochalov V V *et al.* 2013 *Phys. Part. Nucl.* **44** 930; 2016 *Int. J. Mod. Phys.: Conf. Ser.* **40** 1660106
[4] Abramov V V *et al.* 2017 *CDR for SPASCHARM experiment.* v. August 13
[5] Bravar A *et al.* 1996 *Phys. Rev. Lett.* **77** 2626
[6] Soffer J and Törnqvist N A 1992 *Phys. Rev. Lett.* **68** 907
[7] Barni R, Preparata G and Ratcliffe P G 1992 *Phys. Lett.* B **296** 251
[8] Boros C, Zuo-tang L and Ta-chung M 1993 *Phys. Rev. Lett.* **70** 1751; 1995 *Phys. Rev.* D **51** 4867
[9] Bonner B E *et al.* 1988 *Phys. Rev. Lett.* **61** 1918
[10] Adams D L *et al.* 1991 *Phys. Lett.* B **261** 201
[11] Adams D L *et al.* 1996 *Phys. Rev.* D **53** 4747

IOP Conf. Series: Journal of Physics: Conf. Series **938** (2017) 012015 doi:10.1088/1742-6596/938/1/012015

COMPASS results on DVCS and exclusive π^0 production

Andrzej Sandacz[1]
(on behalf of the COMPASS Collaboration)
Division of High Energy Physics, National Centre for Nuclear Research, PL 00-681 Warsaw, Poland
E-mail: Andrzej.Sandacz@ncbj.gov.pl

Abstract. The high energy polarised muon beam available at CERN, with positive or negative charge, makes COMPASS a unique place for studies of General Parton Distributions (GPDs). The dedicated COMPASS GPD program started in 2012 with commissioning of a new long liquid hydrogen target and new detectors, such as the large recoil particle detector CAMERA and the large-angle electromagnetic calorimeter. It was followed by a short pilot 'DVCS run'. The first results on DVCS and exclusive π^0 channels have been obtained. From the sum of cross sections measured with positive and negative beam polarities the pure DVCS cross-section and its dependence on the squared four-momentum transfer t have been extracted. It allowed a first estimate of the transverse size of the partonic distribution of the nucleon $\sqrt{\langle r_\perp^2 \rangle}$ in a range of intermediate x_{Bj} with a mean x_{Bj} value of 0.056. We report also on the absolute value of the cross section for exclusive π^0 production and its dependence on t and azimuthal angle ϕ between the scattering plane and π^0 production plane. These measurements are aiming to constrain the GPDs, in particular chiral-odd (transversity) GPDs. The full data taking for the GPD program approved within the COMPASS-II proposal continues in 2016 and 2017.

1. Introduction

The theoretical framework of General Parton Distributions (GPDs) [1, 2, 3] provides a novel description of the nucleon's partonic structure and contains a wealth of new information. In particular, GPDs embody both the nucleon form factors and Parton Distribution Functions (PDF), such as parton number, helicity and transversity densities. Very importantly, GPDs provide a description of the nucleon as an extended object, sometimes referred to as a 3-dimensional 'nucleon tomography' [4], which correlates (transverse) spatial and (longitudinal) momentum degrees of freedom. The GPDs also allow access to such a fundamental property of the nucleon as the orbital angular momentum of quarks [2]. Like ordinary PDFs, GPDs describe the structure of the nucleon independently of the specific reaction by which the nucleon is probed, i.e. they are expected to be universal. The mapping of the nucleon GPDs requires a comprehensive program of measuring various hard processes in a broad kinematic range, in particular Deeply Virtual Compton Scattering (DVCS). In addition, Hard Exclusive Meson Production (HEMP) is expected to add an independent and complementary information. A

[1] Work supported by the Polish NCN Grant 2015/18/M/ST2/00550.

IOP Conf. Series: Journal of Physics: Conf. Series **938** (2017) 012015 doi:10.1088/1742-6596/938/1/012015

comprehensive experimental overview of DVCS is presented in Ref. [5] and results of a recent phenomenological studies and predictions for DVCS can be found in Ref. [6].

The GPD part of COMPASS-II proposal [7] is devoted to measurements of both DVCS and HEMP with polarised μ^+ and μ^- beams and a liquid hydrogen target. The new detectors, the 4-m long recoil proton detector CAMERA and (a central part of) the new large-angle electromagnetic calorimeter ECAL0, which are essential for measurements of exclusive processes, were constructed and incorporated into the COMPASS set-up in 2012. The commissioning of these new detectors was done in 2012 and was followed by a 4-week pilot run. The results on DVCS and exclusive π^0 production, which are presented in this contribution, have been obtained from this pilot run. The dedicated data taking for the GPD program, with the complete ECAL0, is continuing in 2016-2017.

2. Experimental set-up

COMPASS is a fixed-target experiment situated at the high-intensity M2 beam-line of the CERN SPS. It can deliver either hadron or naturally polarised muon beams of a given charge in the energy range between 50 and 280 GeV. For the GPD program the data are collected with the muon beam nominal energy of 160 GeV and with muons of both polarities: μ^+ (μ^-) with polarisation of about -0.8 (+0.8).

The COMPASS apparatus [8] consists of a two-stage forward spectrometer comprising various tracking detectors, electromagnetic and hadronic calorimeters, muon identification detectors, and a ring imagining Cherenkov counter grouped around two dipole magnets SM1 and SM2. For the GPD program the set-up was complemented by installing a large proton recoil detector CAMERA around the new 2.5 m long liquid hydrogen target and a new large-angle electromagnetic calorimeter ECAL0. The 4-meter long CAMERA consists of two concentric barrels, each of 24 scintillator slats with a read out at both ends. The recoil proton detection is based on the ToF measurement between the two barrels. The calorimeter ECAL0 that is situated just downstream of CAMERA allows to extend the accessible kinematic domain for DVCS and exclusive π^0 production towards higher x_{Bj}, improve hermeticity for detection of exclusive events and reduce the background to single-photon events production, which originates from π^0 and other decays.

By adding these detectors COMPASS has been converted into a facility measuring exclusive reactions within kinematic domain from $x \sim 0.01$ to ~ 0.15, which cannot be explored at any other existing or planned facility in the near future.

3. Event selection

For brevity, in this section we discuss in more detail only the selections for the single photon channel. As for the exclusive π^0 sample the analysis is analogous, we will just only mention those issues that are specific for the π^0 channel.

The event selection exploits the overconstrained kinematics for the measured exclusive process $\mu p \rightarrow \mu' p' \gamma$. Events with a single outgoing muon and a single neutral cluster above the DVCS thresholds for the electromagnetic calorimeters are selected first. The muon and photon are then combined with all possible proton candidates reconstructed in CAMERA. The combinatorial ambiguity is resolved by applying cuts on variables that are sensitive to the exclusivity of events such as: coplanarity

$$\Delta\phi = \phi^{proton}_{meas} - \phi^{proton}_{spec} \quad , \tag{1}$$

the transverse momentum balance

$$\Delta p_T = p^{proton}_{T,meas} - p^{proton}_{T,spec} \quad , \tag{2}$$

as well as the four-momentum balance

$$M_X^2 = (k + p - k' - q' - p'_{meas})^2 \quad . \tag{3}$$

Here, the symbols k, p, k', q' and p'_{meas} denote the four-momenta of the incident muon, target proton, scattered muon, produced photon and recoil proton, respectively. The subscript 'meas' denotes a quantity measured by the recoil proton detector, while 'spec' denotes a quantity derived from the spectrometer measurements only, when assuming exclusivity, i.e. $p'_{spec} = k+p-k'-q'$. In addition, the measured longitudinal hit position z_A in the inner barrel of the recoil detector is compared to the one obtained from an interpolation between the interaction vertex and the hit in the outer barrel, z_{interp}. This reverse vertex pointing is encoded in the variable $\Delta z_A = z_{interp} - z_A$. The distributions of exclusivity variables are shown in Fig. 1

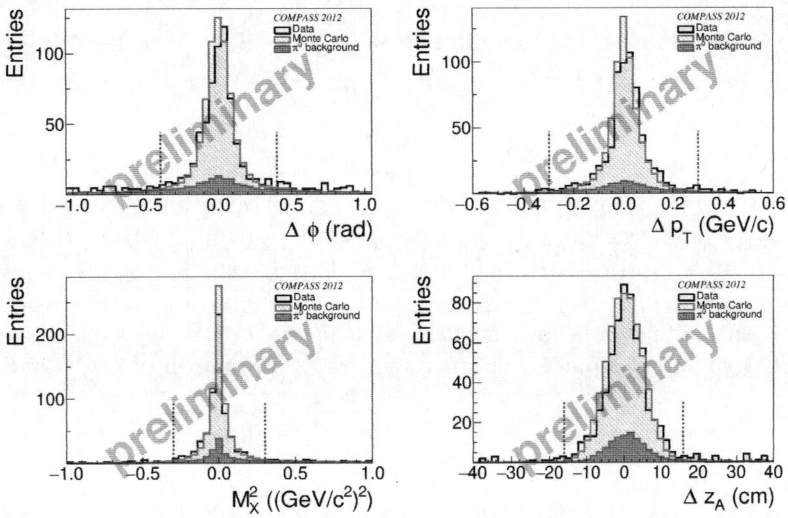

Figure 1: Distributions of exclusivity variables the single photon sample. The variables $\Delta\phi$, Δp_T and M_X^2 are defined by Eqs. (1-3), while Δz_A encodes the reverse vertex pointing. The whole Monte Carlo estimate is displayed in red, while in grey only contamination from exclusive π^0 production is shown. The blue dotted lines indicate the applied cuts.

The background is further reduced by applying a kinematic constrained fit. The input to the fit is supplied with all measured quantities, their uncertainties and correlations. The fit performs a chi-squared minimisation using constraints on the energy and momentum conservation. In the following the kinematic variables determined by the kinematic fit are used since the fit offers their most precise determination.

For selection of the sample of exclusive π^0 production, $\mu p \to \mu' p' \pi^0$, an analogous procedure is used as described for the DVCS sample. Now, the variable q' corresponds to the four-momentum of π^0 instead of single photon. In addition to the cuts on the four exclusivity variables introduced above, a cut on the invariant mass of two-photon system is applied as shown in Fig. 2. For exclusive π^0 sample an additional constraint is included in the kinematic constrained fit. In addition to the conservation of energy and momentum, the invariant mass of two photons is required to be constrained by the PDG π^0 mass.

IOP Conf. Series: Journal of Physics: Conf. Series **938** (2017) 012015 doi:10.1088/1742-6596/938/1/012015

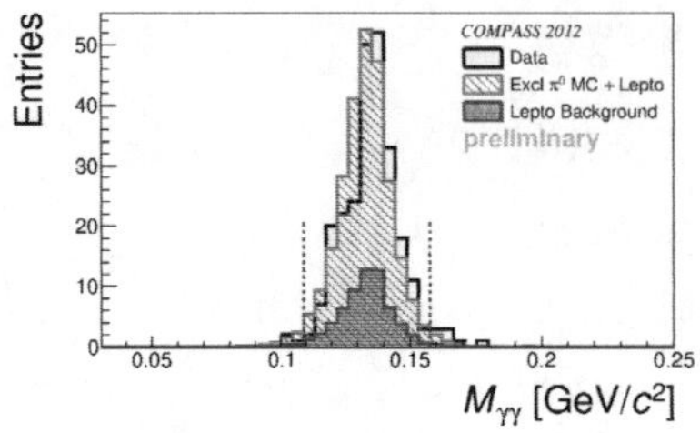

Figure 2: Invariant mass of the $\gamma\gamma$ system for exclusive π^0 sample. The blue dotted lines indicate the applied cuts.

4. Results on DVCS

In this analysis we determine the differential cross section

$$\frac{\mathrm{d}^4\sigma^{\mu p \to \mu' p' \gamma}}{\mathrm{d}Q^2 \mathrm{d}\nu \mathrm{d}|t| \mathrm{d}\phi} \tag{4}$$

for exclusive production of single photon. Here, Q^2 and ν denote the virtual photon virtuality and its energy in the lab. system, t is the square of the four-momentum transfer between the target and recoil protons and ϕ is the azimuthal angle between the lepton scattering plane and the photon production plane. In order to extract the differential cross section as a function of Q^2, ν, t and ϕ the data has been corrected for acceptance effects using bin-wise acceptance correction factors. As an example, acceptance factors as a function of t, Q^2 and ν are shown in Fig. 3.

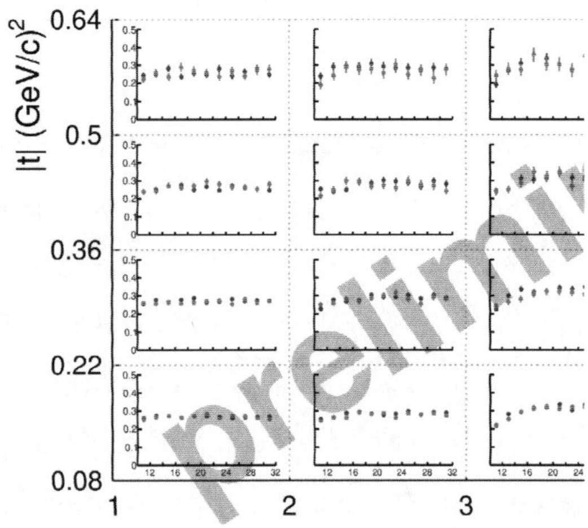

Figure 3: Acceptance correction factors for the single photon sample as a function of t, Q^2 and ν.

The Bethe Heitler contribution has been subtracted from the data for each of the bins individually. A major background source for single photon production is the production of π^0 mesons, where one of the decay photons is undetected. The contamination arising from π^0 production was estimated and subtracted from the data inside each bin using LEPTO 6.5.1

together with an exclusive π^0 Monte Carlo based on a parameterisation from Goloskokov-Kroll model [9].

For brevity the differential cross section of Eq. (4) will be denoted in the following by $\mathrm{d}\sigma$, while the signs of muon charge and its helicity will be indicated by $+(-)$ and $\leftarrow$ $(\rightarrow)$, respectively. The sum of cross sections for μ^+ and μ^- incident beams

$$\mathcal{S}_{CS,U} = \mathrm{d}\sigma^{\overleftarrow{+}} + \mathrm{d}\sigma^{\overrightarrow{-}} = 2(\mathrm{d}\sigma^{BH} + \mathrm{d}\sigma^{DVCS}_{unpol} + e_\mu P_\mu \mathrm{Im}I) \tag{5}$$

allows us to access the unpolarised DVCS cross section after subtracting the Bethe Heitler term. Furthermore, using the notation of Ref. [10] for

$$\mathrm{d}\sigma^{DVCS}_{unpol} = \frac{e^6}{y^2Q^2}(c_0^{DVCS} + c_1^{DVCS}\cos\phi + c_2^{DVCS}\cos 2\phi) \tag{6}$$

and

$$\mathrm{Im}I = \frac{e^6}{x_{Bj}y^3tP_1(\phi)P_2(\phi)}(s_1^I\sin\phi + s_2^I\sin 2\phi) \tag{7}$$

one observes that upon integration of $\mathcal{S}_{CS,U}$ over ϕ the contributions of all terms cancel, except c_0^{DVCS}. The latter corresponds to the dominant leading-twist and helicity-conserving contribution to the DVCS cross section.

The goal of this measurement is to determine the t-dependence of the unpolarised (phi-integrated) DVCS cross section in virtual photon-proton scattering. The γ^*p cross section is derived from the measured muon-proton cross section by taking into account the transverse virtual photon flux Γ

$$\frac{\mathrm{d}^3\sigma^{\mu p}}{\mathrm{d}Q^2\mathrm{d}\nu\mathrm{d}t} = \Gamma\left(\frac{\mathrm{d}\sigma_T^{\gamma^*p}}{\mathrm{d}t} + \epsilon\frac{\mathrm{d}\sigma_L^{\gamma^*p}}{\mathrm{d}t}\right), \tag{8}$$

where according to the Hand's convention

$$\Gamma = \frac{\alpha_{em}(1-x_{Bj})}{2\pi Q^2 y E_\mu}\left[y^2(1-\frac{2m_\mu^2}{Q^2}) + \frac{2}{1+Q^2/\nu^2}(1-y-\frac{Q^2}{4E_\mu^2})\right] \tag{9}$$

and the subscript T (L) refers to transverse (longitudinal) virtual photons. Since the extracted DVCS cross section is proportional to the helicity-conserving term c_0^{DVCS}, the term containing σ_L in Eq. (8) is discarded for the present analysis.

The values of the differential cross section

$$\frac{\mathrm{d}\sigma_T^{\gamma^*p\to\gamma p}}{\mathrm{d}|t|} \tag{10}$$

for four bins of $|t|$ are shown in Fig. 4 (left). They have been fitted with an exponential function in order to determine the t-slope parameter B. The value of B is equal to $(4.31 \pm 0.62\ ^{+0.09}_{-0.25})(\mathrm{GeV}/c)^{-2}$ inside the kinematic range that is indicated in the legend of Fig. 4 (left), and which corresponds to $\langle x_{Bj}\rangle = 0.056$, $\langle Q^2\rangle = 1.8$ $(\mathrm{GeV}/c)^2$ and $\langle W\rangle = 5.8$ GeV. The extracted value of B allows us to estimate the transverse size of partonic distibution in the nucleon $r_\perp$ using the relation

$$\langle r_\perp^2(x_{Bj})\rangle = 2B(x_{Bj}), \tag{11}$$

which for our measurement results in $\sqrt{\langle r_\perp^2\rangle} = (0.578 \pm 0.042 \pm^{+0.006}_{-0.018})$ fm at a mean x_{Bj} value of 0.056.

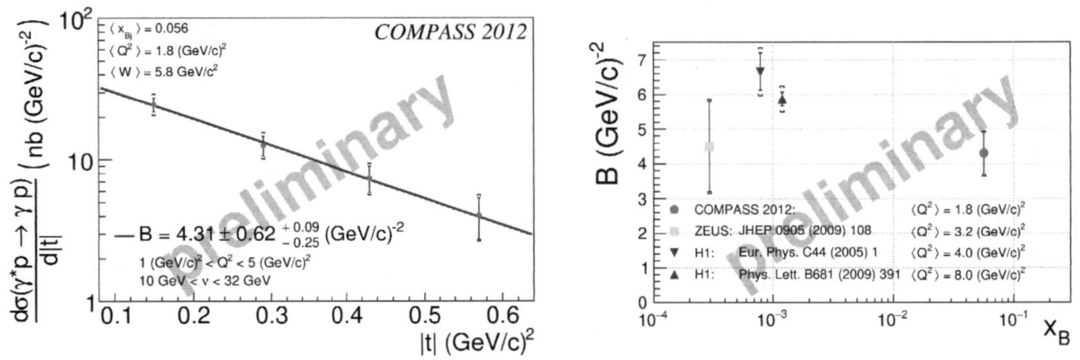

Figure 4: **Left:** The differential DVCS cross section as a function of $|t|$. **Right:** COMPASS results for the t-slope parameter compared to previous results from HERA.

The main systematic error of the cross section is due to uncertainties of absolute normalisation and determination of the amount of π^0 background. In case of parameter B the error from the absolute normalisation plays a minor role, while the main systematic uncertainty originates from the normalisation of the π^0 background.

In Fig. 4 (right) the values of parameter B from COMPASS and HERA [11, 12] are plotted as a function of x_{Bj}. As the slope B is related to the transverse size of the partonic distribution (see Eq. (10)), one may conclude that these model-independent results indicate a decrease of the transverse size of the proton with increasing x_{Bj}.

5. Results on exclusive π^0 production

The analysis of the exclusive π^0 sample mostly follows the lines of the DVCS analysis. A complete simulation of the experimental set-up is used to determine acceptance of the experiment. In order to determine background originating from non-exclusive events, the data are modeled with Monte Carlo in both kinematic regions where either the signal or the background dominates. Here, LEPTO 6.5.1 is used for a non-exclusive part. For the signal contribution a dedicated Monte Carlo generator [13] was used with the cross section parameterisation from the Goloskokov-Kroll model. The background is subtracted bin-wise from the data. Other sources of background, like misidentified exclusive $\omega \to \gamma\gamma\gamma$ events, where one photon is lost, are found to be negligible.

First, we extract the four-fold differential μp cross sections separately for μ^+ and μ^- data. Then the unpolarised cross section is obtained as the weighted mean of the two. In this way the dependence on the beam polarisation cancels. In order to convert μp cross section into $\gamma^* p$ cross section we divide the former one by the tranasverse virtual photon flux (cf. Eq. (9)).

The systematic uncertainty in this measurement is mostly due to scaling, which can be decomposed into effects depending on the absolute normalisation and background estimate, as well as to the threshold effects and kinematic uncertainties.

The unpolarised $\gamma^* p$ cross section for exclusive π^0 production reads

$$\frac{\mathrm{d}^2\sigma^{\gamma^* p}}{\mathrm{d}t\mathrm{d}\phi} = \frac{1}{2\pi}\left[\frac{\mathrm{d}\sigma_T}{\mathrm{d}t} + \epsilon\,\frac{\mathrm{d}\sigma_L}{\mathrm{d}t} + \epsilon\cos(2\phi)\frac{\mathrm{d}\sigma_{TT}}{\mathrm{d}t} + \sqrt{2\epsilon(1+\epsilon)}\cos(\phi)\frac{\mathrm{d}\sigma_{LT}}{\mathrm{d}t}\right], \tag{12}$$

where σ_T, σ_L, σ_{TT}, σ_{LT} are the structure functions, ϵ is the virtual photon polarisation parameter and ϕ represents the azimuthal angle between lepton scattering plane and π^0 production plane. Here, the subscript $T(L)$ denotes the contribution from transversely (longitudinally) polarised virtual photons, while the subscripts TT and LT denote the contributions from the interference between transversely-transversely and longitudinally-transversely polarised photons.

After integration over ϕ, we extract the cross section in bins of $|t|$, which is displayed in Fig. 5 (left). Note, that it decreases exponentially with increasing $|t|$ at values of $|t|$ larger than about 0.25 $(\text{GeV}/c)^2$, while at smaller $|t|$ the t-dependence becomes weaker. Being the first measurement at energies significantly higher than previous results, it is worth to compare it to the model prediction. In Fig. 5 (right) the COMPASS result is compared to the prediction of the Goloskokov-Kroll model [14]. The blue curve represents the cross section from the model as a function of t, while the blue points correspond to the cross section averaged over $|t|$ bins of the data. The mean cross sections for the full t-range are compared at the rightmost part of the plot. We observe that the magnitude of the predicted cross section overshoots our measurement by approximately a factor of two.

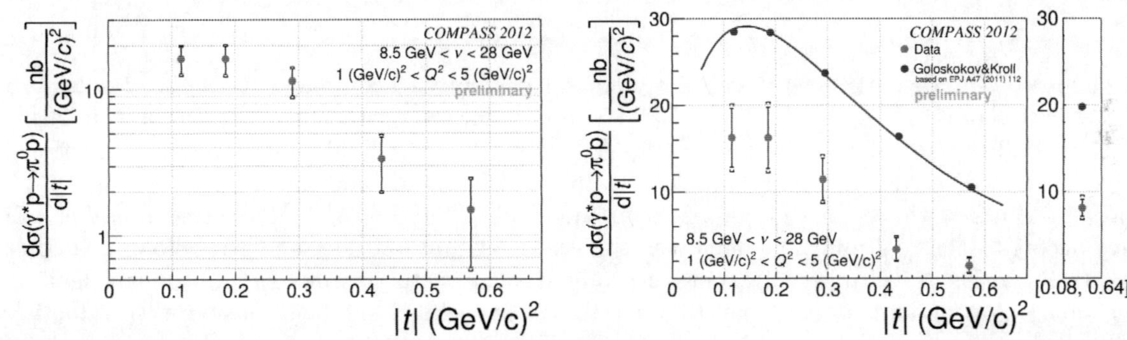

Figure 5: **Left:** The exclusive π^0 cross section as a function of $|t|$ (note vertical log scale). **Right:** COMPASS results compared to predictions of the GPD model of Ref. [14].

Furthermore, we extract the cross section as a function of ϕ in one bin of t and eight bins of ϕ of equal width. The red dots in Fig. 6 (left) show the measured cross section for each bin and a fit of the function given by Eq. (11), which is represented by the red curve. The values of the fitted parameters are indicated in the legend of the plot. We observe a large contribution from σ_{TT} and a smaller positive one from σ_{LT}. This indicates a significant role of transversely polarised photons in exclusive π^0 production.

The comparison to the prediction of the Goloskokov-Kroll model is shown in Fig.6 (right). Apart from the discrepancy in the magnitude of cross sections mentioned before, here we observe also different shapes of ϕ-dependence for the measurement and the model prediction. It indicates that relative contributions of the interference terms σ_{TT} and σ_{LT} are different for the measurement and the model.

According to Refs. [14, 15] the structure functions for exclusive pseudo-scalar meson production, which appear in Eq. (11), depend on GPDs $\tilde{H}^q$, $\tilde{E}^q$, H_T^q and $\bar{E}_T^q = 2\tilde{H}_T^q + E_T^q$. For π^0 production a large contribution from transversely polarised virtual photons is expected, which is mainly generated by the chiral-odd ('transversity') GPDs $\bar{E}_T^q$. It manifests itself in a large contribution from σ_{TT} and a dip in the differential cross section $\mathrm{d}\sigma/\mathrm{d}t$ as $|t|$ decreases to zero. These features are in a qualitative agreement with our results, although the predicted values of cross section overshoot the measured ones by a factor of about 2.

6. Outlook

In 2016 and 2017 the COMPASS collaboration is continuing to take data on the GPD program.

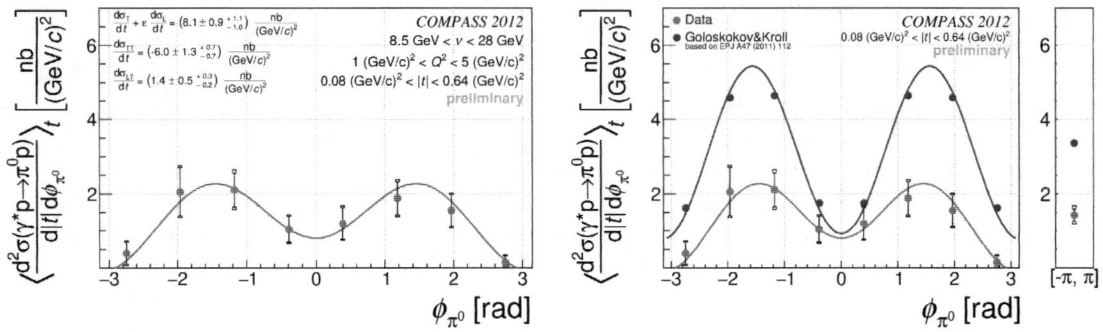

Figure 6: **Left:** The exclusive π^0 cross section as a function of ϕ. **Right:** COMPASS results compared to predictions of the GPD model of Ref. [14].

An expected increase of the statistic is by a factor of about 15 compared to the 2012 data. An investigation of GPDs with DVCS and HEMP on unpolarised protons using the data from 2016-2017 will allow us to determine the x_{Bj}-dependence of t-slope of the differential cross sections. That is related to the spatial transverse distribution of partons and the 'nucleon tomography'. Measurements of the beam charge and spin sum and difference of single-γ cross sections will give access to the real and imaginary parts of the DVCS amplitude, and will allow to further constrain GPDs H. Studies of exclusive production of vector mesons (ρ^0, ω, ϕ) will lead to the quark flavour and gluon separation for GPDs H, while that of exclusive π^0 production will provide constraints on on chiral-odd GPDs. The projected accuracies for measurements of various observables within the GPD program are presented in the COMPASS-II proposal [7].

References

[1] D. Mueller *et al*, Fortsch. Phys. **42** (1994) 101.
[2] X. Ji, Phys. Rev. Lett. **78** (1997) 610; Phys. Rev. D **55** (1997) 7114.
[3] A.V. Radyushkin, Phys. Lett. B **385** (1996) 333; Phys. Rev. D **56** (1997) 5524.
[4] M. Burkardt, Phys. Rev. D **62** (2000) 071503; erratum-ibid. D **66** (2002) 119903; Int. J. Mod. Phys. A **18** (2003) 173; Phys. Lett. B **595** (2004) 245.
[5] N. d'Hose, S. Niccolai and A. Rostomyan, Eur. Phys. J. A **52** (2016) 157.
[6] K, Kumerički, S. Liuti and H. Moutarde, Eur. Phys. J. A **52** (2016) 151.
[7] F. Gautheron *et al.* [COMPASS Collaboration], CERN-SPSC-2010-014, SPSC-P-340, May 2010.
[8] COMPASS Collaboration, P. Abbon *et al.*, Nucl. Instr. Meth. A **577** (2007) 455.
[9] S.V. Goloskokov and P. Kroll, Eur. Phys. J. C **65** (2010) 137.
[10] A.V. Belitsky, D. Mueller and A. Kirchner, Nucl. Phys. B **629** (2002 173.
[11] ZEUS Collaboration, S. Chekanov *et al.*, JHEP **0905** (2009) 108.
[12] H1 Collaboration, F.D. Aaron *et al.* Phys. Lett. B **681** (2009) 391; A. Aktas *et al.* Eur. Phys. J. C **44** (2005) 1.
[13] A. Sandacz and P. Sznajder, arXiv:1207.0333 [hep-ph].
[14] S.V. Goloskokov and P. Kroll, Eur. Phys. J. A **47** (2011) 112.
[15] CLAS Collaboration, I. Bedlinsky *et al.*, Phys. Rev. C **90** (2014) 025205.

Polarization observables in few nucleon systems with CLAS

Nicholas Zachariou, for the CLAS collaboration

University of Edinburgh, James Clerk Maxwell Building, Peter Guthrie Tait Road, Edinburgh EH9 3FD, UK

E-mail: nick.zachariou@ed.ac.uk

Abstract. The CEBAF Large Acceptance Spectrometer (CLAS), housed in Hall-B at the Thomas Jefferson National Accelerator Facility provides us with the experimental tools to study strongly-interacting matter and its dynamics in the transition from hadronic to partonic degrees of freedom in nuclear interactions. In this paper we discuss the progress made in understanding the relevant degrees of freedom using polarisation observables of deuteron photodisintegration in the few-GeV photon-energy region. We also address progress made in studying the interaction between Hyperons and Nucleons via polarisation observables, utilising high-statistics experiments that provided us with the large data samples needed to study final-state interactions, as well as perform detailed studies on initial-state effects. The polarisation observables presented here provide us with unique experimental tools to study the underlying dynamics of both initial and final-state interactions, as well as the information needed to disentangle signal from background contributions.

1. Introduction

Photoproduction experiments, using polarized beams incident on few-nucleon systems, provide us with unique experimental tools to 1) study the underlying dynamics of strongly interacting matter and investigate the relevant degrees of freedom, and 2) study initial- and final-state effects. Hadron interaction at low and high energies has been extensively studied and it is currently well understood in terms of hadronic and partonic degrees of freedom, respectively. What governs this interaction at medium energies has been, however, a long standing problem in nuclear physics. This transition region has been extensively studied over the past 20 years in deuteron photodisintegration, which provides us with the simplest two-body system. Here we present recent polarization measurements on deuteron photodisintegration, that provide stringent constraints on the underlying dynamics.

Understanding initial- and final-state effects allows us to reduce model-dependent interpretation of experimentally accessible phenomena on bound nucleon targets, as well as study in detail interactions that are otherwise experimentally difficult to perform. Understanding the hyperon-nucleon (YN) interaction is crucial in obtaining a comprehensive picture of the strong interaction. On one hand, the nucleon-nucleon interaction has been extensively studied for many decades and a good understanding of this system at energies beyond the pion production threshold exists. On the other hand, direct studies of interactions between other members of the octet – namely the hyperons – remain scarce and inaccurate due to the challenges imposed by the short-lived hyperon. Reliable and precise data on the interaction between hyperons and nucleons will allow us to address the “Hyperon Puzzle”, which reflects how theoretical models cannot predict the role of hyperons in neutron stars in a manner that is consistent with the most recent observations of massive neutron starts [1]. Here we present a new approach that utilizes final-state interactions (FSI) in exclusive hyperon photoproduction reactions to obtain a clear insight of the dynamics of the YN interaction.

The studies presented here were performed using data collected at the Jefferson Lab (JLab) using the CEBAF Large acceptance Spectrometer (CLAS) [2] from experiment E06-103 [3]. During this experimental run, data were collected using a circularly and linearly polarized tagged photon beam incident on a 40 cm long liquid deuterium target.

2. Hall B and the CLAS detector

The Continuous Beam Accelerator Facility (CEBAF) at JLab provided the continuous electron wave beams up to 6 GeV used for the nuclear and particle physics scientific program [4]. The injector was able to produce polarized beams with polarizations > 80%, which allowed polarization experiments to be performed. The beam was simultaneously delivered to three end stations (Halls A, B, and C) to conduct complementary experiments. The CLAS detector, which was the main scientific instrument for experiments performed in Hall B until 2012, provided a nearly hermetic acceptance. The CLAS detector was split into six sectors – each comprising of identical mass spectrometers – by six kidney-shaped superconducting coils, which produced a toroidal magnetic field primarily in the ϕ direction. Each sector was composed of three layers of drift chambers to measure charged-particle trajectories, scintillators to measure time-of-flight, a Čerenkov counter to separated electrons from π^-s, and electromagnetic calorimeters primarily used for photon and neutron identification (see left panel of Fig. 1). This system allowed for an efficient charged-particle tracking between 8° and 144° polar angles and a neutral particle detection between 8° and 70° polar angles with momentum and angular resolutions of $\sim 0.5\%$ and 2 mrad, respectively. These detector characteristics allowed for a reliable particle identification between protons and pions below 3.5 GeV/c, and between pions and kaons below 2 GeV/c. The entire detector system has since been upgraded for the new scientific programme at JLab and will receive beams with energies up to 10.6 GeV [5].

The tagger spectrometer, also housed in Hall B, allowed tagged real-photon experiments to be performed, with photon energies between 20% and 95% of the incident electron energy E_0. The right panel of Fig. 1 shows a schematic of the photon tagger. The tagger hodoscope design provided an energy resolution of $0.001E_0$ and a timing resolution of 110 ps. Circularly and linearly polarized photon beams were generated with the use of amorphous and crystal radiators placed upstream of the tagger, with polarizations reaching 80% and 75% respectively. These capabilities were utilised in a plethora of experiments performed in Hall B and here we report recent measurements on the study of few nucleon systems to obtain detailed understanding of the underlying dynamics in transition region and the hyperon-nucleon interaction.

3. Studies of the transition region

Studies of the transition region were initially employed searching for experimentally accessible phenomena, such as dimensional scaling and hadron helicity conservation, as an attempt to search and identify the onset of quark-gluon dynamics. Historically, these dimensional scaling rules have been derived first in the framework of perturbative QCD (pQCD) and predict a scaling of the energy dependence of the invariant cross section, $d\sigma/dt \propto s^{-n+2}$, where n is the number of elementary fields involved in the interaction, and s and t are the usual Mandelstam variables [6]. Extensive experimental efforts in many reactions were performed to search for the onset of scaling. CLAS data from experiment E93-017 allowed a detailed study of the two-body deuteron photodisintegration differential cross section at photon energies between 0.5 and 3.0 GeV. The results were consistent with the power-law dependence $\propto s^{-11}$ above proton transverse

IOP Conf. Series: Journal of Physics: Conf. Series **938** (2017) 012016   doi:10.1088/1742-6596/938/1/012016

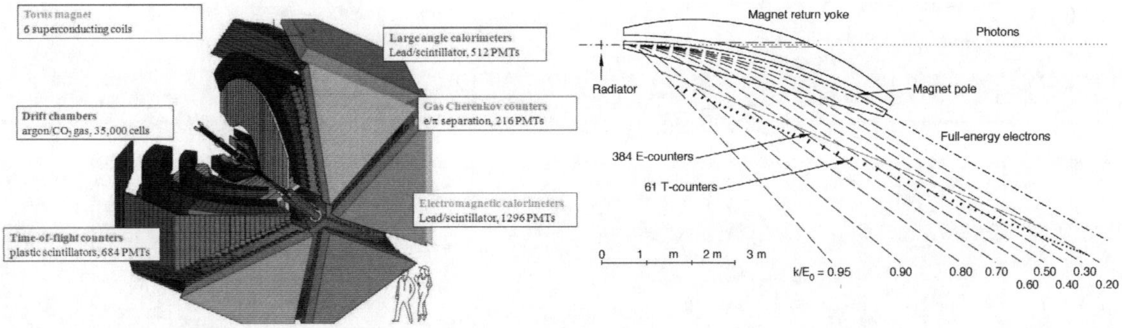

Figure 1. Left: schematic CLAS detector. The kidney-shaped superconducting coils are shown in yellow, drift chambers in blue, Čerenkov counters in magenta, time-of-flight scintillators in red, and electromagnetic calorimeters in green. Right: Hall-B photon tagging system located upstream of the CLAS detector.

momenta of $p_\perp = 1.1$ GeV/c [7]. It was thus, initially claimed that the quark-gluon regime is reached above these energies.

More recent theoretical efforts, however, using the correspondence between string theories in Anti-de-Sitter space-time and conformal field theories in physical space time, have predicted the same cross-section scaling in a very different dynamical regime [8]. This was done by using the scale invariance of the interaction between hadron constituents at very large distance scales in the so-called "conformal window", where the effective coupling is large but constant. Therefore, the onset of dimensional scaling alone, does not provide with direct evidence of quark-gluon dynamics. In this respect, polarization observables provide us with the tools needed to probe the underlying dynamics in the few-GeV region. Several theoretical models have been developed in an attempt to describe these dynamics. The main models for deuteron photodisintegration are based on nonperturtbative phenomenological approaches. The two models that provide us with predictions for polarization observables, are the Hard Rescattering Mechanism (HRM) [9], which is described as a phenomenological extension of pQCD, and the Quark- Gluon-String-Model (QGSM) [10], which is a purely non-perturbative partonic model. Both of these models describe the available cross-section data with the same degree of success [11], however predictions on the beam-spin asymmetry, Σ, which is more sensitive to the reaction mechanism, differ by about 40% between the two models [12].

Experiment E06-103 allowed a precise determination of Σ for a wide range of proton angles ($30° < \theta_p^{c.m.} < 145°$) and photon energies from 1.1 to 2.3 GeV [13]. The results from this study are shown in Fig. 2, with the left panel showing the energy dependence of Σ at proton *c.m.* angle around 90° and the right panel showing the angular dependence for six photon energy bins 200-MeV wide from 1.1 to 2.3 GeV.

The energy dependence of the measured Σ values shows a clear transition from lower to higher values around 1.7 GeV, that might be described by the HRM model. Neither model predicts, however, the magnitude of the observable. The angular dependence of Σ shows rich structures especially at lower photon energies. Such rich structures are predicted by the QGSM; however, none of the models is able to adequately describe the angular dependence of the CLAS data. The polarization results from CLAS provide stringent constraints for the available theoretical models, as they significantly extend the existing database to a much broader kinematic range, with significantly improved precision compared to previous measurements. Theoretical efforts are needed to better describe our data, and obtain a clear understanding of hadron dynamics in the few-GeV region.

IOP Conf. Series: Journal of Physics: Conf. Series **938** (2017) 012016 doi:10.1088/1742-6596/938/1/012016

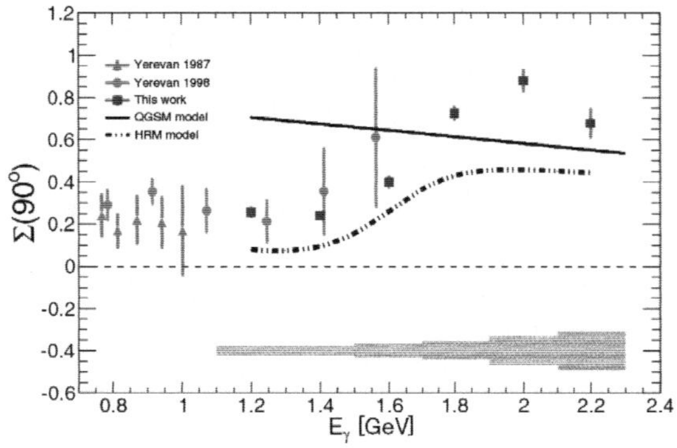

Figure 2. Recent published data from CLAS on the beam-spin asymmetry of the reaction $\gamma d \to pn$ [13]. The left panel shows the energy dependence of the 90° data (blue squares) compared to previous measurements from Yerevan. Theoretical predictions from QGSM and HRM are shown with solid and dashed line respectively. The right panel shows the angular dependence for six photon-energy bins 200-MeV wide.

4. Studies of the hyperon-nucleon interaction and initial state effects

Information on the YN interaction was accessed through final-state interactions (FSI) in the reaction $\gamma d \to K^+\Lambda n$ using data from experiment E06-103. Figure 3 illustrates the four main mechanisms that contribute to our reaction. These includes the quasi-free hyperon photoproduction off the bound proton (panel a) – which dominates the cross-section, and three FSI: the pion-mediate reaction (panel b), the kaon rescattering (panel c), and the hyperon-nucleon rescattering reaction (panel d). The large acceptance of CLAS and the tagged photon beam allowed for kinematically selecting events in which FSI effects dominate, permitting the extraction of information on the dynamics of YN interaction in a model-dependent way. The FSI contributions were enhanced by selecting events in which the momentum of the final-state neutron was inconsistent with the typical Fermi motion. This corresponded to a cut of the spectator momentum above 200 MeV/c. While this selection significantly reduced contributions from the quasi-free mechanism to the FSI sample, the relative contributions between the three mechanisms remained unknown.

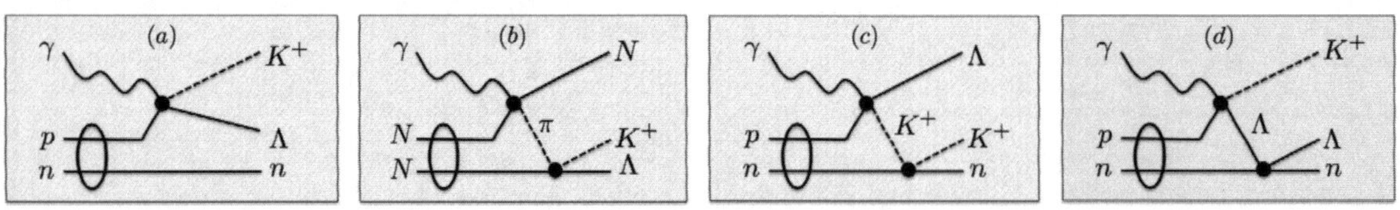

Figure 3. Four main mechanisms that contribute to the reaction $\gamma d \to K^+\Lambda n$ according to theoretical models [14, 15]: (a) quasi-free Λ photoproduction on the proton; (b) pion mediated production; (c) K^+ rescattering on spectator neutron; (d) Λ rescattering on spectator neutron.

Polarization observables allow us to obtain estimates of the contributions from the three FSI mechanisms in a relatively model-independent way, or even identify kinematical regimes where the YN mechanism dominates. The self-analyzing nature of the Λ hyperon allows a simultaneous determination of several polarization observables, utilizing both the linearly and circularly polarized photon beam. Equation (1) shows the polarization observables accessible to our experiment

$$\frac{d\sigma}{d\Omega} = \left(\frac{d\sigma}{d\Omega}\right)_0 [1 - P_{lin}\Sigma\cos 2\phi + \alpha\cos\theta_x(-P_{lin}O_x\sin 2\phi - P_{circ}C_x)$$
$$-\alpha\cos\theta_y(-P_y + P_{lin}T\cos 2\phi) - \alpha\cos\theta_z(P_{lin}O_z\sin 2\phi + P_{circ}Cz)\ldots], \quad (1)$$

where $\left(\frac{d\sigma}{d\Omega}\right)_0$ is the cross section obtained from an unpolarised photon beam, P_{lin} and P_{circ} are the degree of linear and circular photon polarisation, respectively, and Σ, O_x, O_z, C_x, C_z, P_y, and T are the polarisation observables. The kaon azimuthal angle ϕ is measured from the photon-polarization plane and $\cos\theta_{x,y,z}$ are the direction cosines of the Λ decay products in the Λ rest frame. The self-analysing power of Λ, which allows the determination of the Λ polarisation by studying the distribution of its decay products, is denoted by α. The beam-spin asymmetry can be obtained using the azimuthal distribution of the kaon or the hyperon. In a pure quasi-free data sample the two polarization observables Σ_{K^+} and Σ_Λ are consistent as the two particles fall in the same plane. However, FSI effects dilute the magnitude of Σ. Specifically, events that originate from the pion-mediate reaction are expected to significantly dilute the beam-spin asymmetry Σ_{K^+} and Σ_Λ, whereas kaon rescattering samples will dilute Σ_{K^+} but leave Σ_Λ unchanged. In a very similar manner YN samples will dilute Σ_Λ but leave Σ_{K^+} unchanged. Similar arguments
hows
-free
imes

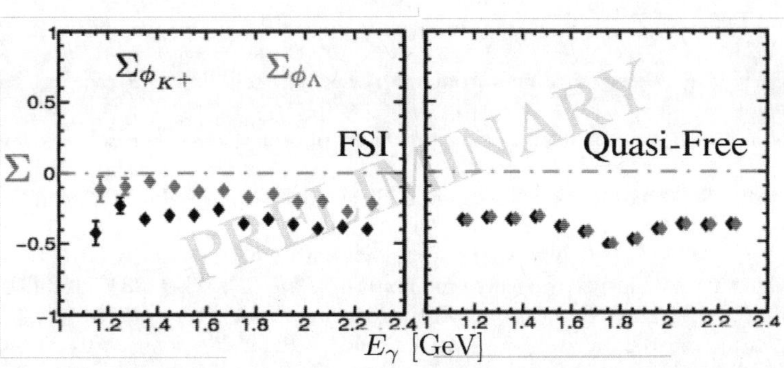

Figure 4. Beam-spin asymmetry Σ determined using the kaon (black points) and the Hyperor (red points) azimuthal distribution for an FSI dominated sample (left panel) and a a quasi-free dominate sample (right panel).

where the Σ_{K^+} determined in the FSI data is consistent with the observable determined in the quasi-free data, reflect regimes in which the YN interaction dominates. The determination of a large set of polarization observables in such kinematical regimes places very stringent constraints on the underlying dynamics of the YN interaction and allows us to tune the existing and largely unconstraint YN potentials [14, 16]. The available statistics from CLAS allows us to obtain these set of observables and their kinematical dependence for FSI, not only for the one-fold differential observables as shown here, but differential up to three kinematic variables, providing an unprecedented insight on the bare YN interaction.

It has been shown recently that exclusive meson photoproduction off the neutron is mportant for the research of the excited nucleon spectrum and the search for missing resonances, as some resonances are predicted to couple strongly to the neutron. However, since no free neutron targets exist, bound neutron targets, the simplest of which is deuterium, are typically employed. The information extracted from such targets is modified by both initial and final state effects. The method discussed above allows us to establish the contributions of FSI. Initial state effects, can be determined by studying the dependence of the observables with the momentum of the target nucleon. Studies performed with CLAS indicated that such effects can be well understood in a model independent way [17]. The left panel of Fig. 5 shows the evolution of Σ for K^+ photoproduction off the bound proton, with the target proton momentum, and the right panel shows the results obtained from a polynomial extrapolation (red points) as compared to results from a free-proton target (black points). It has been shown that selecting a quasi-free sample (in which the target nucleon momentum is less than 150 MeV/c), the determination of observables by neglecting any target-nucleon momentum dependence, leads to biased estimates compared to ones that are determined by extrapolation to the free-nucleon point of 0 MeV/c. The extrapolated results are consistent with the free-proton data, providing confidence of the

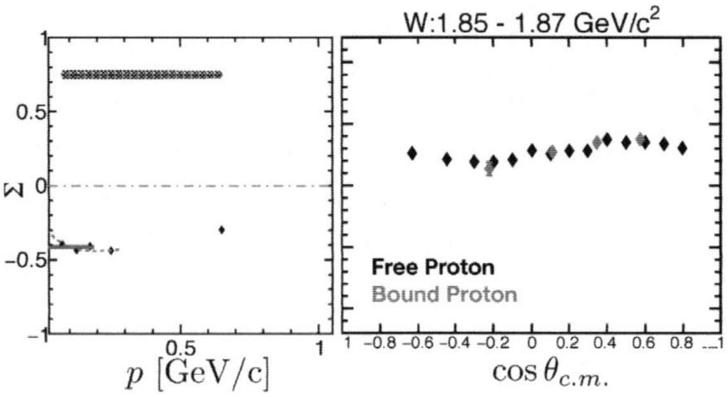

Figure 5. Evolution of Σ with target proton momentum determined using exclusive kaon photoproduction off the bound proton (left panel) and a comparison of results obtained through extrapolation with results from a free proton target.

References

[1] Lonardoni D, Lovato A, Gandolfi S and Pederiva F (2015) Hyperon Puzzle: Hints from quantum monte carlo calculations. *Phys. Rev. Lett.* **114** 092301-5

[2] Mecking B A et. al. (2003) The CEBAF large acceptance spectrometer (CLAS). *Nucl. Instr. Meth.* A **503** 513-53

[3] Nadel-Turonski P N et. al., Kaon production on the deuteron using polarized photons. Jefferson Lab PAC30 Proposal PR-06-103 (2006)

[4] Leemann C W, Douglas D R and Krafft G A(2001), The Continuous Electron beam Accelerator Facility: CEBAF at the Jefferson Laboratory. *Annu. Rev. Nucl. Part. Sci.* **51** 413-50

[5] Burkert V D (2008), CLAS12 and its initial Science Program at the Jefferson Lab Upgrade. arXiv:0810.4718

[6] Brodsky S J and Farrar G (1973), Scaling laws at large transverse momentum. *Phys. Rev. Lett.* **31** 1153-6

[7] Rossi P et al. (2005), Onset of asymptotic scaling in deuteron photodisintegration.*Phys. Rev. Lett.* **94** 012301

[8] Polchinski J and Strassler M J (2002), Hard rescattering and gauge/string duality. *Phys. Rev. Lett.***88** 031601-4

[9] Sargsian M (2004) Polarisation observables in Hard rescattering mechanism of deuteron photodisintegration. *Phys. Lett.* B **587** 41-51

[10] Grishina V Y et. al. (2004) Forward-backward angle asymmetry and polarisation observables in higj-energy deuteron photodisintegration. *Eur. Phys. J.* A **19** 117-123

[11] Mirazita M et. al. (2004) Complete anglular distribution of two-body photodisintegration between 0.5 and 3.0 GeV. *Phys. Rev.* C **70** 014005-12

[12] Avakian R et. al. (2008) Determination of the azimuthal asymmetry in deuteron photodisintegration by linearly polarised photons at E_γ = 1.1 - 2.3 GeV. CLAS Approved Analysis, CAA-NP07-01

[13] Zachariou N et. al., Determination of the beam-spin asymmetry of deuteron photodisintegration in the energy region E_γ = 1.1 - 2.3 GeV. *Phys. Rev. C* **91** 055202-16

[14] Salam A et al. (2006), K^0photoproduction on the deuteron and the extraction of the elementary amplitude. *Phys. Rev. C* **74** 044004-8

[15] Laget J M (2007), Pentaquark, cusp, and rescattering in single kaon photoproduction off deuterium. *Phys. Rev. C* **75** 014002-7

[16] Miyagawa K et al. (2006), Polarization observables in exclusive kaon photoproduction on the deuteron. *Phys. Rev. C* **74** 034002-11

[17] Ilieva Y (2017), Studies of final-state Interactions via helicity asymmetries in exclusive pseudoscalar meson photoproduction off deuteron. *Few-Body Syst.* **58** 124

Source of polarized ions for the JINR accelerator complex

A S Belov[2], D E Donets[1], V V Fimushkin[1], A D Kovalenko[1], L V Kutuzova[1], Yu V Prokofichev[1], V B Shutov[1], A V Turbabin[2], and V N Zubets[2]

[1]Joint Institute for Nuclear Research, Joliot-Curie 6, 141980 Dubna, Moscow region, Russia
[2]Institute for Nuclear Research of RAS, prospect 60letiya Oktyabrya, 7A, 117312, Moscow, Russia

E-mail: belov@inr.ru

Abstract. The JINR atomic beam type polarized ion source is described. Results of tests of the plasma ionizer with a storage cell and of tuning of high frequency transition units are presented. The source was installed in a linac injector hall of NUCLOTRON in May 2016. The source has been commissioned and used in the NUCLOTRON runs in 2016 and February – March 2017. Polarized and unpolarized deuteron beams were produced as well as polarized protons for acceleration in the NUCLOTRON. Polarized deuteron beam with pulsed current up to 2 mA has been produced. Deuteron beam polarization of 0.6-0.9 of theoretical values for different modes of high frequency transition units operation has been measured with the NUCLOTRON ring internal polarimeter for the accelerated deuteron and proton beams.

1. Introduction

The high intensity pulsed source of polarized ions (SPI) has been developed at JINR in collaboration with INR RAS for injection of polarized deuterons and protons into the Nuclotron and future collider of heavy and light ions NICA [1]. The SPI is an atomic beam-type polarized ion source with a charge-exchange plasma ionizer and a storage cell in an ionization region [2-4]. Many parts of the polarized ion source CIPIOS from IUCF [5] are used in the project.

General description of the source is given in chapter 2 of the paper. Results of tests of the source ionizer and tuning of the high frequency transition (HFT) units are presented in chapter 3. Commissioning of the SPI is described in chapter 4.

2. General description of the SPI

A schematic diagram of the SPI is shown in figure 1. The source consists of an atomic beam apparatus, the plasma charge-exchange ionizer and a system for polarized ion beam transport and spin turn to vertical direction.

2.1. Atomic beam apparatus

Polarized deuterium or hydrogen atoms with thermal energy are produced by the atomic beam apparatus consisting from a pulsed radio frequency (RF) discharge dissociator, sextupole magnets and HFT units. For definiteness we describe further generation of polarized atomic deuterium beam and of polarized deuterons unless otherwise noted.

Dissociation of molecular deuterium in RF discharge is used for production of atomic deuterium as in a conventional atomic beam type source. Molecular deuterium is injected by a pulsed electromagnetic gas valve into the dissociator tube where pulsed RF discharge is induced. Deuterium molecules dissociate into atoms in the RF discharge plasma in collisions with plasma electrons. Atomic deuterium flows into vacuum out the dissociator tube through 100 mm long 5mm internal diameter Pyrex channel which ends with sonic nozzle of 2 mm in diameter. Walls of the channel are cooled to ~80K with the cryocooler (model 350 Cryodyne refrigerator system).

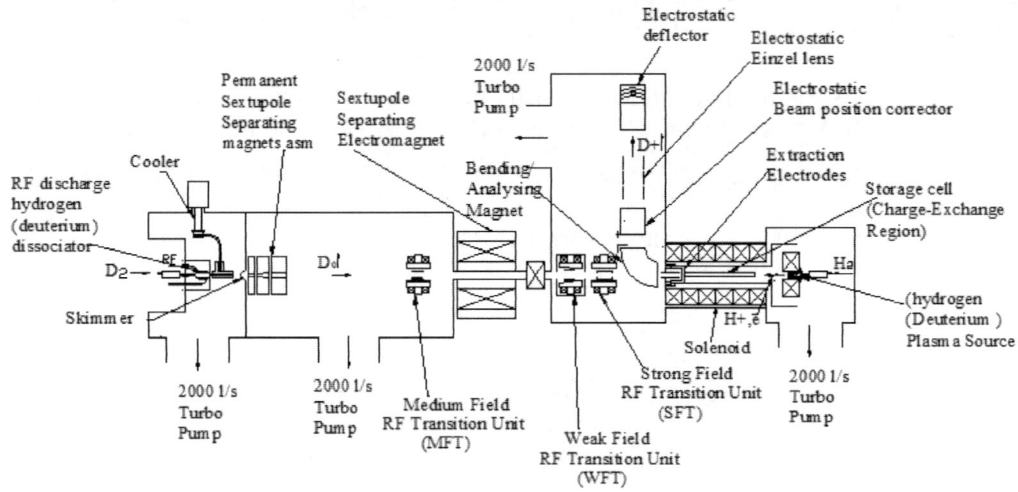

Figure 1. Schematic diagram of the SPI.

The sextupole magnets system consists of assembly of three permanent sextupoles and one electromagnet sextupole. The sextupole magnets parameters are presented in ref. [5]. The atomic beam passes through the sextupole magnets where atoms in spin states with $m_j = 1/2$ are focused and those in spin state with $m_j = -1/2$ are defocused. Thus, after passage the sextupole magnets the atoms become polarized.

Nuclear polarization of deuterium atoms being ionized in strong magnetic field is increased with system of the HFT units which includes a medium field transition unit (MFT) (installed between the assembly of permanent sextupole magnets and electromagnet sextupole), a weak field (WFT) and a strong field (SFT) transition units which are installed downstream the electromagnet sextupole. For hydrogen atoms alternate turning on of the WFT ($1 \rightarrow 3$) and the SFT ($2 \rightarrow 4$) will lead to switching between theoretical values of proton vector polarization -1 and +1. For polarized deuterium use of MFT, WFT and SFT will allow switching deuteron vector polarization between +1 and -1 and tensor polarization between +1 and -2 [5,6]. Many another deuteron polarization states can be produced using different combinations of the HFT. The sextupole magnets and the HFT units were parts of the CIPIOS designed at IUCF [5].

2.2. Charge-exchange plasma ionizer with the storage cell

The polarized atomic deuterium beam produced by the atomic beam apparatus is injected into the storage cell installed inside a solenoid of the charge-exchange plasma ionizer. The atomic beam has duration of up to 3 ms determined by the pulse duration of the RF discharge dissociator. During the pulse time polarized deuterium atoms are injected and stored in the cell. Then hydrogen plasma jet generated by a plasma arc-discharge source is injected into the storage cell in direction opposite to the deuterium atomic beam through the orifice of 3 mm in diameter at end of the storage cell. Polarized deuterons are produced in the storage cell via charge-exchange collisions between polarized deuterium atoms and unpolarized protons:

$$D^0\uparrow + H^+ \rightarrow D^+\uparrow + H^0 \qquad (1)$$

For production of polarized protons the next reaction is used:

$$H^0\uparrow + D^+ \rightarrow H^+\uparrow + D^0 \qquad (2)$$

In this case the atomic beam apparatus produces the polarized atomic hydrogen beam and the plasma source produces deuterium plasma.

Cross-section of the charge-exchange reactions (1) and (2) increases with decrease of relative energy of colliding particles and reach of $5\cdot10^{-15}$ cm^2 at energy of incident particles of ~10 eV (typical for a gas discharge plasma).

Radial confinement of the low energy polarized ions formed in the charge-exchange region is provided by magnetic field of the ionizer solenoid.

The polarized ions move then slowly under influence of weak electric fields in plasma in direction toward an extraction electrode system where they are accelerated to energy of up to 25 keV together with unpolarized plasma ions. The three electrode accel-deccel quasi Pierce system is used for the ion beam extraction and formation.

2.3. Beam transport and spin rotator system

The 25 keV ion beams extracted pass through the 90° bending magnet in which the polarized ion beam is separated from unpolarized ions. The unpolarized ion beam current is recorded using the ion beam collector downstream the bending magnet. The polarized beam comes out the magnet in vertical direction, passes through the electrostatic einzel lens and then is deflected by the 90° electrostatic deflector into the horizontal plane to x direction. Direction of ions spin remains unchanged during passage the deflector. At the source exit the polarized ion beam passes through a solenoid which is used to rotate the deuterons or protons spin to vertical direction. By this way it is possible to ensure at the source exit vertical direction of spin for polarized protons and deuterons.

3. Tests of the SPI with the storage cell and tuning of the HFT units

3.1. Tests of the ionizer with the storage cell

After tests of the SPI with free polarized atomic deuterium beam in the plasma ionizer described in [7] the storage cell has been installed into the plasma ionizer. The storage cell allows increasing density of polarized atoms in the charge-exchange region one order of magnitude in comparison with free polarized atomic beam. However, even in pulsed mode of operation unpolarized ion current which can be transported through the storage cell in noiseless mode is restricted to ~50 mA [4]. This method has been tested at INR RAS [3] and the polarized proton beam with peak current of 11 mA and polarization of 80% has been obtained from the source installed at the INR RAS test bench [4].

Initial tests of the SPI with the storage cell in the ionizer region resulted to low intensity of the polarized deuteron beam of ~0.4 mA. It has been found that this polarized deuteron beam low intensity was determined in part by low efficiency of collection and extraction of polarized ions from charge-exchange region inside the storage cell. The collection efficiency has been increased by increase of magnetic field in the charge-exchange region up to 2.4 kG and increase of internal diameter and shape of the storage cell. Initially the storage cell had cylindrical shape of the internal surface with diameter of 15 mm and length of 250 mm (as in ref. [4]). Now the storage cell has conical shape with internal diameter changing from 16 mm to 20 mm and length of 200 mm. The end of the cell directed to the plasma source is closed by diaphragm of 3 mm in diameter through which hydrogen plasma in injected into the storage cell. The cell is manufactured from aluminum alloy.

With the storage cell of this geometry the polarized deuteron beam with peak current of 2 mA has been obtained from the source (difference of the ion current recorded downstream the analyzing/bending magnet with the polarized atomic deuterium beam "on" and "off"). Unpolarized pulsed proton beam extracted simultaneously with the polarized deuterons was about 40 mA as well as beam of H_2^+ ions of 2 mA peak. The beam of H_2^+ ions is not separated from the polarized deuteron beam in the analyzing/bending magnet and in the electrostatic deflector of the SPI. But after acceleration in a linac and passage through a stripping target unpolarized protons formed due to striping of the H_2^+ ions are deflected from the polarized deuteron beam entirely in transport line bending magnets. Thus, the unpolarized H_2^+ ions do not dilute polarized deuteron beam injected into the NUCLOTRON.

3.2. Measurements of efficiencies of the HFT with a Breit – Rabi polarimeter

HFT of the PSI were tuned in their operational position with use of a Breit – Rabi polarimeter.

A schematic diagram of the ABS components and the polarimeter is shown in figure 2.

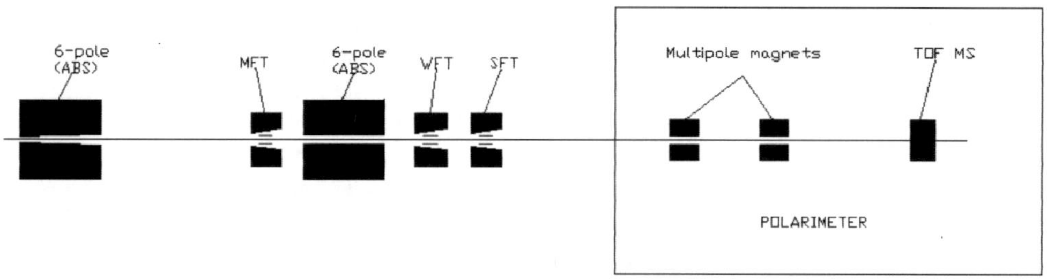

Figure 2. Schematic diagram of the ABS with the Breit – Rabi polarimeter.

Two permanent multipole magnets are used in the polarimeter. Their parameters were determined using race-tracing calculations. The first magnet is a quadrupole with aperture of 17mm and length of 140 mm. The second magnet is a sextupole with aperture of 18mm and length of 125mm. The maximum magnetic field of both magnets is 1.6 T. These magnets were installed into a vacuum chamber inside of the ionizer solenoid instead of the storage cell. The time of-flight mass spectrometer (TOF MS) was installed into a vacuum chamber of the plasma source and was used to measure the polarized atomic beam intensity (the plasma source was removed from the chamber during the tuning procedure).

If one of the HFT is "on", then respective part of the atoms undergoing transition between spin states (with change of their electron spin state in strong magnetic field from $m_j = 1/2$ to $m_j = -1/2$) will be defocused by the multipole magnets of the polarimeter. For deuterium atomic beam this leads to decrease of its intensity measured by the TOF MS by one third. Simultaneous activation of two HFT working with 100% efficiency leads to decrease of intensity by two third. Measurement of relative change of density of the polarized atomic beam by the TOF MS with HFT "off" and "on" will allow one to determine efficiencies of the HFT of the ABS.

Example of tuning of the HFT units with the Breit-Rabi polarimeter is shown in figure 3. The TOF MS signal is shown vs. current in the static coil of the SFT. The SFT RF cavity was powered with 397 MHz. Changing of magnetic field results to inducing of different transitions between hfs of deuterium atoms passing through the SFT unit. Transitions 2-6, 2-5+3-6 and 3-5 are recorded and identified using magnetic field value in the SFT magnet (calibrated for the static coil current) and used frequency of the SFT cavity. A width of the each transition curve is determined by gradient of magnetic field provided by a gradient coil of the SFT. We use here common notation of spin states of deuterium atoms. The RF cavity of the SFT produces oscillating magnetic field which is parallel to static magnetic field in the region where polarized atoms pass through the cavity. So, it is designed to induce 2-6 and 3-5 transitions (σ transitions, $\Delta m_F = 0$). 2-5 and 3-6 transitions (π transitions, $\Delta m_F = \pm 1$) should not be induced because it is necessary to have RF magnetic field to be perpendicular to the static magnetic field for these transitions. However, transitions 2-5 and 3-6 are induced also as can be seen in the figure 3 which can be explained by relatively big aperture of the TOF MS (27 mm) recording atoms passing the RF cavity far from the SFT axis where there is transversal RF field component to the static magnetic field.

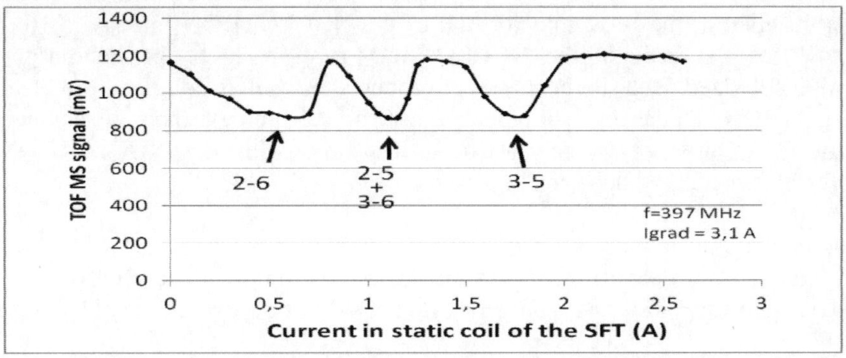

Figure 3. TOF MS signal vs. current in static coil of the SFT (only the SFT is turned on). Transitions between spin states of deuterium atoms 2-6, 3-5, and 2-5+3-6 are shown.

Measured efficiencies of another HFT units and their combinations are shown in table 1.

Table 1. The measured efficiencies of the HFT.

	SFT 2-6	SFT 3-5	MFT 3-4 + SFT 2-6	MFT 1-4 +SFT 2-6	MFT 1-4 +SFT 3-5	WFT 1-4	MFT 3-4+ WFT 1,2 -3,4
Efficiency	0.87±0.06	0.83±0,06	0,91±0.03	0.93±0.03	0.92±0.03	0.93±0.06	0.62±0,07

The efficiences were calculated taking into account not complete (97%) defocusing of atoms undergoing the transition with the polarimater multipole magnets into the TOF MS aperture (27 mm). The defocusing factor by the magnets of the polarimeter was estimated using MC simulation of tracking of polarized atoms through ABS and the polarimeter system. As shown in the table 1 the efficencies of the most HFT are around 90% except WFT 1,2 – 3,4. We plan to rebuild the WFT and improve its efficency.

4. Commissioning of the SPI

The SPI has been moved to the linac hall in May 2016. It was commissioned and used in the NUCLOTRON runs in 2016 and February – March 2017. In these runs the SPI worked in mode with polarized deuterons during 900 hrs and with unpolarized deuterons during 600 hrs. Also there was short test run during of 40 hrs with polarized protons.

Measurements of deuteron polarization made with an internal NUCLOTRON polarimeter [8] resulted to deuteron polarization of 0.6-0.88 of theoretically maximum values. Tensor deuteron polarization of 0.88±0.049; -1.47±0.031 (theoretically maximum +1, -2) has been measured with additional tune of the HFTs with the internal polarimeter. Measurements show long term stability of polarization. This demonstrates the lack of depolarization in the plasma ionizer storage cell which was used for the first time for polarized ion source working during continuous accelerator runs. Proton beam polarization of -0,354±0,022 has been measured [9] which is close to expected polarization of -0.39 with present transport line between the linac and the NUCLOTRON ring [10].

5. Conclusions

The SPI of JINR has been tested with the storage cell installed into the charge-exchange region of the plasma ionizer. Polarized deuteron beam with pulsed current up to 2 mA has been produced after study of the storage cell operation. The HFT units were tuned with using of the Breit-Rabi polarimeter. The SPI was installed in the linac injector hall of the NUCLOTRON in May 2016. The source has been commissioned and used in the NUCLOTRON runs in 2016 and February – March 2017. Polarized and unpolarized deuteron beams were produced as well as polarized protons for acceleration in the NUCLOTRON. Deuteron beam polarization of 0.6-0.88 of theoretical values for different modes of the HFT units operation has been measured with the NUCLOTRON ring internal polarimeter for the accelerated deuteron and proton beams.

Acknowledgements

We greatly acknowledge delivering of parts of polarized ion source CIPIOS from IUCF. We are grateful to accelerator division staff members who helped in the source assembly, commissioning and operation.

References

[1] Trubnikov G, Agapov N, Brovko O, Butenko A, Donets E, Eliseev A, Fimushkin V, Gorbachev E, Govorov A, Ivanov E *et al.* 2013 *Proc. of the 4th Int. Particle Accelerator Conf. IPAC2013 (Shanghai, China, 2013)* TUPFI009 p 1343

[2] Belov A S, Esin S K, Kubalov S A, Kuzik V E, Stepanov A A and Yakushev V P 1987 *Nucl. Instrum. and Meth. In Phys. Res.* **A255** 442

[3] Belov A S, Esin S K, Netchaeva L P, Klenov V S, Turbabin A V and Vasil'ev G A 1998 *Proc. Int. Workshop on Polarized Gas Targets and Polarized Beams* (*Urbana IL, USA, 1997)* ed. Roy J. Holt and M. A. Miller, AIP Conf. Proc. **421** p 362

[4] Belov A S, Esin S K, Netchaeva L P, Turbabin A V and Vasil'ev G A 1999 *Proc. Int. Symp. on High Energy Spin Physics (Protvino, Russia*, 1998) ed. N. E. Tyurin *et al*, (Singapore: World Scientific) p 22

[5] Derenchuk V P and Belov A S 2001 *Proc. 2001 Particle Accelerator Conf* (*Chicago, USA)* WPAH008 p 2093

[6] Fimushkin V V, Kovalenko A D, Kutuzova L V, Prokofichev Yu V, Belov A S, Turbabin A V and Zubets V N 2014 *Proc. 20th Int. Symp. on High Energy Spin Physics* (*Dubna, Russia, 2012) PEPAN* **45** part 1 p 297

[7] Fimushkin V V, Kovalenko A D, Kutuzova L V, Prokofichev Yu V, Belov A S, Turbabin A V and Zubets V N, XVI Workshop on High Energy Spin Physics (D-SPIN2015) IOP Publishing Journal of Physics: Conference Series 678 (2016) 012058 doi:10.1088/1742-6596/678/1/012058

[8] Skhomenko Y et al., Measurement of the deuteron beam polarization at internal target at Nuclotron for DSS experiment, this conference.

[9] Ladygin V P et al., First results on measurements of the proton beam polarization at internal target at Nuclotron, this conference.

[10] Filatov Yu N, Butenko A V, Kovalenko A D, Mikhaylov V A, Kondratenko A M, Kondratenko M A, Acceleration of polarized protons up to 3.4 GeV/c in the NUCLOTRON JINR, this conference.

IOP Conf. Series: Journal of Physics: Conf. Series **938** (2017) 012018 doi:10.1088/1742-6596/938/1/012018

Acceleration of Polarized Protons up to 3.4 GeV/c in the Nuclotron at JINR

A D Kovalenko[1], A V Butenko[1], V A Mikhaylov[1], M A Kondratenko[2], A M Kondratenko[2] and Yu N Filatov[1,3]

[1]Joint Institute for Nuclear Research, 141980, Dubna, Russia
[2]Science and Technique Laboratory "Zaryad", 630090, Novosibirsk, Russia
[3]Moscow Institute of Physics and Technology, 141700, Dolgoprudny, Russia

[1]kovalen@dubna.ru

Abstract. To preserve proton polarization in the Nuclotron up to 13.5 GeV/c, it is enough to use a partial solenoid snake with maximal field integral of 25 Tm that allows one to eliminate crossings of the most dangerous intrinsic and integer spin resonances. The insertion of weak field integral is sufficient to preserve the proton polarization up to 3.4 GeV/c. This momentum corresponds to the first intrinsic resonance. To preserve polarization during crossings of five integer spin resonances, it is possible to choose crossing rates that correspond to either the fast or the slow resonance crossings. Another possibility is a deliberate increasing of the resonance strength. To eliminate depolarization during protons injection into the Nuclotron, a scheme of matching of the polarization with a vertical direction is presented. During the run in February-March 2017, the three measurements of the proton polarization at kinetic energies of 0.5 GeV, 1 GeV and 2 GeV were made that allow one to obtain the integer spin resonances strengths.

1. Preservation of the proton polarization by the partial solenoidal snakes

The beam depolarization during acceleration of the protons at the Nuclotron occurs due to crossing of a large number of spin resonances. A full solenoidal snake allows one to suppress resonant depolarization of the proton beam and requires a longitudinal magnetic field integral of about 50 T·m at the maximum momentum of 13.5 GeV/c [1]. To preserve proton polarization, it is enough to use a partial snake that allows one to eliminate crossings of the most dangerous intrinsic and integer spin resonances by the proper choice of the betatron tunes. In the paper [2] it was shown that a solenoidal 50%-snake, which requires a field integral of 25 T·m at the maximum momentum and rotates the spin at 90 degrees around the longitudinal direction, allows one to preserve proton polarization and still leaves a lot of freedom in choosing the betatron tunes. To eliminate a series of integer resonances, it is sufficient to use a partial snake with a small field integral. Thus, the proton polarization can be preserved without usage of solenoids with large field integrals up to the energy corresponding to the first intrinsic resonance. Spin tracking simulation had demonstrated that the 5% solenoid snake preserves proton polarization at the Nuclotron up to 3.4 GeV/c. The 5%-snake eliminates crossing of five integer spin resonances and requires the field integral of 0.65 T·m at the momentum of 3.4 GeV/c [2]. The value of the longitudinal field integral can be reduced further, but then it is necessary to take into account the synchrotron modulation of the energy. The rough estimations show that the weak solenoid with field integral of 0.06 T·m eliminates polarization losses during protons acceleration up to 3.4 GeV/c.

2. Preservation of the proton polarization without additional magnetic insertions

To preserve the polarization during protons acceleration without insertion of additional magnetic fields into the Nuclotron lattice it is sufficient to provide either the fast or the slow spin resonance crossings. To choose an appropriate field ramp rate at the moments of the crossing it requires detailed information on the values of the integer spin resonances strengths.

Such information can be obtained from the experimental data received in the run with the polarized protons (February-March 2017), when the following three proton polarizations at the Nuclotron

$$P_1 = (34.5 \pm 2.4)\%, \qquad P_2 = (12 \pm 1)\%\,, \qquad P_3 = -(9 \pm 1)\%$$

were measured at the beam kinetic energies of 0.5 GeV, 1 GeV and 2 GeV correspondently [3,4].

Figure 1 shows the location of the polarization measurement points, as well as the location of five points of the integer spin resonances $\gamma G = k$ (γ is a relativistic factor, G is a proton anomalous part of gyromagnetic ratio).

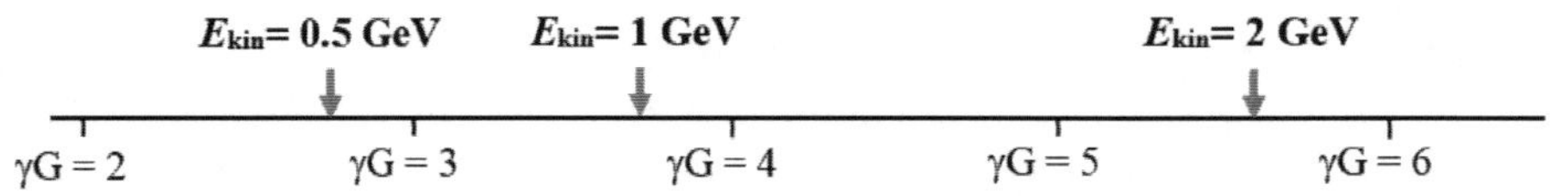

Figure 1. Location of the polarization measurement points (red arrows) and of integer spin resonances points.

Analysis of the experimental data shows that the loss of the beam polarization occurs for two reasons. The first reason is a matching of the proton polarization to vertical direction during beam injection to the Nuclotron ring (see the next section). The maximum initial polarization at the beam injection to the Nuclotron could not exceed the value of about 40% due to the polarization mismatching.

The spin resonances crossings are the second reason of the depolarization. The polarizations before and after crossing of the resonance region are connected by Froissart-Stora formula [5]

$$P_{after} = \left[2\exp\left(-\frac{\pi\omega^2}{2\,|\varepsilon'|}\right) - 1\right] P_{before}.$$

Here ω is the spin resonance strength, ε' is resonance crossing rate which is proportional to the field ramp rate dB/dt. In the experiments the field ramp rate was of 0.6 T/s and the corresponded resonance crossing rate was about of $\varepsilon' \approx 10^{-6}$. During fast resonance crossing ($\varepsilon' \gg \omega^2$) the polarization remains practically unchanged. At slow (adiabatic) resonance crossing ($\varepsilon' \ll \omega^2$) the polarization makes reversal and changes its sign. The polarization loss occurs at intermediate resonance crossing ($\varepsilon' \sim \omega^2$). The resonance strength corresponding to the intermediate crossing with a full beam depolarization is about of $5.7 \cdot 10^{-4}$ at the field ramp rate of 0.6 T/s.

The integer spin resonances occur due to manufacturing and alignment errors of the Nuclotron magnetic elements. The closed orbit distortion gives the main contribution to the resonance strength. Here we use a model of random quadrupole shifts. The accuracy of statistical model is determined by the number of the Nuclotron quadrupoles N_{quad} and in our calculations is about of $1/\sqrt{N_{quad}} \approx 10\%$.

The initial polarization P_{inj} at the Nuclotron injection and the rms deviation of the quadrupoles vertical shifts $\sigma_{\Delta y}$ are used as the fitting parameters. The best fitting to the measured polarizations occurs when parameters are equal to $P_{inj} \approx 39.2\%$ and $\sigma_{\Delta y} \approx 0.86$ mm. In this case our model gives the next results

$$P_1^* = 34\%, \qquad P_2^* = 12.2\%\,, \qquad P_3^* = -8.3\%,$$

which are in good agreement with the measured polarizations. The integer spin resonances strengths obtained from experimental data are presented in table 1. Also table 1 shows the depolarization degree at crossing the resonance with P_{before} = 100% and the field ramp rates corresponding to the fast and slow crossing with 5% loss of the polarization after crossing of the resonance.

Table 1. The integer spin resonances strengths obtained from the experimental data.

γG	2	3	4	5	6
E_{kin}, GeV	0.108	0.632	1.155	1.679	2.202
ω, 10^{-4}	2.09	4.96	10.8	23.6	57.5
D, %	13	63	33	≈0	≈0
$(dB/dt)_{fast}$, T/s	1.6	9.1	43	206	1220
$(dB/dt)_{slow}$, T/s	0.01	0.06	0.29	1.39	8.28

From table 1 it follows that the first resonance $\gamma G = 2$ is crossed rather fast with 13% polarization loss relative to the initial value of the polarization P_{inj}. The main depolarization occurs at crossing of $\gamma G = 3$ and $\gamma G = 4$ resonances. The spin resonances $\gamma G = 5$ and $\gamma G = 6$ are crossed adiabatically without polarization losses. During each adiabatic crossing the polarization is flipped.

The decreasing field ramp rate by half provides the adiabatic crossing of the $\gamma G = 4$ resonance with a 5% relative depolarization.

To reduce polarization losses down to 5% at the each $\gamma G = 2$ and $\gamma G = 3$ resonances it sufficient to increase the field ramp rate in the resonance regions up to 1.6 T/s and 9.1 T/s correspondently. The time of crossing the resonance region is about of a few milliseconds. Another possibility to reduce polarization losses is the decreasing the field ramp rate down to 0.01 T/s and 0.06 T/s. But at such a significant decrease of field ramp rate it is necessary to take into account the high order spin resonances and the synchrotron modulation of the energy.

The usage of combined spin resonance crossing techniques can be very efficient. A deliberate increase of the integer resonance strengths by means of the correcting dipoles allows one to reduce requirement on decreasing of the field ramp rate. In principal, it is possible to efficiently increase the resonances strength by correcting dipoles leaving acceptable closed orbit distortion. The influence of correcting dipoles on spin dynamics increases with increasing the beam energy. It becomes real to provide adiabatic crossing of all integer spin resonances by using correcting dipoles.

3. Matching the polarization direction at the injection of protons to the Nuclotron ring

The scheme of the beam polarization transportation from Source of Polarized Ions (SPI) [6] to the Nuclotron is shown in figure 2. At the SPI exit the polarization is directed vertically and keeps its orientation after passing the LU-20 linac. After passing the injection channel the polarization changes vertical direction. The proton polarization after beam injection to the Nuclotron ring is determined by the angle α between the polarization and vertical direction and equals to $\cos\alpha$ at the 100% polarization from SPI.

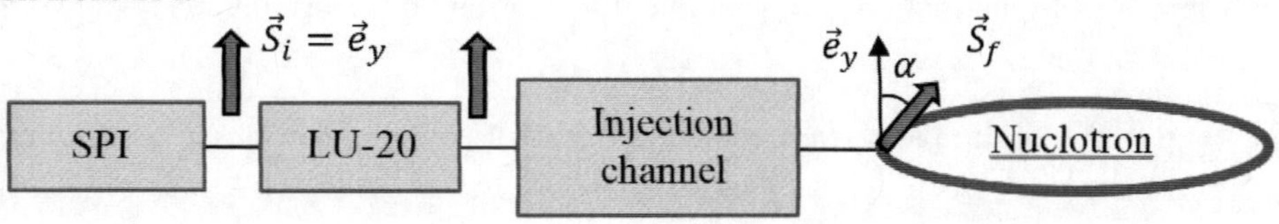

Figure 2. The transportation of the beam polarization from SPI to the Nuclotron ring.

The injection channel consists of three dipole magnets with different field directions (see table 2 and figure 3). The change of polarization direction after passing the injection channel occurs due to rotations around magnet's field directions. Calculation shows that the initial polarization P_{inj} at protons injection to the Nuclotron is of 39.2% if the polarization is directed vertically at the SPI exit. Additional rotation of the polarization could take place in the SPI solenoid, which is placed at the SPI exit. In this case calculation shows that the value of P_{inj} could not exceed of 44.2%.

Table 2. Dipole magnet in the injection channel.

Dipole	K_{x1} **(radial)**	K_y **(vertical)**	K_{x2} **(radial)**
Length, m	1.0	2.4	0.425
Orbital angle, rad	0.272	0.944	–0.158

To match proton polarization with vertical direction at the injection to the Nuclotron it is sufficient to insert two solenoids into the injection channel as shown in figure 3.

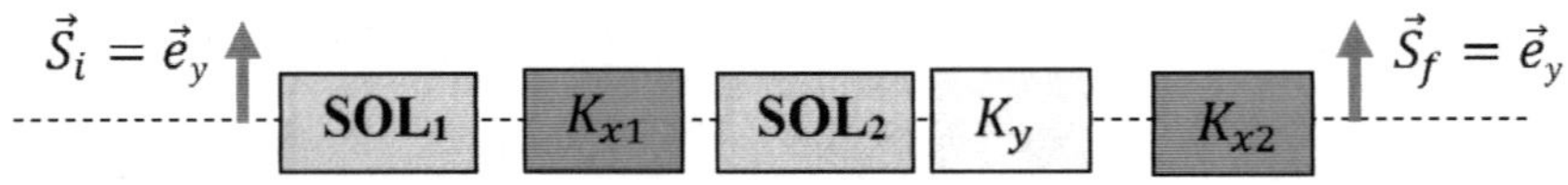

Figure 3. Placement of two solenoids into the injection channel for the polarization matching.

Table 3 shows values of the solenoidal field's integrals to obtain the proton polarization directed either upward or downward the vertical. Thus, the presented matching scheme allows one to make spin reversal during the beam injection into the Nuclotron.

Table 3. The vertical up and down polarization directions at proton injection to the Nuclotron.

$(S_y)_f$	$(B_{sol}L)_1$, T·m	$(B_{sol}L)_2$, T·m
+1	0.249	-0.207
-1	0.115	0.259

Conclusion

Let us briefly summarize the main results:

- the weak solenoid with field integral of 0.06 T·m eliminates polarization losses during protons acceleration up to 3.4 GeV/c,
- the information on the spin resonance strengths obtained from the experimental data allows one to accelerate protons up to 3.4 GeV/c with the polarization loss less than 15 % without insertion of additional magnetic fields into the Nuclotron lattice.

Analysis of the measured polarizations allows one to draw the following conclusions:

- essential depolarization about of 60% occurs at injection protons into the Nuclotron ring,
- the proton polarization at the exit of the source of polarized ions (SPI) was about $(90 \pm 10)\%$,
- rms deviation of random quadrupole vertical shifts $\sigma_{\Delta y}$ of about 1 mm causes the closed orbit distortion with maximum deviation is about ± 2 cm,
- measurements of polarization give valuable information on the orbital and spin parameters.

Acknowledgement

We would like to thank all the participants of the seminar at Veksler and Baldin Laboratory of High Energy Physics (JINR, Dubna) for the useful discussions of some questions, considered in this paper.

References

[1] Kovalenko A D *et al* 2014 *PEPAN Lett* **45** 1 p 260
[2] Filatov Yu N *et al* 2017 *Proc. of IPAC'17 Copenhagen Denmark* TUPVA112
[3] Ladugin V *et al* 2017 *First results on measurements of the proton beam* DSPIN-17 Dubna
[4] Shindin R *et al* 2017 *New F3-polarimeter at the 54-th seance at Nuclotron* DSPIN-17 Dubna
[5] Froissart M, Stora R 1960 *Nuclear Instruments and Methods* **7** 3 pp 297-305
[6] Belov A *et al* 2017 *Source of polarized deuterons and protons for Nuclotron-M* DSPIN-17

IOP Conf. Series: Journal of Physics: Conf. Series **938** (2017) 012019 doi:10.1088/1742-6596/938/1/012019

Online polarimetry of the Nuclotron internal deuteron and proton beams

A Yu Isupov

Laboratory of High Energy Physics, Joint Institute for Nuclear Research, Dubna 141980, Russia

E-mail: isupov@moonhe.jinr.ru

Dedicated to the bright memory of Prof. L.S.Zolin

Abstract. The spin studies at Nuclotron require fast and precise determination of the deuteron and proton beam polarization. For these purposes new powerful VME–based data acquisition (DAQ) system has been designed for the Deuteron Spin Structure setup placed at the Nuclotron Internal Target Station. The DAQ system is built using the *netgraph*–based data acquisition and processing framework *ngdp*. The software dealing with VME hardware is a set of *netgraph* nodes in form of the loadable kernel modules, so works in the operating system kernel context. The specific for current implementation nodes and user context utilities are described. The online events representation by *ROOT* classes allows us to generalize code for histograms filling and polarization calculations. The DAQ system was successfully used during 53rd and 54th Nuclotron runs, and their suitability for online polarimetry is demonstrated.

1. Introduction

The precise online polarization measurements of the deuteron and proton beam in the Nuclotron ring are required for many projects of the LHEP research program. The Deuteron Spin Structure (DSS) setup [1], [2], [3], [4] placed at the Nuclotron Internal Target Station (ITS) [5] studies the polarization observables in the *dp* elastic scattering for a several years [6]. Vector and tensor polarizations of the deuteron and proton beams from the LHEP–JINR polarized ions source (SPI) [7], as well as A_y, A_{yy}, A_{xx} analyzing powers of the reactions have been successfully measured in the beam energy range from 135 MeV/nucleon to 900 MeV/nucleon. These measurements were done using the VME–based data acquisition (DAQ) system which will be described in present paper.

Through the presented text the file and software package names are highlighted as *italic text*, C and other languages constructions — as `typewriter text`. Reference to the manual page named "qwerty" in the 9th section is printed as ***qwerty(9)***. All mentioned trademarks are properties of their respective owners.

2. Dealing with VME hardware

The heart of DSS DAQ system — *ngdp* graph to deal with VME hardware and manage the produced data streams — we can see in Fig.1. This picture was automatically generated by the `dot` command of ***ngctl(8)*** utility from graph was really built, and manually simplified by stripping out both nodes non–essential for understanding and details of too technical nature.

IOP Conf. Series: Journal of Physics: Conf. Series **938** (2017) 012019 doi:10.1088/1742-6596/938/1/012019

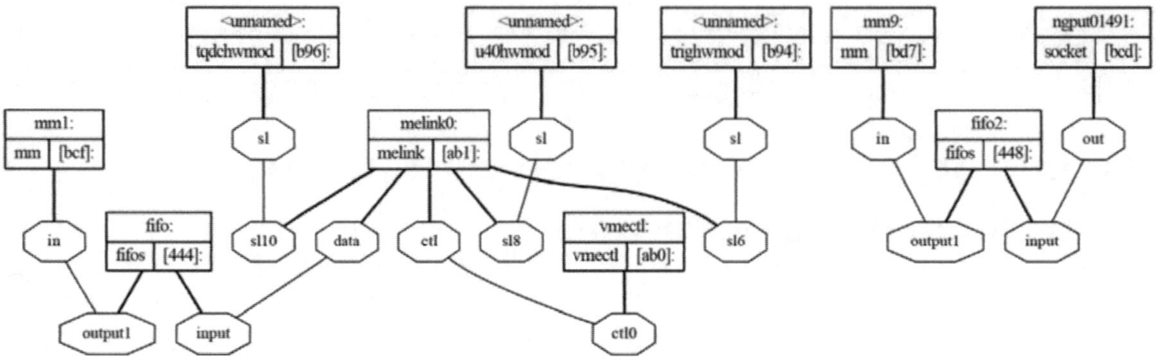

Figure 1. The simplified *netgraph* graph used by DSS DAQ system to deal with VME hardware and manage the produced data streams. Rectangles are nodes with: name (up), type (left), ID (right); octagons are hooks named within.

The *ngdp* graph like any ***netgraph(4)*** graph works in the operating system (OS) kernel context and contains the nodes (rectangles in Fig.1) connected by graph edges, which produced by pairs of so called hooks (octagons in Fig.1). The data messages flow along the edges. For reasons to use the *netgraph* package see [8]. The nodes are implemented in form of the loadable kernel modules (KLD), which approach simplifies and speeds up the debugging essentially. The data processing code execution in the kernel context allows us to handle data obtained from the VME master in the fastest possible way [8].

The VME hardware handling scheme is as follows. Each kind of VME master (with embedded CPU or adapter card for computer interconnection) and VME hardware module is represented by corresponding *netgraph* node type — VME driver and handler of VME hardware module (lets name it VME node below), respectively. Each node type should be instantiated as many times as needed to reflect an existing quantity of corresponding hardware units.

The VME driver ***ng_melink(4)*** is a KLD module (named `melink0` in Fig.1) intended to handle M–Link FVME/FVME2 VME master hardware [9]. So the VME driver contains UNIX hardware driver with `probe()`, `attach()`, `detach()` procedures, interrupt handler, kernel thread(s) ***kthread(9)*** to serve a requests ***queue(3)*** and process data, and is a *netgraph* node simultaneously.

The VME node is a KLD module in the *netgraph* style intended to deal with some VME hardware module by mediation of VME driver. Currently we support: the trigger TTCM/FVME2TMWR modules [9] by the ***ng_trighwmod(4)*** node type, the trigger U40 modules [9] by the ***ng_u40hwmod(4)*** node type, and the TQDC16 time– and charge–to–digital converter modules [9] by the ***ng_tqdchwmod(4)*** node type.

Interconnections in the DAQ graph (see Fig.1) are following: VME driver in `<M>`th crate is connected by own `ctl` hook to `ctl<M>` hook of ***ng_vmectl(4)*** node, while VME node in `<N>`th slot — by own `sl` hook to `sl<N>` hook of VME driver of corresponding VME crate.

The full data flow scheme of the DSS DAQ system is as follows:

$$
\begin{array}{l}
\qquad\qquad\qquad\qquad\quad \nearrow \texttt{ng\_mm} \not\rightarrow \textit{ngget} \mid \textit{writer} \\
\texttt{ng\_melink} \rightarrow \texttt{ng\_fifos} \rightarrow \texttt{ng\_mm} \not\rightarrow \textit{ngget} \mid \textit{b2r} \mid \textit{ngput} \not\rightarrow \texttt{ng\_socket} \rightarrow \\
\qquad\qquad\qquad\qquad\quad \searrow \texttt{ng\_mm} \not\rightarrow \textit{ngget} \mid \textit{b2r}\ \text{(EoB counts)}
\end{array}
$$

$$
\begin{array}{l}
\qquad\qquad\qquad \nearrow \texttt{ng\_mm} \not\rightarrow \textit{ngget} \mid \textit{r2h}\ \ --- \rightarrow \textit{histGUI} \\
\rightarrow \texttt{ng\_fifos} \qquad\qquad\qquad\qquad\qquad \backslash --- \rightarrow \textit{histGUI} \\
\qquad\qquad\qquad \searrow \texttt{ng\_mm} \not\rightarrow \textit{ngget} \mid \textit{r2h}\ \text{(polar. calc.)} \quad \cdots
\end{array}
$$

IOP Conf. Series: Journal of Physics: Conf. Series **938** (2017) 012019 doi:10.1088/1742-6596/938/1/012019

where:

$\rightarrow$	***packet(9)*** [10] data stream in the kernel context (along graph edges);
$\not\rightarrow$	context boundary crossing by the data stream;
$\mid$	***packet(3)*** [10] data stream in the user context (so called pipe);
$--\rightarrow$	socket connection (here `TServerSocket` / `TSocket` from *ROOT* framework [11]);
$\cdots$	number of the same elements.

The *netgraph* (kernel context) part of this scheme we can see in Fig.1.

The `melink0` VME driver node (see Fig.1) is the source of data packets stream which supplied to the ***ng_fifos(4)*** node [12] named `fifo` through the pair of hooks `data-input`. The `fifo` provides own copy of the obtained data stream for each of its consumers (in Fig.1 for simplicity only one of them connected through pair of hooks `output1-in` is depicted). In the DSS DAQ the `fifo` feeds the ***b2r(1)*** for *ROOT* events production [12] by the full (or possibly downscaled) data stream of trigger events (as well as all Begin–of–Burst (BoB) and End–of–Burst (EoB) events), while the ***b2r(1)*** for EoB counts viewing — by the EoB events only. (Here the Burst means accelerator spill.) The ***ng_mm(4)*** node [12] is used as replacement for the *netgraph* standard ***ng_socket(4)*** node due to more handful bufferization and better performance.

The ***b2r(1)*** directs *ROOT* events to `fifo2` node through the pair of hooks `out-input` (see Fig.1 right part) with mediation of ***ngput(1)*** utility and ***ng_socket(4)*** node. The `fifo2` feeds both the ***r2h(1)*** for histograms filling [12] and ***r2h(1)*** for polarization calculations (see section 3.1) by the *ROOT* representation of trigger events (possibly with downscaling) as well as all BoB and EoB events (in Fig.1 for simplicity only one of consumers connected through pair of hooks `output1-in` is depicted).

The ***b2r(1)*** and ***r2h(1)*** are user context utilities because C++ code and *ROOT* libraries could not be linked into OS kernel. So data streams cross the context boundary using ***ng_mm(4)*** or ***ng_socket(4)*** nodes and ***ngget(1)*** or ***ngput(1)*** utilities.

High level control over DAQ graph is performed by special ***ng_vmectl(4)*** control node type, which intended to issue commands (in the form of *netgraph* control messages) to VME drivers and VME nodes. ***ng_vmectl(4)*** obtains responses to commands as control messages from VME nodes and as data messages from VME drivers, so supplies the next command in time.

3. User context utilities

Lets describe only software utilities specific for DSS DAQ system. The *ngdp* generic ones used here are: ***writer(1)*** [10] of packet stream to files on hard disk, ***ngget(1)*** packet stream extractor from *netgraph* graph, ***ngput(1)*** packet stream injector to *netgraph* graph, ***b2r(1)*** (binary–to–*ROOT*) converter, ***r2h(1)*** (*ROOT*–to–histograms) converter, and ***histGUI(1)*** standalone client for ***r2h(1)*** (see [12]). According to scheme of events representation by *ROOT* classes [12] the specific classes for trigger, BoB, EoB events as well as required helper classes are designed. These classes are linked into ***b2r(1)*** and ***r2h(1)*** as well as provided in the *libElinpol.so* shared library form to be used by the express–offline *ROOT* scripts. The EoB events viewer (so called dumper *b2r_dump*) is a specially compiled form of the ***b2r(1)*** converter.

3.1. Polarization calculator

The specially compiled ***r2h(1)*** converter supports additional entities `Calcvp` and `Calctp` in the configuration file format ***r2h.conf(5)*** [12]. This allows it to calculate polarizations instead of *ROOT* histograms filling. Currently the polarization calculator produces textual output of calculation results at each EoB event arrival. In the future these results could be output in the HTML form to be included into Web–based representation scheme described in [13] and [14].

All used arm scaler counts of the DSS setup polarimetry part (see [2]) are derived from raw data by so called universal calculation mechanism (***cell(3)***, see [12]). This allows us to organize polarization calculations very flexibly without recompilation of the ***r2h(1)*** executable itself. We not explain too huge details of these counts preparation (at least counts should be normalized to number of bursts), however such procedure could be revised easily and quickly. For example, at beam energy change we can operatively replace the scintillation counters combination producing any arm counts, as well as the analyzing power values (which should be already known from some previous measurements). Also the data cuts in terms of 1D–ranges on the time and/or amplitude 1D– and 2D–histograms applied during the counts preparation could be revised.

3.2. ***sv(1)*** *supervisor utility*

For the human operator convenience the ***sv(1)*** supervisor utility with GUI which expected to be self–explanatory is implemented. To perform base DAQ commands like start/stop run, load software components, etc. the supervisor relies on configuration file *linpolsv.mk* in the so called makefile format (see ***make(1)*** for syntax description). This file contains named targets (i.e. `continue`, `pause`, `init`, etc.) and for each of them it provides some actions in terms of shell commands, while the supervisor simply executes something like `make -f linpolsv.mk continue`. To perform VME hardware configuration the supervisor uses the ***netgraph(3)*** application program interface (API) to send the *netgraph* control messages to VME nodes. The configuration is represented to human operator by additional windows and is stored in per hardware module files of ***termcap(5)*** format. Also to load the whole hardware module(s) configuration from the file(s) the special command–string utility ***hwmod_conf(1)*** is implemented, which shares sources with supervisor.

4. Conclusions

The explained DSS DAQ system was successfully used during 53rd (December 2016) and 54th (March 2017) Nuclotron runs. The A_y, A_{yy}, A_{xx} analyzing powers of the dp elastic scattering and vector and tensor polarizations of the deuteron beam from the SPI [7] have been successfully obtained using the designed express–offline software in the *ROOT* scripts form. The first measurements of the 500 MeV proton beam polarization were performed during 54th run also. In the future the polarizations could be determined online by the polarization calculator utility and their results could be integrated into Web–based representation scheme described in [13] and [14]. So the DSS DAQ system is suitable for online polarimetry.

Acknowledgments

The work has been supported in part by the RFBR under grant 16-02-00203a.

References

[1] Ladygin V P, Gurchin Y V, Piyadin S M, Terekhin A A, Isupov A Y *et al.* 2014 *Few Body Syst.* **55** 709–712
[2] Kurilkin P K, Ladygin V P, Uesaka T *et al.* 2011 *Nucl.Instr.and Meth.in Phys.Res.* **A** 45–51
[3] Gurchin Y V *et al.* 2011 *Phys.Part.Nucl.Lett.* **8** 950–958
[4] Piyadin S M *et al.* 2011 *Phys.Part.Nucl.Lett.* **8** 107–113
[5] Isupov A Y, Krasnov V A, Ladygin V P, Piyadin S M and Reznikov S G 2013 *Nucl. Instr. and Meth. in Phys. Res.* **A** 127–134
[6] Kurilkin P K *et al.* 2012 *Phys.Lett.B* **B** 61–65
[7] Fimushkin V V *et al.* 2016 *J. Phys. Conf. Ser.* **678** 012058–012062
[8] Isupov A Y 2010 *arXiv:1004.4474 [physics.ins-det]*
[9] `http://afi.jinr.ru/`
[10] Gritsaj K I and Isupov A Y 2001 *JINR Communications* **E10–2001–116** 1–19
[11] Brun R and Rademakers F 1996 *Proc. of the AIHENP'96 Workshop* (Nucl.Instr.and Meth.in Phys.Res. (1997) vol **A**(389)) (Lausanne, Switzerland) pp 81–86 See also `http://root.cern.ch/`
[12] Isupov A Y 2010 *arXiv:1004.4482 [physics.ins-det]*
[13] Isupov A Y 2005 *Czech. J. Phys. Suppl.* **A55** A407–A414
[14] Isupov A Y 2006 *Czech. J. Phys. Suppl.* **C56** C385–C392

Algorithm for Calculation of Longitudinal Target Polarization

Yu Kiselev[a] , N Doshita[b], K Kondo[b], T Iwata[b], G Nukazuka [b]

[a]JINR, VBLPHE, 141980 Dubna, Moscow Reg. Russia, E-mail: yury.kiselev@cern.ch

[b]Department of Physics, Faculty of Science, Yamagata University, 990-8560, Yamagata, Japan

Abstract. We suggest using the Lagrange polynomial interpolation to calculate the longitudinal polarization of the large COMPASS target with a length of 2x55 cm. The algorithm does not change the polarizations measured by the NMR sensors. In addition, it allows taking into account the influence of the microwave field on the nuclear polarization at the resonator boundaries. As a result, measurements of polarization at several points of a long target and knowledge of the cavity design allow one to obtain a more explicit analytical expression for the target polarization. The work is performed in the COMPASS collaboration at CERN.

1. Introduction

Polarization of long targets produced by the method of dynamic nuclear polarization (DNP) depends on the inhomogeneity of the bias field, temperature of the material, and parameters of the microwave (MW) field. In these targets, the polarization is measured by several NMR probes. To reduce the inductive coupling between the probes, their number is limited, and the axial angles between the NMR coils are rotated by $\pi/2$. It is clear that between probes the polarization should be calculated by interpolation of local measurements. Here we consider the Lagrange interpolation that not only allows calculating polarization and its errors at any place along the target but also helps estimate the efficiency of the target design.

2. Lagrange interpolation of the long target polarization

In its canonical form, the Lagrange interpolation formula [1] of the *n*-th order approximates the functions $p_k(x_k)$ by the polynomial of the *n*-th degree $P(x)$. Here p_k=(p_0, p_1,... p_n) are polarization values at $n+1$ probe locations x_k= (x_0, x_1, ... x_n) and the function $P(x)$ has the form:

$$P(x) = \frac{(x-x_1)(x-x_2)\dots(x-x_n)}{(x_0-x_1)(x_0-x_2)\dots(x_0-x_n)}p_0 + \frac{(x-x_0)(x-x_2)\dots(x-x_n)}{(x_1-x_0)(x_1-x_2)\dots(x_1-x_n)}p_1 + \dots +$$

$$+\frac{(x-x_0)(x-x_1)\dots(x-x_{n-1})}{(x_n-x_0)(x_n-x_1)\dots(x_n-x_{n-1})}p_n. \qquad (1)$$

Let the coordinate of a probe be $x=x_1$, then $P(x_1) = p_1(x_1)$ in (1), because the fraction at $p_1$equals to one and all other terms of the sum are zeroed. Thus, at $x=x_1$ and similarly for all other coordinates of the probes, the polarizations exactly coincides with the measured values. Hence, the Lagrange algorithm does not distort the measured polarizations. Apart from the data of the direct NMR measurements, the indirect data following from the Maxwell equations can be additionally used in (1). For example, since the polarizing effect is produced by the radial magnetic components of the MW field (TM-modes), which have the maximum at the ends of the cavity walls, the polarization produced over there must be no lower than those measured by the nearest NMR probe. Of course, this condition is fulfilled when inhomogeneity of the bias field falls within the allowable limits and the stopper-insulator effectively suppresses MW power penetration. Let us consider the helpful example.

Figure 1 shows coordinates of probes in the upstream and downstream COMPASS cells. The targets are located in the center of the cylindrical MW cavities (not shown in the figure) along the bias field. The MW stopper-isolator installed between the cells allows independent polarization of each target. Ten NMR coils of the NMR probes p_1, p_2, p_3, p_4, p_5 and p_6, p_7, p_8, p_9, p_{10} (Fig. 1) are connected by coaxial cables to ten measuring units (Q-meters) kept at room temperature.
Following the above logic, for the upstream cell one can assume $p_{up0} \cong p_1$ and $p_{up5} \cong p_5$, where p_{up0} and p_{up5} are the nuclear polarizations at x=0.0 and x_{up5}, respectively (Fig. 1). Similarly, for the downstream cell we can put pd6 $\cong p_6$ and $p_{d10} \cong p_{10}$ (Fig. 1).

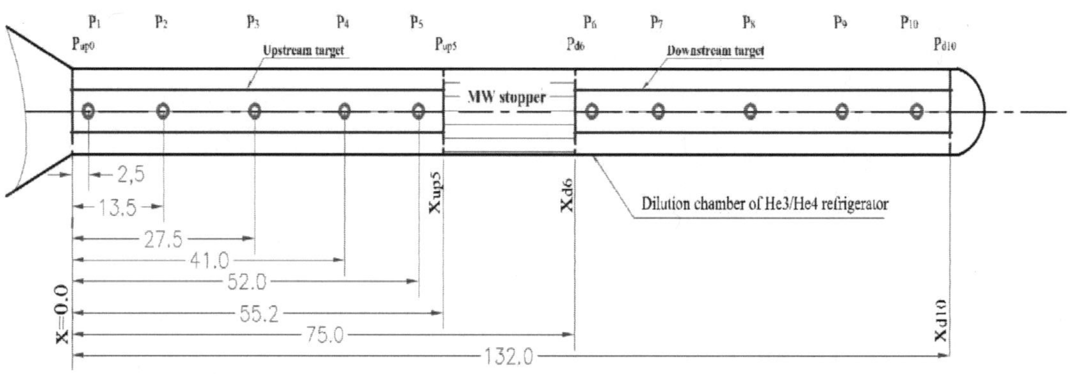

Fig. 1. The two targets are built into the dilution chamber of the ^{3}He/^{4}He refrigerator. The MW stopper isolates the MW field allowing independent DNP in targets. The circles show location of NMR probes: their numerical positions and polarizations

Now, formula (1) for the upstream target takes the form:

$$\mathrm{Pup}(x) = P0 + P1 + P2 + P3 + P4 + P5 + \mathrm{Pend} \quad (2)$$

where each summand describes the contribution of the probe polarizations at the point x.

$$\begin{aligned} P0 &= (x - x_1)(x - x_2)(x - x_3)(x - x_4)(x - x_5)\,(x - x_{\mathrm{end}}) \cdot p_0/C\,0, \\ P1 &= (x - x_0)(x - x_2)(x - x_3)(x - x_4)(x - x_5)\,(x - x_{\mathrm{end}}) \cdot p_1/C\,1, \\ P2 &= (x - x_0)(x - x_1)(x - x_3)(x - x_4)(x - x_5)\,(x - x_{\mathrm{end}}) \cdot p_2/C\,2, \\ P3 &= (x - x_0)(x - x_1)(x - x_2)(x - x_4)(x - x_5)\,(x - x_{\mathrm{end}}) \cdot p_3/C\,3 \\ P4 &= (x - x_0)(x - x_1)(x - x_2)(x - x_3)(x - x_5)\,(x - x_{\mathrm{end}}) \cdot p_4/C\,4, \\ P5 &= (x - x_0)(x - x_1)(x - x_2)(x - x_3)(x - x_4)\,(x - x_{\mathrm{end}}) \cdot p_5/C\,5, \\ \mathrm{Pend} &= (x - x_0)(x - x_1)(x - x_2)(x - x_3)(x - x_4)\,(x - x_5) \cdot p_{\mathrm{end}}/\mathrm{Cend}. \end{aligned} \quad (3)$$

As follows from (1), the constants $C0$...Cend are

$$C0 = (x_0 - x_1)(x_0 - x_2)(x_0 - x_3)(x_0 - x_4)(x_0 - x_5)(x_0 - x_{\mathrm{end}}),$$

$$
\begin{aligned}
C1 &= (x_1 - x_0)(x_1 - x_2)(x_1 - x_3)(x_1 - x_4)(x_1 - x_5)(x_1 - x),\\
C2 &= (x_2 - x_0)(x_2 - x_1)(x_2 - x_3)(x_2 - x_4)(x_2 - x_5)(x_2 - x_{\text{end}}),\\
C3 &= (x_3 - x_0)(x_3 - x_1)(x_3 - x_2)(x_3 - x_4)(x_3 - x_5)(x_3 - x_{\text{end}}),\\
C4 &= (x_4 - x_0)(x_4 - x_1)(x_4 - x_2)(x_4 - x_3)(x_4 - x_5)(x_4 - x_{\text{end}}),\\
C5 &= (x_5 - x_0)(x_5 - x_1)(x_5 - x_2)(x_5 - x_3)(x_5 - x_4)(x_5 - x_{\text{end}}),\\
\text{Cend} &= (x_{\text{end}} - x_0)(x_{\text{end}} - x_1)(x_{\text{end}} - x_2)(x_{\text{end}} - x_3)(x_{\text{end}} - x_4)(x_{\text{end}} - x_5).
\end{aligned}
\tag{4}
$$

In the RUN_2015, the upstream sensors had the following local coordinates (Fig. 1): x_0=0.1, x_1=2.5, x_2=13.5, x_3=27.5, x_4=41.0, x_5=52.0, and x_{end} =55.2 cm. Substitution into (4) yields

$$C_0 = 1.03064 \cdot 10^8, \tag{5}$$

It is convenient to present probe coordinates together with measured polarizations as a list {{x_0,p_0}, {x_1,p_1},…,{x_{end},p_{end}}}. In this notation, the upstream list in Units {cm,%} is

{{0.1,84.1},{2.5,84.1},{13.5,79.0},{27.5,82.1},{41.0,81.4},{52.0,86.8},{55.2,86.8}}. (6)

A similar the downstream list was obtained as {{75.0,-72.7}, {78.0,-72.7}, {88.5,-71.1}, {103.-81.3},{116.5,-82.5},{127.5,-81.2},{132.0,-81.2}}. (7)

The interpolation functions of both upstream (6) and downstream (7) are shown in Fig. 2.

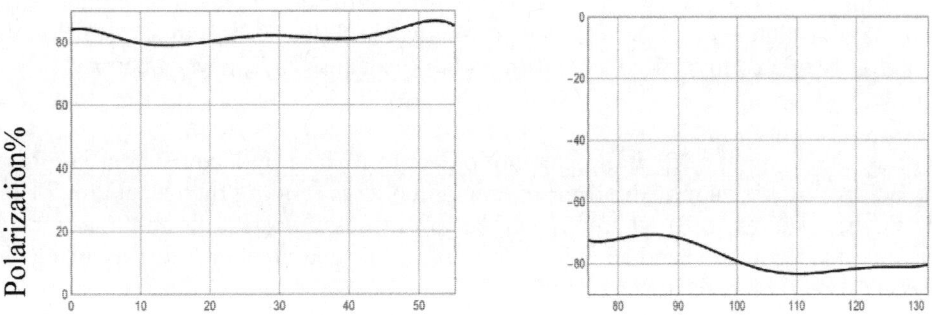

Fig .2. Upstream (left) and downstream (right) cell polarizations shown along their length X in [cm]. The stopper place is located at the coordinates 55.2< X <75.0 cm.

In the simplest application, the Lagrange algorithm allows a set of digital polarization data (6) and (7) to be replaced by a convenient graphical interface, which facilitates visualization of the DNP regime and allows refining the calculation of the polarization in long targets.

3. Results

Figure 2 shows that the targets were polarized oppositely, therefore MW penetration through the stopper could affect their inhomogeneity. We study this effect using calculations of average polarizations with their standard deviations in the form

$$\langle P\rangle_{up} = \frac{1}{N}\sum_{i=1}^{N} P_{up}(x_i) \pm \langle\sigma\rangle_{up}, \text{ where } \langle\sigma\rangle_{up} = \sqrt{\frac{1}{N}\sum_{i=1}^{N}\left(P_{up}(x_i) - \langle P\rangle_{up}\right)^2}, \tag{8}$$

The same formulas for the downstream are denoted by the symbol «*down*». If x_i is scanned within coordinate lists (6) and (7) i.e. N=7 we obtained

$$\langle P\rangle_{up} = \frac{1}{7}\sum_{i=1}^{7} P_{up}(x_i) = +83.5\% \pm 2.9\%, \ \langle P\rangle_{down} = \frac{1}{7}\sum_{i=1}^{7} P_{down}(x_i) = -77.5\% \pm 5.1\%. \tag{9}$$

IOP Conf. Series: Journal of Physics: Conf. Series **938** (2017) 012020 doi:10.1088/1742-6596/938/1/012020

The refined values obtained with the Lagrange functions are,

$$\langle P \rangle_{up} = \frac{1}{55.}\sum_{x_i=0.1}^{55.1} P_{up}(x_i) = +82.1\% \pm 2.2\%, \quad 0.1<X<55.1 \text{ cm} \quad (N=55), \tag{10}$$

$$\langle P \rangle_{down} = \frac{1}{57}\sum_{x_i=75}^{132} P_{down}(x_i) = -78.0\% \pm 4.8\%, \quad 75.<X<132 \text{ cm } (N=57). \tag{11}$$

Here the calculations are retained in the discrete form, and the coordinate is stepped at 1 cm. For comparison, in the interval 75<X<100 cm (Fig. 2), we have

$$\langle P \rangle_{down} = \frac{1}{25}\sum_{x_i=75}^{100} P_{down}(x_i) = -73.2\% \pm 2.5\%, \qquad 75<X<100 \text{ cm } (N=25). \tag{12}$$

While for $100 \le x_i \le 132$ cm, where MW leakage is suppressed by absorption in the initial part of the downstream target, the absolute value of polarization becomes almost 9% higher

$$\langle P \rangle_{down} = \frac{1}{32}\sum_{100}^{132} P_{down}(x_i) = -81.9\% \pm 1.1\%, \quad 100 <X<132 \text{ cm } (N=25). \tag{13}$$

It can be seen that the average upstream polarization (10) almost exactly equals to the polarization at the end of the downstream target (13), which indicate the absence of significant temperature gradients in a long dilution chamber and a high enough homogeneous microwave field. It becomes clear that the heterogeneity (12) is closely related to a microwave penetration (leakage) through MW stopper.

4. Conclusion

The Lagrange algorithm allows a set of digital polarization data to replace by a convenient graphical interface, which facilitates visualization of the DNP regime and allows refining the calculation of the polarization in long targets. The use of the Lagrange algorithm makes it possible to more accurately estimate the efficiency of microwave irradiation and the cryogenic equipment of targets using the averaged magnitude and the structure of the polarization.

5. References

[1] G Korn and T Korn 1961 *Mathematical Handbook for Scientists and Engineers.* Ch. 20.5-2 (New York Toronto London)

Impact of the Tilted Detector Solenoid on the Ion Polarization at JLEIC*

A M Kondratenko[1], M A Kondratenko[1], Yu N Filatov[2], Ya S Derbenev[3], F Lin[3], V S Morozov[3] and Y Zhang[3]

[1]Science and Technique Laboratory "Zaryad", 630090, Novosibirsk, Russia
[2]Moscow Institute of Physics and Technology, 141700 Dolgoprudny, Russia
[3]Jefferson Lab,12000 Jefferson Avenue, Newport News, VA 23606, USA

kondratenkom@mail.ru

Abstract. Jefferson Lab Electron Ion Collider (JLEIC) is a figure-8 collider "transparent" to the spin. This allows one to control the ion polarization using a universal 3D spin rotator based on weak solenoids. Besides the 3D spin rotator, a coherent effect on the spin is produced by a detector solenoid together with the dipole correctors and anti-solenoids compensating betatron oscillation coupling. The 4 m long detector solenoid is positioned along a straight section of the electron ring and makes a 50 mrad horizontal angle with a straight section of the ion ring. Such a large crossing angle is needed for a quick separation of the two colliding beams near the interaction point to make sufficient space for placement of interaction region magnets and to avoid parasitic collisions of shortly-spaced 476 MHz electron and ion bunches. We present a numerical analysis of the detector solenoid effect on the proton and deuteron polarizations. We demonstrate that the effect of the detector solenoid on the proton and deuteron polarizations can be compensated globally using an additional 3D rotator located anywhere in the ring.

1. Spin Transparency of the JLEIC Collider

Jefferson Lab Electron Ion Collider (JLEIC) is designed to provide high polarizations of both ion and electron colliding beams. The figure-8 JLEIC collider is "transparent to the spin". In such a collider, effect on the spin of one arc is compensated by the other arc. Thus, effect of "strong" arc fields on the spin is reduced to zero. Any spin direction repeats itself after each particle turn, i.e. the collider has no preferred spin direction [1-3].

Colliders transparent to the spin offer a unique opportunity to efficiently control the ion polarization using small magnetic field integrals. In such a collider, any small perturbation has a strong effect on the beam polarization. To stabilize the spin direction, one must introduce a 3D spin rotator based on 'weak' solenoids, which "shifts" the spin tune by a small value ν and sets the necessary orientation of the polarization. The "weak" solenoids have essentially no effect on the beam's orbital characteristics.

For polarization stability, one must ensure that the spin tune ν induced by the 3D spin rotator significantly exceeds the strength of the zero-integer spin resonance: $\nu \gg \omega$. Our calculations show

* This material is based upon work supported by the U.S. Department of Energy, Office of Science, Office of Nuclear Physics under contracts DE-AC05-06OR23177 and DE-AC02-06CH11357.

that, for stable control of the ion polarization, it is sufficient to use 3D spin rotators, which allow one to induce spin tunes of 10^{-2} for protons and 10^{-4} for deuterons.

The resonance strength consists of two parts: a coherent part ω_{coh} arising due to additional transverse and longitudinal fields on the beam trajectory deviating from the design orbit and an incoherent part ω_{emitt} associated with the particles' betatron and synchrotron oscillations (beam emittances). In practice, the coherent part significantly exceeds the incoherent one: $\omega_{coh} \gg \omega_{emitt}$. The coherent part does not cause beam depolarization and only results in a simultaneous rotation of all particles' spins. In principle, the direction and magnitude of the coherent part of the resonance strength can be measured and compensated by an additional 3D spin rotator. To preserve the polarization, it is then sufficient to satisfy a weaker condition: $\nu \gg \omega_{emitt}$.

Experiments require insertion of additional magnetic elements into the collider lattice. When integrating an experimental insertion into the lattice, one must make sure that the insertion does not violate the spin transparency of the collider. This is related to the fact that, besides the 3D spin rotators, an additional coherent effect on the spin can be caused by the magnetic elements of the insertion.

Let us next consider influence of the detector solenoid on the ion polarization.

2. Crab crossing scheme of the JLEIC collider

Effect of the detector solenoid on the ion polarization is determined by specific details of the interaction region design of the JLEIC collider. To obtain a high luminosity of the colliding electron and ion bunches, the collider employs a crab crossing scheme illustrated in Fig. 1.

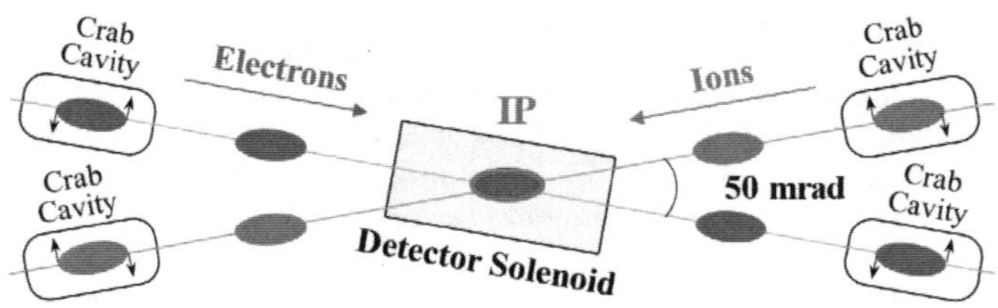

Figure 1. Schematic of the detector solenoid placement in the JLEIC ion collider ring.

The JLEIC interaction region design has a crossing angle of 50 mrad for a rapid separation of the colliding beams near the IP in order to create sufficient space for placement of interaction region magnets, eliminate harmful parasitic collisions under a high (476 MHz) bunch repetition frequency and enable detection of outgoing small-angle particles. To avoid luminosity loss associated with non-head-on collisions, following the success of the KEK B-factory, a scheme of crab crossing [4] utilizing SRF crab cavities is employed for restoring head-on collisions [5,6]. A 4 m detector solenoid is aligned with a straight section of the electron ring and makes a 50 mrad horizontal angle with a straight section of the ion ring.

3. Calculation of the proton and deuteron spin dynamics in the JLEIC ion collider ring including the detector solenoid

The detector solenoid in the JLEIC collider is oriented at an angle of 50 mrad to the ion trajectory and therefore has both transverse and longitudinal magnetic field components along the trajectory. Figure 2 shows a scheme of the detector solenoid placement in the JLEIC ion collider ring [7,8]. The interaction point divides the tilted detector solenoid into two parts with a length ratio of $1.6\text{ m} : 2.4\text{ m} = 2 : 3$ and lies at the crossing point of the electron and ion straight sections when the solenoid is off. When the field of the detector solenoid is on, the ions are affected not only by the longitudinal field component but by the radial field component as well, which shifts the ions vertically away from the interaction point. To stabilize the interaction point and correct the ion orbit at the exit and entrance of the Final Focusing Quadrupole (FFQ) triplets, there are a pair of dipole correctors on each side of the detector solenoid: the first and second correctors are directly to the left of the solenoid

IOP Conf. Series: Journal of Physics: Conf. Series **938** (2017) 012021 doi:10.1088/1742-6596/938/1/012021

and the third and fourth correctors on the right-hand side of the solenoid are separated by a detector dipole with vertical field. Each dipole corrector has vertical K_{yi} and radial K_{xi} components of the transverse field.

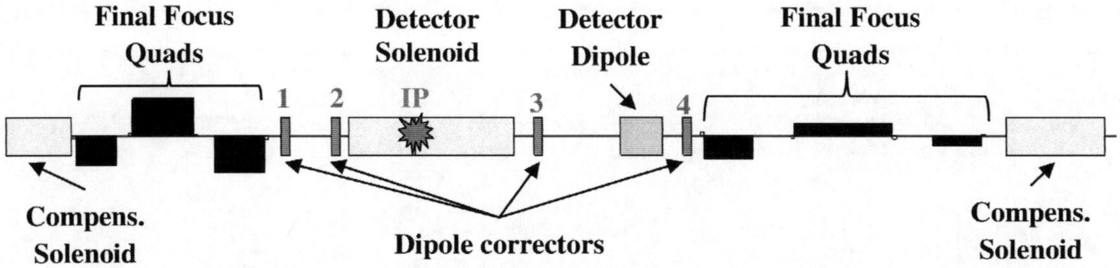

Figure 2. Schematic of the detector solenoid placement in the JLEIC ion collider ring.

To compensate betatron oscillation coupling introduced by the detector solenoid, 1.6 m and 2.4 m anti-solenoids are placed on the two sides of the detector solenoid. The axes of the anti-solenoids are aligned with the axes of the ion beam lines.

The ion orbital and spin motions in the tilted detector solenoid are calculated using the linear approximation. Its accuracy depends on the choice of the reference orbit. If one uses the straight trajectory, which particles follow when the field of the detector solenoid is off, then the deviation of a particle trajectory from the reference orbit is mainly determined by the excursion of the closed orbit. It is proportional to the crab-crossing angle α_{crab} and, for a solenoid of a length L_{sol} is about:

$$\Delta r \sim \alpha_{crab} K_{sol} L_{sol}^2.$$

The deviation of a trajectory can reach values of up to a few mm. If one considers that a particle trajectory in a tilted solenoid is a spiral and uses it as a reference orbit then the deviation of a particle trajectory from the reference orbit is determined by the betatron beam size σ, which is a few μm. In this case, the accuracy of calculations in the linear approximation improves by 2-3 orders of magnitude. Later in our calculations, we use the spiral trajectory as the reference orbit.

Figure 3 shows the vertical excursion of the ion beam trajectory caused the detector solenoid field of 3 T at the beam momentum of 100 GeV/c. As we can see, the new orbit is completely restored at the exit and entrance of the FFQ triplets. To stabilize the interaction point, the transverse components of the kickers in units of the magnetic rigidity must take the following values:

$K_{x1} \approx -3.3 \cdot 10^{-3}\mathrm{m}^{-1}$, $K_{y1} \approx 2.9 \cdot 10^{-5}\,\mathrm{m}^{-1}$, $K_{x2} \approx 6.9 \cdot 10^{-3}\,\mathrm{m}^{-1}$, $K_{y2} \approx -5.5 \cdot 10^{-5}\,\mathrm{m}^{-1}$, $K_{x3} \approx 7.9 \cdot 10^{-3}\,\mathrm{m}^{-1}$, $K_{y3} \approx 9.2 \cdot 10^{-5}\,\mathrm{m}^{-1}$, $K_{x4} \approx -3.1 \cdot 10^{-3}\,\mathrm{m}^{-1}$, $K_{y4} \approx -4. \cdot 10^{-5}\,\mathrm{m}^{-1}$.

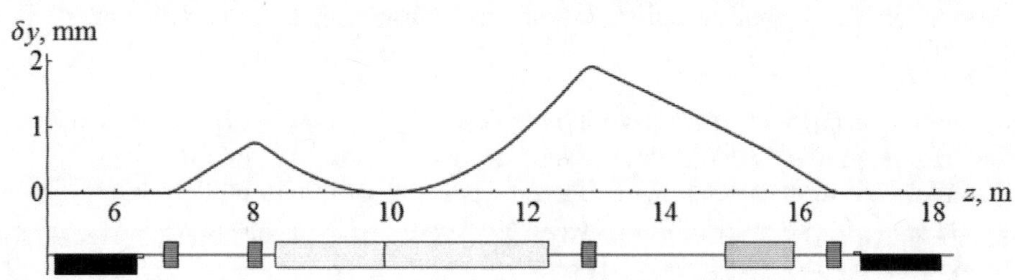

Figure 3. Vertical deviation of the ion beam from the axis of the ion straight section. The field of the detector solenoid is 3 T. The beam momentum is 100 GeV/c.

Besides the 3D spin rotator, a coherent effect on the spin is produced by a detector solenoid together with the dipole correctors and anti-solenoids compensating betatron oscillation coupling. Figure 4 shows the influence of the detector solenoid insertion on the proton and deuteron spin dynamics as a function of momentum. The calculations assume that the beam orbit is stable and corresponds to the beam parameters shown in Fig. 3.

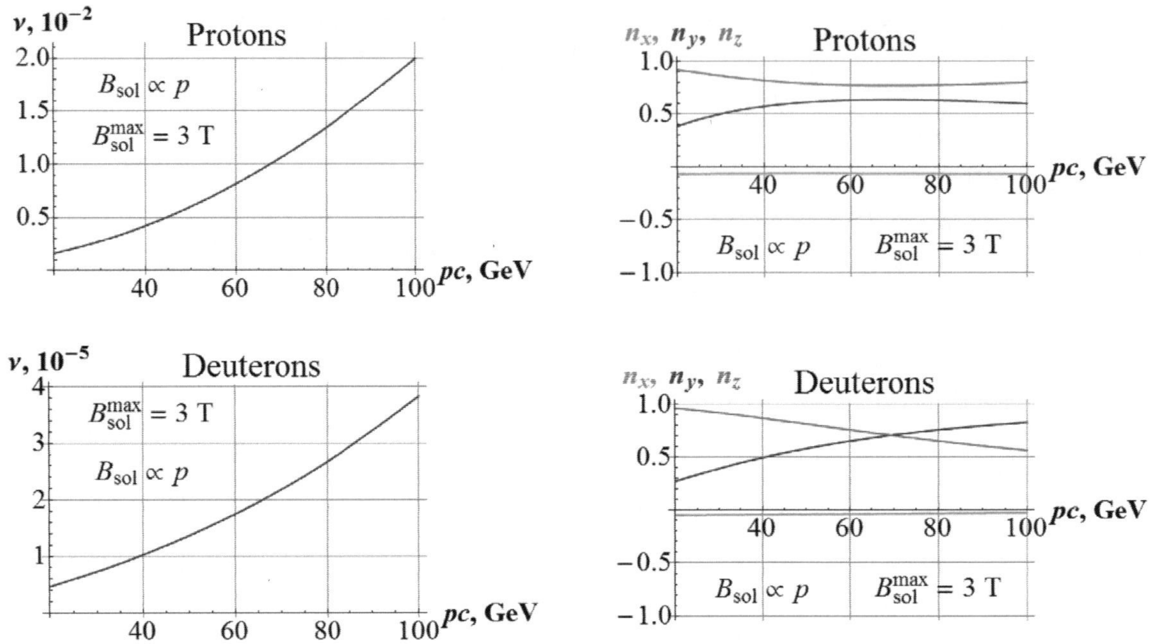

Figure 4. Momentum dependence of the spin tune and $\vec{n}$ – axis components induced by insertion of the detector solenoid for protons (top) and deuterons (bottom). In these calculations, the solenoid field scales proportionally to the beam momentum and reaches 3 T at the maximum beam momentum of 100 GeV/c.

4. Conclusions

The calculations presented above show that the spin tune induced by the detector solenoid insertion does not exceed the value of $2 \cdot 10^{-2}$ for protons and $4 \cdot 10^{-5}$ for deuterons when the field of the detector solenoid changes from 0 to 3 T in the whole momentum range of the JLEIC collider. Influence of the detector solenoid insertion on the proton and deuteron polarizations can be compensated using an additional 3D rotator, which can be placed in any available space in the collider. The above analysis shows that the influence of a tilted detector solenoid on the ion polarization depends on locations of the correcting dipoles and compensating anti-solenoids. In principle, it is possible to optimize the scheme for stabilization of the bunch collision point on the basis of correcting dipoles and anti-solenoids so that the spin transparency condition is automatically satisfied.

References

[1] Abeyratne S *et al* 2015 *MEIC Design Summary* arXiv:1504.07961 [physics.acc-ph]
[2] Kondratenko A M *et al* 2015 *in Proc IPAC'15* paper TUPWI029 p 2301
[3] Kondratenko A M *et al* 2017 *in Proc IPAC'17* paper WEPIK038 p 3014
[4] Fukuma H *et al* 1995 *KEKB B–Factory Design Report* KEK–Report–95–7
[5] Palmer R 1988 *SLAC Report No. SLAC-PUB-4707*
[6] Oide K and Yokoya K 1989 *Phys. Rev.* **A 40** p 315
[7] Wei G H *et al* 2016 *in Proc. of IPAC'16 Busan Korea* paper WEPMW015 p 2454
[8] Wei G H *et al* 2016 *in Proc. of NAPAC'16 Chicago IL* paper MOA4CO03 p 57

Measurement of the deuteron beam polarization at internal target at Nuclotron for DSS experiment

Ya T Skhomenko[1,2], V P Ladygin[1], Yu V Gurchin[1], A Yu Isupov[1], M Janek[3], J-T Karachuk[1,4], A N Khrenov[1], P K Kurilkin[1], A N Livanov[1], S M Piyadin[1], S G Reznikov[1], A A Terekhin[1], A V Tishevsky[1], A V Averyanov[1], A S Belov[5], E V Chernykh[1], D Enache[4], V V Fimushkin[1], D O Krivenkov[1]

[1] Joint Institute for Nuclear Research, Dubna, Russia
[2] Belgorod State National Research University, Belgorod, Russia
[3] Physics Department, University of Žilina, Žilina, Slovakia
[4] National Institute for R&D in Electrical Engineering ICPE-CA, Bukharest, Romania
[5] Institute of Nuclear Physics, Moscow, Russia

E-mail: skhomenko@jinr.ru

Abstract. The current deuteron beam polarimetry at Nuclotron is provided by the Internal Target polarimeter based on the use of the asymmetry in *dp*- elastic scattering at large angles in the cms at 270 MeV. The upgraded deuteron beam polarimeter has been used obtain the vector and tensor polarization during 2016/2017 runs for the DSS experimental program. The polarimeter has been used also for tuning of the polarized ion source parameters for 6 different spin modes.

1. Introduction

The study of the spin structure of two-nucleon and three-nucleon short-range correlations via the measurements of the polarization observables in the deuteron induced reactions is main goal of the DSS project at Nuclotron[1, 2, 3]. The high precision polarimetry of the deuteron and proton beams is important for these investigations.

The goal of the present article is to report new results on the measurements of the vector and tensor beam polarizations using upgraded polarimeter based on the asymmetry measurements in *dp*- elastic scattering at 270 MeV [4] at the Internal Target Station (ITS) at Nuclotron. These measurements were performed during the DSS experiment on the study of the vector A_y, tensor A_{yy} and A_{xx} analyzing powers in *dp*- elastic scattering at large transverse momenta [5].

2. Deuteron polarimeter at ITS

Efficient polarimetry can be achieved even at relatively low beam intensity with using a thin solid target inside the inner ring of the accelerator. The luminosity can be increased significantly due to multiple beam passage through the interaction point and the use of a correctly configured internal target trajectory. Therefore, the internal beam polarimeter with a very thin target may have approximately the same efficiency as the extracted beams polarimeters.

The polarimeter based on the use of *dp*- elastic scattering at large angles ($\theta_{\rm cm} \geq 60°$) at 270 MeV [4], where precise data on analyzing powers [6]-[8] exist, has been developed at internal target station (ITS) at Nuclotron[9]. The accuracy of the determination of the deuteron beam polarization achieved with this method is better than 2% because of the values of the analyzing powers were obtained for the polarized deuteron beam, which absolute polarization had been calibrated via the $^{12}C(d,\alpha)^{10}B^*[2^+]$ reaction[8].

Deuteron beam polarimeter [4] is placed in the Nuclotron ring. It consists of a spherical scattering chamber and system change targets that can be set six different targets. A detector support with 39 mounted plastic scintillation counters is placed downstream the ITS spherical chamber. Each plastic scintillation counter was coupled to a photo-multiplier tube Hamamatsu H7416MOD. Eight proton detectors were installed for left , as well as for right and up, but due to space limitation – only four for down. The angular span of one proton detector was 2° in the laboratory frame, which corresponds to ~4° in the cms. Three deuteron detectors are placed at scattering angles of deuterons coinciding kinematically with the protons. Only one deuteron detector can cover the solid angle corresponding to four proton detectors placed down. In addition, one pair of detectors is placed to register two protons from quasi-elastic $p-p$ scattering at θ_{pp}=90° in the cms in the horizontal plane. The scattered deuterons and recoil protons at 270 MeV were detected in kinematic coincidence over the cms angular range of 65–135° at eight different angles, defined by the positions of the proton detectors.

The VME (Versa Module Eurocard) based data acquisition system is used for the data taking from scintillation detectors [10]. The signals from the detectors are fed in 16-channel TQDC-16 charge-time-digital converters via commutator bar. TQDC-16 module allows to measure the amplitude and time appearance of the signal simultaneously. The hardware of the DSS VME system consists of 4 TQDC-16 modules, trigger module TTCM and VME controller. There is a possibility to tune the first-level trigger using logic of trigger and TQDC-16 modules. This system has been significantly upgraded recently [11].

Newly developed multichannel high-voltage power supply system (Wiener MPod) [12] is used to provide the power for about 70 scintillation detectors based on Hamamatsu photomultipliers.

3. Experiment at ITS

The internal target station (ITS) setup is well suited for study of the energy dependence of polarization observables for the deuteron-proton elastic scattering and deuteron breakup reaction with the detection of two protons at large scattering angles. For these purposes the CH_2-target of 10 μm thick is used for the measurements. The yield from carbon content of the CH_2-target is estimated in separate measurements using several twisted 8μm carbon wires. The measurements were performed using internal target station at Nuclotron [9] with new control and data acquisition system [13].

New source of polarized ions (SPI) [14] has been used to provide polarized deuteron beam. In the current experiment the spin modes with the maximal ideal values of (P_z,P_{zz})= (0,0), (-1/3,-1) and (-1/3, +1) were used. The deuteron beam polarization has been measured at 270 MeV [4].

The DSS data taking was separated on 3 parts: November 2016, December 2016 and February 2017. The deuteron beam polarization measurements were performed at 270 MeV before each energy studied in the range of 400-1800 MeV [5]. The beam polarization measurements were performed also in the end of each part of the experiment. The polarimeter [4] has been used to tune other spin modes of SPI.

The software for data analysis was developed using the ROOT package in C++. The *dp*-elastic scattering events at 270 MeV were selected using correlation of the energy losses and time-of-flight difference for deuteron and proton detectors. The measurements were performed using CH_2 target only. The carbon contamination was measured to be less than 0.5%.

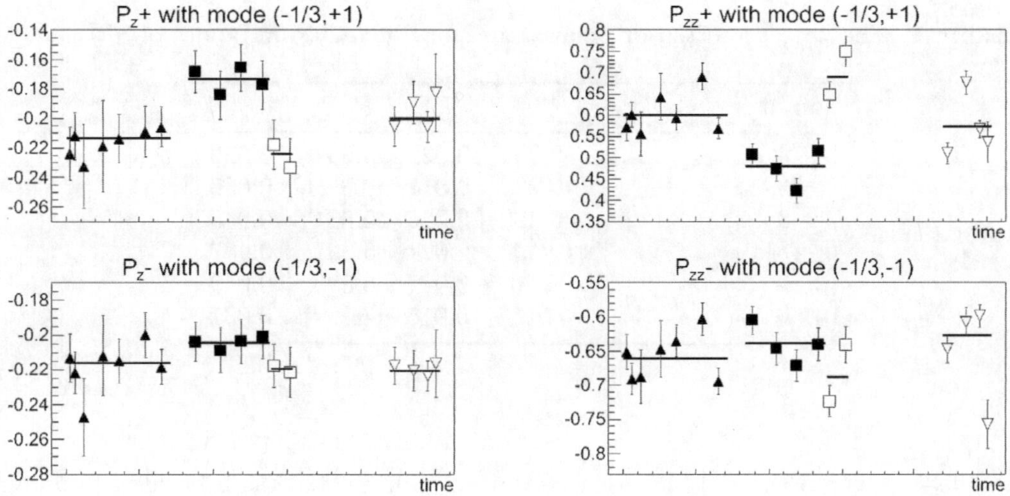

Figure 1. Polarizations values P_z and P_{zz} for spin modes (P_z,P_{zz})= (-1/3,-1) and (-1/3, +1) during the runs in 2016 — 2017 yy.

The precise data on the deuteron analyzing powers at 270 MeV [6]-[8] were used to get the polarization values at several angles [4]. Assuming that the Y-axis is a symmetry axis (β=90°, φ =0°) one can calculate P_z and P_{zz} using the normalized *dp*-elastic scattering events and analyzing powers known [4]. The values of the beam polarization for different spin have been obtained as weighted averages for 8 scattering angles for *dp*- elastic scattering in the horizontal plane only. The typical values of the beam polarization were ~65-75% from the ideal values.

4. The results of the polarization measurements

The vector and tensor polarizations were measured seven, six and four times in the parts at November-2016, at December-2016 and at February-2017, respectively. The values have small statistical and systematics errors. They are rather stable within each part of the experiment. The exception is the December-2016 part, when the physics program was separated on two parts by the tuning of the SPI for pure tensor modes (P_z,P_{zz})= (0,-2) and (0,+1) during 8 hours. This is the reason why December-2016 part is divided by sets of the measurements. The polarization values were approximated by the constants for all four sets of the data. The results of the measurements and approximation are presented in Fig. 1. All the results are within two standard deviations from these constants. One can see that the beam polarization values are quite stable within more than 200 hour of the SPI operation. On the other hand, SPI demonstrates good reproducibility of the polarization values for different sets of the data after long interruptions. The typical values of the vector and tensor components of the beam polarization for the spin modes (-1/3,-1) and (-1/3,+1) are given in the table 1.

The polarimeter [4] has been used also to tune the SPI [14] operation for pure tensor spin modes (0,-2) and (0,+1), for pure vector spin mode (+2/3,0) and for the spin mode (-2/3,+1) with both vector and tensor components. The preliminary results are presented in table 1. One can see, that the typical values of the beam polarization were ~65-75% from the ideal values for all 6 spin modes of SPI.

Table 1. The vector and tensor polarization for different spin modes of SPI [14].

Spin mode(P_z, P_{zz})	P_z	dP_z	P_{zz}	dP_{zz}
(-1/3,+1)	-0.254	0.022	0.637	0.039
(-1/3,-1)	-0.223	0.017	-0.621	0.030
(-2/3,+1)	-0.489	0.026	0.631	0.045
(+2/3,0)	0.427	0.021	0.061	0.037
(0,+1)	0.030	0.027	0.880	0.049
(0,-2)	0.046	0.015	-1.469	0.031

5. Conclusion

The upgraded version of the 270 MeV deuteron beam polarimeter has been used to obtain the vector and tensor polarization during 2016/2017 runs.

The long-term stability of the vector and tensor components of the beam polarization has been demonstrated for the spin modes (-1/3,+1) and (-1/3,-1) of SPI.

The polarimeter has been used for tuning of the polarized ion source parameters for 6 different spin modes.

Acknowledgments

The authors thank the Nuclotron staff for providing good conditions of the experiment. They thank V.B. Shutov for the help with the SPI [14] tuning. They express the gratitude to S.N. Bazylev, V.I. Maximenkova, I.V. Slepnev, V.M. Slepnev and A.V. Shutov for the help during the preparation of the detector and DAQ system. The work has been supported in part by the RFBR under grant $N^0$16-02-00203a, JINR- Slovak Republic and JINR-Romania scientific cooperation programs in 2016-2017.

References

[1] Ladygin V P *et al.* 2014 *Phys.Part.Nucl.* **45** 327.
[2] Ladygin V P *et al.* 2014 *Few Body Syst.* **55** 709.
[3] Janek M *et al.* 2017 *Few Body Syst.* **58** 40.
[4] Kurilkin P K *et al.* 2011 *Nucl.Instr.Meth. in Phys.Res. A* **642** 45.
[5] Ladygin V P *et al.* *talk at this Conference.*
[6] Sekiguchi K *et al.* 2002 *Phys.Rev.* C **65** 034003.
[7] Sekiguchi K *et al.* 2004 *Phys.Rev.* C **70** 014001
[8] Suda K *et al.* 2007 *Nucl.Instr.Meth. in Phys.Res.* A **572** 745.
[9] Malakhov A I *et al.* 2000 *Nucl.Instrum.Meth. in Phys.Res. A* **440** 320.
[10] http://afi.jinr.ru
[11] Isupov A Yu *talk at this Conference.*
[12] Skhomenko Ya T *et al.* 2016 *Scientific Statements of Belgorod State University. Series: Mathematics and Physics* **43** 115.
[13] Isupov A Yu *et al.* 2013 *Nucl.Instrum.Meth. in Phys.Res. A* **698** 127.
[14] Fimushkin V V *et al.* 2016 *J.Phys.Conf.Ser.* **678** 012058.

IOP Conf. Series: Journal of Physics: Conf. Series **938** (2017) 012023 doi:10.1088/1742-6596/938/1/012023

Design of magnetic system to produce intense beam of polarized molecules of H_2 and D_2

A V Yurchenko*[1], D M Nikolenko[2], I A Rachek[2], Yu V Shestakov[1,2], D K Toporkov[1,2] and A V Zorin[1,2]
[1] Novosibirsk State University, 630090 Novosibirsk, Russia
[2] Budker Institute of Nuclear Physics, 630090 Novosibirsk, Russia

E-mail: arseniy-1995@mail.ru

Abstract. A magnetic-separating system is designed to produce polarized molecular high-density beams of H_2/D_2. The distribution of the magnetic field inside the aperture of the multipole magnet was calculated using the Mermaid software package. The calculation showed that the characteristic value of the magnetic field is 40 kGs, the field gradient is about 60 kGs/cm. A numerical calculation of the trajectories of the motion of molecules with different spin projections in this magnetic system is performed. The article discusses the possibility of using the magnetic system designed for the creation of a high-intensity source of polarized molecules. The expected intensity of this source is calculated. The expected flux of molecules focused in the receiver tube is $3.5 \cdot 10^{16}$ mol/s for the hydrogen molecule and $2.0 \cdot 10^{15}$ mol/s for the deuterium molecule.

1. Introduction

The existing sources of polarized hydrogen and deuterium atoms make it possible to obtain beams with an intensity of about 10^{17} atoms/s. A detailed review of the existing sources of polarized hydrogen atoms is given in [1, 2]. These sources are based on the spin-separation method. This principle can also be used for the production of polarized molecules. Thus, hydrogen molecules having different projections of the nuclear magnetic moment can be spatially separated in an inhomogeneous magnetic field, which was first demonstrated in Ref. [3].

We previously demonstrated [4, 5, 6] the production of polarized molecules of hydrogen/deuterium using the existing atomic beam source of polarized deuterium atoms [7]. In this source, superconducting sextupole magnets were used for the spatial separation of molecules with different nuclear spin projections. The experimentally measured flux of polarized molecules with a negative nuclear spin projection focused into the receiving detector tube was found to be $Q_{\text{tube } H_2} = 1.8 \cdot 10^{12}$ mol/s for the hydrogen molecules and $Q_{\text{tube } D_2} = 3.8 \cdot 10^{11}$ mol/s for the deuterium molecules.

These intensities can be increased by orders of magnitude if all linear dimensions of the experimental setup and magnets will be up-scaled, for example, by a factor of 10 [8]. This article discusses the possibility of creating this type of molecular source and its expected intensity.

---

* To whom any correspondence should be addressed.

2. Simulation of the spin separating magnet.

In a new source, a 60-pole magnet for spin separation of molecules should be used with the inner bore of 400 mm. An increase in the number of poles of the magnet will make it possible to obtain the larger gradient of the magnetic field, with the same value of the magnetic field at the poles, and, thereby, to increase the deflection force.

The geometry of the 1/60 part of such a developed magnet is shown in Figure 1. The area of the coil with current is S_{coil}=3.92 cm^2, the current density passed through the coil can reach j=100 kA/cm^2, the total current, respectively, is up to I=392 kA.

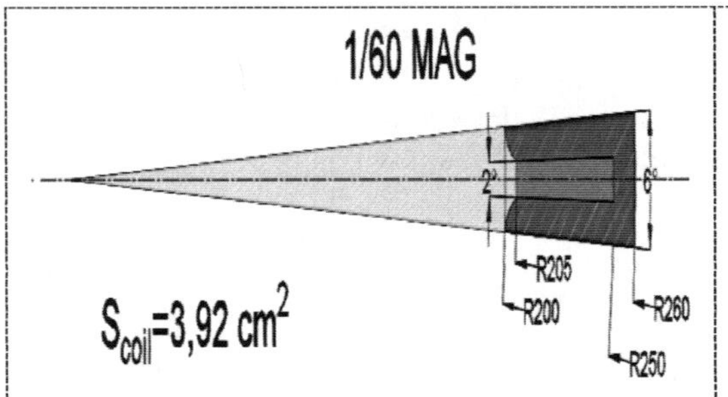

Figure 1. The geometry of a 60-pole magnet. Dimensions are in (mm). The following areas are highlighted: aperture of vacuum chamber – blue color, coil with current – red color, iron heel – green color

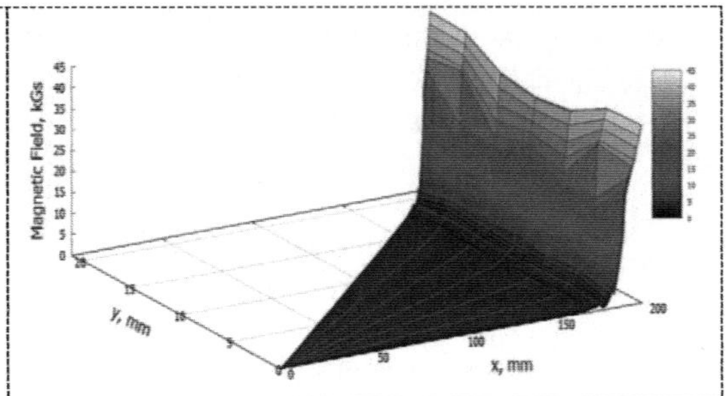

Figure 2. The magnetic field distribution $B(r,\varphi)$ in the vacuum chamber region, created by a 1/60 part of the magnet. The distribution is obtained as a result of computer simulation

The magnetic field distribution B(r,φ) in the inner aperture of the magnet was calculated by Mermaid software package (developed by BINP). The distribution of B(r,φ) is shown in Figure 2. It can be seen that the field and, consequently, its gradient is maximal near the pole of the magnet.

The simulation showed that the characteristic value of the magnetic field at the magnet pole at r=200 mm is B=40 kGs, the field gradient near the pole is about ∇B=60 kGs/cm.

3. Prototype of the experimental setup

For the spatial separation of molecules with a specific projection of the nuclear spin, the following geometry of the installation prototype was proposed, which is shown in Figure 3. Molecules with different projections of the nuclear spin are emitted from the nozzle of the source, whose internal diameter is 380 mm, the outer diameter is 400 mm (width of the nozzle, respectively, is 10 mm; the nozzle area is 122 cm^2, which is approximately 10^3 greater than the area of the nozzle used before). The distance from the nozzle to the entrance of the focusing 60-pole magnet is 1000 mm. The length of the magnet is 2000 mm. At the entrance of the magnet a disk is placed with a diameter of 380 mm, forming an annular diaphragm, an outer diameter is 400 mm and a width is 10 mm. With such a geometry the molecules are moving in the magnet in the vicinity of the poles, where the field gradient and, consequently, the deflecting force, are maximal. The magnet has a linearly converging aperture, such that the output diameter of the magnet is 370 mm. At a distance of 500 mm from the end of the magnet, along with the axis of the installation, a receiver for molecules is placed, with a diameter of 330 mm. It contains a turbopump for compressing a polarized gas. In such a geometry the direct ballistic flux of molecules cannot reach a receiver.

To determine the fraction of molecules with different spin projections that move out of the magnet from the total number of molecules from the source nozzle, a Monte Carlo computer program was developed to simulate the motion of neutral particles in installations with cylindrical geometry in the presence of an inhomogeneous magnetic field. The program takes into account of Maxwell's particle velocity dispersion, the flow distribution from the nozzle according to the cosine law, the dependence of the magnetic moment for each quantum state of the molecule on the magnitude of the magnetic field, the unit coefficient of adhesion of molecules to the surface of the vacuum chamber. In the simulation, in-beam scattering is not taken into account. The magnetic field is given as a table of numbers from Mermaid simulation. The trajectories were calculated for a hydrogen molecule with a nozzle temperature T=8K and for a deuterium molecule with T=11K (below this temperature molecular deuterium condensates).

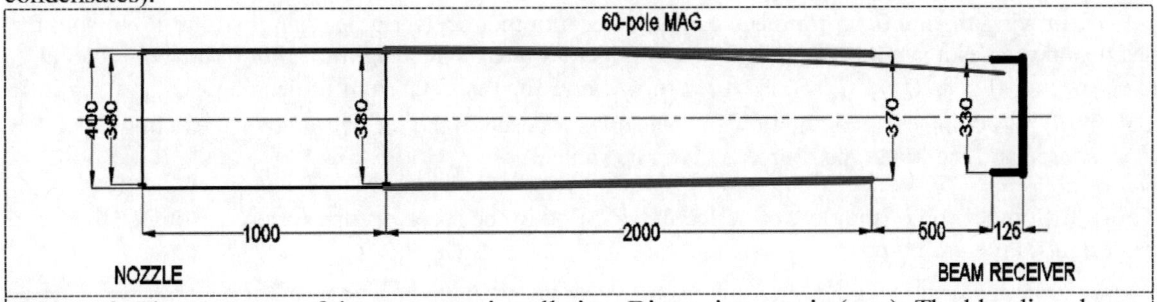

Figure 3. The geometry of the prototype installation. Dimensions are in (mm). The blue line shows one of the possible trajectories of molecules, which, being focused by the magnetic field, will enter the molecular receiver

4. Computer simulation of molecules motion

The results of computer simulation of the movement of molecules in the prototype of installation are given below.

Table 1 contains simulation results for the ratio of the number of molecules with different spin projections at the exit from the magnet to the total number of molecules from the source nozzle and the polarization. Here we use the notation: $N_{positive}$ is the number of molecules with a positive effective magnetic moment (negative nuclear spin projection m_I=-1, focusing), N_{zero} is the number of molecules with zero spin projection (free or ballistic motion, m_I=0), $N_{negative}$ is the number of molecules with negative effective magnetic moment (positive nuclear spin projection m_I=+1, defocusing), N_{all} is total number of molecules with all possible spin projections.

Table 1. Simulation results: the ratio of the number of molecules with different projections of nuclear spin and polarization **at the exit from the magnet** to the total number of molecules from the nozzle

molecule	$\frac{N_{positive}}{N_{nozzle}}$	$\frac{N_{zero}}{N_{nozzle}}$	$\frac{N_{negative}}{N_{nozzle}}$	$\frac{N_{all}}{N_{nozzle}}$	P_z
H_2	$6.31\cdot10^{-5}$	$7.66\cdot10^{-6}$	$6.78\cdot10^{-8}$	$7.08\cdot10^{-5}$	-0.88
D_2	$1.87\cdot10^{-5}$	$6.04\cdot10^{-6}$	$2.44\cdot10^{-6}$	$2.72\cdot10^{-5}$	-0.59

In Table 2, the same ratios are given, only for molecules that fall into the receiver.

Table 2. Simulation results: the ratio of the number of molecules with different projections of nuclear spin and polarization **in the receiver** to the total number of molecules from the nozzle

molecule	$\frac{N_{positive}}{N_{nozzle}}$	$\frac{N_{zero}}{N_{nozzle}}$	$\frac{N_{negative}}{N_{nozzle}}$	$\frac{N_{all}}{N_{nozzle}}$	P_z
H_2	$4.60\cdot10^{-5}$	0	0	$4.60\cdot10^{-5}$	-1.0
D_2	$3.83\cdot10^{-6}$	0	0	$3.83\cdot10^{-6}$	-1.0

From the analysis of the tables, it can be seen that the geometry of the projected installation is chosen in such a way that only molecules with a negative projection of the nuclear spin m_I =-1 enter the receiver tube. It is clear that the aperture of the magnet was chosen to converging, so that the molecules with the projection of the spin m_I=0,+1 hit the walls of the vacuum chamber and pumped out. It can be seen from Table 1 that the fraction of these molecules at the exit of the magnet is the order of magnitude smaller than the molecules with m_I=-1.

Knowing the fraction of molecules that enter into the receiver tube from the total flow from the nozzle, one can estimate the expected flow of molecules into the receiver tube. For hydrogen molecules, the expected flux is $Q_{\mathrm{tube}\ H_2} = 3.5 \cdot 10^{16}\ \mathrm{mol/s}$ and for the deuterium molecule is $Q_{\mathrm{tube}\ D_2} = 2.0 \cdot 10^{15}\ \mathrm{mol/s}$. A comparison of the flow of molecules focused in the receiver between existing BINP SB RAS source and the prototype source is given in Table 3.

Table 3. Comparison of the flow of molecules focused in the receiver tube for an existing (BINP) and projected source prototype

molecule	existing source in the BINP (experiment)	prototype source (simulation)
H_2	$1.8\cdot10^{12}$ mol/s	$3.5\cdot10^{16}$ mol/s
D_2	$3.8\cdot10^{11}$ mol/s	$2.0\cdot10^{15}$ mol/s

It can be seen that the gain when using a prototype source is about 10^4 for a hydrogen molecule and 10^3 for a deuterium molecule.

5. Conclusion

The magnet system of the new source to produce an intense beam of polarized molecules of hydrogen isotopes has been suggested. The field of a 60-pole magnet and the trajectory of molecules in this large installation are calculated. The calculations have shown that the intensity of the developed molecular source may exceed available at BINP source by more than 10^4 for the hydrogen molecules and more than 10^3 for the deuterium molecules.

The resulting high-intensity polarized molecular beams can be used for future accelerator experiments and as a possible fuel for fusion reactors [9].

This work was done with the support of the Russian Science Foundation under the joint with Deutsche Forschungsgemeinschaft grant RSF 16-42-01009 and DFG BU 2227/1-1.

6. References

[1] D. Toporkov, in XVth International Workshop on Polarized Sources, Targets, and Polarimetry September 9–13, 2013. Charlottesville, Virginia, USA, p. 064 (PoS, PSTP 2013)
[2] E. Steffens and W. Haeberl, Rep. Prog. Phys.66, 1887 (2003)
[3] R. Frisch, O. Stern, Z. Phys.85, 4 (1933)
[4] D. K. Toporkov et al., JETP Letters, 2017, Vol. **105**, No. **5**, pp. 289–291.
[5] D. K. Toporkov et al., Nucl. Instrum. Methods A **868**, 15 (2017) http://dx.doi.org/10.1016/j.nima.2017.06.038
[6] Yu. Shestakov et al., Proceedings of this Conference
[7] L. Isaeva et al., Nucl. Instrum. Methods A **411**, 201 (1998)
[8] Yu. Shestakov et al., in Proceedings of 13th International Symposium on High Energy Spin Physics, p. 415 (Protvino, Russia, 1998)
[9] G. Ciullo, R. Engels, M. Buscher, A. Vasilyev. Editors. Nuclear Fusion with Polarized Fuel. Springer Proceedings in Physics, Volume 187, (2016)

IOP Conf. Series: Journal of Physics: Conf. Series **938** (2017) 012024 doi:10.1088/1742-6596/938/1/012024

Studies of the General Parton Distributions.

Sergey Goloskokov

Bogoliubov Laboratory of Theoretical Physics, Joint Institute for Nuclear Research, Dubna 141980, Moscow region, Russia

E-mail: goloskkv@theor.jinr.ru

Abstract.
We discuss possibility to study Generalized Parton Distributions (GPSs) induced processes using polarized beams at NICA. We show that important information on GPDs structure can be obtained at NICA in exclusive meson production and in Drell-Yan (D-Y) process that determined by the double GPDs contribution.

1. Introduction

In this report we consider what information about the GPDs can be obtained at NICA. This analysis is done here on the basis of handbag approach which was successfully applied to light meson leptoproduction [1]. At large photon virtuality Q^2 the meson photoproduction amplitude factorizes [2] into hard meson production off partons which has a perturbative nature and GPDs. The cross sections and spin observables of light meson leptoproduction obtained in GPD model were found to be in good agreement [1, 3, 4] with HERMES, COMPASS, HERA and CLAS data. This means that information on GPDs can be obtained from the leptoproduction reactions.

In what follows we discuss a possibility to study GPDs induced processes using polarized proton beams at NICA. It is proposed that the NICA [5] detectors will have a detailed determination of the final particles. In this case it is possible to detect an exclusive reaction like $pp \to ppM$ where M is a some meson.

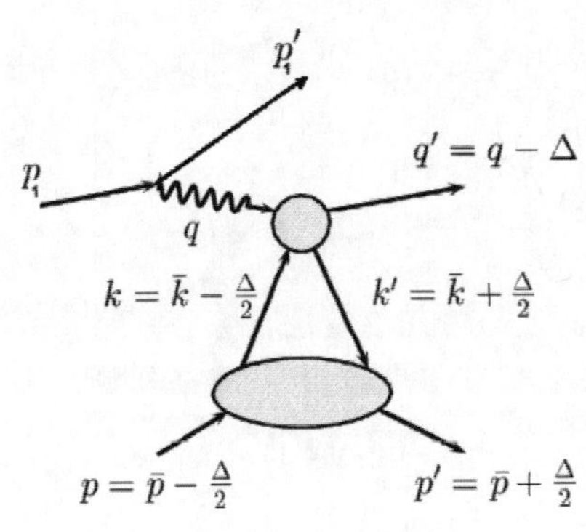

Figure 1. Meson production in exclusive proton-proton reaction. In the final state one detect two protons and meson.

The essential contribution in this reaction is following. One of the proton radiates a hard photon with virtuality Q^2 which interacts with the other proton. In the final state of the

IOP Conf. Series: Journal of Physics: Conf. Series **938** (2017) 012024 doi:10.1088/1742-6596/938/1/012024

exclusive process we should observe meson M and two protons. The photon virtuality Q^2 can be determined from analysis of the final proton kinematics as in the case of the meson leptoproduction. At large Q^2 other contributions might be rather small. The proposed reaction is similar to the corresponding process in the lepton-proton reaction.

At NICA we shall have the photon-proton interaction energy $W \sim 5-12$GeV. This W range is closed to HERMES and COMPASS energies. Generally NICA can scan this energy interval. At these energies the spin effects in different meson production can be studied [1, 3]. It seems that a more precise information on the spin asymmetries can be obtained at NICA with respect to HERMES and COMPASS experiments.

2. Handbag approach. Light meson production.

In what follows we consider the light meson leptoproduction reaction. Results for physical observables at NICA should be similar.

Within the handbag approach the meson production amplitude off proton can be presented at high photon virtuality Q^2 in the factorized form [2] as a convolution of the hard subprocess amplitude $\mathcal{H}^a_{\mu'+,\mu+}$ and GPDs H

$$M_{\mu'+,\mu+} \propto \int_{-1}^{1} dx \mathcal{H}^a_{\mu'+,\mu+} H^a(x,\xi,t) =< H > . \tag{1}$$

Here a is a flavor factor and μ and μ' are helicities of the photon and produced meson.

GPDs contain information on the hadron structure. In the forward limit $t = 0$ and zero skewness $\xi = 0$ GPDs are equivalent to ordinary Parton Distribution Functions (PDFs). With the help of sum rules GPDs are connected with hadron form factors, and information on the parton angular momenta can be extracted. We estimate GPDs by using the double distribution representation [6] which connects GPDs with PDFs. The PDFs parameterizations are obtained, e.g, from the CTEQ6 analysis [7], or from the nucleon form factor studies [8].

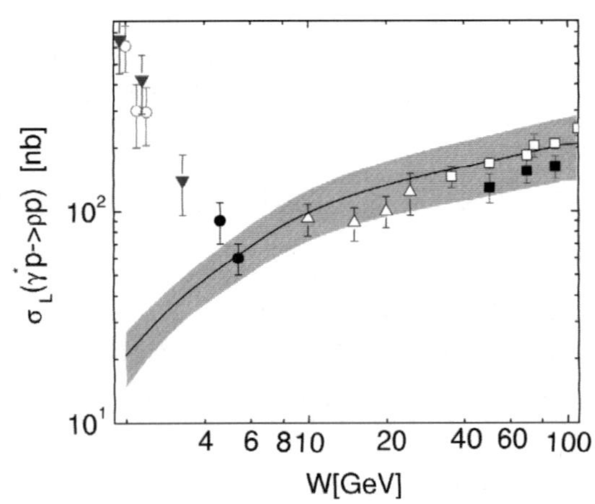

Figure 2. The longitudinal cross section for ρ^0 at $Q^2 = 4.0\,\mathrm{GeV}^2$. Data: HERMES (solid circle), ZEUS (open square), H1 (solid square), E665 (open triangle), open circles- CLAS, CORNEL -solid triangle.

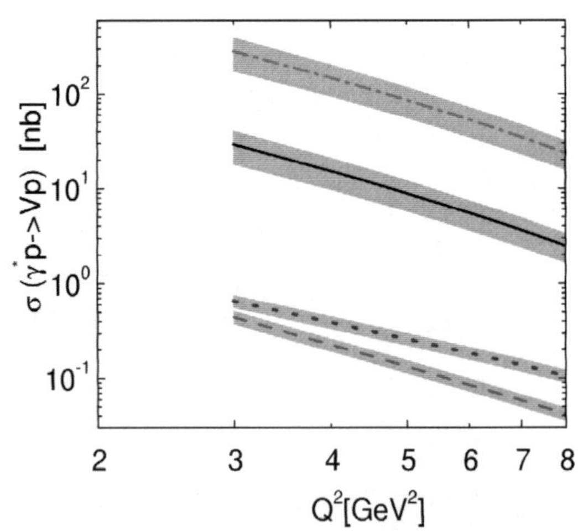

Figure 3. Hierarchy of meson production cross section at $W = 10$GeV. Dot-dashed line ρ^0, full line ω, dotted line ρ^+, dashed line K^{*0}

In Fig 2 we show the longitudinal cross section of the ρ^0 production at wide energy range [1]. It can be seen that the model results are consistent with experiment for $W > 4$GeV.

Cross section behavior at lower energies is not yet understood. Because NICA energy is closed to $W \sim 10$GeV, we show in Fig 3 cross section of various meson production at this energy [3]. Information on GPDs H can be obtained from the meson production cross section. From analyses of different meson production one can extract GPDs H for the valence quarks, gluon and sea quarks.

The proton helicity flip amplitude is a convolution of the subprocess amplitude $\mathcal{H}^a_{\mu'+,\mu+}$ and the GPDs E

$$M_{\mu'-,\mu+} \propto \frac{\sqrt{-t}}{2m} \int_{-1}^{1} d\overline{x}\, \mathcal{H}^a_{\mu'+,\mu+}\, E^a(\overline{x},\xi,t) =< E > . \tag{2}$$

This amplitude is essential in the case of transversely polarized proton beam. The information on the GPD E can be obtained from $A_{UT}^{\sin(\phi-\phi_s)}$ asymmetry which is determined as the interference of $< E >$ and $< H >$ contributions

$$A_{UT}^{\sin(\phi-\phi_s)} \propto \frac{\mathrm{Im} < E >^* < H >}{|< H >|^2}. \tag{3}$$

Our model results for A_{UT} asymmetry at COMPASS energy [9] are shown in Fig 4 and indicate a good agreement with data [10]. Study of A_{UT} asymmetry for different meson production will give possibility for flavor separation of the E GPDs. Predictions for such asymmetries integrated over t' [3] are shown in Fig 5.

Note that at HERMES and COMPASS energies A_{UT} asymmetries were obtained with sufficiently large error at two energies $W \sim 5$GeV and $W \sim 8$GeV respectively. At NICA there is a possibility to scan energy range $W \sim 5 - 12$GeV. This will give an important information on the energy dependencies of GPDs.

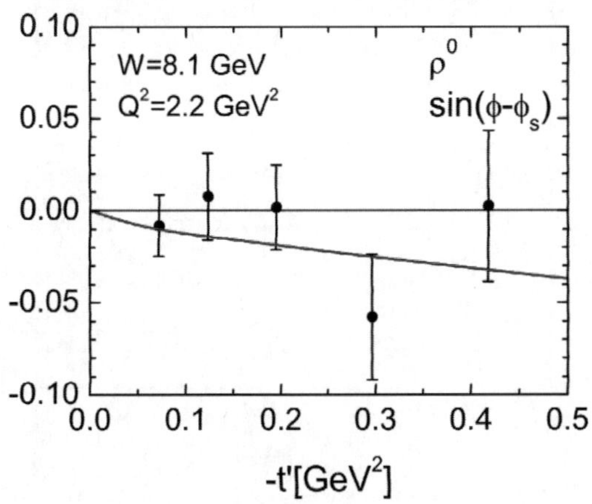

Figure 4. Model results for A_{UT} asymmetry at COMPASS energy W = 8GeV. The data [10] are shown.

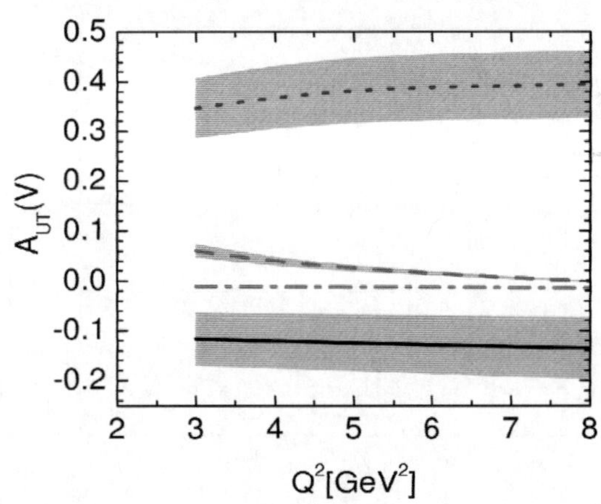

Figure 5. Hierarchy of A_{UT} asymmetry for various meson production for W = 10GeV. Dot-dashed line ρ^0, full line ω, dotted line ρ^+, dashed line K^{*0}.

GPDs H and E contribute to the Ji sum rule [11], which determine the proton angular momenta

$$\int x dx (H^a(x,\xi,0) + E^a(x,\xi,0)) = 2J^a. \tag{4}$$

The new information on the E GPDs will shed a light to solution of so called Spin Crises.

3. Pseudoscalar meson production. Transversity effects.

The leading twist contributions are not sufficient to describe the experimental results on pseudoscalar meson (PM) production at low Q^2. It was found that the transversity effects which have twist-3 nature and caused by the GPDs H_T and $\bar{E}_T$ are essential in this process [4]. The transversity effects were studied for the first time in the SIDIS reactions [12]. In meson production transversity contributions are essential in the $M_{0\pm,++}$ amplitudes. These amplitudes are determined by the transversity GPDs which contribute together with the subprocess $\mathcal{H}_{0-,++}$ amplitude. The twist-3 pion wave function [4] is an ingredient of the $\gamma_T^* \to \pi^0$ transition amplitude $\mathcal{H}_{0-,++}$

$$M_{0-,++} \propto \int_{-1}^{1} d\bar{x} \mathcal{H}_{0-,++}(\bar{x},...)\, H_T. \tag{5}$$

The amplitude $M_{0+,++}$ has the similar form but it is determined by the GPDs $\bar{E}_T$ and contains the kinematical factor $\sqrt{-t'}/(4m)$ like the corresponding term in (2).

We present here our results on the π^0 production based on the handbag approach [4]. In calculation, we use the leading twist contributions determined by the GPDs $\tilde{H}$ and $\tilde{E}$ and the transversity effects from the $M_{0\pm,++}$ amplitudes which are essential at low Q^2. In Fig. 6 we present the model results together with CLAS experimental data [13]. Note that at non-zero momentum transfer the $\bar{E}_T$ contribution becomes important and we find essential correlation between σ_T and σ_{TT} which is observed experimentally. It can be seen from Fig. 7 a relative smallness of the leading-twist σ_L cross section (full line) with respect to σ_T determined mainly by the transversity H_T and $\bar{E}_T$ effects (dashed line) at moderate $Q^2 \sim 3\text{GeV}^2$. The smallness of σ_L with respect to σ_T was confirmed at JLAB [14]. Similar effects were found in the pion-induced D-Y process [15].

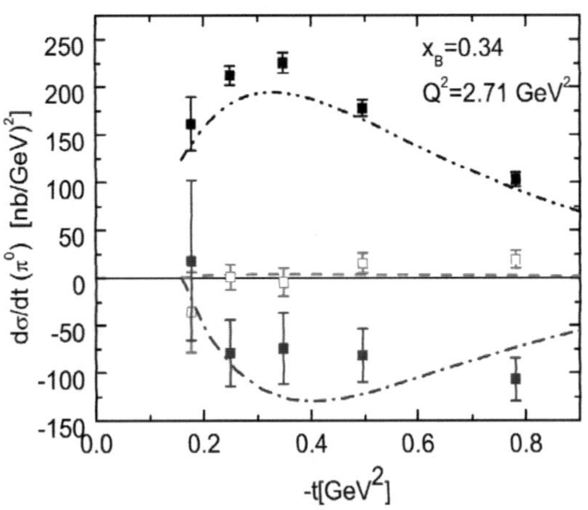

Figure 6. π^0 production at CLAS energy range together with the data [13]. Dashed-dot-dotted line- $\sigma = \sigma_T + \epsilon\sigma_L$, dashed line-$\sigma_{LT}$, dashed-dotted line- σ_{TT}.

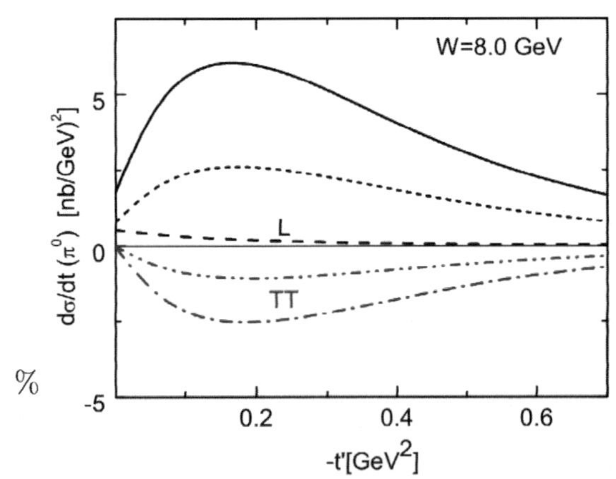

Figure 7. π^0 production at COMPASS energy range. Full line- unseparated cross section at $Q^2 = 3\text{GeV}^2$ -and short dashed at $Q^2 = 5\text{GeV}^2$; dashed dotted- σ_{TT} at $Q^2 = 3\text{GeV}^2$, dashed dotted- dotted at $Q^2 = 5\text{GeV}^2$. Dashed line σ_L at $Q^2 = 3\text{GeV}^2$.

The transversity effects at NICA can be investigated if one will consider the π_0 PM in the final state. It was shown here that the transversity GPDs generate the large transverse cross section which exceed essentially the leading twist longitudinal cross section at moderate Q^2. This important result [4] can be tested at NICA. The π_0 cross section was studied at JLAB

IOP Conf. Series: Journal of Physics: Conf. Series **938** (2017) 012024 doi:10.1088/1742-6596/938/1/012024

energies only. Some preliminary data were obtained at COMPASS. NICA experiments permits one to get more information on the transversity GPDs. Really one can analyse at NICA π_0 production cross section at different W and Q^2. This can essentially improve our knowledge of the transversity GPDs.

4. Drell-Yan process. Two GPDs contribution.

The exclusive D-Y production in the proton-proton reaction determined by two GPDs contribution was proposed in [16]. This gives possibility to study GPDs at NICA within the exclusive reaction where in the final state we observe two protons and the lepton pair. An example of such process that contains gluon and quark contribution is shown in Fig. 8. Investigation of the cross section determined by two-GPDs effects is in progress in collaboration with O. Teryaev and P. Kroll. It is shown that the gluon and quark sea GPDs lead to the cross section which does not decrease with energy. Thus, this reaction can be studied at NICA and at higher energies, at LHC e.g. We estimate cross section integrated over s_1 and s_2 energies (Fig. 8) and momenta transfer t_1 and t_2 at $\sqrt{s} = 24$GeV. We find that $d\sigma/dQ^2 \sim 20$pb/GeV^2 for $Q^2 = 5$GeV^2. This result is smaller with respect to the cross section of inclusive D-Y reaction because the contribution Fig. 8 is a part of inclusive process.

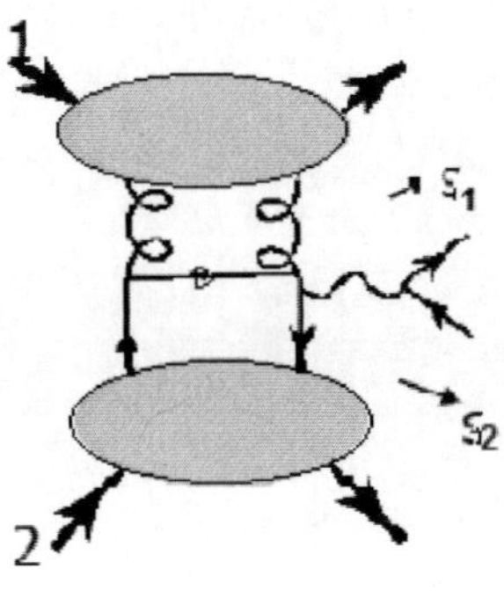

Figure 8. Drell-Yan process with gluon and quark GPDs

5. Conclusion

In this report we discussed the possibility to study exclusive meson productions at NICA to collect our knowledge on GPDs. It is shown that information on the GPDs H can be obtained from analysis of exclusive meson production data. The polarized proton beams give possibility to study the GPDs E from A_{UT} asymmetries. Note that the data on different meson production will give possibility to perform flavor separation of GPDs. Study of the E GPDs contribution to the Ji sum rules should give a new information on the parton contributions to the proton spin.

Transversity effects can be investigated in the PM production. Essential role here plays π_0 reaction. From analysis of π^0 cross section the transversity GPDs H_T and $\bar{E}_T$ can be extracted. As a result information from JLAB and COMPASS experiments can be improved. The real possibility to test transversity effects at different W and Q^2 at NICA is very important. This will give information on dependencies of the transversity GPDs on ξ and Q^2. Study of exclusive D-Y reactions determined by the double GPDs contribution, will give important information about structure of this process.

Thus we can conclude that NICA experiments with the polarized beams can essentially improve our knowledge on the GPDs.

The work was supported in part by the Heisenberg-Landau program.

References

[1] Goloskokov S V and Kroll P 2005 *Euro. Phys. J.* C **42** 281;
Goloskokov S V and Kroll P 2007 *Euro. Phys. J.* C **50** 829;
Goloskokov S V and Kroll P 2008 *Euro. Phys. J.* C **53** 367
[2] Ji X 1997 *Phys. Rev.* D **55** 7114;
Radyushkin A V 1996 *Phys. Lett.* B **380** 417;
Collins J C et al. 1997 *Phys. Rev.* D **56** 2982
[3] Goloskokov S V and Kroll P 2009 *Euro. Phys. J.* C **59** 809
[4] Goloskokov S V and Kroll P 2011 *Euro. Phys. J.* A **47** 112
[5] V. Kekelidze et al. 2016 *EPJ Web Conf.* **126** 02015.
[6] Musatov I V and Radyushkin A V 2000 *Phys. Rev.* D **61** 074027
[7] Pumplin J et al. 2002 *JHEP* **0207** 012
[8] Diehl M and Kroll P 2013 *Euro. Phys. J.* C **73** 2397
[9] Goloskokov S V and Kroll P 2014 *Euro. Phys. J.* C **74** 2725
[10] Adolph C, et al. (COMPASS Collaboration) *Phys.Lett.* B **731** 19
[11] Ji X D 1997 *Phys. Rev. Lett* **78** 610
[12] Anselmino M, Boglione M, D'Alesio U, Kotzinian A, Murgia F, Prokudin A and Melis S 2009 *Nucl. Phys. Proc. Suppl.* **191** 98
[13] Bedlinskiy I et al. *(CLAS Collab.)* 2012 *Phys. Rev. Lett.* **109** 112001
[14] Defurne, M et al. *(Jeff. Lab Hall A Collab.)* 2016 *Phys. Rev. Lett.* **117** 262001
[15] Goloskokov S V and Kroll P 2015 *Phys. Lett* B **748** 323
[16] Teryaev O. 2012 *PoS QNP* 055; Pivovarov A A and Teryaev O V 2015 *AIP Conf. Proc.* 1654 070008

Spin Transparency Mode in the NICA Collider with Solenoid Siberian Snakes for Proton and Deuteron Beam

A D Kovalenko[1], **A V Butenko**[1], **V A Mikhaylov**[1], **M A Kondratenko**[2], **A M Kondratenko**[2] **and Yu N Filatov**[1,3]

[1]Joint Institute for Nuclear Research, 141980, Dubna, Russia
[2]Science and Technique Laboratory "Zaryad", 630090, Novosibirsk, Russia
[3]Moscow Institute of Physics and Technology, 141700, Dolgoprudny, Russia

[1]kovalen@dubna.ru

Abstract. Two solenoid Siberian Snakes are required to obtain ion polarization in spin transparency mode of the NICA collider. The snake solenoids with a total field integral of 2×50 T·m are placed into the straight sections of the NICA collider. It allows one to control polarization of protons and deuterons up to 13.5 GeV/c and 4 GeV/c respectively. The snakes introduce a strong betatron oscillation coupling. The calculations of orbital parameters of proton and deuteron beams in the NICA collider with solenoid Snakes are presented.

1. Ion colliders in the "Spin transparency" mode

The NICA (Nuclotron-based Ion Collider fAcility) collider at the Joint Institute for Nuclear Research (Dubna, Russia) is designed to provide high polarization of both protons and deuterons colliding beams. To control ion polarization the NICA collider is set up in a "spin transparency" mode [1]. "Spin transparency" means that total effect of all magnetic fields along design orbit on the spin is reduced to zero. Thus, any spin direction repeats after a particle turn. Particles are in the region of a zero-integer spin resonance and the spin tune is zero. Colliders transparent to the spin offer a unique opportunity to efficiently control the ion polarization using small quasi-stationary magnetic field.

The natural example of a collider transparent to the spin is the Jefferson Lab Electron Ion Collider (JLEIC) with figure-8 shaped rings [2]. The conventional NICA collider becomes "transparent to the spin" due to introduction of two identical solenoidal snakes into its opposite straight sections.

Let us demonstrate the main features of the ion polarization control in the "spin transparency" mode comparing the JLEIC and NICA colliders with the RICH collider [3]. Table 1 shows a general information and ability to use these colliders in the "spin transparency" mode. The spin tune at RHIC is equal to one half and does not depend on energy as well, but there is no "spin transparency".

Table 1. Spin transparency mode at colliders with polarized ions.

Collider	Momentum range, GeV/c	Colliding particles	Spin Tune	Spin Transparency
RHIC (BNL)	25-250	*pp*	1/2	–
JLEIC (JLAB) (figure-8)	8-100	*eN*	0	+
NICA (JINR)	2.5-13.5	*NN*	0	+

The independence of the spin tune on the beam energy solves a problem of the polarization preservation during beam acceleration. In colliders, a crucial task is not only preservation of the polarization during the energy ramp but also its control at the interaction points during experiments. Table 2 shows information about ion polarization control at the considering colliders.

Table 2. Ion Polarization Control.

Collider	Spin Rotators based on	Polarization Direction at IP	Spin Flipping	
			Reversal Time	**Orbital Parameters**
RHIC (BNL)	'strong' magnetic fields	Any directions (*without deuterons*)	A few min	Change
JLEIC (JLAB)	'weak' solenoids	Any directions (*any particles*: *p, d, He*^{3}, ...)	A few ms	Do not change
NICA (JINR)	'weak' solenoids	Any directions (*any particles*: *p, d, He*^{3}, ...)	A few ms	Do not change

At RHIC polarization is controlled using a pair of spin rotators with "strong" fields, which provide the required spin orientation at an interaction point and then return the spin to its original direction. This process changes not only the polarization direction but the trajectories of the beam particles as well. This causes a change in the orbital characteristics of the beam. There are betatron tune shifts, changes in dispersion, beta-functions, luminosity, etc. Accounting for the changes in the beam orbital characteristics becomes a problem when doing experiments with polarized beams in colliders.

In colliders with spin transparency, any small perturbation has a strong effect on the beam polarization. To stabilize the spin direction, one must introduce spin rotators based on "weak" fields into the collider's lattice, which "shift" the spin tune by a small value ($\nu \ll 1$) and set the necessary orientation of the polarization. The required "weak" field integrals are significantly lower than the field integrals used in the RHIC spin rotators and are limited by the strength of the zero-integer spin resonance w_0: $\nu \gg w_0$. What especially stands out is the possibility of using weak solenoids, which do not impact the closed orbit at all. Spin rotators based on "weak" solenoids have essentially no effect on the beam's orbital characteristics during adjustment of the beam polarization direction.

There is a possibility to implement a spin flipping system based on the spin rotators. Spin flipping system allows one to make spin reversal during. At RHIC spin reversal time is determined by the time of strong field variation and is about of few minutes. At JLEIC and NICA in spin transparency mode spin reversal time is about few milliseconds. Due to orbital parameters become unchanged during spin manipulation in spin transparency mode the luminosity of colliding bunches does not change at JLEIC and NICA. Thus, spin transparency mode allow one to carry out high precision experiments with polarized ions.

Let us emphasize that the ion polarization control scheme in spin transparency mode is universal for all particle species including both protons and deuterons, while obtaining longitudinal deuteron polarization at RHIC presents a serious problem.

2. The NICA collider in spin transparency mode

The scheme of polarization control is based on the use of two types of solenoids, namely, strong-field solenoids and weak-field solenoids (see figure 1). The strong-field solenoids are used for Siberian snakes, which create the "spin transparency" mode in the NICA collider. Two solenoidal snakes are inserted in the opposite straight sections of the racetrack-shaped collider. The snakes are divided symmetrically onto two parts by MPD and SPD setups. Such snake design allows one to control the beam polarization in vertical plane of SPD and MPD detectors while the polarization leis in the plane of the collider in the NICA arcs.

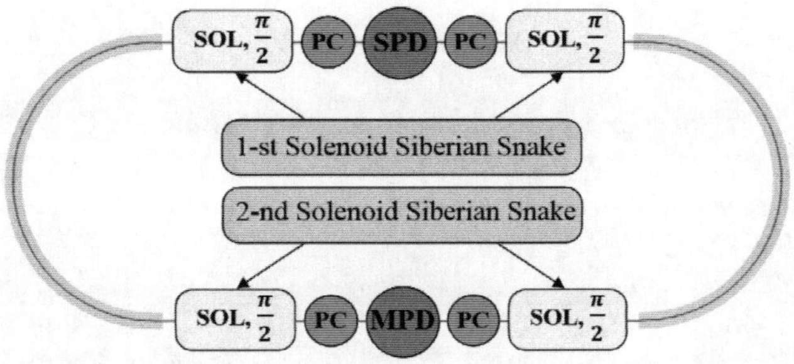

Figure 1. Polarization control scheme at the NICA collider in spin transparency mode.

The weak-field solenoids with small integral of the longitudinal field are used directly for the control of the beam polarization of any particle species (*p*, *d*, ^3He, …) in any orbit place by means of polarization control insertions (PC insertions) [4]. For example, it is possible to arrange both vertical and longitudinal directions of the beam polarization in MPD and SPD detectors. One can obtain any spin direction in the collider arcs, which is necessary for the matching of the polarizations during the beam injection from Nuclotron. One can provide the spin reversals in the detectors during an experiment (spin flipping systems) [5]

3. Orbit matching of the snake solenoids to the NICA collider lattice

The PC insertions weak solenoids do not influence on the collider orbital characteristics. On the contrary, the snake solenoids introduce a strong coupling of the betatron oscillations. The snake solenoid fields are proportional to the momentum and the collider optics remains unchanged during the beam acceleration.

The total solenoid field integral required for the snake can be obtained in parts by inserting solenoids into available spaces of the NICA straight section. We choose design of the one snake which consist of eight solenoids of 1.2 m length each. At maximum momentum of 13.5 GeV/c the solenoid field is 6 T. It allows one to control polarization in the total momentum range for protons and up to 4 GeV/c for deuterons. Figure 2 shows placement of the snake and PC insertion solenoids into the collider half straight section. Here SOL are snake solenoids, PC are polarization control solenoids, FFQ are final focus quadrupoles, VB are arc's vertical-field bending magnets and RB are radial-field bending magnets, which provide beams collisions at interaction point IP.

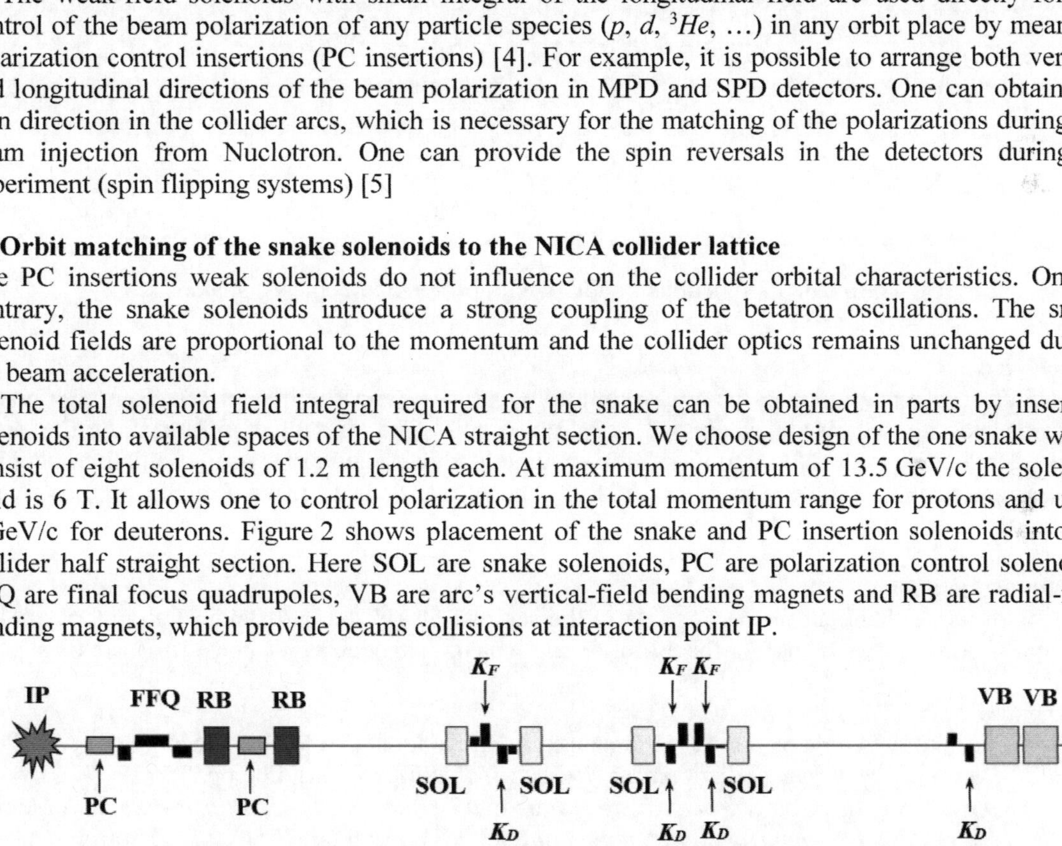

Figure 2. Placement of the snake solenoids in the NICA half straight.

Two families of focusing K_F and defocusing K_D quadrupoles are used for matching the snake solenoids to the collider optics. Additional quadrupoles for compensation of the coupling of the betatron oscillations are not used in our matching scheme [6].

Let us give an example of the snake solenoids matching, when the values of the β- functions at interact points of MPD and SPD detectors remain the same as for the case of NICA without snakes. The β-functions for NICA with proton and deuteron snakes are shown in figures 3. The deviation δK_F and δK_D of gradients K_F and K_D in unit of the magnetic rigidity are indicted on the graphs. Proton snakes practically do not change beam's orbital parameters. Deuteron snakes increase values of β-functions in the collider arcs which can be compensated by optimization of the collider straights sections lattice.

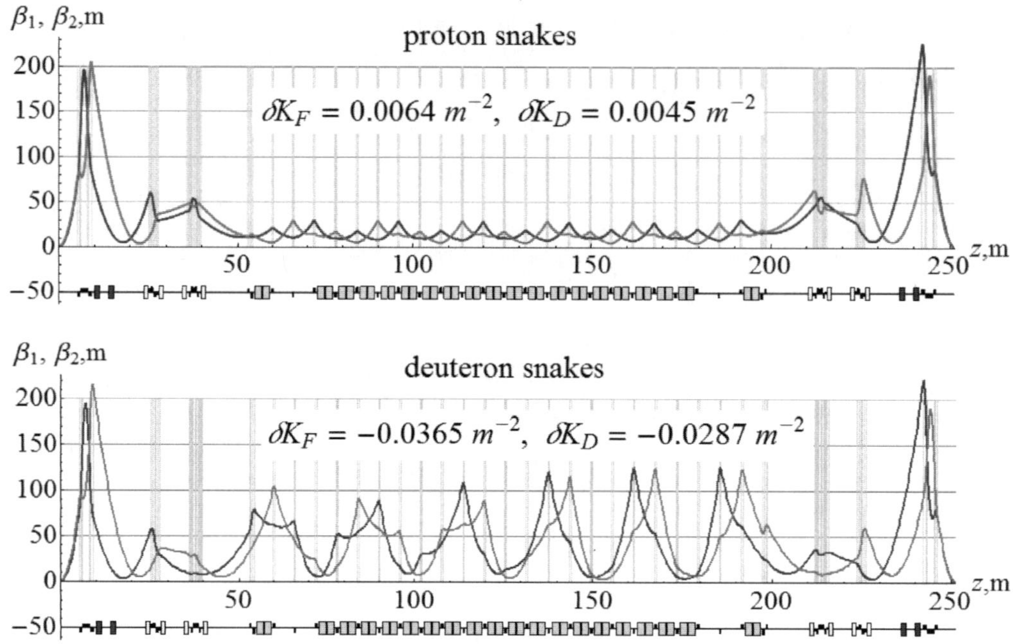

Figure 3. β – functions in NICA with proton and deuteron snakes.

Conclusion

The spin transparency mode in the NICA collider opens wide opportunities to carry out high quality experiments with polarized proton and deuteron beams. The presented examples of the snake solenoids matching to the NICA collider lattice without compensation of betatron oscillations coupling demonstrate that the spin transparency mode in the NICA collider looks feasible for proton and deuteron beams in the real collider lattice.

Acknowledgement

We would like to thank all the participants of the seminar at Veksler and Baldin Laboratory of High Energy Physics (JINR, Dubna) for the useful discussions of some questions, considered in this paper.

References

[1] Kovalenko A D, Filatov Yu N, Kondratenko A M, Kondratenko M A and Mikhaylov V A 2014 *Polarized deuterons and protons at NICA@JINR, PEPAN Lett.* **45** 1 pp 321–22

[2] Abeyratne S *et al* 2012, *Science Requirements and Conceptual Design for a Polarized Medium Energy Electron-Ion Collider at Jefferson Lab*, arXiv:1209.0757 [physics.acc-ph]

[3] Harrison M, Peggs S and Roser T 2002 *The RHIC Accelerator*, *Annual Review of Nuclear and Particle Science* **52** pp 425–469

[4] Filatov Yu N, Kovalenko A D, Kondratenko A M, Kondratenko M A and Mikhaylov V A 2013 *Proton And Deuteron Polarization Control In NICA Collider Using Small Solenoids*, *Proc.XV Workshop on HESP DSPIN 2013* Dubna pp 351-54

[5] Derbenev Ya S, Kondratenko A M, Kondratenko M A, Morozov V S and Filatov Yu N 2011 *Spin-Flipping Systems For Storage Rings Proc. XIV Workshop on HESP DSPIN-11, Dubna,* pp 377-84

[6] Kondratenko M A, Kondratenko A M, Kovalenko A D, Mikhaylov V A and Filatov Yu N 2013 *Orbital Parameters Of Proton Beam In Nuclotron With Solenoid Siberian Snake*, *Proc.XV Workshop on HESP DSPIN 2013* Dubna pp 371-74

A possible layout of the Spin Physics Detector with toroid magnet.

A.P. Nagaytsev[1]

Joint Institute for Nuclear Research, Dubna, Moscow region, Russia

Abstract. The Spin Physics Detector project for carrying out experiments at the 2-nd interaction point of the NICA collider is under preparation. The design of the collider allows reaching collision energy in the c.m.s. as high as $\sqrt{s}$= 26 GeV for polarized proton-proton collisions and $\sqrt{s}$ = 12 GeV for polarized deuteron-deuteron collisons with a luminosity of up to 10^{32} cm^2s^{-1} (for protons) and 10^{31} cm^2s^{-1} for deuterons. Such a high luminosity of polarized beams interactions opens unique possibilities to investigate a variety of polarization phenomena including those related to the nucleon spin structure. A proposal for the experimental set-up based on a toroid type magnet is presented.

The main physical tasks proposed for spin program at NICA (see Ref.1) can be divided into four blocks shown in Fig.1.

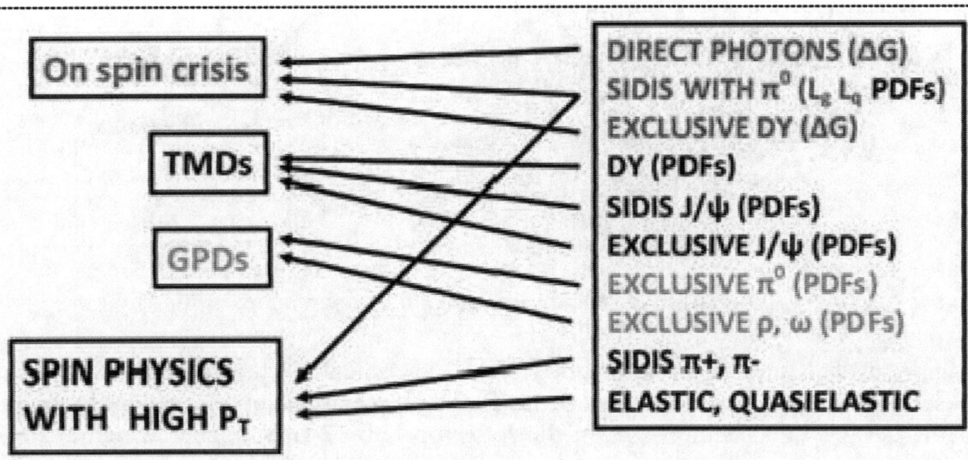

Figure 1. The main physical tasks proposed for spin program at NICA.

[1] Alexander Nagaytsev, nagajcev@gmail.com; phone +7 49621 65124.

IOP Conf. Series: Journal of Physics: Conf. Series **938** (2017) 012026 doi:10.1088/1742-6596/938/1/012026

Preliminary considerations of the event topologies (see Ref.1) for measurements shown in Fig.1 required the following characteristics of the experimental setup: 4π angular region around the beam, high-precision (better than 50 μm) and fast vertex detector, a tracking system that provides high accuracy (~200 μm) along the track, fast DAQ, minimum of material, measurement of neutral (π^0 etc) secondary particles, identification of charged particles with efficiency close to 100%, fast and modern trigger system, modularity and access to the elements of the setup, which will allow to upgrade and modify detectors for new research. SPD has to be equipped with the following sub-detectors: vertex detectors, tracking detectors, electromagnetic calorimeters, hadron and muon detectors and trigger system. Some of them must be in the magnetic field for which there are two options: either toroid or solenoid type.

A toroid magnet provides a field free region around the beam pipe and, due to that, does not influence on the beam trajectories and polarizations. It can consist of 8 superconducting coils symmetrically placed around the beam axis (see Fig. 2). A support ring upstream (downstream) of the coils hosts the supply lines for electric power and for liquid helium. At the downstream end, a hexagonal plate compensates the magnetic forces to hold the coils in place. The field lines of ideal toroid magnet are always perpendicular to the particles originating from the beam intersection point. Since the field intensity increases inversely proportional to the radial distance, greater bending power is available for particles scattering at smaller angles and having higher momenta. These properties help to design a compact spectrometer that keeps the investment costs for the detector tolerable. The toroid magnet requires insertion of the coils into the tracking volume occupying a part of the azimuthal acceptance. Preliminary studies show that the use of superconducting coils, made by the Nb_3Sn-Copper core surrounded by a winding of aluminum for support and cooling, allows one to reach an azimuthal detector acceptance of about 85%. For SPD measurements there is a disadvantage of the toroid solution related to a high non-uniformity of acceptance as a function of azimuthal angle in the laboratory frame and the some loss of detector acceptance.

Figure 2. Possible view of SPD toroid magnet.

One example of the use of a toroid magnet for an experimental setup is the TORUS magnet for CLAS12 (see Ref.2). The torus magnet for the CLAS12 spectrometer is a 3.6-T superconducting magnet being designed and built as part of the Jefferson Lab 12-GeV upgrade (Fig.3). The magnet consists of six coil case (enclosed in a vacuum-impregnated coil pack) assemblies mounted to a cold central hub. The coil pack consists of a 117-turn double-pancake winding wrapped with two layers of 0.635-mm-thick copper cooling sheets. The coil case assembly is cooled by supercritical helium at 4.6 K. The details are given in Ref.2.

IOP Conf. Series: Journal of Physics: Conf. Series **938** (2017) 012026 doi:10.1088/1742-6596/938/1/012026

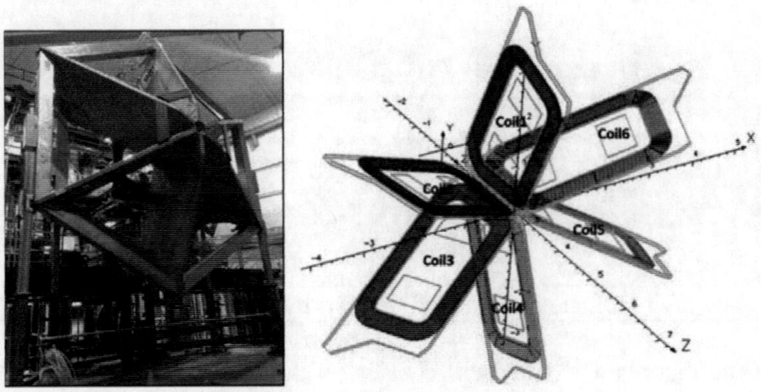

Figure 3. The views of TORUS magnet for CLAS12 (left) and superconducting magnet system (right).

The preliminary design of SPD with toroid magnet system is shown in Fig.4. The main parts of this scheme described below. They are shown in Fig.4 in red- toroid magnet, in yellow – electromagnetic calorimeter and in blue – muon system (RS).

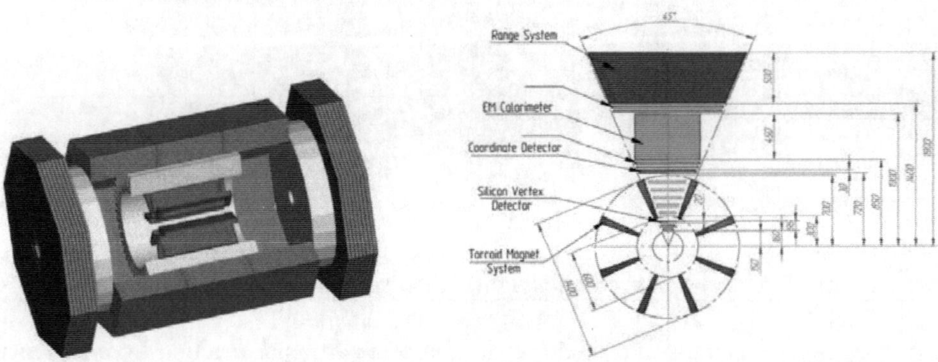

Figure 4. The view of SPD (left) and view of the barrel sector (right). Details are given in text.

The first coordinate system shown in Fig.4 (right panel) is vertex silicon detector. The most obvious technology for the vertex detector is a silicon one. Preliminary one can consider the vertex system approved for the MPD (see Ref.3). This detector can be located outside the beam pipe and can consist of several layers of double-sided silicon strips which provide a precise vertex reconstruction and tracking of the particles before they reach the general SPD tracking system and reject the secondary decay vertexes.

The coordinate system before calorimeter can be produced with straw tubes (Fig.5). This choice is based to the following properties of the straw tubes: the minimum of material for the tracks of the secondary particles ($X_0 \sim 0.1$), the time (~ 200-300 ns) and spatial (~100 μm) resolutions, provided the expected particle rates (DAQ rates ~ 100 Khz); developed production sites in JINR, Dubna. The details can be found in Refs.4.

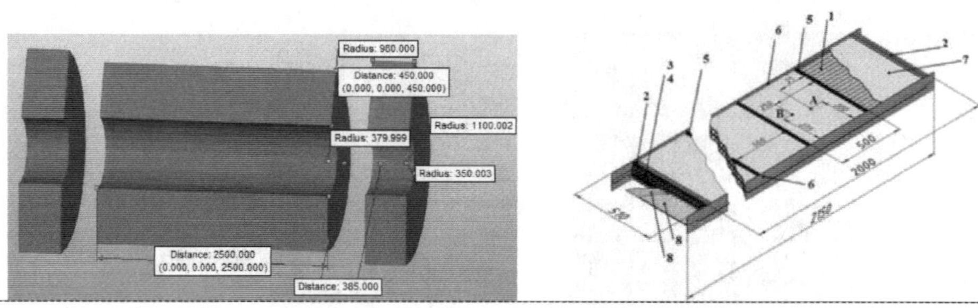

Figure 5. The view of coordinate system (left) and straw module (right).

The electromagnetic calorimeter (EM) will consists of "shashlyk" modules with the application of new readout techniques based on AMPD technology. The latest version of the electromagnetic calorimeter (ECAL) module, developed at JINR for the COMPASS-II experiment at CERN (see Fig.6), [5], can be a good candidate for ECAL in the barrel and end-cap parts of SPD. The modules can have an area of $4x4cm^2$ and a length of 30-40 cm (see Ref.5). The expected energy resolution can be $\sigma(E)/E = (5 - 8)\%/ E$. The calorimeter also can be used for the triggering of DY electrons.

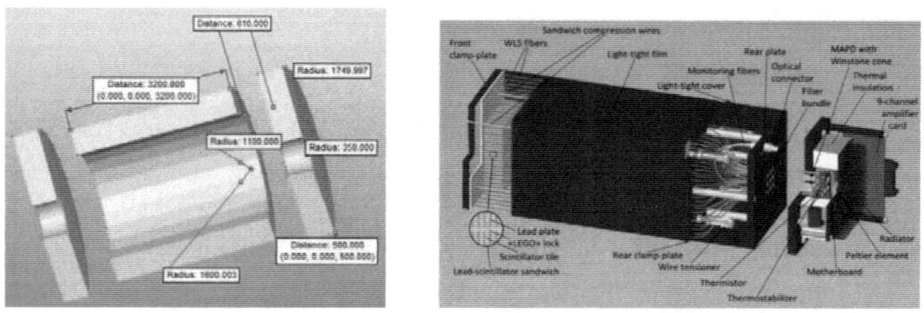

Figure 6. The view of ECAL(left) and module (right) Details are given in text.

The main task of the trigger system is to provide separation of a particular reaction from all reactions occurred in collisions. Each of them will be pre-scaled with: two muons (or electrons/positrons) in final states, various types of hadrons in final states (π+/-, K, p,...), photons (π^0, ω, η...), other reactions. RPC are proposed to be used as main trigger detector [see Ref. 6]. Also hodoscopes of scintillating counters can be used for triggering. They can be located before and after RS (or mounted in the last layers of RS) and before RPC. The view of detectors for trigger system is given in Fig.7.

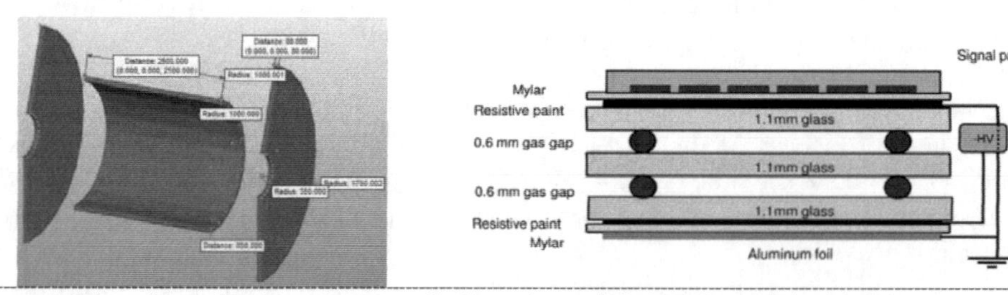

Figure 7. The view of trigger system (left) and possible RPC module (right).

IOP Conf. Series: Journal of Physics: Conf. Series **938** (2017) 012026 doi:10.1088/1742-6596/938/1/012026

The system of mini-drift layers with Fe layers called by Range System (RS) is the main muon system of the detector (see Fig.8). One can provide the clean (> 99%) muon identification for muons with momenta more than 1 GeV. The combination of responses from EM calorimeter and RS can be used for the identification of pions and protons in wide energy range. This system was taken as main muon detector for PANDA experiment (see Ref.7)

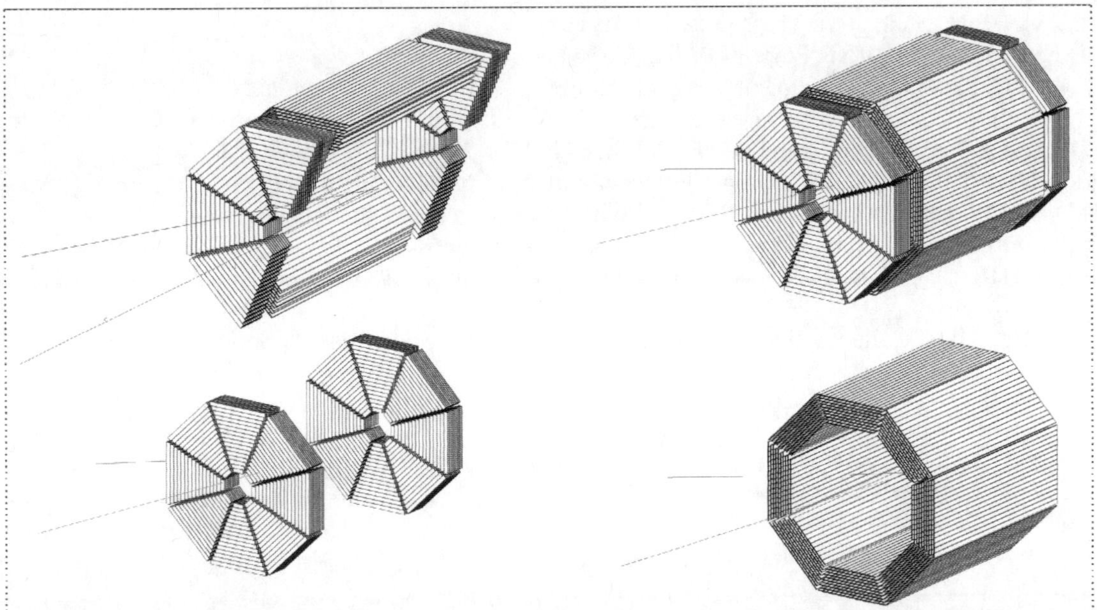

Figure 8. The views of Range System in SPD with muon tracks from DY process, obtained with PYTHIA generator.

The final version of the SPD will be defined after detailed Monte-Carlo simulations and consideration of requirements for other spin effects studies. Some preliminary pictures of experimental set-up with Drell-Yan and SIDIS events are shown in Figure 9.

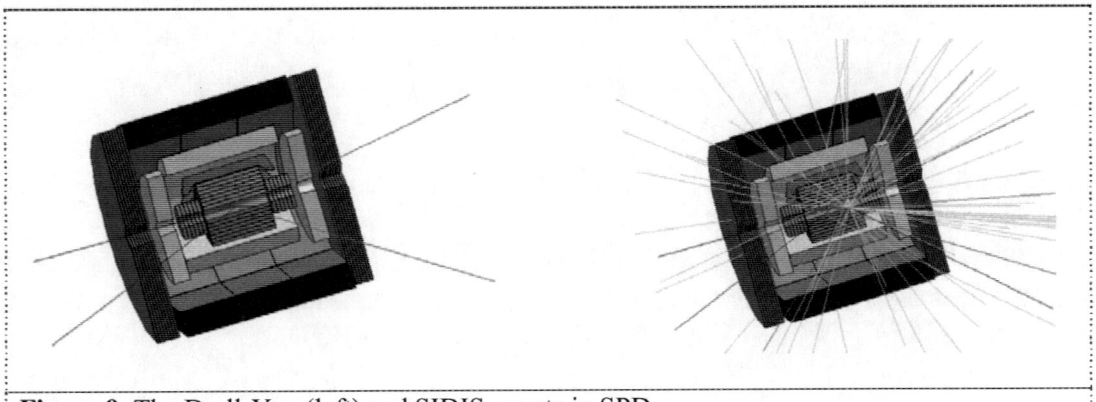

Figure 9. The Drell-Yan (left) and SIDIS events in SPD.

References:

[1] A.V.Efremov et al., EPJ Web Conf. 85 (2015) 02039
[2] C.Luongo et al.,IEEE Trans.Appl.Supercond. 26 (2016) no.4, 4500105
[3] V.Golovatyuk et al., Eur.Phys.J. A52 (2016) no.8, 212
[4] V.A.Baranov et al, Instrum.Exp.Tech. 55 (2012) 26-28 ,
Bazylev S.N. et al., JINR Preprint P13-2010-60,
Davkov K.I. et al., JINR Preprint P13-2012-93
NA-62 Collaboration, Technical Design Document, NA62-10-07, December 2010
[5] N. Anfimof et al., COMPASS Note 2011-2; N. Anfimov, talk at the International Workshop "ADVANCED STUDIES INSTITUTE SYMMETRIES AND SPIN", Prague, July 2013.
[6] RPC: R. Santonico and R. Cardarelli, Development of resistive plate counters, NIM A187(1981)377; R. Santonico and R. Cardarelli, Progress in resistive plate counters, NIM, A263(1988)20; ATLAS muon spectrometer technical design report. CERN/LHCC 9722; ATLAS TDR 10, CERN, 1997; CMS muon technical design report. CMS TDR 3, CERN/LHCC 9732, 1997.
[7] FAIR/PANDA Collaboration, Technical Design Report - Muon System, September 2012.

IOP Conf. Series: Journal of Physics: Conf. Series **938** (2017) 012027 doi:10.1088/1742-6596/938/1/012027

Fast polarimetry of the colliding proton beams based on the elastic *pp* analyzer using the *NICA* detectors

V I Sharov
Veksler and Baldin Laboratory of High Energy Physics, Joint Institute for Nuclear Research, Dubna, Russia
E-mail: vsharov41@mail.ru

Abstract. It is shown that the existing data on analyzing power A_n of the elastic *pp* scattering could be successfully applied for polarimetry of the colliding proton beams using the *NICA* detectors. Performed calculations of the count rates of the elastic events have revealed that the polarimeter based on using A_n for elastic *pp* will have a high polarization measurement velocity.

1. Introduction.

The study of spin effects on modern colliders [1] is a complex methodical work. For the full reliability of the experimental results it is required to know the polarization values of colliding beams in their intersection points. The report analyzes opportunities of using the existing A_n set [2] of elastic *pp* data for the polarimetry of the *NICA* collider [1] polarized proton beams.

An example of using the existing $A_n(pp)$ and $d\sigma/d\Omega_{CM}(pp)$ data sets [2] for polarimetry of the proton beams at energies of the *NICA* collider is presented. It is shown that polarimetry with the elastic *pp* process as polarization analyzer is possible only near the lower boundary of the *NICA* energy range at $T_{p,CM} \approx 2\ GeV$. The event count rates have been estimated at the statistics accumulation during the beam polarization p_Z measurements for a number of energy values of the colliding proton beams. It is shown that this polarimeter will provide fast polarization measurement approximately $\simeq 20$ minutes.

2. The coordinate systems used for colliding beam polarimetry.

For the polarimetry, we have applied a convenient and often used projectile helicity frame (PHF) - $P(x_P, y_P, z_P)$, see Figure 1, in which we consider the polarizations p_i and analyzing powers A_i, A_{ij}. The coordinate unit vectors of this frame are defined as:

$$\mathbf{z}_P = \mathbf{k}_{in}/|\mathbf{k}_{in}|, \qquad \mathbf{y}_P = \mathbf{n} = \mathbf{k}_{in} \times \mathbf{k}_{out}/|\mathbf{k}_{in} \times \mathbf{k}_{out}|, \qquad \mathbf{x}_P = \mathbf{n} \times \mathbf{z}_P = \mathbf{y}_P \times \mathbf{z}_P. \qquad (2.1)$$

Under this choice of the coordinate system polarization observables A_i, A_{ij} satisfy the following relations for the parity conserving interaction:

$$A_x = A_z = A_{xy} = A_{yz} = 0. \qquad (2.2)$$

For the polarized beams produced by an ion source, a second coordinate system of interest for the projectile is that with its **Z** axis along the axis of symmetry **S** of the beam polarization.

IOP Conf. Series: Journal of Physics: Conf. Series **938** (2017) 012027 doi:10.1088/1742-6596/938/1/012027

In the PHF, the spin quantization axis, **S**, of the colliding polarized beam is described by angles β, ϕ, where β is the angle between **S** and $\mathbf{k}_{in}$ and ϕ - the angle between the projection of **S** on x_P, y_P plane and the $\mathbf{y}_P$ -axis:

$$cos\ \beta = \mathbf{S} \cdot \mathbf{k}_{in}, \qquad cos\ \phi = (\mathbf{S} \times \mathbf{k}_{in}) \cdot (\mathbf{y}_P \times \mathbf{k}_{in}). \tag{2.3}$$

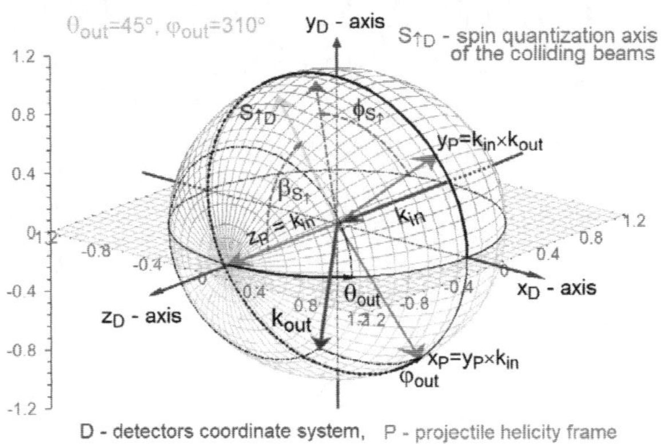

Figure 1. The coordinate systems for the secondary particles detected by the collider detectors located around the intersection point of the colliding beams for its polarimetry: $D_c(x_D, y_D, z_D)$ - right-hand Cartesian coordinate system, $D_s(r, \theta, \varphi)$ - spherical coordinates are preferable since the angular distribution of the scattered particles is axially symmetrical and $P(x_P, y_P, z_P)$ - projectile helicity frame (PHF).

The case of the $\mathbf{S}\|\mathbf{y}_D$ is actually used, when the vector $\mathbf{S} \perp \mathbf{k}_{in}$, and the angle $\beta = 90°$, and $\phi = \varphi$. But the angle ϕ is measured from the axis $\mathbf{y}_P$, and not from the $\mathbf{x}_P$ axis to save a long-standing convention $\phi = 0$ for scattering to the left of the beam with spin up.

3. Spin formalism for elastic *pp* scattering.

In the PHF, taking into account the relations (2.2), the general expression for the cross sections of elastic collision of two polarized protons is the following: (see ref. [3])

$$I = I_o[1 + p_y^B A_y^B + p_y^R A_y^R + p_x^B p_x^R C_{xx} + p_z^B p_x^R C_{zx} + p_y^B p_y^R C_{yy} + p_z^B p_z^R C_{zz} + p_x^B p_z^R C_{xz}], \tag{3.1}$$

where the indices B and R refer to the parameters of blue and red colliding beams, respectively. This expression includes the polarization $p_y^{B,R}$ and the analyzing power $A_y^{B,R}$ of each proton, as well as the spin correlation coefficients C_{ij}.

To separate polarimetry of the B or R colliding beams, in the expression (3.1) we need to isolate a member $p_y^B A_y^B$ or $p_y^R A_y^R$ by means of all usual methods to resolve differences in the efficiencies of the detectors. The individual term in equation (3.1) can be distinguished by selecting combinations of the polarized - unpolarized beam, flip the orientation of the spins of the colliding particles, by special arrangement of detectors, etc. For example, to find the p^B polarization of the blue beam, it is necessary to collide it with the unpolarized red beam $p^R = 0$ and vice versa. Then the expression (3.1) is reduced to the following:

$$I = I_o[1 + p_y^B A_y^B] \tag{3.2}$$

and allows one to determine p_y^B at known A_y^B and vice versa. In the PHF polarization p_y^B of the colliding beam is associated with p_Z by means of the angles β and ϕ as:

$$p_y = p_Z\ sin\ \beta\ cos\ \phi = p_Z\ cos\ \phi, \tag{3.3}$$

where p_Z is the beam polarization in the ion source system.

IOP Conf. Series: Journal of Physics: Conf. Series **938** (2017) 012027 doi:10.1088/1742-6596/938/1/012027

4. How to measure the colliding proton beams polarization.

To measure the value and sign of the beam polarization p_Z, it is necessary to use two detectors of the scattered proton - left (Lf) and right (Rg), placing them in diametrically opposite areas near $\phi = 0$, where $p_{yP}^{Lf} = p_Z\,|cos\,\phi|$ and near $\phi = 180°$ where $p_{yP}^{Rg} = -p_Z\,|cos\,\phi|$. The current value of the proton beam polarization p_Z from the polarized ion source is given by the following expression:

$$p_Z = [N_{Lf}^{pol}/N_{Lf}^{o} - N_{Rg}^{pol}/N_{Rg}^{o}]/(2\,|cos\,\phi|\,A_n(T_p,\theta)), \tag{4.1}$$

where N_{Lf}^{pol} and N_{Lf}^{o} are the output of the elastic pp process with scattering to the left and N_{Rg}^{pol} and N_{Rg}^{o} are the output with scattering to the right polarized (*pol*) and unpolarized (*o*) beams, respectively. In these measurements it is necessary to use an unpolarized "target" beam.

In this formula, the current value of the angle ϕ in PHF is calculated by using equation (2.3) and known coordinates of the vectors $\mathbf{k}_{in}, \mathbf{y}_P$ and $\mathbf{S}$ in the system of detectors $D_c(x,y,z)$ or $D_s(r,\theta,\varphi)$. The values of the analysing power $A_n(T_p,\theta)$ of the elastic *pp* scattering at angle θ with the energy of the colliding protons T_p are taken from the $A_n(T_p,\theta)$ data set in the comp. [2].

5. Examples of existing $A_n(T_p,\theta)$ data in the CM ($D_{c,s}$) system.

Figure 2 shows the experimental data of $A_n(T_p,\theta)$ depending on the angle $\theta_{p,CM}$.

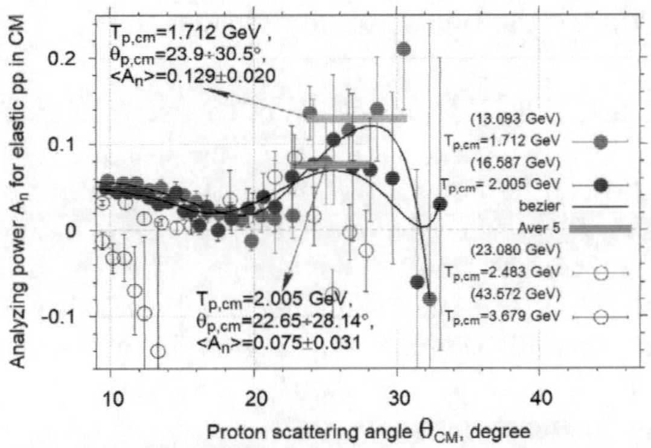

Figure 2. Experimental A_n data [2] at the energies of colliding protons $T_{p,CM}$ = 1.7 GeV and 2,0 GeV($T_{p,Lab}$ = 13.1 GeV and 16.6 GeV) depending on the angle $\theta_{p,CM}$ of the elastically scattered proton. The legend indicates the values of the energies $T_{p,Lab}$ (the upper number) and $T_{p,CM}$ (the bottom number). The solid curves are the Bezier approximation. The average $\overline{A_n}$ over shown intervals are indicated by arrows.

It is seen that at energies of $T_{p,CM} > 2.0\ GeV$, the A_n angular dependences are small and do not have the explicit extremum, which can be used for the polarimetry. We have also seen

Table 1. The parameters for several sets of experimental data [2] on angular dependencies $A_n(\theta_{CM})$ for several values of the colliding proton energies $T_{p,CM}$.

$T_{p,Col}$, GeV	$T_{p,FT}$, GeV	$\theta_{p,CM}^{o}$ min	$\theta_{p,CM}^{o}$ max	$-t$, $(GeV/c)^2$ min	$-t$, $(GeV/c)^2$ max	$\overline{A_n}$	$d\overline{A_n}$
1.106	7.027	37.89	43.87	1.39	1.84	0.154	0.008
1.505	10.849	29.05	36.49	1.28	2.00	0.146	0.006
1.712	13.093	23.90	30.50	1.05	1.70	0.129	0.020
2.005	16.587	22.65	28.14	1.20	1.84	0.075	0.031
2.483	23.080	18.34	24.18	1.10	1.90	0.042	0.010

that in the range $20° \leq \theta \leq 40°$ of the working angle of the collider detectors the average A_n has a significant value $\overline{A_n} \approx 0.08$ and acceptable precision $d\overline{A_n} \approx 0.03$ only for the energies of the colliding protons $T_{p.CM} \approx 2\ GeV$. Therefore polarimetry with the elastic *pp* process is possible only near the lower boundary of the $NICA$ energy range.

IOP Conf. Series: Journal of Physics: Conf. Series **938** (2017) 012027 doi:10.1088/1742-6596/938/1/012027

6. Estimations of the elastic pp output in the CM ($D_{c,s}$) system.

According to the equations (4.1) the statistical accuracy of the polarization p_Z measurements is determined by the value of accumulated statistics with polarized and unpolarized beams for scattering to the left and right: N^o_{Lf}, N^{pol}_{Lf}, N^o_{Rg}, N^{pol}_{Rg}. To estimate the count rate ($R = L \cdot \sigma_{int}$) of the elastic proton scattering in a given angular intervals $\Delta\theta$, $\Delta\varphi$ at a given energy $T_{p,Col}$, one needs to determine the aperture cross section:

$$\Delta\sigma(T_p, \Delta\theta, \Delta\varphi) = \int_{\Delta\varphi} \int_{\Delta\theta} d\sigma/d\Omega(T_p, \theta) sin(\theta) d\theta d\varphi. \quad (6.1)$$

The angular $\theta_{p,CM}$ dependence of the $d\sigma/d\Omega_{CM}$ at given$T_{p,CM}$ energy can be determined from the existing data sets (see [2]) for the $\theta_{p,CM}$ dependencies of differential cross sections $d\sigma/d\Omega_{CM}(T_{p,CM}, \theta_{p,CM})$ for elastic pp scattering. The experimental $d\sigma/d\Omega_{CM}$ data for a number of $T_{p,CM}$ energies have been approximated by the exponential type dependencies $d\sigma/d\Omega_{CM} = C \cdot exp(-b \cdot \theta)$ with the parameter $C[cm^2]$ and a slope of b. With this approximation the aperture cross section $\Delta\sigma$ is defined as follows:

$$\Delta\sigma = \Delta\varphi \cdot C \int_{\theta_L}^{\theta_H} \exp(-b \cdot \theta) \sin\theta d\theta, \quad (6.2)$$

where $\Delta\varphi$ - is detector aperture in φ, and θ_L and θ_H are the border of integration.

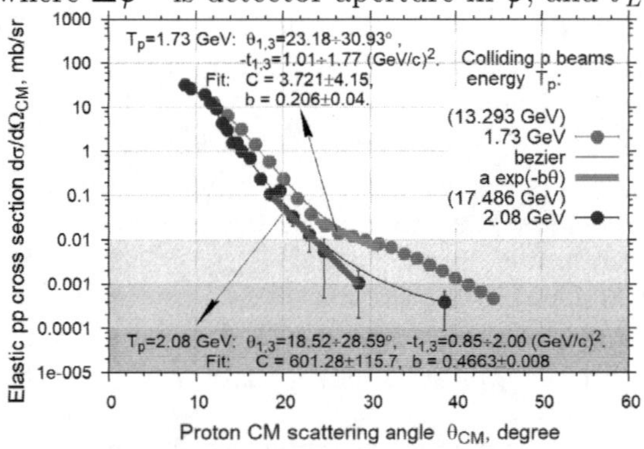

Figure 3. Fitting results of the experimental $d\sigma/d\Omega_{CM}$ data [2]. For the fit, we have chosed the $d\sigma/d\Omega_{CM}$ sets with the values of the $T_{p,CM}$ energies close to the $T_{p,CM}$ values for which A_n data are available. The region of the $\theta_{p,CM}$ angles for fitting has been selected to include the $\Delta\theta_{p,CM}$ region in which we are going to use the $A_n(\theta_{p,CM})$ values for the polarimetry purpose.

Figure 3 shows the $d\sigma/d\Omega_{CM}(T_{p,CM}, \theta_{CM})$ fitting results for $T_{p.CM} = 1.730$ and $2.080 GeV$.

Table 2. The results of $d\sigma/d\Omega_{CM}(T_{p,CM}, \theta_{p,CM})$ data fitting and calculation results of the aperture cross sections $\Delta\sigma$ and event count rate R.

$T_{p,Col}$, GeV	$\theta^o_{p,CM}$		C, mb/sr	b	$\Delta\sigma$, mb	R, s^{-1}	$\overline{A_n}$	$\Delta t, min$	
	θ_L	θ_H						10%	5%
0.999	37.9	43.9	127.3	0.195	0.280	280	0.154	0.4	1.6
1.128	37.9	43.9	583.2	0.254	0.119	119	0.154	1.0	3.7
1.528	29.1	36.5	0.47	0.113	0.074	74	0.146	1.8	6.5
1.730	23.9	30.5	3.72	0.206	0.068	68	0.129	2.5	9.0
2.080	22.7	28.1	601.3	0.466	0.020	20	0.075	24.0	92.0
2.483	18.3	24.2	6.047	0.295	0.041	41	0.042	36.0	144.0

Table 2 presents the results of $d\sigma/d\Omega_{CM}(T_{p,CM}, \theta_{p,CM})$ data fitting - the parameters C and b, the values of aperture cross sections $\Delta\sigma$, event count rate R and necessary time Δt for statistics accumulation to get the measured p_Z result with the statistical precision of $\delta = 10\%$ and $\delta = 5\%$. In the $\Delta\sigma$ and R calculations it is supposed that the azimuthal $\Delta\varphi$ - capture of the polarimeter detectors in the right and left shoulders is equal to 90°. It has been also assumed that the luminosity L of the colliding proton beams is equal to $L = 1 \times 10^{30}\ cm^{-2}sec^{-1}$.

The results of the R calculations, presented in table 2, show a high count rate of elastic *pp* events -tens events per second at the energies of the colliding protons $T_{p,CM} \approx 2\ GeV$, that provides a short period of time of statistics accumulation $\sim 20\ minutes$ for obtaining p_Z results with statistical precision of 10%.

The opportunities of registering and selecting of elastic *pp* events by the collider detectors was considered by us earlier [4]. In this case, we have used the information about the expected parameters of the *NICA* colliding *pp* beams and detectors surrounding the point of their intersection from [5, 6].

7. Conclusion

Existing data sets [2] on the analysing power A_n of elastic *pp* scattering have been analyzed. It has been shown that in the working range of the proton scattering angle $20° \leq \theta \leq 40°$ the average $\overline{A_n}$ values are significant $\overline{A_n} \approx 0.08$ and have acceptable accuracy of $dA \sim 0.03$ only for the energies of the colliding protons: $T_{p,CM} \approx 2\ GeV$. Therefore polarimetry with the elastic *pp* process is possible only near the lower boundary of the energy range of the *NICA* collider.

The calculations of the count rates of the elastic events have been performed and it is shown that the polarimeter using A_n for elastic *pp* as polarization analyzer will have a high measurement velocity. A single polarization measurement will require $\sim 20\ minutes$.

Acknowledgment
The author is grateful to his colleague E.V. Chernykh for the helpful discussions. The author would also like to thank I.I. Strakovsky from the George Washington University, who was always ready to help us in providing some of the data for elastic *NN* scattering.

References

[1] *NICA - Nuclotron - based Ion Collider fAcility - a Base for Fundavental Research and Innovation Development*, JINR, Dubna 2014, http://nica.jinr.ru/files/Booklet-NICA.pdf
[2] Bystricky J and Lehar F 1978 *Nucleon-Nucleon Scattering Data* Nr. **11-1**, Part I-II Ed. by H. Behrens and G. Ebel Karlsruhe Fachinformationszentrum; 1981 *Nucleon-Nucleon Scattering Data. Summary Tables* Nr. **11-2,3**
[3] Ohlsen G G 1972 Polarization transfer and spin correlation experiments in nuclear physics *Rep. Prog. Phys.* **35** 717-801.
[4] Sharov V I 2017 *Selection of the elastic scattering events in interactions of the NICA colliding proton (deuteron) beams* In Proceedings of Baldin ISHEPP XXIII Dubna September 19-24 2016 *EPJ Web of Conferences* **138** 11002
[5] Savin I A, Efremov A V et al. 2014 *Spin Physics Experiments at NICA SPD with polarized proton and deuteron beams. Letter of Intent* Dubna
[6] Kovalenko A D 2013 *Polarized deuterons and protons at NICA: The disign concept development* Proc. of "SPIN-13" Dubna JINR **E1,2-2014-12** 345-350

IOP Conf. Series: Journal of Physics: Conf. Series **938** (2017) 012028 doi:10.1088/1742-6596/938/1/012028

Spin shifts scale of gravitational interaction, opening new path to particle physics

Alexander Burinskii

Theor. Phys. Lab., NSI, Russian Academy of Sciences, B. Tulskaya 52 Moscow 115191, Russia

E-mail: bur@ibrae.ac.ru

Abstract. We argue that spin deforms space along with mass, and the great spin/mass ratio of the elementary particles shifts gravitational interaction from Planck to Compton scale, opening new way to unification of gravity with quantum theory, based on the supersymmetric bag model.

1. Introduction

During last century the weakness of gravitational interaction based on the Planck scale of gravitational interaction was considered as indisputable truth. It was also supported by Schwarzschild solution, which distorts space in zone $r_g \sim 2m$, commensurable with other interactions only at Planck scale. As is well known, until all of the projects were based on this concept and failed, and *"... a realistic model of elementary particles still appears to be a distant dream ...* ", J. Schwarz [arXiv:1201.0981]. About half century ago, Kerr obtained a new solution for a spinning black hole (BH) [1], and Carter noticed [2], that the Kerr-Newman (KN) solution (charged version of the Kerr solution) has gyromagnetic ratio $g = 2$, i.e. the same as Dirac electron [2, 3]. In addition, it was noticed in [4, 5, 6], that the KN solution with parameters of an electron had influence on space not on the expected Planck scale, but changed really topology of space on the Compton distances, which are on 22 orders higher than Planck scale. In particular, the horizons disappeared and there appeared a naked singular ring deforming space topologically. The zone of space distortion becomes depending on spin/mass ratio $a = J/m$, showing that spin deforms space along with mass, which should be taken into account by estimations of gravitational influence.

The action of spin on space is connected with gravitational effect of "frame-dragging" [7], which is a local deformation of the light cone system. Contrary to centrally symmetric "frame-dragging" in spherical black hole solutions, in the Kerr and KN solutions frame-dragging forms a vortex, which was called as Lense-Thirring effect. One can say that *spin is gravitating,* but its action on the metric is different. Contrary to Schwarzschild solution, which distorts space in zone $r_g \sim 2m$, the Kerr solution with angular momentum J and mass m distorts space in zone $r_K \sim J/m$. This becomes especially essential for particle physics, because J/m ratio of particles is giant, about $\approx 10^{20} - 10^{22}$. [1] and just underestimations of the role of spin led to failure of superstring project and others modern theoretical constructions.

[1] We use the 'natural' (Planck) units $\hbar = c = G = 1$, in which Planck's mass, length and energy are $M_P = l_P = E_P = 1$. In this system the energy equivalent to quantum spin $\hbar = 1$ is $E_P = 1$, while the typical masses of particles are about $\sim 10^{-20}$.

IOP Conf. Series: Journal of Physics: Conf. Series **938** (2017) 012028 doi:10.1088/1742-6596/938/1/012028

Account of spin shifts scale of gravitational interaction from Planck to Compton scale, just to zone of action of quantum theory, leading to *new concept*: gravity and quantum theory become equally strong and should be considered on equal footings. Rank of gravity is raised to theory which lies in the hearth of particle physics along with quantum theory.

We present here a model which is based on this new concept. This is a nonperturbative bag-like solution base on supersymmetric Higgs (or Landau-Ginzburg) field model, [8, 9, 10].

2. Supersymmetric bag model compatible with spinning Kerr-Newman solution

Metric of the KN solution in the Kerr-Schild form is [3]

$$g_{\mu\nu} = \eta_{\mu\nu} + 2Hk_\mu k_\nu, \tag{1}$$

where $\eta_{\mu\nu}$ is metric of auxiliary Minkowski space M^4, (signature $(-+++)$), where scalar function

$$H = \frac{mr - e^2/2}{r^2 + a^2\cos^2\theta}, \tag{2}$$

is given in oblate spheroidal coordinates r and θ, determined by transformations [3],

$$x + iy = (r + ia)\exp\{i\phi_K\}\sin\theta, \quad z = r\cos\theta, \quad \rho = r - t. \tag{3}$$

The null field $k_\mu(x)$, $(k_\mu k^\mu = 0)$ determines the lightlike directions of 'frame-dragging', which takes vortex form Fig.1.,Fig.2. For ultra-high spin of spinning particles, $a = J/m >> m$, 'frame-dragging' becomes so strong that black hole horizons disappear, opening a naked singular ring at $r = 0, \cos\theta = 0$. The congruence k^-_μ, which goes to the disk $r = 0$ as 'in-going', intersects disk and turns into 'out-going' k^+_μ, which is extended analytically on the same background, forming a second sheet of the metric. The space becomes two-sheeted, with two different metrics on the same background, $g^\pm_{\mu\nu} = \eta_{\mu\nu} + 2Hk^\pm_\mu k^\pm_\nu$. The Kerr singular ring $r = 0$, $\cos\theta = 0$ of the Compton radius $a = \hbar/2m$ breaks space topologically, playing the role of a branch line.

Remarkable features of Kerr space led to study of KN solution as a classical model of electron, mainly in two directions:

A) W. Israel [4] first suggested to truncate negative sheet of Kerr space along disk $r = 0$ and put at the disk a distribution of matter in accord with right hand side of the Einstein equations. He got a very strange matter distribution without a clear physical interpretation, however, he found an analogue with a quantum process of mass renormalization and first declared that this disk-like source of Compton radius can be considered as a classical model of electron. Very important consequent calculation by Hamity [11] showed that the matter distribution of the source took diagonal form in the corotating coordinate system, and thus, the source of Kerr solution turned into a relativistically rotating disk, boundary of which is moving with speed of light. Later on López generalized this model to model of a rotating bubble [6].

B) Another stringy version of the Kerr source was developed in our works with Prof. D. Ivanenko [5, 12], where Kerr singular ring was compared with Nielsen-Olesen (NO) vortex string model in superconductor [13]. It was shown that Kerr's singular ring can be considered as a lightlike string which can bear the electromagnetic and spinor traveling waves. Much later, such pp-wave strings were considered as solitonic string solutions to low energy string theory [14, 15].

Bag model merges directions A) and B). In the both these directions of investigation source of the Kerr geometry created a gravitational vortex of the Compton size, and there appeared the question: Why this vortex and singular ring were not observed experimentally, and electron looks like a point? Answer follows from mechanism of Lorentz contraction [16]. *For external observer Kerr's vortex shrinks to point by Lorentz contraction.*

The suggested in [8, 9, 10] supersymmetric bag model solves problem of unification with gravity by separation of their zones of influence. Quantum theory requires flat space, at least in

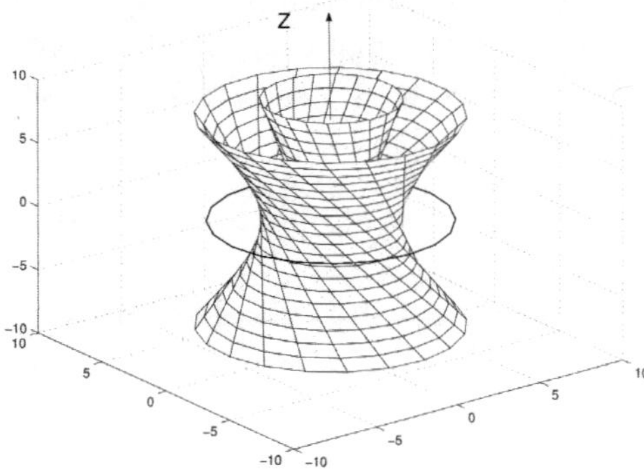

Figure 1. The lightlike Kerr Null congruence k^μ determines space-dragging caused by mass and spin.

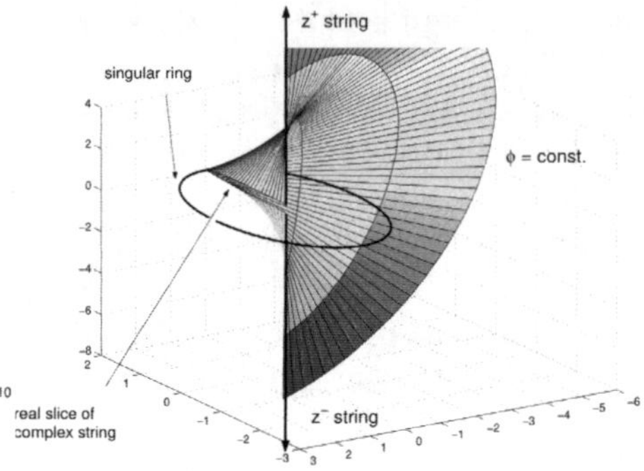

Figure 2. Rotational dragging of vector-potential A^μ along border of bag creates a quantum Wilson loop.

the Compton zone, while spinning Kerr's gravity spoils space topologically. Supersymmetry expels gravitational field from Compton zone of spinning particle, similar to expulsion of electromagnetic field from superconductor. Supersymmetric bag model realizes such expulsion of gravity and electromagnetic field, forming tree zones: flat quantum interior **(I)**, external zone with exact KN solution **(E)**, and a thin zone of transition **(R)**, see Fig.3.

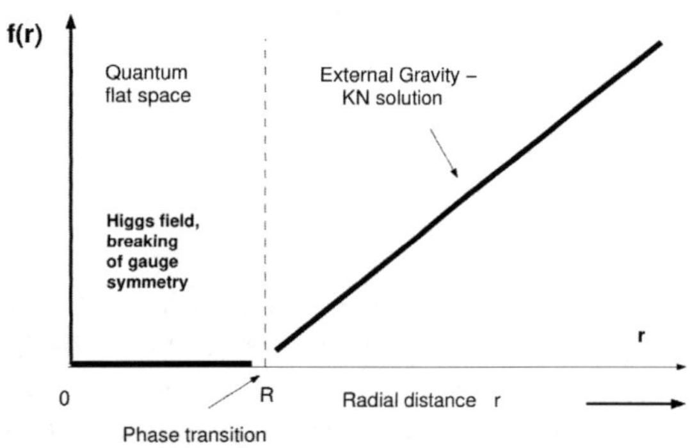

Figure 3. Separation of the flat Quantum interior **(I)** from zone of KN Gravity **(E)** by domain wall phase transition **(R)**.

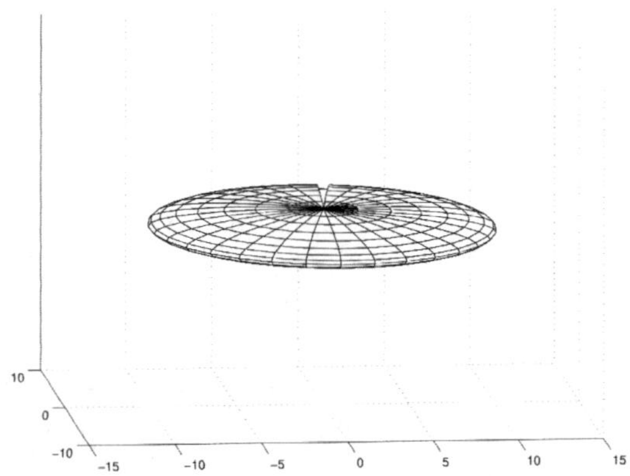

Figure 4. Shape of the KN bag.

Surface (R) is defined by the continuous transition of KN solution to Minkowski interior of the bag, (C. López [6]). According (1) and (2), zone **(R)** corresponds to $H_{KN}(r) = 0$, which gives $r = R = e^2/2m$, and (3) shows that domain wall boundary has an ellipsoidal form, the disk with radius $r_c = \sqrt{R^2 + a^2}$ and thickness R, see Fig.4.

To satisfy **(I)**,**(E)**,**(R)**, it is natural to use Higgs mechanism of symmetry breaking which is used in many nonperturbative electroweak models and in the MIT and SLAC bag models [17, 18]. The corresponding Lagrangian is also known as Landau-Ginzburg (LG) field model for superconducting phase transitions and used in famous Nielsen-Olesen (NO) model of vortex string in superconductor, [13], $\mathcal{L}_{NO} = -\frac{1}{4}F_{\mu\nu}F^{\mu\nu} - \frac{1}{2}(\mathcal{D}_\mu\Phi)(\mathcal{D}^\mu\Phi)^* - V(|\Phi|)$, where $\mathcal{D}_\mu = \nabla_\mu + ieA_\mu$ are $U(1)$ covariant derivatives, $F_{\mu\nu} = A_{\mu,\nu} - A_{\nu,\mu}$ the corresponding field strength. Usual potential $V = \lambda(\Phi^\dagger\Phi - \eta^2)^2$, is inappropriate, since it breaks gauge symmetry in external zone **(E)**. To satisfy conditions **(I)**,**(E)**,**(R)** one should use supersymmetric LG model with three Higgs-like fields, [19] $(H, Z, \Sigma) \equiv (\Phi_1, \Phi_2, \Phi_3)$. Corresponding Lagrangian differs from $\mathcal{L}_{NO}$ only by summation over fields $\Phi_i, i = 1,2,3$, while the potential V is to be formed from a superpotential $W(\Phi_i), as\ V(r) = \sum_i F_i F_i^*, \quad F_i = \partial W/\partial\Phi_i \equiv \partial_i W$, where $W(\Phi_i, \bar{\Phi}_i) = Z(\Sigma\bar{\Sigma} - \eta^2) + (Z + \mu)H\bar{H}$, The conditions $F_i = \partial_i W = 0$ determine two vacuum states with $V = 0$: internal-**(I)**, $r < R - \delta$, condensation of Higgs field, $|H| = \eta$, and external-**(E)**, where $r > R + \delta$ and $H = 0$, separated by domain wall with $V > 0$, – zone **(R)**, in correspondence with **(I)**,**(E)**,**(R)**.

Supersymmetric potential concentrates Higgs field in zone **(I)**, creating superconducting vacuum state inside bag, where we can use the Lagrangian $\mathcal{L}_{NO}$ leading to equations

$$\partial_\nu \partial^\nu A_\mu = J_\mu = e|H|^2(\chi,_\mu + eA_\mu), \qquad (4)$$

where χ is oscillating phase of Higgs field $H = |H|e^{i\chi(t,\phi)}$. Vector potential of the KN solution concentrates in equatorial plane near the Kerr singular ring and forms a circular string which is similar to Nielsen-Olesen vortex string in superconductor, but placed on the boundary of superconducting disk at $r = R = e^2/2m$ where it takes maximal value $A^*_\mu dx^\mu = -Re\ \frac{2m}{e}(dt - ad\phi)$ and forms a longitudinal lightlike field dragged in t, ϕ direction by the Kerr congruence near the ring, see Fig. 4. It forms a stringy Wilson loop which is compensated by phase of Higgs field, $\chi,_\mu + eA_\mu = 0 \Rightarrow J_\mu = 0$ inside the disk, leading to $\chi = 2m(t + a\phi)$. Thus, current J_μ is expelled to surface layer of the superconducting disk. Periodicity of the phase $\chi(\phi)$ leads *quantization of angular momentum,* [8, 9, 10], $J = n/2, \quad n = 1,2,3,...$

By nature bag models are assumed to be soft and flexible [22, 20, 21]. Under excitations bags are deformed and take the shape of a stringy flux-tube joining the quark-antiquark pair [17, 20], or toroidal string [18, 21, 22]. In the KN gravitational field, bag takes ellipsoidal shape, see Fig.4, and acquires circular string at the sharp boundary, closely to Kerr singular ring [9]. This string bears lightlike traveling wave and creates circulating singular pole, reproducing zitterbewegung of a naked electron. On the other hand, this singular pole can be treated as a quark-antiquark pair, playing the role of end-points of the closed circular string, by analogy with the string models and with hadron bag models [9, 23].

3. References

[1] Kerr R P 1963 *Phys. Rev. Lett.* **11** 237
[2] Carter B 1968 *Phys. Rev.* **174** 1559
[3] Debney G C, Kerr R P and Schild A 1969 *J. Math. Phys.* **10** 1842
[4] Israel W 1970 *Phys. Rev.* D **2** 64
[5] Burinskii A Ya 1974 *Sov. Phys. JETP* **39** 193
[6] López C A 1984 *Phys. Rev.* D **30** 313
[7] Misner Ch W, Thorne K S and Wheeler J A 1973 *Gravitation* v.3, San Francisco: W. Freeman
[8] Burinskii A 2015 *JETP (Zh. Eksp. Teor. Fiz.)* **148**(8) 228, [arXiv:1505.03439]
[9] Burinskii A 2015 *JETP (Zh. Eksp. Teor. Fiz.)* **148** (11) 937 [arXiv:1706.02979]
[10] Burinskii A 2016 *Phys. Lett.* B **754** 99
[11] Hamity V 1976 *Phys. Lett.* **A 56** 77
[12] Ivanenko D D and Burinskii A Ya 1975 *Izv. Vuz. Fiz.* **5** 135
[13] Nielsen H B and Olesen P 1973 *Nucl. Phys.* B **61** 45
[14] Burinskii A 1995 *Phys. Rev.* **D 52** 5826 [arXiv:hep-th/9504139]
[15] Dabholkar A et. al. 1996 *Nucl. Phys.* **B 474** 85
[16] Punsly B 1985 *J. Math. Phys.* **26** 1728
[17] Chodos A. et al. 1974 *Phys. Rev.* D **9** 3471
[18] Bardeen W. A. at al. 1974 *Phys. Rev.* D **11** 1094
[19] Morris J R 1996 *Phys. Rev.* **D 53** 2078 [arXiv:hep-ph/9511293]
[20] Johnson K and Thorn C B 1976 *Phys. Rev.* **D 13** 1934
[21] Tye S-H H 1976 *Phys. Rev.* **D 13**, 3416
[22] Giles R C and Tye S-H H 1976 *Phys. Rev.* **D 13** 1690
[23] Burinskii A 2017 [arXiv:1701.01025]

IOP Conf. Series: Journal of Physics: Conf. Series **938** (2017) 012029 doi:10.1088/1742-6596/938/1/012029

About long-range interaction of spheroidal solitons in scalar field nonlinear model

Alexander A. Chernitskii

[1] Department of Mathematics
St. Petersburg State Chemical Pharmaceutical Academy
Prof. Popov str. 14, St. Petersburg, 197022, Russia
[2] A. Friedmann Laboratory for Theoretical Physics
St. Petersburg, Russia

E-mail: AAChernitskii@mail.ru

Abstract. The nonlinear scalar field model of space-time film (Born – Infeld type nonlinear scalar field model) is considered. Its spherically symmetrical solution is obtained. This solution gives the class of moving solitary solutions or solitons with the Lorentz transformation. We consider the distant interaction between such spheroidal solitons or spherons. This interaction is caused by the nonlinearity of the model. Starting from the static configuration with two spherons we show that the interaction under investigation is similar to electromagnetic one.

1. Introduction

We consider the nonlinear scalar field model that can be called the extremal space-time film one [1, 2]. The investigation of this field model is interesting from the various point of view.

The variational formulation of this field model is similar to the appropriate formulation for Born – Infeld nonlinear electrodynamics [3, 4]. What is more, the scalar nonlinear equation under consideration has the same spherically symmetrical solution that we have in Born – Infeld electrodynamics for zero component of the electromagnetic potential and which is sometimes called the Born's electron.

The Lorentz transformations give the appropriate moving solitary solutions or solitons. Such solution will be called spheroidal soliton or spheron.

As it is known the nonlinearity of the field model account for the interaction between the solitons. The appropriate methods for investigation for BI electrodynamics [5, 4] can be applied to the scalar field model under consideration.

Here we show that the interaction between the spherons in the scalar field model under consideration looks like the electromagnetic one.

2. Extremal space-time film and energy-momentum conservation law

The equation of the model under consideration has the following form in orthogonal coordinates

$$\frac{\partial \Upsilon^{\mu}}{\partial x^{\mu}} = 0\,, \qquad \Upsilon^{\mu} \doteqdot \frac{\Phi^{\mu}}{\pounds}\,, \quad \Phi_{\nu} \doteqdot \frac{\partial \Phi}{\partial x^{\nu}}\,, \quad \pounds = \sqrt{\left|1 + \chi^2\, \underline{\mathrm{m}}^{\mu\nu}\, \Phi_{\mu}\, \Phi_{\nu}\right|}\,, \tag{1}$$

IOP Conf. Series: Journal of Physics: Conf. Series **938** (2017) 012029 doi:10.1088/1742-6596/938/1/012029

where Φ is the scalar real field function, χ is the dimensional constant, $\mathfrak{m}_{\mu\nu}$ are the components of the Minkowski metric. The Greek indices take values $\{0,1,2,3\}$.

Customary method gives the following differential conservation law for energy-momentum tensor in the area outside of singularities:

$$\frac{\partial \mathbb{T}^{\mu\nu}}{\partial x^{\mu}} = 0\,, \qquad \mathbb{T}^{\mu\nu} = \frac{1}{4\pi}\left(\frac{\Phi^{\mu}\,\Phi^{\nu}}{\pounds} - \frac{\mathfrak{m}^{\mu\nu}}{\chi^2}\,(\pounds - 1)\right) . \tag{2}$$

Let us consider the integral from the energy-momentum differential conservation law (2) over the four-volume $\mathbb{V}$ including the three-volume $\not{\mathbb{V}}$ and the time interval $\left[\boldsymbol{x}^0 - \Delta x^0/2\,,\ \boldsymbol{x}^0 + \Delta x^0/2\right]$. Integration by parts in this integral gives the appropriate integral conservation law

$$\int\limits_{\mathbb{V}} \frac{\partial \mathbb{T}^{\mu\nu}}{\partial x^{\nu}}\,\mathrm{d}x^0\mathrm{d}x^1\mathrm{d}x^2\mathrm{d}x^3 = 0 \qquad \Longrightarrow \qquad \int\limits_{\Sigma} \mathbb{T}^{\mu\nu}\,\mathrm{d}\Sigma_{\nu} = 0\,, \tag{3}$$

where Σ is the three-dimensional boundary hypersurface for the four-volume $\mathbb{V}$, $\mathrm{d}\Sigma_{\nu}$ are components of the outer hypersurface element four-vector such that $\mathrm{d}\Sigma_0 = \pm \mathrm{d}x^1\mathrm{d}x^2\mathrm{d}x^3$, $\mathrm{d}\Sigma_1 = \pm \mathrm{d}x^2\mathrm{d}x^3\mathrm{d}x^0$, $\mathrm{d}\Sigma_2 = \pm \mathrm{d}x^3\mathrm{d}x^1\mathrm{d}x^0$, $\mathrm{d}\Sigma_3 = \pm \mathrm{d}x^1\mathrm{d}x^2\mathrm{d}x^0$.

Let us divide the three-dimensional closed hypersurface Σ into two unclosed hypersurfaces in some coordinate system such that $\Sigma = \overset{P}{\Sigma} \cup \overset{F}{\Sigma}$:

$$\overset{P}{\Sigma} = \not{\mathbb{V}}\Big|_{x^0=\boldsymbol{x}^0-\Delta x^0/2} \cup\ \not{\mathbb{V}}\Big|_{x^0=\boldsymbol{x}^0+\Delta x^0/2}\,, \qquad \overset{F}{\Sigma} = \sigma \cup \left[\boldsymbol{x}^0 - \Delta x^0/2\,,\ \boldsymbol{x}^0 + \Delta x^0/2\right] . \tag{4}$$

where σ is the closed two-surface bounding the three-volume $\not{\mathbb{V}}$, $\overset{P}{\Sigma}$ will be called the momentum hypersurface and $\overset{F}{\Sigma}$ will be called the force hypersurface. The three-dimensional sections for these hypersurfaces are shown on Fig. 1.

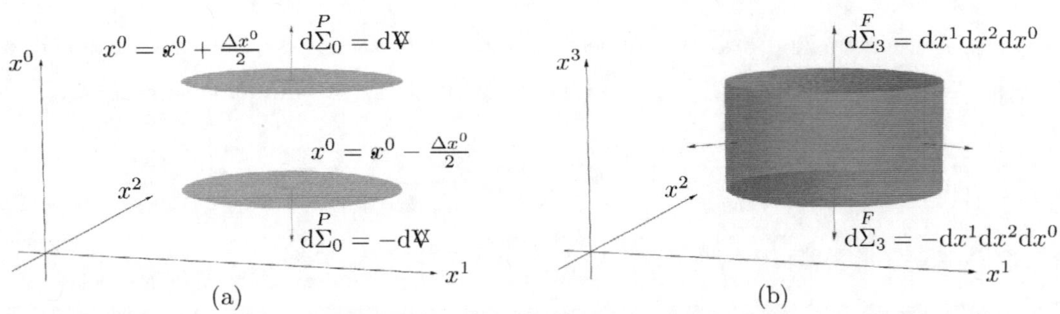

Figure 1. Section for $\overset{P}{\Sigma}$ in three-dimensional space $\{x^0, x^1, x^2\}$ (a) and section for $\overset{F}{\Sigma}$ in three-dimensional space $\{x^1, x^2, x^3\}$ (b).

Thus we have the obtained integral conservation law in the following integral force dynamic law form:

$$\Delta \mathbb{P}^{\mu}_{\not{\mathbb{V}}} = \bar{\Delta}\mathbb{F}^{\mu}_{\sigma}\,, \qquad \Delta \mathbb{P}^{\mu}_{\not{\mathbb{V}}} \doteqdot \mathbb{P}^{\mu}_{\not{\mathbb{V}}}\Bigg|^{x^0=\boldsymbol{x}^0+\frac{\Delta x^0}{2}}_{x^0=\boldsymbol{x}^0-\frac{\Delta x^0}{2}}\,, \qquad \bar{\Delta}\mathbb{F}^{\mu}_{\sigma} \doteqdot \int_{\boldsymbol{x}^0-\frac{\Delta x^0}{2}}^{\boldsymbol{x}^0+\frac{\Delta x^0}{2}} \mathbb{F}^{\mu}_{\sigma}\,\mathrm{d}x^0\,, \tag{5}$$

where $\mathbb{P}^{\mu}_{\mathbb{V}}$ is the momentum of field into the three-dimensional volume $\mathbb{V}$, $\bar{\Delta}\mathbb{F}^{\mu}_{\sigma}$ is the integral force on the two-dimensional surface σ. The momentum and the force are defined as follows:

$$\mathbb{P}^{\mu}_{\mathbb{V}} \doteqdot \int\limits_{\mathbb{V}} \mathcal{T}^{\mu 0}\, \mathrm{d}\mathbb{V} \ , \qquad \mathbb{F}^{\mu}_{\sigma} \doteqdot - \int\limits_{\sigma} \mathcal{T}^{\mu i}\, \mathrm{d}\sigma_i \ . \tag{6}$$

where $\mathrm{d}\mathbb{V} = \pm \mathrm{d}x^1 \mathrm{d}x^2 \mathrm{d}x^3$, $\{\mathrm{d}\sigma_i\} = \{\pm \mathrm{d}x^2 \mathrm{d}x^3, \pm \mathrm{d}x^1 \mathrm{d}x^3, \pm \mathrm{d}x^1 \mathrm{d}x^2\}$ are the outer normal components of the two-dimensional surface σ bounded the three-dimensional volume $\mathbb{V}$.

3. Spheroidal soliton

In the case when the field function depends on the radial coordinate r only we have the following simplest solution

$$\frac{\partial \Phi}{\partial r} = - \frac{q}{\sqrt{r^4 \pm \bar{r}^4}} \ , \quad \bar{r} \doteqdot \sqrt{|q\,\chi|} \quad \Longrightarrow \quad \Phi = \frac{q}{r}\, F_{\frac{1}{4};\frac{1}{2};\frac{5}{4}} \left(\mp \frac{\bar{r}^4}{r^4} \right) \ , \tag{7}$$

where $F_{\alpha;\beta;\gamma}(z)$ is hypergeometric function, top and bottom sings are correspond to different signatures of metric in the model action: $\{+,-,-,-\}$ (top) and $\{-,+,+,+\}$ (bottom).

The obtained solution (7) give birth to the class of the moving soliton solutions with the Lorentz transformations:

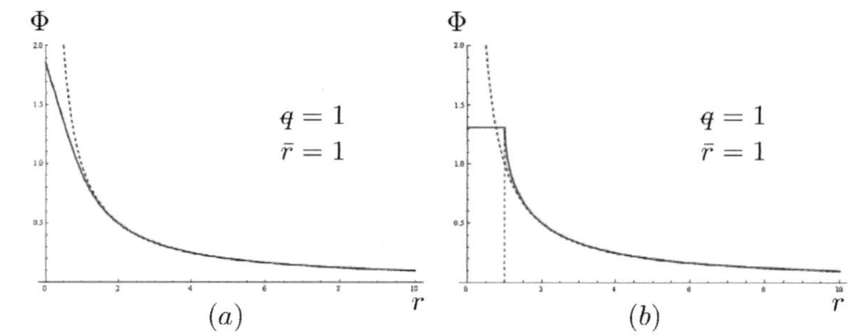

Figure 2. The field function $\Phi(r)$ for the solution with two metric signatures: $\{+,-,-,-\}$ (a) and $\{-,+,+,+\}$ (b). The dashed lines show the function $1/r$.

$$\underset{\cdot}{x}^{\mu} = L^{\mu}_{\cdot\nu}\, x^{\nu} \ , \quad x^{\mu} = L^{\cdot\mu}_{\nu}\, \underset{\cdot}{x}^{\nu} \ , \quad L^{\mu}_{\cdot\nu}\, L^{\cdot\nu}_{\rho} = \delta^{\mu}_{\rho} \ . \tag{8}$$

Here $\{\underset{\cdot}{x}^{\mu}\}$ is the intrinsic coordinate system of soliton, thus we make the substitution $r \to \underset{\cdot}{r} = \sqrt{\underset{\cdot}{x}^i\, \underset{\cdot}{x}^i}$ in (7). The components $L^{\mu}_{\cdot\nu}$ (the point designates that the index is second) are

$$L^{0}_{\cdot 0} = u^0 = \frac{1}{\sqrt{1 - \mathbf{v}^2}} \ , \quad L^{0}_{\cdot i} = L^{i}_{\cdot 0} = -u^i = -\frac{v^i}{\sqrt{1 - \mathbf{v}^2}} \ , \quad L^{i}_{\cdot j} = \delta^{i}_{j} + \frac{u^i\, u^j}{1 + u^0} \ ,$$
$$\mathbf{v}^2 \doteqdot \left(v^1\right)^2 + \left(v^2\right)^2 + \left(v^3\right)^2 \ , \quad \left| \mathfrak{m}_{\mu\nu}\, u^{\mu}\, u^{\nu} \right| = 1 \ , \tag{9}$$

where v^i are the components of three-dimensional velocity of the soliton, u^{μ} are the components of its four-dimensional velocity.

Energy-momentum four-vector components for spheroidal soliton are

$$\mathbb{P}^{\mu} = L^{\cdot\mu}_{0}\, \underset{\cdot}{\mathbb{E}} = u^{\mu}\, \mathsf{m} \ , \tag{10}$$

where m the energy of the soliton in its intrinsic coordinate system. Here it should be noted that the transformation (10) is true if only we consider the moving tree-volume $\mathbb{V}$ in definition (6) and transformation of its element as the zero component of four-vector $\mathrm{d}\overset{P}{\Sigma}_0$.

The spheroidal soliton under consideration will be called spheron for short.

4. Interacting spherons and electromagnetism

Let us consider the sum of the moving spherons which are distant from each other as the first approximation to an unknown multisoliton solution. But we assume that the velocities of these solitons can depend on time. To investigate the trajectories of the spherons in this first approximation we will use the integral conservation law for momentum (5). This method was used for the similar problem in nonlinear electrodynamics [5, 4].

The calculation gives that the interaction of two spheron at rest fully conforms with the electrostatic interaction of two charged point particles. Then we can transform this static configuration to a moving reference frame with the Lorentz transformations (8). The appropriate transformation for acceleration or force is known [6, 7]. Thus for this case we obtain also the magnetic component of the force in the moving reference frame.

Let us consider the internal tensor structure of the four-vector of integral force $\bar{\Delta}\mathbb{F}^{\mu}_{\sigma}$ (5).

Assuming that the time interval is vanishingly small ($\Delta x^0 \to 0$) and designating the symbols related to intrinsic coordinate system of the two spherons with point under letter we can write

$$\mathrm{d}\overset{n}{\underset{\cdot}{\mathbb{P}}}{}^{i} = \overset{n}{\underset{\cdot}{\mathbb{F}}}{}^{i}\,\mathrm{d}\underset{\cdot}{x}^{0}\,, \quad \mathrm{d}\overset{n}{\underset{\cdot}{\mathbb{P}}}{}^{0} = 0\,. \tag{11}$$

Here we must consider $\mathrm{d}\underset{\cdot}{x}^0$ as zero component for four-vector with component of trajectory differentials:

$$\mathrm{d}\underset{\cdot}{x}^{\mu} = \{\mathrm{d}\underset{\cdot}{x}^0, 0, 0, 0\} \quad \Longrightarrow \quad \mathrm{d}\mathbf{x}^{\mu} = \{\mathrm{d}x^0, \mathrm{d}\mathbf{x}^1, \mathrm{d}\mathbf{x}^2, \mathrm{d}\mathbf{x}^3\}\,. \tag{12}$$

Thus to obtain the right tensor law of motion we must consider the electrostatic force components $\overset{n}{\underset{\cdot}{\mathbb{F}}}{}^{i}$ as components of a second-rank tensor such that

$$\mathrm{d}\overset{n}{\mathbb{P}}{}^{\mu} = \overset{n}{q}\,\overset{n}{F}{}^{\mu}_{.\nu}\,\mathrm{d}\mathbf{x}^{\nu}\,, \quad \overset{n}{\underset{\cdot}{F}}{}^{0}_{.0} = 0\,, \quad \overset{n}{q}\,\overset{n}{\underset{\cdot}{F}}{}^{i}_{.0} = \overset{n}{\underset{\cdot}{\mathbb{F}}}{}^{i} \quad \Longrightarrow \quad \mathsf{m}\,\mathrm{d}\overset{n}{\mathbf{u}}{}^{\mu} = \overset{n}{q}\,\overset{n}{F}{}^{\mu}_{.\nu}\,\overset{n}{\mathbf{u}}{}^{\nu}\,\mathrm{d}\overset{n}{\mathbf{s}}\,, \tag{13}$$

where $\mathrm{d}\overset{n}{\mathbf{s}}{}^2 \doteqdot \mathrm{d}x^0\mathrm{d}x^0 - \mathrm{d}\overset{n}{\mathbf{x}}{}^i\mathrm{d}\overset{n}{\mathbf{x}}{}^i$ is invariant parameter of trajectory for the n-th spheron.

Thus we have obtained that the dynamical equation for interacting spheron looks like the equation of motion for point charged particle in electromagnetic field. But here we based on the static configuration of spherons where the electrostatic interaction was obtained. In a conventional manner for electrostatic interaction we can introduce the appropriate electrostatic potential satisfying the Laplace's equation with a charge density as the source. The appropriate relativistic generalization gives the four-vector electromagnetic potential components satisfying the wave equations with the appropriate four-vector current density as the source. It is obvious that the rotating configuration with the spherons has the angular momentum or spin the dynamics of which is investigated with the help of the appropriate integral conservation law.

5. Conclusions

Thus we have the similarity between the interacting spherons of the scalar field and the classical electrodynamics of the point charged particles. But because the starting configuration of spherons is statical we must keep in reserve the investigation of additional effects for the interacting spherons with large relative velocities.

References

[1] Chernitskii A A 2016 *Journal of Physics: Conference Serie* **678** 012016 (*Preprint* `http://stacks.iop.or`)
[2] Chernitskii A A 2015 (*Preprint* `1506.09137`)
[3] Chernitskii A A 2004 *Encyclopedia of Nonlinear Science* ed Scott A (New York and London: Routledge) pp 67–69 (*Preprint* `hep-th/0509087`)
[4] Chernitskii A A 2012 *Nonlinear electrodynamics: singular solitons and their interactions (in Russian)* (Saint-Petersburg: ENGECON)
[5] Chernitskii A A 1999 *J. High Energy Phys.* **1999** Paper 10, 1–34 (*Preprint* `hep-th/9911093`)
[6] Einstein A 1905 *Ann. Phys.* 891–921
[7] Pauli W 1958 *Theory of relativity* (Pergamon Press Ltd.)

IOP Conf. Series: Journal of Physics: Conf. Series **938** (2017) 012030 doi:10.1088/1742-6596/938/1/012030

Constituent quark masses in Poincaré-invariant quantum mechanics

Viktor Andreev
Francisk Skorina Gomel State University, Theoretical Physics Department, Belarus.
E-mail: vik.andreev@gsu.by

Vadzim Haurysh
Sukhoi State Technical University of Gomel, Technical Mechanics Department, Belarus.
E-mail: mez0n@inbox.ru

Abstract. The masses of the quarks in the Poincaré-invariant quantum mechanics are the constituent masses. Even in this framework it is possible to obtain an estimate of the constituent quark masses from the Ward identity for the axial current ant the current quark masses.

1. Introduction

It's well-known that quark masses are free parameters of composite quark models and quantum field theories, based on effective QCD-motivated lagrangians (χ-PT model, models, based on sum rules etc.). There are two types of quark masses that do not coincide in magnitude: the mass of the current quark, estimated in processes with a significant transfer of the 4-momentum square, and constituent mass, which includes the mass of the gluon field around the quark and is estimated from the mass of hadrons and their quark composition (see, for example, [1]). The search for a connection between the current and the constituent masses of quarks in the framework of QCD is an actual and complex task.

In this paper, using the idea of the identity for the pseudoscalar density [2], we propose a procedure for fixing the values of the constituent masses of quarks with the use of current masses and calculating the parameters of the model, based on the point form of Poincaré-invariant quantum mechanics [3] (PIQM), using experimental data on electroweak decays of pseudoscalar and vector mesons.

2. Leptonic decay constants in PIQM

The scheme for solving the problem in PIQM is as follows: at the first stage a basis of direct product of two particles without interaction is constructed. In the case of a system of two particles with the masses m_q and m_Q and, respectively, with 4-momentums $p_1 = \left(\omega_{m_q}(\mathrm{p}_1), \mathbf{p}_1\right)$ and $p_2 = \left(\omega_{m_Q}(\mathrm{p}_2), \mathbf{p}_2\right)$ basis

$$|\mathbf{p_1}, \lambda_1\rangle\, |\mathbf{p_2}, \lambda_2\rangle \equiv |\mathbf{p_1}, \lambda_1, \mathbf{p_2}, \lambda_2\rangle \tag{1}$$

defines a reducible representation of the Poincaré group.

In the second stage, using the Clebsch-Gordan decomposition for the Poincaré group (see [4],[5]) a basis of an irreducible representation, that characterizes the entire system, is constructed. To do this, we introduce a full momentum

$$\mathbf{P} = \mathbf{p_1} + \mathbf{p_2} \tag{2}$$

and the relative momentum $\mathbf{k}$ of particles [6]. The basis of the two-particle irreducible representation is defined by the quantum numbers of the total momentum (2), total angular momentum J with a projection μ and effective mass of noninteracting particles

$$M_0 = \omega_{m_Q}(\mathrm{k}) + \omega_{m_q}(\mathrm{k}) \ , \ \ \omega_m(\mathrm{k}) = \sqrt{\mathrm{k}^2 + m^2}, \tag{3}$$

where $\mathrm{k} = |\mathbf{k}|$, and two additional numbers that remove the degeneracy of this basis.

As a result, the meson with momentum $\mathbf{Q}$, mass M, spin J and it's projection μ is defined as the direct product of the state vectors of free particles (quarks) with wave function $\Phi^J_{\ell s}(\mathrm{k})$:

$$|\mathbf{Q}, J\mu, M\rangle = \int d\mathbf{k}\sqrt{\frac{\omega_{m_q}(\mathrm{p}_1)\,\omega_{m_Q}(\mathrm{p}_2)\,M_0}{\omega_{m_q}(\mathrm{k})\,\omega_{m_Q}(\mathrm{k})\,\omega_{M_0}(\mathrm{P})}}\times$$

$$\times\sum_{\lambda_1\lambda_2}\sum_{\nu_1\nu_2}\Omega\left\{{\ell \atop \nu_1,}{s \atop \nu_2,}{J \atop \mu}\right\}(\theta_k,\phi_k)\,\Phi^J_{\ell s}(\mathrm{k})\,D^{1/2}_{\lambda_1,\nu_1}(\mathbf{n}_{W_1})\,D^{1/2}_{\lambda_2,\nu_2}(\mathbf{n}_{W_2})\,|\mathbf{p}_1,\lambda_1,\mathbf{p}_2,\lambda_2\rangle\,, \tag{4}$$

where

$$\Omega\left\{{\ell \atop \nu_1,}{s \atop \nu_2,}{J \atop \mu}\right\}(\theta_k,\phi_k) = \mathbf{C}\left\{{s_1 \atop \nu_1,}{s_2 \atop \nu_2,}{s \atop \lambda}\right\}\mathbf{C}\left\{{\ell \atop m,}{s \atop \lambda,}{J \atop \mu}\right\}Y_{\ell\, m}(\theta_k,\phi_k) =$$

$$= \mathbf{C}\left\{{s_1 \atop \nu_1,}{s_2 \atop \nu_2,}{s \atop \nu_1+\nu_2}\right\}\mathbf{C}\left\{{\ell \atop \mu-(\nu_1+\nu_2),}{s \atop \nu_1+\nu_2,}{J \atop \mu}\right\}Y_{\ell\,\mu-(\nu_1+\nu_2)}(\theta_k,\phi_k)\ . \tag{5}$$

In Eq.(5) the functions $\mathbf{C}\left\{{s_1 \atop \nu_1,}{s_2 \atop \nu_2,}{s \atop \lambda}\right\}$, $\mathbf{C}\left\{{\ell \atop m,}{s \atop \lambda,}{j \atop \mu}\right\}$ are Clebsch-Gordan coefficients of $SU(2)$ group, $Y_{lm}(\theta_k,\phi_k)$ are the spherical functions and $D(\mathbf{n}_W)$ is Wigner D-function.

Wave function (WF) $\Phi^J_{\ell s}(\mathrm{k})$ taking into account the number of quark colors N_c is normalized by the condition

$$\sum_{\ell,s}\int_0^\infty d\mathrm{k}\mathrm{k}^2\left|\Phi^J_{\ell s}(\mathrm{k})\right|^2 = N_c. \tag{6}$$

The decay constants $P(q\bar{Q}) \to \ell + \nu_\ell$ and $\ell \to V(q\bar{Q}) + \bar{\nu}_\ell$ pseudoscalar and vector mesons with masses M_P and M_V, after removing the element of the Kobayashi-Maskawa matrices, are determined by the expressions:

$$\left\langle 0\left|\hat{J}^\mu_P(0)\right|\mathbf{P}, M_P\right\rangle = \frac{\mathrm{i}}{(2\pi)^{3/2}}\frac{1}{\sqrt{2\omega_{M_P}(\mathrm{P})}}P^\mu f_P\,, \tag{7}$$

$$\left\langle 0\left|\hat{J}^\mu_V(0)\right|\mathbf{P}, 1\lambda_V, M_V\right\rangle = \frac{\mathrm{i}}{(2\pi)^{3/2}}\frac{\varepsilon^\mu(\lambda_V)}{\sqrt{2\omega_{M_V}(\mathrm{P})}}M_V f_V\,. \tag{8}$$

Using relations (4), (7) and (8) matrix element of the quark currents one can obtain the integral representations of leptonic decays constants of pseudoscalar and vector mesons [7, 8]:

$$f_I(m_q, m_Q, \beta^I_{qQ}) = \sqrt{\frac{3}{2}\frac{1}{\pi}}\int_0^\infty d\mathrm{k}\mathrm{k}^2\Phi(\mathrm{k},\beta^I_{qQ})\sqrt{\frac{W^+_{m_q}(\mathrm{k})\,W^+_{m_Q}(\mathrm{k})}{M_0\omega_{m_q}(\mathrm{k})\,\omega_{m_Q}(\mathrm{k})}}\times$$

$$\times\left(1 + a_I\frac{\mathrm{k}^2}{W^+_{m_q}(\mathrm{k})\,W^+_{m_Q}(\mathrm{k})}\right)\ ,\quad I = P,V;\ \ a_P = -1, a_V = 1/3\,, \tag{9}$$

where

$$W^{\pm}_{m}\,(\mathrm{k}) = \omega_m\,(\mathrm{k}) \pm m\;. \tag{10}$$

Note, that the integral representation (9) match with the expressions for the decay constants of pseudoscalar and vector mesons in the instant and front-forms of PIQM [9, 10].

For fixing the parameters we use the constant of pseudoscalar density, which determined by relation (see e.g. [2, 11]):

$$\langle 0\,|\bar{Q}\gamma_5 q|\,\mathbf{P}, M_P\rangle = -\frac{\mathrm{i}}{(2\pi)^{3/2}}\frac{g_P}{\sqrt{2\,\omega_{M_P}\,(\mathrm{P})}}\;. \tag{11}$$

where axial current $\hat{J}^{\alpha}_{P}(x) = \bar{Q}(x)\gamma^{\alpha}\gamma_5 q(x)$ and pseudoscalar density $j^5(x) = \mathrm{i}\bar{Q}(x)\gamma_5 q(x)$ are related by [2]

$$\partial_\alpha \hat{J}^{\alpha}_{P}(x) = (\hat{m}_Q + \hat{m}_q) j^5(x). \tag{12}$$

The equations (7), (11) and (12) leads to the fact, that the constants f_P and g_P are related by [2, 11, 12]:

$$(\hat{m}_q + \hat{m}_Q)\, g_P = M_P^2 f_P\;. \tag{13}$$

where $\hat{m}_{q,Q}$ – current masses of quarks. The Eq. (13) is used, as usual, for u and d-quarks (sometimes and for s-quark [2]). After the calculations we obtain the integral representation constants g_P of pseudoscalar meson

$$g_P\left(m_q, m_Q, \beta^P_{qQ}\right) = \sqrt{\frac{3}{2}\frac{1}{\pi}}\int\limits_0^\infty \mathrm{d}\mathrm{k}\,\mathrm{k}^2 \Phi\left(\mathrm{k}, \beta^P_{qQ}\right)\sqrt{\frac{M_0}{\omega_{m_q}(\mathrm{k})\,\omega_{m_Q}(\mathrm{k})}} \times$$
$$\times \left(\sqrt{W^{+}_{m_q}(\mathrm{k})\, W^{+}_{m_Q}(\mathrm{k})} + \sqrt{W^{-}_{m_q}(\mathrm{k})\, W^{-}_{m_Q}(\mathrm{k})}\right). \tag{14}$$

3. Numerical results and discussions

The values of the constituent masses of light quarks (u, d and s) and the parameters of the wave function β^I_{qQ} can be fixed by means of the experimental values of the decay constants [13]

$$f^{(\mathrm{exp})}_{\pi^+} = (130.50 \pm 0.01 \pm 0.03 \pm 0.13)\ \mathrm{MeV},\ f^{(\mathrm{exp})}_{K^+} = (155.72 \pm 0.17 \pm 0.45 \pm 0.16)\ \mathrm{MeV} \tag{15}$$

and the lattice calculations of the quark current masses [13, 14, 15]:

$$\hat{m}_u = \left(2.2^{+0.6}_{-0.4}\right)\ \mathrm{MeV},\quad \hat{m}_d = \left(4.7^{+0.5}_{-0.4}\right)\ \mathrm{MeV},$$
$$\frac{(\hat{m}_u + \hat{m}_d)}{2} = \left(3.7^{+0.7}_{-0.3}\right)\ \mathrm{MeV},\quad \hat{m}_s = \left(96^{+8}_{-4}\right)\ \mathrm{MeV}. \tag{16}$$

Wave function $\Phi(\mathrm{k}, \beta^I_{qQ})$ is treated as a trial wave function with a parameter β^I_{qQ}, which can be chosen for the ground state in one of the species:

$$\Phi^{\mathrm{os}}(\mathrm{k}, \beta) = N_{os}\exp\left[-\frac{\mathrm{k}^2}{2\beta^2}\right],\quad N_{os} = \frac{2}{\pi^{1/4}\beta^{3/2}}\,, \tag{17}$$

$$\Phi^{\mathrm{coul}}(\mathrm{k}, \beta) = \frac{N_{coul}}{(1 + \mathrm{k}^2/\beta^2)^2}\,,\quad N_{coul} = 4\sqrt{\frac{2}{\pi\beta^3}}\,, \tag{18}$$

$$\Phi^{\mathrm{pl}}(\mathrm{k}, \beta) = \frac{N_{pl}}{(1 + \mathrm{k}^2/\beta^2)^3}\,,\quad N_{pl} = 16\sqrt{\frac{2}{7\pi\beta^3}}\;. \tag{19}$$

Table 1. The values of the constituent masses of u,d and s-quarks and parameters of the trial WF β^P_{qQ}

WF	m_u, MeV	$\bar{m}_{ud}$, MeV	$\bar{m}_s$, MeV	β^P_{ud}, MeV	β^P_{us}, MeV
Φ^{os} – (17)	218.3 ± 4.2	219.6 ± 4.2	426.2 ± 59.7	370.8 ± 9.3	373.2 ± 20.9
Φ^{coul} – (18)	—	—	—	—	—
Φ^{pl} – (19)	235.2 ± 14.7	233.9 ± 14.7	—	562.1 ± 28.2	—

Assuming that the scale of the violation of isotopic symmetry for the constituent u and d-quarks is the same as for current, i.e. $m_d - m_u = \hat{m}_d - \hat{m}_u = 2.5$ MeV from the system of equations

$$\begin{cases} f_P(m_q, m_Q, \beta^P_{qQ}) = f_P^{(\exp)} , \\ (\hat{m}_q + \hat{m}_Q)\, g_P(m_q, m_Q, \beta^P_{qQ}) = f_P^{(\exp)} M_P^2, \end{cases} \tag{20}$$

for pseudoscalar $\pi^\pm$ and $K^\pm$-mesons one can find the values of the constituent masses of the quarks and the parameters of the wave functions for u, d and s-quarks (see table 1). Mark "–" means no solution to the system (20) within the error.

Values for the test oscillator (Gaussian) WF correlate with the model parameters [16] with linear confinement potential: $m_u = 220$ MeV, $m_s = 450$ MeV, $\beta^P_{ud} = 365.9$ MeV and $\beta^P_{us} = 388.6$ MeV. Also, constituent masses $m_u = 220$ MeV и $m_s = 419$ MeV are used in the Mock-model [17].

We note that this technique is sensitive to the choice of the trial wave function. So for the Coulomb WF (18) in general one can not find a solution of the system of equations (20). Wave function (19) is only suitable for u, d-quarks (using (16) for s-quark is problematic because of the strong violation of the chiral symmetry).

Using the values G_F, $|V_{ud}|$, $|V_{us}|$, m_τ, τ_τ, M_ρ и M_{K^*} from [13] find, that experimental values $f_V^{(\exp)}$ are:

$$f_{\rho^+}^{(\exp)} = (209.3 \pm 1.5) \text{ MeV}, \quad f_{K^{*+}}^{(\exp)} = (205.3 \pm 6.0) \text{ MeV} . \tag{21}$$

From the equation $f_V(m_q, m_Q, \beta^V_{qQ}) = f_V^{(\exp)}$ we find the values of the parameters β^V_{ud} and β^V_{us} for oscillator WF (17):

$$\beta^V_{ud} = (310.9 \pm 2.2) \text{ MeV}, \quad \beta^V_{us} = (314.3 \pm 83.6) \text{ MeV} . \tag{22}$$

4. Conclusion and remarks

The paper is dedicated to calculation the parameters of a model, based on the point form of Poincaré-invariant quantum mechanics. The obtained values of the constituent masses of quarks and parameters of the wave functions correlate with the values, obtained in models based on the light-front and instant form of dynamics. Authors note, that analysis of model parameters, depending on the type of trial wave function, also have been done.

5. Acknowledgments

We are grateful to the organizational and material support to the organizers of the Conference DSPIN'17. This work was supported by the Belarusian Republican Foundation for Basic Research.

IOP Conf. Series: Journal of Physics: Conf. Series **938** (2017) 012030 doi:10.1088/1742-6596/938/1/012030

References

[1] Watson A 2004 The Quantum Quark
[2] Jaus W 2003 *Phys.Rev.* **D67**. – 094010
[3] Dirac Paul A M 1949 *Rev. Mod. Phys.* **21** 392–399
[4] Shirokov Yu 1958 *ZhETF* **34**, № 4 1005–1011
[5] Werle J 1970 Relativistic theory of reactions
[6] Keister B 1991 *Adv. Nucl. Phys.* **20** 225–479
[7] Andreev V 2011 *Physics of Particles and Nuclei Letters* **8**, № 4. 347–355
[8] Andreev V 2013 (*Preprint* 1305.4266)
[9] Krutov A 2016 *Phys.Rev.* **D93**, № 3 036007
[10] Jaus W 1991 *Phys. Rev.* **D44** 2851–2859 Lucha W 2006 *Phys. Rev.* **D74** 054004
[11] Lucha W 2006 *Phys. Rev.* **D74** 054004
[12] Leutwyler H *Nato Sci.* **B363** 149–164
[13] Patrignani C et al 2017 *Rev. of Particle Physics* **C40**, № 10 100001.
[14] Carrasco N 2014 *Nucl. Phys.* **B887** 19–68
[15] Hoelbling C et al. 2016 *Phys. Rev. Lett.* **117** , № 8 082001
[16] Choi H M 1999 *Phys. Rev.* **D59** 034001
[17] Godfrey S 1985 *Phys. Rev.* **D32** 189–231

Quantum scattering beyond the plane-wave approximation

Dmitry Karlovets
Tomsk State University, Lenina 36, 634050 Tomsk, Russia

Abstract. While a plane-wave approximation in high-energy physics works well in a majority of practical cases, it becomes inapplicable for scattering of the vortex particles carrying orbital angular momentum, of Airy beams, of the so-called Schrödinger cat states, and their generalizations. Such quantum states of photons, electrons and neutrons have been generated experimentally in recent years, opening up new perspectives in quantum optics, electron microscopy, particle physics, and so forth. Here we discuss the non-plane-wave effects in scattering brought about by the novel quantum numbers of these wave packets. For the well-focused electrons of intermediate energies, already available at electron microscopes, the corresponding contribution can surpass that of the radiative corrections. Moreover, collisions of the cat-like superpositions of such focused beams with atoms allow one to probe effects of the quantum interference, which have never played any role in particle scattering.

1. Introduction

When studying scattering of particles, one usually treats their states as the eigenstates of a momentum operator, that is, as the plane waves, $\hat{\mathbf{p}}|\mathbf{p}\rangle = \mathbf{p}|\mathbf{p}\rangle$. The only objection one can raise is that the plane waves, being completely delocalized in space and time, do not exist in Nature. That is why the well-normalized wave packets are oftentimes considered first and a limit of the infinitely narrow ones in momentum space is taken when passing to the cross section. And yet the plane-wave approximation successfully describes the overwhelming majority of the processes studied at modern colliders. On the other hand, a proper discussion of its limits of applicability in the scattering problems seems to be absent in literature, which is a bit disturbing, especially when striving for higher-order radiative corrections. There are also important examples from the collider physics, atomic physics, neutrino physics, etc., in which the plane-wave model becomes inapplicable – see, for instance, [1, 2, 3, 4, 5, 6] and references therein.

In mathematical physics, a lot of non-plane-wave solutions for the Dirac equation are well known [7]. Yet it was only in the 1990s that the first photons possessing a well-defined orbital angular momentum (OAM) were generated [8]. Such a twisted photon, $|\omega, k_z, \kappa, \ell, \lambda\rangle$, is a (paraxial) monochromatic state with the OAM ℓ, a helicity λ, a longitudinal momentum k_z, and an absolute value of the transverse momentum κ. During the last several years, 200-300-keV electrons and cold neutrons carrying the OAM were also generated [9, 10, 11, 12], along with the so-called Airy photons and electrons [13, 14] and their different generalizations. Possible applications of these states to the hadronic physics have been discussed in Refs.[5, 15, 16].

Here we study how scattering processes with these particles can be described within an S-matrix formalism. A non-plane-wave contribution to the cross section turns out to be attenuated by a ratio of the Compton wave length, $\lambda_c = \hbar/(mc)$, to the beam width σ_x, λ_c/σ_x, which can reach the values of $\lambda_c/\sigma_x \sim 10^{-3}$ for electrons, but stays several orders of magnitude smaller for available hadronic beams. Thus for the well-focused electrons of intermediate energies, the corresponding contribution can even surpass that of the radiative corrections, as the former is attenuated as $f(s,t)\,\lambda_c/\sigma_x$, being no longer negligible when $f(s,t)$, which depends on a phase ζ_{fi} of a scattering amplitude M_{fi}, is not vanishing. Moreover, colliding coherent superpositions of the wave packets with atoms, one can get out of the paraxial approximation when the beams are focused to an atomic size, which is already feasible. The units with $\hbar = c = 1$ are used.

2. Qualitative considerations

Let us consider a process of $2 \to N$ scattering. Staying first within the common plane-wave model, the quantum effects arise because of recoil, because of such intrinsic degrees of freedom as, for instance, spin, and because of the radiative corrections. The latter are oftentimes said to be the "genuinely quantum" phenomena, and in a perturbative regime their contribution is attenuated by $\alpha \ll 1$ (in QED, $\alpha = 1/137$). Going beyond the plane-wave approximation, however, we get yet another source of the genuinely quantum corrections, which contribute already to the tree level. These are evoked by the wave nature of particles, with their de Broglie wave length, $\lambda_{dB} = 2\pi/p$, being a characteristic parameter. The plane-wave approximation is usually said to be applicable when the wave packets describing particles are narrow in the momentum space, that is,

$$\frac{\sigma_p}{|\langle \mathbf{p} \rangle|} \sim \frac{\lambda_{dB}}{\sigma_x} \ll 1, \tag{1}$$

where $\sigma_p \sim 1/\sigma_x$ is a packet's width and $\langle \mathbf{p} \rangle$ is its mean momentum. This inequality is manifestly not Lorentz invariant. Let σ_x be a packet's width in a plane perpendicular to the direction of a boost, which is made along a collision axis for definiteness. Then a Lorentz-invariant generalization of (1) would be

$$\frac{\sigma_p}{m} \sim \frac{\lambda_c}{\sigma_x} \ll 1, \tag{2}$$

with λ_c instead of λ_{dB}. Note that in a more general case of arbitrary boost, one should deal with a tensor σ_{ij} instead of the scalar σ_p. As a result, the small parameter of the problem no longer preserves its simple form (2) [5].

For ultrarelativistic energies, $|\langle \mathbf{p} \rangle| \gg m$, the de Broglie wave length is much smaller than the Compton one, which does not mean, however, that the wave nature of particles does not reveal itself in scattering. As we demonstrate below, in fact it does and the non-plane-wave contribution to the (Lorentz invariant) cross section is attenuated by the parameter (2) and it does not decrease with the increase of the energy. The smallness of (2) also necessitates the applicability of a one-particle theory with no negative-energy solutions.

3. Corrections to the plane-wave cross section

The plane-wave cross section $d\sigma^{(pw)}$ depends on an absolute value of the amplitude $|M_{fi}(\mathbf{p}_i)|^2$ with $\mathbf{p}_i$ being the incoming momenta. When dealing with the wave packets instead, one gets

$$|M_{fi}(\mathbf{p}_i)|^2 \to M_{fi}(\mathbf{p}_i)M^*_{fi}(\mathbf{p}'_i) \text{ or } M_{fi}(\mathbf{p}_i + \mathbf{k}_i/2)M^*_{fi}(\mathbf{p}_i - \mathbf{k}_i/2) \tag{3}$$

where $\mathbf{p}_i$ and $\mathbf{p}'_i$ share the same mean value $\langle \mathbf{p}_i \rangle$, but the mean value of $\mathbf{k}_i$ is vanishing. Then, presenting the complex amplitude as $M_{fi} = |M_{fi}| \exp\{i\zeta_{fi}\}$, we obtain

$$M_{fi}(\mathbf{p}_i + \mathbf{k}_i/2)M^*_{fi}(\mathbf{p}_i - \mathbf{k}_i/2) = \left(|M_{fi}(\mathbf{p}_i)|^2 + \mathcal{O}(k_i^2)\right) \exp\left\{i\mathbf{k}_i \frac{\partial \zeta_{fi}}{\partial \mathbf{p}_i} + \mathcal{O}(k_i^3)\right\}, \tag{4}$$

where the neglected terms are negligible when Ineq.(2) holds true, akin to the paraxial or WKB approximation. Hence, the first correction to the plane-wave cross section is linearly attenuated by λ_c/σ_x and depends on derivatives of the phase ζ_{fi},

$$\frac{d\sigma^{(1)}}{d\sigma^{(pw)}} = \frac{\lambda_c}{\sigma_x}\left(f_1(s,t)\frac{\partial\zeta_{fi}}{\partial s} + f_2(s,t)\frac{\partial\zeta_{fi}}{\partial t}\right), \tag{5}$$

with s, t being the Mandelstam variables. It is these functions, f_1 and f_2, that depend on such quantum numbers of the incoming particles as, for instance, the OAM or the Airy phase.

One can infer from (4) that these functions do not vanish only when the in-state can be characterized with a polar vector $\mathbf{b}_i$: $\mathbf{k}_i \to \mathbf{b}_i$. In other words, the initial system of particles as a whole must possess a finite *dipole moment.* The simplest way to realize this is to collide two beams head on, but at a finite impact parameter. An alternative would be to collide two non-symmetric wave packets like Airy beams or vortex electrons with a center of the ring slightly shifted from that of the beam. The latter vortex beams can be shown to possess a finite uncertainty of the OAM. This finite uncertainty plays a crucial role, as the pure Bessel states are orthogonal and, hence, neither overlap nor interfere, exactly like the plane waves. It is a finite overlap of the incoming packets that results in a non-vanishing $d\sigma^{(1)}$ from (5), and that is why the mutual disposition of the packets (in either x- or p-space) comes into play [5].

Next, such a dipole moment is obviously a potential source of *an azimuthal asymmetry* in scattering, as the functions f_1, f_2 can now depend on the products $\mathbf{b}\cdot\mathbf{p}_f$ with $\mathbf{p}_f$ being a particle's final momentum. The asymmetry can reveal itself, say, in ee, e^+e^-, pp, or $p\bar{p}$ collisions (see also [15, 16]), and it can be measured, for instance, by taking two electron beams, each focused to a spot of 0.1 nm, as in the Refs.[17, 18], and colliding them slightly off-center at an impact parameter of $b \sim \sigma_x$. The order of the predicted asymmetry, very roughly, is

$$\frac{d\sigma^{(1)}}{d\sigma^{(pw)}} \sim \frac{\lambda_c}{\sigma_x} \sim 10^{-4} - 10^{-3}. \tag{6}$$

The more detailed calculations support this estimate [5].

4. Going beyond the WKB regime

The paraxial approximation fails to work when $\sigma_x \sim \lambda_c$, and the one-particle theory itself becomes inapplicable. In order to understand better what effects are lost when (2) holds true, let us recall that the RHS of (3) represents just a density matrix of the process,

$$M_{fi}(\mathbf{p}_i + \mathbf{k}_i/2)M^*_{fi}(\mathbf{p}_i - \mathbf{k}_i/2) \to \rho_{fi}(\mathbf{p}_i, \mathbf{k}_i), \tag{7}$$

a Fourier transform of which is called a Wigner function,

$$n_{fi}(\mathbf{r}_i, \mathbf{p}_i) = \int \frac{d^3k}{(2\pi)^3}\, e^{i\mathbf{k}_i\mathbf{r}_i}\rho_{fi}(\mathbf{p}_i, \mathbf{k}_i), \tag{8}$$

studied, for instance, in [19]. Clearly the expansion (4) leaves this Wigner function *positive everywhere*, that is, quasi-classical.

The same holds for the incoming wave packets. If a wave function of an in-state is $\psi(\mathbf{p}_i)$, a square of the matrix element (7) depends on $\psi(\mathbf{p}_i + \mathbf{k}_i/2)\psi^*(\mathbf{p}_i - \mathbf{k}_i/2) \to \rho(\mathbf{p}_i, \mathbf{k}_i)$ with ρ being a particle's density matrix. In the perturbative regime (2), an expansion analogous to (4) would also leave the corresponding Wigner functions positive everywhere. It has a clear interpretation: the quasi-classical states with the non-negative Wigner functions are nearly coherent, whereas those with the not-everywhere positive functions experience *destructive self-interference.* For a single packet, the negative regions become noticeable only when it is focused to a spot comparable to λ_c. From this one may deduce that it is practically impossible to get out of the WKB regime.

There are, however, possibilities to go beyond the paraxial approximation even without such a tight focusing. Indeed, the non-classical states of light, such as a so-called Schrödinger cat state, are known to reveal a high degree of non-classicality even if they are not focused below a shot-noise limit. Taking instead of a single (Gaussian) packet a quantum superposition of them, which may not be Gaussian because of the quantum interference, we get a cat-like system with a Wigner function whose negative regions become noticeable at a much larger scale than λ_c.

The next step is to collide such nonclassical states with the composite "particles" like atoms or molecules having an effective size of $a \gg \lambda_c$. Then the parameter (2) is to be changed to a/σ_x, which can be of the order of unity, even if (2) still holds. For a hydrogen atom, a is just a Bohr radius, $a \approx 0.053$ nm. As beams of the electrons with $\sigma_x \sim 0.1$ nm have been already obtained [17, 18], one can probe the highly nonclassical regime of scattering, in which negative values of the Wigner functions give noticeable contribution to the cross section. The simplest effect here would also be an azimuthal asymmetry in elastic scattering of a Schrödinger cat state off an atom, which would be of the order of $a/\sigma_x \sim 10^{-1}$ (see details in [6]).

5. Conclusion

The non-plane-wave effects in quantum scattering are no longer vanishing for the focused beams of electron microscopes, which allows one to probe a contribution of the Coulomb phase of the scattering amplitude to the cross section. These effects can even surpass those of the radiative corrections, and they become only moderately attenuated beyond the paraxial approximation, which is already feasible with current technologies. For hadronic beams, however, the corresponding contribution seems to be several orders of magnitude smaller. On the other hand, quantum interference can also play a role in describing final jets in hadronic collisions [16], which is yet to be studied in more detail.

Acknowledgments

I am grateful to V. G. Bagrov, I. F. Ginzburg, I. P. Ivanov, P. O. Kazinski, P. Kruit, A. Di Piazza, V. G. Serbo, O. V. Teryaev, J. Verbeeck, and A. S. Zhevlakov for useful discussions and criticism. This work is supported by the Russian Science Foundation (project No. 17-72-20013).

References

[1] Kotkin G L, Serbo V G and Schiller A 1992, *Int. J. Mod. Phys.* A **07**, 4707
[2] Akhmedov E Kh, Kopp J *J. High Energy Phys.* JHEP04(2010)008
[3] Melnikov K, Kotkin G L, Serbo V G, 1996 *Phys. Rev. D* **54** 3289
[4] Ivanov I P, Serbo V G 2011 *Phys. Rev. A* **84** 033804
[5] Karlovets D V *J. High Energy Phys.* JHEP03(2017)049
[6] Karlovets D V, Serbo V G 2017 *Phys. Rev. Lett.* **119** 173601
[7] Bagrov V G, Gitman D M 1990 *Exact solutions of relativistic wave equations* (Dordrecht Netherlands: Kluwer Academic Publ.)
[8] *Twisted photons. Applications of light with orbital angular momentum* 2011 ed J P Torres and L Torner (New York: WILEY-VCH)
[9] Uchida M, Tonomura A 2010 *Nature* **464** 737
[10] Verbeeck J, Tian H, Schlattschneider P 2010 *Nature* **467** 301
[11] McMorran B J, Agrawal A, Anderson I M, et al 2011 *Science* **331** 192
[12] Clark C W, Barankov R, Huber M G, et al 2015 *Nature* **525** 504
[13] Siviloglou G A, Broky J, Dogariu A and Christodoulides D N 2007 *Phys. Rev. Lett.* **99** 213901
[14] Voloch-Bloch N, Lereah Y, Lilach Y, et al 2013 *Nature* **494** 331
[15] Ivanov I P, Seipt D, Surzhykov A, Fritzsche S 2016 *Phys. Rev. D* **94** 076001
[16] Dodonov V V, Dremin I M, Man'ko O V, et al 1998 *J. Russ. Laser Res.* **19** 427
[17] Verbeeck J, Schattschneider P, Lazar S, et al 2011 *Appl. Phys. Lett.* **99** 203109
[18] Pohl D, Schneider S, Zeiger P, et al 2017 *Sci. Rep.* **7** 934

[19] Vall A N, Perevalova I A, Polyakov M V and Soldatenko O N 2011 *Russ. Phys. J.* **54** 47

IOP Conf. Series: Journal of Physics: Conf. Series **938** (2017) 012032 doi:10.1088/1742-6596/938/1/012032

Supersymmetric formulation of classical, quantum and stochastic dynamics

Nugzar Makhaldiani
Joint Institute for Nuclear Research
Dubna, Moscow Region, Russia
E-mail: mnv@jinr.ru

Abstract. q-deformed algebra and its fermion, boson and intermediate limits; universal stochastic and supersymmetric dynamics; hamiltonization of the dynamics and spin dynamics in external fields considered.

-Why supersymmetry is So universal? - Because it describes an universal connection of a system to its environment. Supermathematics unifies discrete and continual aspects of mathematics. It unifies fermionic and bosonic systems, most fundamental of which are fermionic and bosonic oscillators with the following algebra for creation - a^+ and annihilation-a operators

$$aa^+ - qa^+a = 1 \tag{1}$$

where $q = -1$ for fermion and $q = 1$ for boson. Naively the parameter q interpolates fermion into boson, but only for fermions there are finite dimensional matrix representations, e.g. the simplest one as 2×2 matrixes with only nonzero elements $a_{21} = 1$ and $a^+_{12} = 1$. The known matrix representation for boson extends to general value of q by change: $n \to n_q = (q^n - 1)/(q - 1)$ and reduce in the fermion case to the infinite dimensional matrix representation with the simplest 2×2 representation matrixes on the diagonal.

In a coordinate representation the algebra (1) presents as

$$a^+ = z,\ af(z) = \partial_q f(z) = \frac{f(qz) - f(z)}{(q-1)z} = \frac{q^\delta - 1}{(q-1)z} f(z),\ \delta = z\frac{d}{dz} = z\partial \tag{2}$$

To the bosonic case correspond $q \to 1$, $\partial_1 = \partial$. To the fermion case: $q \to -1$, $\partial_{-1} f(z) = (f(z) - f(-z))/(2z)$. In this form it is possible to realize a small deviation from boson and fermion, a small deviation from the Pauli exclusion principle. There is also an intermediate region: $q = \varepsilon$, $\partial_\varepsilon f(z) = (f(z) - f(0))/z + O(\varepsilon)$. When $q = 2, 3, 5, ...29...137...-$ prime number, the q-analysis reduce to the p-adic one, [1].

Quantum, classical and stochastic dynamics and supersymmetry. After formulation of the mathematical framework of quantum mechanics (QM), operatorial formulation of QM, Koopman and von Neumann gave operatorial approach to classical Hamiltonian mechanics [2], [3]. After Wiener introduction of the functional integrals, Dirac and Feynman gave formal functional integral formulation of the quantum theory [4]. Gozzi invented functional integral formulation of the classical theory [5]. All stochastic and deterministic differential equations, describing all natural and engineered dynamical systems, possess a topological supersymmetry. Its spontaneous breakdown could be interpreted as the stochastic generalization of deterministic chaos. This conclusion stems from the fact that such phenomenon encompasses features that are traditionally associated with chaotic dynamics such as non-integrability, positive topological entropy, sensitivity to initial conditions. Spontaneous topological symmetry breaking can be considered as the most general definition of continuous-time dynamical chaos. For supersymmetric gauge theories stochastic quantization appears to have one definite advantage: since a gauge fixing term is unnecessary, supersymmetry will not be broken at any step. This holds both for the Abelian and non-Abelian case. It appears at the moment as if stochastic regularization is the only viable candidate for a regularization scheme which manifestly conserves both supersymmetry, chiral symmetry and gauge invariance. However, supersymmetry is related to stochastic quantization also at a much deeper level. As an example, even purely scalar field theories will, when quantized stochastically, display a 'hidden' supersymmetry. This issue, is intimately connected with the existence of a so-called 'Nicolai map' for supersymmetric field theories [6]. Parisi-Sourlas 'dimensional reduction' of scalar field theories in external random fields [7], is closely related to both supersymmetry and stochastic quantization. This becomes apparent when one establishes the connection to the Nicolai map. The phenomenon of dynamical 'dimensional reduction' was first noted within the context of critical phenomena associated with spin systems in random external fields. Systems very close to such a situation can in fact be created and studied in the laboratory.

As an example, let us consider the Langevin equation associated with a point particle being subjected to random background noise. This corresponds to the very real physical problem of the Brownian motion of a (classical) particle in a heat bath. The Langevin equation for the particle reads

$$\frac{dx}{dt} \equiv \dot{x} = -\frac{\delta S}{\delta x} + \eta(t), \tag{3}$$

where x represents the space coordinate of the particle. Expectation values are, as usual, evaluated as the path integral

$$< x(t_1)...\, x(t_n) >= \int d\eta\; x(t_1)...\, x(t_n) \exp(-\frac{1}{4}\int dt \eta(t)^2) \tag{4}$$

over a Gaussian noise, i.e.

$$< \eta(t_1)\eta(t_2) >= 2\delta(t_1 - t_2). \tag{5}$$

We now attempt to make a change of variables: $\eta \to x$. This involves the Jacobian

$$det(\delta\eta(t)/\delta x(t')) = det((d/dt + V')\delta(t - t')),\; V = \delta S/\delta x. \tag{6}$$

For partition function Z, we have

$$\begin{aligned} Z &= \int d\eta \exp(-\frac{1}{4}\int dt\eta(t)^2) \\ &= \int d\eta dx det(d/dt + V')\delta(\dot{x} + V - \eta(t)) \exp(-\frac{1}{4}\int dt\eta(t)^2) \\ &= \int dx det(d/dt + V') \exp(-\frac{1}{4}\int dt(\dot{x} + V)^2) \\ &= \int dx d\psi d\bar{\psi} \exp(-S),\; S = \int dt(\frac{1}{4}(\dot{x} + V)^2 - \bar{\psi}(d/dt + V')\psi). \end{aligned} \tag{7}$$

This system is recognized as Witten's example of supersymmetric quantum mechanics. So, purely classical (although stochastic) problem of Brownian motion is completely equivalent to a (supersymmetric) quantum mechanical problem!

Let us consider a general dynamical system described by the following system of the ordinary differential equations

$$\dot{x}_n = v_n(x), \;\; 1 \leq n \leq N, \tag{8}$$

where $\dot{x}_n$ stands for the total derivative with respect to the parameter t.

Let us consider the following Lagrangian

$$L = (\dot{x}_n - v_n(x))\psi_n \tag{9}$$

and the corresponding equations of motion

$$\dot{x}_n = v_n(x), \dot{\psi}_n = -\frac{\partial v_m}{\partial x_n}\psi_m. \tag{10}$$

The system (10) extends the general system (8) by linear equation for the variables ψ. The extended system can be put in the Hamiltonian form [8]

$$\dot{x}_n = \{x_n, H_1\}_1, \dot{\psi}_n = \{\psi_n, H_1\}_1, \tag{11}$$

where first level (order) Hamiltonian is

$$H_1 = v_n(x)\psi_n \tag{12}$$

and (first level) bracket is defined as

$$\{A, B\}_1 = A(\frac{\overleftarrow{\partial}}{\partial x_n}\frac{\overrightarrow{\partial}}{\partial \psi_n} - \frac{\overleftarrow{\partial}}{\partial \psi_n}\frac{\overrightarrow{\partial}}{\partial x_n})B. \tag{13}$$

The quasi-classical description of the motion of a relativistic (nonradiating) point particle with spin in accelerators and storage rings includes the equations of orbit motion

$$\begin{aligned} &\dot{x}_n = f_n(x), \; f_n(x) = \varepsilon_{nm}\partial_m H, \; n, m = 1, 2, ..., 6; \\ &x_n = q_n, \; x_{n+3} = p_n, \; \varepsilon_{n,n+3} = 1, \; n = 1, 2, 3; \\ &H = e\Phi + c\sqrt{\wp^2 + m^2c^2}, \; \wp_n = p_n - \frac{e}{c}A_n \end{aligned} \tag{14}$$

and Thomas-BMT equations [9, 10] of classical spin motion

$$\begin{aligned} &\dot{s}_n = \varepsilon_{nmk}\Omega_m s_k = \{H_1, H_2, s_n\}, \; \{A, B, C\} = \varepsilon_{nmk}\partial_n A \partial_m B \partial_k C, H_1 = \Omega \cdot s, \; H_2 = s^2, \\ &\Omega_n = \frac{-e}{m\gamma c}((1 + k\gamma)B_n - k\frac{(B \cdot \wp)\wp_n}{m^2c^2(1+\gamma)} + \frac{1 + k(1+\gamma)}{mc(1+\gamma)}\varepsilon_{nmk}E_m\wp_k) \end{aligned} \tag{15}$$

where, parameters e and m are the charge and the rest mass of the particle, c is the velocity of light, $k = (g-2)/2$ quantifies the anomalous spin g factor, γ is the Lorentz factor, p_n are components of the kinetic momentum vector, E_n and B_n are the electric and magnetic fields, and A_n and Φ are the vector and scalar potentials;

$$B_n = \varepsilon_{nmk}\partial_m A_k, \; E_n = -\partial_n\Phi - \frac{1}{c}\dot{A}_n, \gamma = \frac{H - e\Phi}{mc^2} = \sqrt{1 + \frac{\wp^2}{m^2c^2}} \tag{16}$$

The spin motion equations we put in the Nambu-Poisson form [11]. Hamiltonization of this dynamical system we will put in the ground of the optimal control theory of the accelerator.

Electric dipole moments (EDM) are one of the keys to understand the origin of our Universe. CP and P violation entail nonvanishing P and T violating electric dipole moments of elementary particles $\vec{d} = d\vec{s}$. Non-vanishing EDMs give rise to the precession of the spin of a particle in an electric field. In the rest frame of a particle

$$\dot{s}_n = \varepsilon_{nmk}(\Omega_m s_k + d_m E_k), \ \Omega_m = -\mu B_m, \tag{17}$$

where in terms of the lab frame fields

$$B_n = \gamma(B^l_n - \varepsilon_{nmk}\beta_m E^l_k), E_n = \gamma(E^l_n + \varepsilon_{nmk}\beta_m B^l_k) \tag{18}$$

Now we can apply the Hamiltonization and optimal control theory methods to this dynamical system.

References

[1] Makhaldiani N 2011 *Fractal Calculus (H) and some Applications* *Physics of Particles and Nuclei Letters* **8** 325

[2] Koopman B O 1931 *Hamiltonian systems and transformations in Hilbert space* *Proc. Nat. Acad. Sci. USA* **17** 315

[3] von Neumann J 1932 *Zur operatorenmethode in der klassichen mechanik* *Ann.Math.* **33** 587

[4] Feynman R P Hibbs R A 1965 *Quantum mechanics and path-integrals* (McGraw-Hill, Inc.)

[5] Gozzi E 2001 *Functional Techniques in Classical Mechanics* *Nucl.Phys.Proc.Suppl.* **104** 243

[6] Nicolai H 1980 *Phys. Lett. B* **89** 341; *Nucl. Phys. B* **176** 419

[7] Parisi G and Sourlas N 1979 *Phys. Rev. Left.* **43** 744; *Nucl. Phys. B* **206** 321

[8] Makhaldiani N and Voskresenskaya O 1997 *On the correspondence between the dynamics with odd and even brackets and generalized Nambu's mechanics, JINR Communications* **E2-97-418**

[9] Thomas L H 1927 *Philos. Mag.* **3** 1

[10] Bargmann V Michel L Telegdi V L 1959 *Precession of the polarization of particles moving in a homogeneous electromagnetic field Phys. Rev. Lett.* **2(10)** 435

[11] Makhaldiani N 2016 *Renormdynamics, Discrete Dynamics and Spin* *Journal of Physics: conference series* **678** 012029

IOP Conf. Series: Journal of Physics: Conf. Series **938** (2017) 012033 doi:10.1088/1742-6596/938/1/012033

Impenetrable barriers for positrons in neighbourhood of superheavy nuclei with Z>118

V P Neznamov[1]
Federal state unitary enterprise Russian Federal Nuclear Center – All-Russian Research Institute of Experimental Physics (FSUE "RFNC-VNIIEF"), 37 Mira pr., Sarov, Nizhny Novgorod region, 607188, Russia

E-mail: vpneznamov@vniief.ru

Abstract. Analysis of quantum mechanical motion of charged half-spin particles in the repulsive Coulomb field results in that an impenetrable potential barrier not explored earlier was found. For a particle at rest with a reduced mass m, the barrier radius is equal to half classical radius: the barrier radius decreases with increase in the particle energy. For the stable and quasi-stable nuclei with $Z>118$, presence of an impenetrable barrier as β^+-decay leads to the existence of "traps" for positrons in the neighbourhood of nuclei and as $Z_{cr}\simeq 170$ (with emission of electron-positron pairs by vacuum) leads to the existence of a quasi-constant source of annihilation quanta.

1. The equations for fermions in the Coulomb field

1.1. The Dirac equations in the Coulomb field
Here and below, we will generally use the system of units $\hbar=c=1$.
Bispinor wave function for stationary states.

$$\Psi(\mathbf{r},t)=\begin{pmatrix}\varphi(\mathbf{r})\\ \chi(\mathbf{r})\end{pmatrix}e^{-iEt}. \tag{1}$$

E is the energy of the Dirac particle.
In spherical coordinates (r,θ,φ)

$$\begin{aligned}\varphi(\mathbf{r})&=e^{im_\varphi\varphi}\xi(\theta)F(r),\\ \chi(\mathbf{r})&=e^{im_\varphi\varphi}(-i)\sigma^3\xi(\theta)G(r).\end{aligned} \tag{2}$$

In (2) $F(r),G(r)$ are the radial wave functions; m_φ is the azimuthal quantum number; σ^k are the Pauli matrices; the spinor $\xi(\theta)=\begin{pmatrix}{}_{-1/2}Y(\theta)\\ {}_{+1/2}Y(\theta)\end{pmatrix}$, where ${}_{-1/2}Y(\theta)$, ${}_{+1/2}Y(\theta)$ are spherical harmonics for spin ½.

[1] vpneznamov@vniief.ru

The dimensionless variables: $\rho = r/l_c$; $\varepsilon = E/m$; the Coulomb field $V(\rho) = \pm Z\alpha/\rho$; $l_c = 1/m$ is the Compton wavelength of the Dirac particle; $\alpha = e^2/\hbar c$ is the constant of the fine structure; Z is the number of protons in atomic nucleus.

The separation constant $\kappa = \mp 1, \mp 2 ... = \begin{cases} -(l+1), & j = l+1/2 \\ l, & j = l-1/2 \end{cases}$; j, l are the quantum numbers of the angular and orbital momenta of the Dirac particle.

The Dirac equations for components of bispinor $\Psi(\mathbf{r})$

$$\begin{aligned} &(\varepsilon+1-V(\rho))\chi(\rho,\theta,\varphi) = \boldsymbol{\sigma}\mathbf{p}\varphi(\rho,\theta,\varphi), \\ &-(\varepsilon-1-V(\rho))\chi(\rho,\theta,\varphi) = \boldsymbol{\sigma}\mathbf{p}\chi(\rho,\theta,\varphi). \end{aligned} \tag{3}$$

After separation of variables: The Dirac equations for radial wave functions are

$$\begin{aligned} &(\varepsilon+1-V)G(\rho) = \frac{dF}{d\rho} + \frac{1+\kappa}{\rho}F, \\ &-(\varepsilon-1-V)F(\rho) = \frac{dG}{d\rho} + \frac{1-\kappa}{\rho}G, \end{aligned} \tag{4}$$

1.2. Second-order equation for spinor $\varphi(\mathbf{r})$

$$\begin{aligned} &\left((\varepsilon - V(\rho))^2 - \mathbf{p}^2 - ie\boldsymbol{\alpha}\mathbf{E}\right)\Psi(\rho,\theta,\varphi) = 0 \text{ [1]}, \\ &\alpha^k = \begin{pmatrix} 0 & \sigma^k \\ \sigma^k & 0 \end{pmatrix}; \; e\mathbf{E} = -\nabla V, \end{aligned} \tag{5}$$

From equations (3)

$$\chi = \frac{1}{\varepsilon+1-V}\boldsymbol{\sigma}\mathbf{p}\varphi, \tag{6}$$

and

$$\left((\varepsilon-V)^2 - \mathbf{p}^2 - \frac{1}{\varepsilon+1-V} ie(\boldsymbol{\sigma}\mathbf{E})(\boldsymbol{\sigma}\mathbf{p})\right)\varphi(\rho,\theta,\varphi) = 0. \tag{7}$$

Self-adjoint second-order equation can be represented in the form

$$\begin{aligned} &\frac{1}{\rho^2}\frac{d}{d\rho}\left(\rho^2 \frac{d}{d\rho}\right)\psi(\rho) + \left((\varepsilon-V)^2 - 1 - \frac{\kappa(\kappa+1)}{\rho^2} - \right. \\ &\left. -\frac{3}{4}\frac{\left(\frac{dV}{d\rho}\right)^2}{(\varepsilon+1-V)^2} - \frac{1}{2}\frac{\frac{d^2V}{d\rho^2}}{(\varepsilon+1-V)} + \frac{1}{(\varepsilon+1-V)}\frac{1}{\rho}\frac{dV}{d\rho}\kappa\right)\psi(\rho) = 0. \end{aligned} \tag{8}$$

For self-conjugacy in (7), a non-unitary transformation is implemented [2]

$$\psi(\rho) = \frac{\rho}{\sqrt{\varepsilon+1-V}}F(\rho) \tag{9}$$

In case of the repulsive Coulomb field $V(\rho) = Z\alpha/\rho$, there is a value $\rho = \rho_{cl}$ at which $(\varepsilon+1-V(\rho_{cl})) = 0$. At that

$$\rho_{cl} = Z\alpha/(\varepsilon+1) \tag{10}$$

IOP Conf. Series: Journal of Physics: Conf. Series **938** (2017) 012033 doi:10.1088/1742-6596/938/1/012033

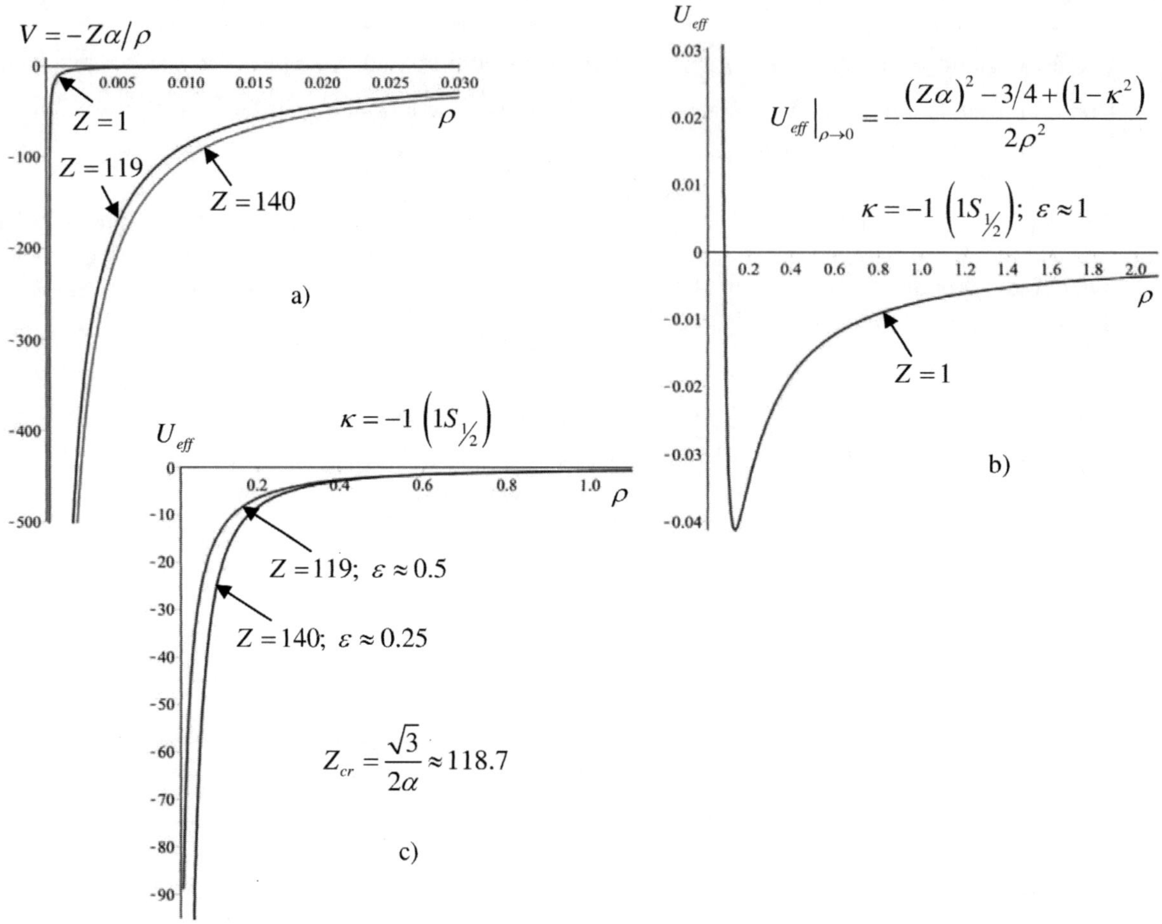

Figure 1. a) the attracting Coulomb field $(V=-Z\alpha/\rho)$; the effective potentials of second-order equation (8): b) as $Z<Z_{cr}$, c) as $Z>Z_{cr}$.

In dimensional units

$$r_{cl}=Z\frac{e^2}{mc^2}\bigg/\left(1+\frac{E}{mc^2}\right). \tag{11}$$

For $\psi(\rho)=f(\rho)/\rho$ equation (8) is

$$\frac{d^2 f}{d\rho^2}+2\left(E_{Schr}-U_{eff}(\rho)\right)f=0, \tag{12}$$

where

$$E_{Schr}=\frac{1}{2}\left(\varepsilon^2-1\right) \tag{13}$$

and

$$U_{eff}\big|_{\rho\to\rho_{cl}}=\frac{3}{8}\frac{1}{(\rho-\rho_{cl})^2}. \tag{14}$$

Barrier (14) is quantum mechanically impenetrable. If $f\big|_{\rho\to\rho_{cl}}=(\rho-\rho_{cl})^s\sum_k f_k(\rho-\rho_{cl})^k$ then from the inditial equation for (12) $s_1=3/2;\ s_2=-1/2$. Solution $s_2=-1/2$ is unphysical because of square-nonintegrability of the wave function $f(\rho)$. Then $f\big|_{\rho\to\rho_{cl}} = f_0(\rho-\rho_{cl})^{3/2}$ and in accordance with (9) $F\big|_{\rho\to\rho_{cl}} = C(\rho-\rho_{cl})^2$; $F\big|_{\rho\to\rho_{cl}}$, $\left.\frac{dF}{d\rho}\right|_{\rho\to\rho_{cl}} = 0$; $G\big|_{\rho\to\rho_{cl}} = C_1$, $\left.\frac{dG}{d\rho}\right|_{\rho\to\rho_{cl}} = C_2$.

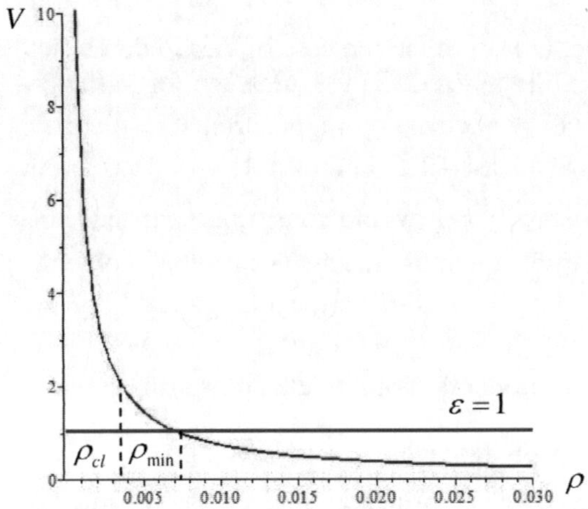

a) $V(\rho)=Z\alpha/\rho$ at $Z=1$ (for $\varepsilon=1$, $\rho_{\min}=1/137$, $\rho_{cl}^{e}=1/274$)

$F(\rho_{cl})=0$; $\left.\frac{dF}{d\rho}\right|_{\rho=\rho_{cl}}=0$

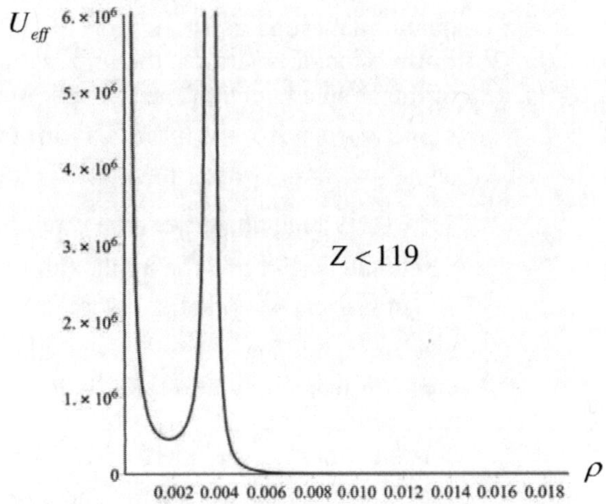

$U_{eff}(\rho)$

b) Z<Z$_{cr}$ ($Z=1$, $\kappa=-1$, $\varepsilon=1$, $\rho_{cl}=0.00365$)

$f(\rho_{cl})=0$

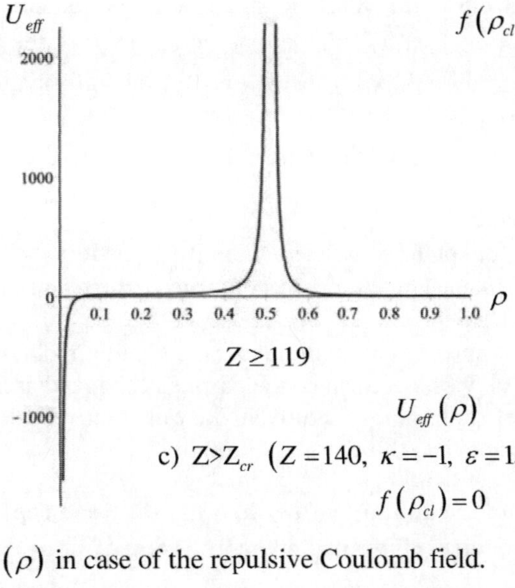

Z ≥ 119

$U_{eff}(\rho)$

c) Z>Z$_{cr}$ ($Z=140$, $\kappa=-1$, $\varepsilon=1$, $\rho_{cl}=0.51$)

$f(\rho_{cl})=0$

Figure 2. The curves $V(\rho)$ and $U_{eff}(\rho)$ in case of the repulsive Coulomb field.

Consequences of existence of impenetrable barrier

1. The existence of an impenetrable barrier in an electron motion in the repulsive Coulomb field, in terms of the scattering theory, means the existence of a well-defined impact parameter p

$$p = r_{cl}. \tag{15}$$

When probing the internal structure of an electron in all possible experiments with electron-electron scattering, the impossibility to penetrate towards radii $r < r_{cl}$ should be recorded.

In this case, the value r_{cl} decreases with the growth of the electron energy:

IOP Conf. Series: Journal of Physics: Conf. Series 938 (2017) 012033 doi:10.1088/1742-6596/938/1/012033

$$r_{cl} \approx \frac{Zr_e}{E} m_e c^2 \quad \text{at } E >> m_e c^2. \tag{16}$$

In (16), r_e is classical radius of electron $\left(r_e = e^2/mc^2\right)$.

Let us emphasize that this conclusion holds true only within the applicability of one-particle quantum mechanics.

2. For motion of positrons, the impenetrable barrier (14) exists in the case of the Coulomb field of the atomic nuclei of the Periodic system. For fully ionized stable atoms, the Ze variation range is from $Z=1$ (Hydrogen ion) to $Z=92$ (U^{238} nucleus). For a positron at rest in the center-of-mass system, the barrier radius varies from $1.4\cdot 10^{-13}$ cm $(Z=1)$ to $1.3\cdot 10^{-11}$ cm $(Z=92)$. Such change can be establish experimentally that would confirm the existence of discussed barrier and the applicability of one-particle quantum mechanics for this problem.
   At subsequent hypothetical increase Z up to $Z = Z_{scr} = (170-177)$, an energy level of electron ground state reaches the upper limit of the lower continuum $E=-m$. Then, in case of electron-free S-states on K-shell, a spontaneous production of two electron-positron pairs are occurred [3].
   The electrons occupy the free K-shell, the atomic charge decreases by two units $((Z_{scr}-2)e)$, positrons leave the atomic system where, in principal, they can be recorded. However, taking into account an impenetrable barrier (14), the positrons will be in potential well within the barrier with $r = r_{cl}$ (see Fig. 2). Hereafter the positrons annihilate with electrons of the K-shell with emission of quanta and the system reverts to the initial state $(Z = Z_{scr})$ with further event recurrence. For the outer world, the source of the annihilation of electrons with positrons will be the sole information source for the atomic system with $Z = Z_{scr}$.
3. The existence of the impenetrable barrier indirectly revealed the relationship between the spin and the charge of particles.
   Indeed:
   - for spinless particles, there is no barrier;
   - when signs of charges the spin particle and the Coulomb potential are opposite, there is no barrier;
   - when the spin particle and the Coulomb potential have like charges, the barrier is present, and arrangement of the barrier is changed at change of particle energy.

Future models of the substructure should quantitatively describe specified above the relationship between the spin and the charge of particles.

2. Second-order equation and Foldy-Wouthuysen representation

For the spinor $u(\mathbf{r})$, the equation has the form of

$$\left[(E-eA_0)^2 - (\mathbf{p}-e\mathbf{A})^2 - m^2 + e\boldsymbol{\sigma}\mathbf{H} - \frac{1}{E-eA_0+m} ie\boldsymbol{\sigma}\mathbf{E}\boldsymbol{\sigma}(\mathbf{p}-e\mathbf{A})\right]u(\mathbf{r}) = 0. \tag{17}$$

In case of stationary states, the electromagnetic potentials $A_0(\mathbf{r}), A^k(\mathbf{r})$ are independent of time. Let $A_0(\mathbf{r})=0,\ A^k(\mathbf{r}) \neq 0$. Then equation (17) is self- adjoint and

$$Eu(\mathbf{r}) = \left(\pm\sqrt{m^2 + (\mathbf{p}-e\mathbf{A})^2 + e\boldsymbol{\sigma}\mathbf{H}}\right)u(\mathbf{r}). \tag{18}$$

Then, let us consider the case of $A^k(\mathbf{r})=0, A_0(\mathbf{r})\neq 0$. In this case, the latter addendum in (17) is a nonself-adjoint operator. Let us perform the non-unitary transformation of the similarity of the equation (17) and spinor $u(\mathbf{r})$

$$\Phi(\mathbf{r})=gu(\mathbf{r}), \tag{19}$$

where

$$g=(E-eA_0+m)^{-1/2}. \tag{20}$$

As a result, equation (17) is reduced to the form:

$$g\left[(E-eA_0)^2-\mathbf{p}^2-m^2-\frac{1}{(E-eA_0+m)}i\boldsymbol{\sigma}\mathbf{E}\boldsymbol{\sigma}\mathbf{p}\right]g^{-1}\Phi(\mathbf{r})=0 \tag{21}$$

and, finally, to

$$\left[(E-eA_0)^2-\mathbf{p}^2-m^2-\frac{3}{4}\frac{1}{(E-eA_0+m)}\mathbf{E}^2+\frac{1}{2}\frac{1}{E-eA_0+m}\mathrm{div}\mathbf{E}+\frac{1}{E-eA_0+m}\boldsymbol{\sigma}(\mathbf{E}\times\mathbf{p})\right]\Phi(\mathbf{r})=0. \tag{22}$$

Equation (18) with $A^k(\mathbf{r})\neq 0, A_0(\mathbf{r})=0$, written actually in the Hamiltonian form and can be easily compared with the Dirac equation in the Foldy-Wouthuysen representation (FW) [4].
In case of $A^k(\mathbf{r})=0, A_0(\mathbf{r})\neq 0$, it is impossible to write equations (17), (22) in the closed Hamiltonian form. It can be done only by the method of successive approximations.

2.1. Foldy-Wouthuysen chiral representation

If we present bispinor as

$$\psi(\mathbf{r},t)=\begin{pmatrix}\psi_L(\mathbf{r})\\ \psi_R(\mathbf{r})\end{pmatrix}e^{-iEt}, \tag{23}$$

then the addendum βm in the Dirac equation mixes spinors $\psi_L(\mathbf{r}),\psi_R(\mathbf{r})$ and the Dirac equation with the non-zero mass of fermions does not possess chiral symmetry.
Conversely, in second-order equation there are no addendums mixing spinors ψ_L,ψ_R and it can be written as

$$\left[(p_0-eA_0)^2-(\mathbf{p}-e\mathbf{A}_0)^2-m^2+\boldsymbol{\sigma}\mathbf{H}-ie\boldsymbol{\sigma}\mathbf{E}\right]\psi_L=0, \tag{24}$$

$$\left[(p_0-eA_0)^2-(\mathbf{p}-e\mathbf{A}_0)^2-m^2+\boldsymbol{\sigma}\mathbf{H}+ie\boldsymbol{\sigma}\mathbf{E}\right]\psi_R=0. \tag{25}$$

Equations (24), (25) are chiral-symmetric. Presence or absence of fermion mass will have no effect on chiral symmetry (24), (25). Noteworthy that equations (24), (25) allow obtaining a closed expression for the Hamiltonian FW in chiral representation

$$H_{FW}=eA_0+\gamma_5\sqrt{m^2+(\mathbf{p}-e\mathbf{A})^2+e\boldsymbol{\sigma}\mathbf{H}-ie\boldsymbol{\alpha}\mathbf{E}}. \tag{26}$$

3. References

[1] Dirac P A M 1958 *The Principles of Quantum Mechanics.* Forth Edition (Oxford at the Clarendon Press).
[2] Zeldovich Ya B and Popov V S 1971 *Usp. Fiz. Nauk* **105**, 403; 1972 *Sov. Phys. Usp.* **14**, 673.
[3] Greiner W, Mueller B and Rafelski J 1985 *Quantum Electrodynamics of Strong Fields* (Berlin-Heidelberg: Spring-Verlag).
[4] Case K M 1954 *Phys. Rev.* **95**, 1323.

IOP Conf. Series: Journal of Physics: Conf. Series **938** (2017) 012034 doi:10.1088/1742-6596/938/1/012034

EMERGENT SPIN AND NEW PHYSICS

Ivanhoe B.Pestov †

Bogoliubov Laboratory of Theoretical Physics, Joint Institute for Nuclear Research, 141980 Dubna, Moscow Region, Russia
† E-mail: pestov@theor.jinr.ru

Abstract

In quantum mechanics and particle physics, spin is considered as an intrinsic form of orbital angular momentum. Our goal here is to demonstrate that the well known fundamental idea of Kronig, Uhlenbek and Goudsmit can be presented as an intrinsic form of symmetric top rotation motion. It will be shown that internal symmetry of intrinsic top can be realized on a set of the simplest geometrical quantities, which themselves do not exhibit this property. That is why an intrinsic form of top angular momentum will be called emergent spin. The concept of emergent spin is more general than the concept of spin and, hence, it can be interesting in terms of discussion of the possible ways to look for a physics beyond the Standard Model.

A. On the fundamental (field - theoretical) level the idea of spin as the other being of orbital angular momentum was realized in the form of the Pauli equation and the Dirac equation. In the contemporary physics the Dirac equation is considered as a fundamental equation connecting the electron's spin with its electromagnetic properties. The Pauli-Dirac spin obeys commutation relations analogous to those of the orbital angular momentum: $[S_j,\, S_k] = i\varepsilon_{jkl}S_l, \quad S_1^2 + S_2^2 + S_3^2 = s(s+1) = \frac{3}{4}$.

Emergent spin obeys commutation relations analogous to those of the symmetric top angular momentum [1]:

$$[S_j,\, S_k] = i\varepsilon_{jkl}S_l, \quad S_1^2 + S_2^2 + S_3^2 = s(s+1) = \frac{3}{4}, \tag{1}$$

$$[\tilde{S}_j,\, \tilde{S}_k] = i\varepsilon_{jkl}\tilde{S}_l, \quad \tilde{S}_1^2 + \tilde{S}_2^2 + \tilde{S}_3^2 = \frac{3}{4}, \quad [S_j,\, \tilde{S}_k] = 0. \tag{2}$$

Being realized this Lie algebra of internal symmetry defines intrinsic top. A field which carries emergent spin and appropriate internal symmetry will be called the emergent spin field or simply the spin field.

B. A set of the simplest (irreducible) geometrical quantities on the four-dimensional smooth manifold consists from scalar field $a(x)$, covariant vector field $a_i(x)$, and antisymmetric covariant tensor fields $a_{ij}(x)$, $a_{ijk}(x)$ and $a_{ijkl}(x)$, which we arrange as a tuple

$$\mathbf{A} = (a, a_i, a_{ij}, a_{ijk}, a_{ijkl}), \quad i, j, k, l = 1, 2, 3, 4. \tag{3}$$

IOP Conf. Series: Journal of Physics: Conf. Series **938** (2017) 012034 doi:10.1088/1742-6596/938/1/012034

The set of the tuples in question can be considered as a linear space of the dimension 2^4. With the general definition of linear space we have nothing to do with preferred dimension of this space. However, it will be shown that from the geometrical and group-theoretical points of view the dimension 2^4 (in general 2^n) is singled out.

Let us exhibit a backbone property that gives possibility to interpret 2^4- tuple (3) as a system. We consider linear transformations $L : \overline{\mathbf{A}} = L\mathbf{A}$ which are defined by a set of totally antisymmetrical tensor fields of the type (p,q), $T^{i_1\cdots i_p}_{j_1\cdots j_q}$, $\quad (p, q = 0, 1, 2, 3, , 4)$. We put $\overline{\mathbf{A}} = (\overline{a}, \overline{a}_i, \overline{a}_{ij}, \overline{a}_{ijk}, \overline{a}_{ijkl})$, $\quad (i, j, k, l = 1, 2, 3, 4)$ and have

$$\overline{a} = Ta + T^m a_m + \frac{1}{2!}T^{mn}a_{mn} + \frac{1}{3!}T^{mnp}a_{mnp} + \frac{1}{4!}T^{mnpq}a_{mnpq},$$

$$\overline{a}_i = T_i a + T_i^m a_m + \frac{1}{2!}T_i^{mn}a_{mn} + \frac{1}{3!}T_i^{mnp}a_{mnp} + \frac{1}{4!}T_i^{mnpq}a_{mnpq},$$

$$\overline{a}_{ij} = T_{ij} a + T_{ij}^m a_m + \frac{1}{2!}T_{ij}^{mn}a_{mn} + \frac{1}{3!}T_{ij}^{mnp}a_{mnp} + \frac{1}{4!}T_{ij}^{mnpq}a_{mnpq},$$

$$\overline{a}_{ijk} = T_{ijk} a + T_{ijk}^m a_m + \frac{1}{2!}T_{ijk}^{mn}a_{mn} + \frac{1}{3!}T_{ijk}^{mnp}a_{mnp} + \frac{1}{4!}T_{ijk}^{mnpq}a_{mnpq},$$

$$\overline{a}_{ijkl} = T_{ijkl} a + T_{ijk}^m a_m + \frac{1}{2!}T_{ijk}^{mn}a_{mn} + \frac{1}{3!}T_{ijk}^{mnp}a_{mnp} + \frac{1}{4!}T_{ijk}^{mnpq}a_{mnpq}.$$

The identical transformation E is defined by the following natural conditions $E^{i_1\cdots i_p}_{j_1\cdots j_q} = 0, \;$ if $\; p \neq q, \quad E^{i_1\cdots i_p}_{j_1\cdots j_p} = \delta^{i_1\cdots i_p}_{j_1\cdots j_p}$, where $\delta^{i_1\cdots i_p}_{j_1\cdots j_p}$ is the generalized Kronecker delta.

The so defined local internal transformations form the group $GL(2^4, \mathbf{R})$ which allows one to consider the 2^4- tuple (3) as a unit. The hidden and underlying properties of this group can be opened if we consider the space of tuples on the Riemann manifold with metric $g_{ij}(x)$, which is called upon to demonstrate a deep connection between the concept of intrinsic top and gravity.

We will construct a natural general covariant basis in the Lie algebra $gl(2^4, \mathbf{R})$ of $GL(2^4, \mathbf{R})$ and describe emergent spin as a bipolar structure on the group in question. The bipolar structure means that the two dual sets of commuting operators exist which define the Lie algebra of the group in question and its two dual subgroups S and $\widetilde{S}$. The transformations of these subgroup commute with each other. These results express in the mathematical form that the tuple in question can be put in correspondence to the intrinsic top and that emergent spin field $\mathbf{A}$ is the space of the two-valued representation of the dual groups S and $\widetilde{S}$.

To this end, let us consider in the linear space of tuples (3) the natural algebraic operators $\overline{\mathbf{A}} = E_{\mathbf{v}}\mathbf{A}$ and $\overline{\mathbf{A}} = I_{\mathbf{v}}\mathbf{A}$ defined by the vector field v^i as follows:

$$E_{\mathbf{v}} : \overline{\mathbf{A}} = E_{\mathbf{v}}\mathbf{A} = (0,\ v_i a,\ v_{[i}a_{j]},\ v_{[i}a_{jk]},\ v_{[i}a_{jkl]}),$$

$$I_{\mathbf{v}} : \overline{\mathbf{A}} = I_{\mathbf{v}}\mathbf{A} = (v^m a_m,\ v^m a_{mi},\ v^m a_{mij},\ v^m a_{mijk},\ 0),$$

where the square brackets $[\cdots]$ denote the process of alternation and $v_i = g_{ij}v^j$. For any vector fields v^i and w^i we have $I_{\mathbf{v}}E_{\mathbf{w}} + E_{\mathbf{w}}I_{\mathbf{v}} = (\mathbf{v}|\mathbf{w})\cdot E$, where $(\mathbf{v}|\mathbf{w}) = g_{ij}v^i w^j$ is scalar product. We mention also the evident relations $E_{\mathbf{v}}E_{\mathbf{w}}+E_{\mathbf{w}}E_{\mathbf{v}}=0, \quad I_{\mathbf{v}}I_{\mathbf{w}}+I_{\mathbf{w}}I_{\mathbf{v}}=0$. To complete, let us introduce the numerical diagonal operator Z that is defined by the conditions $Z^{i_1\cdots i_p}_{j_1\cdots j_q}=0, \quad \text{if} \quad p\neq q, \quad Z^{i_1\cdots i_p}_{j_1\cdots j_p}=(-1)^p\delta^{i_1\cdots i_p}_{j_1\cdots j_p}$. From the definition of Z, we immediately have the following relations: $E_{\mathbf{v}}Z+ZE_{\mathbf{v}}=0, \quad I_{\mathbf{v}}Z+ZI_{\mathbf{v}}=0, \quad Z^2=E$.

Now we introduce the fundamental operators:

$$Q_{\mathbf{v}}=E_{\mathbf{v}}-I_{\mathbf{v}}, \quad \tilde{Q}_{\mathbf{v}}=(E_{\mathbf{v}}+I_{\mathbf{v}})Z, \tag{4}$$

which define a bipolar structure on the group $GL(2^4,\mathbf{R})$ and at the same time a concept of intrinsic top.

From the definition (4), it follows that

$$Q_{\mathbf{v}}Q_{\mathbf{w}}+Q_{\mathbf{w}}Q_{\mathbf{v}}=-2(\mathbf{v}|\mathbf{w})\cdot E, \quad \tilde{Q}_{\mathbf{v}}\tilde{Q}_{\mathbf{w}}+\tilde{Q}_{\mathbf{w}}\tilde{Q}_{\mathbf{v}}=-2(\mathbf{v}|\mathbf{w})\cdot E,$$

and hence $Q^2_{\mathbf{v}}=\tilde{Q}^2_{\mathbf{v}}=-(\mathbf{v}|\mathbf{v})\cdot E$. From definition (4), we also deduce another important relation

$$\tilde{Q}_{\mathbf{v}}Q_{\mathbf{w}}=Q_{\mathbf{w}}\tilde{Q}_{\mathbf{v}},$$

that fulfills at any vector fields $\mathbf{v}$ and $\mathbf{w}$.

Further, let us introduce the dual operators

$$Q_{\mathbf{v}\wedge\mathbf{w}}=\frac{1}{2!}(Q_{\mathbf{v}}Q_{\mathbf{w}}-Q_{\mathbf{w}}Q_{\mathbf{v}}), \quad \text{and} \quad \tilde{Q}_{\mathbf{v}\wedge\mathbf{w}}=\frac{1}{2!}(\tilde{Q}_{\mathbf{v}}\tilde{Q}_{\mathbf{w}}-\tilde{Q}_{\mathbf{w}}\tilde{Q}_{\mathbf{v}}).$$

The last formula can be generalized as follows. We take four linear independent vector fields $\mathbf{u},\mathbf{v},\mathbf{w},\mathbf{z}$ and introduce the operators that are defined as an alternated product of the operators $Q_{\mathbf{u}}$, $Q_{\mathbf{v}}$, $Q_{\mathbf{w}}$, $Q_{\mathbf{z}}$ and $\tilde{Q}_{\mathbf{u}}$, $\tilde{Q}_{\mathbf{v}}$, $\tilde{Q}_{\mathbf{w}}$, $\tilde{Q}_{\mathbf{z}}$. We see that the operators $Q_{\mathbf{u}\wedge\mathbf{v}\cdots\wedge\mathbf{z}}$ and $\tilde{Q}_{\mathbf{u}\wedge\mathbf{v}\cdots\wedge\mathbf{z}}$ commute with each other and all their possible products put together the general covariant basis in the Lie algebra $gl(2^4,\,\mathbf{R})$ of the group $GL(2^4,\,\mathbf{R})$). The total number of these operators is equal to $2^4\cdot 2^4$ (including the identical operator E). Thus, in the space in question the transformations of the group $GL(2^4,\mathbf{R})$ can be represented in terms of the so called infinitesimal operators.

Now let us consider three orthogonal and unit vector fields $\mathbf{u}$, $\mathbf{v}$ and $\mathbf{w}$, $(\mathbf{u}|\mathbf{v})=(\mathbf{u}|\mathbf{w})=(\mathbf{v}|\mathbf{w})=0$, $(\mathbf{u}|\mathbf{u})=(\mathbf{v}|\mathbf{v})=(\mathbf{w}|\mathbf{w})=1$ and put

$$S_1=\frac{i}{2}Q_{\mathbf{v}\wedge\mathbf{w}}, \quad S_2=\frac{i}{2}Q_{\mathbf{w}\wedge\mathbf{u}}, \quad S_3=\frac{i}{2}Q_{\mathbf{u}\wedge\mathbf{v}},$$

$$\tilde{S}_1=\frac{i}{2}\tilde{Q}_{\mathbf{v}\wedge\mathbf{w}}, \quad \tilde{S}_2=\frac{i}{2}\tilde{Q}_{\mathbf{w}\wedge\mathbf{u}}, \quad \tilde{S}_3=\frac{i}{2}\tilde{Q}_{\mathbf{u}\wedge\mathbf{v}}.$$

One can verify that for the operators, so defined, the relations (1) and (2) are fulfilled and, hence, intrinsic rotational motion really exists, when we consider space of emergent spin fields.

C. Thus, the generalized intrinsic rotational motion (the generalized intrinsic top) emerges as the bipolar structure of the group $GL(2^4,\mathbf{R})$ in question and two sets of the commuting operators $Q_{\mathbf{v}},\cdots,Q_{\mathbf{v}\wedge\mathbf{w}\cdots\wedge\mathbf{z}}$ and $\tilde{Q}_{\mathbf{v}},\cdots,\tilde{Q}_{\mathbf{v}\wedge\mathbf{w}\cdots\wedge\mathbf{z}}$ realize this structure.

The bipolar structure gives a natural way of derivation of equations of the emergent spin field [2]-[6].

Equations for the free complex and real spin fields formally have the same form :

$$\Pi\mathbf{\Psi} + \frac{1}{2}\varphi Q_t\mathbf{\Psi} = m\mathbf{\Psi}, \quad \Pi\mathbf{A} + \frac{1}{2}\varphi Q_t\mathbf{A} = m\mathbf{A},$$

but there is a principal difference between the Lagrangians of these equations. The Lagrangian of the complex emergent spin field is invariant with respect to the transformations

$$\overline{\mathbf{\Psi}} = exp(\alpha Q_e)\mathbf{\Psi}, \quad \overline{\mathbf{\Psi}} = exp(\beta Q_\nu)\mathbf{\Psi},$$

where Q_e is the operator of electrical charge and Q_ν is the operator of the neutrino charge. The operators Q_e and Q_ν anticommute $Q_eQ_\nu + Q_\nu Q_e = 0, \quad Q_e^2 = Q_\nu^2 = -E$. But it is remarkable that the Lagrangian of the real spin field is invariant only with respect to the transformations $\overline{\mathbf{A}} = exp(\alpha Q_e)\mathbf{A}$, generated by the operator of the electric charge. We see that in one case the spin field carries only the electrical charge but in the second case it carries the electrical charge and neutrino charge. The operator of the neutrino charge Q_ν is connected with the parity nonconservation and, hence, it is defined by the orientation of the Riemann manifold. The orientation is the antisymmetrical tensor e_{ijkl} ("element of volume") normalized as $e_{1234} = \sqrt{g}$, where $g = \mathrm{Det}(g_{ij})$. The operator of the neutrino charge Q_ν is defined by the operator $\widetilde{H}$: $(\widetilde{H}\mathbf{A}) = (\frac{1}{4!}e_{mnpq}a^{mnpq}, -\frac{1}{3!}e_{imnp}a^{mnp}, -\frac{1}{2!}e_{ijmn}a^{mn}, e_{ijkm}a^m, e_{ijkl}a)$. For more details see the works sited above.

D. Conclusion. Since searches of new physics is not successful till now we need to have new approaches to the fundamental physics as a whole. Spin is already a great unifying principle in the theoretical physics, but, as it is demonstrated, its potential is far from being exhausted. Concept of emergent spin (intrinsic top) provides new and complete explanation of spin and, hence, theory of elementary particles as well.

References

[1] L.C.Biedenharn and J.D.Louck, *Angular Momentum in Quantum Physics*, Addison-Wesley Publishing Company, Inc, 1981.

[2] I.B. Pestov, Theor. and Math. Phys. **34(1)**, 48 (1978).

[3] I.B. Pestov, Physics of Atomic Nuclei, Vol. **74**, n.7, 1084 (2011); Preprint No. E2-2008-93, JINR (Dubna, 2008).

[4] I.B. Pestov. Physics of Particles and Nuclei, Vol. **44**, No.3, 442 (2013).

[5] I.B. Pestov, J.Phys.Math.6:135 (2015). doi: 10.4172/2090-0902.1000135.

[6] I.B. Pestov, Eur.Phys.J. Web of Conf. **138**, 08008 (2017). doi: 10.1051/epjconf/201713808008.

IOP Conf. Series: Journal of Physics: Conf. Series **938** (2017) 012035 doi:10.1088/1742-6596/938/1/012035

Polarized hydrogen/deuterium molecules

Yu V Shestakov*[1,2], **D M Nikolenko**[1], **I A Rachek**[1], **R Sh Sadykov**[1], **D K Toporkov**[1,2], **A V Yurchenko**[2], **S A Zevakov**[1]
[1] *Budker Institute of Nuclear Physics, 630090 Novosibirsk, Russia*
[2] *Novosibirsk State University, 630090 Novosibirsk, Russia*

E-mail: Yu.V.Shestakov@inp.nsk.su

Abstract. The prototype of a polarized molecular hydrogen/deuterium source which is based on the classical Stern-Gerlach separation scheme has been tested at the Budker Institute of Nuclear Physics (BINP), Novosibirsk. It consists of the circular slit nozzle cooled down to 6.5 K and the two superconducting sextupole magnets. The flux of polarized hydrogen molecules of $3 \cdot 10^{12}$ mol/s was measured for a total gas flow through the nozzle of $5 \cdot 10^{-2}$ Torr·l/s. The obtained results will be used to develop a much more intense source of polarized molecules.

1. Introduction

A polarized deuterium gas target is used in the electron scattering experiments at the VEPP-3 storage ring, BINP [1]. The principal feature of the polarized gas target at VEPP-3 is the use of the Cryogenic Atomic Beam Source (CABS) to feed a storage cell [2]. As it's shown in ref. [3] the intensity of the CABS is limited by an intra beam scattering.

In 1998 the proposal to obtain a high density polarized hydrogen gas target was published [4]. The idea was to use a ring-shaped slit nozzle to form the directed molecular beam passing near the poles of superconducting multipole magnet where the field gradient is maximal. This can ensure a good separation of the molecules with different projections of nuclear spin. Moreover in this configuration the gas density of molecular beam can be reduced compared to a density of beam formed by the ordinary conical nozzle using in the CABS. Hence an intra beam scattering which depends on beam density can also be reduced allowing to obtain a more intense source of polarized molecules. To increase the directivity of the molecular flow the ring-shaped nozzle has to consist of the capillaries.

In 2016 the joint Russian-German project “Towards a Molecular Source for Polarized Deuterium Fuel in Nuclear Fusion Research and other Applications” was supported by RSF-DFG cooperation. This support was provided due to the growing interest in using of polarized fuel for thermonuclear fusion to significantly increase the energy production [5]. The joint project involves the study of the possibility to obtain the nuclear spin polarized molecules of hydrogen isotopes through the implementing the above proposal. To do this, the existing CABS was modified.

2. Modified CABS and the results of computer simulations

Fig. 1 shows the schematic view of the experimental setup developed to transform the existing Cryogenic Atomic Beam Source to a polarized molecular beam source. The dissociator was replaced by the inlet tube. It was coupled to the ring-shaped nozzle with inner and outer diameters of 41.7 and

* To whom any correspondence should be addressed.

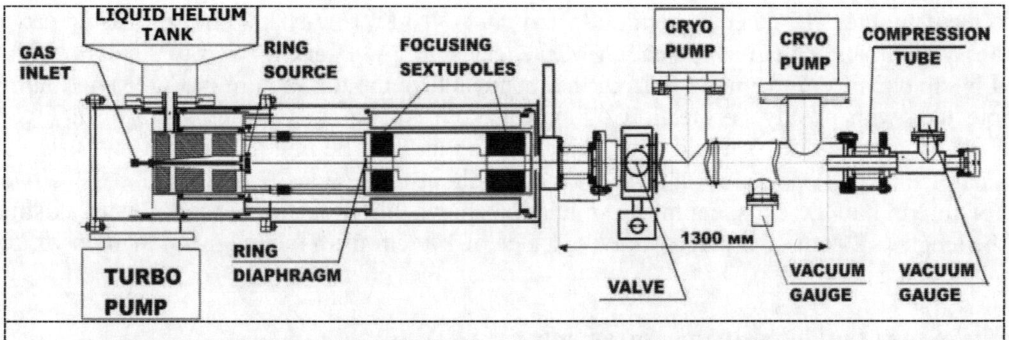

Figure 1. Schematic view of the experimental setup.

41.9 mm respectively (slit width is 0.1 mm). Nozzle was set coaxially with the sextupole magnets of the CABS. Ring diaphragm with inner and outer diameters of 40 and 42 mm was placed at the entrance of the focusing sextupoles. In this geometry the molecules are passing near the poles of magnets, where the gradient of magnetic field is maximal (up to 32 kG/cm). The inner surface of superconducting magnets has helium temperature and was used as a cryopump to get a high vacuum. Additionally the two cryopumps were installed to achieve a suitable vacuum condition along the beam of molecules coming into the compression tube (CT). A vacuum gauge was set in the volume of CT to determine the pressure change when the current through the magnets was turned on and off. The conductance of the CT and the sensitivity of the vacuum gauge to the given gas are known. Therefore it is possible to determine the flux of molecules injected into the CT.
The computer simulations were performed to find the optimal distance from the sextupole magnets to the compression tube in order to obtain the maximum flow of nuclear polarized H_2 entering the CT.

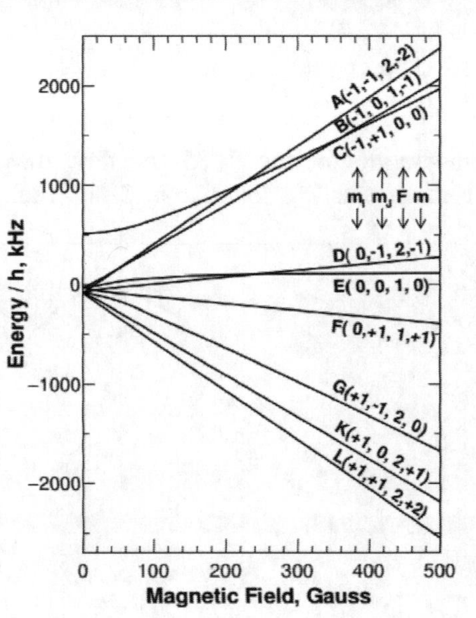

Figure 2. Energies of H_2 states versus a magnetic field. F=I+J, $m=m_I+m_J$ – total angular momentum and its projection.

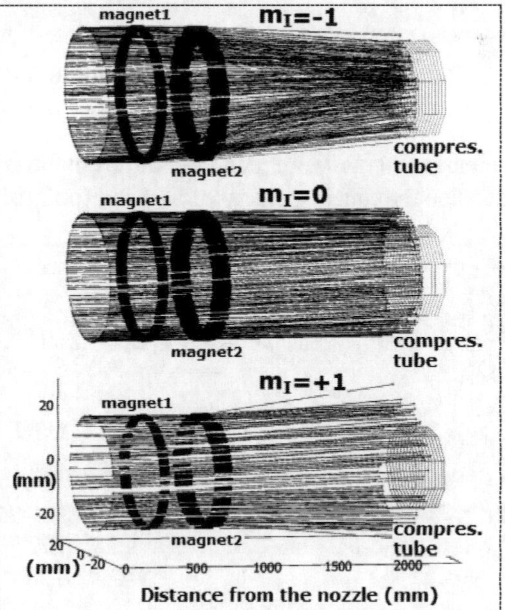

Figure 3. Result of computer simulation of H_2 molecules trajectories for different projections of nuclear spin m_I.

The Zeeman structure of H_2 is shown in Fig. 2. Each energy state is defined by a nuclear spin projection m_I and projection of an orbital angular momentum m_J. Fig. 3 presents the result of ray tracing calculation for H_2 molecules emitted from cold (~10 K) ring-shaped nozzle. As on can see the molecules which have negative projection of nuclear spin (A, B and C energy states in Fig. 2) are focused by a nonuniform magnetic field of the magnets (see the top picture in Fig. 3) and hit the CT. The molecules with positive projection of nuclear spin (G, K and L energy states in Fig. 2) are defocused and miss the CT (see the bottom picture in Fig. 3). The molecules with m_I=0 (D, E and F energy states in Fig. 2) pass through the magnets without deflections and also miss the CT (see the middle picture in Fig. 3). Thus, computer simulation shows that if one takes the CT with a diameter of 3 cm (the length is 20 cm) and places it at a distance of 130 cm from the magnets, then only molecules with m_I = -1 can enter the CT.

3. Results of the beam intensity measurements

In this section the results of the measurements with prototype of polarized hydrogen/deuterium molecules source based on existing CABS are presented. Fig. 4 shows the behavior of the pressure in the CT for a H_2 molecular beam when the current through the coils of magnets was turned on and off.

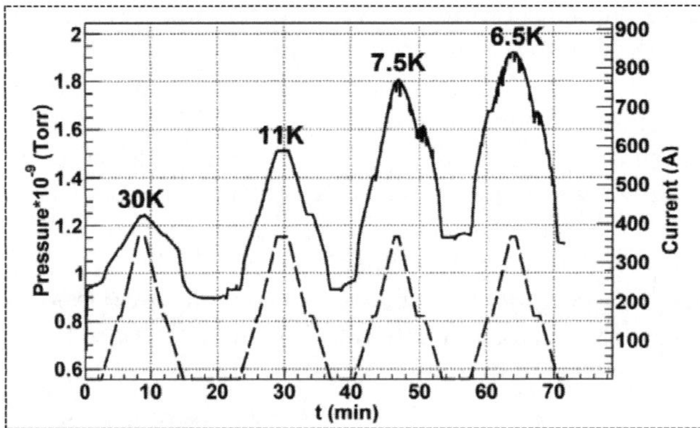

Figure 4. Pressure in the CT (solid curve) and current through the coils of the magnets (dashed curve).

The measurements were done at different nozzle temperature for the fixed gas flow through the nozzle. The maximal intensity of focused (polarized) H_2 molecules was found to be $3\cdot10^{12}$ mol./sec.

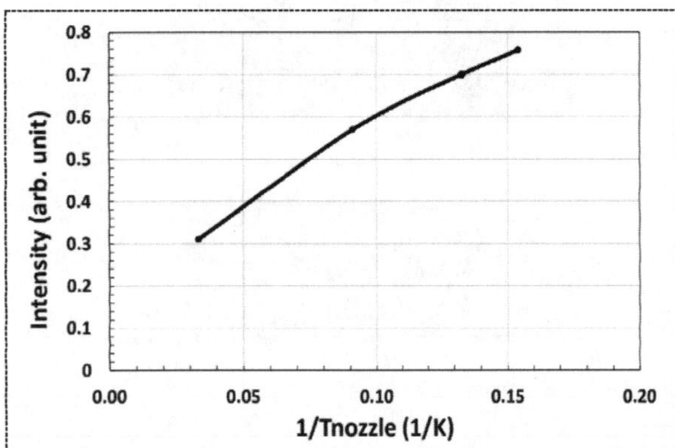

Figure 5. Intensity of polarized H_2 entering the CT as a function of the nozzle temperature.

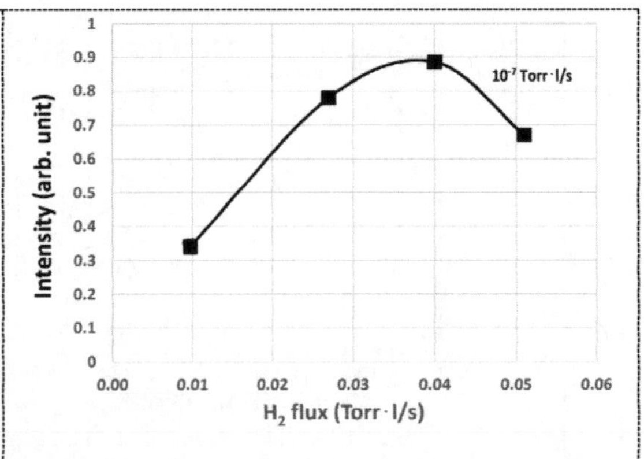

Figure 6. Intensity of polarized H_2 entering the CT versus a total flux through the nozzle.

The measured intensity of polarized H_2 beam as a function of the nozzle temperature is presented in Fig. 5. The measurements were carried out for the fixed total gas flow through the nozzle of $4\cdot10^{-2}$

Torr·l/sec. The computer simulation gives the inverse relation between intensity of polarized molecules and nozzle temperature. As one can see in Fig. 5 the nozzle temperature goes down from 30 to 6.5 K (more than 4 times) but the intensity increases only by the factor 2. This indicates the beam intensity attenuation as the nozzle temperature decreases. The other picture that shows the molecular beam attenuation is presented in Fig. 6. It demonstrates the intensity of polarized H_2 entering the CT versus a gas flow through the nozzle for the fixed temperature of the nozzle.
One probable explanation is an attenuation by a residual gas. A rough estimate shows this is quite possible under given vacuum conditions. But for a full understanding, more research is needed to determine the H_2-H_2 scattering cross section for a low gas temperature (~ 10 K). The other probable reasons for decreasing of the beam intensity are the changing of the picking factor of the gas flow and clusters production into the beam under the pressure rise in the nozzle.

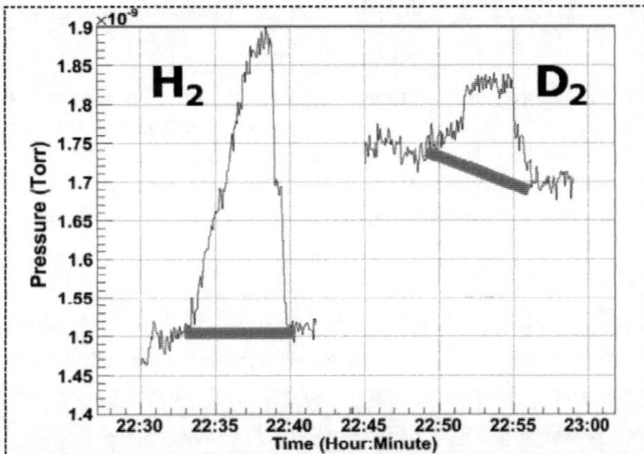

Figure 7. Pressure in the CT (solid curve) and current through the coils of the magnets

The possibility to achieve polarized deuterium molecules was tested. Fig. 7 shows the comparison of focusing efficiency for H_2 and D_2. The changes in the pressure presented in Fig. 7 are proportional to the intensity of polarized molecules. It's clear (see Fig. 7) that the intensity of polarized D_2 molecules is about 5 times smaller than the intensity of H_2 molecules. The reason is that the magnetic moments of D_2 are much smaller than those of H_2. Moreover the geometry of the molecular source prototype was not optimized to get the highest flux of polarized deuterium entering the compressing tube.

4. Outlook

The immediate prospects for the future include:

- full understanding the reasons for decreasing of the molecular beam intensity;
- the block of capillaries has to be used for obtaining a more directional beam from the nozzle;
- measurement of the polarization of molecules using the Lamb-shift polarimeter;
- development of the novel high intensity source of polarized hydrogen/deuterium.

This work was supported by the Russian Science Foundation under the joint with Deutsche Forschungsgemeinschaft grant RSF 16-42-01009 and DFG BU 2227/1-1.

References

[1] Nikolenko D M et al. 2010 *Phys. Atom. Nucl.,* **73, No 8,** 1322
[2] Dyug M V et al. 2002 *Nucl. Instr. Meth.* **A 495,** 8
[3] Dmitri Toporkov 2007 *Proc. of 12th Int. Work. on Pol. Ion Sour., Targ. and Polar. (Upton)* 135
[4] Shestakov Yu V et al. 1998 *Proc. of 13th Int. Sym. on High Ener. Spin Physics (Protvino)* 415
[5] Giuseppe Ciullo 2016 *Springer Proc. in Physics. Nuclear Fusion with Polarized Fuel.* **187,** 1

Empirical analysis of particle masses from PDG-2016

Sergey Sukhoruchkin

Petersburg Nuclear Physics Institute named by B.P. Konstantinov
of NRC Kurchatov Institute, 188300 Gatchina, Russia

E-mail: `sukhoruchkin_si@pnpi.nrcki.ru`

Abstract.
Scalar field is responsible for particle mass generation within the Standard Model. We show here the specific role of scalar and vector field masses among masses of other particles. It is described as a common "tuning effect" in particle masses with the parameters $\delta = 16m_e$, $m_e/3$ derived directly from CODATA relation between masses of nucleons and the electron m_e. Masses of both leptons could be expressed as QED correction ($\alpha/2\pi$) of boson masses M_H, M_Z.

It was noticed by Y. Nambu [1] that the Standard Model "is theoretically unsatisfactory because a) the unification of forces is only partially realized; and b) there are too many input parameters, especially concerning the masses, which are not explained." We consider here an electron–based SM–development [2-6], derived from the fact that nucleons and the electron are stable particles, which determine the mass of the universe and are in the ratio (between their masses) very accurately evaluated in CODATA review as m_n/m_e=1838.6836605(11) [7]. We take into account a place of the nucleon mass between masses of other particles [2-6,8], the discreteness ("tuning effect") with the period $\delta = 16m_e$ close to doubled pion mass splitting.

Observed distribution of particle mass differences (Fig.1 [2]) contains maxima at integer values of δ, namely, k=2 (17 MeV), k=6 (48 MeV), k=13 (104 MeV = m_μ) and k=17 (142 MeV = $m_{\pi^\pm}$). Presence of a doublet at 1671-1687 MeV = $12m_\pi$ and the maximum at 3370 MeV = 24m$_\pi$ means the existence of long-range correlations with $m_\pi = 17\delta$ which corresponds to n=115 for the nucleon mass found by Y. Nambu [9]. Shift of neutron mass from $115{\cdot}16m_e$ - m_e accounts δm_n = 161.6491(6) keV equal to 1/8 of nucleon mass splitting δm_N=1293.3322(4) keV.

Exactly integer ratio δm_N : δm_n =8.00086(3)≈8×1.0001(1) allows representation:

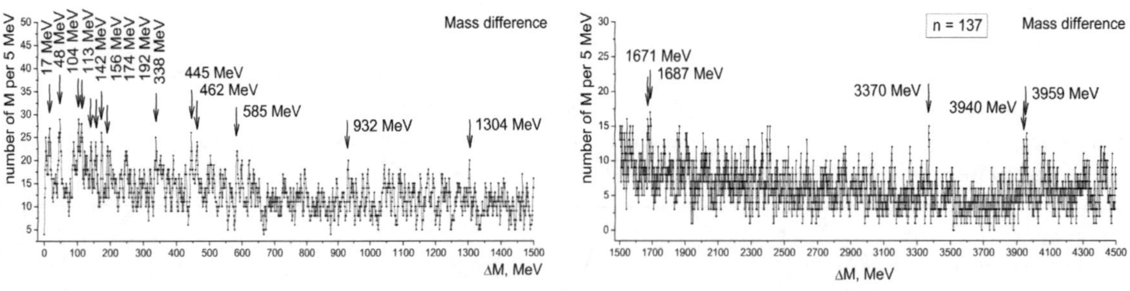

Figure 1. Distribution of differences between particle masses ΔM in regions 0-1500, 1500-4600 MeV.

$$m_n = 115 \cdot 16 m_e - m_e - \delta m_N/8 \qquad\qquad m_p = 115 \cdot 16 m_e - m_e - 9\delta m_N/8 \qquad (1)$$

The value of the shift is close to the QED radiative correction $\alpha/2\pi$=115.95$\cdot 10^{-5}$ to the pion mass $\delta m_n/m_{\pi^\pm}$ = 115.82 $\cdot\, 10^{-5}$ and coincides with the parameter Δ^{TF}=161 keV found in excitations of nuclei in which (according to T. Otsuka and I. Tanihata [10,11]) the one–pion exchange dynamics plays an important role. In Fig. 2 and Table 1 and 2 both parameters of the fine structure 161 keV=17δ' and 170 keV=18δ' observed in excitations of nuclei with valence nucleons [3,5] are presented as integers of the period δ=9.5 keV (integers n=13,14,16). In CODATA relation the shift of -m_e = -3$\cdot m_e/3$ could be assigned to each of three quarks in the nucleon [2-6]. Parameters $\delta m_n = \delta m_N/8$ and $m_e/3$ are considered as the fine structure.

Table 1. Comparison of excitations E^* (in keV) in Z = 51 A-odd nuclei with integer numbers of Δ^{TF}=161 keV=(δm_N=1293 keV)/8 and close to δm_N intervals in 0^+-2^+-1^+ levels of ^{116}Sn (right, boxed).

^{A}Z	^{123}Sb	^{125}Sb	^{127}Sb	^{129}Sb	^{131}Sb	^{133}Sb	^{119}Sb	^{116}Sn	^{116}Sn
(N-70)/2	1	2	3	4	5	6		N=	66
E^*, keV	160.3	332.1	491.2	645.2	798.4	962.0	644	1294	1292
E^*-$\frac{\delta m_N}{8}$	-1	-9	+7	-1	-10	-7	-2	1	-1
n$\frac{\delta k_N}{8}$	161	323	484	646	808	969	646	1293	1293

An appearance of the same fine structure intervals 161 keV and 170 keV in nucleon mass representation and in nuclear data could be connected with the general character of all these QCD-based parameters and with position of the nucleon mass (in nuclear medium, m_N^{nucl} ≈932 MeV, Fig.3) among other above discussed stable mass interval ($m_\mu, f_\pi, m_\pi, \Delta M_\Delta$).

Distinguish stability of intervals ΔM with n=13 (m_μ) and n=17 (m_π) in Fig. 1 as an extension of the discreteness with stable splitting at 2δ and 6δ can be considered with the analogy between

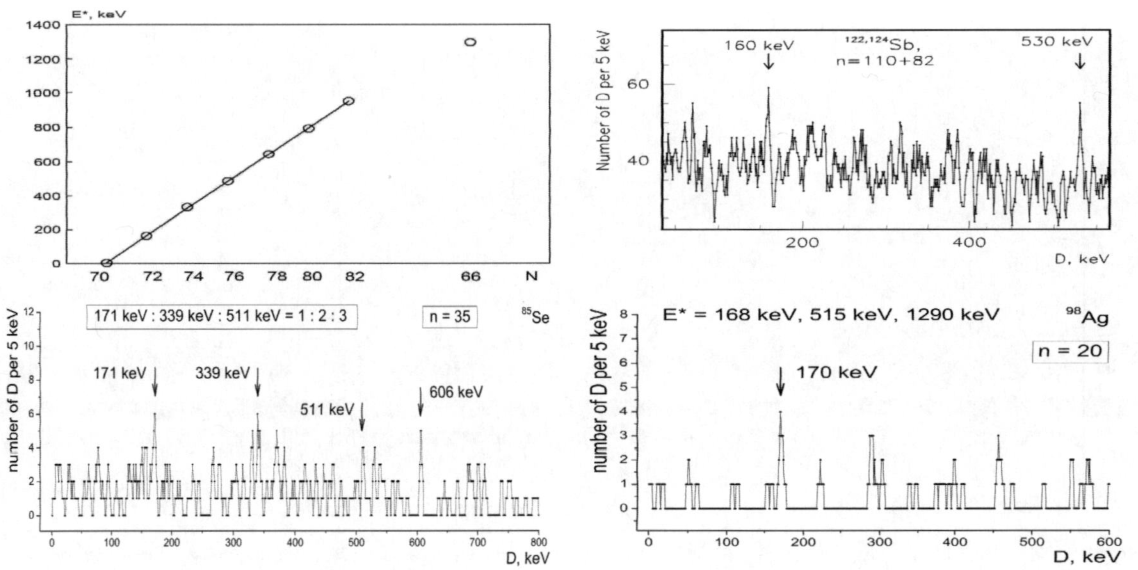

Figure 2. *Top left*: Linear trend in E^* of ^{odd}Sb with a slope Δ^{TF}=161 keV=$\delta m_N/8$ (Table 1). A point at right (N=66) corresponds to stable E^* in ^{116}Sn (equidistant E^* J^π=0^+, 2^+, 1^+ see Table 1, right, boxed).
Top right: Sum D-distribution in ^{122,124}Sb. Maxima at 160 keV=17δ' and 530 keV=4×14δ'.
Bottom left: D-distribution in ^{85}Se (N=51); maxima at k×18δ', similar to E^* in ^{101,103}Sn (N=51,53, Table 2, left).
Bottom right: D-distribution in ^{98}Ag (N=51) with E^*=168, 515, 1290 keV, (Table 2, right).

Table 2. Excitations (in keV) in nuclei with Z=50,34,47, close to integers of $18\delta'$=170 keV=$m_e/3$. Integer numbers of Δ^{TF}=$17\delta'$=161 keV=$\delta m_N/8$ and D in ^{85}Se are given in the center.

Z	50	50	50		34			34	47			47
N	51	53	83		51			50	51			50
^AZ	^{101}Sn	^{103}Sn	^{133}Sn		^{85}Se			^{84}Se	^{98}Ag			^{97}Ag
E^*	170	168	854	1363	170	339	511	1455	167.8	515	1291	1290
$2J^\pi$	7^+	$(7)^+$	3^-	3^-	D	D	D	2+	(3+)	2,3+	1,3+	13^+
$18\delta'$	170	170	851	1362	170	340	511	$(9/8)\delta m_N$	170	511	δm_N	δm_N

the lepton ratio $L = 207 \approx (m_\mu/m_e)$ and the number of fermions in the central field N,ferm. shown in the 1-st line of Table 3 [2-6]. Interval $\delta = 16m_e$ was taken from CODATA relation.

Relations 1:2:17, 1:3 and 1:12:24 between positions of maxima in Fig. 1, the muon and pion mass presentation (as 13×16-1 and 17×16+1) and maxima at ΔM=1687 MeV (n=12×17) and ΔM=3370 MeV (n=24×17) have traces of symmetry–motivated origin of the mass discreteness.

Table 3. Comparison of ratios between masses m_μ/M_Z, $f_\pi/(2/3)m_t$ and $\Delta M_\Delta/M_H$, QED parameter $\alpha/2\pi$ and numbers of fermions in the central field (central line, boxed in the bottom line is the hole configuration in $1p$ shell). One asterisk: configuration $1s^4_{1/2}$, $1p^8_{3/2}$, $1p_{1/2}$; two asterisks - configuration: new principal quantum number; three asterisks - configuration – $1s^4_{1/2}$, $1p^8_{3/2,1/2}$.

N ferm.	N = 1	N=16	16·13-1	16·16	16·17+1	16·18
Part./param.	m_e/M_q	δ	m_μ/M_Z	f_π/M'_H	$m_{\pi^\pm}$	$\Delta M_\Delta/M_H$
Ratio	$115.9\cdot10^{-5}$		$115.87\cdot10^{-5}$	$114\cdot10^{-5}$		$117\cdot10^{-5}$
Comments			hole in $1p$, *	filled shells, ***	(**)	

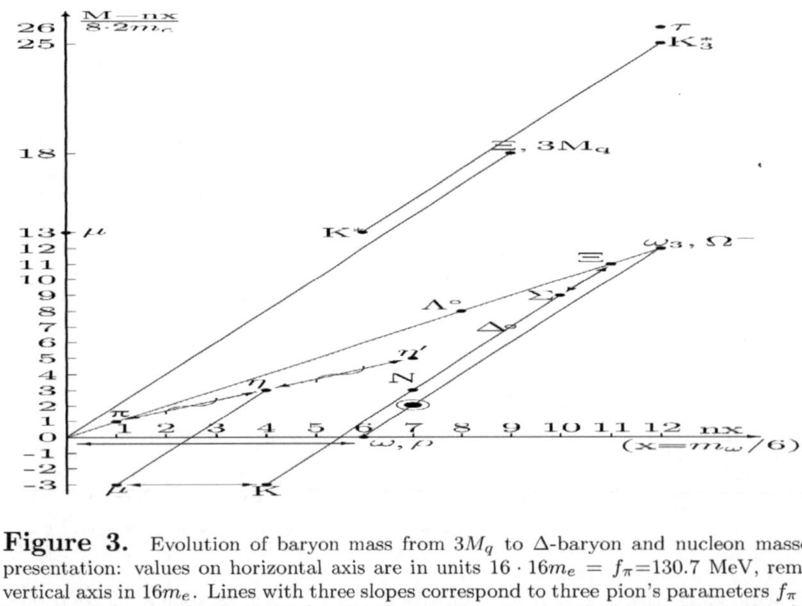

Figure 3. Evolution of baryon mass from $3M_q$ to Δ-baryon and nucleon masses is shown here in two-dimensional presentation: values on horizontal axis are in units $16 \cdot 16m_e$ = f_π=130.7 MeV, remainders – M_i-n($16 \cdot 16m_e$) are along vertical axis in $16m_e$. Lines with three slopes correspond to three pion's parameters $f_\pi = 16\delta$, $m_{\pi^\pm} = 17\delta$ and $\Delta M_\Delta = 18\delta$. Line with m_π=140 MeV=$f_\pi+\delta$ (n=16+1) goes through masses of Λ, Ξ, Ω hyperons (=$8m_\pi$, $11m_\pi$, $12m_\pi$). Stable interval in pseudoscalar mesons $m_{\eta'}$-m_η=m_η-$m^\pm_\pi$ (crossed arrows), is close to M^Δ_q=410 MeV=50δ. Nucleon mass in nuclear medium (m^{nucl}_N, circled point) is close to sum ΔM_Δ+$6f_\pi$ which corresponds to important role of pion's parameters f_π, ΔM_Δ.

Table 4. Presentation of parameters of tuning effect in particle masses (three top sections) and in nuclear data (bottom, sections marked X = 0, 1 at left) by the expression $n{\cdot}16m_e(\alpha/2\pi)^X M$ with QED correction $\alpha/2\pi$ (α=137^{-1}) [6].

X	M	n = 1	n = 13	n = 16	n = 17	n = 18	Comments
-1	3/2			m_t=172.0			
GeV	1	$16M_q=\delta^\circ$	M_Z=91.2	M_H=115 [10]		M_H=126	Part.
	1/2	(m_b-M_q)		M^{L3}=58 [9]			masses
0	1	$16m_e=2m_d-2m_e$	m_μ=106	f_π=130.7	m_π-m_e	ΔM_Δ=147	NRCQM
MeV	3			$M''_q=m_\rho/2$		M_q=441=ΔE_B	param.
1	1	$16m_e=\delta=8\varepsilon_\circ$			$k\delta$-m_n-m_e=	170 = $m_e/3$	Part.
					=161.651(6)		masses
keV	8				δm_N=1293.3		CODATA
1	1	$9.5=\delta'=8\varepsilon'$	123	152	Δ^{TF}=161	170 (Sn)	Nuclear
keV	3				484 (E^*)	512 (Pd)	data

In this work, we considered additional indirect confirmations (Tables 1,2) of the fine structure in CODATA relations (parameters 161 keV=$\delta m_N/8$ and 170 keV=$m_e/3$). The first parameter can be considered together with the mass of the pion due to a closeness of their ratio to $\alpha/2\pi$ which is shown in Tables 3,4 and was discussed in connection with the gravitation in [12-14].

We draw attention to the grouping effect in the mass differences at the value 932 MeV=$6f_\pi+\Delta M_\Delta$ close to the nucleon mass in nuclear media. Presence of similar effect at $M_q=3\Delta M_\Delta$, at $9M_q$, $10M_q$, $12m_\pi$ and $24m_\pi$ (Fig. 1) reflects the stability of intervals connected with the pion ($f_\pi, m_\pi, \Delta M_\Delta$, n=16,17,18). Small splitting within such groupings allowed to confirm general character of mass discreteness connected directly with the electron rest mass $\delta = 16m_e$ (as a cumulative effect with the value $m_e/3$). Interconnection between m_e and QED correction $\alpha/2\pi$ was suggested by R. Feynman, V. Belokurov and D. Shirkov. Presence of a single dynamics of mass generation which follows from exact CODATA relation and includes boson masses (Tables 3,4) allows to construct the "new physics" with existing SM–parameters.

References
[1] Nambu Y 1998 *Nucl. Phys. A* **629** 3c
[2] Sukhoruchkin S I *Nucl. Part. Phys. Proc.* (in press)
[3] Sukhoruchkin S I 2017 *Nucl. Part. Phys. Proc.* **282-284** 189
[4] Sukhoruchkin S I PoS(EPS-HEP2017)746
[5] Sukhoruchkin S I 2017 *Discreteness in particle masses* (LAP) ISBN 978-620-2-02258-3
[6] Sukhoruchkin S I 2016 *J. Phys.: Conf. Series* **678** 012052
[7] Mohr P, Newell D B, Taylor B N 2016 *Rev. Mod. Phys.* **88** 035009
[8] Patrignani C *et al.* 2016 (Particle Data Group) *Chin. Phys.* C **40** 100001
[9] Nambu Y 1952 *Progr. Theor. Phys.* **7** 595
[10] Otsuka T, Suzuku T, Utsino Y 2008 *Nucl. Phys.* A **805** 127c
[11] Tanihata I 2010 *Mod. Phys. Lett.* A **25** 1886 and private communication
[12] Sukhoruchkin S I 2016 PoS(ICHEP2016)1198
[13] Sukhoruchkin S I 1990 *Program and Theses* 40th Meet. Nucl. Spectroscopy (Leningrad, Nauka) 147
[14] Sukhoruchkin S I 1994 *Symmetry Methods in Physics* JINR E2-94-347 vol. 2, 528

IOP Conf. Series: Journal of Physics: Conf. Series **938** (2017) 012037 doi:10.1088/1742-6596/938/1/012037

Statistical analysis of 2D patterns and its application to astrometry

Petr Zavada and Karel Píška

Abstract. A general statistical procedure for analysis of finite 2D patterns, inspired by analysis of heavy-ion data, is developed. The method is verified in the study of publicly available data obtained by the Gaia-ESA mission. We prove that the procedure can be sensitive to the limits of accuracy of measurement, but it can also clearly identify the real physical effects on the large background of random distributions. As an example, the method confirms presence of binary and ternary star systems in the studied data. At the same time the possibility of statistical detection of gravitational microlensing effect is discussed.

Institute of Physics AS CR, Na Slovance 2, CZ-182 21 Prague 8, Czech Republic

E-mail: zavada@fzu.cz

1. Introduction

The motivation of the present study was to modify and generalize the known method for analyzing anisotropic flow in relativistic nuclear collisions [1, 2], which is effectively applied in the current experiments [3] (and citations therein). The method is based on the use of the Fourier expansion of azimuthal distributions of produced particles and allows us to obtain important information on mechanism of nuclear collisions. However, mathematical frame of this method is more general and corresponding formalism can be after minor modifications used even for quite different kind of analysis. Our present idea is focused primarily on astrometry. Recently, some similarity between spiral structures in galactic patterns and heavy ion collisions has been discussed in [4]. However, our approach is different. The astrometric data are decomposed into the set of limited, finite patterns whose parameters are statistically analyzed. The input data are taken from the Gaia survey [5, 6].

The detailed description of Fourier analysis and its application to astrometry is given in the extended version of this report [7]. In the same paper we have described also a complementary statistical method for analysis of the patterns of stars, which deals with angular distances and is important for identification of binary and ternary star systems. The results obtained by application of both methods are presented and discussed in the next section. Sec. 2.1 is devoted to the discussion on gravitational microlensing effect and conditions of its observation.

2. Application to astrometry

The simulations described in [7] have been compared with the data of the recent Gaia catalogue DR1 [6]. We present the results of our analysis from the regions marked in Fig.1.

First, we checked the Fourier analysis, the results are shown in Fig.2. The upper part corresponds to a dense field C and very similar results can be obtained over other regions at the galactic plane. This result can be compared with the Monte-Carlo calculation in [7]. From the

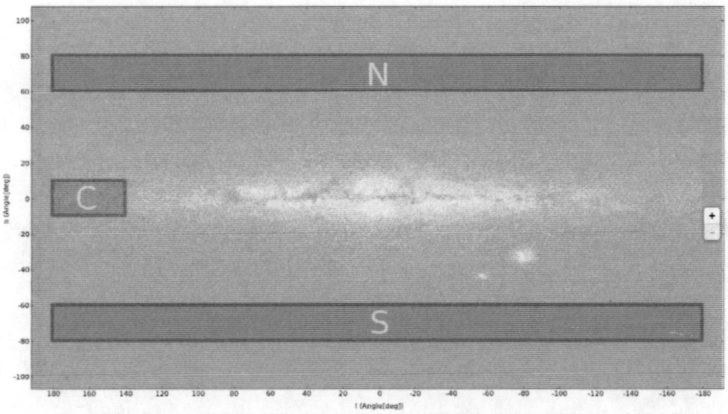

	area: $l \times b$	δ[deg]	$\langle M \rangle$	N_e
C	$\langle 140, 180 \rangle \times \langle -10, 10 \rangle$	0.005	2.60	6962671
N	$\langle -180, 180 \rangle \times \langle 60, 80 \rangle$	0.02	3.26	1437067
S	$\langle -180, 180 \rangle \times \langle -80, -60 \rangle$	0.02	3.30	1448647

Figure 1: Analyzed regions in the GAIA survey, where δ is angular radius of the events, M is their multiplicity and N_e is number of events. Analysis is done only for events $2 \le M \le 80$.

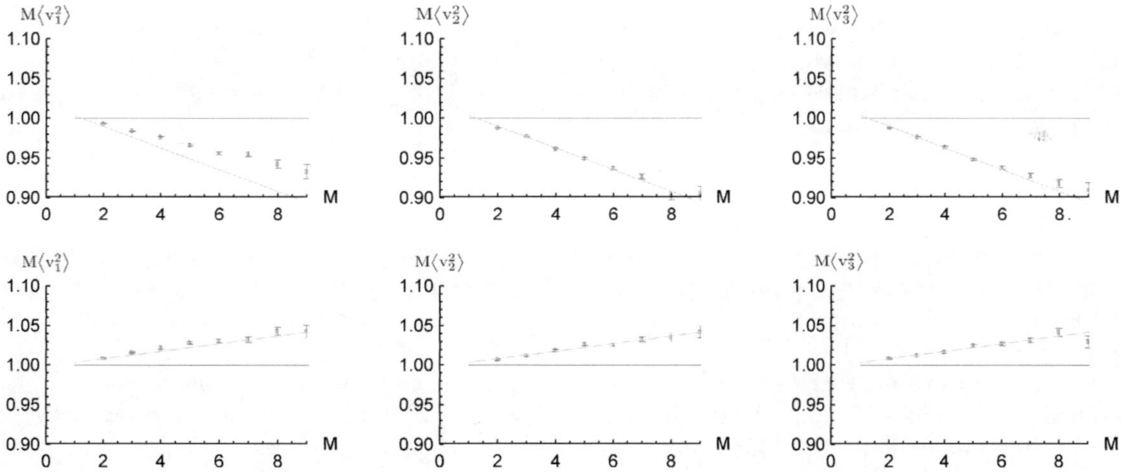

Figure 2: Dependence of Fourier coefficients $\langle v_1^2 \rangle, \langle v_2^2 \rangle, \langle v_3^2 \rangle$ on multiplicity M for the events in the area C (upper panels) and N&S (lower panels). The green line is a linear fit of data taken in $\langle v_2^2 \rangle$ panel.

negative slope of the line one can estimate the radius of the empty spots. The result $M \langle v_i^2 \rangle > 1$ for the sparse field N&S suggests the presence of clustering.

To analyze obtained results in more detail, we applied the complementary method on angular distances. The results are presented in the upper part of Fig.3 and can be compared with the corresponding Monte-Carlo simulation in [7]. In fact we observe a 2D representation of the effect reported in [8] (Sec. 4.4.1., Fig.17), which is due to reduced resolution of two sources at angular distances $d \lesssim 3''$. In the same figure we observe also a peak at small separation $d \lesssim 1''$. In the

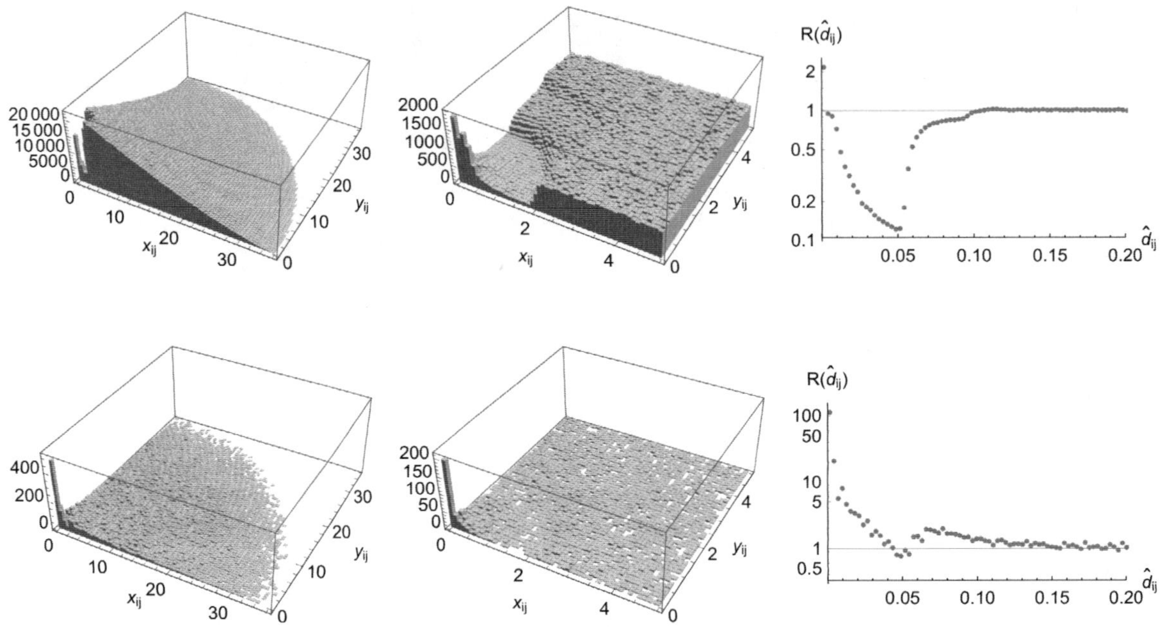

Figure 3: Distributions of angular distances in the region C for $G \leq 15$ mag (lower panels) and for all G (upper panels). The left and central panels are 3D plots, where the unit of x, y is $1''$. The right panels are the ratios of measured distribution of relative distances $\hat{d}_{ij}$ to the interpolation of corresponding random Monte-Carlo data.

cited paper such peak is observed only in sparse field and is interpreted as a presence of binary stars. A simulation described in the same paper suggests that reduced efficiency at $d \lesssim 3''$ correlates with fainter magnitudes G in the field. This can be in agreement with the lower part of Fig.3, where the sources with $G > 15$ mag are excluded and as a result the resolution dip is reduced.

Similar analysis for region N&S is demonstrated in Fig.4. Obviously, in the sparse field the resolution is improved. Apart of the pronounced peak at $d \lesssim 1''$, we observe a clear excess of pairs separated by $d \lesssim 8''$. The whole excess, giving one broader peak for $G \leq 15$ mag can represent binaries.

The next results are related to the ternary star systems. We compare random Monte-Carlo simulation in [7] with the corresponding data from the field N&S in Fig.5. Again, we observe an excess in the region of small ternary distances, roughly $\hat{d}_\alpha \lesssim 0.1$ or equivalently $d_\alpha \lesssim 15''$. The excess is broader and less dependent on G than for binaries. It can represent the bounded ternary star systems.

2.1. Gravitational microlensing

The principle of the gravitational microlensing effect is explained in [9]. For a small angular distance β between two stars the light beams from the more distant one ($\mathbf{S}_2$) and passing the gravitational field of the star on the way ($\mathbf{S}_1$), can reach the observer $\mathbf{O}$ by two ways. The

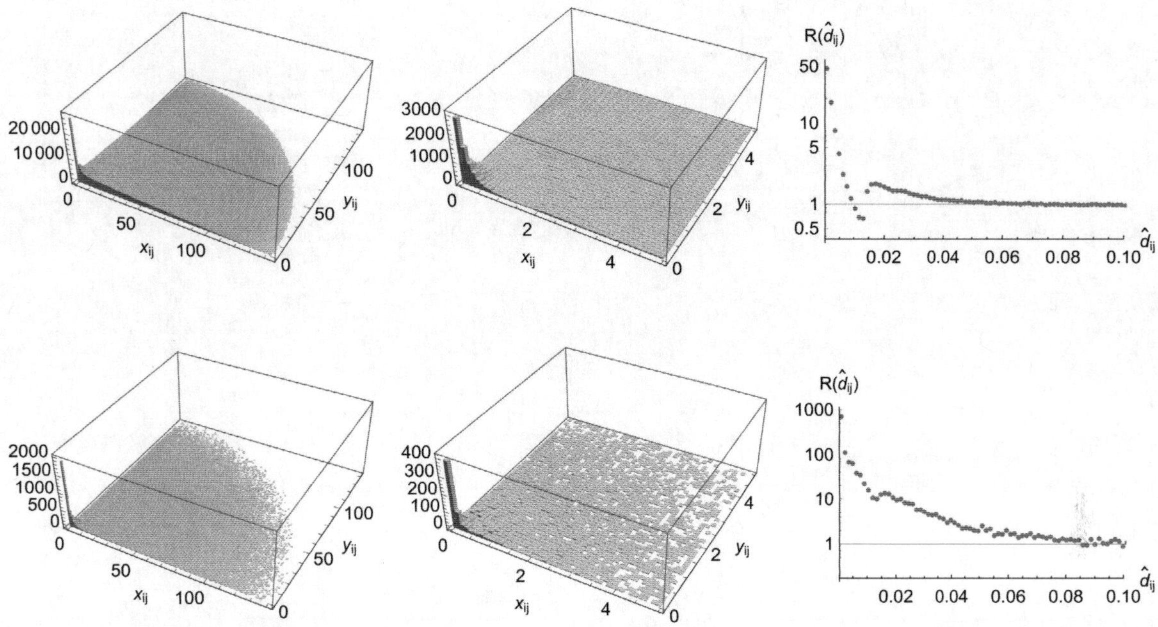

Figure 4: The same as in the previous figure but for region N&S.

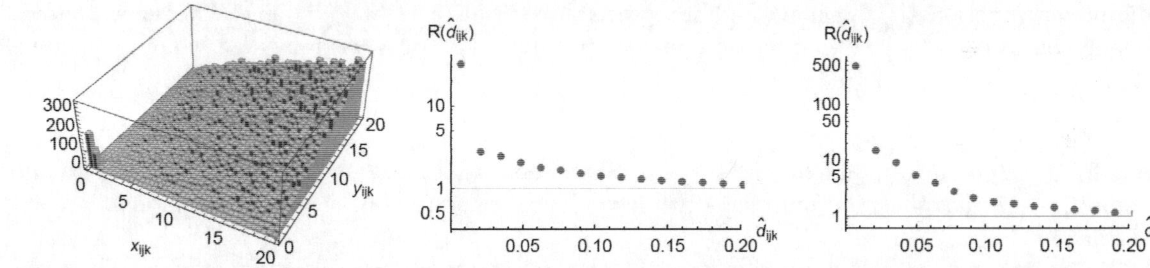

Figure 5: 3D plot of angular ternary distances in the region S&N for all G (left) and ratios of measured distribution of relative distances $\hat{d}_{ijk}$ to the interpolation of corresponding random Monte-Carlo data for all G (centre) and $G \leq 18$ mag (right).

angles seen by the observer are

$$\theta_{12} = \frac{\beta \pm \sqrt{4\alpha_0^2 + \beta^2}}{2}; \qquad \alpha_0 = \sqrt{\frac{4\varkappa M}{c^2}\frac{D_{ds}}{D_d D_s}}, \tag{1}$$

where $\varkappa$ is gravitational constant, M is the mass of the star $\mathbf{S}_1$ and D_d, D_s, D_{ds} are distances $\mathbf{OS}_1, \mathbf{OS}_2, \mathbf{S}_1\mathbf{S}_2$. For $\beta = 0$ the Einstein ring with angular radius $\theta = \alpha_0$ is created. Apparently

$$|\theta_2| \leq \alpha_0 \leq \theta_1. \tag{2}$$

For observation of the effect it is important that angular separation β of the pair is close to α_0. And for α_0 estimate the distance D_d is critical. For example $D_d \approx 10 - 10^2$l.y., $D_s \approx 2D$

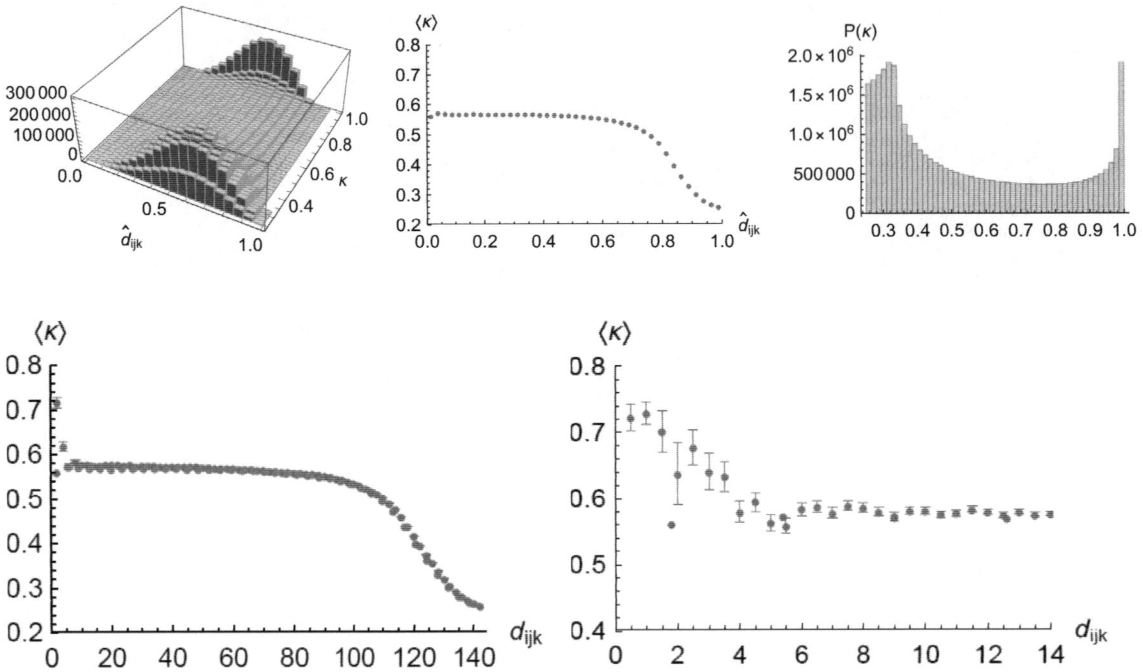

Figure 6: The Monte-Carlo simulation (upper panels) of distribution $P(\hat{d}_{ijk}, \kappa)$ (left), the alignment function $\langle\kappa\rangle$ (centre) and integrated distribution $P(\kappa)$ (right). The lower panels display on two scales the comparison of the $\langle\kappa\rangle$ from the data region N&S (blue) with the Monte-Carlo simulation (red).

and $M \approx M_S$ give roughly $\alpha_0 \approx 20$ mas. Such separation is probably beyond present Gaia resolution. In fact the minimum separation we have registered in any of the regions C,N,S is 59 mas.

There can be the following signature of the gravitational microlensing effect. The light sources $\mathbf{S}_2(\theta_2), \mathbf{S}_1, \mathbf{S}_2(\theta_1)$ as seen by the observer, should have a small separation d_{ijk} (or $\hat{d}_{ijk}$) and should be aligned in a line, or make a narrow triangle within the errors of measurement. So, as a measure of the alignment we define the parameter κ:

$$\kappa = \frac{1}{3}\left(\cos^2\eta_1 + \cos^2\eta_2 + \cos^2\eta_3\right), \tag{3}$$

where η_i are angles in the observed triangle. For $\kappa = 1$ there is maximum alignment (e.g. $\eta_1, \eta_2, \eta_3 = 0, 0, \pi$) and for $\eta_1 = \eta_2 = \eta_3 = \pi/3$ we have minimum $\kappa = 1/4$. In the upper panels of Fig.6 we have shown results of the related Monte-Carlo simulation. In the lower part we have shown its comparison with the data from the field N&S (note rescaled variable d_{ijk} instead of $\hat{d}_{ijk}$). It is important, that the shape of (normalized) Monte-Carlo distribution $P(\hat{d}_{ijk}, \kappa)$ does not depend on the multiplicity of events ≥ 3. The same holds for the distribution $P(\kappa)$ and the dependence $\langle\kappa\rangle$ on d_{ijk}. One can observe a perfect agreement with the data for $d_{ijk} \gtrsim 5''$. On contrary, for the smaller distances among the three sources there is a clear excess of the alignment. Can there be a connection between this effect and the gravitational microlensing - image splitting? Or, more probably, is the excess only another form of the distortion of measurement at small separations? One should add that the conditions for observation of the aligned triplet described above can be rather strict. For $\beta < \alpha_0$ the sources $\mathbf{S}_2(\theta_2), \mathbf{S}_2(\theta_1)$ are getting strongly magnified, so the source $\mathbf{S}_1$ may not be resolved. On the other hand, brightness of $\mathbf{S}_2(\theta_2)$ falls rapidly for $\beta > \alpha_0$ [9].

3. Summary and conclusion

We have proposed the statistical method for analysis of the finite 2D patterns. Each pattern (event) involves random points located inside the disc with fixed radius. We have demonstrated that the method can identify tiny deviations from purely random distributions, for example a tendency to clustering or anti-clustering.

The method has been applied to the analysis of astrometric data obtained by the Gaia mission. In this case the events are sets of the stars observed inside discs of small fixed angular radius in the galactic reference frame, which cover chosen region of the sky. In parallel the statistics of random events (star patterns) is generated by the Monte-Carlo code. Comparison of relevant distributions obtained from the data and Monte-Carlo events proves the presence of binary and ternary star systems in the Gaia survey.

A special attention has been paid to the discussion on possibility of detection of the gravitational microlensing and image splitting effect. Our present conclusion is that the statistical method suggested above can be useful for detection of this effect, however further effort is needed to extract reliable results from the data.

Acknowledgments

This work has made use of data from the European Space Agency (ESA) mission *Gaia* (`https://www.cosmos.esa.int/gaia`), processed by the *Gaia* Data Processing and Analysis Consortium (DPAC, `https://www.cosmos.esa.int/web/gaia/dpac/consortium`). Funding for the DPAC has been provided by national institutions, in particular the institutions participating in the *Gaia* Multilateral Agreement. Further, we are grateful to J.Grygar for deep interest and many valuable comments, D.Heyrovsky and A.F. Zakharov for illuminating discussions on gravitational microlensing effect and O.Teryaev for very useful discussions and inspiring comments.

References

[1] S. Voloshin and Y. Zhang, Z. Phys. C **70**, 665 (1996)
[2] A. M. Poskanzer and S. A. Voloshin, Phys. Rev. C **58**, 1671 (1998)
[3] J. Adam *et al.* [ALICE Collaboration], Phys. Rev. Lett. **116**, no. 13, 132302 (2016)
[4] A. Rustamov and J. N. Rustamov, arXiv:1602.01812 [hep-ph].
[5] Gaia Collaboration, T. Prusti, J. H. J. de Bruijne, A. G. A. Brown, A. Vallenari, C. Babusiaux, C. A. L. Bailer-Jones, U. Bastian, M. Biermann, D. W. Evans and et al. (2016b) The Gaia mission. A&A 595, pp. A1.
[6] Gaia Collaboration, A. G. A. Brown, A. Vallenari, T. Prusti, J. H. J. de Bruijne, F. Mignard, R. Drimmel, C. Babusiaux, C. A. L. Bailer-Jones, U. Bastian and et al. (2016a) Gaia Data Release 1. Summary of the astrometric, photometric, and survey properties. A&A 595, pp. A2.
[7] P. Zavada and K. Píška, arXiv:1708.06560 [astro-ph.IM].
[8] Gaia Collaboration, F. Arenou et al. Gaia Data Release 1. Catalogue validation. A&A 599 (2017), pp. A50.
[9] P. Schneider, J. Ehlers and E.E. Falco, *Gravitational Lenses*, Springer-Verlag 1992.

On the A-dependence of the neutron single-spin asymmetry in pA-collisions

V V Abramov

NRC "Kurchatov" Institute IHEP, Protvino 142281, Moscow region, Russia

E-mail: Victor.Abramov@ihep.ru

Abstract. Preliminary results of the PHENIX experiment on the single-spin asymmetry (A_N) for a neutron production in proton-nucleus collisions are discussed. The asymmetry A_N changes its sign and increases three-fold as the transition from p + p to p + Au interactions occurs. We call this phenomena "anomalous A-dependence of A_N". This unusual behavior is explained in the framework of the chromomagnetic polarization of quarks (CPQ) model. Dependence of A_N on centrality of collisions and the "ridge" effect in heavy ions collisions are discussed.

1. Introduction

Recently, the PHENIX experiment reported about the observation of an anomalous A_N dependence on the atomic weight A of the target [1]. The data are measured in the p+p, p+Al and p+Au collisions at 200 GeV in the c.m. frame. Neutrons are produced in the fragmentation region of polarized protons at a small transverse momentum: $p_T \approx 0.108$ GeV/c, $x_F \approx 0.54$ (see Fig. 1). The data [1] (solid points) and the CPQ model [2] predictions of $A_N(A)$ for the inclusive

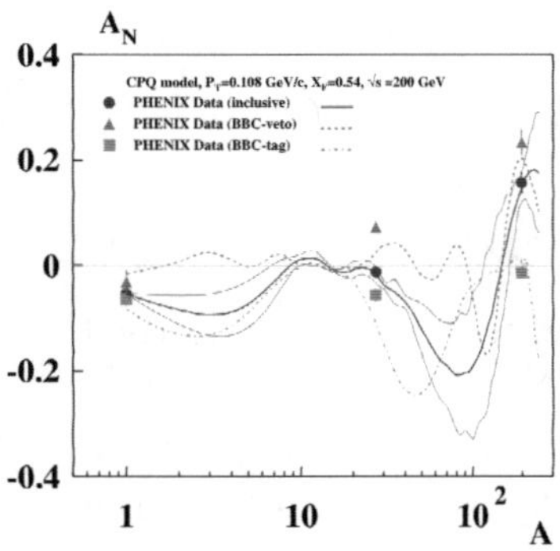

Figure 1. The $A_N(A)$ data [1] for the reaction $p^{\uparrow} + p(A) \to n + X$ and the CPQ model calculations.

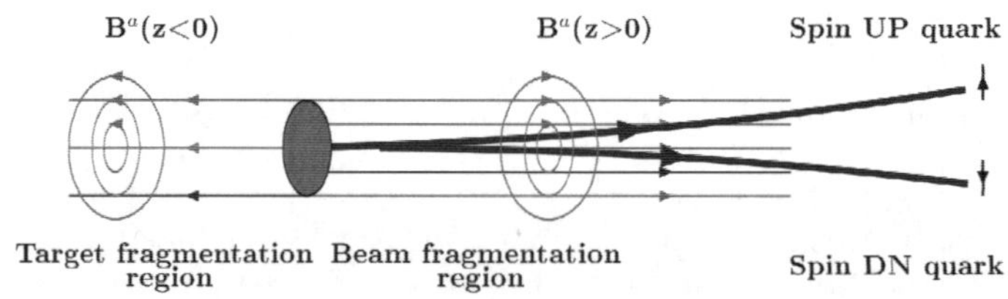

Figure 2. Chromo-magnetic polarization of quarks model (microscopic Stern-Gerlach apparatus). The circular transverse chromomagnetic field $\mathbf{B^a}$ is created by the relativistic moving spectator quarks and antiquarks. The probe quark gets an additional p_T kick to the left (right) when its spin is directed upwards (down). In general, the field direction and its value are different for the beam and target fragmentation regions.

trigger (solid curve) are shown in Fig. 1. Thin lines show estimates of one standard deviation for the CPQ model calculations. When the trigger prohibits any signal in the beam-beam counters

(BBC-veto trigger), the values of $A_N(A)$ are shifted upwards (triangles in Fig. 1). The signal requirement in the BBC counters (BBC-tag trigger) shifts $A_N(A)$ down (squares in Fig. 1).

2. Chromo-magnetic polarization of quarks model

A longitudinal chromoelectric field $\mathbf{E^a}$ and a circular transverse chromomagnetic field $\mathbf{B^a}$ are generated in the interaction region. Single-spin asymmetry is due to an additional p_T kick, caused by the Stern-Gerlach force that is acting on the quark dipole color moment in the inhomogeneous transverse circular chromomagnetic field (see Fig. 2). The precession of the quark spin in the chromomagnetic field changes the Stern-Gerlach force and leads to an oscillation of A_N as a function of its arguments in the case of a high $\mathbf{B^a}$ value. The asymmetry A_N is given by

$$A_N = C(\sqrt{s})F(p_T, A)[G(\phi_A) - \sigma G(\phi_B)], \tag{1}$$

where the function $G(\phi)$ takes into account the quark spin precession and is given by

$$G(\phi) = (1 - \cos\phi)/\phi + \epsilon \cdot \phi. \tag{2}$$

The quark spin precesses in the color field, as is described by the FT-BMT equation [3], updated for the case of chromomagnetic field and color quarks [2]. The parameter $\epsilon = -0.00497 \pm 0.00009$

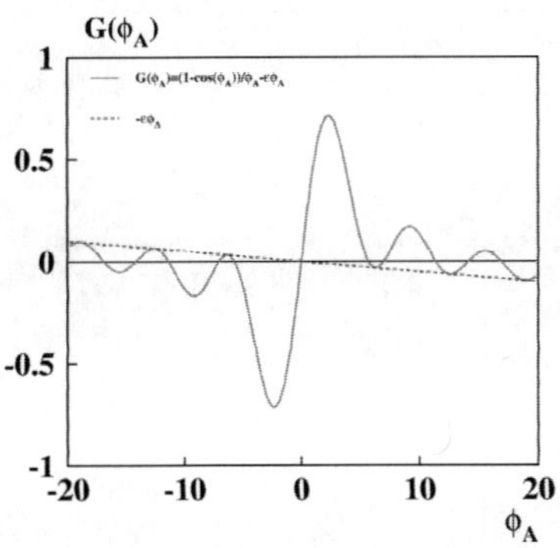

Figure 3. The function $G(\phi_A)$ in (1) and the linear term -$\epsilon\phi_A$.

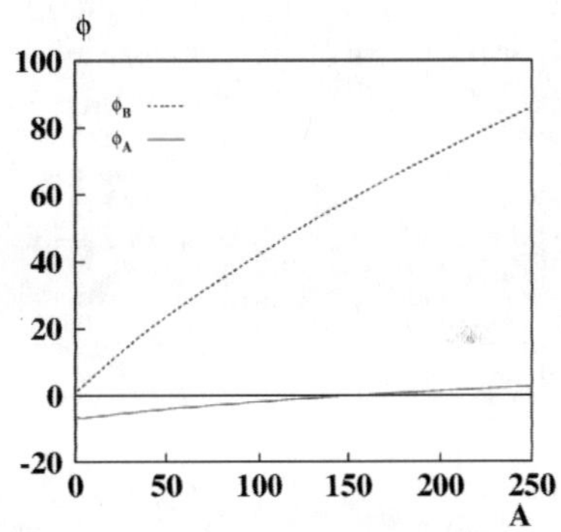

Figure 4. The dependence of ϕ_A and ϕ_B on A.

in (2) and the other phenomenological parameters are determined from the global fit of the polarization data for the 81 inclusive reactions with 3465 data points [2, 4].

The main role in the origin of the anomalous A-dependence of A_N for the neutron production by polarized protons plays the term $[G(\phi_A) - \sigma G(\phi_B)]$ in (1), where $\sigma = 0.178$, ϕ_A and ϕ_B are the "integral angles of quark spin-precession" in the fragmentation regions of a projectile A and target B, respectively. The function $G(\phi_A)$, shown in Fig. 3, is an antisymmetric oscillating function of ϕ_A. The function $G(\phi_A)$ has the minimum at $\phi_A \approx -2.3$ and the maximum at $\phi_A \approx 2.3$. The peaks of $G(\phi_A)$ transform into the corresponding extrema of A_N, as a function of A. The spin-precession angles ϕ_A and ϕ_B for the inclusive reaction $p^\uparrow + p(A) \to n + X$ depend on the target atomic mass A, as is shown in Fig. 4. In particular, $\phi_A \approx -2.3$ at A=86 and $\phi_A \approx 2.3$ at A=215, where the corresponding extrema of $A_N(A)$ are expected.

The contribution (relative weight) ν of spectator quark or antiquark to the effective color field is determined by the color factors: $\nu(qq) \equiv \lambda = -0.1363 \pm 0.0003$ and $\nu(q\bar{q}) = 1$ [2, 4, 5]. It was found that for the qq-pair the antitriplet color state is dominant and for the case of the $q\bar{q}$-pair the singlet color state dominates in the two quark interactions. One of the quarks in the pair above is a spectator and the other is an active probe quark from the detected neutron.

The contributions of spectator quarks and antiquarks to the effective color field, in the fragmentation regions of colliding particles A and B (ν_A and ν_B) are given by

$$\nu_A = 1 + \lambda - 3B_{eff}\tau\lambda, \tag{3}$$

$$\nu_B = 1 + \lambda + 3(B_{eff} - 1)\lambda - 3\tau\lambda, \tag{4}$$

where B_{eff} is the effective number of nucleons in target nuclei, where $\tau = 0.0267 \pm 0.0005$. The spin-precession angles ϕ_A and ϕ_B are proportional to ν_A and ν_B, respectively. In the CPQ model, the effective number of nucleons, participating in the creation of a color field in $p + A_2$ collisions, is given by

$$B_{eff} = A_2(1 - (1 - (A_b/A_2)^{2/3})^{3/2}, \tag{5}$$

where

$$A_b = A_0 + (A_2 - A_0)exp(-W_0/(f_a\sqrt{s}))exp[-(p_T/p_m)^{a_f}], \tag{6}$$

where $f_a = A_2^{1/6}$, $A_2 \equiv A$ - the target atomic mass [2]. The transverse radius of the field $\mathbf{B^a}$ is $R_b \approx 1.2A_b^{1/3}$ fm. The CPQ model parameters are: $A_0 = 0.3084 \pm 0.0012$, $W_0 = 275.6 \pm 1.3$ GeV, $p_m = 0.152 \pm 0.038$ GeV/c, and $a_f = 3.09 \pm 0.05$. For large $\sqrt{s} > W_0/f_a$ and small $p_T \leq p_m$, the second term in (6) dominates, $A_b \approx A$, and the anomalous A-dependence of A_N is expected. At small $p_T \leq p_m$, all nucleons in the target nuclei are a source of spectator quarks due to the uncertainty relation principle. These additional spectator quarks increase the effective color field and cause the anomalous behaviour of $A_N(A)$.

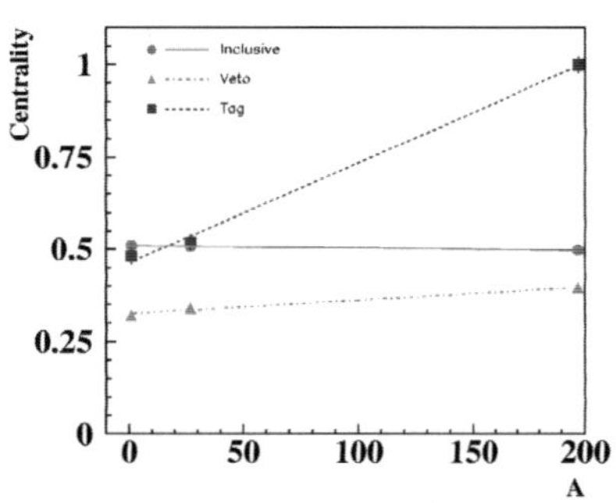

Figure 5. The A-dependence of centrality for the inclusive, BBC-veto and BBC-tag triggers.

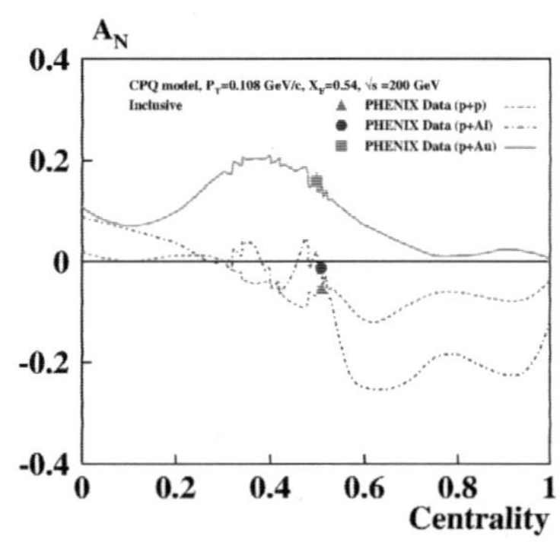

Figure 6. The dependence of A_N on the collision centrality for the p+p, p+Al and p+Au collisions.

3. Comparison of the $\mathbf{A_N(A)}$ data and the CPQ model calculations

The results of the $A_N(A)$ measurements [1] and the CPQ model predictions are shown in Fig. 1 for the three types of triggers. For the inclusive trigger, the data are well reproduced using the CPQ model and, in addition, two minima are predicted at A=3 and A=86, respectively. The letter correspons to the main minimum of $G(\phi_A)$ in (1). The large positive A_N on the Au target is due to the main maximum of $G(\phi_A)$ at A=215. In the case of not inclusive triggers, the selected samples have centrality of collision that differs from the mean value (0.5 for the inclusive one). The A-dependence of centrality is shown in Fig. 5 for the three types of triggers, disscussed above. For the inclusive trigger, the centrality is close to 0.5, as expected. For the BBC-veto trigger, the centrality of collisions is much less than 0.5 and since in this case the color field is stronger, the asymmetry A_N is shifted upwards. The BBC-tag trigger selects centrality that is close to 1 for A=215 and corresponds to a very peripheral collisions. The color field is much weaker for peripheral collisions due to a smaller number of spectator quarks produced in the target fragmentation. As a result, the asymmetry A_N is shifted downwards for large A.

The dependence of A_N on centrality is shown in Fig. 6 for the cases of p+p, p+Al and p+Au collisions. As is seen in Fig. 6, the asymmetry $A_N(A)$ has a strong dependence on centrality for the Au target that explains the observed trigger dependence of A_N.

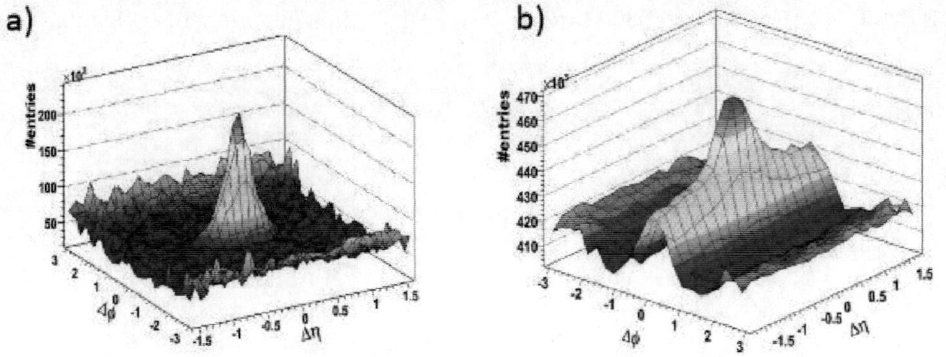

Figure 7. The two particle correlation function for the d+Au (left) and Au+Au (right) central collisions at $\sqrt{s} = 200$ GeV measured by the STAR experiment [6].

4. The origin of the "ridge" effect in the framework of the CPQ model

To evaluate the dependence of the chromomagnetic field on centrality, the so called "ridge" effect in heavy ions collisions [6] was analyzed in the framework of the CPQ model. The "ridge" phenomenon means the long-range correlations of two produced hadrons that depend on difference of their pseudorapidities ($\Delta\eta$). Pseudorapidity is defined as $\eta = -\ln\tan(\theta/2)$, where θ - polar angle. The difference in pseudorapidities of two hadrons is larger, on average, for more central collisions. At the same time, the difference in azimuthal angles ($\Delta\phi$) of two hadrons is narrow and has no strong dependence on centrality. In Fig. 7 is shown an example of the "ridge" structure measured in the STAR experiment [6]. For the central Au+Au collisions the width of the peak in the $\Delta\eta$ direction is much larger than along the $\Delta\phi$ axes.

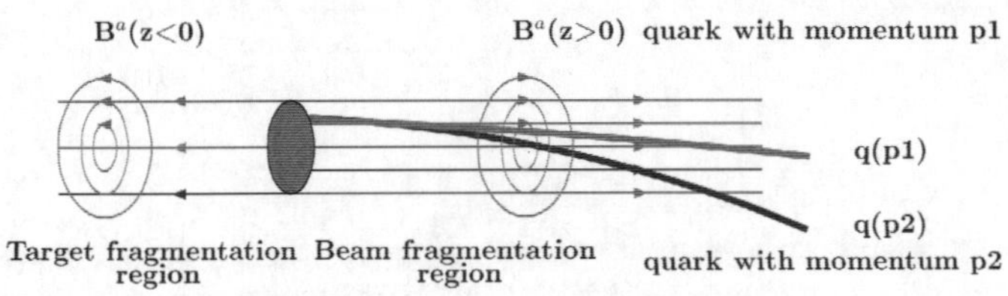

Figure 8. Schematic view of the interaction region as a microscopic color-magnetic lens, smearing a jet of quarks and antiquarks in a radial direction due to the Lorentz force.

This behavior is explained in the CPQ model as a result of the Lorentz force acting on a color charge in a circular transverse chromomagnetic field. In the CPQ model the interaction region can be considered as a microscopic color-magnetic lens, shown schematically in Fig. 8. The polar angle of each probe quark changes after passing through the field and the width $\Delta\eta$ of the difference in the pseudorapidity of the two quarks increases, in agreement with the data

IOP Conf. Series: Journal of Physics: Conf. Series **938** (2017) 012038 doi:10.1088/1742-6596/938/1/012038

behavior. The width of azimuth angle difference $\Delta\phi$ does not change significantly due the fact that the Lorentz force is directed radially to (or from) the center of the collision zone. The "ridge" effect should decrease when the momentum (or p_T) increases, in accordance with the data. Additional contribution to $\Delta\eta$ comes from the difference of Lorentz force for the quark and antiquark (which have opposite color charges). The dependence of the effective chromomagnetic field B^a, normalized to its value for the inclusive trigger, on the centrality u_c is found as follows:

$$U(u_c) = B^a(u_c)/B^a(0.5) = c_1 - c_2(u_c + u_c^2) + c_3(u_c^3 + u_c^4 - u_c^5), \quad (7)$$

where c1 = 6.1184, c2 = 9.5397, c3 = 13.0326, and $0 < u_c < 1$ is the centrality of collisions.

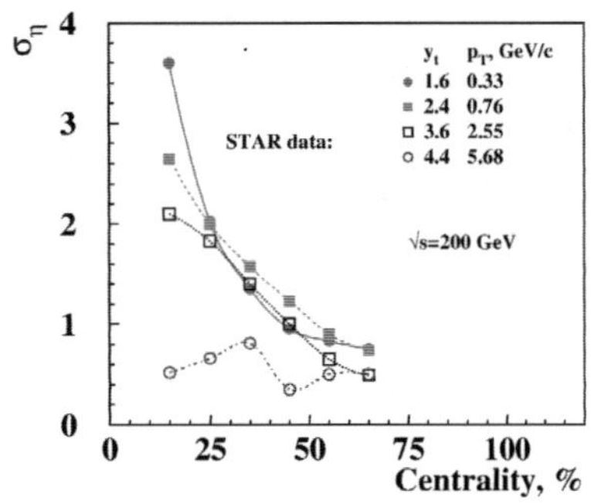

Figure 9. The dependence of σ_η on centrality. The STAR data are measured at $\sqrt{s} = 200$ GeV [7].

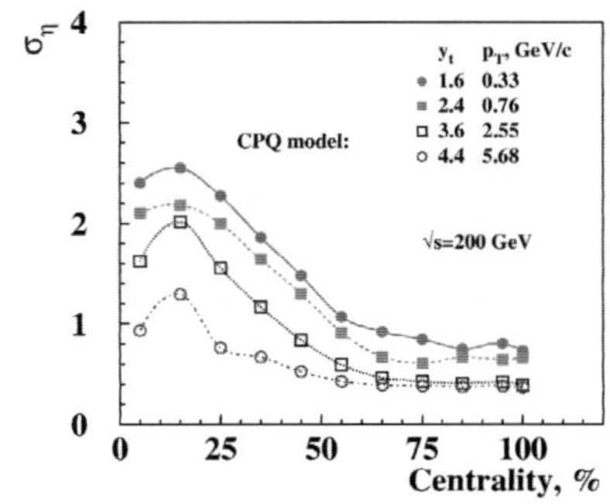

Figure 10. The dependence of σ_η on centrality. The CPQ model calculations are at $\sqrt{s} = 200$ GeV.

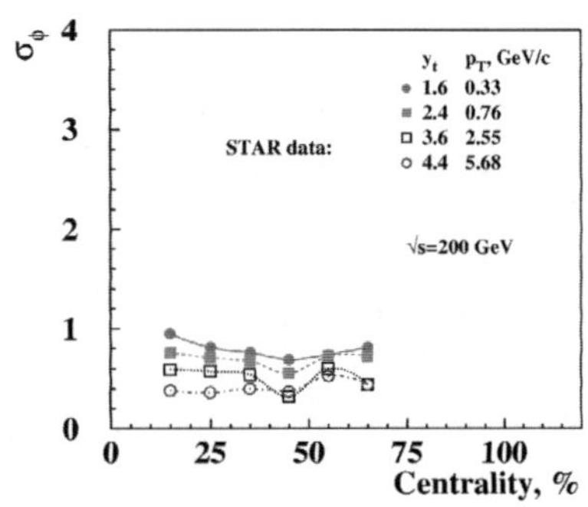

Figure 11. The dependence of σ_ϕ on centrality. The STAR data are measured at $\sqrt{s} = 200$ GeV [7].

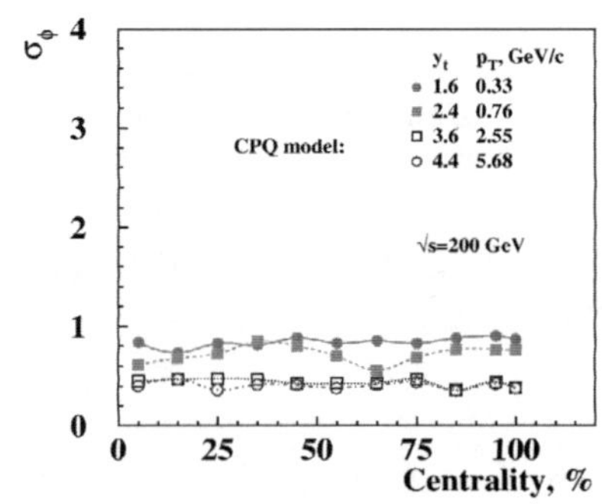

Figure 12. The dependence of σ_ϕ on centrality. The CPQ model calculations are at $\sqrt{s} = 200$ GeV.

To compare the "ridge" data and the CPQ model calculations the projections of two particle correlations on $\Delta\eta$ and $\Delta\phi$ axis were fitted by the Gaussian distribution. The dependence of the fit parameter σ_η on centrality for the data at $\sqrt{s} = 200$ GeV [7] is shown in Fig. 9. The width of the peak (σ_η) is decreasing with the rise of centrality value and p_T of hadrons. This behavior is consistent with the model predictions, shown in Fig. 10. In Figs. 11 and 12 are shown the data [7] and the model estimates for the parameter σ_ϕ, as a function of centrality. There is no significant dependence of σ_ϕ on centrality but the dependence on p_T is still essential.

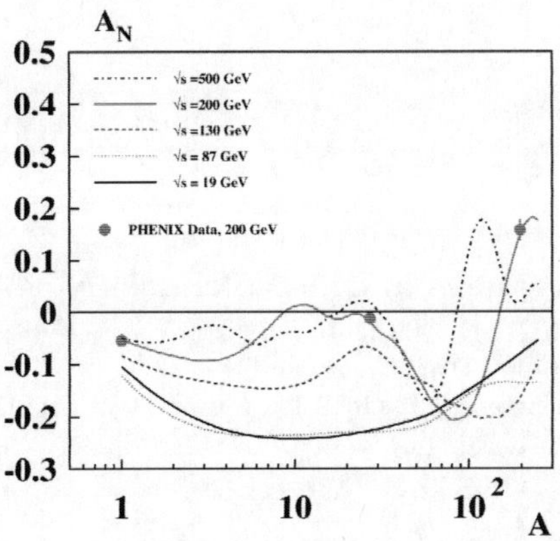

Figure 13. The dependence of A_N on A for p_T = 0.108 GeV/c, x_F = 0.54, and the inclusive trigger.

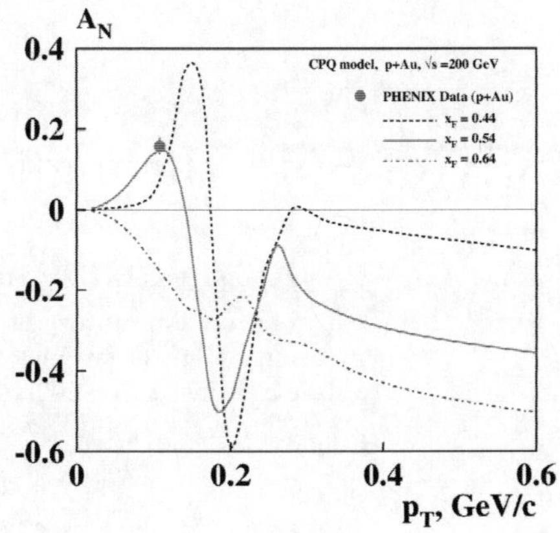

Figure 14. The dependence of A_N on p_T for $\sqrt{s}$ = 200 GeV, Au target, and a few values of x_F – 0.44, 0.54, and 0.64.

5. Calculations of $A_N(A)$ and $A_N(p_T)$

The data [1] and calculations of $A_N(A)$ for different energies are shown in Fig. 13. The positions of maxima and minima of $A_N(A)$ are moving to lower A values with the rise of energy. For $\sqrt{s} \leq 100$ GeV the sign of $A_N(A)$ is negative for all values of A. The dependences of A_N on p_T for $\sqrt{s} = 200$ GeV, Au target, and $x_F = 0.44$, 0.54 or 0.64 are shown in Fig. 14. The oscillating shape of $A_N(p_T)$ is due to the quark spin precession in the effective chromomagnetic field.

Conclusions and outlook

1) The anomalous A_N dependence on A for the production of neutrons in pA-collisions is due to the Stern-Gerlach force action and the quark spin-precession in the effective chromomagnetic field. This is expected for $p_T \leq 0.3$ GeV/c, $x_F = 0.6$ and $\sqrt{s} \geq 100$ GeV.
2) The model of chromo-magnetic polarization of quarks predicts not only a large positive peak in the $A_N(A)$ dependence at $A \approx 215$ but also negative peaks at A=3 and A=86.
3) The dependence of $A_N(A)$ on the trigger type can be caused by the dependence of the color field on the collision centrality.
4) The "ridge" effect in the two hadron correlations that was observed in the heavy ions collisions can be explained in the framework of the CPQ model.
5) A similar anomalous A-dependence of $A_N(A)$ can exist for the other inclusive reactions at low $p_T \leq 0.3$ GeV/c. This can be tested in the experiments at IHEP, JINR and BNL.

Acknowledgments

This work was supported in part by the RFBR Contract (Grant No. 15-02-01669).

References

[1] Aidala C et al 2017 *Preprint* arXiv:1703.10941
[2] Abramov V V 2009 *Phys. At. Nucl.* **72** 1872-88
[3] Bargman V, Michel L and Telegdy V 1959 *Phys. Rev. Lett.* **2** 435
[4] Abramov V V 2016 *J. Phys.: Conf. Series* **678** 012039
[5] Abramov V V 2014 *Phys. Part. Nucl.* **45** 62-65
[6] Abelev B *et al* 2009 *Phys. Rev. C* **80** 064912
[7] Kettler D (STAR Collaboration) 2011 *J. Phys. Conf. Ser.* **270** 012058

IOP Conf. Series: Journal of Physics: Conf. Series **938** (2017) 012039 doi:10.1088/1742-6596/938/1/012039

Study of Spin through Gluon Poles

I V Anikin[1], **L Szymanowski**[2], **O V Teryaev**[1] **and N Volchanskiy**[1,3]
1 Bogoliubov Laboratory of Theoretical Physics, JINR, 141980 Dubna, Russia
2 National Centre for Nuclear Research (NCBJ), 00-999 Warsaw, Poland
3 Research Institute of Physics, Southern Federal University, 344090 Rostov-on-Don, Russia
E-mail: anikin@theor.jinr.ru

Abstract. Based on the use of contour gauge and collinear factorization, we propose a new set of single spin asymmetry which can be measured in polarized Drell-Yan process by SPD@NICA. We stress that all of discussed single spin asymmetries exist owing to the gluon poles manifesting in the twist-3 or twist-2⊗twist-3 parton distributions related to the transverse-polarized Drell-Yan process.

1. Introduction

The studies of hadron structure are based on the investigations of both semi-inclusive and exclusive processes which can be described by transverse momentum dependent (TMDs) and generalized parton distributions (GPDs). The duality and matching between different factorization regimes are one of importances for the coherent QCD description of hadron structure. We discuss the manifestation of such effects in the pion-nucleon Drell-Yan process at large x_F provided pions are described by wave functions and distribution amplitudes rather than parton distributions.

We calculate the gauge invariant Drell-Yan-like hadron tensors. In connection with new COMPASS results, we predict the new single spin asymmetry which probes gluon poles together with chiral-odd and time-odd functions. The relevant pion production as a particular case of Drell-Yan-like process has been discussed. For the meson-induced Drell-Yan process, we model an analog of the twist three distribution function, which is a collinear function in inclusive channel, by means of two non-collinear distribution amplitudes which are associated with exclusive channel. This modelling demonstrates the fundamental duality between different factorization regimes.

2. Drell-Yan-like processes: new result for pion-nucleon Drell-Yan process

We study the process which has the following schematic representation, see Fig.1

$$A^{(\uparrow\downarrow)}(p_1) + B(p_2) \to \gamma^*(q) + X(P_X) \to \ell(l_1) + \bar{\ell}(l_2) + X(P_X), \tag{1}$$

where A denotes the (transverse)polarized nucleon, B stands for an arbitrary hadron and $l_1 + l_2 = q$ has a large mass squared ($q^2 = Q^2$). The single transverse spin asymmetry (SSA) under our consideration is given by

$$\mathcal{A} = \frac{d\sigma^{(\uparrow)} - d\sigma^{(\downarrow)}}{d\sigma^{(\uparrow)} + d\sigma^{(\downarrow)}}, \quad \frac{d\sigma^{(\uparrow\downarrow)}}{d^4q d\Omega} = \frac{\alpha_{em}^2}{2jq^4}\mathcal{L}_{\mu\nu} H_{\mu\nu}\,, \tag{2}$$

IOP Conf. Series: Journal of Physics: Conf. Series **938** (2017) 012039 doi:10.1088/1742-6596/938/1/012039

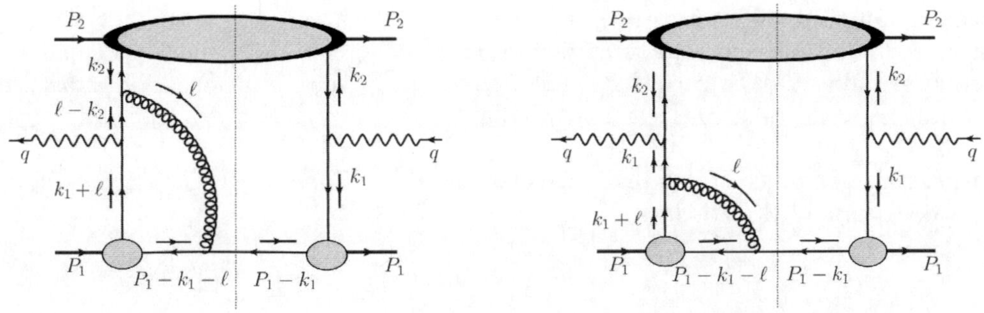

Figure 1. The Feynman diagrams which contribute to the polarized Drell-Yan hadron tensor: the left panel corresponds to the standard diagram while the right panel – to the nonstandard diagram. Note that the standard and nonstandard diagrams differ by the hard parts [1].

where $\mathcal{L}_{\mu\nu}$ is a lepton tensor, and $H_{\mu\nu}$ - the QED gauge invariant hadron tensor which corresponds to the difference between direct and mirror channels in the regime of $x_F \to 1$.

In Fig.1, the upper blob with the light-cone minus dominant direction is given by the following hadron matrix element:

$$\langle P_2, S_\perp | \left[\bar{\psi}\, \sigma^{-\perp}\, \psi\right]^{\text{tw-2}} | S_\perp, P_2\rangle \overset{\mathcal{F}}{\sim} \varepsilon^{-\perp S^\perp P_2}\, \bar{h}_1(y) \tag{3}$$

and the lower blob with the light-cone plus dominant direction corresponds to

$$\langle P_1, S | \left[\bar{\psi}\, \gamma^+\, A^\alpha_\perp\, \psi\right]^{\text{tw-3}} | S, P_1\rangle\Big|_{\text{DY}} \Longrightarrow \tag{4}$$
$$\mathcal{D}^{\alpha\beta}\, \langle 0 | \left[\bar{\psi}\gamma^+(\gamma_5)\psi\right]^{\text{tw-2}} | S, P_1\rangle\, \gamma_\beta\, \langle P_1, S | \left[\bar{\psi}\sigma^{-+}(\gamma_5)\psi\right]^{\text{tw-3}} | 0\rangle\Big|_{\pi N\text{-DY}}$$

which finally leads to (see, [2–6] for details)

$$\widetilde{B}(x_1, x_2) = \frac{1}{2}\frac{\Phi^{\text{tw-3}}_{(1)}(x_1) \overset{\vec{\mathbf{k}}_1^\perp}{\circledast} \Phi^{\text{tw-2}}_{(2)}(x_2)}{x_2 - x_1 - i\epsilon}\,. \tag{5}$$

where $-i\epsilon$ stems from the contour gauge: $[z^-, -\infty^-]_{A^+} = 1$.

For the chiral-odd contributions, within our framework, we predict a new SSA which reads

$$\mathcal{A}_T = \frac{S_\perp}{Q}\, \frac{D_{[\sin 2\theta]}\, \sin\phi_S B^{\sin\phi_S}_{UT}}{\bar{f}_1(y_B)\, H_1(x_B)}, \; D_{[\sin 2\theta]} = \frac{\sin 2\theta}{1 + \cos^2\theta} \tag{6}$$

where

$$B^{\sin\phi_S}_{UT} = 2\bar{h}_1(y_B)\, \Phi^{\text{tw-3}}_{(1)}(x_B) \overset{\vec{\mathbf{k}}_1^\perp}{\circledast} \Phi^{\text{tw-2}}_{(2)}(x_B) \tag{7}$$

and $\bar{f}_1(y_B)$, $H_1(x_B)$ emanate from the unpolarized cross-section and they parameterize the following matrix elements. The result is presented in terms of [7]. We want to emphasize that the predicted SSA (6) exists due to the complexity of $\widetilde{B}(x_1, x_2)$-function (5), *i.e.* $\Im\text{m}\widetilde{B}(x_1, x_2) \neq 0$.

From comparison with recent COMPASS results on pion-nucleon Drell-Yan process, we can see that our $\mathcal{A}_T$ is completely new one. Indeed, the leading twist Sivers asymmetry $A^{\sin\phi_S}_{UT}$ which formally stands at the similar tensor combination $\varepsilon^{l_1^\perp S^\perp P_1 P_2}$, appears only together with the depolarization factor $D_{[1+\cos^2\theta]}$. In its turn, the higher twist asymmetries $A^{\sin(\phi_S \pm \phi)}_{UT}$ at $D_{[\sin 2\theta]}$ correspond to the different tensor structures, $\varepsilon^{q S^\perp P_1^\perp P_2} \sim \sin(\phi_S \pm \phi)$.

3. Factorization theorem, in a nutshell

Since the main our tool is the factorization procedure, let us outline the principal items of factorization we adhere. With a help of factorization, a given amplitude of hadron tensor of any hard processes can be estimated, not calculated, by the well-defined procedure within the corresponding asymptotical regime. As a result of this estimation, we arrive at the well-known factorization theorem which states that the short (hard) and long (soft) distance dynamics can be separated out provided Q^2 is large, *i.e.*

$$\mathcal{W}_{\mu\nu...} = \int d^4k_1 d^4k_2 ... \,\mathrm{tr}\big[\Phi_1(k_1) E_{\mu\nu...}(k_1, k_2, ...) \Phi_2(k_2)...\big] \stackrel{Q^2\to\infty}{\Longrightarrow}$$

$$\int dx dy ... \,\mathrm{tr}\big[\Phi_1(y) E_{\mu\nu}(x, y, ...)\, \Phi_2(x)...\big] + \mathcal{O}(1/Q^2) \tag{8}$$

where $E_{\mu\nu...}(k_i)$ implies the products of propagators multiplying by the momentum conservation δ-function, and

$$E_{\mu\nu}(x, y, ...|\mathbf{k}_i^T = 0), \quad \Phi(k) \stackrel{\mathcal{F}}{=} \langle \mathcal{O}^{(\bar\psi,\psi,A)}(z, 0)\rangle\,, \tag{9}$$

$$\Phi(x_i) = \int dk_i^+\, \delta(x_i - k_i^+ \cdot n) \int dk_i^-\, d^2\mathbf{k}_i^T\, \Phi(k_i)\,. \tag{10}$$

Besides, both the hard and soft parts should be independent of each other, UV- and IR-renormalizable and, finally, parton distributions must possess the universality property. We notice that the transverse momentum integral in (10) gives us the possibility for the study of TMDs. In the other words, the TMDs can contribute through the corresponding moments. However, if the nontrivial $\mathbf{k}_i^T$-dependence has been kept in $E_{\mu\nu}(x, y, ...|\mathbf{k}_i^T)$, the factorization theorem can be in question. It can be happened, for example, in case we want to restore the $\mathbf{k}^T$-dependent Wilson line in the correlators $\Phi(k_i)$.

4. Contour gauge and complexity of $B(x_1, x_2)$-functions

We show that the contour gauges for gluon fields play the crucial role for our study. The contour gauge belongs to the class of non-local gauges that depends on the path connecting two points in the correlators. It turns out, in the cases which we consider, the prescriptions for the gluonic poles in the twist 3 correlators, see (5), are dictated by the contour gauge $[x, \pm\infty] = 1$. Indeed, one can easily check that the representation with $x_1 - x_2 + i\epsilon$ in the denominator belongs to the gauge $[x, -\infty] = 1$, while the representation with $x_1 - x_2 - i\epsilon$ belongs to the gauge $[+\infty, x] = 1$.

Thus, we find that the nonstandard new terms, which exist in the case of the complex twist-3 B-function with the corresponding prescriptions, do contribute to the hadron tensor exactly as the standard term known previously. This is another important result of our work.

In conclusion of this section, let us briefly discuss the main items of the contour gauge conception. To describe this class of gauges, we adhere the geometrical interpretation of gluons where the gluon field is a connection of the principal fiber bundle $\mathcal{P}(\mathbb{R}^4, G, \pi)$. Here, $\mathbb{R}^4$ is the base where the principal fiber bundle is determined, G denotes the group defined on the given fiber and π is a transformation of the base $\mathbb{R}^4$ into the fiber bundle $\mathcal{P}$). The corresponding path-dependent Wilson factor (or line), representing each element $\mathbf{g}(x)$ of the fiber, defines the gauge-transformed field:

$$A^{\mathbf{g}}_\mu(x) = \mathbf{g}^{-1}(x) A_\mu(x) \mathbf{g}(x) + \frac{i}{g}\mathbf{g}^{-1}(x)\partial_\mu \mathbf{g}(x). \tag{11}$$

The set of these fields for all $\mathbf{g}(x)$ forms the gauge orbit. In order to quantize a system of the gauge fields, one chooses the only element of each orbit. In contrast to the standard way, we first fix an arbitrary point $(x_0, \mathbf{g}(x_0))$ in the fiber. Then, we define two directions: one of them in the base, the other in the fiber. The direction in the base $\mathbb{R}^4$ is the tangent vector of a curve which goes through the given point x_0. At the same time, the direction in the fiber can be uniquely determined as the tangent subspace related to the parallel transition. Due to this procedure, we uniquely define the point in the fiber bundle.

Having solved the parallel transport equation defined on the fiber as

$$\frac{dx_\alpha(v)}{dv}\mathcal{D}_\alpha \mathbf{g}(x(v)) = 0\,, \tag{12}$$

we find the solution in terms of the Wilson line:

$$\mathbf{g}(x) = Pexp\Big\{ig \int\limits_{\mathbb{P}(x_0,x)} d\omega \cdot A(\omega)\Big\}\mathbf{g}(x_0), \tag{13}$$

where the points x_0 and x are connected by the path $\mathbb{P}$. The starting point x_0 is usually fixed. Here, $\mathbf{g}(x_0)$ is chosen to be equal to unity. Note that the fixing of $\mathbf{g}(x)$ ensures a unique choice of the element in the orbit. We now insert (13) into (11) and can see that the field $A^{\mathbf{g}}_\mu(x)$ is completely determined by the form of the path which connects the starting and final points. Moreover, using (11) and (13), one obtains the property:

$$Pexp\Big\{ig \int\limits_{x_0}^{x} d\omega \cdot A^{\mathbf{g}}(\omega)\Big\} = \mathbf{g}^{-1}(x) Pexp\Big\{ig \int\limits_{x_0}^{x} d\omega \cdot A(\omega)\Big\}\mathbf{g}(x_0). \tag{14}$$

Then, if we insert (13) into (11), we arrive at

$$A^{\mathbf{g}}_\mu(x) = \int\limits_{\mathbb{P}(x_0,x)} dz_\alpha \frac{\partial z_\beta}{\partial x_\mu}\, \mathbf{g}^{-1}(z)\, G_{\alpha\beta}(z|A)\, \mathbf{g}(z) = \int\limits_{\mathbb{P}(x_0,x)} dz_\alpha \frac{\partial z_\beta}{\partial x_\mu}\, G_{\alpha\beta}(z|A^{\mathbf{g}}), \tag{15}$$

where

$$G_{\alpha\beta}(z|A^{\mathbf{g}}) \equiv G^{\mathbf{g}}_{\alpha\beta}(z) = \mathbf{g}^{-1}(z)\, G_{\alpha\beta}(z|A)\, \mathbf{g}(z) \tag{16}$$

The contour gauge condition demands that $\mathbf{g}(x)$ is equal to unity for all x belonging to the base, *i.e.*

$$[x,\, x_0] \stackrel{def}{=} Pexp\Big\{ig \int\limits_{x_0}^{x} d\omega \cdot A(\omega)\Big\} = 1, \;\; \forall x \in \mathbb{R}^4. \tag{17}$$

Therefore, within the contour gauge, the field $A^{\mathbf{g}}_\mu$ (see, (15)) becomes

$$A^{c.g.}_\mu(x) = \int\limits_{\mathbb{P}(x_0,x)} dz_\alpha \frac{\partial z_\beta}{\partial x_\mu}\, G_{\alpha\beta}(z|A^{c.g.}), \tag{18}$$

i.e. the gluon field $A^{\mathbf{g}}_\mu$ is a linear functional of the tensor $G_{\mu\nu}$.

5. Conclusions

To conclude we derive the gauge invariant meson-induced DY hadron tensor with the essential spin transversity and primordial transverse momenta. Our calculation includes both the standard-like, which is well-known, and nonstandard-like diagram, which is first found in [4], contributions. The latter contribution plays a crucial role for the gauge invariance.

We study the case of pion distribution amplitudes which has been projected onto the chiral-odd combination. The latter singles out the chiral-odd parton distribution inside nucleons. The chiral-odd tensor combinations are very relevant for the future experiments implemented by COMPASS [7]. We predict a new single transverse spin asymmetry which can be measured experimentally, for example, by SPD@NICA. The latter asymmetry can eventually be treated as a signal for the transverse "primordial" momentum dependence which probes both gluon poles and time-odd functions.

Acknowledgments

We thank A V Efremov, L Motyka and A P Nagaytsev for useful discussions. The work by I.V.A. was partially supported by the Bogoliubov-Infeld Program and Heisenberg-Landau Program HL-2017. L.Sz. is partly supported by grant No 2015/17/B/ST2/01838 from the National Science Center in Poland.

References

[1] I V Anikin and O V Teryaev 2009 *Phys. Part. Nucl. Lett.* **6** 3
[2] I V Anikin, L Szymanowski, O V Teryaev and N Volchanskiy 2017 *Phys. Rev.* D **95** 111501
[3] I V Anikin, I O Cherednikov and O V Teryaev 2017 *Phys. Rev.* D **95** 034032
[4] I V Anikin and O V Teryaev 2010 *Phys. Lett.* B **690** 519
[5] I V Anikin and O V Teryaev 2015 *Eur. Phys. J.* C **75** 184
[6] I V Anikin and O V Teryaev 2015 *Phys. Lett.* B **751** 495
[7] M Aghasyan *et al.* [COMPASS Collaboration] 2017 arXiv:1704.00488 [hep-ex].

IOP Conf. Series: Journal of Physics: Conf. Series **938** (2017) 012040 doi:10.1088/1742-6596/938/1/012040

Tests for the extraction of Boer-Mulders functions

Ekaterina Christova
Institute for Nuclear Research and Nuclear Energy, Bulgarian Academy of Sciences, Sofia, Bulgaria
E-mail: echristo@inrne.bas.bg

Elliot Leader
Imperial College, London, UK
E-mail: e.leader@imperial.ac.uk

Michail Stoilov
Institute for Nuclear Research and Nuclear Energy, Bulgarian Academy of Sciences, Sofia, Bulgaria
E-mail: mstoilov@inrne.bas.bg

Abstract. At present, the Boer-Mulders (BM) functions are extracted from asymmetry data using the simplifying assumption of their proportionality to the Sivers functions for each quark flavour. Here we present two independent tests for this assumption. We subject COMPASS data on semi-inclusive deep inelastic scattering on the $\langle\cos\phi_h\rangle$, $\langle\cos 2\phi_h\rangle$ and Sivers asymmetries to these tests. Our analysis shows that the tests are satisfied with the available data if the proportionality constant is the same for all quark flavours, which does not correspond to the flavour dependence used in existing analyses. This suggests that the published information on the BM functions may be unreliable.

The $\langle\cos\phi_h\rangle$ and $\langle\cos 2\phi_h\rangle$ asymmetries receive contributions also from the, in principle, calculable Cahn effect. We succeed in extracting the Cahn contributions from experiment (we believe for the first time) and compare with their calculated values, with interesting implications.

1. Introduction

At present it is already recognized that the collinear picture of the parton model, according to which quark momenta are parallel to proton momentum, is a rather rough approximation for the nucleon structure – quarks have also transverse momentum. This leads especially to a completely new type – T-odd, parton densities (pdf's). Here we shall focus on two of them – the Boer-Mulders (BM) and Sivers transverse momentum dependent (TMD) parton densities.

In present analyses [1] it is assumed that the BM functions are proportional to the Sivers functions. This much simplifies the analysis, but it is clearly model dependent – a different assumption would lead to different BM functions.

Here we suggest two independent tests for the above assumption using only measurable quantities – relations between the $\langle\cos\phi_h\rangle$, $\langle\cos 2\phi_h\rangle$ and Sivers asymmetries. We work with the so-called difference asymmetries i.e. the difference between the production of particles and their anti-particles. We then utilise COMPASS data on semi-inclusive deep inelastic scattering (SIDIS) on a deuteron target in the formulated tests. Further, as the $\langle \cos\phi_h \rangle$ and $\langle \cos 2\phi_h \rangle$ azimuthal asymmetries receive contributions from both the BM and Cahn effects, we are able to extract information on the Cahn effect from experiment - as far as we know for the first time. The details are presented in [2].

2. BM and Sivers functions

The BM parton densities $\Delta^N f_{q^\uparrow/p}(x_B, k_\perp)$ describe the distribution of transversely polarized quarks $q^\uparrow$ in an unpolarized proton [3]. The Sivers parton densities $\Delta^N f^{Siv}_{q/p^\uparrow}(x_B, k_\perp)$ describe the distribution of unpolarized quarks in a transversely polarized proton $p^\uparrow$ [4].

For the BM and Sivers functions, as well as for all TMD functions that enter our cross sections, we use the standard parametrization. For SIDIS on deuterium target only the sum of the valence quarks, $Q_V \equiv u_V + d_V$, enters and we have [5]:

$$\Delta f_J^{Q_V}(x_B, k_\perp, Q^2) = \Delta f_J^{Q_V}(x_B, Q^2)\, \sqrt{2e}\, \frac{k_\perp}{M_J}\, \frac{e^{-k_\perp^2/\langle k_\perp^2\rangle_J}}{\pi \langle k_\perp^2\rangle_J}, \qquad J = BM, Siv \tag{1}$$

with

$$\Delta f_J^{Q_V}(x_B, Q^2) = 2\mathcal{N}_J^{Q_V}(x_B)\, Q_V(x_B, Q^2), \qquad \langle k_\perp^2\rangle_J = \frac{\langle k_\perp^2\rangle\, M_J^2}{\langle k_\perp^2\rangle + M_J^2}. \tag{2}$$

Here the $\mathcal{N}_J^{Q_V}(x_B)$ are unknown functions, and M_J, or equivalently $\langle k_\perp^2\rangle_J$, are unknown parameters.

3. The transverse quark momenta

Further in our considerations the parameters $\langle k_\perp^2\rangle$ and $\langle p_\perp^2\rangle$, which appear in the unpolarized cross sections that normalize all TMD asymmetries will enter. They are interpreted as the average transverse quark and hadron momenta and are determined from multiplicities. At present the obtained values are rather controversial:

• From old measurements we have:
1) $\langle k_\perp^2\rangle \approx 0.25\, GeV^2$ and $\langle p_\perp^2\rangle \approx 0.20\, GeV^2$ [6], from the EMC and FNAL data;
2) $\langle k_\perp^2\rangle = 0.18\, GeV^2$ and $\langle p_\perp^2\rangle = 0.20\, GeV^2$ [7].

• The more recent data from HERMES and COMPASS separately, gives quite different values [8]:
3) $\langle k_\perp^2\rangle = 0.57 \pm 0.08\, GeV^2$ and $\langle p_\perp^2\rangle = 0.12 \pm 0.01\, GeV^2$, extracted from HERMES data
4) $\langle k_\perp^2\rangle = 0.61 \pm 0.20\, GeV^2$ and $\langle p_\perp^2\rangle = 0.19 \pm 0.02\, GeV^2$, extracted from COMPASS data.

Further we shall comment on this controversial situation, since the Cahn effect, which we extract from data, is calculable, and depends sensitively on $\langle k_\perp^2\rangle$ and $\langle p_\perp^2\rangle$.

4. The difference asymmetries

We consider the production of charged hadrons $h^\pm$ in SIDIS of charged leptons on an unpolarized and a transversely polarized deuteron target:

$$l + d \to l' + h^\pm + X, \qquad l + d^\uparrow \to l' + h^\pm + X \tag{3}$$

We work with the so called difference asymmetries:

$$A^{h^+ - h^-} \equiv \frac{\Delta\sigma^{h^+} - \Delta\sigma^{h^-}}{\sigma^{h^+} - \sigma^{h^-}}. \tag{4}$$

where σ^{h^+,h^-} and $\Delta\sigma^{h^+,h^-}$ are the unpolarized and polarized SIDIS cross sections, respectively.

The difference asymmetries do not present a new measurement – they are expressed in terms of the usual asymmetries A^{h^+,h^-} and the ratio of the corresponding multiplicities r [9]:

$$A^{h^+-h^-} = \frac{1}{1-r}\left(A^{h^+} - rA^{h^-}\right), \qquad A^{h^+} = \frac{\Delta\sigma^{h^+}}{\sigma^{h^+}}, \quad A^{h^-} = \frac{\Delta\sigma^{h^-}}{\sigma^{h^-}}, \quad r = \frac{\sigma^{h^-}}{\sigma^{h^+}}. \tag{5}$$

As shown in ref.[5], the advantage of using the difference asymmetries is that, based only on charge conjugation (C) and isospin (SU(2)) invariance of the strong interactions, they are expressed purely in terms of the best known valence-quark distributions and fragmentation functions; sea-quark and gluon distributions do not enter. For a deuteron target there is the additional simplification – independently of the final hadron, only the sum of the valence-quark distributions enters.

5. The azimuthal asymmetries

Δf^{BM} and Δf^{Siv} are measured via the dependence on the azimuthal angle ϕ_h of the final hadron:

$$\begin{aligned} d\sigma^h(x,z,Q^2,P_T,\phi) &= d\sigma_0^h \left\{1 + A^h_{\cos\phi}\cos\phi_h + A^h_{\cos 2\phi}\cos 2\phi_h + ... \right. \\ &\left. + S_T\left[A^h_{Siv}\sin(\phi_s-\phi_h) + ..\right]\right\} \end{aligned} \tag{6}$$

where $d\sigma_0^h$ is the unpolarized, ϕ_h-independent cross section, $d\sigma_0^h \propto f_q(x)\otimes D_q^h(z)$.

The asymmetries $A^h_{\cos\phi}$ & $A^h_{\cos 2\phi}$ receive contributions from both Δf_{BM} and the Cahn effect – $A_{\cos\phi}$ receives a subleading $1/Q$-contribution from both BM and Cahn effects:

$$1)\ A^h_{\cos\phi} \simeq \frac{1}{Q}\sum_q\left[\Delta f^{BM}_{q\uparrow/p}\otimes\Delta D^h_{q\uparrow} + f_q\otimes D^h_q\right] \tag{7}$$

while $A^h_{\cos 2\phi}$ receives a leading BM-contribution and twist-4 $1/Q^2$-contribution from Cahn effect:

$$2)\ A^h_{\cos 2\phi} \simeq \sum_q\left[\Delta f^{BM}_{q\uparrow/p}\otimes\Delta D^h_{q\uparrow} + \frac{1}{Q^2}f_q\otimes D^h_q\right] \tag{8}$$

The Sivers asymmetry A_{Siv} is induced by Δf_q^{Siv}:

$$3)\ A^h_{Siv} \simeq \sum_q \Delta f_q^{Siv}\otimes D^h_q \tag{9}$$

6. Tests for the relation between the BM and Sivers functions on a deuteron target

As on a deuterium target only the sum of the valence-quarks Q_V enters, in contrast to the currently used assumption of proportionality between BM and Sivers functions for each quark flavour, we assume the relation:

$$\Delta f^{Q_V}_{BM}(x,k_\perp,Q^2) = \lambda_{Q_V}\,\Delta f^{Q_V}_{Siv}(x,k_\perp,Q^2), \qquad Q_V = u_V + d_V \tag{10}$$

where λ_{Q_V} is a constant.

This assumption leads to relations between the BM induced contributions in $A^h_{\cos\phi}$ or $A^h_{\cos 2\phi}$ and the Sivers asymmetries. They are particularly simple and present predictive tests for (10) when the Q^2-evolution of the collinear pdf's and fragmentation functions (FFs) can be neglected. Here we present these tests.

Taking into account both the BM and Cahn contributions to the unpolarized cross section, the relation (10) leads to the following relations between the x_B-dependent $A^h_{\cos\phi}$ or $A^h_{\cos 2\phi}$ and Sivers asymmetries on deuterium target [2]:

$$A^{h-\bar{h}}_{\cos\phi,d}(x_B) - \Phi(x_B)\, C^h_{\widetilde{BM}}\, A^{h-\bar{h}}_{Siv,d}(x_B) \;=\; \Phi(x_B)\, C^h_{Cahn}, \tag{11}$$

$$A^{h-\bar{h}}_{\cos 2\phi,d}(x_B) - \hat{\Phi}(x_B)\, \hat{C}^h_{\widetilde{BM}}\, A^{h-\bar{h}}_{Siv,d}(x_B) \;=\; \frac{MM_d}{\langle Q\rangle^2}\, \hat{\Phi}(x_B)\, \hat{C}^h_{Cahn}, \quad h=\pi^+, K^+, h^+ \tag{12}$$

Here the functions $\Phi(x_B)$ and $\hat{\Phi}(x_B)$ are completely fixed by kinematics:

$$\Phi(x_B) = \frac{\sqrt{\pi}\,(2-\bar{y})\sqrt{1-\bar{y}}}{\langle Q\rangle\,[1+(1-\bar{y})^2]}, \quad \hat{\Phi}(x_B) = \frac{2\,(1-\bar{y})}{[1+(1-\bar{y})^2]}, \quad \bar{y} = \frac{\langle Q\rangle^2}{2M_d E\, x_B}, \tag{13}$$

$\langle Q\rangle^2$ is some mean value of Q^2 for each x_B-bin, M_d is the mass of the deuterium target. C^h_{Cahn}, $\hat{C}^h_{Cahn}$ and $C^h_{\widetilde{BM}}$, $\hat{C}^h_{\widetilde{BM}}$ are constants, determined entirely by the collinear and Collins FFs, the explicit expressions are given in ref.[2].

Relations (11) and (12), in which C^h_i, respectively $\hat{C}^h_i$ are parameters, present:

1) two independent direct tests of the assumed relation (10) between the BM and Sivers functions, in which only measurable quantities enter, and no knowledge about the TMD functions is required and,

2) two independent ways for extracting the Cahn contribution from data.

7. Tests using the COMPASS data for $h^\pm$ production on a deuterium target

Here we use relations (11) and (12) between the difference asymmetries to test assumption (10) between the TMD functions, using the COMPASS data on deuterons for production of charged hadrons $h^\pm$ for the angular distributions $A^{h^\pm}_{\cos\phi,d}(x_B)$ and $A^{h^\pm}_{\cos 2\phi,d}(x_B)$ [10], and the single-spin Sivers asymmetry data $A^{h^\pm}_{Siv,d}(x_B)$ [11]. We proceed in 3 steps:

1) Via Eq.(5), we form the difference asymmetries $A^{h^+-h^-}_J$, $J = \cos\phi, \cos 2\phi, Siv$ from the measured asymmetries $A^{h^+}_j$ and $A^{h^-}_j$ for positive and negative charged hadron production, r is given in [9]. The obtained asymmetries are presented on Fig.1. As seen from the Figure, the Sivers asymmetry $A^{h-\bar{h}}_{Siv,d}(x_B)$ is determined with large relative errors and is close to 0. This suggests that $C^h_{\widetilde{BM}}$ and $\hat{C}^h_{\widetilde{BM}}$ may be poorly determined. The same arguments hold for $A^{h^+-h^-}_{\cos 2\phi,d}(x_B)$ and we expect tests with $A^{h^+-h^-}_{\cos\phi,d}(x_B)$ to give more precise information.

2) We choose the Q^2 interval where the Q^2-dependence of the collinear pdf's and FFs can be neglected. In the COMPASS kinematics to each value of $\langle Q^2\rangle$ corresponds one definite value of $\langle x_B\rangle$, thus fixing the Q^2 interval we fix also the x_B-interval. Using the available CTEQ parametrizations for the pdf's and the parametrization in [12] for FFs we see that aside from the small values of $Q^2 < 1.5\,GeV^2$, the Q^2-dependence is weak. This implies that Eqs (11,12) can be applied for $x_B \geq 0.014$.

3) Finally, we fit the parameters in Eqs (11,12) from the data, using χ^2-analysis with linear interpolation of the data. There are two ways to utilize (11) and (12), we shall follow both of them:

($\mathcal{A}$) We consider both C^h_{Cahn} and $C^h_{\widetilde{BM}}$ (respectively $\hat{C}^h_{Cahn}$ and $\hat{C}^h_{\widetilde{BM}}$), as fitted parameters. Our analysis showed that for both tests, Eqs (11,12), we obtain a good fit in almost the same kinematic interval $x_B \gtrsim 0.014$, the results are presented on Fig.2.

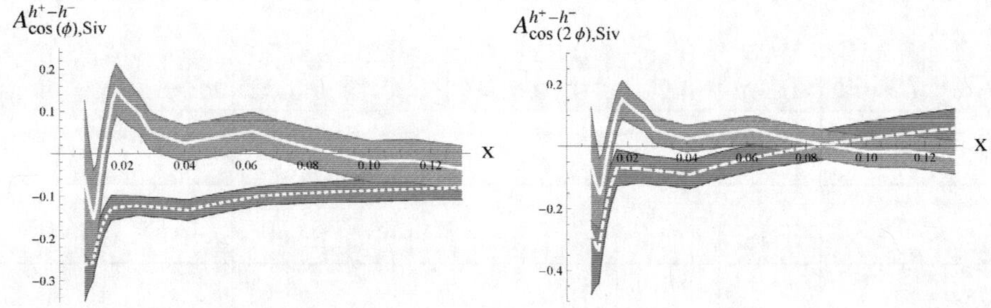

a b

Figure 1. The difference asymmetries with their statistical errors : a) $A^{h^+-h^-}_{Siv,d}(x_B)$ (solid line) and $A^{h^+-h^-}_{\cos\phi,d}(x_B)$ (dashed line) b) $A^{h^+-h^-}_{Siv,d}(x_B)$ (solid line) and $A^{h^+-h^-}_{\cos 2\phi,d}(x_B)$ (dashed line)

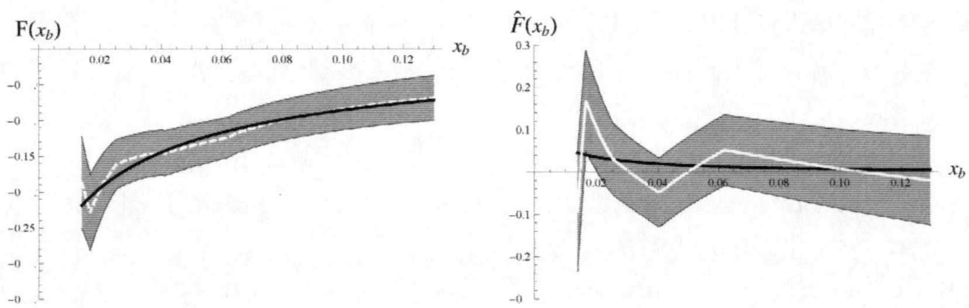

a b

Figure 2. The fits for $x_B \geq 0.014$: a) of Eq.(11)), the dashed white line is for $F(x_B) \equiv A^{h^+-h^-}_{\cos\phi,d}(x_B) - C^h_{\widetilde{BM}}\,\Phi(x_B) A^{h-\bar{h}}_{Siv,d}(x_B)$, the black line is $C^h_{Cahn}\,\Phi(x_B)$; b) of Eq.(12)), the white line is $\hat{F}(x_B) \equiv A^{h^+-h^-}_{\cos 2\phi,d}(x_B) - \hat{C}^h_{\widetilde{BM}}\,\hat{\Phi}(x_B)\, A^{h^+-h^-}_{Siv,d}(x_B)$, the black one is $MM_d\,\hat{\Phi}(x_B)\hat{C}^h_{Cahn}/\langle Q\rangle^2$

($\mathcal{B}$) In the second approach, first we calculate the Cahn constants, C^h_{Cahn} or $\hat{C}^h_{Cahn}$, using their explicit expressions. For example, for C^h_{Cahn} we have:

$$C^h_{Cahn} \quad = \quad -\langle k_\perp^2\rangle \frac{\int dz_h\, z_h [D^h_{q_V}(z_h)]/\sqrt{\langle P_T^2\rangle}}{\int dz_h\, [D^h_{q_V}(z_h)]}, \quad \langle P_T^2\rangle = \langle p_\perp^2\rangle + z_h^2\langle k_\perp^2\rangle \tag{14}$$

They depend on the FFs and on the parameters $\langle k_\perp^2\rangle$ and $\langle p_\perp^2\rangle$ which, as discussed in Section (3), vary considerably. Then we fit the same data with just one parameter $C^h_{\widetilde{BM}}$, respectively $\hat{C}^h_{\widetilde{BM}}$. Consequently, the main interest in this approach is to compare the calculated Cahn constants with the parameters as in ($\mathcal{A}$).

The calculated values for C^h_{Cahn} and $\hat{C}^h_{Cahn}$ for the different values of $\langle k_\perp^2\rangle$, $\langle p_\perp^2\rangle$, and the fitted parameter $C^h_{\widetilde{BM}}$ and $\hat{C}^h_{\widetilde{BM}}$ are presented in Tables 1. For comparison, in the last column the corresponding fitted values from approach $\mathcal{A}$ are also presented. The errors in the parameters are calculated using Monte Carlo simulation.

To the best of our knowledge this is the first time that the Cahn contribution has been determined from data and it is intriguing that its value is in agreement with a calculated value based on the early values of the Gaussian parameters $\langle k_\perp^2\rangle = 0.18$, GeV^2 and $\langle k_\perp^2\rangle = 0.25$, GeV^2 and completely disagree with the later values.

IOP Conf. Series: Journal of Physics: Conf. Series **938** (2017) 012040 doi:10.1088/1742-6596/938/1/012040

$\langle k_\perp^2 \rangle$ [GeV^2]	0.25	0.18	0.57 ± 0.08	0.61 ± 0.20	$(\mathcal{A})$
$\langle p_\perp^2 \rangle$ [GeV^2]	0.20	0.20	0.12 ± 0.01	0.19 ± 0.02	
C^h_{Cahn}	-0.21	-0.16	-0.49 ± 0.05	-0.4 ± 0.1	-0.167 ± 0.043
$C^h_{\widetilde{BM}}$	1.43	0.44	13 ± 2	11 ± 4	0.55 ± 0.88
$\hat{C}^h_{Cahn}$	0.077	0.044	0.41 ± 0.08	0.32 ± 0.15	0.083 ± 0.22
$\hat{C}^h_{\widetilde{BM}}$	-1.86	-1.59	-4.9 ± 0.8	-4.0 ± 1.5	-1.6 ± 1.6

Table 1. C^h_{Cahn} and $\hat{C}^h_{Cahn}$ are calculated using FFs from LSS [12], $C^h_{\widetilde{BM}}$ and $\hat{C}^h_{\widetilde{BM}}$ are fitted. Their values are compared to those of $(\mathcal{A})$.

8. Conclusions

We have performed two independent tests of the assumption that the proportionality between the BM and Sivers functions holds for the sum of the valence-quark TMD distributions, using the COMPASS data on the asymmetries $A^{h^+-h^-}_{\cos\phi,d}(x_B)$, $A^{h^+-h^-}_{\cos 2\phi,d}(x_B)$ and $A^{h^+-h^-}_{Siv,d}(x_B)$. Both tests are consistent with this assumption in the same kinematic interval $x_B = [0.014, 0.13]$.

However, in the published extractions of the BM functions [1] the assumption made is $\Delta f^q_{BM} = \lambda_q \Delta f^q_{Siv}$ for each quark separately. It would agree with our result only if $\lambda_u = \lambda_{\bar{u}} = \lambda_d = \lambda_{\bar{d}} = \lambda_{Q_V}$, which does not correspond to the values obtained in [1]. This suggests that the published BM functions may be unreliable.

We have also determined the kinematical Cahn contribution, both directly from a fit to the data (as far as we know for the first time) and from a calculation. The calculated values are very sensitive to the average transverse momentum-squared $\langle k_\perp^2 \rangle$. Surprisingly, the calculated values agree with the extracted ones only for the old experimental values $\langle k_\perp^2 \rangle \approx 0.18$ GeV^2 and $\langle k_\perp^2 \rangle \approx 0.25$ GeV^2 and completely disagree with the much bigger present-day values.

Acknowledgments

E.Ch. and M.S. acknowledge the support of Grant D-08-17 of the Bulgarian Science Foundation. E. L. is grateful to the Leverhulme Trust for an Emeritus Fellowship and E.Ch. to a collaborative Grant with JINR, Dubna.

References

[1] Vincenzo Barone, Stefano Melis and Alexej Prokudin, Phys. Rev. **D 81**, 114026 (2010)
[2] E. Christova, E. Leader and M. Stoilov, arXiv:1705.10613
[3] D. Boer and P.J.Mulders, Phys. Rev. **D 57** 5780 (1998)
[4] D.Sivers, Phys. Rev. **D41** (1990) 83; **43** 261 (1991)
[5] E. Christova and E. Leader, Phys. Rev. **D92** (2015) 114004
[6] M. Anselmino et al., Phys. Rev. **D71**, 074006 (2005)
[7] F.Giordano, report No DESY-THESIS-2008-030
[8] M. Anselmino et al., JHEP **04** (2014) 005
[9] COMPASS, M. Alekseev et al, Phys.Lett **B660** (2008) 458
[10] COMPASS collab. C. Adolph et al., Nucl. Phys. **B886** (2014) 1046-1077
[11] COMPASS collab. E.S.Ageev et al, Nucl. Phys. **B 765** (2007) 31; A. Martin, Czech. J. Phys. **56** (2006) F33
[12] E. Leader, A.V. Sidorov and D. Stamenov, Proceedings of DSPIN-13, Dubna, Russia (2013), arXiv:1312.5200.

IOP Conf. Series: Journal of Physics: Conf. Series **938** (2017) 012041 doi:10.1088/1742-6596/938/1/012041

Global extraction of partonic transverse momentum distributions from semi-inclusive deep inelastic scattering and Drell-Yan data

Filippo Delcarro*[1,2], Alessandro Bacchetta[1,2], Cristian Pisano[1,2], Marco Radici[2], Andrea Signori[3]

[1] Dipartimento di Fisica, Università di Pavia INFN Pavia, via Bassi 6,I-27100 Pavia
[2] INFN, Sezione di Pavia, via Bassi 6, I-27100 Pavia
[3] Theory Center, Thomas Jefferson National Accelerator Facility, 12000 Jefferson Avenue, Newport News, VA 23606, USA

E-mail: filippo.delcarro@pv.infn.it, alessandro.bacchetta@unipv.it, cristian.pisano@unipv.it, marco.radici@pv.infn.it, asignori@jlab.org

Abstract.
We discuss a first attempt at a global extraction of unpolarized partonic transverse momentum dependent (TMD) distributions and fragmentation functions from a simultaneous fit of data measured in semi-inclusive deep-inelastic scattering, Drell–Yan and Z boson production. This analysis is performed in the low transverse momentum region, at leading order in perturbative QCD. To connect data at different scales, we use TMD evolution at next-to-leading logarithmic accuracy.

1. Introduction

Parton distribution functions describe the internal structure of the nucleon in terms of its elementary constituents, quarks and gluons. Transverse momentum dependent distributions distributions and fragmentations functions (TMDs) carry fundamental information on the intrinsic motion of partons and the correlation between the nucleons spins and momenta, including however also the dependence on transverse momentum components $k_\perp^2$ and thus providing a full three-dimensional picture of hadrons in momentum space. Similarly to the more common collinear distributions, TMDs are not purely perturbative quantities, they cannot be easily computed from first principles but have to be extracted from experimental measurements to have a complete determination.
Many observables in hadronic hard scattering experiments are related to PDFs and FFs, in a way specified by factorization theorems (see, e.g., Refs. [1, 2]). These theorems also illustrate the universality of PDFs and FFs (i.e., the fact that they are the same in different processes) and their evolution equations, that connect the different values that the distributions assume when the hard scale of the process change. Availability of measurements of different processes in different experiments makes it possible to test the reliability of factorization theorems and extract PDFs and FFs through so-called global fits.

*Speaker

In this contribution to the proceedings, based on [3] to which we refer for details, we focus only on the unpolarized TMD PDF $f_1^q(x, k_\perp^2)$ and the unpolarized TMD FF $D_1^{q\to h}(z, P_{hT}^2)$, with a flavor independent analysis. We take into consideration three kinds of processes: semi-inclusive DIS, and Drell–Yan processes (DY) with the production of virtual photons and Z bosons.

2. Formalism

2.1. Semi-inclusive DIS

In one-particle SIDIS, a lepton ℓ with momentum l scatters off a hadron target N with mass M and momentum P. In the final state, the scattered lepton momentum l' is measured together with one hadron h with mass M_h and momentum P_h. The available data refer to SIDIS hadron multiplicities; in the single-photon-exchange approximation and taking into account all powers of the form $\alpha_S^n L^{2n} \approx 1$ (Leading Logarithms –LL) and $\alpha_S^n L^n \approx 1$ (Next-to-Leading Logarithms – NNL), the multiplicities can be written as (see Ref. [4] for details):

$$m_N^h(x, z, |\boldsymbol{P}_{hT}|, Q^2) = \frac{d\sigma_N^h/(dx dz d|\boldsymbol{P}_{hT}| dQ^2)}{d\sigma_{DIS}/(dx dQ^2)} \simeq \frac{2\pi|\boldsymbol{P}_{hT}| F_{UU,T}(x, z, \boldsymbol{P}_{hT}^2), Q^2)}{F_T(x, Q^2)} \tag{1}$$

where x,y, z and γ are the usual kinematic variables, $\boldsymbol{P}_{hT}$ is the component of $\boldsymbol{P}_h$ transverse to $\boldsymbol{q}$, $Q^2 = -q^2 = -(l-l')^2)$ and ϵ is the polarization ratio of the virtual photon.
The semi-inclusive cross section can be factorized in terms of TMDs only in the kinematic limits $M^2 \ll Q^2$ and $\boldsymbol{P}_T^2 \ll Q^2$. Moreover, in the present analysis, we will take into account all powers of the form $\alpha_S^n L^{2n} \approx 1$ (Leading Logarithms –LL) and $\alpha_S^n L^n \approx 1$ (Next-to-Leading Logarithms – NNL). With this accuracy, only the contribution $F_{UU,T}$ is relevant in our study. This function can be expressed in terms of TMD PDFs and FFs using the factorized formula for SIDIS [5]:

$$\begin{aligned} F_{UU,T}(x, z, \boldsymbol{P}_{hT}^2, Q^2) = & \sum_a \mathcal{H}_{UU,T}^a(Q^2; \mu^2) \times \int d\boldsymbol{k}_\perp \, d\boldsymbol{P}_\perp \, f_1^a(x, \boldsymbol{k}_\perp^2; \mu^2) \, D_1^{a\to h}(z, \boldsymbol{P}_\perp^2; \mu^2) \\ & \cdot \delta(z\boldsymbol{k}_\perp - \boldsymbol{P}_{hT} + \boldsymbol{P}_\perp) + Y_{UU,T}(Q^2, \boldsymbol{P}_{hT}^2) + \mathcal{O}(M^2/Q^2) \, . \end{aligned} \tag{2}$$

Here, $\mathcal{H}_{UU,T}$ is the hard scattering part; the term $Y_{UU,T}$ is introduced to ensure a matching to the perturbative fixed-order calculations at higher transverse momenta. In this analysis we neglect this term, leaving a detailed treatment of the matching to the high $P_{hT}^2 \approx Q^2$ region to future investigations.
In order to apply TMD evolution equations, we need to calculate the Fourier transform of the part of Eq. (2) involving TMDs. The structure function thus reduces to

$$F_{UU,T}(x, z, \boldsymbol{P}_{hT}^2, Q^2) \approx 2\pi \sum_a \int_0^\infty d\xi_T \, \xi_T \cdot J_0(\xi_T |\boldsymbol{P}_{hT}|/z) \tilde{f}_1^a(x, \xi_T^2; Q^2) \tilde{D}_1^{a\to h}(z, \xi_T^2; Q^2) \tag{3}$$

where we introduced the Fourier transforms $\tilde{f}_1^a(x, \xi_T; Q^2)$ and $\tilde{D}_1^{a\to h}(z, \xi_T; Q^2)$ of the TMD PDF and FF, respectively.

2.2. Drell–Yan processes

In a Drell–Yan process, two hadrons A and B with momenta P_A and P_B collide at a center-of-mass energy squared $s = (P_A + P_B)^2$ and produce a virtual photon or a Z boson plus hadrons. The boson then decays into a lepton-antilepton pair. The experimental cross sections can be analyzed in terms of structure functions [6, 7]:

$$\frac{d\sigma}{dQ^2 \, dq_T^2 \, d\eta} = \sigma_0^{\gamma,Z} \left(F_{UU}^1 + \frac{1}{2} F_{UU}^2 \right). \tag{4}$$

IOP Conf. Series: Journal of Physics: Conf. Series **938** (2017) 012041 doi:10.1088/1742-6596/938/1/012041

where $\sigma_0^{\gamma,Z}$ are the elementary cross sections for the process. Similarly to the SIDIS case, in the kinematic limit $q_T^2 \ll Q^2$ and neglecting the hadron masses the structure function F_{UU}^2 can be neglected. With the choices discussed for the SIDIS case, the structure function F_{UU}^1 can be expressed as a Fourier transform:

$$F_{UU}^1(x_A, x_B, \boldsymbol{q}_T^2, Q^2) = \sum_a \mathcal{H}_{UU}^{1a} \int_0^\infty \frac{d\xi_T}{2\pi} \xi_T \, J_0(\xi_T|\boldsymbol{q}_T|) \cdot \tilde{f}_1^a(x_A, \xi_T; \mu^2) \, \tilde{f}_1^{\bar{a}}(x_B, \xi_T; \mu^2). \quad (5)$$

2.3. TMDs and their evolution

Following the formalism of Refs. [2, 8], the unpolarized TMDs at LO in configuration space for a parton flavor a at a certain scale μ^2 can be written as

$$\widetilde{f}_1^a(x, \xi_T^2; Q^2) = f_1^a(x; \mu_b^2) \, e^{S(\mu_b^2, Q^2)} \cdot e^{g_K(\xi_T)\ln(Q^2/Q_0^2)} \, \widetilde{f}_{1\text{NP}}^a(x, \xi_T^2) \quad (6)$$

$$\widetilde{D}_1^{a\to h}(z\xi_T^2; Q^2) = D_1^{a\to h}(z; \mu_b^2) \, e^{S(\mu_b^2, Q^2)} \cdot e^{g_K(\xi_T)\ln(Q^2/Q_0^2)} \, \widetilde{f}_{1\text{NP}}^{a\to h}(z, \xi_T^2) \quad (7)$$

where $f_1^a(x; \mu_b^2)$ and $D_1^{a\to h}(z; \mu_b^2)$ are the usual collinear distribution and fragmentation functions, evaluated at the initial energy scale μ_b, which is chosen such that at $Q_0 = 1$ GeV there are no evolution effects.

The Sudakov exponent S can be generally written as

$$S(\mu_b^2, \mu^2) = -\frac{1}{2} \int_{\mu_b^2}^{\mu^2} \frac{dk_T^2}{k_T^2} [A(\alpha_s(k_T)) \ln\left(\frac{Q^2}{k_T^2}\right) + B(\alpha_s(k_T))]. \quad (8)$$

Following Refs. [9, 10, 11], for the nonperturbative Sudakov factor we make the traditional choice $g_K(\xi_T) = -g_2\xi_T^2/2$ with g_2 a free parameter. Finally, we parametrize the intrinsic nonperturbative parts of the TMDs as the normalized linear combination of a Gaussian and a weighted Gaussian:

$$\widetilde{f}_{1\text{NP}}^a(x, \xi_T^2) = \frac{1}{2\pi} e^{-g_{1a}\frac{\xi_T^2}{4}} \left(1 - \frac{\lambda g_{1a}^2}{1 + \lambda g_{1a}} \cdot \frac{\xi_T^2}{4}\right), \quad (9)$$

$$\widetilde{D}_{1\text{NP}}^{a\to h}(z, \xi_T^2) = \frac{g_{3a\to h} e^{-g_{3a\to h}\frac{\xi_T^2}{4}} + (\lambda_F/z^2) \, g_{4a\to h}^2 \left(1 - g_{4a\to h}\frac{\xi_T^2}{4}\right) e^{-g_{4a\to h}\frac{\xi_T^2}{4}}}{2\pi z^2 \left(g_{3a\to h} + (\lambda_F/z^2) \, g_{3a\to h}\right)}. \quad (10)$$

Based on the analyses of Refs. [12, 13], we consider that the Gaussian width of the TMDs depends on their fractional momentum, according to

$$g_1(x) = N_1 \frac{(1-x)^\alpha \, x^\sigma}{(1-\hat{x})^\alpha \, \hat{x}^\sigma}, \qquad g_{3,4}(z) = N_{3,4} \frac{(z^\beta + \delta) \, (1-z)^\gamma}{(\hat{z}^\beta + \delta) \, (1-\hat{z})^\gamma}, \quad (11)$$

where α, σ, $N_1 \equiv g_1(\hat{x})$ with $\hat{x} = 0.1$ and β, γ, δ,$N_{3,4} \equiv g_{3,4}(\hat{z})$ with $\hat{z} = 0.5$ are free parameters.

3. Data analysis and results

In our work we included a wide range of measurements taken from semi-inclusive DIS [14, 15], Drell-Yan at low energy [16, 17] and Z boson production [18, 19, 20, 21].

To avoid issues relative to known errors in the normalization of Compass data, we divided every data point by the value of the first point of their bin, excluding them from the d.o.f counting. The application of the TMD formalism to SIDIS depends on the capability of identifying the current fragmentation region, we identify it by operating a cut on z only, namely $0.2 < z < 0.74$.

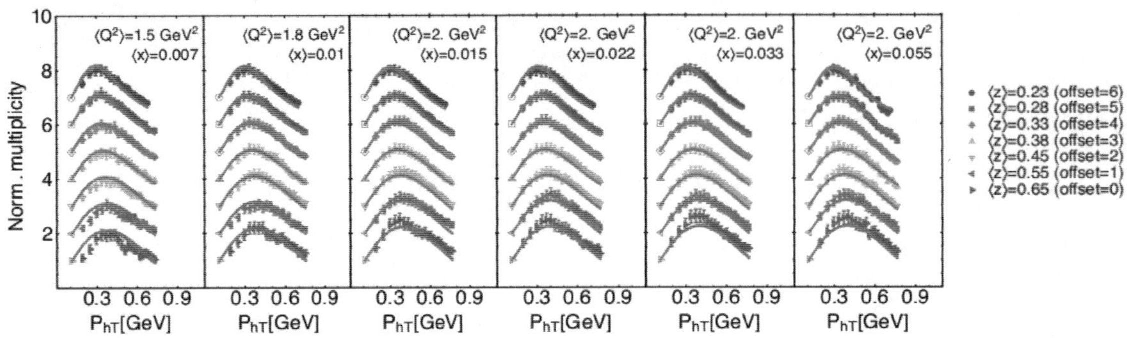

Figure 1: HERMES multiplicities for the process $ep \to ep + X$ as a function of the transverse momentum of the detected hadron P_{hT} at different $\langle x\rangle$, $\langle z\rangle$, $\langle Q^2\rangle$ bins. Each $\langle z\rangle$ bin has been shifted for clarity by an offset.

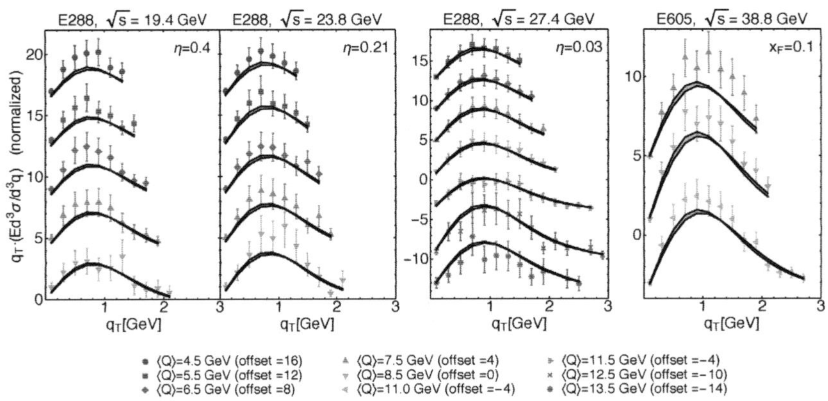

Figure 2: DY cross section as a function of the qT of the virtual photon for different values of $\sqrt{s}$ and

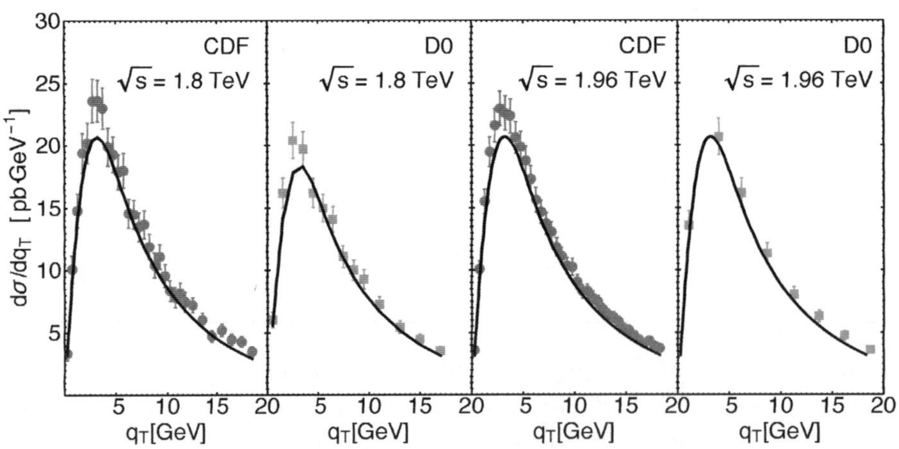

Figure 3: Cross section for the process $\bar{p}p \to ZX$ as a function of the transverse momentum q_T of the Z boson, for different energies of the CDF and D0 experiments at the Tevatron.

Another requirement for the applicability is the presence of two separate scales in the process, Q^2 and P_{hT}^2, which should satisfy the condition $P_{hT}^2/z^2 \ll Q^2$. We implement this condition by imposing $Q^2 > 1.4$ GeV^2 and restricting our fit to the small transverse momentum region. With those kinematic cuts we have 6252 points from COMPASS, 1514 from HERMES, 203 from Drell-Yan experiments and another 90 from Z boson production, for a total of 8059 data points. Our fit is based on the replica methodology, a Monte Carlo approach that consists in creating $\mathcal{M}$ replicas of the data points, by shifting independently each data point i by a Gaussian noise with the same variance as the measurement. A minimization procedure is applied to each replica separately, using MINUIT.

The overall quality of our fit is good, with a global χ^2/d.o.f. $= 1.55 \pm 0.05$. Uncertainties are computed as the 68% confidence level (C.L.) from the replica methodology. The number of degrees of freedom (d.o.f.) is given by the number of data points analyzed (8059), reduced by the number of free parameters in the error function (i.e. 11 free parameters).
With the adopted kinematic cuts, we included regions where TMD factorization could be questioned, but we checked that our results describe very well the regions where TMD factorization is supposed to hold. The χ^2/d.o.f. can be improved up to 1.02 restricting the kinematic cuts, without changing the parameters. In Tab. 1 the values obtained for the fit parameters are presented.
For SIDIS at HERMES off a proton, most of the contribution to the χ^2 comes from events with a π^+ in the final state and the bins with the worst agreement are at low Q^2. The main reason for the large χ^2 at HERMES is probably a normalization difference. SIDIS at COMPASS involves scattering off deuteron only, $D \to h^\pm$, and we identify $h \equiv \pi$. The quality of the agreement between theory and COMPASS data is better than in the case of pion production at HERMES, mainly because our fit is essentially driven by the COMPASS data and the observable that we fit in this case is a normalized multiplicity.
In regard to the low energy Drell–Yan data we observe that they have large error bands. This is why their χ^2 values are rather low compared to the other data sets. The agreement is also good for Z boson production. The statistics from Run-II is higher, which generates smaller experimental uncertainties and higher χ^2, especially for the CDF experiment.
For illustration, in Fig. 1 a selection of our results is compared with the COMPASS experimental data for the production of positively charged hadrons at different $\langle x\rangle$, $\langle z\rangle$ bins, as a function of the transverse momentum of the final hadron P_{hT}. The bands are computed as the 68% confidence level envelope of the full sets of fit curves from all the 200 replicas.
Results from DY and Z-boson productions are presented in Figs. 2 and 3, respectively.

4. Conclusions

In this work we demonstrated for the first time that it is possible to perform a simultaneous fit of unpolarized TMD PDFs and FFs to data of SIDIS, Drell–Yan and Z boson production at small transverse momentum collected by different experiments.

Using a replica methodology we extracted unpolarized TMDs using 8059 data points with 11 free parameters, obtaining an average χ^2/d.o.f. of 1.55 ± 0.05. Most of the discrepancies between experimental data and theory comes from the normalization and not from the transverse momentum shape. This issue could probably be addressed in future studies adopting a framework with an higher order of accuracy. This analysis could be further improved by considering different functional forms for the nonperturbative elements.

Acknowledgments
This research has received funding from the European Research Council (ERC) under the European Unions Horizon 2020 research and innovation programme (grant agreement No. 647981, 3DSPIN)

Table 1: 68% confidence intervals of best-fit values for TMD parameters at $Q = 1$ GeV.

TMD PDFs	g_1 [GeV^2]	α	σ		λ [GeV^{-2}]	
	0.28 ± 0.06	2.95 ± 0.05	0.17 ± 0.02		0.86 ± 0.78	
TMD FFs	g_3 [GeV^2]	β	δ	γ	λ_F [GeV^{-2}]	g_4 [GeV^2]
	0.21 ± 0.02	1.65 ± 0.49	2.28 ± 0.46	0.14 ± 0.07	5.50 ± 1.23	0.13 ± 0.01

References

[1] Collins J C, Soper D E and Sterman G F 1989 *Adv. Ser. Direct. High Energy Phys.* **5** 1–91 (*Preprint* hep-ph/0409313)
[2] Collins J 2013 *Foundations of perturbative QCD* (Cambridge University Press) ISBN 9781107645257, 9781107645257, 9780521855334, 9781139097826 URL http://www.cambridge.org/de/knowledge/isbn/item5756723
[3] Bacchetta A, Delcarro F, Pisano C, Radici M and Signori A 2017 *Journal of High Energy Physics* **2017** 81 ISSN 1029-8479 URL https://doi.org/10.1007/JHEP06(2017)081
[4] Bacchetta A, Diehl M, Goeke K, Metz A, Mulders P J and Schlegel M 2007 *JHEP* **02** 093 (*Preprint* hep-ph/0611265)
[5] Collins J C and Soper D E 1981 *Nucl. Phys.* **B193** 381 [Erratum: Nucl. Phys.B213,545(1983)]
[6] Boer D and Vogelsang W 2006 *Phys. Rev.* **D74** 014004 (*Preprint* hep-ph/0604177)
[7] Arnold S, Metz A and Schlegel M 2009 *Phys. Rev.* **D79** 034005 (*Preprint* 0809.2262)
[8] Aybat S M and Rogers T C 2011 *Phys. Rev.* **D83** 114042 (*Preprint* 1101.5057)
[9] Nadolsky P M, Stump D R and Yuan C P 2000 *Phys. Rev.* **D61** 014003 [Erratum: Phys. Rev.D64,059903(2001)] (*Preprint* hep-ph/9906280)
[10] Landry F, Brock R, Nadolsky P M and Yuan C P 2003 *Phys. Rev.* **D67** 073016 (*Preprint* hep-ph/0212159)
[11] Konychev A V and Nadolsky P M 2006 *Phys. Lett.* **B633** 710–714 (*Preprint* hep-ph/0506225)
[12] Signori A, Bacchetta A, Radici M and Schnell G 2013 *JHEP* **11** 194 (*Preprint* 1309.3507)
[13] Bacchetta A, Echevarria M G, Mulders P J G, Radici M and Signori A 2015 *JHEP* **11** 076 (*Preprint* 1508.00402)
[14] Airapetian A *et al.* (HERMES) 2013 *Phys. Rev.* **D87** 074029 (*Preprint* 1212.5407)
[15] Adolph C *et al.* (COMPASS) 2013 *Eur. Phys. J.* **C73** 2531 [Erratum: Eur. Phys. J.C75,no.2,94(2015)] (*Preprint* 1305.7317)
[16] Ito A S *et al.* 1981 *Phys. Rev.* **D23** 604–633
[17] Moreno G *et al.* 1991 *Phys. Rev.* **D43** 2815–2836
[18] Affolder T *et al.* (CDF) 2000 *Phys. Rev. Lett.* **84** 845–850 (*Preprint* hep-ex/0001021)
[19] Abbott B *et al.* (D0) 2000 *Phys. Rev.* **D61** 032004 (*Preprint* hep-ex/9907009)
[20] Aaltonen T *et al.* (CDF) 2012 *Phys. Rev.* **D86** 052010 (*Preprint* 1207.7138)
[21] Abazov V M *et al.* (D0) 2008 *Phys. Rev. Lett.* **100** 102002 (*Preprint* 0712.0803)

IOP Conf. Series: Journal of Physics: Conf. Series **938** (2017) 012042 doi:10.1088/1742-6596/938/1/012042

The contribution of pseudoscalar and axial-vector mesons to hyperfine structure of muonic hydrogen

A. E. Dorokhov
Joint Institute of Nuclear Research, BLTP, 141980, Moscow region, Dubna, Russia
E-mail: dorokhov@theor.jinr.ru

N. I. Kochelev
Institute of Modern Physics of Chinese Academy of Sciences, 730000, Lanzhou, China
Joint Institute of Nuclear Research, BLTP, 141980, Moscow region, Dubna, Russia
E-mail: nikkochelev@mail.ru

A. P. Martynenko
Samara University, 443086, Samara, Russia
E-mail: a.p.martynenko@samsu.ru

F.A. Martynenko
Samara University, 443086, Samara, Russia
E-mail: f.a.martynenko@gmail.com

A. E. Radzhabov
Institute of Modern Physics of Chinese Academy of Sciences, 730000, Lanzhou, China
Matrosov Institute for System Dynamics and Control Theory SB RAS, 664033, Irkutsk, Russia
E-mail: aradzh@icc.ru

R.N. Faustov
Institute of Informatics in Education, FRC CSC RAS, 119333, Moscow, Russia

Abstract. In the framework of the quasipotential method in quantum electrodynamics we calculate the contribution of light pseudoscalar (PS) and axial-vector (AV) mesons to the interaction operator of a muon and a proton in muonic hydrogen atom. The coupling of mesons with the muon is via two-photon intermediate state. The parametrization of the transition form factor of two photons into PS and AV mesons, based on the experimental data on the transition form factors and QCD asymptotics, is used. Numerical estimates of the contributions to the hyperfine structure of the spectrum of the S and P levels are presented. It is shown that such contribution to the hyperfine splitting in muonic hydrogen is rather important for a comparison with precise experimental data.

1. Proton radius puzzle

Precise investigation of the Lamb shift (LS) and hyperfine structure (HFS) of light muonic atoms is a fundamental problem for testing the Standard model and establishing the exact values of its parameters, as well as searching for effects of new physics. Recently, the CREMA (Charge Radius Experiments with Muonic Atoms) Collaboration from PSI by using the laser spectroscopy method measured with unprecedented accuracy the transition frequencies between the 2P and 2S states in muonic hydrogen (μp) [1, 2]

$$\nu_t\left(2\mathrm{P}_{3/2}^{F=2} - 2\mathrm{S}_{1/2}^{F=1}\right) = 49881.35(65)\ \mathrm{GHz}, \qquad h\nu_t = \Delta E\left(2\mathrm{P}_{3/2}^{F=2} - 2\mathrm{S}_{1/2}^{F=1}\right) \tag{1}$$

$$\nu_s\left(2\mathrm{P}_{3/2}^{F=1} - 2\mathrm{S}_{1/2}^{F=0}\right) = 54611.16(1.05)\ \mathrm{GHz}, \qquad h\nu_s = \Delta E\left(2\mathrm{P}_{3/2}^{F=1} - 2\mathrm{S}_{1/2}^{F=0}\right). \tag{2}$$

From these two transition measurements, both the Lamb shift ΔE_{LS} and the 2S-HFS ΔE_{HFS} was determined independently with the result [2]

$$\Delta E_{LS}^{\mathrm{exp}}\left(2\mathrm{P}_{1/2} - 2\mathrm{S}_{1/2}\right) = 202.3706(23)\ \mathrm{meV}, \tag{3}$$

$$\Delta E_{HFS}^{\mathrm{exp}}\left(2\mathrm{S}_{1/2}^{F=1} - 2\mathrm{S}_{1/2}^{F=0}\right) = 22.8089(51)\ \mathrm{meV}. \tag{4}$$

From theory side these quantities calculated in the frame work of bound-state QED are expressed in terms of the charge r_E^2 and Zemach r_Z radii of the proton as [3]

$$\Delta E_{\mathrm{LS}}^{\mathrm{th}} = 206.0668(25) - 5.2275(10)\ \mathrm{r}_E^2\ \mathrm{meV}, \tag{5}$$

$$\Delta E_{\mathrm{HFS}}^{\mathrm{th}} = 22.9843(30) - 0.1621(10)\ \mathrm{r}_Z\ \mathrm{meV}. \tag{6}$$

Remind that the charge RMS radius r_E is defined via the normalized proton charge distribution ρ_{E}, while the Zemach radius is correlated with the proton magnetic moment distribution ρ_{M} as

$$\mathrm{r}_{\mathrm{E}}^2 = \int d^3r r^2 \rho_{\mathrm{E}}(r), \qquad \mathrm{r}_Z = \int d^3r \int d^3r' r' \rho_{\mathrm{E}}(r)\, \rho_{\mathrm{M}}(r - r'). \tag{7}$$

The first term on the right side of (5) accounts for radiative, relativistic, and recoil effects, while the first term on the right side of (6) is the Fermi energy arising from the interaction between the muon and the proton magnetic moment and different corrections to it. It is also important, that like for the problem of the anomalous magnetic moments of leptons [4, 5], the coefficients in front of the radius terms in (5), (6) are much stronger enhanced for the muonic system relative to the electronic system, and thus much more sensitive in extraction of the radii parameters from the experimental data.

The comparison of (5),(6) with (3),(4) provides [2, 3]

$$\mathrm{r}_{\mathrm{E}}^{\mathrm{CREMA}} = 0.84087(39)\ \ \mathrm{fm}, \qquad \mathrm{r}_{\mathrm{Z}}^{\mathrm{CREMA}} = 1.082(37)\ \ \mathrm{fm}. \tag{8}$$

This should be compared with the values recommended by CODATA [6] and based on the electron-proton scattering and electronic hydrogen spectroscopy

$$\mathrm{r}_{\mathrm{E}}^{\mathrm{CODATA}} = 0.8751(61)\ \ \mathrm{fm}. \tag{9}$$

Thus we see two basic results of the laser spectroscopy experiment for μp. First, from the μp spectroscopy the value of r_{E} is determined with precision 10 times higher than from electronic data. Second, and more striking, that there is the large discrepancy between $\mathrm{r}_{\mathrm{E}}^{\mathrm{CREMA}}$ and $\mathrm{r}_{\mathrm{E}}^{\mathrm{CODATA}}$ at the level of 5.6 σ (or the muonic hydrogen value is 4% smaller than the CODATA value). This is so-called the **proton size puzzle**, not explained up to now! Later on, the similar problem was detected for the deutron radius [7].

At present, several experimental groups plan to measure HFS of various muonic atoms with more high precision [8, 9, 10]. This will make it possible to better understand the existing "puzzle" of the proton charge radius, to check the Standard Model with greater accuracy and, possibly, to reveal the source of previously unaccounted interactions between the particles forming the bound state in QED. One of the ways of overcoming the crisis situation arises from a deeper theoretical analysis of the fine and hyperfine structure of muonic atom spectrum, with the verification of previously calculated contributions and the more accurate construction of the particle interaction operator in quantum field theory, the calculation of new corrections whose value for muonic atoms can increase substantially in comparison with electronic atoms. The expected results will allow to get also a new very important information about the forces which are responsible for the structure of atoms.

2. Light meson exchange contributions to HFS of μp

From the theory side it is urgently needed to study the possible effects of exchanges between muon and proton which can contribute to HFS of μp. Some of such effects was considered in recent papers [11, 12, 13, 14, 15]. Below, we discuss the effects of exchanges between muon and proton which can contribute to HFS of μp coming from the light pseudoscalar (PS) and axial-vector (AV) meson exchanges between muon and proton induced by meson coupling to muon through two photons (see Fig. 1).

The leading contribution to HFS of μp is coming from one-photon exchange and has the following form [16, 17, 18]:

$$\Delta V_B^{hfs} = \frac{8\pi\alpha\mu_p}{3m_\mu m_p}(\mathbf{S}_p\mathbf{S}_\mu)\delta(\mathbf{r}) - \frac{\alpha\mu_p(1+a_\mu)}{m_\mu m_p r^3}\left[(\mathbf{S}_p\mathbf{S}_\mu) - 3(\mathbf{S}_p\mathbf{n})(\mathbf{S}_p\mathbf{n})\right] + \frac{\alpha\mu_p}{m_\mu m_p r^3}\left[1 + \frac{m_\mu}{m_p} - \frac{m_\mu}{2m_p\mu_p}\right](\mathbf{L}\mathbf{S}_p) \quad (10)$$

where m_μ, $\mathbf{S}_\mu$ and m_p, $\mathbf{S}_p$ are masses and spins of muon and proton, correspondingly, μ_p is the proton magnetic moment. The potential (10) gives the main contribution of order α^4 to the HFS of muonic atom. Precision calculation of the HFS of the spectrum, which is necessary for comparison with experimental data, requires the consideration of various corrections accounting for the vacuum polarization, nuclear structure and recoil, and relativistic effects [16, 20, 19, 21, 22, 23].

We calculate further the contribution to HFS coming from the pion [1] and axial-vector $f_1(1285)$, $a_1(1260)$ and $f_1(1420)$ meson exchanges shown in Fig. 1. The effective vertices of the interaction of the PS and AV mesons and virtual photons can be expressed in terms of the transition form factors as follows:

$$V^{\mu\nu}(k_1,k_2) = i\varepsilon^{\mu\nu\alpha\beta}k_{1\alpha}k_{2\beta}\frac{\alpha}{\pi F_\pi}F_{\pi^0\gamma^*\gamma^*}(k_1^2,k_2^2), \quad (11)$$

$$T^{\mu\nu\alpha} = 8\pi i\alpha\varepsilon_{\mu\nu\alpha\tau}k^\tau k^2 F^{(0)}_{AV\gamma^*\gamma^*}(k_1^2,k_2^2), \quad (12)$$

where k_1, k_2 are four-momenta of virtual photons. For small values of the relative momenta of particles in the initial and final states and small value of transfer momentum t between muon and proton, the transition amplitude takes a simple form

$$T^{\mu\nu\alpha} = 8\pi i\alpha\varepsilon_{\mu\nu\alpha\tau}k^\tau k^2 F^{(0)}_{AV\gamma^*\gamma^*}(t^2,k^2,k^2), \quad (13)$$

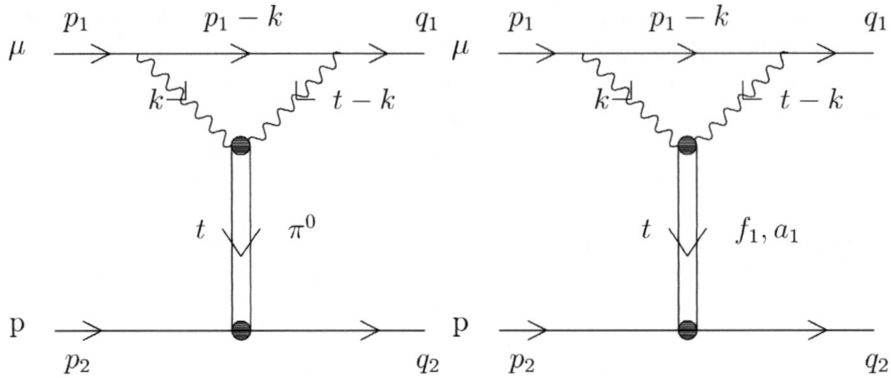

Figure 1. Muon-proton interaction induced by mesonic exchange.

where $k = k_1 = -k_2$.

The final result for the HFS potential is equal to

$$\Delta V_{PS}^{hfs}(\mathbf{t}) = \frac{\alpha^2}{6\pi^2}\frac{g_p}{m_p F_\pi}\frac{\mathbf{t}^2}{\mathbf{t}^2 + m_\pi^2}\mathcal{A}(\mathbf{t}^2), \tag{14}$$

$$\Delta V_{AV}^{hfs}(\mathbf{t}) = -\frac{32\alpha^2 g_{AVPP} F_{AV\gamma\gamma}(0,0)}{3\pi^2(\mathbf{t}^2 + M_A^2)} I(\frac{m_\mu}{\Lambda_A}). \tag{15}$$

where

$$\mathcal{A}(\mathbf{t}^2) = \frac{2}{\pi^2 \mathbf{t}^2}\int \frac{id^4k[t^2k^2 - (tk)^2]}{k^2(k-t)^2(k^2 - 2kp_1)} F_{PS\gamma^*\gamma^*}(k^2, (k-t)^2), \tag{16}$$

$$I(\frac{m_\mu}{\Lambda_A}) = \int \frac{id^4k(2k^2 + k_0^2)}{k^2(k^2 - 2m_\mu k_0)} F_{AV\gamma^*\gamma^*}(k^2, k^2). \tag{17}$$

The integral $\mathcal{A}$ in (16) is well studied in connection to the problem of interpretation [24, 25] of the KTeV (FermiLab) data on the pion decay into e^+e^- pair. In order to fix the transition form factors in the most model independent way we used corresponding data from CLEO [26] and L3 [27, 28, 29] collaborations for the PS and AV mesons, respectively (see for details [11, 12])[2]. We would like to point out, that one can expect the important contribution of the AV exchange to spin dependent part of muon-proton interaction because the exchange particle has the spin one. Furthermore, it is also well known that in the channel with quantum number 1^{++} axial anomaly effects can play an important role and, in particularly, these effects might be considered as a cornerstone to solve so-called "proton spin crisis" [31, 32]. The other phenomenological input, the meson-nucleon couplings, were determined by using the Regge approach analysis of the deep-inelastic scattering, f_1 and a_1 trajectories contributing to the polarization of quarks in the nucleon [33, 34].

Calculating the matrix elements with wave functions of 1S , 2S and 2P 1/2 states, we obtain the corresponding contributions to the HFS spectrum

$$\Delta E^{hfs}(1S) = \frac{\mu^3\alpha^5 g_A}{6F_\pi^2\pi^3}\left\{\mathcal{A}(0)\frac{4W(1+\frac{W}{m_\pi})}{m_\pi(1+\frac{2W}{m_\pi})^2} - \frac{1}{\pi}\int_0^\infty \frac{ds}{s} Im\mathcal{A}(s)\times\right. \tag{18}$$

[1] The contribution of the η and η' mesons is negligible [11]

[2] In recent paper [30]within NJL model the estimation for form factor $F^{0(NJL)}_{f_1(1285),\gamma^*\gamma^*} \approx 0.276 GeV^{-2}$ was obtained. This value is in the agreement with L3 data which was used in our paper [12] (see Table 1.)

$$\left[1+\frac{1}{4W^2(s-m_\pi^2)}\left(\frac{m_\pi^4}{(1+\frac{m_\pi}{2W})^2}-\frac{s^2}{(1+\frac{\sqrt{s}}{2W})^2}\right)\right]\Bigg\},$$

$$\Delta E^{hfs}(2S)=\frac{\mu^3\alpha^5 g_A}{48F_\pi^2\pi^3}\Bigg\{\mathcal{A}(0)\frac{W(8+11\frac{W}{m_\pi}+8\frac{W^2}{m_\pi^2}+2\frac{W^3}{m_\pi^3})}{2m_\pi(1+\frac{W}{m_\pi})^4}-\frac{1}{\pi}\int_0^\infty\frac{ds}{s}Im\mathcal{A}(s)\times \tag{19}$$

$$\left[1+\frac{1}{(s-m_\pi^2)}\left(\frac{m_\pi^2(2+\frac{W^2}{m_\pi^2})}{2(1+\frac{W}{m_\pi})^4}-\frac{s(2+\frac{W^2}{s})}{2(1+\frac{W}{\sqrt{s}})^4}\right)\right]\Bigg\},$$

$$\Delta E^{hfs}_{AV}(1S)=\frac{32\alpha^5\mu^3 g_{AVPP}F^{(0)}_{AV\gamma^*\gamma^*}(0,0,0)}{3M_A^2\pi^3\left(1+\frac{2W}{M_A}\right)^2}I\left(\frac{m_\mu}{\Lambda_A}\right), \tag{20}$$

$$\Delta E^{hfs}_{AV}(2S)=\frac{2\alpha^5\mu^3 g_{AVPP}F^{(0)}_{AV\gamma^*\gamma^*}(0,0,0)\left(2+\frac{W^2}{M_A^2}\right)}{3M_A^2\pi^3\left(1+\frac{W}{M_A}\right)^4}I\left(\frac{m_\mu}{\Lambda_A}\right), \tag{21}$$

where $W=\mu\alpha$ and μ is reduced mass.

Table 1. Pion and axial-vector mesons exchanges contribution to HFS of muonic hydrogen.

mesons	$I^G(J^{PC})$	Λ_A in MeV	$F^{(0)}_{AV\gamma^*\gamma^*}(0,0)$ in GeV^{-2}	$\Delta E^{hfs}(1S)$ in meV	$\Delta E^{hfs}(2S)$ in meV
$f_1(1285)$	$0^+(1^{++})$	1040	0.266	-0.0093 ± 0.0033	-0.0012 ± 0.0004
$a_1(1260)$	$1^-(1^{++})$	1040	0.591	-0.0437 ± 0.0175	-0.0055 ± 0.0022
$f_1(1420)$	$0^+(1^{++})$	926	0.193	-0.0013 ± 0.0008	-0.0002 ± 0.0001
π^0	$1^-(0^{-+})$	776		-0.0017 ± 0.0001	-0.0002 ± 0.00002
Sum				-0.0560 ± 0.0178	-0.0071 ± 0.0024

In Table 1 our results for contribution of the pion and AV mesons exchanges to HFS are presented. For the case of $2S$ state the summary contribution from pion and AV meson exchanges is equal to (-0.0071) meV, which is quite important to obtain the total value of HFS with high precision.

3. Conclusion

A new important contribution to the muon-nucleon interaction is discovered. It is coming from effective pion and AV mesons exchanges induced by anomalous meson vertices with two photon state. The contribution of this exchange to the HFS of μp is calculated in framework of the quasipotential method in QED and the use of the technique of projection operators on the states of two particles with a definite spin. It is shown that this contribution is rather large and should be taking into account for the interpretation of the new data on HFS in this atom.

As has been mentioned the CREMA Collaboration measured two transition frequencies in muonic hydrogen for the 2S triplet state ($2P^{F=2}_{3/2}-2S^{F=1}_{1/2}$) and for 2S singlet state ($2P^{F=1}_{3/2}-2S^{F=0}_{1/2}$) [2]. From these measurements it is possible to extract the value of HFS for $2S$ level. Obtained value (4) allows to get the value of the Zemach radius (8) with accuracy 3.4 % with help of relation (6). This is in the agreement with another numerical values $r_Z=1.086(12)$ fm [35], $r_Z=1.045(4)$ fm [36], $r_Z=1.047(16)$ fm [37], $r_Z=1.037(16)$ fm [38] obtained from electron-proton scattering and from H and muonium spectroscopy. We should emphasize that the changing of the theoretical value of HFS on 0.001 meV leads to the changing of the Zemach radius on 0.006 fm. Therefore, our contribution coming from pion and AV meson exchange leads to the new value of this radius

$$r_Z = 1.040(37) \text{ fm}, \tag{22}$$

which is smaller in the comparison with most listed results but still agree with them within errorbars.

The work is supported by Russian Science Foundation (grant No. RSF 15-12-10009) (A.E.D.), the Chinese Academy of Sciences visiting professorship for senior international scientists (grants No. 2013T2J0011) (N.I.K.) and President's international fellowship initiative (Grant No. 2017VMA0045) (A.E.R.), Russian Foundation for Basic Research (grant No. 16-02-00554) (A.P.M., F.A.M.).

References

[1] R. Pohl, A. Antognini, F. Nez et al., *Nature* **466**, 213 (2010).
[2] A. Antognini et al., *Science* **339**, 417 (2013).
[3] A. Antognini et al., *Ann. Phys. (NY)* **331**, 127 (2013).
[4] A. E. Dorokhov, A. E. Radzhabov, A. S. Zhevlakov, *Eur. Phys. J.* C **72**, 2227 (2012).
[5] A. E. Dorokhov, A. E. Radzhabov, A. S. Zhevlakov, *JETP Lett.* **100,** 133 (2014).
[6] P. J. Mohr, D. B. Newell, and B. N. Taylor (CODATA Recommended Values of the Fundamental Physical Constants: 2014) *Rev. Mod. Phys.* **88**, 035009 (2016).
[7] R. Pohl, F. Nez, L. M. P. Fernandes et al., *Science* **353**, 669 (2016).
[8] Y. Ma et al., *Int. J. Mod. Phys. Conf. Ser.* **40**, 1660046 (2016).
[9] A. Adamczak et al. [FAMU Collaboration], *JINST* **11**, no. 05, P05007 (2016).
[10] R. Pohl [CREMA Collaboration], *J. Phys. Soc. Jap.* **85**, 091003 (2016).
[11] A. E. Dorokhov, N. I. Kochelev, A. P. Martynenko, F. A. Martynenko, and R. N. Faustov, *Phys. Part. Nucl. Lett.* **14,** 857 (2017) ; arXiv:1704.07702 [hep-ph].
[12] A. E. Dorokhov, N. I. Kochelev, A. P. Martynenko, F. A. Martynenko, and A. E. Radzhabov, arXiv:1707.04138 [hep-ph].
[13] H. Q. Zhou, H. R. Pang, *Phys. Rev.* A **92**, 032512 (2015).
[14] F. Hagelstein, V. Pascalutsa, *PoS* CD15 077 (2016).
[15] N. T. Huong, E. Kou, B. Moussallam, *Phys. Rev.* D **93**, 114005 (2016).
[16] A. P. Martynenko and R. N. Faustov, *J. Exp. Theor. Phys.* **98**, 39 (2004).
[17] A. P. Martynenko and R.N. Faustov, *J. Exp. Theor. Phys.* **88**, 672 (1999).
[18] A. A. Krutov, A. P. Martynenko, F. A. Martynenko and O. S. Sukhorukova, *Phys. Rev.* A **94**, 062505 (2016).
[19] M. I. Eides, H. Grotch, and V. A. Shelyuto, Theory of Light Hydrogenic Bound States, Springer Tracts in Modern Physics, V. 222 (Springer, Berlin, Heidelbeg, New York, 2007).
[20] K. Pachucki, *Phys. Rev.* A **53**, 2092 (1996).
[21] E. Borie, *Ann. Phys.* **327**, 733 (2012).
[22] R. N. Faustov and A. P. Martynenko, *Phys. Lett.* B **541**, 135 (2002).
[23] A. P. Martynenko, F. A. Martynenko, and R. N. Faustov, *J. Exp. Theor. Phys.* **124**, 895 (2017).
[24] A. E. Dorokhov and M. A. Ivanov, *Phys. Rev.* D **75**, 114007 (2007).
[25] A. E. Dorokhov, M. A. Ivanov and S. G. Kovalenko, *Phys. Lett.* B **677**, 145 (2009).
[26] J. Gronberg *et al.*[CLEO], *Phys. Rev.* **D57**, 33 (1998).
[27] P. Achard et al. [L3 Collaboration], *Phys. Lett.* B **526**, 269 (2002).
[28] P. Achard et al. [L3 Collaboration], *JHEP* 03, 018 (2007).
[29] H. Aihara et al., *Phys. Rev.* D **38**, 1 (1988).
[30] A. A. Osipov, A. A. Pivovarov and M. K. Volkov, Phys. Rev. D **96** no.5, 054012 (2017)
[31] A.E. Dorokhov, N.I. Kochelev and Yu.A. Zubov, *Int. J. Mod. Phys.* A **8** (1993) 603.
[32] M. Anselmino, A. Efremov and E. Leader, *Phys. Rept.* **261**, 1 (1995); Erratum: [*Phys. Rept.* **281**, 399 (1997)]
[33] N. I. Kochelev and Y. Oh, *Phys. Rev.* D **85**, 016012 (2012).
[34] N. I. Kochelev, D.-P. Min, Y. Oh, V. Vento, and A. V. Vinnikov, *Phys. Rev.* D **61**, 094008 (2000).
[35] J. Friar and I. Sick, *Phys. Lett.* B **579**, 285 (2004).
[36] M. O. Distler, J. C. Bernauer, and T. Walcher, *Phys. Lett.* B **696**, 343 (2011).
[37] A. V. Volotka, V. M. Shabaev, G. Plunien, and G. Soff, *Eur. Phys. J.* D **33**, 23 (2005).
[38] A. Dupays, A. Beswick, B. Lepetit, C. Rizzo, and D. Bakalov, *Phys. Rev.* A **68**, 052503 (2003).

IOP Conf. Series: Journal of Physics: Conf. Series **938** (2017) 012043 doi:10.1088/1742-6596/938/1/012043

Neutrino spin oscillations in curved space-time under the influence of external fields

Maxim Dvornikov

Pushkov Institute of Terrestrial Magnetism, Ionosphere and Radiowave Propagation (IZMIRAN), 108840 Troitsk, Moscow, Russia

Physics Faculty, National Research Tomsk State University, 36 Lenin Avenue, 634050 Tomsk, Russia

E-mail: maxdvo@izmiran.ru

Abstract. We study neutrino spin oscillations in background matter under the influence of strong electromagnetic and gravitational fields. The neutrino spin evolution is treated quasiclassically. We derive the effective Hamiltonian governing spin oscillations of a neutrino moving in the vicinity of a black hole and interacting with a relativistic magnetized accretion disk around a black hole. Applications for the studies of spin oscillations of astrophysical neutrinos are considered.

1. Introduction

Neutrinos play an important role in the evolution of various astrophysical objects like neutron stars, supernovæ, quasars, etc. In these environments, neutrinos are subject to the interaction not only with a background matter and magnetic fields, but also with a strong gravitational field. The study of astrophysical neutrinos propagating in these backgrounds becomes important owing to the recent development of the multi-messenger approach in astronomy and astrophysics, when a correlation between a neutrino signal and, e.g., the detection of gravitational waves is searched.

Neutrinos are known to be mainly left-polarized particles in the standard model. However, the interaction with external fields can change their polarization and neutrinos become invisible for a terrestrial detector. This process is known as neutrino spin oscillations. Thus, the interaction of astrophysical neutrinos with external fields, which include gravity, can significantly modify the predictions of the multi-messenger approach.

In the paper, we review our results on the description of neutrino spin oscillations in non-trivial gravitational backgrounds under the influence of various external fields. In section 2, we derive the most general quasiclassical equation for the description of the neutrino spin evolution in these external fields. In sections 3 and 4, we apply our results for the studies of neutrino spin oscillations in the vicinity of Schwarzschild and Kerr black holes (BH). Finally, in section 5, we summarize the obtained results.

2. Covariant description of the neutrino spin evolution

In this section we shall construct the quasiclassical approach for the description of the spin evolution of a Dirac neutrino moving in a background matter under the influence of electromagnetic and gravitational fields. In particular, we generalize the neutrino spin evolution equation in matter and electromagnetic fields to include the effects of the nontrivial geometry of space-time.

The equation describing the neutrino spin evolution in matter under the influence of an external electromagnetic field in Minkowski space-time was derived in [1]. The most straightforward generalization of this equation to include a nontrivial gravitational background has the form [2],

$$\frac{\mathrm{D}S^\mu}{\mathrm{D}\tau} = 2\mu\left(F^{\mu\nu}S_\nu - U^\mu U_\nu F^{\nu\lambda}S_\lambda\right) + \sqrt{2}G_\mathrm{F}E^{\mu\nu\lambda\rho}G_\nu U_\lambda S_\rho. \tag{1}$$

where S^μ and U^μ are the four vectors of the spin and the velocity of a neutrino, $F_{\mu\nu}$ is the skew-symmetric tensor of the electromagnetic field, $E^{\mu\nu\lambda\rho} = \varepsilon^{\mu\nu\lambda\rho}/\sqrt{-g}$ is the antisymmetric tensor is curved space-time, $g = \det(g_{\mu\nu})$ is the determinant of the metric tensor $g_{\mu\nu}$, μ is the neutrino magnetic moment, and $G_\mathrm{F} = 1.17 \times 10^{-5}\,\mathrm{GeV}^{-2}$ is the Fermi constant. The covariant derivative $\mathrm{D}/\mathrm{D}\tau$ along the world line in equation (1) is taken with respect to the interval τ. Equation (1) should be completed by the geodesic equation for the evolution of the four velocity, $\mathrm{D}U^\mu/\mathrm{D}\tau = 0$.

The neutrino interaction with a background matter is characterized by the four vector G^μ, which is a linear combination of hydrodynamic currents, J_f^μ, and polarizations, Λ_f^μ, of background fermions of the type f,

$$G^\mu = \sum_f \left(q_f^{(1)}J_f^\mu + q_f^{(2)}\Lambda_f^\mu\right). \tag{2}$$

We shall express J_f^μ using the invariant number density n_f, i.e. the density in the rest frame of fermions, and the four velocity U_f^μ of background fermions: $J_f^\mu = n_f U_f^\mu$. The explicit expression of Λ_f^μ is given in [3]. The coefficients $q_f^{(1,2)}$ can be found in [1].

Then we introduce a locally Minkowskian frame by considering the vierbein $e^a{}_\mu$, $a = 0, \dots, 3$, satisfying the relations, $g_{\mu\nu} = e^a{}_\mu e^b{}_\nu \eta_{ab}$ and $\eta_{ab} = e_a{}^\mu e_a{}^\nu g_{\mu\nu}$, where $e_a{}^\mu$ is the inverse vierbein and $\eta_{ab} = \mathrm{diag}(1, -1, -1, -1)$ is the metric tensor in the Minkowski space-time. Decomposing all the quantities in equation (1) in the vierbein basis, we get the following equation for the neutrino spin $s^a = e^a{}_\mu S^\mu$:

$$\frac{\mathrm{d}s^a}{\mathrm{d}t} = \frac{1}{\gamma}\left[G^{ab}s_b + 2\mu\left(f^{ab}s_b - u^a u_b f^{bc}s_c\right) + \sqrt{2}G_\mathrm{F}\varepsilon^{abcd}g_b u_c s_d\right], \tag{3}$$

where $G^{ab} = \eta^{ac}\eta^{bd}\gamma_{cde}u^e$ is the antisymmetric tensor accounting for the gravitational interaction of neutrinos, $\gamma_{abc} = \eta_{ad}e^d{}_{\mu;\nu}e_b{}^\mu e_c{}^\nu$ are the Ricci rotation coefficients (the semicolon stays for the covariant derivative), $\gamma = U^0$, $g^a = e^a{}_\mu G^\mu = (g^0, \mathbf{g})$ is the effective potential of the matter interaction in the vierbein frame, $f_{ab} = e_a{}^\mu e_b{}^\nu F_{\mu\nu} = (\mathbf{e}, \mathbf{b})$ is the electromagnetic field tensor in the vierbein frame, and $u^a = e^a{}_\mu U^\mu = (u^0, \mathbf{u})$ is the neutrino velocity in the vierbein basis. We should also reformulate the dynamics of U^μ in the vierbein frame. One can obtain the following evolution equation for u^a:

$$\frac{\mathrm{d}u^a}{\mathrm{d}t} = \frac{1}{\gamma}G^{ab}u_b. \tag{4}$$

The details of the derivation of equations (3) and (4) can be found in [2, 4].

The three vector of the neutrino polarization $\boldsymbol{\zeta}$ is defined as

$$s^a = \left((\boldsymbol{\zeta}\cdot\mathbf{u}), \boldsymbol{\zeta} + \frac{\mathbf{u}(\boldsymbol{\zeta}\cdot\mathbf{u})}{1 + u^0}\right). \tag{5}$$

Using equations (3) and (5) one finds the evolution equation for $\boldsymbol{\zeta}$ as

$$\frac{\mathrm{d}\boldsymbol{\zeta}}{\mathrm{d}t} = \frac{2}{\gamma}[\boldsymbol{\zeta} \times \mathbf{G}], \tag{6}$$

where

$$\mathbf{G} = \frac{1}{2}\left[\mathbf{b}_g + \frac{1}{1+u^0}(\mathbf{e}_g \times \mathbf{u})\right] + \frac{G_\mathrm{F}}{\sqrt{2}}\left[\mathbf{u}\left(g^0 - \frac{(\mathbf{gu})}{1+u^0}\right) - \mathbf{g}\right] + \mu\left[u^0\mathbf{b} - \frac{\mathbf{u}(\mathbf{ub})}{1+u^0} + (\mathbf{e}\times\mathbf{u})\right]. \tag{7}$$

Here $\mathbf{e}_g$ and $\mathbf{b}_g$ are the components of the tensor G_{ab}: $G_{ab} = (\mathbf{e}_g, \mathbf{b}_g)$. To derive equation (7) we use the fact that $u_a u^a = 1$.

3. Neutrino spin oscillations in Schwarzschild metric

Let us apply general equations (6) and (7) to describe the neutrino spin evolution in the vicinity of a non-rotating BH. In this section, we shall omit the influence of the background matter and the electromagnetic field on neutrino oscillations.

When we study the gravitational field of a non-rotating BH, the interval is known to be expressed with help of the Schwarzschild metric,

$$d\tau^2 = A^2 dt^2 - A^{-2}dr^2 - r^2(d\theta^2 + \sin^2\theta d\phi^2), \tag{8}$$

where $A = \sqrt{1 - r_g/r}$, $r_g = 2M$ is the Schwarzschild radius, and M is the BH mass. In Eq. (8) we use the spherical coordinates (r, θ, ϕ).

The neutrino spin precession is determined by the vector $\boldsymbol{\Omega} = \mathbf{G}/\gamma$. The components of this vector can be found on the basis of equation (7) (see also [4]),

$$\Omega_1 = \frac{1}{2}v_\phi \cos\theta, \quad \Omega_2 = v_\phi \sin\theta \frac{1}{2}\left(-A + \frac{\gamma}{(1+\gamma A)}\frac{r_g}{2r}\right), \quad \Omega_3 = v_\theta \frac{1}{2}\left(A - \frac{\gamma}{(1+\gamma A)}\frac{r_g}{2r}\right), \tag{9}$$

where $\mathbf{v} = (v_r, v_\theta, v_\phi)$ are the components of the world velocity.

Let us discuss a neutrino orbiting BH. For simplicity we consider only circular orbits with the radius R. We may restrict ourselves to the consideration of the orbits lying only in the equatorial plane ($\theta = \pi/2$ or equivalently $v_\theta = 0$) because of the spherical symmetry of the gravitational field. In this case $\Omega_1 = \Omega_3 = 0$ in equation (9). The expressions for the neutrino angular velocity and γ have the form,

$$v_\phi = \frac{d\phi}{dt} = \sqrt{\frac{r_g}{2R^3}}, \quad \gamma^{-1} = \frac{d\tau}{dt} = \sqrt{1 - \frac{3r_g}{2R}}. \tag{10}$$

It should be noted that the vierbein four velocity now takes the form, $u^a = (\gamma A, 0, 0, \gamma v_\phi r)$. One can verify that $u^a u_a = 1$ with help of equation (10). We also mention that equation (4) is also identically satisfied since $G^{ab}u_b = 0$. Therefore a neutrino has a constant four velocity with respect to the vierbein frame.

Using equation (9), we can rewrite the remaining nonzero component of $\boldsymbol{\Omega}$ in the more simple form, $\Omega_2 = -v_\phi/2\gamma$. We suppose that initially a neutrino is left-handed, i.e. its initial spin vector is antiparallel to the particle's velocity. According to equations (6) and (9), the neutrino spin rotates around the second axis. Therefore we can derive the effective Schrödinger equation for neutrino spin oscillations in the gravitational field of a non-rotating BH,

$$\mathrm{i}\frac{\mathrm{d}\nu}{\mathrm{d}t} = H_\mathrm{eff}\nu, \quad H_\mathrm{eff} = -(\boldsymbol{\sigma}\cdot\boldsymbol{\Omega}) = \begin{pmatrix} 0 & -\mathrm{i}\Omega_2 \\ \mathrm{i}\Omega_2 & 0 \end{pmatrix}, \tag{11}$$

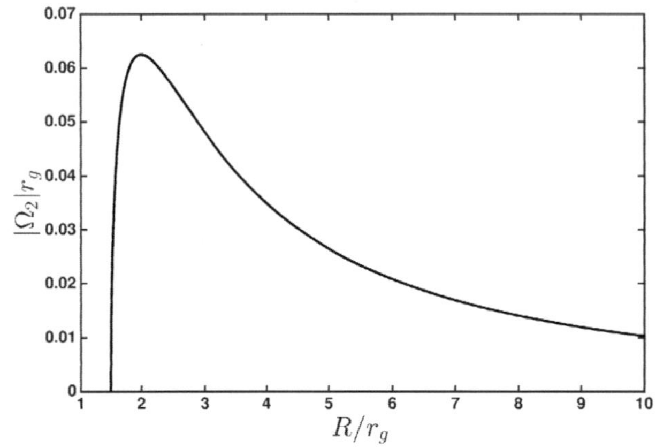

Figure 1. Neutrino spin oscillations frequency versus the radius of the neutrino orbit.

where $\nu^{\mathrm{T}} = (\nu_{\mathrm{R}}, \nu_{\mathrm{L}})$ is the effective neutrino wave function, H_{eff} is the effective Hamiltonian, and $\boldsymbol{\sigma} = (\sigma_1, \sigma_2, \sigma_3)$ are the Pauli matrices.

Using equation (11), we obtain the expression for the transition probability for neutrino spin oscillations, $P(t) = \sin^2(\Omega_2 t)$. One can see that there is a full mixing in our case, and thus, the neutrino transition probability can achieve a unit value. Let us plot the frequency of neutrino spin oscillations versus the radius of the orbit. It is possible to see in figure 1 that $|\Omega_2| = 0$ at $R = 1.5 r_g$ and $|\Omega_2| \to 0$ at $R \to \infty$. One can also conclude that $|\Omega_2|$ has its maximal value, which is equal to $6.25 \times 10^{-2} r_g^{-1}$, at $R = 2r_g$. Let us evaluate the number of revolutions N, that a neutrino should make, necessary for the total spin flip. Using equation (10) we can find that $N = \gamma = 2$ at $R = 2r_g$. It is interesting to evaluate the characteristic period of the neutrino spin oscillations, i.e. the oscillations length. For $M = 10 M_\odot$ at $R = 2r_g$ we get for $T = \pi/|\Omega_2| \approx 4.94 \times 10^{-3}$ s.

4. Neutrino propagation in the vicinity of a rotating black hole

Now let us describe neutrino spin oscillations in matter and a magnetic field when a particle moves in the vicinity of a rotating BH. As in section 3, we shall consider circular neutrino motion and derive the effective Schrödinger equation as well as find the transition probability for spin oscillations.

The space-time of a rotating BH can be described by the Kerr metric. In Boyer-Lindquist coordinates $x^\mu = (t, r, \theta, \phi)$, this metric has the form,

$$\mathrm{d}\tau^2 = \left(1 - \frac{r r_g}{\Sigma}\right) \mathrm{d}t^2 + 2\frac{r r_g a \sin^2\theta}{\Sigma} \mathrm{d}t\mathrm{d}\phi - \frac{\Sigma}{\Delta}\mathrm{d}r^2 - \Sigma \mathrm{d}\theta^2 - \frac{\Xi}{\Sigma}\sin^2\theta \mathrm{d}\phi^2, \tag{12}$$

where

$$\Delta = r^2 - r r_g + a^2, \quad \Sigma = r^2 + a^2\cos^2\theta, \quad \Xi = \left(r^2 + a^2\right)\Sigma + r r_g a^2 \sin^2\theta. \tag{13}$$

The angular momentum of BH is $J = Ma$. Coordinates x^μ in equations (12) and (13) correspond to the world frame rather than to the vierbein frame.

We shall discuss BH surrounded by matter forming an accretion disk in equatorial plane. In this case only U_f^0 and U_f^ϕ are nonzero. Moreover we shall consider a stationary accretion. By the symmetry reasons, the quantities n_f, U_f^0, and U_f^ϕ can be functions of r and θ only. Assuming that the temperature of background fermions is sufficiently high, we can take that the background matter is unpolarized. In this case, the effective neutrino interaction with matter is characterized by the four vector,

$$g^a = n_{\text{eff}} \left(rU_f^0 \sqrt{\frac{\Delta}{\Xi_0}}, 0, 0, \frac{U_f^\phi \Xi_0 - arr_g U_f^0}{r\sqrt{\Xi_0}} \right), \quad n_{\text{eff}} = \sum_f q_f^{(1)} n_f, \tag{14}$$

where $\Xi_0 = \Xi(\theta = \pi/2) = r^4 + a^2 r(r + r_g)$. In equation (14) we assume that all types of background fermions have the same four velocity.

Now let us specify the structure of the electromagnetic field. We suggest that, at relatively great distance from BH, there is a constant and uniform magnetic field parallel to the rotation axis of BH and having the strength B_0. This magnetic field can be created by the plasma motion in the accretion disk. The structure of the axially symmetric electromagnetic field in a curved space-time having such an asymptotics was studied in [5], where one can find the nonzero components of the vector potential. On the basis of the explicit form of the vector potential (see, e.g., [2]), the electric and magnetic field strengths can be calculated as $F_{\mu\nu} = \partial_\mu A_\nu - \partial_\nu A_\mu$.

As in the case of Schwarzschild metric, studied in section 3, here one can show that the four velocity in the vierbein frame u^a is constant if we study circular neutrino orbits. Using this fact, we can reformulate the neutrino spin dynamics using the effective Schrödinger equation (11). The vector $\mathbf{\Omega}$, which defines the neutrino spin precession, has the following nonzero components:

$$\begin{aligned}
\Omega_2 &= \frac{1}{2\gamma r_g} \left\{ \mp \frac{1}{\sqrt{2} x^{3/2}} - \mu B_0 r_g \frac{2\sqrt{2}x^2(x-1) \pm \alpha\sqrt{x}(2x-1) + \sqrt{2}\alpha^2}{x^{3/2}\sqrt{2x^3 - 3x^2 \pm 2\sqrt{2}\alpha x^{3/2}}} \right\}, \\
\Omega_3 &= \frac{G_{\text{F}} n_{\text{eff}}}{\sqrt{2}\gamma} \frac{\pm U_f^0 - r_g U_f^\phi (\sqrt{2}x^{3/2} \pm \alpha)}{\sqrt{2x^3 - 3x^2 \pm 2\sqrt{2}\alpha x^{3/2}}},
\end{aligned} \tag{15}$$

where $x = r/r_g$ and $\alpha = a/r_g$. The upper sign refers to the direct orbits, i.e. when a neutrino corotates with BH, and the lower sign corresponds to retrograde ones (a particle counter-rotates).

Supposing that B_0 and n_{eff} do not depend on time, we can solve the Schrödinger equation analytically. Assuming that initially only left-polarized neutrinos are present, i.e. $\nu^{\text{T}}(0) = (0, 1)$, we get the probability to detect a right-polarized neutrino in the form,

$$P(t) = P_{\text{max}} \sin^2 \left(\frac{\pi}{L_{\text{osc}}} t \right), \quad P_{\text{max}} = \frac{\Omega_2^2}{\Omega_2^2 + \Omega_3^2}, \quad L_{\text{osc}} = \frac{\pi}{\sqrt{\Omega_2^2 + \Omega_3^2}}. \tag{16}$$

Here P_{max} is the amplitude and L_{osc} is the length of oscillations .

Considering the situation when $n_{\text{eff}} = 0$ and $B_0 = 0$, i.e. only the gravitational interaction is accounted for, we present the dependence $\pi r_g / L_{\text{osc}} = |\Omega_2| r_g$ versus $x = r/r_g$ for different values of α in figure 2. It can be seen that for retrograde orbits the maximal value of $|\Omega_2|$ decreases with α, whereas for direct orbits this value increases with α.

If we discuss the maximal possible angular momentum of BH corresponding to $\alpha = 1/2$, we get from figure 2 that for a direct neutrino orbit with $r \approx r_g$, the frequency of neutrino spin oscillations reaches its maximal value $\Omega_2^+ \approx 0.12 r_g^{-1}$. For a retrograde orbit with $r \approx 2.60 r_g$ we get the maximal frequency $\Omega_2^- \approx 0.05 r_g^{-1}$. Considering BH with $M = 10 M_\odot$, we obtain the corresponding oscillations lengths, $L_{\text{osc}}^+ \approx 7.85 \times 10^7\,\text{cm}$ and $L_{\text{osc}}^- \approx 1.88 \times 10^8\,\text{cm}$. On the basis of these oscillations lengths one gets the times of the neutrino spin-flip, $T_{\text{sf}} = L_{\text{osc}}/2$, as $T_{\text{sf}}^+ \approx 1.31 \times 10^{-3}\,\text{s}$ and $T_{\text{sf}}^- \approx 3.13 \times 10^{-3}\,\text{s}$.

Note that for $\alpha = 1/2$ the minimal possible radius of a stable retrograde orbit is equal to $r_{\text{min}} = 4.5 r_g$. Therefore the obtained Ω_2^- corresponds to an unstable orbit. In case of a direct orbit $r_{\text{min}} \to r_g/2$. In figure 2, we get that in this case $\Omega_2 = 0$.

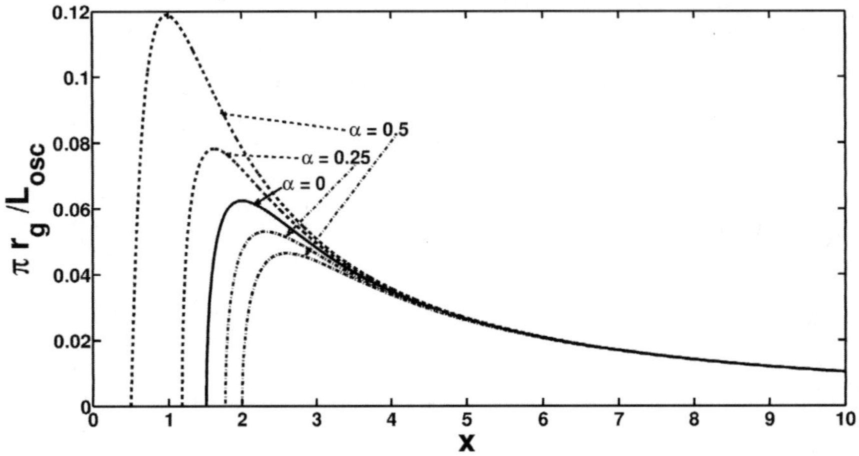

Figure 2. The function $\pi r_g/L_{\mathrm{osc}} = |\Omega_2| r_g$ versus $x = r/r_g$ for $\alpha = 0$, 0.25, and 0.5. The case $\alpha = 0$ (solid line) corresponds to a nonrotating BH, described by the Schwarzschild metric studied in section 3. For $\alpha \neq 0$, dashed lines are depicted for direct orbits and dash-dotted lines for retrograde ones.

5. Conclusion

In this work, we developed the quasiclassical approach for the description of neutrino spin oscillations in background matter under the influence of electromagnetic and gravitational fields. As applications of the obtained results, we considered spin evolution of neutrinos orbiting Schwarzschild and Kerr BH. We also discussed the possibility of the neutrino interaction with an ultrarelativistic magnetized accretion disk around BH.

Recently, the dynamics of the fermion spin in gravitational fields was studied in [6]. In that work, the spin evolution was based on the analysis of the Dirac equation in curved space-time. Although the results of [6] are analogous to that of our works [2, 4], where we used the quasiclassical approximation, we can describe the neutrino spin evolution in an arbitrary gravitational field without referring to the Dirac equation. This fact makes the approach in [2, 4] to be more attractive for possible applications.

I am thankful to the organizers of DSPIN-17 for the invitation and a financial support, as well as to the Tomsk State University Competitiveness Improvement Program and RFBR (research project No. 15-02-00293) for a partial support.

References

[1] Dvornikov M and Studenikin A 2002 Neutrino spin evolution in presence of general external fields *J. High Energy Phys.* JHEP09(2002)016 (*Preprint* hep-ph/0202113)
[2] Dvornikov M 2013 Neutrino spin oscillations in matter under the influence of gravitational and electromagnetic fields *J. Cosmol. Astropart. Phys.* JCAP06(2013)015 (*Preprint* arXiv:1306.2659)
[3] Lobanov A E and Studenikin A I 2001 Neutrino oscillations in moving and polarized matter under the influence of electromagnetic fields *Phys. Lett.* B **515** 94–98
[4] Dvornikov M 2006 Neutrino spin oscillations in gravitational fields *Int. J. Mod. Phys.* D **15** 1017–1034 (*Preprint* hep-ph/0601095)
[5] Wald R M 1974 Black hole in a uniform magnetic field *Phys. Rev.* D **10** 1680–1685
[6] Obukhov Yu N, Silenko A J and Teryaev O V 2017 General treatment of quantum and classical spinning particles in external fields *Preprint* arXiv:1708.05601

IOP Conf. Series: Journal of Physics: Conf. Series **938** (2017) 012044 doi:10.1088/1742-6596/938/1/012044

Intrinsic mass scale in QCD factorization

B I Ermolaev[1] and S I Troyan[2]
[1] Ioffee Physico-Technical Institute, 194021 St.Petersburg, Russia
[2] St.Petersburg Institute of Nuclear Physics, 188300 Gatchina, Russia

E-mail: ermolaev@mail.cern.ch

Abstract. We argue for existence of an intrinsic mass scale in QCD factorization and present a possible origin of it. Values of this scale are within the Non-Perturbative QCD mass range. This scale has nothing to do with the the factorization scales: their values are within the perturbative mass range and dependence on them vanishes in the factorization convolutions. We show that the intrinsic mass scale plays the key part in reduction of K_T Factorization to Collinear Factorization: such a reduction can be done only if dependence of the non-perturbative inputs in K_T Factorization on invariant energy has a sharp-peaked form, so the intrinsic mass scale is associated with location of the peak(s). We also present models where the intrinsic scale is generated by such peaks.

1. Introduction

The concept of QCD factorization suggests that description of hadronic reactions at high energies can be divided into perturbative and non-perturbative stages. It can apply to any high-energy process involving hadrons. Among the simplest applications, there are such constructions as the DIS structure functions, parton distribution in hadrons, etc. Basically, both Single-Parton and Multi-Parton Scattering scenarios contribute to those processes but in the present paper we focus on the simplest and at the same time most popular scenario of the Single-Parton Scattering albeit our results can easily be generalized to the Multi-Parton Scattering scenario. In order to be specific, we consider the gluon distribution D in hadrons though our conclusions are valid for all hadronic reactions. The Optical theorem relates D to the color singlet amplitude of the elastic gluon-hadron scattering in the forward kinematics with $t = 0$. Factorization of D into two blobs is illustrated by the graphs consisting of two blobs connected by two gluon lines in all available forms of QCD factorization (see Ref. [5]). Being related to the gluon distribution D, A is mostly essential for us, so throughout the paper we will ignore the other parts. The upper blob corresponds to the perturbative amplitude $A^{(pert)}$ of scattering the gluon with momentum q off the intermediate gluon with momentum k. The lowest blob corresponds to the scattering of the intermediate gluon off the hadron with momentum p. It is addressed as non-perturbative and is defined through various models and fits. In general, the gluon distributions in the hadrons depend on the kinematic variables w, q^2, with $w = 2pq$, and on the gluon and hadron spins, S_g and S_h respectively: $D = D(w, q^2, S_g, S_h)$. For the sake of simplicity, we skip writing the spin dependence. In Collinear Factorization[1], K_T-Factorization [2] and Basic Factorization [3] D can be represented respectively as

$$D \approx D_{col}(x, q^2/\mu^2) = D_{col}^{(pert)}(x/\beta, q^2, \mu_F^2) \otimes \phi_{col}(\beta, \mu_F^2) , \quad (1)$$

$$D \approx D_{KT}(x,q^2/\mu^2) = D_{KT}^{(pert)}(x/\beta, q^2, k_\perp^2, \mu_F'^2) \otimes \Phi_{KT}(\beta, k_\perp^2, \mu_F'^2),$$
$$D \approx D_B(x,q^2/\mu^2) = D_B^{(pert)}(x/\beta, q^2, k^2) \otimes \Phi_B(\alpha, k^2, \mu_B^2) \;.$$

We introduced a mass μ in the l.h.s. of Eq. (1) to make dimensionless the q^2-dependence, which is necessary because D is dimensionless. The notations $D_{col}^{(pert)}, D_{KT}^{(pert)}$ and $D_B^{(pert)}$ stand for perturbative contributions while ϕ_{col}, Φ_{KT} and Φ_B are called the integrated, unintegrated and totally unintegrated gluon distributions respectively. Whereas ϕ_{col} and Φ_{KT} include both perturbative and non-perturbative contributions, the totally unintegrated gluon distribution Φ_B contains non-perturbative contributions only. The symbols $\otimes$ in Eq. (1) refer to integrations which are different for different forms of factorization. In Collinear Factorization $\otimes$ means the integration over the longitudinal fraction β of momentum k. In K_T-Factorization the symbol $\otimes$ means the three-dimensional integration: over β and over the two-dimensional transverse momentum $k_\perp$. In Basic Factorization $\otimes$ denotes the four-dimensional integration: besides the integrations over β and $k_\perp$, it involves integration over the second longitudinal variable, α. The variables $\alpha, \beta, k_\perp$ are related to the Sudakov parametrization[4] (or the light-cone representation) of momentum k:

$$k = -\alpha q' + \beta p' + k_\perp \;, \tag{2}$$

where the light-cone momenta p', q' are made of the external momenta p and q: $p = p' + x_p q'$, $q = q' - x_q p'$, $x_p = p^2/w$, $x_q = q^2/w$, $w = 2p'q' \approx 2pq$. The parameters μ_F and μ_F' are called the factorization scale in Collinear and K_T- Factorizations respectively. As these forms of factorization were introduced independently of each other, μ_F can coincide with μ_F' or differ from it. The perturbative components of Eq. (1), $D_{col}^{(pert)}$ and $D_{KT}^{(pert)}$, explicitly depend on the factorization scales. The distributions ϕ_{col} and Φ_{KT} depend on the factorization scales through phenomenological numerical factors. The QCD factorization concept states that dependence of ϕ_{col} (Φ_{KT}) on the factorization scales is exactly compensated in the factorization convolutions by the inverse dependence of $D_{col}^{(pert)}$ ($D_{KT}^{(pert)}$) which means that the convolutions in Eqs (1) are free of the scales μ_F, μ_F'. This may be interpreted as if Eqs (1) were free of any mass scales at all, which would be totally unacceptable. Indeed, all l.h.s. of Eq. (1) need a mass scale μ to arrange for the dimensionless way of dependence on q^2. On the other hand, an origin of μ is unknown. In practice, solution to this contradiction is very simple: as ϕ_{col} depends on μ_F implicitly, the fact of total vanishing of μ_F in D_{col} is ignored, so the scale μ_F is used in the same context as μ: $D = D(x, q^2/\mu_F^2)$. Obviously, this is absolutely incorrect.

In the present paper we prove that there is an intrinsic scale in all forms of factorization. It does not vanish in factorization convolutions. We argue that this scale is generated by the distribution Φ_B in Basic Factorization, where values of this scale are $\sim \Lambda_{QCD}$ and then it evolves to the domain of non-perturbative QCD.

Our paper is organized as follows: In Sect. 2 we study the intrinsic scale in the framework of Basic Factorization. In Sect. 3we reduce Basic Factorization to K_T-Factorization and then reduce K_T-Factorization to Collinear Factorization. By doing so, we find a possible source of the intrinsic mass scale and study its properties in a model-independent way. In Sect. 4 we suggest a simple model which naturally generates the intrinsic mass scale. Sect. 5 is for our concluding remarks.

2. Basic facts about Basic Factorization

Basic Factorization (BF) suggested in Ref. [3] is the most recent and the most general of the known forms of QCD factorization. It has been considered in the Single-Parton Scattering approximation only but it can easily be extended to the Multi-Parton Scattering. As this form of factorization is much less known than Collinear and K_T-Factorizations, we briefly remind below its essence. The expressions for A and D for both polarized and unpolarized hadrons are obtained in Ref. [5]. For instance, in the case of unpolarized hadrons the gluon-hadron scattering amplitude A is

$$A = -\imath \int \frac{d\beta}{\beta} dk_\perp^2 d\alpha A^{(pert)}(s_2, q^2, k^2) \frac{w}{k^2 k^2} \widetilde{M}(s_1, k^2), \tag{3}$$

where, in terms of the Sudakov variables, (see Eq. (2))

$$s_1 = (p-k)^2 \approx w\alpha + k^2 + p^2 \,, \qquad s_2 = (q+k)^2 \approx w\beta + q^2 + k^2 \,, \qquad k^2 = -w\alpha\beta - k_\perp^2 \,. \tag{4}$$

The factors k^2 in the denominator correspond to propagators of the active gluons. The notation $A^{(pert)}$ stands for the perturbative gluon-gluon scattering amplitude and $\widetilde{M}$ is the non-perturbative input. The renormalization makes the perturbative amplitude $A^{(pert)}$ be free of ultraviolet (UV) divergences whereas infrared (IR) divergences are regulated by virtualities of the external momenta q and k providing that $q^2 \neq 0, k^2 \neq 0$. However, even with regulated $A^{(pert)}$, the integrand of Eq. (3) can be divergent at $k^2 = 0$ because the integration runs over the whole phase space. Similarly, integration over α can yield divergent result at large $|\alpha|$. In Ref. [3, 5] we proved that those singularities are killed when the input $\widetilde{M}$ satisfies the following requirements (see Ref. [5] for detail):

$$\widetilde{M} \sim \left(k^2\right)^{1+\eta}, \quad \widetilde{M} \sim \alpha^{-\kappa}, \tag{5}$$

with $\eta > 0$, and $\kappa > 0$. These restrictions are valid for the cases of both polarized and unpolarized hadrons. They can be regarded as criteria of validity for all models of the inputs. Applying the Optical theorem to Eq. (3), we obtain the gluon parton distribution D_B in Basic Factorization. In Ref. [3, 5] we showed that Basic Factorization can be reduced to K_T- Factorization. To this end, let us notice that when integration over α in Eq. (3) has been performed, the resulting convolution contains integrations over β and $k_\perp$ and therefore it looks as a convolution in K_T- Factorization. However, the problem is that in order to reduce Eq. (3) to the one of K_T- Factorization, one should take $A^{(pert)}$ out of integral over α, which cannot be done straightforwardly because $A^{(pert)}(s_2, q^2, k^2)$ depends on α through $k^2 = -w\alpha\beta - k_\perp^2$. It is possible only approximately if the integration runs over subregion

$$|\alpha| \ll k_\perp^2/(w\beta) \,. \tag{6}$$

Then $k^2 \approx -k_\perp^2$, so the only α- dependent factor in the integrand of Eq. (3) is M. Integrating it over the region of Eq. (6), we arrive at the gluon-hadron scattering amplitude A_{KT} in K_T-Factorization:

$$A_{KT} = -\imath \int \frac{d\beta}{\beta} dk_\perp^2 A^{(pert)}(s_2, q^2, k_\perp^2) \frac{1}{k_\perp^2} M_{KT}(\beta, k_\perp^2), \tag{7}$$

where the input M_{KT} is obtained from $\widetilde{M}$:

$$M_{KT}(\zeta, k_\perp^2) = \left(w/k_\perp^2\right) \int_{-\zeta}^{\zeta} d\alpha \widetilde{M}(s_1, k_\perp^2). \tag{8}$$

The notation ζ in Eq. (8) stands for the invariant energy of M_{KT}:

$$\zeta = \xi k_\perp^2/\beta, \tag{9}$$

with $0 < \xi \ll 1$ while invariant sub-energies $s_{1,2}$ are defined in Eq. (4). M_{KT} is the non-perturbative input to the gluon-hadron scattering amplitude in K_T Factorization. According to

IOP Conf. Series: Journal of Physics: Conf. Series **938** (2017) 012044 doi:10.1088/1742-6596/938/1/012044

Eq. (5), $M_{KT} \sim (k_\perp^2)^\eta$ at small $k_\perp^2$. applying the Optical theorem to Eq. (7), we arrive at the gluon distribution in K_T Factorization:

$$D_{KT} = \int \frac{d\beta}{\beta} \frac{dk_\perp^2}{k_\perp^2} D_{KT}^{(pert)}(x/\beta, q^2/\zeta) \Phi_{KT}(\zeta, k_\perp^2), \tag{10}$$

where $D^{(pert)}$ is the perturbative contribution and Φ_{KT} is the non-perturbative input. Our analysis demonstrates that considering the gluon-hadron scattering in Basic Factorization and reducing it to K_T Factorizations do not involve any mass scale, so the problem of origin of the intrinsic mass scale still remains unclear. However, we show in the next Sect. that existence of the intrinsic mass scale plays the key role, when K_T-Factorization is reduced to Collinear one.

3. Reducing K_T- Factorization to Collinear Factorization

Collinear Factorization involves one-dimension integration over β, so at first sight, transition to it from Eq. (10) can be done with straightforward integration of Eq. (10) over $k_\perp$. However such a procedure involves integrating both $\Phi_{KT}(\zeta, k_\perp^2)$ and $D_{KT}^{(pert)}$ and therefore a result would be a medley of them instead of the product, which destroys the convolutional form of Collinear Factorization. Therefore, one should find a way to integrate out the $k_\perp$-dependence in Eq. (10) and at the same time to preserve the pattern of Eq. (1)

The only approximate way to do so, is to presume that the ζ-dependence of $\Phi_{KT}(\zeta, k_\perp^2)$ has a sharp-peaked form. Generally speaking, the number of the peaks can be unlimited, their widths and heights can be different. In what follows we will address the peaks as resonances. The suggestion of having one or several resonances in $\Phi_{KT}(\zeta, k_\perp^2)$ is well grounded. Indeed, after the active quark(s) has been emitted off the hadron, the remaining ensemble of spectators becomes unstable, so it can likely be described in terms of resonances. As Φ_{KT} is altogether non-perturbative, all μ_r should be within the non-perturbative domain ($\mu_r \sim \Lambda_{QCD}$). As soon as we accept the peaked form of $\Phi_{KT}(\zeta, k_\perp^2)$, we can integrate over ζ the input Φ_{KT} only, arriving at the explicit expression for the gluon-hadron scattering amplitude $A_{(col)}$ in Collinear Factorization. It consists of the resonance and background contributions:

$$A_{col} = A_R + A_B \ . \tag{11}$$

The resonance contribution A_R is

$$A_R(x, q^2/\mu^2) \approx \sum_n \int \frac{d\beta}{\beta} A^{(pert)}(x/\beta, \mu_n^2)\, \varphi_R^{(n)}(\beta, \mu_n^2) \ , \tag{12}$$

with μ_n^2 being the location of the n_{th} maximum and $\mu^2 = \max \mu_n^2/\xi \gg \mu_n^2$. The input φ_R in Eq. (12) is

$$\varphi_R(\beta, \mu_n^2) \approx \int_{\Omega_n} d\zeta\, \Phi_{KT}(\zeta, k_\perp^2) \ , \tag{13}$$

where the integration region Ω_n is located around the maximum of the n_{th} peak. Apparently, the regions Ω_n in Eq. (13) are not well-defined and they are only a part of the total integration region. So besides the resonance contribution A_R, there are additional non-factorized contributions which we interpret as background contribution A_B to the amplitude A. Their impact strongly depend on the specific expressions for Φ_{KT}.

We have shown that reduction of K_T Factorization down to Collinear Factorization requires existence of at least one mass scale located in the non-perturbative domain. We call such scales μ_n the intrinsic mass scales. These scales have a physical meaning: Φ_{KT} has the maximums at $\zeta = \mu_n^2$. Below we present a simple model for Ψ_{KT}.

IOP Conf. Series: Journal of Physics: Conf. Series **938** (2017) 012044 doi:10.1088/1742-6596/938/1/012044

4. Minimal Resonance Model for K_T Factorization

Resonance Model was presented in Refs. [5, 6]. This model is based on the following observation: After the active quark has been emitted off the hadron, the remaining set of spectators pick up a color and thereby it becomes unstable. This observation guides us to model Φ_{KT} through interference of several resonances. The case of the single resonance we call Minimal Resonance Model (MRM):

$$\Phi_{KT}(\zeta, k_\perp^2) \approx \Phi_{MRM}(\zeta, k_\perp^2) = R_{KT}(k_\perp^2) \frac{\Gamma_1}{(\zeta - \mu_1^2)^2 + \Gamma_1^2} \,. \tag{14}$$

R_{KT} should decrease at large $k_\perp^2$, which is often achieved through exponential factors, see e.g. Refs. [7, 8, 9]. Being motivated by these models, we choose R_{KT} as follows: $R_{KT} = N\left(k_\perp^2\right)^\eta e^{-k_\perp^2/k_0^2}$, with N being a constant and k_0^2 being a parameter. In order to get the sharp peak in Eq. (14) we presume that $\Gamma_1 \ll \mu_1^2$, which is standard for the Breit-Wigner expressions. Replacing Φ_{KT} in Eq. (10) by Φ_{MRM} and integrating over ζ, we obtain the gluon distribution in Collinear Factorization in the form of Eq. (11)

$$A_{col}(x, q^2/\mu^2) = A_R^{(1)} + A_B^{(1)} \tag{15}$$

but with different resonance and background contributions. The resonance contribution $A_R^{(1)}$ looks very simple:

$$A_R^{(1)}(x, q^2/\mu^2) = \int_x^1 \frac{d\beta}{\beta} A^{(pert)}(x/\beta, q^2/\mu_1^2)\, \varphi_R^{(1)}(\beta, \mu_1^2) \,, \tag{16}$$

where $\mu^2 = \mu_1^2/\xi \gg \mu_1^2$ and

$$\varphi_R^{(1)} = \pi N \left(\frac{\beta\mu_1^2}{k_0^2}\right)^\eta e^{-\beta\mu_1^2/k_0^2} \approx \left[\pi N \left(\frac{\mu_1^2}{k_0^2}\right)^\eta\right] \beta^\eta \left(1 - \lambda\beta\right), \tag{17}$$

with $\lambda = \mu_1^2/k_0^2$, while the background contribution $A_B^{(1)}$ is given by a more involved expression:

$$A_B^{(1)} = \sum_{n=1}^{\infty} \frac{N}{n!}\left(\frac{\Gamma_1}{\mu_1^2}\right)^n \int_{-\mu_1^2/\Gamma}^{q^2/\Gamma} \frac{dt\, t^n}{1+t^2} \int_x^1 \frac{d\beta}{\beta}\left[\frac{\partial^n}{\partial y^n}\left(\frac{A^{(pert)}(x/\beta, q^2/(y\mu_1^2))}{y}\right)\right]_{y=1} \left(\frac{\beta\mu_1^2}{k_0^2}\right)^\eta e^{-\beta\mu_1^2/k_0^2} \,. \tag{18}$$

Let us discuss these results. Eqs. (17,18) contain the parameters μ_1^2 and k_0^2. Their values are widely different: μ_1^2 is in the non-perturbative domain, $\mu_1 \sim \Lambda_{QCD}$ whereas k_0^2 of Refs. [7, 8, 9] is much greater. Therefore, the exponential factor in Eqs. (17,18) can be either neglected or approximated by few terms of the power expansion. The parameter μ in the l.h.s. of Eq. (15,16) is generated by the intrinsic scale μ_1 and its value can easily be within the domain of Perturbative QCD. We stress that μ, μ_1 have nothing in common with the conventional factorization scale μ_F. Then, the factors C_n in Eq. (18) are $\sim \ln^n(q^2/\Gamma_1)$ and $\ln^n(\mu_1^2/\Gamma_1)$. Despite they grow with increase of n, the power factors $(\Gamma/\mu_1^2)^n$ decrease at the same time much faster, so the series of Eq. (18) can be approximated by only few first terms.

5. Conclusion

In present paper we have considered the problem of origin of the intrinsic mass scale in QCD factorization. To begin with, we demonstrated that all available forms of QCD factorization implicitly need a mass scale. Then we demonstrated that reduction of K_T Factorization down to Collinear Factorization is possible only if the inputs Φ_{KT} depend on the invariant energy in the form of sharp peaks. so that locations of the peaks introduce a new mass scale, intrinsic scale. Its value should be in domain of Non-perturbative QCD ($\sim \Lambda_{QCD}$). This scale does not disappear in factorization convolutions and it generates a new scale in domain of Perturbative QCD, which is responsible for dimensionless way of the q^2-dependence. We model the peaked structure of Φ_{KT} by resonances of the Breit-Wigner type. As the simplest case, we present a model with the single resonance and represent the factorization convolutions in Collinear Factorization as the sum of the resonance contribution and background.

Acknowledgments

B.I. Ermolaev is grateful to Organizing Committee of the workshop DSPIN-2017 for financial support of his participation.

References

[1] Efremov A V and Radyushkin A V 1976 *Phys. Lett.* B **63** 449
Efremov A V and Radyushkin A V 1980 *Teor. Mat. Fiz.* **42** 147
Efremov A V and Radyushkin A V 1980 *Theor. Math.Phys.* **44** 573
Efremov A V and Radyushkin A V 1980 *Teor. Mat. Fiz.* **44 17**
Efremov A V and Radyushkin A V 1977 *Lett. Nuovo Cim.* **19** 83
Sterman G and Weinberg S 1977 *Phys. Rev. Lett.* **39** 1436
Libby S and Sterman G 1978 *Phys. Rev.* D **18** 3252
Sterman G 1978 *Phys. Rev.* D **17** 2773
Sterman G 1978 *Phys. Rev.* D **17** 2789
Amati D, Petronzio R and Veneziano G 1978 *Nucl. Phys.* B **140** 54
Amati D, Petronzio R and Veneziano G 1978 *Nucl. Phys.* B **146** 29
Brodsky S J and Lepage G P 1979 *Phys. Lett.* B **87** 359
Brodsky S J and Lepage G P 1980 *Phys. Rev.* D **22** 2157
Collins J C and Soper D E 1981 *Nucl. Phys.* B **193** 381
Collins J C and Soper D E 1982 *Nucl. Phys.* B **194** 445
Collins J C, Soper D E and Sterman G 1985 *Nucl. Phys.* B **250** 199
Efremov A V and Radyushkin A V 2009 *Mod. Phys. Lett.* A **24** 2803 (Report JINR E2-80-521)
Politzer H D 1977 *Phys. Lett.* B **70** 430
Ellis R K, Georgy H, Machacek M, Politzer H D and Ross G G 1978 *Phys. Lett.* B **78** 281
Ellis R K, Georgy H, Machacek M, Politzer H D and Ross G G 1979 *Nucl. Phys.* B **152** 285
[2] Catani S, Ciafaloni M and Hautmann F 1990 *Phys. Lett.* B **242** 97
Catani S, Ciafaloni M and Hautmann F 1991 *Nucl. Phys.* B **366** 135
Collins J C and Ellis R K 1991 *Nucl.Phys.* B **360** 3
[3] Ermolaev B I, Greco M and Troyan S I 2011 *Eur. Phys. J.* C **71** 1750
Ermolaev B I, Greco M and Troyan S I 2012 *Eur. Phys. J.* C **72** 1953
[4] Sudakov V V 1956 *Sov. Phys. JETP* **3** 65
[5] Ermolaev B I and Troyan S I 2017 *Eur. Phys. J.* C bf 77 167
[6] Ermolaev B I, Greco M and Troyan S I 2015 *Eur. Phys. J.* C **75** 7, 306
[7] Golec-Biernat K and Wusthoff M 1999 *Phys. Rev.* D **60** 114023 (*Preprint* arXiv:hep-ph/9903358)
[8] Jung H 2004 Un-integrated PDFs in CCFM *Preprint* arXiv:hep-ph/0411287
[9] Pumplin J 2006 *Phys. Rev.* D **73** 114015

IOP Conf. Series: Journal of Physics: Conf. Series **938** (2017) 012045 doi:10.1088/1742-6596/938/1/012045

Magnetic Polarizability of Virtual $(s\bar{s})$ and $(c\bar{c})$ Pairs in the Nucleon

Peter Filip

Institute of Physics, Slovak Academy of Sciences, Dúbravská cesta 9, Bratislava 845 11, SK

E-mail: peter.filip@savba.sk

Abstract. We suggest ${}^3P_0$ quantum state of virtual $(s\bar{s})$ pairs in the nucleon can be polarised by the internal fields permeating the volume of the nucleon (proton or neutron). Due to the quadratic Zeeman interaction, ${}^3P_0$ wavefunction of virtual $(q\bar{q})$ pairs acquires the admixture of ${}^1P_{10}$ quantum state in the magnetic field, which generates the antiparallel polarization of s and $\bar{s}$ quarks (in the nucleon). Considering the internal magnetic fields of neutron and proton (originating from their measured magnetic dipole moments), we suggest the induced s-quark polarization in the neutron to be of the oposite direction compared to the proton case. We mention the influence of the internal chromo-magnetic fields on the quantum state of $(q\bar{q})$ pairs in the nucleon and we discuss also the expected behaviour of virtual $(c\bar{c})$ pairs.

1. Introduction

Internal structure of the nucleon is an important and challenging question of modern physics. Due to remaining uncertainities in the spin (angular momentum) structure of Proton (only 1/3 of the expected part of the proton spin was found [1] to be directly related to the spin orientations of quarks), the intensive experimetal efforts continue with the goal to determine the spin contribution of gluons (ΔG), and quark orbital angular momentum L_q. The spin of virtual $(s\bar{s})$ pairs may also contribute to the total angular momentum $(J = \hbar/2)$ of proton.

The presence of virtual $(s\bar{s})$ pairs in the nucleon is naturally expected within Parton model, and the experimental data from lepton-proton scattering as well as the observed OZI rule violations in $\bar{p}p$ annihilation [2] suggest the polarization of virtual $(s\bar{s})$ pairs [3] in the nucleon.

In this contribution we consider a polarization of ${}^3P_0$ $(s\bar{s})$ quantum state, which may occur due to the interaction of quark magnetic (or chromo-magnetic) moments with the oriented internal electromagnetic or gluonic fields in the nucleon.

2. ${}^3P_0$ quantum state in external fields

Quantum state of virtual $(s\bar{s})$ component in the proton is not fully fixed by theoretical arguments, and several possibilities (e.g. ${}^1S_0$, ${}^3S_1$, ${}^3P_0$ or ${}^1P_1$ type of wavefunctions) exist [3]. Experimental data indicate that ${}^1S_0$ and ${}^3S_1$ configurations are unlikely [3], while ${}^3P_0$ state of $(s\bar{s})$ pair with vacuum quantum numbers $J^{PC} = 0^{++}$ appears to be a reasonable assumption.

Bound state of real $(c\bar{c})$ quark-antiquark pair in ${}^3P_0$ configuration (0^{++}) is observed as χ_{c0} meson [4], and the properties of ${}^3P_0$ state in Positronium (e^+e^-) are well understood [5]. If we presume the response of (e^+e^-) and $(q\bar{q})$ bound systems (in ${}^3P_0$ state) to the external field to be similar, the existing knowledge about the Positronium $^3P_0$ state behavior in external fields [5] can be used to make the predictions regarding the state of virtual $(s\bar{s})$ pairs in the nucleon.

2.1. Internal magnetic fields in the nucleon

Neutron and proton have anomalous magnetic moments ($\mu_n = -1.91\mu_N$ and $\mu_p = 2.79\mu_N$) due to their nontrivial internal structure. Using SU(6) symmetric constituent quark $|qqq\rangle$ wavefunction, the nucleon magnetic moment is [4]: $\mu_N = (4\mu_{q_a} - \mu_{q_b})/3$. Constituent quark magnetic moments then are: $\mu_u = 1.85\mu_N$ and $\mu_d = -0.97\mu_N$. Hyperons $\Sigma^\pm$ and $\Xi^{0,-}$ also agree with such a simple ansatz (within 20% precision) and $\mu_s \approx -0.61\mu_N$ value of s-quark magnetic moment agrees well also with the magnetic moments of Λ^0 and Ω^- ($3\mu_s$) baryons [4].

Magnetic moments (μ_p, μ_n) of proton and neutron are the source of dipole magnetic field $B_{\vec{\mu}}$, which is generated within the interior (MIT bag volume) of the nucleon. Compton length of the proton ($\lambda_c = h/m_p c = 1.3\,\mathrm{fm}$) is comparable to the proton radius (size), and one can imply a dipole field $B_{\vec{\mu}}$ does penetrate the interior (partonic matter) of the nucleon. Direction of internal magnetic field $B_{\vec{\mu}}$ within the proton (or neutron) volume is parallel (or antiparallel) to the spin orientation, and its magnitude is approximately $|B_{\vec{\mu}}| \approx 10^{13}\,\mathrm{T} = 10^{17}$Gauss. This estimate can be obtained at semi-classical level, requiring the electric current I inside the coil of radius $R_p \approx 0.82$ fm to generate the magnetic dipole of size $\mu = \pi R_p^2 I \approx 2\mu_N$. Magnetic field inside such a theoretically considered coil [6] then is: $B = \mu_0 I/2R_p \approx 10^{13}\,\mathrm{T}$ (here $\mu_0 = 4\pi \times 10^{-7} N/A^2$).

2.2. Magnetic polarizability of $^3P_0$ state

Let us consider the wavefunction of bound fermion-antifermion $(f\bar{f})$ pair in $^3P_0$ quantum state with $J^{PC} = 0^{++}$ quantum numbers. Using the Positronium notation [7] we have

$$^3P_{00} = \frac{1}{\sqrt{3}}(Y_{11}\chi_{1-1} + Y_{1-1}\chi_{11} - Y_{10}\chi_{10}) \tag{1}$$

where Y_{lm} are orbital (spherical harmonics) functions, and χ_{s,s_z} denote the spin part of $^3P_{00}$ wave function. Now we shall assume that quantum state of the virtual $(s\bar{s})$ pairs in the nucleon has a similar (the same) quantum structure as given by Eq.(1). Using the uncoupled spin eigenstates $\chi_{1-1} = |\downarrow\downarrow\rangle$, and $\chi_{1+1} = |\uparrow\uparrow\rangle$, and $\chi_{10} = (|\downarrow\uparrow\rangle + |\uparrow\downarrow\rangle)/\sqrt{2}$, one has

$$^3P_{00}[s\bar{s}] = \frac{1}{\sqrt{3}}\Big[Y_{11}|\downarrow\downarrow\rangle + Y_{1-1}|\uparrow\uparrow\rangle - Y_{10}\frac{|\downarrow\uparrow\rangle + |\uparrow\downarrow\rangle}{\sqrt{2}}\Big] \tag{2}$$

The probability of antiparallel $s, \bar{s}$ spins ($\uparrow\downarrow$ or $\downarrow\uparrow$) is 1/3, and parallel orienations ($\uparrow\uparrow$) and ($\downarrow\downarrow$) occur with the same probability (1/3). For the net helicities of s and $\bar{s}$ quarks in such $^3P_{00}$ state we have $\Delta s = s^\uparrow - s^\downarrow = 0$ and $\Delta\bar{s} = \bar{s}^\uparrow - \bar{s}^\downarrow = 0$, and $L_s + L_{\bar{s}} = 0$ also applies.

It has been suggested [8] that component $Y_{11}|\downarrow\downarrow\rangle$ in $^3P_{00}[s\bar{s}]$ wave function (2) is enhanced in the proton, due to strong attraction of the valence u and virtual $\bar{s}$ quarks in the pseudoscalar channel, which results in the polarization of strange quarks in the nucleon. Magnetic fields can also modify the spin structure of $^3P_{00}$ state. In the Positronium [7] case, $^3P_{00}$ wavefunction acquires the admixture of $^1P_{10}$ state. Eigenstates $^3\tilde{P}_{00}, {}^1\tilde{P}_{10}, {}^3\tilde{P}_{20}$ in small magnetic fields are [7]:

$$^3\tilde{P}_{00}(B) \approx {}^3P_{00} + \varepsilon_B {}^1P_{10}/\sqrt{3} \tag{3}$$

$$^1\tilde{P}_{10}(B) \approx {}^1P_{10} - \varepsilon_B {}^3P_{00}/\sqrt{3} - \varepsilon'_B {}^3P_{20}\sqrt{2/3} \tag{4}$$

$$^3\tilde{P}_{20}(B) \approx {}^3P_{20} + \quad + \varepsilon'_B {}^1P_{10}\sqrt{2/3} \tag{5}$$

where admixture parameters $\varepsilon_B = \frac{2\mu_e B}{\Delta E_{10}}$ and $\varepsilon'_B = \frac{2\mu_e B}{\Delta E_{21}}$ depend on the magnetic field B, and on eigenenergy differences $\Delta E_{10}, \Delta E_{21}$. Strange quarks also have their magnetic moment μ_s, and therefore, a similar behavior of $^3P_{00}[s\bar{s}]$ wavefunction in external B field can be anticipated.

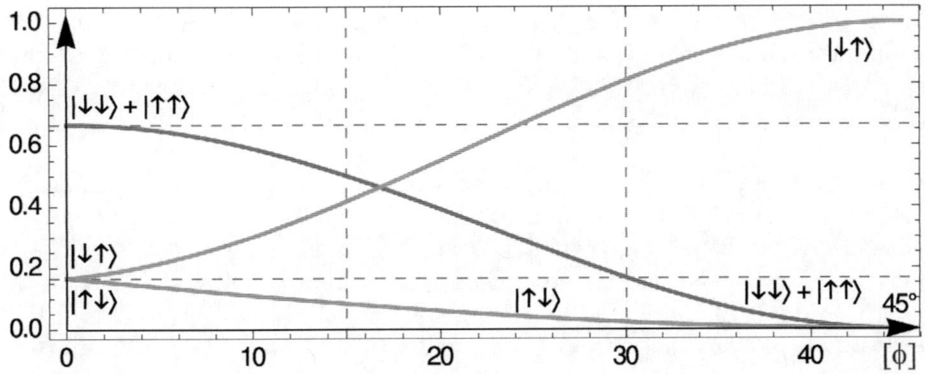

Figure 1. Probability of the spin components in $^3\tilde{P}_{00}$ state in the magnetic field, as a function of the mixing angles (value $\phi=45°$ corresponds to Paschen-Back limit of very large fields: $B\to\infty$).

Using $^1P_{10}=Y_{10}(|\downarrow\uparrow\rangle-|\uparrow\downarrow\rangle)/\sqrt{2}$, the virtual $^3\tilde{P}_{00}$ state of $(s\bar{s})$ pair in *small* magnetic fields is:

$$^3\tilde{P}_{00}[s\bar{s}](B)\approx\frac{1}{\sqrt{3}}\Big[Y_{11}|\downarrow\downarrow\rangle+Y_{1-1}|\uparrow\uparrow\rangle\Big]-\frac{Y_{10}}{\sqrt{3}}\Big(|\downarrow\uparrow\rangle\frac{1+\varepsilon_B}{\sqrt{2}}+|\uparrow\downarrow\rangle\frac{1-\varepsilon_B}{\sqrt{2}}\Big) \tag{6}$$

Relative imbalance of $\uparrow\downarrow$ and $\downarrow\uparrow$ spin configurations in the wavefunction (6) generates the induced magnetic moment of $^3\tilde{P}_{00}[s\bar{s}](B)$ state. Such behavior corresponds to magnetic polarizability [9].

Energy of the Positronium $^3\tilde{P}_{00}$ state decreases in the magnetic field [5], which corresponds to $\uparrow_{\mu_{e^-}}\uparrow_{\mu_{e^+}}$ orientation of e^-, e^+ magnetic moments. In Paschen-Back limit (see Fig. 1), component $|\downarrow\uparrow\rangle$ becomes dominant in $^3P_{00}$ state. Therefore, one can expect enhanced probability of $\uparrow_{\mu_s}\uparrow_{\mu_{\bar{s}}}$ configuration $|\downarrow\uparrow\rangle$ for the virtual $(s\bar{s})$ pairs in the polarized proton as well. The behavior of $^3\tilde{P}_{00}$ Positronium wavefunction in the electric field is well known [7], and our observation (analysis) gives: No polarization of $s,\bar{s}$ spins occurs, if $^3\tilde{P}_{00}$ state is exposed to the electric fields.

3. Virtual $s\bar{s}$ and $c\bar{c}$ pairs in the proton and neutron

If other effects (e.g. pseudoscalar QCD interaction [8]) do not determine the polarization of $(q\bar{q})$ pairs in the nucleon, and the polarizability of virtual $(s\bar{s})$ pairs by the internal magnetic field is indeed significant, the following qualitative predictions can be inferred from Eq.(6):

- Net spin polarization of s quarks in proton is *antiparallel* to the proton spin.
- Net spin polarization of s quarks in the neutron is *parallel* to the spin of the neutron.
- Polarization of $\bar{s}$ quarks in the nucleon favors to be opposite to the s quark polarization.
- The polarization of c quarks in the nucleon is opposite to the polarization of s quarks.
- The polarization of $\bar{c}$ (and $\bar{u}$) quarks is parallel to the polarization of s quarks.

The first two statements originate from the internal magnetic dipole field $B_{\vec{\mu}}$ orientation relative to the spin of proton and neutron, and the last two statements come from the fact, that c quarks have positive charge $(2/3)e$, and thus their magnetic moment is parallel to their spin (contrary to the case of negatively charged s quarks). The third statement applies only if $\uparrow\uparrow$ and $\downarrow\downarrow$ components of the function (6) have the same probability (QCD interaction [8] is inactive).

4. Magnetic moments of virtual quarks and magnitude of the effect

Magnetic polarizability of $(q\bar{q})$ pair depends on the quark magnetic moment $|\mu_q|$. The observed magnetic moments of baryons [4] require μ_u=1.85, μ_d=-0.97, μ_s=-0.61 (in μ_N units), for the *constituent* quarks u, d, s. This agrees with Dirac equation (demanding $\mu_q{=}\hbar Q/2m_q^*$), if constituent quark masses are: $m_u^*{\approx}338$, $m_d^*{\approx}322$, $m_s^*{\approx}510$ (in MeV units). One can estimate the magnetic moment of c quark to be $\mu_c{=}{+}0.4\mu_N$ (a significant value) using $m_c^* \approx 1.5\,\mathrm{MeV}$.

However, for virtual $(q\bar{q})$ quarks, the constituent masses are unjustifiably large and one should use smaller (e.g. *current* quark [4]) masses: $\tilde{m}_u{=}\,2.3$, $\tilde{m}_d{=}4.8$, $\tilde{m}_s{=}\,95$ and $\tilde{m}_c{=}1.27$ (MeV). In such a case, magnetic moments $\tilde{\mu}_q$ of light virtual quarks are scaled up according to $(m_q^*/\tilde{m}_q)$ ratios, which gives factors 135×, 67×, and 5× for u, d, s quarks. The *current* quark magnetic moments thus are: $\tilde{\mu}_u{=}{+}250\mu_N$, $\tilde{\mu}_d{=}{-}65\mu_N$, $\tilde{\mu}_s{=}{-}3\mu_N$ (and $\tilde{\mu}_c{=}{+}0.47\mu_N$).

Using $\tilde{\mu}_{\bar{s}}{=}\,3\mu_N$ for *current* quark $\bar{s}$, one has $|\tilde{\mu}_s B| \approx 1\,\mathrm{MeV}$ for $B \approx 10^{13}\,\mathrm{T}$ in the nucleon ($\mu_N{=}3.15 \times 10^{-14}\,\mathrm{MeV/T}$). Assuming hyperfine splitting $\Delta E_{10}^{s\bar{s}} \approx 100\,\mathrm{MeV}$, for ${}^3\tilde{P}_{00}$ and ${}^1\tilde{P}_{10}$ states of virtual $(s\bar{s})$ pair, mixing parameter $\varepsilon_B = 2\tilde{\mu}_s B/\Delta E_{10}^{s\bar{s}} = 0.019$. From Eq.(6) we obtain

$$\Delta s = s{\uparrow} - s{\downarrow} = [(1-\varepsilon_B)^2 - (1+\varepsilon_B)^2]/6 = -0.0126 \quad ; \qquad \Delta\bar{s} = \bar{s}{\uparrow} - \bar{s}{\downarrow} = +0.0126 \qquad (7)$$

for proton, and $(\Delta s{+}\Delta\bar{s}){=}0$. For virtual c quarks we have $\Delta c \approx +0.002$, using $|\tilde{\mu}_s/\tilde{\mu}_c| \approx 6.4$, $\Delta E_{10}^{c\bar{c}} \approx 100\,\mathrm{MeV}$. One can also expect $\Delta\bar{d} > \Delta\bar{s} > 0$ and $\Delta\bar{u} < 0$, for the virtual anti-quarks.

Parallel orientation of magnetic moments $\uparrow_{\mu_s}\uparrow_{\mu_{\bar{s}}}$ is more probable than the opposite $(\downarrow^{\mu_s}\downarrow^{\mu_{\bar{s}}})$ one in the polarized ${}^3\tilde{P}_{00}$ state. Induced magnetic moment $\langle\mu\rangle_{s\bar{s}}$ of virtual $(s\bar{s})$ pairs can thus contribute [10] to the magnetic moment of proton, while the overall spin as well as total (and orbital) angular momentum of virtual $(s\bar{s})$ pairs remain to be zero.

5. Summary and conclusions

We have considered a simple model of ${}^3\tilde{P}_{00}[q\bar{q}]$ quantum state polarization of virtual $(s\bar{s})$ and $(c\bar{c})$ pairs in the nucleon, occuring due to the internal magnetic field $B \approx 10^{13}\,\mathrm{T}$, which originates from the nucleon magnetic moment. Our estimates are based on the analogy with the behavior of ${}^3\tilde{P}_{00}$ state of Positronium (e^+e^-) in the magnetic field [7]. The angular momentum of virtual $(s\bar{s})$ pairs remains zero, while s quarks are polarized oppositely (and $\bar{s}$ parallel) to the proton spin. The estimated polarizations are: $-\Delta\bar{s} = \Delta s \approx -0.013$, and $\Delta c = -\Delta\bar{c} \approx +0.002$.

However, quarks interact also via strong interactions, and QCD effects [8] can be larger than the electromagnetic interactions we have considered here. Moreover, internal volume of the polarized nucleon presumably contains an oriented chromo-magnetic fields B^a [11]. The interaction $H_{QCD}^{int} = -\mu_q^a B^a$ of such gluonic fields B^a with chromo-magnetic moments μ_q^a of virtual quarks may induce a stronger polarization of ${}^3\tilde{P}_{00}[q\bar{q}]$ state. Our inferences about the $(s\bar{s})$ polarization magnitude, the polarization orientation of $(c\bar{c})$ pairs, and the opposite polarization of $\bar{s}, \bar{d}$, and $\bar{u}$ quarks in neutron (relative to proton case) should thus be taken with caution.

Acknowledgments

Author is indebted to the organizers of DSPIN 2017 conference for the invitation to participate in this valuable scientific meeting, and to JINR, for the financial support and kind hospitality.

References

[1] Adams D, et al. (SMC Collaboration) 1997 *Physical Review D* **56** 5330
[2] Amsler C 1998 *Review of Modern Physics* **70** 1293
[3] Ellis J, Karliner M, Kharzeev D E and Sapozhnikov M G 2000 *Nuclear Physics A* **673** 256
[4] Patrignani C, et al. (Particle Data Group) 2016 *Chinese Physics C* **40** 100001
[5] Curry S M 1973 *Physical Review A* **7** 447
[6] Filip P 2014 *Proc. of DSPIN-2013 Conference,* October 8-12, 2013, Dubna, 53-56
[7] Dermer C D and Weisheit J C 1989 *Physical Review A* **40** 5526
[8] Alberg M, Ellis J and Kharzeev D E 1995 *Physics Letters B* **356** 113
[9] Petrunkin V A 1981 *Soviet Journal of Particles and Nuclei* **12** 278
[10] Young R D 2017 *Nature* **544** 419
[11] Barone V, Calarco T and Drago A 1998 *Physics Letters B* **431** 405

Infrared models for the Bjorken sum rule in the APT approach

I.R. Gabdrakhmanov[a], **O.V. Teryaev**[a,b], **and V.L. Khandramai**[c]
[a]Veksler and Baldin Laboratory of High Energy Physics, JINR, Dubna, Russia
[b]Bogoliubov Laboratory of Theoretical Physics, JINR, Dubna, Russia
[c]Gomel State Technical University, Gomel, Belarus

E-mail: ilnur@jinr.ru

Abstract. We consider continuation of the Bjorken sum rule (BSR) in the low Q^2 region. QCD coupling is 'frozen' with the help of 'Massive' APT modification. For higher twist corrections we apply spectral representation with several particular spectral physically motivated functions. Analytic restrictions are imposed stemming from Gerasimov-Drell-Hearn sum rule and the finiteness of respective spin-dependent cross-sections. Results are compared with the current experimental data.

1. Introduction

In the framework of the Quantum Chromodynamics (QCD) at the high energy scale observable quantities are being successfully extracted (and experimentally confirmed) using perturbative expansion due to the asymptotic freedom. On the over hand low energy QCD behavior is not well understood yet, The standard perturbative approach fails to describe low energy behavior due to Landau singularity of the coupling $\alpha_S(Q^2)$ with Q^2 of the order of $\Lambda^2_{QCD} \simeq 1\ GeV^2$.

Several QCD modifications has been proposed to improve coupling low energy behavior, one of them is Analytic Perturbation Theory (APT) [1] (for a review see [2, 3]). The APT using analyticity and causality converts power expansion in the perturbative BSR corrections $\Delta_{\rm Bj}(Q^2)$ to the nonpower one. The APT has been successfully applied to describe polarized $\Gamma_1^{p-n}(Q^2)$ data down to few hundred MeV [4].

The effective gluon mass as infrared regulator ($Q^2 \to Q^2 + M^2_{gl}$) was introduced by Cornwall [5] and also by Simonov [6]. The 'Massive' APT (MPT) recently proposed by D.V. Shirkov [7, 8] combined it with the APT scheme and extended consideration to the higher orders up to NNNLO.

In this work we apply MPT approach with spectral HT resummation from [9] with the use of Breit-Wigner (BW) distribution [10] to analyze lowest polarized moment $\Gamma_1^{p-n}(Q^2)$ up to 4-loop level and compare it with the low-Q^2 JLab data.

2. Fitting procedure details

We base on non-perturbative HT resummation proposed in [9]:

$$\sum_{n=1}^{\infty} \mu^{p-n}_{2n+2}\left(\frac{M^2_{HT}}{Q^2}\right)^n = \int_{-\infty}^{\infty} \frac{f(x)M^2}{Q^2 - xM^2}dx, \tag{1}$$

Delta-like spectral density was analyzed in [11]. Here we use Breit-Wigner like spectral density instead:

$$\Delta_{HT}(Q^2) = \int_{-\infty}^{\infty} \frac{1}{\pi} \frac{M_{HT}\,\sigma}{(y+M_{HT}^2)^2 + M_{HT}^2\,\sigma^2} \frac{M_{HT}^2}{Q^2-y} dy, \tag{2}$$

that leads to HT term:

$$\Delta_{HT}(Q^2) = \frac{M_{HT}^2(M_{HT}^2+Q^2)}{(M_{HT}^2+Q^2)^2 + M_{HT}^2\,\sigma^2} \tag{3}$$

There are 4 free parameters: "glueball mass" M_{gl}^2, M_{HT}^2, μ_4^{p-n} and BW width σ.
By the use of condition:

$$\Gamma_1^{p-n}(Q^2=0) = 0, \tag{4}$$

determined by the finiteness of cross section in the real photon limit one can eliminate μ_4^{p-n}.
Similarly to [12, 13] basing our $\Gamma_1^{p-n}(Q^2)$ IR analysis on Gerasimov-Drell-Hearn [14, 15] and Burkhardt-Cottingham [16] sum rules the equation

$$\frac{d}{dQ^2}\Gamma_1^{p-n}(Q^2=0) = \frac{-(\mu_p-1)^2+\mu_n^2}{8M^2}, \tag{5}$$

implies additional restriction and leaves 2 free parameters. If we fix resonance width where will be only one free parameter to fit.

We consider 4 orders in coupling constant from the leading order up to NNNLO in order to trace sensitivity and stability of our numerical calculation. For simplicity we will limit ourselves by the leading order RG equation for higher order analytic couplings calculation.

3. $\sigma = \rho_0$ width case

One important case motivated by VDM is $\sigma = \Gamma_{\rho_0}$, width of ρ_0 meson. It is worth to mention that in the LO and NLO model exhibits weak dependence on variable parameters such as M_{gl}^2 (or in other words has infinite 1 σ confidence interval) as can be seen in the Fig. 1 while the numerical detail are given in the table 1.

In the Fig. 1 one can see fitting results at the each of the four loops respectively. Contour plot represents χ^2 dependence on the $(M_{gl}^2 \times m_{HT}^2)$-plane, thin green line corresponds to the $m_{HT}^2(M_{gl}^2)$ dependence as an additional constraint, the red dot on it points to the best fit, while the blue dots show the edges of 1 *sigma* dispersion.

Table 1. Fitting fixed $\sigma = \Gamma_\rho$

Order	M_{gl}^2, GeV^2	m_{HT}^2, GeV^2	μ_4, GeV	χ^2
LO	$0.789 + \infty - 0.354$	$0.694 + 0.088 - 0.011$	$-0.180 + 0.012 - 0.010$	8.133
NLO	$0.915 + \infty - 0.276$	$0.681 + 0.121 - 0.017$	$-0.163 + 0.017 - 0.020$	8.634
N^2LO	$1.226 + 0.847 - 0.177$	$0.679 + 0.218 - 0.058$	$-0.125 + 0.021 - 0.037$	9.071
N^3LO	$0.509 + 0.043 - 0.044$	$0.245 + 0.309 - 0.202$	$-0.059 + 0.003 - 0.024$	0.782

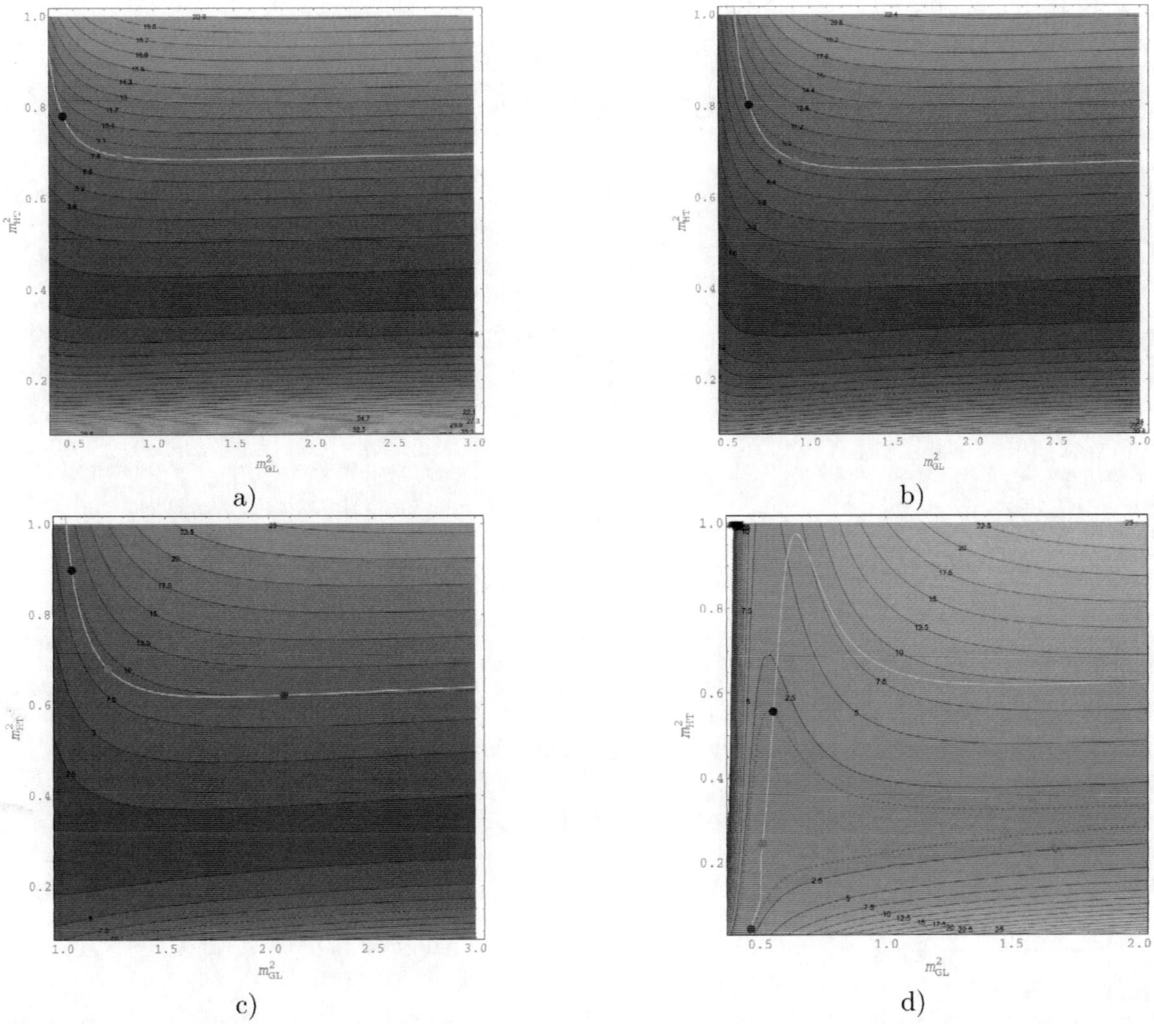

Figure 1. Contour plots on χ^2 in the LO, NLO, NNLO, NNNLO respectively. Green line - $m^2_{HT}(M^2_{gl})$. Red dot - found fit. Blue dots - edges for 1 σ dispersion

4. On the choice of σ

Let us now return to general evaluation of applicability of the finite width model. On the Fig. 2 the $\chi^2(\sigma)$ dependence for N^3LO order is illustrated. As one can see best fit parameters demonstrate stable values and almost do not change up to $\sigma \approx 0.2$, what at one side admits possibility of the ρ_0 meson exchange. On the other side tell us that the Dirac δ function dispersion approximation for spectral function is also applicable and gives reasonable results. Although it manifests less stable results for m^2_{HT} giving in the N^2LO deviation at the order of magnitude lower. In order to reveal stabilty of the results and improve precision it is important to derive exact analytic constant in the each order of consideration.

It is worth noting some qualitative characteristics of the $m^2_{HT}(M^2_{gl})$ dependence. The character of the dependence qualitatively changes with the σ as one can see in the Fig 1. Nevertheless in the LO, NLO and NNLO it has an independent on σ singularity. Which in the

LO for example is determined by

$$4g_A M_p^2 + 3(\mu_p - 1)M_{gl}^2 \pi \log^2\left(\frac{M_{gl}^2}{\Lambda_{PTLO}^2}\right)\beta_0 = 0 \tag{6}$$

which gives $M_{gl\ pole}^2 \approx 0.225$ in the LO as well as $M_{gl\ pole}^2 \approx 0.432$, $M_{gl\ pole}^2 \approx 0.229$ in the NLO and NNLO respectively.

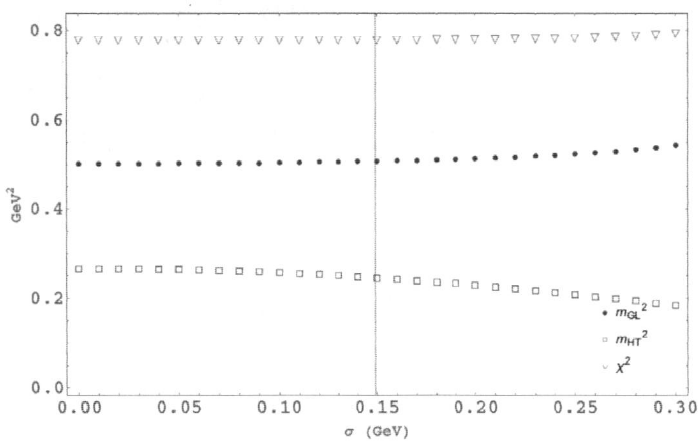

Figure 2. Best fit dependence on σ in the N^3LO. Vertical red line corresponds to ρ_0 width.

5. Acknowledgments

We are indebted to S.V. Mikhailov, O.P. Solovtsova and D. Strozik-Kotlorz for the discussion and essential remarks.

References

[1] Solovtsov I L and Shirkov D V 1999 *Theor. Math. Phys.* **120** 1220–1244 [Teor. Mat. Fiz.120,482(1999)] (*Preprint* hep-ph/9909305)
[2] Shirkov D V and Solovtsov I L 2007 *Theor. Math. Phys.* **150** 132–152 (*Preprint* hep-ph/0611229)
[3] Stefanis N G 2013 *Phys. Part. Nucl.* **44** 494–509 (*Preprint* 0902.4805)
[4] Khandramai V L, Pasechnik R S, Shirkov D V, Solovtsova O P and Teryaev O V 2012 *Phys. Lett.* **B706** 340–344 (*Preprint* 1106.6352)
[5] Cornwall J M 1982 *Phys. Rev.* **D26** 1453
[6] Simonov Yu A 1993 *JETP Lett.* **57** 525–529 [Pisma Zh. Eksp. Teor. Fiz.57,513(1993)]
[7] Shirkov D V 2013 *Phys. Part. Nucl. Lett.* **10** 186–192 (*Preprint* 1208.2103)
[8] Shirkov D V 2015 *Phys. Part. Nucl. Lett.* **12** 205–213
[9] Teryaev O 2013 *Nucl. Phys. Proc. Suppl.* **245** 195–198 (*Preprint* 1309.1985)
[10] Breit G and Wigner E 1936 *Phys. Rev.* **49** 519–531
[11] Khandramai V L, Teryaev O V and Gabdrakhmanov I R 2016 *J. Phys. Conf. Ser.* **678** 012018
[12] Soffer J and Teryaev O 1993 *Phys. Rev. Lett.* **70** 3373–3375
[13] Soffer J and Teryaev O V 1995 *Phys. Rev.* **D51** 25–31 (*Preprint* hep-ph/9405228)
[14] Gerasimov S B 1966 *Sov. J. Nucl. Phys.* **2** 430–433 [Yad. Fiz.2,598(1965)]
[15] Drell S D and Hearn A C 1966 *Phys. Rev. Lett.* **16** 908–911
[16] Burkhardt H and Cottingham W N 1970 *Annals Phys.* **56** 453–463

IOP Conf. Series: Journal of Physics: Conf. Series **938** (2017) 012047 doi:10.1088/1742-6596/938/1/012047

Infrared Confinement: Mass-Scale Dependence of the Strong Effective Coupling, Leptonic Decay Constants and Spectrum of Mesons, the Lowest Glueball Mass

Gurjav Ganbold

Bogoliubov Laboratory of Theoretical Physics, JINR, Joliot-Curie 6, 141980 Dubna, Russia;
Institute of Physics and Technology, Enkh Taivan ave. 54b, 13330 Ulaanbaatar, Mongolia

E-mail: ganbold@theor.jinr.ru

Abstract. Properties of two-particle bound states composed of spin-zero and spin-half constituents have been investigated within a relativistic field model based on new analytically (infrared) confined propagators. A new mass scale dependence of the strong effective coupling has been obtained. By using this dependence, new estimates on the mass spectrum and leptonic decay constants of mesons have been performed with reasonable accuracy. Independent and new estimates have been obtained for the mass and radius of the lowest (scalar) glueball.

1. Introduction

QCD predicts a dependence of the physical coupling g under changes of energy scale Q (or, distance $\sim 1/Q$). This dependence $\alpha_s(Q) \doteq g^2/(4\pi)$ is described theoretically by the renormalization group equations and determined experimentally at relatively high energies $Q > 1$ GeV [1]. Meanwhile, understanding of a number of phenomena such as quark confinement, hadronization etc., requires a correct description of hadron dynamics in the infrared (IR) region below $Q \sim 1$ GeV. Particularly, many quantities in particle physics are affected by the IR behavior of α_s. However, the long-distance behavior of α_s has not been well defined yet, it needs to be more specified. The correct description of QCD effective coupling in the IR regime remains one of the actual problems in particle physics.

2. Model

Let's consider a QCD-inspired relativistic field model with Lagrangian [2]:

$$\mathcal{L} = -\frac{1}{4}\left(\partial^\mu \mathcal{A}^A_\nu - \partial^\nu \mathcal{A}^A_\mu - g f^{ABC}\mathcal{A}^B_\mu \mathcal{A}^C_\nu\right)^2 + \left(\bar{q}^a_f \left[\gamma_\alpha \partial^\alpha - m_f\right]^{ab} q^b_f\right) + g\left(\bar{q}^a_f \left[\Gamma^\alpha_C \mathcal{A}^C_\alpha\right]^{ab} q^b_f\right), \quad (1)$$

where $\mathcal{A}^C_\alpha$ is the gluon and q^a_f is a quark field of flavor f with mass $m_f = \{m_{ud}, m_s, m_c, m_b\}$ and $\Gamma^\alpha_C = i\gamma_\alpha t^C$. For the spectra of quark-antiquark and di-gluon bound states we solve Bethe-Salpeter type equations obtained in [3, 4]. By omitting intermediate stages of calculation (see for details [5]) we rewrite the master equation determining the meson mass as follows:

$$1 = \alpha_s \cdot \lambda_{JJ}(M^2_J, m_1, m_2) = \alpha_s \frac{16\pi C_J}{9} \int \frac{d^4k}{(2\pi)^4} \iint dx dy\; e^{-ik(x-y)}\, U_{\mathcal{N}}(x)\sqrt{D(x)D(y)}\, U_{\mathcal{N}}(y)$$

$$\cdot \mathrm{Tr}\left[O_J \tilde{S}_{m_1}\left(\hat{k}+\xi_1\hat{p}\right) O_J \tilde{S}_{m_2}\left(\hat{k}-\xi_2\hat{p}\right)\right]|_{-p^2=M_J^2}, \qquad (2)$$

where $C_J = \{1, 1, 1/2, -1/2\}$, $\xi_i = m_i/(m_1+m_2)$, $O_J = \{I, i\gamma_5, i\gamma_\mu, \gamma_5\gamma_\mu\}$ and the gluon ($D(x)$) and quark propagator ($\tilde{S}_{m_1}(\hat{p})$) are represented in Euclidean space.

Note, the polarization kernel $\lambda_{JJ'}(-p^2)$ has to be diagonalized on a complete system of orthonormal functions $\{U_\mathcal{N}\}$, where $\mathcal{N} = \{n, l, \mu, ...\}$ is a set of quantum numbers. The solution of Eq.(2) is nothing else but the solution of the corresponding ladder Bethe-Salpeter equation.

Ultraviolet divergences in the model have been removed by renormalization of wave function and charge, but infrared singularities remain in Eq.(2) because of integration over variable k. To avoid the appearance of the singularities in the mass formula, we follow theoretical predictions in favor of an IR-finite behavior of the gluon propagator [6, 7] and introduce a new scheme of infrared cutoffs on the limits of scale integrations for the propagators as follows:

$$D(x) = \frac{1}{4\pi^2 x^2} \to \int\limits_{\Lambda/2}^{\infty} ds \exp[-sx^2], \quad \tilde{S}_{m_f}(\hat{p}) = \frac{1}{-i\hat{p}+m_f} \to (i\hat{p}+m_f)\int\limits_0^{1/\Lambda} dt \exp[-t(p^2+m_f^2)], \qquad (3)$$

where Λ is the mass scale of the IR confinement domain. These propagators are entire analytic functions in the Euclidean space. Note, another type of IR confinement applied to whole 'quark-antiquark' loop was applied in [8, 9]. The analytic confinement disappears as $\Lambda \to 0$.

3. Effective Strong Coupling in the IR Region

The meson mass M_J is defined by Eq.(2) at given α_s, quantum numbers $\mathcal{N}$ and constituent quark masses $\{m_1, m_2\}$. And vice versa, α_s can be estimated for given mass.

The QCD coupling may feature an IR-finite behavior (e.g., in [10, 11]). To study this, we consider M_J as an appropriate energy-scale for α_s and choose a special case $m_1 = m_2 \sim M/2$ in the ground-state. Then, a new effective (mass-dependent) strong coupling $\alpha_s(M)$ in time-like domain may be defined by:

$$\alpha_s(M) = -1/\lambda(M, M/2, M/2). \qquad (4)$$

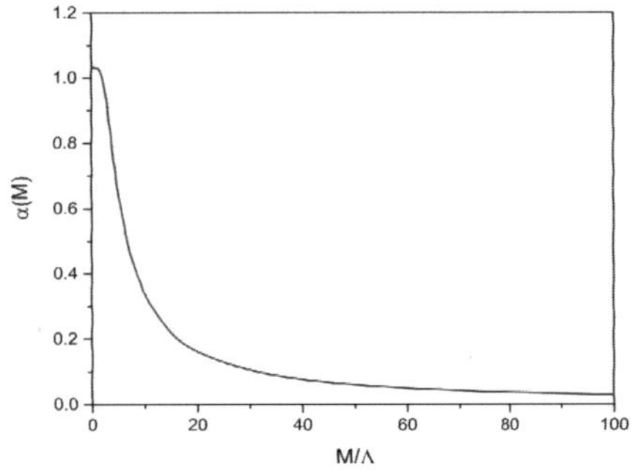

Figure 1. Dependence of the effective strong coupling α_s on relative mass scale M/Λ.

A new variational upper bound $\alpha(M)$ to $\alpha_s(M)$ is shown in Fig.1. The slope of $\alpha(M/\Lambda)$ depends on $\Lambda > 0$, but the newly revealed origin $\alpha(\mathbf{0}) = \mathbf{1.032}$ (or, $\alpha(\mathbf{0})/\pi = \mathbf{0.328}$) remains unchanged. This upper bound to the IR-fixed value $\alpha(0)$ is in a reasonable agreement with often quoted estimates [12]:

$$\alpha_s^0/\pi \simeq 0.19 \div 0.25 \quad [13], \qquad \alpha_s^0/\pi \simeq 0.265 \quad [14], \qquad \alpha_s^0/\pi \simeq 0.26 \quad [15].$$

4. Lowest Glueball State

The existence of glueballs, the bound states of gluons, is predicted by QCD because of the self-interaction of gluons. Most known experimental signatures for glueballs are an enhanced production in gluon-rich channels of radiative decays and some decay branching fractions incompatible with $(q\bar{q})$ states. There are predictions expecting glueball-like states in the mass range $M_G \sim 1.5 \div 5.0$ GeV with spin $J = 0, 1, 2, 3$ [16].

Below we consider a two-gluon scalar bound state with $J^{PC} = 0^{++}$. By omitting details of intermediate calculations (similar to those represented in the previous section) we define the scalar glueball mass $M_{0^{++}}$ from equation:

$$1 - \frac{8\,\alpha}{3\pi} \int dz\, e^{izp}\, \Pi_G(z) = 0\,, \qquad p^2 = -M^2_{0^{++}}\,, \tag{5}$$

where $\Pi_G(z)$ is the self-energy (polarization) function of the scalar glueball.

Particularly, for $\Lambda = 236$ MeV and $\alpha(M_G)$ defined from Eq.(4) we obtain new estimates:

$$M_{0^{++}} = 1739 \text{ MeV}\,, \qquad \alpha(M_{0^{++}}) = 0.451\,. \tag{6}$$

The new value of $M_{0^{++}}$ is in reasonable agreement with other predictions [2, 16, 17, 18, 19].

We also estimate the scalar glueball 'radius' (r_G) which leads to a new value:

$$r_{0^{++}} \cdot M_{0^{++}} = 4.41 \tag{7}$$

that is in reasonable agreement with data in [16].

5. Meson Spectrum

The dependence of meson mass on α_s and other model parameters is defined by Eq.(2). Note, the kernel function $\lambda_{\mathcal{N}}$ is real and finite, that allows us to evaluate a variational solution to M_J.

Particularly, for the obtained $\Lambda = 236\, MeV$ and $\alpha(M)$ defined in Eq.(4) we derive meson mass formula Eq.(2) by fitting the conventional meson masses with adjustable constituent quark masses $\{m_{ud}, m_s, m_c, m_b\}$. We have fixed a new final set of model parameters (in units of MeV) as follows:

$$\Lambda = 236\,, \quad m_{ud} = 227.6\,, \quad m_s = 420.1\,, \quad m_c = 1521.6\,, \quad m_b = 4757.2\,. \tag{8}$$

and represent in Tab. 1 our new estimates on the pseudoscalar (P) and vector (V) meson masses.

Table 1. Estimated masses of conventional mesons M_P and M_V (in units of MeV) for new model parameters (8) compared to the recent experimental data [16].

$J^{PC} = 0^{-+}$	M_P	Data	$J^{PC} = 1^{--}$	M_V	Data
D	1893.6	1869.62	ρ	774.3	775.26
D_s	2003.7	1968.50	K^*	892.9	891.66
η_c	3032.5	2983.70	D^*	2003.8	2010.29
B	5215.2	5259.26	D^*_s	2084.1	2112.3
B_s	5323.6	5366.77	J/Ψ	3077.6	3096.92
B_c	6297.0	6274.5	B^*	5261.5	5325.2
η_b	9512.5	9398.0	Υ	9526.4	9460.30

Our estimates represented in Tab. 1 are in reasonable agreement with experimental evidences with relative errors less than 1.8 per cent.

IOP Conf. Series: Journal of Physics: Conf. Series **938** (2017) 012047 doi:10.1088/1742-6596/938/1/012047

6. Leptonic Decay Constants

The leptonic decay constants f_J are important quantities in meson physics. The precise knowledge of their values provides more improvement in our understanding of various processes convolving meson decays. To describe effectively the 'sawtooth'-type unsmooth dependence of f_J on meson masses (see Tab.3), we introduce additional parameters R_J into the basis functions as follows: $\tilde{U}_R(k) = \int_0^1 ds\, h(s) \exp\left[-sk^2/R_J^2\right]$, where $h(s)$ is a smooth functions and R_J characterizes the meson 'size' in units of mass scale. Then, by using our model parameters in Eq.(8) along with the optimal meson 'sizes' shown in Tab. 2 we estimate new results on the leptonic decay constants of conventional mesons shown in Tab. 3.

Table 2. Meson 'size' parameters R_J (in units of GeV)

D	D_s	η_c	B	B_s	B_c	η_b	ρ	K^*	D^*	D_s^*	J/Ψ	B^*	Υ
0.93	1.08	1.83	1.73	2.18	3.34	3.80	0.33	0.38	0.78	0.90	2.40	3.34	2.80

Table 3. Estimated leptonic decay constants of conventional mesons f_P and f_V (in MeV) compared to experimental data in [20, 21, 22, 23].

0^{-+}	f_P	Data	Ref.	1^{--}	f_V	Data	Ref.
f_D	207	206.7 ± 8.9	[20]	f_ρ	221	221 ± 1	[20]
f_{D_s}	257	257.5 ± 6.1	[20]	f_{K^*}	217	217 ± 7	[20]
f_{η_c}	238	238 ± 8	[22]	f_{D^*}	245	245 ± 20	[23]
f_B	193	192.8 ± 9.9	[21]	$f_{D_s^*}$	271	272 ± 26	[23]
f_{B_s}	239	238.8 ± 9.5	[21]	$f_{J/\Psi}$	416	415 ± 7	[20]
f_{B_c}	488	489 ± 5	[22]	f_{B^*}	196	196 ± 44	[23]
f_{η_b}	800	801 ± 9	[22]	f_Υ	715	715 ± 5	[20]

References

[1] Bethke S 2009 *Eur. Phys. J.* C **64** 689
[2] Ganbold G 2009 *Phys. Rev.* D **79** 034034
[3] Ganbold G 2010 *Phys. Rev.* D **81** 094008
[4] Ganbold G 2014 *Phys. Part. Nucl.* **45** 10
[5] Ganbold G 2012 *Phys. Part. Nucl.* **43** 79
[6] Fischer C S, Alkofer R and Reinhardt H 2002 *Phys. Rev.* D **65** 094008
[7] Lerche C and von Smekal L 2002 *Phys. Rev.* D **65** 125006
[8] Ganbold G et al. 2015 *J. Phys.: Nucl. Part. Phys.* G **42** 075002
[9] Ganbold G 2017 *Eur. Phys. J.: Web of Conference* **138** 04004
[10] Aguilar A C, Mihara A and Natale A A 2004 *Int. J. Mod. Phys.* A **19** 249
[11] Brodsky S J and de Teramond G F 2004 *Phys. Lett.* B **582** 211
[12] Baldicchi M et al, 2008 *Phys. Rev.* D **77** 034013
[13] Godfrey S and Isgur N 1985 *Phys. Rev.* D **32** 189
[14] Zhang T and Koniuk R 1991 *Phys. Lett.* B **261** 311
[15] Halzen F, Krein G I and Natale A A 1993 *Phys. Rev.* D **47** 295
[16] Patrignani C et al. 2016 *Chin. Phys.* C **40** 100001
[17] Morningstar C J and Peardon M 1999 *Phys. Rev.* D **60** 034509
[18] Chen Y et al. 2006 *Phys. Rev.* D **73** 014516
[19] Gregory E et al. 2012 *J. High Energ. Phys.* **10** 170
[20] Olive K A et al (Particle Data Group Collaboration) 2014 *Chin. Phys.* C **38** 090001
[21] Laiho J, Lunghi E and van de Water R S 2010 *Phys. Rev.* D **81** 034503
[22] Chiu T-W et al (TWQCD Collaboration) 2007 *Phys. Lett.* B **651** 171
[23] Becirevic D et al 1999 *Phys. Rev.* D **60** 074501

IOP Conf. Series: Journal of Physics: Conf. Series **938** (2017) 012048 doi:10.1088/1742-6596/938/1/012048

The role of charge-symmetry invariance in the σ_{-1} "bremsstrahlung-weighted" photo-absorption sum rules for the non-relativistic and relativistic few-body systems

S.B. Gerasimov
BLTPh, JINR, Dubna, Russia
E-mail: gerasb@theor.jinr.ru

Abstract. Within the non-relativistic approach to photo-absorption sum rules for the 3N(4N)-nuclei the new σ_{-1} sum rules proposed which are based on general charge-symmetry (CS) consequences for a pair of the "CS-conjugated" triton and ^3He nuclei and for the "self-conjugated" nucleus ^4He. The numerical deviations from the less general "charge-independent" approach leading to the long-known Foldy-Khokhlow sum rules are estimated. In the relativistic domain the functional inapplicability is demonstrated of the "charge independence" approximation in the derivation of the σ_{-1} - type sum rules in the case of nucleons. With application of the relevant "charge symmetry" approach, the general system of the sum rules is derived and the inevitable isospin-dependent dynamics of the constituent quarks is underlined.

1. Introduction and the non-relativistic σ_{-1} sum rules for few-body nuclei

The NR sum rules σ_{-1} for the s-shell nuclei ($A = Z + N \leq 4$) were earlier obtained from the assumption that the ground-state wave function is symmetric in the space coordinates of nucleons [1],[2] for each nucleus of the isotopic doublet ^3H and ^3He and for ^4He

$$4\pi^2\alpha\frac{NZ}{A-1}(\frac{1}{3})\cdot\langle r_{ch}^2\rangle_{NR} = \int\frac{d\nu}{\nu}\sigma_{E1}(\nu) \tag{1}$$

The present-day situation is that the accuracy of the theoretical results is higher than the experimental uncertainties in the important energy range up to the pion production threshold suggesting that new experimental data would be timely. The main objectives of this Section is a comparison of the Charge Independence (CI) and Charge Symmetry (CS) initial assumptions taken for the derivation of the σ_{-1} sum rules for the "mirror" ^{3}He -^{3}H and the "charge-self-conjugate" ^{4}He nuclei. The validity of the Foldy-Khokhlov sum rules assumes validity of the (nucleon) charge independence (CI) symmetry.

Our objective in this Section is to apply more strong charge symmetry (CS) requirements in derivation of the Foldy-Khokhlov-type sum rules. In accord with the CS-provisions for the mean values of the nucleon radii correlations in the mirror pair ^{3}H- and ^{3}He nuclei and the "self-conjugate" ^{4}He nucleus, we have get that the values of $< D^2 >_{^3H}$, and $\langle r_{ch}^2\rangle_{^3H}$ can be expressed via one and the same linear combination of the introduced and undetermined parameters which is different from the analogues expression for $\langle r^2_{ch}\rangle_{^3He}$ and $< D^2 >_{^3He}$. Taken at face value this means that with our more strict assumption on the underlying symmetry i.e. the CS instead of CI-option, we get only one "survived" sum rule instead of three (for ^{3}H,^{3}He and ^{4}He) earlier derived and often-used Foldy-Khokhlov sum rules:

$$\begin{array}{ccccc} \sigma_{-1}(^3H) & = & \frac{4\pi^2\alpha}{3}\cdot\langle r^2_{pp}\rangle^{^3H}_{ch} & \neq & \sigma_{-1}(^3He) \\ 2.45mb & & 2.42mb \equiv .242f^2 & & 2.54 \pm .09mb \end{array} \qquad (2)$$

(the numerical values of the experimental data are from works [3].

The nuclei charge radii in the above relations and in what follows are understood as corresponding to the case of "point-charge-protons" ones which reached by subtracting from "physical" values standard corrections [4]. The problem of the radii difference in the charge-symmetry-conjugated 3N-nuclei is of similar degree of complexity as the long-discussed binding energy difference of the ^{3}He and ^{3}H: $E_b(^3H)$=8.52MeV and $E_b(^3He)$=7.76MeV. e.g.[5]. For a qualitative understanding of the intended estimation, we resort to dimensional arguments and factors as follows:

$$\begin{array}{ccc} \frac{E_b(^3He)}{E_b(^3H)} & \simeq & \frac{\langle r^2_{ch}\rangle^{-1/2}_{^3He}}{\langle r^2_{ch}\rangle^{-1/2}_{^3H}} \\ 0.911 & & 0.899 \end{array} \qquad (3)$$

The unexpected numerical closeness of two ratios written down on exclusively dimensional ground signals nevertheless that for its full dynamical explanation one should include into consideration not only the Coulomb interaction effects but all collection of the Charge Symmetry Breaking (CSB) - explications that represent on the nuclei and hadronic scales the fundamental current u- and d-quark mass difference. The modified $\sigma_{-1}(^4He)$ sum rule keeping the basic CS-symmetry limitations can be presented in the form

$$\sigma_{-1}(E1)^{^4He} = \frac{16}{9}\pi^2\alpha(\langle r_c{}^2\rangle_{^4He} + \beta_{pp} - \beta_{pn}) \qquad (4)$$

The notations for the $\beta_{NN'}$ through the scalar products of the nucleon radial coordinates were introduced by Foldy [1]. For the CS-self-conjugate $^4He_2$-nucleus all β_{pp}=β_{nn} $\neq$$\beta_{pn}$ can be computed if the explicit form of the ground state radial wave function is available.

The finite difference calculated with the experimental data [3]

$$\beta_{pp} - \beta_{pn} \simeq 0.2fm^2 \qquad (5)$$

corresponds to the transition off the "CI - case" to the more general "CS - case" applied to the photodisintegration of the "self-adjoint" ^4He - nucleus

2. Relativistic Constituent Quark Model and Nucleons

Following formally to the $p_z \to \infty$ techniques derivation of the Cabibbo-Radicati [6] and GDH [7],[8],[9] sum rules we can obtain the relations

$$4\pi^2\alpha(\frac{1}{3} < \vec{D}^2 > -(\frac{\kappa_N}{2m_N})^2) = \int\frac{d\nu}{\nu}\sigma^{res}_{tot}(\nu), \qquad (6)$$

$$4\pi^2\alpha[\frac{1}{3} < \vec{D}^2 >] = \int_{thr}^{\infty}\frac{d\nu}{\nu}(\sigma^{res}_p(\nu) \qquad (7)$$

IOP Conf. Series: Journal of Physics: Conf. Series **938** (2017) 012048 doi:10.1088/1742-6596/938/1/012048

We use the definitions

$$\hat{D} = \int \vec{x}\hat{\rho}(\vec{x})d^3x = \sum_{j=1}^{3} Q_q(j)\vec{d_j} \tag{8}$$

The defined operators $Q_q(j)$ and $\vec{d_j}$ are the electric charges and configuration variables of point-like interacting quarks in the infinite-momentum frame of the bound system. Finally, we relate the electric dipole moment operator correlators, sucessively for the proton, the neutron and the pure "isoscalar(or isovector)-nucleon" part equal for both nucleons and the isovector part of the mean-squared radii operators, which all are sandwiched by the nucleon state vectors in the "infinite - momentum frame", with experimentally measurable data on the resonance parts of the photoabsorption cross sections on the proton and neutron presently known below ~ 2 GeV.

The listed operator mean values are parametrized as follows

$$\begin{aligned}
R_V &= \frac{1}{2}(< r_1^2 >_P - < r_1^2 >_N) = \alpha - \frac{1}{2}\beta \\
J_P &= \frac{1}{3} < \hat{D^2} >_P = \frac{8}{27}\alpha + \frac{1}{27}\beta + \frac{8}{27}\gamma - \frac{8}{27}\delta \\
J_N &= \frac{1}{3} < \hat{D^2} >_N = \frac{2}{27}\alpha + \frac{4}{27}\beta + \frac{2}{27}\gamma - \frac{8}{27}\delta \\
J_V &= \frac{1}{3} < \hat{D^2} >_V = \frac{2}{3}\alpha + \frac{1}{3}\beta + \frac{2}{3}\gamma - \frac{4}{3}\delta
\end{aligned}$$

where $< \vec{d_1}^2 > = < \vec{d_2}^2 > = \alpha$, $< \vec{d_3}^2 > = \beta$, $< \vec{d_1} \cdot \vec{d_2} > = \gamma$, $< \vec{d_1} \cdot \vec{d_3} > = < \vec{d_2} \cdot \vec{d_3} > = \delta$ indices "1" and "2" refer to the like quarks and "3" to the odd quark. Evaluation of the relativistic electric dipole moment fluctuation and the isovector charge radius sum rules for the nucleon was carried out with the available compilation [10] of the resonance pion-photoproduction data on the proton and neutron $A_{1/2}^{P(N)}$ and $A_{3/2}^{P(N)}$ the numerical values we calculated and all integrals over photoexcited nucleon resonances were taken in the narrow resonance approximation, when

$$J_{p(a)}^{res} \simeq \frac{4\pi m_n |A_{3/2(1/2)}^{res}|^2}{m_{res}^2 - m_n^2}, \tag{9}$$

where $m_{n(res)}$ is the nucleon (or resonance)mass Solving the system of the linear equations and evaluating the $R_V, J_{P,N,V}$ with the help of experimentally known partial amplitudes of main photo-excited resonances, we find our final results for the numerical values α, β and the opening angle θ_{12} and θ_{13} between vectors $\vec{d_1}$ and $\vec{d_2}$ and vectors $\vec{d_1}$ and $\vec{d_3}$, lying in the plane transverse to the "infinite momentum $p_z \to \infty$ vector:

$$\begin{aligned}
\alpha^{1/2} = 0.75 \pm 0.06 fm, \quad \beta^{1/2} = 0.77 \pm 0.12 fm \\
\theta_{12} \simeq 120^0, \quad \theta_{13} \simeq \theta_{12} \sim 120^0 \\
< {r_1}^2 >_V = .25 \pm 0.02 fm^2 (exp: .28 fm^2)
\end{aligned}$$

In the limit of the charge independence option: $\alpha = \beta$ and $\gamma = \delta$, when $J_V - 3J_N - J_P = 0$, and putting $J_{P/N} = A_V^2 + A_S^2 \pm 2|A_V||A_S|$, we get that there is no the real-valued solution for, say, A_S through the isovector amplitude A_V. We accept this fact as the failure of the earlier accepted the charge-independence (CI) assumption. Having got the values of the "global" parameters α, β, etc. defined in the infinite momentum frame, we can put all three vectors with the pertinent numerical parameters into the plane transverse to $p_z \to \infty$ momentum and define parameters of the closed contour circumscribing their ends. The resulting curve is either ellipse with very small excentricitet or the circle. This fact is, in our opinion, at variance with rather popular "quark-diquark" models of the nucleon with, presumably, more noticeable asymmetry of the spatial correlations of the *uu*- and *ud* - quark pairs. The isovector Dirac rms of the valence quark distribution could be at stronger deviation from data because of expected importance of the (omitted in our approach) isovector two-pion exchanges entering in full expressions for the nucleon radii

3. The CS "at the work" for the conclusive, partial, resonance photo-excitation description

Summarizing the results of the global approach via the use of the sum rules, we obtain a kind of averaged picture across the spectrum pattern with the presumably dominating influence of the first $\Delta(1236)$-resonance with its magnetic-dipole photo-excitation of the nucleon spin degrees of freedom, which is also hiding explication of the qualitative isospin-depending features of the (mainly, electric-dipole) 2-nd and (the electric-quadrupole) 3-rd resonance photo-excitation regions. The value of effective dipole moments, determining the excitation cross section of the "electric-dipole-type" nucleon resonances, includes apparently the pionic degrees of freedom of the constituent quarks. This is supported by the fundamental principle of the charge symmetry of strong interaction dynamics. The participation of the pion degrees of freedom plays the role in the dynamical formation of the quark-diquark spatially looking nucleon structure. We suggest to illustrate it numerically choosing the experimental data on the resonance photo-absorption data on the $J^P = 3/2^-$ and $J^P = 1/2^-$-resonances,composing the region of the "second" resonance area. Performing the evaluations analogous the earlier presented for all the resonances with the data from the PDG table,we obtain the results differing from earlier cited and which indicated now in parentheses:

$$\alpha^{1/2} = 0.82(0.75)fm, \;\; \beta^{1/2} = 0.53(0.77)fm$$
$$\theta_{12} \simeq 130^0(120^0), \;\; \theta_{13} \simeq 70^0(120^0), \;\; \theta_{23} \sim 160^0(120^0)$$

A pronounced asymmetry of the sides of the triangles Oud and Ouu

$$\frac{C_{ud}^{min}}{C_{uu}} \simeq \frac{0.8fm}{1.5fm} \tag{10}$$

can testify to the validity of the quark-diquark model of the considered resonances though diquarks are not apparently strongly bound. On the other hand, the obvious need to fulfill the charged symmetry requirements implies a significant role of the isospin-dependent quark-quark potentials, that is the role of the interactions caused by the pion exchanges, for instance. Ultimately, however, both factors can be the sides of the same phenomenon.

References

[1] Foldy L L 1957 *Phys. Rev.* **107** 1330;
[2] Khokhlov Yu K 1957 *ZhETF* **32** 124;
[3] Efros V D, Leidemann W, Orlandini G 1997 *Phys. Lett.* B **408** 1;
[4] Ekström A, Jansen G R, Wendt K A et al 2015 *Phys. Rev.* C **91** 00301(R);
[5] Wu Y, Ishikawa S, Sasakawa T 1990 *Phys. Rev. Lett.* **64** 1875;
[6] Cabibbo N, Radicati L A 1966 *Phys. Lett.* **19** 687;
[7] Gerasimov S B 1965 *Sov. J. Nucl. Phys.* **2** 430;
[8] Drell S D, Hearn A C 1966 *Phys. Rev. Lett.* **16** 908;
[9] Hosoda M, Yamamooto K 1966 *Prog. Theor. Phys.* **36** 425;
[10] Particle Data Group (Olive, K A et al.) 2014 *Chin. Phys.* C **38** 090001;

IOP Conf. Series: Journal of Physics: Conf. Series **938** (2017) 012049 doi:10.1088/1742-6596/938/1/012049

Simple criterium for CP conservation in 2HDM

I. F. Ginzburg[1,2]

[1] Sobolev Inst. Mathem., av. Ac. Koptyug, 4, 630090 Novosibirsk, Russia,
[2] Novosibirsk State University, ul. Pirogova, 2, 630090 Novosibirsk, Russia

E-mail: ginzburg@math.nsc.ru

Abstract. We find a set of necessary and sufficient conditions for the CP conservation in the most general 2HDM in terms of observable quantities. This set contains two simple and relatively easily testable conditions instead of the more complex conditions usually discussed.

1. Introduction

The detailed version of this paper was prepared in collaboration with M. Krawczyk [1]. Maria passed away on May 24, 2017. In this report I omit some details, which look me interesting only for participants of Higgs studies.

CP violation is one of the important yet not well understood aspect of the fundamental physics. The modern LHC data [2],[3] allow us to conclude that the observed particle $h(125)$ is a Higgs boson with spin-CP parity 0^{++} — but only under the assumption that this particle has a definite parity. Generally in many models $h(125)$ does not possess a definite parity. In this case the data give no information about $h(125)$ parity [4].

In the Standard Model the CP violation is described by the CKM matrix, but its origin remains unclear. The extension of SM with two Higgs doublets, called the Two Higgs Doublet Model (2HDM), has been introduced in 1974 with the main aim to provide an extra source of CP violation [5]. Later, many variants of 2HDM were considered with different features, providing various physical realizations. Each of these variants is described by Lagrangian with many parameters. The choice of the set of these parameters, describing the same physical reality is ambiguous. In non-minimal models such as 2HDM the current situation, with the properties of the observed Higgs boson resembling those of the SM Higgs (*SM-like scenario* [7] or *alignment limit* [8]), can be described by different non-equivalent sets of parameters.

We consider in the report two types of problems which appear in the study of CP violation in these models[5]F.

(A) Let us have a variant of 2HDM, constructed as a model for the description of some set of phenomena.
How to know whether the CP symmetry is violated or not in this model without detailed calculations of each particular effect?

(B) Consider 2HDM as an approximation to the description of the Nature. How do we know whether the CP violation is described by this approximation, or whether the observed CP violation is an effect of yet another, weaker interaction.
How to check the presence of CP violation in the experiments with particles that appear in 2HDM?

2. 2HDM

The 2HDM describes a system of two spinless isospinor fields ϕ_1, ϕ_2 with hypercharge $Y = 1$. The most general form of the 2HDM potential is

$$\begin{aligned} V =& \frac{\lambda_1}{2}\left(\phi_1^\dagger\phi_1\right)^2 + \frac{\lambda_2}{2}\left(\phi_2^\dagger\phi_2\right)^2 + \lambda_3\left(\phi_1^\dagger\phi_1\right)\left(\phi_2^\dagger\phi_2\right) + \lambda_4\left(\phi_1^\dagger\phi_2\right)\left(\phi_2^\dagger\phi_1\right) \\ &+\left[\frac{\lambda_5}{2}\left(\phi_1^\dagger\phi_2\right)^2 + \lambda_6\left(\phi_1^\dagger\phi_1\right)\left(\phi_1^\dagger\phi_2\right) + \lambda_7\left(\phi_2^\dagger\phi_2\right)\left(\phi_1^\dagger\phi_2\right) + \text{h.c.}\right] \\ &-\frac{m_{11}^2}{2}\left(\phi_1^\dagger\phi_1\right) - \frac{m_{22}^2}{2}\left(\phi_2^\dagger\phi_2\right) - \left[\frac{m_{12}^2}{2}\left(\phi_1^\dagger\phi_2\right) + \text{h.c.}\right]. \end{aligned} \tag{1}$$

The potential parameters are restricted by the requirement that the potential is positive at large quasiclassical values of ϕ_i (*positivity constraints*). We assume also that these parameters are not too big so that one can use estimates based on the lowest non-trivial approximation of the perturbation theory.

After Electroweak Symmetry Breaking the 2HDM contains 3 neutral Higgs bosons $h_a \equiv h_{1,2,3}$ (in general with indefinite CP parity) and the charged Higgs bosons $H^\pm$, with the masses M_a and $M_\pm$, respectively (*the numbering of h_a is independent of the masses M_a*).

2.1. Reparametrization freedom

2HDM describes system of two fields with identical quantum numbers. Therefore, its description in terms of original fields ϕ_i or in terms of their linear superpositions ϕ'_i with corresponding transformation of the parameters of the Lagrangian are equivalent; this statement verbalizes the *reparameterization* (RPa) freedom of the model. We refer to these different choices as different RPa bases. The RPa group consists of RPa transformations $\hat{\mathcal{F}}$ of the form:

$$\begin{pmatrix}\phi'_1\\ \phi'_2\end{pmatrix} = \hat{\mathcal{F}}_{gen}(\theta,\tau,\rho)\begin{pmatrix}\phi_1\\ \phi_2\end{pmatrix}, \quad \hat{\mathcal{F}}_{gen} = e^{i\rho_0}\begin{pmatrix}\cos\theta\, e^{i\rho/2} & \sin\theta\, e^{i(\tau+\rho/2)}\\ -\sin\theta\, e^{-i(\tau+\rho/2)} & \cos\theta\, e^{-i\rho/2}\end{pmatrix}. \tag{2}$$

A subgroup of the RPa group — the *rephasing* group RPh — describes a freedom of adjusting the relative phase of the fields ϕ_i. The parameter ρ_0 describes an overall phase transformation of the fields and can be ignored since it does not affect the parameters of the potential. The parameter ρ describes the RPh symmetry transformation of system.

The $U(1)_{EM}$ symmetry preserving ground state of this system is given by a global minimum of the potential and reads

$$\langle\phi_1\rangle = \begin{pmatrix}0\\ v_1/\sqrt{2}\end{pmatrix}, \quad \langle\phi_2\rangle = \begin{pmatrix}0\\ v_2 e^{i\xi}/\sqrt{2}\end{pmatrix}, \tag{3}$$

with the standard parameterization $v_1 = v\cos\beta$, $v_2 = v\sin\beta$.

2.2. The Higgs basis

We use below the RPa basis with $v_2 = 0$ (the Higgs, or Georgi, basis [9]), in which the 2HDM potential can be written in the form [10] with the change of lowercase letters of (1) to capital:

$$\begin{aligned} V_{HB} =& \frac{\Lambda_1}{2}\left(\Phi_1^\dagger\Phi_1 - \frac{v^2}{2}\right)^2 + \frac{\Lambda_2}{2}\left(\Phi_2^\dagger\Phi_2\right)^2 + \Lambda_3\left(\Phi_1^\dagger\Phi_1 - \frac{v^2}{2}\right)\left(\Phi_2^\dagger\Phi_2\right) + \Lambda_4\left(\Phi_1^\dagger\Phi_2\right)\left(\Phi_2^\dagger\Phi_1\right) \\ &+\left[\frac{\Lambda_5}{2}\left(\Phi_1^\dagger\Phi_2\right)^2 + \Lambda_6\left(\Phi_1^\dagger\Phi_1 - \frac{v^2}{2}\right)\left(\Phi_1^\dagger\Phi_2\right) + \Lambda_7\left(\Phi_2^\dagger\Phi_2\right)\left(\Phi_1^\dagger\Phi_2\right) + \text{h.c.}\right] + M_\pm^2\left(\Phi_2^\dagger\Phi_2\right). \end{aligned} \tag{4}$$

3. Relative couplings

In the discussion below we use the relative values of couplings for each neutral Higgs boson h_a:

$$\chi_a^P = \frac{g_a^P}{g_{\rm SM}^P}, \quad \chi_a^\pm = \frac{g(H^+H^-h_a)}{2M_\pm^2/v}, \quad \chi_a^{H^+W^-} = \frac{g(H^+W^-h_a)}{M_W/v}; \quad (a=1,2,3)\,. \tag{5}$$

The quantities χ_a^P (where $P = V\,(W,Z)$, $q = t, b, ...$, $\ell = \tau, ...$) are the ratios of the couplings of h_a with the fundamental particles P to the corresponding couplings for the would-be SM Higgs boson with mass M_a. The other couplings describe the interactions of h_a with the charged Higgs boson $H^\pm$. Couplings χ_a^V and $\chi_a^\pm$ are real due to Hermiticity of Lagrangian, they are directly measurable. Couplings χ_a^q, χ_a^ℓ and $\chi_a^{H^+W^-}$ are generally complex. Both the real and imaginary parts of the Yukawa couplings χ_a^q and χ_a^ℓ can be measured in principle using distributions of Higgs bosons decay products in $h_a \to \bar{q}q$, $h_a \to \bar{\ell}\ell$. Besides, we have [10]

$$\chi_{ab}^Z \equiv \frac{g(Zh_ah_b)}{M_Z/v} = -\varepsilon_{abc}\chi_c^V\,. \tag{6}$$

There are useful sum rules among these couplings, namely

$$(a)\ \sum_a (\chi_a^V)^2 = 1\,, \quad (b)\ (\chi_a^V)^2 + |\chi_a^{H^+W^-}|^2 = 1\,. \tag{7}$$

The absolute value of the coupling $\chi_a^{H^+W^-}$ is fixed by the sum rule (7 b) and is well measurable. The unitarity of the rotation matrix describing transition from components of fields ϕ_i to the physical Higgs fields h_a allows to obtain the following relations for couplings $\chi_a^{H^+W^-}$ (the factor $e^{i\rho}$ represents the rephasing freedom in the Higgs basis):

$$\begin{aligned}
\chi_1^{H^+W^-} &\equiv \left(\chi_1^{H^-W^+}\right)^* &&= -e^{i\rho}\,\frac{\chi_1^V\chi_2^V - i\chi_3^V}{\sqrt{1-(\chi_2^V)^2}}\,,\\
\chi_2^{H^+W^-} &\equiv \left(\chi_2^{H^-W^+}\right)^* &&= e^{i\rho}\,\sqrt{1-(\chi_2^V)^2}\,,\\
\chi_3^{H^+W^-} &\equiv \left(\chi_3^{H^-W^+}\right)^* &&= -e^{i\rho}\,\frac{\chi_2^V\chi_3^V + i\chi_1^V}{\sqrt{1-(\chi_2^V)^2}}\,.
\end{aligned} \tag{8}$$

Note, that here we discuss couplings which appear in the Lagrangian. Radiative corrections (RC) change these couplings; however in most cases these corrections are small and therefore the corresponding observables differ weakly from those presented in the Lagranian. In this sense we treat the latter ones as being measurable. Therefore the identities (6)-(8) are valid with accuracy to RC.

When describing interactions of the Higgs bosons with the gauge bosons $V = W, Z$, one should distinguish interactions of different Lorentz structure: the vectorial $h_aV_\mu V^\mu$, tensor $h_aV_{\mu\nu}V^{\mu\nu}$ and axial-tensor $h_a\epsilon_{\mu\nu\alpha\beta}V^{\alpha\beta}V^{\mu\nu}$ interactions, with the coupling constants g_a^V, g_{aT}^V and $\tilde{g}_{aT}^V$, respectively (here $V_{\mu\nu} = \partial_\mu V_\nu - \partial_\nu V_\mu$). These interactions can be separated in the experiment from each other by the study of angular correlations in the decays like $h_a \to ZZ \to e^+e^-\mu^+\mu^-$ [11]. In the 2HDM tensor and axial-tensor interactions appear only due to RC (mainly from t-loops). In most cases their contributions to the observed decays rates of h_a are hardly observable. The axial-tensor interaction can imitate CP violation even if it is absent in the model.

IOP Conf. Series: Journal of Physics: Conf. Series **938** (2017) 012049 doi:10.1088/1742-6596/938/1/012049

4. A minimal complete set of observables in 2HDM

In Ref. [10] a *minimal complete* set of *directly measurable quantities (observables)* defining the 2HDM was found ($a = 1, 2, 3$), namely:

$$\begin{array}{l} v.e.v.\ of\ Higgs\ field\ v = 246\ GeV, \quad masses\ of\ Higgs\ bosons\ M_a,\ M_\pm, \\ 2\ (out\ of\ 3)\ couplings\ \chi_a^V,\ \ 3\ couplings\ \chi_a^\pm,\ \ quartic\ coupling\ g(H^+H^-H^+H^-). \end{array} \tag{9}$$

The parameters of the potential in the Higgs basis are expressed through these observables and free parameter ρ (it appears in $\Lambda_{5,6,7}$ via couplings $\chi_a^{H^+W^-}$ given in eq. (8)):

$$\begin{array}{l} \Lambda_1 = \sum\limits_a (\chi_a^V)^2 M_a^2/v^2;\ \ \Lambda_5 = \sum\limits_a (\chi_a^{H^-W^+})^2 M_a^2/v^2;\ \ \Lambda_4 = (\sum\limits_a M_a^2 - M_\pm^2)/v^2 - \Lambda_1; \\ \Lambda_3 = 2\left(M_\pm^2/v^2\right) \sum\limits_a \chi_a^V \chi_a^\pm;\ \ \Lambda_6 = \sum\limits_a \chi_a^V \chi_a^{H^-W^+} M_a^2/v^2; \\ \Lambda_7 = 2\left(M_\pm^2/v^2\right) \sum\limits_a \chi_a^{H^-W^+} \chi_a^\pm;\ \ \Lambda_2 = 2g(H^+H^-H^+H^-)\,. \end{array} \tag{10}$$

The parameters of the potential (1) with the values of $\tan\beta$ and ξ defined in (3) are obtained from parameters (10) with the aid of transformation (2) in the following form

$$\begin{pmatrix}\phi_1\\ \phi_2\end{pmatrix} = \hat{\mathcal{F}}_{HB}\begin{pmatrix}\Phi_1\\ \Phi_2\end{pmatrix}, \quad \hat{\mathcal{F}}_{HB} = \hat{\mathcal{F}}_{gen}(\theta = -\beta, \tau = \xi, -\rho). \tag{11}$$

5. Conditions for a CP conservation

In the Higgs models like 2HDM neutral scalar particles coincide with their antiparticles. In such models one can discuss the P-parity violation, but not C-parity. When in addition we consider fermions, the P-parity violation is transformed to the CP violation (see e.g. [6]).

5.1. Basic facts

The CP symmetry is conserved in the 2HDM and similar models containing spinless bosons if

- *Each observable physical neutral spinless boson has definite P-parity* (12a)
- *There are no P-violating interactions between these bosons.* (12b)

In the 2HDM we denote neutral spinless P-even bosons as h_1, h_2 (they are called often as h and H) and P-odd boson as h_3 (it is called often as A). The condition (12b) means the absence of the interactions ($i, j, k = 1, 2$)

$$h_ih_jh_3, \quad h_3h_3h_3, \quad h_ih_jh_kh_3, \quad h_ih_3h_3h_3. \tag{13}$$

It is well known that in the 2HDM the CP conservation holds if

$$\begin{array}{c} There\ exists\ the\ RPa\ basis\ in\ which: \\ (\mathbf{a})\ all\ parameters\ of\ potential\ are\ real; \quad (\mathbf{b})\ relative\ phase\ (3)\ \xi = 0. \end{array} \tag{14}$$

It is worth mentioning that the condition (14(a)) forbids the explicit CP violation while both conditions (14(a)) and (14(b)) together forbid the spontaneous CP violation.

The statement (12) only describes CP conservation, but does not provide a criterium for CP conservation in the considered model. The description (14) is RPa basis-dependent. Below we discuss both (A) and (B) set-ups of our problem, presented in the Introduction.

5.2. Method of the CP-odd basis-independent invariants
Many authors found the basis independent criterium for CP violation or conservation in terms of parameters of the Higgs potential; this corresponds to problem (A) stated in the Introduction. For this goal they constructed the RPa basis-invariant CP-odd combinations of parameters of the potential and as a condition for CP conservation they demanded vanishing of all these invariants. For 2HDM three such invariants $Im\mathcal{J}_{1,2,3}$ were found in refs. [12, 13].

To solve the corresponding problem (B) the invariants [12, 13] for 2HDM were expressed via measurable quantities in refs. [14]. Note that in this approach there are four CP-odd invariants but one should check vanishing only of two of them (see e.g. [6]). Since the choice of these two invariants is not fixed from beginning, the presented set contains three conditions, instead of necessary two. In our opinion these equations are too complicated and their experimental verification, discussed in [12]-[15], requires very complex procedure.

5.3. A direct criterium for CP conservation
In this approach we start with a description of CP conservation (12) and use only observables.

In the CP conserving 2HDM, all h_a should have definite P-parity. In particular, one of them is P-odd, while two others are P-even (12a). Therefore, the **necessary condition** for a CP conservation is an existence of one neutral Higgs boson (we denote it h_3), which doesn't couple to the CP-even states[1] VV and H^+H^-:

There exists an neutral Higgs boson h_3 for which vectorial coupling $g_3^V \equiv g(h_3VV) = 0$ and $g_3^\pm \equiv g(h_3H^+H^-) = 0\,.$

(15)

Now, one has to check condition (12b). In order to do this, we substitute Eqs. (15), (8) into (10), choosing $\rho = 0$. The choice $\rho = 0$ together with constraint $\chi_3^V = 0$ in (8) makes both $\chi_{1,2}^{H^+W^-}$ real while $\chi_3^{H^+W^-}$ is imaginary. Next, we insert these $\chi_a^{H^+W^-}$ into (10). The parameter Λ_5 contains real negative quantity $(\chi_3^{H^+W^-})^2$ while in Λ_6 and Λ_7 this imaginary $\chi_3^{H^+W^-}$ is multiplied by χ_3^V or $\chi_3^\pm$, which are zeros (15). Hence, with this choice all parameters of potential Λ_a in the Higgs basis are real. Therefore, in view of the statement (14), the CP-symmetry of model is not violated. In particular, the CP violating interactions (13) do not appear. Therefore

the conditions (15) **are necessary and sufficient for establishing CP conservation** in the boson sector of 2HDM (problem B).

The Eqs. (10), after substitution of Eqs. (15), (11), can be treated as a solution of problem (A).

It is useful to note that in view of identity (6), the first condition (15) can be written also in the form $g(Zh_1h_2) = 0$. In some cases, checking this form of condition may be more convenient. (Such form of criterium for CP non-violation was discussed in Ref. [18].)

6. Possibilities for a verification
The verification of the CP conservation requires observation of all scalars of the model. In the SM-like scenario realized in Nature this looks difficult (see e.g. [17]). Moreover, one should check if some measurable quantities are equal to zero. In any case, these measurements cannot be performed with a high accuracy. Nevertheless, one can not hope for a high accuracy in testing CP conservation in the 2HDM.

[1] Earlier we proposed conditions for CP conservation, based only on condition (12a), without checking up of condition (12b), in the form $\prod_a \chi_a^V = 0$, $\prod_a \chi_a^\pm = 0$ [16, 17].

If in future the experiments show, within the definite experimental uncertainties, an agreement with criteria (15), the important problem arises for the further studies, namely *whether one can expect that the more accurate measurements will show violation of CP in our 2HDM, i. e. violation of criteria* (15), *or the observed CP violation is in fact an effect beyond approximation given by 2HDM, and some new weaker interactions should be implemented in the description of Nature.* (This situation is similar to that in atomic physics. Atomic interaction (QED) conserves P-parity. Parity non-conservation appears at the smallest distances due to next level weak interaction. It is observed as a small effect in rare atomic transitions.)

We are thankful to G.C. Branco, H. Haber, I.P. Ivanov, P. Osland, A. Pilaftsis, M. Rebelo, R. Santos for discussions. This work was supported in part by grants RFBR 15-02-05868 and HARMONIA project under contract UMO-2015/18/M/ST2/00518 (2016-2019).

References

[1] I.F. Ginzburg, M.Krawczyk. *Phys. Rev.* **D 96** (2017) 055030; arXiv:1612.04612 [hep-ph]
[2] The CMS Collaboration. *Phys. Rev. Lett.* **110**, 081803 (2013); arXiv:1212.6639v2 [hep-ex], *Phys. Rev.* **D 92**, 012004 (2015); arXiv:1411.3441 [hep-ex];
[3] The ATLAS Collaboration. *Eur. Phys. J.* **C75**, 231 (2015); arXiv:1503.03643; *Eur. Phys. J.* **C**; arXiv:1506.05669 [hep-ex]; The ATLAS and CMS Collaborations, arXiv:1606.02266 [hep-ex]
[4] I.F. Ginzburg, *Journ. of Phys.: Conference Series 678 (2016) 012002, 1-3*, Proc. XVI Workshop on High Energy Spin Physics (D-SPIN2015), IOP Publishing; *What tell us LHC data about Higgs boson parity.* arXiv:1610.08196 [hep-ph]
[5] T.D. Lee, *Phys. Rev.* **D8** (1973) 1226.
[6] J.F. Gunion, H.E. Haber, G. Kane, S. Dawson, *The Higgs Hunter's Guide* (Addison-Wesley, Reading, 1990); G. C. Branco, L. Lavoura and J. P. Silva, *CP Violation*, Int. Ser. Monogr. Phys. **103** (1999) 1; G. C. Branco, P. M. Ferreira, L. Lavoura, M. N. Rebelo, M. Sher and J. P. Silva, Phys. Rept. **516**, 1 (2012) [arXiv:1106.0034 [hep-ph]].
[7] I.F. Ginzburg, M. Krawczyk, P. Osland, 2nd ECFA/DESY Study 1998-2001, 997-1001; hep-ph/9909455; Nucl. Instrum. Meth. **A 472** (2001) 149-154; hep-ph/0101229; AIP Conf. Proc. Series 578 (2001) 304, IFT-2001-2, 304; hep-ph/0101331; Proc. SUSY02 Conference 703-706; hep-ph/0211371
[8] See e.g. J. Bernon, J. F. Gunion, H.E. Haber, Y. Jiang, S. Kraml, *Phys. Rev.* **D 92**, 075004 (2015); **D 93**, 035027 (2016); A. Pilaftsis. *Phys. Rev.* **D 93**, 075012 (2016)
[9] H. Georgi and D. V. Nanopoulos, Phys. Lett. B **82**, 95 (1979).
[10] I.F. Ginzburg, K.A. Kanishev, *Phys. Rev.* **D 92** (2015) 015024; arXiv:1502.06346 [hep-ph].
[11] J.F.Gunion , H.E. Haber and C. Kao, *Phys. Rev.* **D46** (1992) 2907-2917
[12] L. Lavoura, J. P. Silva, *Phys. Rev.* **D 50** (1994) 4619 [hep-ph/9404276]; F. J. Botella, J. P. Silva, *Phys. Rev.* **D 51** (1995) 3870 [hep-ph/9411288].
[13] J. F. Gunion and H. E. Haber, *Phys. Rev.* **D 72** (2005) 095002 [hep-ph/0506227]; S. Davidson, H. E. Haber, *Phys. Rev.* **D72** (2005) 035004 [hep-ph/0504050]; G. C. Branco, M. N. Rebelo, J. I. Silva-Marcos, *Phys. Lett.* **B 614** (2005) 187 [hep-ph/0502118]; H. E. Haber and D. O'Neil, *Phys. Rev.* **D 74**, 015018 (2006) [E: *Phys. Rev.* **D 74**, 059905 (2006)]; *Phys. Rev.* **D 83**, 055017 (2011); D. O'Neil, arXiv:0908.1363 [hep-ph]
[14] B. Grzadkowski, O. M. Ogreid, P. Osland, *Diagnosing CP properties of the 2HDM, JHEP* **1401** (2014) 105 [arXiv:1309.6229 [hep-ph]; arXiv:1409.7265 [hep-ph]; arXiv:1504.06076 [hep-ph]
[15] B.Grzadkowski, M. Ogreid, P. Osland, arXiv:1609.04764 [hep-ph]
[16] I. F. Ginzburg. *JETP Lett.* **99** # 12 (2014) 742-751; arXiv: 1410.0873 [hep-ph]
[17] I.F. Ginzburg, arXiv:1502.07197 [hep-ph]
[18] D. Fontes, J. C. Romao, R. Santos, J. P. Silva, arXiv:1506.06755 [hep-ph]

IOP Conf. Series: Journal of Physics: Conf. Series **938** (2017) 012050 doi:10.1088/1742-6596/938/1/012050

The analytical $\mathcal{O}(a_s^4)$ expression for the polarized Bjorken sum rule in the miniMOM scheme and the consequences for the generalized Crewther relation

A. L. Kataev[1,2], **V. S. Molokoedov**[1,2,3]

[1] Institute for Nuclear Research of the Russian Academy of Sciences (INR), 60th October Anniversary Prospect, 7a, 117312 Moscow, Russia

[2] Moscow Institute of Physics and Technology (MIPT), 141700, Dolgoprudnyy, Russia

[3] Landau Institute for Theoretical Physics of the Academy of Sciences of Russia, 142432, Moscow Region, Russia

E-mail: kataev@ms2.inr.ac.ru, viktor_molokoedov@mail.ru

Abstract. The analytical $\mathcal{O}(a_s^4)$ perturbative QCD expression for the flavour non-singlet contribution to the Bjorken polarized sum rule in the rather applicable at present gauge–dependent miniMOM scheme is obtained. For the considered three values of the gauge parameter, namely $\xi = 0$ (Landau gauge), $\xi = -1$ (anti–Feynman gauge) and $\xi = -3$ (Stefanis–Mikhailov gauge), the scheme-dependent coefficients are considerably smaller than the gauge-independent $\overline{\text{MS}}$ results. It is found that the fundamental property of the factorization of the QCD renormalization group β-function in the generalized Crewther relation, which is valid in the gauge-invariant $\overline{\text{MS}}$ scheme up to $\mathcal{O}(a_s^4)$-level at least, is unexpectedly valid at the same level in the miniMOM-scheme for $\xi = 0$, and for $\xi = -1$ and $\xi = -3$ in part.

1. Introduction

It is known that the hadronic tensor, which enters into the definitions of differential cross-sections of deep inelastic scattering (DIS) processes of charged leptons on nucleons, contains the terms, that are measured in the DIS of polarized leptons on polarized nucleons, namely

$$\begin{aligned} W_{\mu\nu}(p,q,s) &= \frac{1}{4\pi}\int d^4z\, e^{iqz}\langle p,s|[J_\mu^\dagger(z),J_\nu(0)]|p,s\rangle \qquad (1)\\ &= \left(\frac{q_\mu q_\nu}{q^2}-g_{\mu\nu}\right)F_1(x,Q^2)+\frac{1}{(pq)}\left(p_\mu-\frac{(pq)}{q^2}q_\mu\right)\left(p_\nu-\frac{(pq)}{q^2}q_\nu\right)F_2(x,Q^2)\\ &+ i\varepsilon_{\mu\nu\lambda\rho}\frac{q^\lambda}{(pq)}\left(s^\rho(g_1(x,Q^2)+g_2(x,Q^2))-\frac{(sq)}{(pq)}p^\rho g_2(x,Q^2)\right)+\dots\,, \end{aligned}$$

where $J^\mu(z) = \sum_f Q_f\bar{q}_f(z)\gamma^\mu q_f(z)$ is the quark electromagnetic current, p is the four-momentum of nucleon and s is its spin, q is the transferred momentum with $Q^2 = -q^2$, $0 \leq x = Q^2/(2pq) \leq 1$ is the Bjorken variable. In this expression the structure functions $g_1(x,Q^2)$ and $g_2(x,Q^2)$ are the ones which are extracted from differential cross-section of DIS process of polarized charged leptons on polarized nucleons, while $F_1(x,Q^2)$ and $F_2(x,Q^2)$ characterize the DIS processes with unpolarized particles.

The important characteristic of DIS process of polarized charged leptons on polarized nucleons is the Bjorken polarized sum rule, defined as

$$\int_0^1 \Big(g_1^{lp}(x,Q^2) - g_1^{ln}(x,Q^2) \Big) dx = \frac{1}{6}\left|\frac{g_A}{g_V}\right| C_{Bjp}(Q^2) \, , \tag{2}$$

where index l defines the polarized charged lepton (e or μ) in the concrete different experiments. The structure functions $g_1^{lp(n)}(x,Q^2)$ characterize the spin distribution of polarized quarks and gluons inside nucleons. The world average value of the ratio of axial and vector charges of neutron β decay is $g_A/g_V = -1.2723 \pm 0.0023$ [1]. The general theoretical expression for the Bjorken polarized sum rule contains massive-dependent corrections [2], [3] and non-perturbative high-twist $\mathcal{O}(1/Q^2)$–corrections, discussed in [4]. However, in this work we will consider the massless PT expression for the coefficient function $C_{Bjp}(Q^2)$ in the $SU(N_c)$ colour group only, which is known at the $\mathcal{O}(a_s^4)$ level in the $\overline{\rm MS}$-scheme from the results of [5], supplemented the singlet contribution, which appears first at the same level [6] with the numerically small analytical coefficient, evaluated in [7]. Therefore, in general the massless PT expression for $C_{Bjp}(Q^2)$ can be written down as:

$$C_{Bjp}(Q^2) = C_{Bjp}^{NS}(Q^2) + \sum_f Q_f C_{Bjp}^{SI}(Q^2) \, , \tag{3}$$

where

$$C_{Bjp}^{NS}(Q^2) = 1 + \sum_{n=1}^{\infty} \overline{c}_n^{NS} \overline{a}_s^{\,n}(Q^2) \, , \qquad C_{Bjp}^{SI}(Q^2) = \sum_{n=4}^{\infty} \overline{c}_n^{SI} \overline{a}_s^{\,n}(Q^2) \, , \tag{4}$$

where $\overline{a}_s = \overline{\alpha}_s/\pi$ is the strong coupling constant determined in $\overline{\rm MS}$ scheme. Remind that the first, second and third PT contributions to $C_{Bjp}^{NS}(Q^2)$ in the $\overline{\rm MS}$-scheme were analytically calculated in the works [8], [9] and [10] respectively. Note also that for $n_f = 3$ the singlet contribution is identically equal to zero.

It is known that the scheme–dependent PT series for the renormalization–group (RG) invariant quantities are asymptotic ones. The transformation to the special schemes, which minimize the contributions of high-order PT corrections to physical quantities, is one of theoretical approaches for applying them in phenomenology. In [11] it was proposed to achieve this goal by using the special momentum (MOM) subtractions scheme, which in QCD is gauge–dependent. It should be noted, that in the MOM–like schemes the coefficients of the QCD RG β-function are becoming gauge–dependent starting from the two-loop level [12]. The gauge-dependence of the coefficients of both RG-invariant quantities and QCD β-function in various MOM schemes makes the analysis of the PT expansions for the quantities, interesting from experimental point of view, more complicated. However, the interest to the studies of the PT approximations for various QCD characteristics of observable physical processes in the special gauge-dependent MOM schemes became actual again. It is related mainly to the formulation of minimal MOM (mMOM) scheme [13]. Its analog was used in studies of the Functional Renormalization Group Equations [14].

In Section II we remind the main ideas of the definition of mMOM scheme and present the explicit analytical gauge–dependent relation between the QCD coupling constants in the $\overline{\rm MS}$ and MOM schemes with gauge covariant linear parameter ξ, determined in mMOM scheme. The new analytical results for the PT mMOM coefficients of the $\mathcal{O}(a_s^4)$ approximation of the Bjorken polarized sum rule are presented. Their numerical values, specified to the Landau gauge $\xi = 0$, to the anti–Feynman gauge $\xi = -1$ and to the Stefanis–Mikhailov gauge $\xi = -3$, first used in QCD by Stefanis in the work of [15] (see also [16]) to untangle the special features of

IOP Conf. Series: Journal of Physics: Conf. Series **938** (2017) 012050 doi:10.1088/1742-6596/938/1/012050

renormalizations of gauge–invariant definition of QCD quark correlator, formulated with the help of the Wilson lines, and independently applied later on by Mikhailov in [17] (see also [18]) to clarify the renormalization of the gluon fields by the renormalon chain contributions, are compared with the $\overline{\rm MS}$ scheme results.

In Section III we will clarify that the application of the anti–Feynman and Stefanis–Mikhailov gauges are essential for satisfying the fundamental property of the factorization of the gauge–dependent QCD β-function in the $\mathcal{O}(a_s^3)$ approximation in mMOM scheme variant of the QCD generalized Crewther relation (GCR), discovered in [19] in the gauge–independent $\overline{\rm MS}$ scheme, which in the conformal symmetry limit reproduce well-known massless quark-parton result, obtained in [20]. We will also stress that the transformation to the mMOM scheme when the Landau gauge is chosen will not spoil the property of the factorization of the analytical expression for the QCD β-function in the $SU(N_c)$ GCR at the fourth order of PT, studied in the $\overline{\rm MS}$ scheme in [5].

2. The Bjorken polarized sum rule in mMOM scheme

The mMOM scheme was introduced in [13] and was initially formulated as the easiest way to satisfy the common MOM schemes property of the non-renormalization of the gluon–ghost–anti-ghost vertex in the Landau gauge $\xi = 0$ thanks to the equality between the renormalization constant of this vertex in mMOM and $\overline{\rm MS}$ schemes [13]:

$$Z_{cg}^{\rm mMOM} = Z_{cg}^{\overline{\rm MS}} \,. \tag{5}$$

Taking into account this relation (5) and the MOM–like renormalization conditions for gluon and ghost propagators, one can obtain the following relations between strong coupling constant $a_s^{\rm mMOM} \equiv a_s$ in mMOM scheme and $a_s^{\overline{\rm MS}} \equiv \overline{a}_s$ and between gauge parameters $\xi^{\rm mMOM} \equiv \xi$ and $\xi^{\overline{\rm MS}} \equiv \overline{\xi}$ [13], [21], [22]:

$$a_s(\mu^2) = \frac{\overline{a}_s(\mu^2)}{(1+\Delta(\mu^2))(1+\Omega(\mu^2))^2} \,, \qquad \xi(\mu^2) = \overline{\xi}(\mu^2)(1+\Delta(\mu^2)) \,, \tag{6}$$

where gluon and ghost self-energies functions $\Delta(q^2)$ and $\Omega(q^2)$ are evaluated in the $\overline{\rm MS}$ scheme and defined by the higher order corrections to the corresponding propagators. In the arbitrary $SU(N_c)$ colour group their explicit form was analytically calculated in [23] at $\mathcal{O}(a_s^3)$ level and in [22] in the fourth order PT approximation. Note that these $\overline{\rm MS}$ results for self-energies $\Delta(q^2)$ and $\Omega(q^2)$ depend on the gauge parameter starting from one- and two–loop level correspondingly. The expressions (6) allow us to get the following explicit analytical three-loop approximation of the function $\overline{a}_s(a_s, \xi)$:

$$\overline{a}_s = a_s\Big(1 + \delta_1 a_s + \delta_2 a_s^2 + \delta_3 a_s^3 + \mathcal{O}(a_s^4)\Big) \,, \tag{7}$$

$$\delta_1 = \left(-\frac{169}{144} - \frac{1}{8}\xi - \frac{1}{16}\xi^2\right)C_A + \frac{5}{9}T_F n_f \,, \tag{8}$$

$$\begin{aligned}\delta_2 = {} & \left[-\frac{18941}{20736} + \frac{39}{128}\zeta_3 + \left(\frac{889}{2304} - \frac{11}{64}\zeta_3\right)\xi + \left(\frac{203}{2304} + \frac{3}{128}\zeta_3\right)\xi^2 - \frac{3}{256}\xi^3\right]C_A^2 \\ & + \left[-\frac{107}{648} + \frac{\zeta_3}{2} - \frac{5}{36}\xi - \frac{5}{72}\xi^2\right]C_A T_F n_f + \left[\frac{55}{48} - \zeta_3\right]C_F T_F n_f + \frac{25}{81}T_F^2 n_f^2 \,,\end{aligned} \tag{9}$$

$$
\begin{aligned}
\delta_3 &= \Big[-\frac{1935757}{2985984}+\frac{7495}{18432}\zeta_3+\frac{7805}{12288}\zeta_5+\Big(\frac{4877}{36864}-\frac{611}{1536}\zeta_3+\frac{295}{1024}\zeta_5\Big)\xi \\
&+ \Big(\frac{17315}{110592}-\frac{47}{768}\zeta_3+\frac{175}{6144}\zeta_5\Big)\xi^2+\Big(-\frac{233}{4608}+\frac{59}{1536}\zeta_3+\frac{5}{3072}\zeta_5\Big)\xi^3 \\
&+ \Big(-\frac{235}{36864}-\frac{5}{6144}\zeta_3-\frac{35}{12288}\zeta_5\Big)\xi^4\Big]C_A^3+\Big[-\frac{143}{288}-\frac{37}{24}\zeta_3+\frac{5}{2}\zeta_5\Big]C_F^2T_Fn_f \\
&+ \Big[\frac{25547}{20736}+\frac{107}{144}\zeta_3-\frac{5}{4}\zeta_5+\Big(-\frac{55}{128}+\frac{3}{8}\zeta_3\Big)\xi+\Big(-\frac{55}{256}+\frac{3}{16}\zeta_3\Big)\xi^2\Big]C_FC_AT_Fn_f \\
&+ \Big[\frac{199}{31104}-\frac{223}{384}\zeta_3-\frac{5}{6}\zeta_5+\Big(\frac{833}{13824}+\frac{11}{144}\zeta_3\Big)\xi+\Big(-\frac{143}{6912}-\frac{9}{128}\zeta_3\Big)\xi^2-\frac{5}{1152}\xi^3 \\
&+ \frac{5}{2304}\xi^4\Big]C_A^2T_Fn_f+\Big[\frac{1235}{15552}+\frac{13}{36}\zeta_3+\Big(-\frac{19}{216}-\frac{1}{9}\zeta_3\Big)\xi-\frac{25}{432}\xi^2\Big]C_AT_F^2n_f^2 \\
&+ \Big[\frac{1249}{2592}-\frac{11}{18}\zeta_3\Big]C_FT_F^2n_f^2+\frac{125}{729}T_F^3n_f^3\,,
\end{aligned} \tag{10}
$$

where C_F and C_A are the Casimir operators, T_F is the Dynkin index, n_f is the number of active flavours.

In order to obtain values of the coefficients c_n^{NS} of the flavour non-singlet Bjorken function in the mMOM scheme we use the known four–loop $\overline{\text{MS}}$ results for C_{Bjp}^{NS} from [5] (the corresponding coefficients are denoted as $\overline{c}_n^{NS}$), the relation (7) and the renorm–invariant property of the physical quantity C_{Bjp}^{NS}. The relations between the coefficients in mMOM and $\overline{\text{MS}}$ schemes read:

$$
c_1^{NS} = \overline{c}_1^{NS}\,,\quad c_2^{NS}=\overline{c}_2^{NS}+\delta_1\overline{c}_1^{NS}\,,\quad c_3^{NS}=\overline{c}_3^{NS}+2\delta_1\overline{c}_2^{NS}+\delta_2\overline{c}_1^{NS}\,, \tag{11}
$$

$$
c_4^{NS} = \overline{c}_4^{NS}+3\delta_1\overline{c}_3^{NS}+(2\delta_2+\delta_1^2)\overline{c}_2^{NS}+\delta_3\overline{c}_1^{NS}\,. \tag{12}
$$

Using these relations we find the explicit form of coefficients of the Bjorken function in the mMOM scheme at the $\mathcal{O}(a_s^4)$ level:

$$
c_1^{NS} = -\frac{3}{4}C_F\,, \tag{13}
$$

$$
c_2^{NS} = \frac{21}{32}C_F^2+\Big(-\frac{107}{192}+\frac{3}{32}\xi+\frac{3}{64}\xi^2\Big)C_FC_A+\frac{1}{12}C_FT_Fn_f\,, \tag{14}
$$

$$
\begin{aligned}
c_3^{NS} &= -\frac{3}{128}C_F^3+\Big[\frac{1415}{2304}-\frac{11}{12}\zeta_3-\frac{21}{128}\xi-\frac{21}{256}\xi^2\Big]C_F^2C_A+\Big[-\frac{13}{36}+\frac{\zeta_3}{3}\Big]C_F^2T_Fn_f \\
&+ \Big[\frac{13}{9}+\frac{3}{8}\zeta_3-\frac{5}{6}\zeta_5-\frac{1}{48}\xi-\frac{1}{96}\xi^2\Big]C_FC_AT_Fn_f-\frac{5}{24}C_FT_F^2n_f^2+\Big[-\frac{20585}{9216} \\
&- \frac{117}{512}\zeta_3+\frac{55}{24}\zeta_5+\Big(\frac{215}{3072}+\frac{33}{256}\zeta_3\Big)\xi+\Big(\frac{349}{3072}-\frac{9}{512}\zeta_3\Big)\xi^2+\frac{9}{1024}\xi^3\Big]C_FC_A^2\,,
\end{aligned} \tag{15}
$$

$$
\begin{aligned}
c_4^{NS} &= \frac{d_F^{abcd}d_A^{abcd}}{N_c}\bigg[-\frac{3}{16}+\frac{\zeta_3}{4}+\frac{5}{4}\zeta_5\bigg]+n_f\frac{d_F^{abcd}d_F^{abcd}}{N_c}\bigg[\frac{13}{16}+\zeta_3-\frac{5}{2}\zeta_5\bigg] \qquad (16)\\
&+ \bigg[-\frac{4823}{2048}-\frac{3}{8}\zeta_3\bigg]C_F^4+\bigg[-\frac{13307}{18432}-\frac{971}{96}\zeta_3+\frac{1045}{48}\zeta_5+\frac{9}{1024}\xi+\frac{9}{2048}\xi^2\bigg]C_F^3C_A\\
&+ \bigg[\frac{2543485}{221184}+\frac{90169}{6144}\zeta_3-\frac{1375}{144}\zeta_5-\frac{385}{16}\zeta_7+\bigg(-\frac{1339}{12288}+\frac{121}{1024}\zeta_3\bigg)\xi\\
&+ \bigg(-\frac{1117}{6144}+\frac{415}{2048}\zeta_3\bigg)\xi^2-\frac{21}{4096}\xi^3+\frac{21}{8192}\xi^4\bigg]C_F^2C_A^2+\bigg[-\frac{3927799}{442368}+\frac{49763}{73728}\zeta_3\\
&+ \frac{345755}{147456}\zeta_5+\frac{385}{64}\zeta_7-\frac{121}{96}\zeta_3^2+\bigg(\frac{107569}{147456}+\frac{1623}{2048}\zeta_3-\frac{4405}{4096}\zeta_5\bigg)\xi\\
&+ \bigg(\frac{28303}{49152}-\frac{11}{512}\zeta_3-\frac{3695}{8192}\zeta_5\bigg)\xi^2+\bigg(\frac{151}{3072}-\frac{59}{2048}\zeta_3-\frac{5}{4096}\zeta_5\bigg)\xi^3\\
&+ \bigg(-\frac{41}{49152}+\frac{5}{8192}\zeta_3+\frac{35}{16384}\zeta_5\bigg)\xi^4\bigg]C_FC_A^3+\bigg[\frac{317}{144}+\frac{109}{24}\zeta_3-\frac{95}{12}\zeta_5\bigg]C_F^3T_Fn_f\\
&+ \bigg[-\frac{6229}{864}-\frac{1739}{288}\zeta_3+\frac{205}{72}\zeta_5+\frac{35}{4}\zeta_7+\bigg(\frac{13}{96}-\frac{\zeta_3}{8}\bigg)\xi+\bigg(\frac{13}{192}-\frac{\zeta_3}{16}\bigg)\xi^2\bigg]C_F^2C_AT_Fn_f\\
&+ \bigg[\frac{12265}{1728}-\frac{1237}{512}\zeta_3+\frac{15}{16}\zeta_5-\frac{35}{16}\zeta_7+\frac{11}{12}\zeta_3^2+\bigg(-\frac{8257}{18432}-\frac{49}{96}\zeta_3+\frac{5}{16}\zeta_5\bigg)\xi\\
&+ \bigg(-\frac{869}{3072}-\frac{33}{512}\zeta_3+\frac{5}{32}\zeta_5\bigg)\xi^2-\frac{1}{1536}\xi^3+\frac{1}{3072}\xi^4\bigg]C_FC_A^2T_Fn_f\\
&+ \bigg[-\frac{1283}{864}+\frac{85}{72}\zeta_3-\frac{35}{36}\zeta_5-\frac{\zeta_3^2}{6}+\bigg(\frac{11}{192}+\frac{\zeta_3}{12}\bigg)\xi+\frac{5}{128}\xi^2\bigg]C_FC_AT_F^2n_f^2\\
&+ \bigg[\frac{1891}{3456}-\frac{\zeta_3}{36}\bigg]C_F^2T_F^2n_f^2+\frac{5}{72}C_FT_F^3n_f^3\,.
\end{aligned}
$$

Here in expression (16) $d_F^{abcd} = \mathrm{Tr}(t^at^{(b}t^ct^{d)})/6$ and $d_A^{abcd} = \mathrm{Tr}(C^aC^{(b}C^cC^{d)})/6$ are the total symmetric tensors with generators t^a of the fundamental representation and the adjoint representation C^a of the Lie algebra of the $SU(N_c)$ group.

To study the energy behavior of the obtained by us four-loop PT expression for the Bjorken polarized sum rule in the mMOM scheme it is necessary to take into account the gauge-dependent expression of the RG β-function in this scheme at the four-loop level, analytically evaluated in [21], and to define the corresponding QCD scale parameter Λ_{mMOM}. Using the analytical expressions for the next-to-leading order coefficients of C_{Bjp}^{NS} in the $\overline{\text{MS}}$ and mMOM schemes and the concept of the corresponding effective scale parameter [24], [25], we obtain the following gauge- and flavour-dependent relation:

$$
\Lambda_{\text{mMOM}}(\xi, n_f) = \Lambda_{\overline{\text{MS}}}\cdot\exp\bigg(\frac{(169+18\xi+9\xi^2)C_A-80T_Fn_f}{264C_A-96T_Fn_f}\bigg) \qquad (17)
$$

It will be shown in the next Section that the values of the gauge parameters $\xi = -3, -1$ are highlighted by the presence of factorization of the RG β function in mMOM scheme variant of the $\mathcal{O}(a_s^3)$ approximation of the fundamental Generalized Crewther relation for the product of non-singlet Adler and Bjorken functions, while in the Landau gauge $\xi = 0$ this property is true at $\mathcal{O}(a_s^4)$ order. In view of this it is interesting to consider the asymptotic behavior of the PT series for the obtained by us $\mathcal{O}(a_s^4)$ approximations of the flavour non-singlet Bjorken polarized sum rule in mMOM in these three theoretically prominent gauges with the well-known $\overline{\text{MS}}$ scheme results. This will be done in Table 1.

n_f	ξ	**The flavour NS Bjorken function $\mathbf{C^{NS}_{Bjp}}$ in $\overline{\rm MS}$ and mMOM schemes**
3	—	$1-\overline{a}_s-3.583\,\overline{a}_s^2-20.2153\,\overline{a}_s^3-175.74950\,\overline{a}_s^4$
	0	$1-a_s-0.896\,a_s^2+1.4262\,a_s^3-22.96225\,a_s^4$
	-1	$1-a_s-1.083\,a_s^2+0.2312\,a_s^3-31.54404\,a_s^4$
	-3	$1-a_s-0.333\,a_s^2-1.0317\,a_s^3-44.09174\,a_s^4$
4	—	$1-\overline{a}_s-3.250\,\overline{a}_s^2-13.8503\,\overline{a}_s^3-102.40202\,\overline{a}_s^4$
	0	$1-a_s-0.840\,a_s^2+3.0375\,a_s^3-12.34185\,a_s^4$
	-1	$1-a_s-1.028\,a_s^2+1.8633\,a_s^3-18.49192\,a_s^4$
	-3	$1-a_s-0.278\,a_s^2+0.5170\,a_s^3-31.54819\,a_s^4$
5	—	$1-\overline{a}_s-2.917\,\overline{a}_s^2-7.8402\,\overline{a}_s^3-41.95977\,\overline{a}_s^4$
	0	$1-a_s-0.785\,a_s^2+4.5099\,a_s^3-3.61660\,a_s^4$
	-1	$1-a_s-0.972\,a_s^2+3.3566\,a_s^3-7.57175\,a_s^4$
	-3	$1-a_s-0.222\,a_s^2+1.9269\,a_s^3-21.14145\,a_s^4$

Table 1. The behaviour of the PT series for C^{NS}_{Bjp} at the $\mathcal{O}(a_s^4)$ level for $SU(3)$ QCD with $n_f = 3, 4, 5$ active flavours in the $\overline{\rm MS}$ and mMOM schemes for three values of the gauge parameter $\xi = 0, -1, -3$.

These results demonstrate that for all three used gauges the numerical values of the gauge-dependent coefficients, related to the mMOM scheme, are considerably smaller than the ones, obtained in $\overline{\rm MS}$ scheme. Therefore, the convergence of the PT series is much better in mMOM scheme. Note also that unlike $\overline{\rm MS}$ $\mathcal{O}(\overline{a}_s^4)$ expressions for Bjorken function, the series calculated in mMOM scheme as a rule do not obey sign constant behavior character. It is also interesting to note, that on the contrary to the absolute values of the $\mathcal{O}(\overline{a}_s^3)$ coefficients the $\mathcal{O}(\overline{a}_s^4)$ are decreasing both in the $\overline{\rm MS}$ and mMOM-schemes with increasing number of quarks flavors from n_f=3 to $n_f = 4, 5$. The similiar feature is manifesting itself in the process of applications of mMOM-scheme to the analysis of $\mathcal{O}(a_s^4)$-approximations to other physical quantities (see e.g [26], [27], [28]). These facts became noticeable starting from the three–loop level.

It may be of interest to compare these numerical results with the ones obtained in the process of recent analysis of the PT approximations for the Bjorken polarized sum rule [29] where the mMOM was also used.

3. The generalized Crewther relation in mMOM scheme

One of the most well-known manifestations of the consequences of the conformal symmetry in massless quark-parton model is the existence of the Crewther relation [20]:

$$D^{NS}C^{NS}_{Bjp}\Big|_{c-i\ \rm limit} = \mathbb{1}\ . \tag{18}$$

IOP Conf. Series: Journal of Physics: Conf. Series **938** (2017) 012050 doi:10.1088/1742-6596/938/1/012050

The unity in the l.h.s of this equation corresponds to the normalized massless quark–parton result, obtained from the application of the OPE approach to the one-loop axial-vector-vector (AVV) triangle diagram, which defines $\pi^0 \to \gamma\gamma$ decay amplitude. In equation (18) the physical quantity D^{NS} denotes the Born approximation of the Adler function that characterizes the process of one-photon electron–positron annihilation into hadrons and C^{NS}_{Bjp} is the normalized massless Born approximation of the theoretical expression for the Bjorken polarized sum rule, defined in (2) and (3). However, in realistic QCD the conformal symmetry is broken. This leads to the existence of the discovered in [19] generalized Crewther relation, which in $\overline{\rm MS}$ scheme can be written down as:

$$D^{NS}(\overline{a}_s)C^{NS}_{Bjp}(\overline{a}_s) = \mathbb{1} + \Delta_{csb}(\overline{a}_s) \; . \tag{19}$$

The term $\Delta_{csb}(\overline{a}_s)$ in (19) is appearing starting from the second order of PT. In the related to dimensional regularization *gauge–invariant* $\overline{\rm MS}$ scheme, which is commonly used in the multiloop QCD calculations, its expression was first written down in [19] at the $\mathcal{O}(\overline{a}_s^3)$ level in the following factorized form

$$\Delta_{csb} = \left(\frac{\beta(\overline{a}_s)}{\overline{a}_s}\right)\sum_{i\geqslant 1} K_i \overline{a}_s^{\,i} \tag{20}$$

and confirmed at the $\mathcal{O}(\overline{a}_s^4)$ order in [5]. Here coefficient functions K_i depend on monomials of $SU(N_c)$ group structures C_F, C_A and $T_F n_f$. The RG $\beta(\overline{a}_s)$-function is defined as

$$\beta(\overline{a}_s) = \mu^2 \frac{d\overline{a}_s}{d\mu^2} = -\sum_{i\geqslant 0} \beta_i \overline{a}_s^{\,i+2} \; , \tag{21}$$

In this work we will need to know its three-loop approximation only, which was computed in the $\overline{\rm MS}$ scheme in [30] and confirmed later on in [31]. The corresponding expression of the β function in the mMOM scheme can be found in analytical form in [21], [22] and depends on ξ beginning from two–loop level.

It is important now to understand whether the fundamental property of factorization of the conformal anomaly term $(\beta(\overline{a}_s)/\overline{a}_s)$ in (20) is fulfilled in the *gauge–invariant* $\overline{\rm MS}$–like schemes only. To study this problem we will extend the consideration of the representation (20) to the *gauge–dependent* mMOM scheme.

Regardless of the used renormalization schemes the physical quantities D^{NS} and C^{NS}_{Bjp} in (19) obey the property of the renormalization invariance. Using this property and omitting the details of considerations we obtain the following transition relations for the K_i terms in (20) from $\overline{\rm MS}$ scheme to any other renormalization scheme AS:

$$K_1^{AS} = K_1^{\overline{\rm MS}} \, , \qquad K_2^{AS} = K_2^{\overline{\rm MS}} + \left(\frac{\beta_1^{\overline{\rm MS}} - \beta_1^{AS}}{\beta_0} + 2\delta_1^{AS}\right)K_1^{\overline{\rm MS}} \, , \tag{22}$$

$$\begin{aligned} K_3^{AS} &= K_3^{\overline{\rm MS}} + \left(\frac{\beta_1^{\overline{\rm MS}} - \beta_1^{AS}}{\beta_0} + 3\delta_1^{AS}\right)K_2^{\overline{\rm MS}} + \\ &+ \left(\frac{\beta_2^{\overline{\rm MS}} - \beta_2^{AS}}{\beta_0} + \frac{3\beta_1^{\overline{\rm MS}} - 2\beta_1^{AS}}{\beta_0}\delta_1^{AS} - \frac{\beta_1^{\overline{\rm MS}} - \beta_1^{AS}}{\beta_0}\cdot\frac{\beta_1^{AS}}{\beta_0} + 2\delta_2^{AS} + (\delta_1^{AS})^2\right)K_1^{\overline{\rm MS}} \end{aligned} \tag{23}$$

We emphasize that if all the fractions included in these expressions are explicitly divided by QCD β_0–factors, then the property of factorization of the conformal symmetry breaking factor $(\beta(a_s^{AS})/a_s^{AS})$ in AS scheme will be valid at the $\mathcal{O}(a_S^{AS\,3})$ at least. The terms δ_i^{AS} in (22), (23) are the AS–analogies of the δ_i terms in (7). If we fix mMOM scheme instead of AS scheme, we obtain that at $\mathcal{O}(a_s^2)$ level the factorization of $\beta(a_s)/a_s$ function for K_2 term is possible for three distinguished values of the gauge parameter ξ only, namely for $\xi = 0$ (Landau gauge), for $\xi = -1$ (anti–Feynman gauge) and for $\xi = -3$ [15], [17] (Stefanis–Mikhailov gauge). At the $\mathcal{O}(a_s^3)$ approximation the $\beta(a_s)/a_s$ factorization property is valid for the Landau gauge only and the partial factorization holds for anti–Feynman and Stefanis–Mikhailov gauges (when the factorization condition is imposed, then for one of the six possible color monomials the concrete coefficient is not determined). Thus we conclude that total factorization of the conformal anomaly $\beta(a_s)/a_s$ in the GCR is also possible for *gauge–dependent* renormalization schemes. Therefore we rule out the gauge invariance as a cause of the factorization property. Theoretical reasons of these our findings are yet unclear to us. The detailed description of the outlined in this talk our studies are in progress.

4. Acknowledgments

The authours are grateful to S.V. Mikhailov for constructive remarks. One of us (A.K) would like to thank organizers of XVIIth International Workshop on High Energy Spin Physics (DSPIN-17) and A.V. Efremov personally for hospitality in Dubna. His work on studies of the theoretical features of the generalized Crewther relation in QCD is supported by the Russian Science Foundation Grant No. 14-22-00161. The work of V.M related to the calculation and analysis of the non-singlet Bjorken function in mMOM scheme is supported by the Russian Science Foundation grant No. 16-12-10151.

5. References

[1] C. Patrignani *et al.* [Particle Data Group], Chin. Phys. C **40** (2016) no.10, 100001.
[2] O. V. Teryaev and O. L. Veretin, hep-ph/9602362.
[3] J. Blumlein and W. L. van Neerven, Phys. Lett. B **450** (1999) 417
[4] A. L. Kataev, Mod. Phys. Lett. A **20** (2005) 2007
[5] P. A. Baikov, K. G. Chetyrkin and J. H. Kuhn, Phys. Rev. Lett. **104** (2010) 132004
[6] S. A. Larin, Phys. Lett. B **723** (2013) 348
[7] P. A. Baikov, K. G. Chetyrkin and J. H. Kuhn, Nucl. Part. Phys. Proc. **261-262** (2015) 3
[8] J. Kodaira, S. Matsuda, T. Muta, K. Sasaki and T. Uematsu, Phys. Rev. D **20** (1979) 627.
[9] S. G. Gorishny and S. A. Larin, Phys. Lett. B **172** (1986) 109.
[10] S. A. Larin and J. A. M. Vermaseren, Phys. Lett. B **259** (1991) 345.
[11] W. Celmaster and R. J. Gonsalves, Phys. Rev. Lett. **42** (1979) 1435.
[12] O. V. Tarasov and D. V. Shirkov, Sov. J. Nucl. Phys. **51** (1990) 877 [Yad. Fiz. **51** (1990) 1380]
[13] L. von Smekal, K. Maltman and A. Sternbeck, Phys. Lett. B **681** (2009) 336
[14] J. M. Pawlowski, D. F. Litim, S. Nedelko and L. von Smekal, Phys. Rev. Lett. **93** (2004) 152002
[15] N. G. Stefanis, Nuovo Cim. A **83** (1984) 205.
[16] N. G. Stefanis, Acta Phys. Polon. Supp. **6** (2013) 71
[17] S. V. Mikhailov, Phys. Lett. B **431** (1998) 387
[18] S. V. Mikhailov, Phys. Rev. D **62** (2000) 034002
[19] D. J. Broadhurst and A. L. Kataev, Phys. Lett. B **315** (1993) 179
[20] R. J. Crewther, Phys. Rev. Lett. **28** (1972) 1421.
[21] J. A. Gracey, J. Phys. A **46** (2013) 225403
[22] B. Ruijl, T. Ueda, J. A. M. Vermaseren and A. Vogt, JHEP **1706** (2017) 040
[23] K. G. Chetyrkin and A. Retey, hep-ph/0007088.
[24] N. V. Krasnikov, Nucl. Phys. B **192** (1981) 497.
[25] G. Grunberg, Phys. Lett. **95B** (1980) 70 Erratum: [Phys. Lett. **110B** (1982) 501].
[26] J. A. Gracey, Phys. Rev. D **90** (2014) no.9, 094026.
[27] A. L. Kataev and V. S. Molokoedov, Phys. Rev. D **92** (2015) no.5, 054008.
[28] F. Herzog, B. Ruijl, T. Ueda, J. A. M. Vermaseren and A. Vogt, JHEP **1708** (2017) 113.
[29] C. Ayala, G. Cvetic, A. V. Kotikov and B. G. Shaikhatdenov, arXiv:1708.06284 [hep-ph].
[30] O. V. Tarasov, A. A. Vladimirov and A. Y. Zharkov, Phys. Lett. **93B** (1980) 429.
[31] S. A. Larin and J. A. M. Vermaseren, Phys. Lett. B **303** (1993) 334

IOP Conf. Series: Journal of Physics: Conf. Series **938** (2017) 012051 doi:10.1088/1742-6596/938/1/012051

A Monte Carlo code for the fragmentation of polarized quarks

A Kerbizi[1], X Artru[2], Z Belghobsi[3], F Bradamante[1] and A Martin[1]

[1] INFN Sezione di Trieste and Dipartimento di Fisica, Università di Trieste, Via Valerio 2, 34127 Trieste, Italy

[2] Univ. Lyon, Université Lyon 1, CNRS/IN2P3, Institut de Physique Nucléaire de Lyon, 69622 Villeurbanne, France

[3] Laboratoire de Physique Théorique, Faculté des Sciences Exactes et de l'Informatique, Université Mohammed Seddik Ben Yahia, B.P. 98 Ouled Aissa, 18000 Jijel, Algeria

E-mail: albi.kerbizi@ts.infn.it

Abstract. We describe a Monte Carlo code for the fragmentation of polarized quarks into pseudoscalar mesons. The quark jet is generated by iteration of the splitting $q \to h + q'$ where q and q' indicate quarks and h a hadron. The splitting function describing the energy sharing between q' and h is calculated on the basis of the Symmetric Lund Model where the quark spin is introduced through spin matrices as foreseen in the $^3P_0$ mechanism. A complex mass parameter is introduced for the parametrisation of the Collins effect. The results for the Collins analysing power and the comparison with the Collins asymmetries measured by the COMPASS collaboration are presented. For the first time preliminary results on the simulated azimuthal asymmetry due to the Boer-Mulders function are also given.

1. Introduction

Sarting from the recent work of Ref. [1, 2] we have developed a model for the fragmentation of polarized quarks into pseudoscalar mesons and implemented it in a stand alone Monte Carlo program. The polarized splitting function is calculated on the basis of the relativistic string decay as described in the Symmetric Lund Model (SLM) [3], with confinement as a built in feature and the Left-Right (LR) symmetry, namely symmetry under the exchange of the quarks placed at the string endpoints. The spin information is propagated along the fragmentation chain according to a full quantum version of the semi-classical $^3P_0$ mechanism [4].

Our model for the hadronisation of a polarized quark jet and the Monte Carlo program are described in Section 2 and in Section 3 respectively. In Section 4 we show some results on the Collins effect for transversely polarized u quark jets and compare them with the Collins asymmetries measured by the COMPASS Collaboration in SIDIS off a transversely polarized proton target. In Section 5 we describe how the primordial transverse momentum of the fragmenting quark has been introduced and its effect on the Collins asymmetry and in Section 6 we present for the first time results of the correlation between transverse spin and transverse momentum described by the Boer-Mulders function. Finally in Section 7 we draw our conclusions.

2. polarized splitting with the $^3P_0$ mechanism

We simulate with a stand alone Monte Carlo program the jet produced by an initial polarized quark q_1

$$q_1 \to h_1 + q_2,\, q_2 \to h_2 + q_3, \ldots, q_i \to h_i + q_{i+1} \cdots \tag{1}$$

repeating recursively the splitting

$$q \to h(q\bar{q}') + q' \tag{2}$$

where q, q' indicate quarks and h represents the $q\bar{q}'$ meson state. The index i labels the rank of h. k, k' and p are the $4-$momenta of q, q' and h respectively. They are related by the momentum conservation $k = p + k'$. The splitting in eq. (2) is described by the splitting function $F_{q',h,q}(\mathbf{k}_T, \mathbf{k}'_T, Z)$ which gives the probability of producing a hadron h with transverse momentum $\mathbf{p}_T = \mathbf{k}_T - \mathbf{k}'_T$ and longitudinal momentum fraction $Z = p^+/k^+$ (the lightcone coordinates are defined as $p^{\pm} = p^0 \pm p^3$). $\mathbf{k}_T$ and $\mathbf{k}'_T$ are the transverse momenta of q and q' respectively. The reference axis is the jet axis which we take along the momentum of the fragmenting quark q_1. The quark spin is encoded in the 2×2 spin density matrix $\rho(q) = (\mathbf{1} + \sigma\cdot\mathbf{S}_q)/2$ where $\mathbf{S}_q$ is the polarization vector of q.

The input of the model is the transition operator $T_{q',h,q}(k, k')$ which gives the transition amplitude in momentum $\otimes$ spin space for the generic splitting:

$$T_{q',h,q}(k,k') = C_{q',h,q} \times e^{-\frac{b_T}{2}\mathbf{k'}^2_T} \times (1-Z)^{a/2} e^{-\frac{b_L}{2}\frac{\varepsilon_h^2}{Z}} \times [\mu + \sigma_z \sigma\cdot\mathbf{k}'_T] \times \sigma_z \times \hat{u}^{-1/2}(\mathbf{k}_T). \tag{3}$$

It leads to a LR-symmetric model. The factor $C_{q',h,q}$ describes the splitting (2) in flavour space and is determined by the isospin wave function of h. The first exponential is a cutoff in the transverse momentum $\mathbf{k}'_T$ of the quark q' and the factor containing Z comes from the symmetric string fragmentation ($\varepsilon_h^2 = m_h^2 + \mathbf{p}_T^2$ is the hadrons transverse energy squared). The spin term contains the complex mass parameter μ, responsible for the Collins effect and comes from the $^3P_0$ mechanism. σ_z is the pseudoscalar coupling with h. In the last term $\hat{u}(\mathbf{k}_T)$ is a positive definite single quark density matrix in $\mathbf{k}_T\otimes$ spin space given by

$$\hat{u}(\mathbf{k}_T) = \sum_h |C_{q',h,q}|^2 \int d^2\mathbf{k}'_T e^{-b_T {\mathbf{k}'_T}^2} \sigma_z\, [\mu^* + \sigma\cdot\mathbf{k}'_T\sigma_z]\, [\mu + \sigma_z\sigma\cdot\mathbf{k}'_T]\, \sigma_z \frac{dZ}{Z}(1-Z)^a e^{-\frac{b_L \varepsilon_h^2}{Z}}. \tag{4}$$

b_T, b_L, a and μ are the free parameters of the model obtained by fitting the experimental data.

The splitting function $F_{q',h,q}(\mathbf{k}_T, \mathbf{k}'_T, Z)$ is then calculated from eq. (3), up to a normalisation factor, as

$$F_{q',h,q}(\mathbf{k}_T, \mathbf{k}'_T, Z) \propto tr\left(T_{q',h,q}(\mathbf{k}_T, \mathbf{k}'_T, Z)\,\rho(q)\,T^{\dagger}_{q',h,q}(\mathbf{k}_T, \mathbf{k}'_T, Z)\right) \tag{5}$$

where $\rho(q)$ is the spin density matrix of q.

Finally, the spin density matrix of the quark q' is given by

$$\rho(q') = \frac{T_{q',h,q}(\mathbf{k}_T, \mathbf{k}'_T, Z)\,\rho(q)\,T^{\dagger}_{q',h,q}(\mathbf{k}_T, \mathbf{k}'_T, Z)}{tr\left(T_{q',h,q}(\mathbf{k}_T, \mathbf{k}'_T, Z)\,\rho(q)\,T^{\dagger}_{q',h,q}(\mathbf{k}_T, \mathbf{k}'_T, Z)\right)}. \tag{6}$$

Equation (5) and eq. (6) are used to simulate the quark fragmentation.

3. Simulation program

The simulation program starts choosing the flavour (u, d, s), the energy and the spin density matrix $\rho(q_1)$ of the initial quark q_1. The momentum of q_1 defines the jet axis and we assume it to be aligned with the $\hat{\mathbf{z}}$ axis in an orthonormal $(\hat{\mathbf{x}}, \hat{\mathbf{y}}, \hat{\mathbf{z}})$ reference system. The jet is generated iteratively as a sequence of splittings. At each step we choose at random the flavour of a new $q'\bar{q}'$ pair (s quarks are suppressed by the factor $s/u = s/d = 0.33$), identify the hadron $h(q\bar{q}')$ and calculate its momentum generating first Z and then $\mathbf{p}_T$ using eq. (5). The momentum of h is completely built imposing the mass-shell condition. Afterwards the spin density matrix of q' is calculated using eq. (6). This procedure is applied until the exit condition is reached, namely that the remaining energy allows to generate one (not simulated) baryon. η^0 meson production is suppressed according to $n(\eta^0)/n(\pi^0) \simeq 0.57$ [6]. A more detailed description of the simulation code implementing an approximated version of the $^3P_0$ model is given in Ref. [5].

4. Results on the Collins effect

The free parameters of the model have been tuned comparing the simulation results with experimental data on unpolarized hadron multiplicities as function of p_T^2 [7], on unpolarized fragmentation functions [8] and on Collins asymmetries in e^+e^- [9]. The values of the parameters used in these simulations are $a = 0.9$, $b_L = 0.5\,GeV^{-2}$, $b_T = 5.17\,GeV^{-2}$ and $\mu = (0.42 + i0.76)\,GeV$.

4.1. Single hadron Collins asymmetry

In Fig.1 the solid lines show the Monte Carlo prediction for the Collins analysing power $a_P^{C,u\to\pi^\pm}$ for charged pions obtained in simulations of fully transversely polarized (along $\hat{\mathbf{y}}$ axis) u jets, as function of $z_h = E_h/E_{q_1}$ (left plot) and as function of p_T (right plot). The analysing power is calculated in bins of z_h or p_T as $a_P^{C,u\to h} = 2\langle\sin\phi_C\rangle$, ϕ_C being the Collins angle defined as $\phi_C = \phi_h - \phi_{\mathbf{S}_{q_1}}$, where ϕ_h is the azimuth of $\mathbf{p}_T$ and $\phi_{\mathbf{S}_{q_1}} = \pi/2$ is the azimuth of the polarisation vector of q_1. The points represent the Collins asymmetries for charged pions as obtained from COMPASS in SIDIS off transversely polarized protons [10] . The Monte Carlo prediction is rescaled by a factor $\lambda \simeq 0.05$ calculated from the comparison with the Collins asymmetry as function of p_T for π^- measured in COMPASS. In the u-dominance hypothesis, λ gives a rough estimation of the quark transverse polarization inside a transversely polarized proton.

Looking at Fig. 1 we notice that the model reproduces correctly the opposite sign of the Collins asymmetry for oppositely charged pions. It suggests also a mirror symmetry between π^+ and π^- and the analysing power decays linearly with z_h. For large z_h the MC overestimates the π^- data and the linear trend, not seen in the data, is due to the contribution of different effects like depolarisation of the fragmenting quark, the opposite sign of the analysing powers of even and odd rank hadrons and a somewhat stronger analysing power of the second rank hadron with respect to the first rank. The comparison between MC and data as function of p_T, as can be seen from the right plot of Fig. 1, is quite satisfactory. An interesting feature of the simulated analysing power as function of p_T for π^+ is the change of sign at $p_T \sim 0.9\,GeV$ due to the contribution of π^+ generated immediately after a first π^0. The data seem not to exclude this possibility.

4.2. Di-hadron asymmetries

In Fig. 2 we show the comparison between the simulated asymmetry $a_P^{u\to h^+h^-}$ (solid lines) for pairs of charged hadrons in transversely polarized u jets, as in the previous section, and the asymmetries measured by COMPASS [11] (points) in SIDIS off transversely polarized protons. In both simulations and data the asymmetry is estimated as $2\langle\sin\phi_{R_T}\rangle$ where ϕ_{R_T} is the azimuth

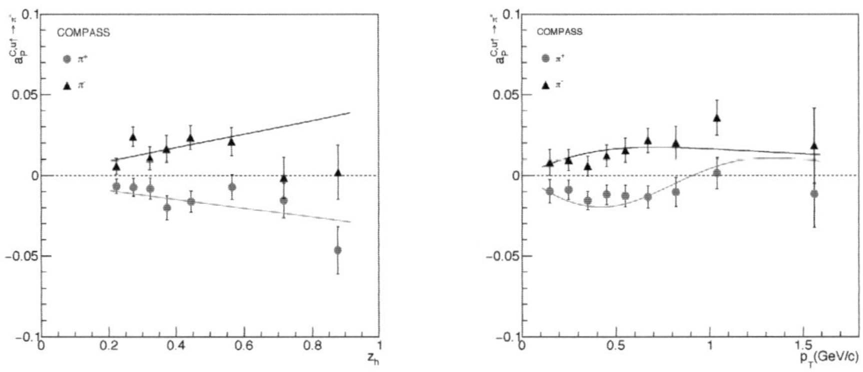

Figure 1. Comparison between the Collins analysing power for charged pions as function of z_h (left plot) and as function of p_T (right plot) predicted by simulations of transversely polarized u quark jets (solid lines) and the Collins asymmetry in SIDIS off transversely polarized protons from COMPASS [10] (points).

of $\mathbf{R}_T = (z_{h_2}\mathbf{p}_{1T} - z_{h_1}\mathbf{p}_{2T})/(z_{h_1} + z_{h_2})$ and 1 (2) label the positive (negative) hadron. The cuts $x_F > 0.1$, $z_h > 0.1$ and $R_T > 0.07\,GeV$ have been applied.

The left plot shows the comparison as function of $z = z_{h_1} + z_{h_2}$ and the right plot as function of their invariant mass M_{inv}. The MC prediction is in satisfactory agreement with data after rescaling it by the same factor λ as for the single Collins asymmetries. Hence the same $^3P_0$ mechanism which produces the Collins asymmetries for single hadrons can explain the di-hadron asymmetry.

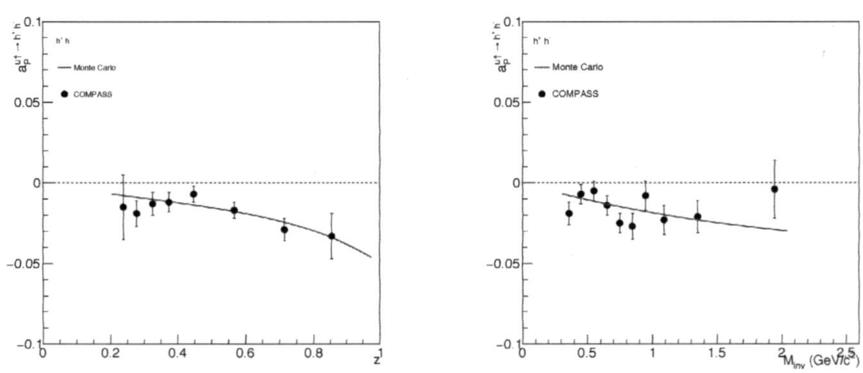

Figure 2. Comparison between the di-hadron analysing power for charged mesons as function of z (left plot) and as function of M_{inv} (right plot) obtained in simulations of transversely polarized u quark jets (solid lines) and the di-hadron asymmetries in SIDIS off transversely polarized protons from COMPASS [11] (points).

5. Effect of the primordial transverse momentum

In simulations we have included the possibility for the initial quark to have some "primordial transverse momentum" $\mathbf{k}_\perp = (k_x, k_y)$, representing for instance the intrinsic transverse momentum of quarks inside the nucleon. It changes the direction of the jet axis and we take it into account by rotating the hadrons accordingly, in the string center of mass frame. The primordial $\mathbf{k}_\perp$ is generated according to the gaussian distribution $\exp(-\mathbf{k}_\perp^2/\langle k_\perp^2\rangle)$. Results for $\langle k_\perp^2\rangle = 0.3\,GeV^2$ show that the smearing of the Collins asymmetries for single hadrons is a small effect and visible at the highest z_h only. Due to the primordial $k_\perp$ the integrated Collins analysing power is reduced by 14% for π^+ and by 6% for π^- . There is no effect on the di-hadron asymmetries, as expected.

6. The Boer-Mulders asymmetry

In this section we present some new results from simulations of quark jets where the primordial transverse momentum and the transverse polarisation of the fragmenting quark are correlated as described by the Boer-Mulders function.

Considering the SIDIS process, in the γ^*-nucleon center of mass frame the distribution of quarks of flavour q and with transverse momentum $\mathbf{k}_\perp$ inside an unpolarized proton moving along $-\hat{\mathbf{z}}$ axis is described by the Transverse Momentum Dependent (TMD) PDF

$$f_{q\uparrow/P}(x,\mathbf{k}_\perp) = \frac{1}{2}\left(f_{q/P}(x,k_\perp^2) + \frac{h_1^{\perp q}(x,k_\perp^2)}{M_P}(\hat{\mathbf{z}}\times\mathbf{k}_\perp)\cdot\hat{\mathbf{S}}_{Tq}\right). \tag{7}$$

where $f_{q/P}$ and $h_1^{\perp q}$ are the unpolarized and the Boer-Mulders TMD PDFs respectively and M is the nucleon mass. The transverse polarization of the quarks is then

$$\mathbf{S}_{Tq} = \frac{h_1^{\perp q}(x,k_\perp^2)}{M_P f_{q/P}(x,k_\perp^2)}(\hat{\mathbf{z}}\times\mathbf{k}_\perp). \tag{8}$$

After the interaction with the γ^*, the polarization vector of the struck (fragmenting) quark is partly depolarized and reflected with respect to the normal of the scattering plane, which we assume here to be the $(\hat{\mathbf{x}},\hat{\mathbf{z}})$ plane. It is

$$\mathbf{S}'_{qT} = D_{NN}\frac{k_\perp h_1^{\perp q}(x,k_\perp^2)}{M_P f_{q/P}(x,k_\perp^2)}(\frac{k_y}{k_\perp},\frac{k_x}{k_\perp}) \tag{9}$$

where D_{NN} is the depolarisation factor.

In our standalone MC, in which only the fragmentation process is described, after having introduced the intrinsic transverse momentum of the initial quark, we consider it to be fully polarized with

$$\mathbf{S}_{Tq_1} = (\frac{k_y}{k_\perp},\frac{k_x}{k_\perp}). \tag{10}$$

In the simulation we first draw the $\mathbf{k}_\perp$ of q_1 as in Section 5 and then calculate the spin density matrix $\rho(q_1) = (\mathbf{1}+\sigma\cdot\mathbf{S}_{Tq_1})/2$ which is taken as input in the generation of the jet in the reference frame where the jet axis lies along the $\hat{\mathbf{z}}$ axis. Afterwards, to take into account the primordial $\mathbf{k}_\perp$, the produced hadrons are rotated accordingly in the string center of mass frame.

The azimuthal distribution of charged hadrons (with $z_h > 0.2$ and $p_T > 0.1\,GeV$) produced in simulations of u jets has a strong modulation, as can be seen in the left plot of Fig. 3. The solid line represents the results of the fit using the function $N(1+a_{\cos\phi_h}^{BM,u\to h}\cos\phi_h + a_{\cos 2\phi_h}^{BM,u\to h}\cos 2\phi_h)$ where N is a normalisation factor. While the amplitude of the $\cos\phi_h$ modulation is compatible with zero, we obtain $a_{\cos 2\phi_h}^{BM,u\to h^+} = 0.120\pm0.003$ for h^+, a factor of 2 smaller than the original (not rescaled by λ) Collins case. In the h^- case it is $a_{\cos 2\phi_h}^{BM,u\to h^-} = -0.084\pm0.004$, a factor of 3 smaller than the original Collins case, of opposite sign with respect to h^+ and somewhat smaller in magnitude. In the right plot of Fig. 3 we show the dependence of $a_{\cos 2\phi_h}^{BM,u\to h^+}$ on p_T which turns out to be similar to the Collins analysing power in shape but with smaller magnitude.

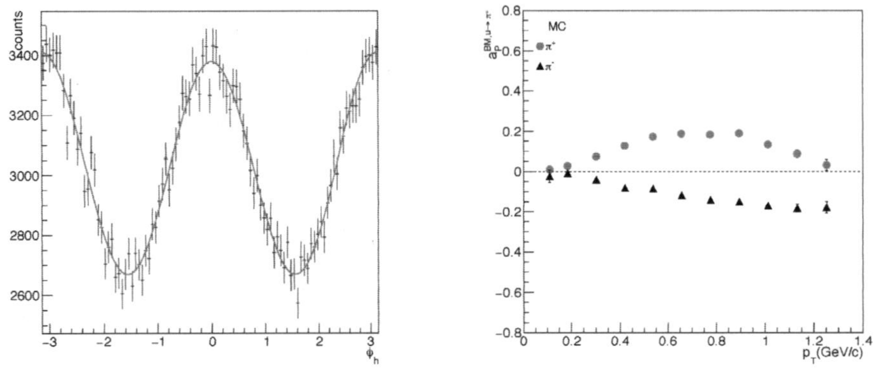

Figure 3. Azimuthal spectrum (left plot) and amplitude of the $\cos 2\phi_h$ modulation as function of p_T (right plot) for positive hadrons obtained in jets of u quarks fully polarized according to the correlation described by the Boer-Mulders function.

7. Conclusions

We have implemented in a stand alone Monte Carlo program a model for the fragmentation of polarized quarks using the Symmetric Lund Model and the $^3P_0$ mechanism.

The kinematical dependence of the Collins analysing power has been studied and the comparison with the experimental Collins asymmetry measured in SIDIS processes is very promising. We find a satisfactory comparison between simulation and experimental data also for the di-hadron asymmetries. The results show that the same mechanism can be responsible for the Collins effect and the di-hadron asymmetry.

The primordial transverse momentum of the fragmenting quarks is also introduced showing that the effect on the single hadron Collins analysing power is weak. Finally we have presented for the first time results on the $\cos 2\phi_h$ azimuthal asymmetry obtained when assuming a correlation between the transverse spin and the transverse momentum of the fragmenting quark as described by the Boer-Mulders function, taken of maximum amplitude.

The work will continue, in order to study the experimental $\cos 2\phi_h$ asymmetries measured by the different experiments.

References

[1] Artru X and Belghobsi Z 2011 String fragmentation model with spinning quarks *XIV Advanced Research Workshop on High Spin Physics (DSPIN-11)* (Dubna, Russia)
[2] Artru X and Belghobsi Z 2013 Theoretical considerations for a jet simulation with spin *XV Advanced Research Workshop on High Energy Spin Physics* ed Efremov A and Goloskokov S (Dubna, Russia) pp 33–40
[3] Andersson B, Gustafson G, Ingelman G and Sjostrand T 1983 *Phys. Rept.* **97** 31–145
[4] Artru X 2009 Recursive fragmentation model with quark spin. Application to quark polarimetry *Proc. of XIII Advanced Research Workshop on High Energy Spin Physics (DSPIN-09) (Dubna, September 1-5, 2009)* ed Efremov A and Goloskokov S (Dubna, JINR.) p 33. (*Preprint* 1001.1061)
[5] Kerbizi A, Artru X, Belghobsi Z, Bradamante F, Martin A and Salah E R 2017 Recursive Monte Carlo code for transversely polarized quark jet *22nd International Symposium on Spin Physics (SPIN 2016) Urbana, IL, USA, September 25-30, 2016* (*Preprint* 1701.08543)
[6] Field R D and Feynman R P 1977 *Phys. Rev. D* **15**(9) 2590–2616
[7] Makke N 2016 Transverse momentum dependent hadron multiplicities at compass *International Journal of Modern Physics: Conference Series* vol 40 (World Scientific) p 1660031
[8] Kniehl B A, Kramer G and Pötter B 2001 *Nuclear Physics B* **597** 337–369
[9] Vossen A *et al.* (Belle) 2011 *Phys. Rev. Lett.* **107** 072004 (*Preprint* 1104.2425)
[10] Adolph C *et al.* (COMPASS) 2015 *Phys. Lett.* **B744** 250–259 (*Preprint* 1408.4405)
[11] Adolph C *et al.* (COMPASS) 2014 *Phys. Lett.* **B736** 124–131 (*Preprint* 1401.7873)

IOP Conf. Series: Journal of Physics: Conf. Series **938** (2017) 012052 doi:10.1088/1742-6596/938/1/012052

Contributions of electromagnetic and strong anomalies to the $\eta(\eta') \to \gamma\gamma$ decays

S. Khlebtsov,[1] Y. Klopot,[2,3] A. Oganesian,[1,2] and O.V. Teryaev,[2]

[1] Institute of Theoretical and Experimental Physics, 117218, Moscow, Russia
[2] Joint Institute for Nuclear Research, 141980, Dubna, Russia
[3] Bogolyubov Institute for Theoretical Physics, 03143, Kiev, Ukraine

E-mail: klopot@theor.jinr.ru

Abstract. We study contributions of electromagnetic and strong axial anomalies in the radiative decays of η and η'. Applying a dispersive approach, we derive the anomaly sum rule for a singlet axial current where both electromagnetic and strong parts of the axial exist. A low energy theorem is generalized for a case of mixing states and is applied to evaluate a subtraction part of the strong anomaly. We found a relatively small contribution of the strong anomaly to the two-photon decay amplitudes of the η and η' mesons.

1. Introduction.

An important property of QCD is the presence of an axial (chiral) anomaly – the violation of the axial symmetry of the classical Lagrangian in the quantum theory. A two-photon decay of the π^0 is a well-known example of the process that happens mainly due to the axial anomaly. In fact, it is the pion decay problem that led to the discovery of the axial anomaly [1, 2] and triggered further studies that revealed a tremendous role of the anomalies in the quantum field theory. Precise measurements of the pion decay width [3, 4] provide a valuable test for theoretical approaches at low energies (e.g. Chiral perturbation theory) as the usual perturbative QCD methods encounter difficulties in this region due to the confinement. Two-photon decays of the η and η' mesons is another important example of the processes which are governed by the axial anomaly and are closely related to the problems of mixing and chiral symmetry breaking.

Besides the case of real photons, axial anomaly plays an important role in the processes involving off-shell photons. In particular, the transitions $\gamma\gamma^* \to \pi^0(\eta, \eta')$ are known to be connected with the axial anomaly. One of the methods to study these processes is based on the anomaly sum rules (ASRs) [6, 7] – the result of the dispersive treatment of the axial anomaly [5]. The anomaly sum rules approach was developed for the study of the π^0 [8] as well as the η and η' [9, ?, 10, 11, 12] transition form factors in the space-like and time-like regions of the photon virtuality. Recent experimental progress in measurements of the π^0, η, η' meson transition form factors has led to extensive theoretical studies [13].

The η and η' mesons, which we deal in this work with, manifest significant mixing, i.e. they are not pure eigenstates of the 8th (octet) component of the octet of axial currents $J^{(8)}_{\mu5} = \frac{1}{\sqrt{6}}(\bar{u}\gamma_\mu\gamma_5 u + \bar{d}\gamma_\mu\gamma_5 d - 2\bar{s}\gamma_\mu\gamma_5 s)$ and singlet axial current $J^{(0)}_{\mu5} = \frac{1}{\sqrt{3}}(\bar{u}\gamma_\mu\gamma_5 u + \bar{d}\gamma_\mu\gamma_5 d + \bar{s}\gamma_\mu\gamma_5 s)$ respectively. This fact is reflected in the non-vanishing off-diagonal terms of the mixing matrix formed by the decay constants f^i_M $(i = 8, 0; M = \eta, \eta')$,

$$\langle 0|J^{(i)}_{\mu 5}(0)|M(p)\rangle = ip_\mu f^i_M. \tag{1}$$

The 8th component of the octet of axial currents acquires an electromagnetic anomaly term

$$\partial^\mu J^{(8)}_{\mu 5} = \frac{2i}{\sqrt{6}}(m_u\overline{u}\gamma_5 u + m_d\overline{d}\gamma_5 d - 2m_s\overline{s}\gamma_5 s) + \frac{e^2}{8\pi^2}C^{(8)}N_c F\tilde{F}, \tag{2}$$

while the singlet axial current additionally acquires a strong (gluon) anomaly term

$$\partial^\mu J^{(0)}_{\mu 5} = \frac{2i}{\sqrt{3}}(m_u\overline{u}\gamma_5 u + m_d\overline{d}\gamma_5 d + m_s\overline{s}\gamma_5 s) + \frac{e^2}{8\pi^2}C^{(0)}N_c F\tilde{F} + \frac{\sqrt{3}\alpha_s}{4\pi}G\widetilde{G}, \tag{3}$$

where F and G are electromagnetic and gluon strength tensors respectively, $\widetilde{F}$ and $\widetilde{G}$ are their duals, and charge factors $C^{(a)}$ are

$$\begin{aligned} C^{(8)} &= \frac{1}{\sqrt{6}}(e_u^2 + e_d^2 - 2e_s^2) = \frac{1}{3\sqrt{6}}, \\ C^{(0)} &= \frac{1}{\sqrt{3}}(e_u^2 + e_d^2 + e_s^2) = \frac{2}{3\sqrt{3}}. \end{aligned} \tag{4}$$

The distinctive feature of the octet of axial currents – absence of the strong component of the axial anomaly – significantly simplifies the study of the anomaly-related processes corresponding to with these currents, leading to exact ASRs in the cases of the 3rd and the 8th components. On the other hand, η and η' two-photon processes are related to the singlet axial current as well. This gives us a good opportunity to study the role of electromagnetic and strong anomalies in these processes. In order to do this, basing on a dispersive representation of axial anomaly, we develop the ASR for the singlet axial current.

2. Dispersive approach to axial anomaly for the singlet axial current

Let us outline the derivation of the anomaly sum in the singlet channel of the axial current. Consider the triangle graph amplitude, composed of the axial current $J_{\alpha 5}$ with momentum $p = k + q$ and two vector currents with momenta k and q

$$\int d^4x d^4y e^{(ikx+iqy)}\langle 0|T\{J_{\alpha 5}(0)J_\mu(x)J_\nu(y)\}|0\rangle = e^2 T_{\alpha\mu\nu}(k,q). \tag{5}$$

This amplitude can be decomposed [14] (see also [15, 16]) as

$$\begin{aligned} T_{\alpha\mu\nu}(k,q) = F_1\, \varepsilon_{\alpha\mu\nu\rho}k^\rho + F_2\, \varepsilon_{\alpha\mu\nu\rho}q^\rho + \; F_3\, k_\nu\varepsilon_{\alpha\mu\rho\sigma}k^\rho q^\sigma + \\ F_4\, q_\nu\varepsilon_{\alpha\mu\rho\sigma}k^\rho q^\sigma + \; F_5\, k_\mu\varepsilon_{\alpha\nu\rho\sigma}k^\rho q^\sigma + F_6\, q_\mu\varepsilon_{\alpha\nu\rho\sigma}k^\rho q^\sigma, \end{aligned} \tag{6}$$

where the coefficients $F_j = F_j(p^2, k^2, q^2; m^2)$, $j = 1, \ldots, 6$ are the Lorentz invariant amplitudes constrained by current conservation and Bose symmetry. Note that the latter includes the interchange $\mu \leftrightarrow \nu$, $k \leftrightarrow q$ in the tensor structures and $k^2 \leftrightarrow q^2$ in the arguments of the scalar functions F_j.

An anomalous axial Ward identity for $T_{\alpha\mu\nu}(k,q)$ for the singlet axial current $J^{(0)}_{\mu 5}(p)$ and photons $\gamma(k, \epsilon^{(k)})$, $\gamma(q, \epsilon^{(q)})$ reads

IOP Conf. Series: Journal of Physics: Conf. Series **938** (2017) 012052 doi:10.1088/1742-6596/938/1/012052

$$p_\alpha T^{\alpha\mu\nu} = 2mG\epsilon^{\mu\nu\rho\sigma}k_\rho q_\sigma + \frac{C_0 N_c}{2\pi^2}\epsilon^{\mu\nu\rho\sigma}k_\rho q_\sigma + N(p^2,q^2,k^2)\epsilon^{\mu\nu\rho\sigma}k_\rho q_\sigma, \tag{7}$$

where

$$\langle 0|\frac{1}{\sqrt{3}}\sum_q m_q\bar{q}\gamma_5 q|\gamma\gamma\rangle = 2mG\epsilon^{\mu\nu\rho\sigma}k_\rho q_\sigma, \tag{8}$$

$$\langle 0|\frac{\sqrt{3}\alpha_s}{4\pi}G\tilde{G}|\gamma\gamma\rangle = e^2 N(p^2,k^2,q^2)\epsilon^{\mu\nu\rho\sigma}k_\mu q_\nu \epsilon_\rho^{(k)}\epsilon_\sigma^{(q)}, \tag{9}$$

$$\langle 0|F\tilde{F}|\gamma\gamma\rangle = 2\epsilon^{\mu\nu\rho\sigma}k_\mu q_\nu \epsilon_\rho^{(k)}\epsilon_\sigma^{(q)}. \tag{10}$$

We have introduced here the form factors G and N, while the transition $F\tilde{F}$–$\gamma\gamma$ is point-like up to QED corrections.

In the kinematical configuration with one real photon ($k^2 = 0$) which we consider in the rest of this section, the above anomalous Ward identity can be rewritten in terms of form factors F_j, G, N as follows ($N(p^2,q^2) \equiv N(p^2,q^2,k^2=0)$),

$$(q^2 - p^2)F_3 - q^2 F_4 = 2mG + \frac{C_0 N_c}{2\pi^2} + N(p^2,q^2). \tag{11}$$

We can write the form factors G, F_3, F_4 as dispersive integrals without subtractions: in the case of isovector and octet channels (free from gluon anomaly) it can be shown explicitly [7]. For the singlet current it can be shown using dimensional arguments. At the same time, one cannot claim the existence of unsubtracted dispersion integral for the form factor N. We rewrite it in the form with one subtraction,

$$N(p^2,q^2) = N(0,q^2) + p^2 R(p^2,q^2), \tag{12}$$

where the new form factor R can be written as an unsubtracted dispersive integral. By taking the imaginary part of (11) w.r.t. p^2 (s in the complex plane), dividing the obtained equation by $(s-p^2)$ and integrating it over $s \in [0,+\infty)$, using the dispersive relations for the form factors F_3, F_4, G, R, in the end we arrive at

$$(q^2 - p^2)F_3 - \frac{1}{\pi}\int_0^\infty ImF_3 ds - q^2 F_4 = 2mG + p^2 R + \frac{1}{\pi}\int_0^\infty ImR ds. \tag{13}$$

Comparing (13) with (11) we can write down the anomaly sum rule for the singlet current:

$$\frac{1}{\pi}\int_0^\infty ImF_3 ds = \frac{C_0 N_c}{2\pi^2} + N(0,q^2) - \frac{1}{\pi}\int_0^\infty ImR(s,q^2)ds. \tag{14}$$

Hereafter, we limit our consideration to the case of real photons ($k^2 = q^2 = 0$).

3. Low-energy theorem generalized for mixing states

The form factor N in the ASR (14) represents the strong anomaly and is related to the matrix element $\langle 0|G\tilde{G}|\gamma\gamma\rangle$. Rigorous QCD calculation of it is not known yet because of difficulties of confinement. Despite this, it is possible to estimate it in the limit $p^\mu = 0$. The idea (see, e.g., [17]) is as follows. *Supposing* that there are no massless particles in the singlet channel in the chiral limit (i.e. no η meson contribution in the singlet channel), as the η' meson remains massive, one must get $\lim_{p\to 0} p^\mu\langle 0|J_{\mu 5}(p)|\gamma\gamma\rangle = 0$. This corresponds to $\langle 0|\partial^\mu J_{\mu 5}|\gamma\gamma\rangle = 0$, and therefore, one immediately relates the matrix elements $\langle 0|G\tilde{G}|\gamma\gamma\rangle$ and $\langle 0|F\tilde{F}|\gamma\gamma\rangle$ in the considered limits using the expressions for the divergence of the singlet axial current in the chiral limit ($m_q = 0$). In

IOP Conf. Series: Journal of Physics: Conf. Series **938** (2017) 012052 doi:10.1088/1742-6596/938/1/012052

reality, the η meson has a significant contribution to the singlet channel (because of the $\eta - \eta'$ mixing) spoiling this low-energy theorem.

Nevertheless, we can follow the line of reasoning of the theorem for a specifically constructed current with no contribution of the states yielding the poles in the chiral limit (namely, η in our approximation). Requiring $\langle 0|J^{(x)}_{\mu 5}|\eta\rangle = 0$ for the linear combination of $J^{(8)}_{\mu 5}$ and $J^{(0)}_{\mu 5}$ we obtain the current that is suitable for the theorem,

$$J^{(x)}_{\mu 5} = b(J^{(8)}_{\mu 5} - \frac{f^8_\eta}{f^0_\eta} J^{(0)}_{\mu 5}), \tag{15}$$

where b is an (arbitrary) constant, and the decay constants f^i_M are defined as the currents' projections onto meson states ($i = 8, 0; M = \eta, \eta'$),

$$\langle 0|J^{(i)}_{\mu 5}(0)|M(p)\rangle = ip_\mu f^i_M. \tag{16}$$

So, for this current we can conclude that even in the chiral limit $\lim_{p\to 0}\langle 0|\partial_\mu J^{(x)}_{\mu 5}(p)|\gamma\gamma\rangle = 0$, and therefore, using (2) and (3) in the chiral limit, at $p^\mu = 0$ we obtain the following relation between the matrix elements of $G\tilde{G}$ and $F\tilde{F}$:

$$\langle 0|\frac{\sqrt{3}\alpha_s}{4\pi} G\tilde{G}|\gamma\gamma\rangle = \frac{N_c}{f^8_\eta}(f^0_\eta C^{(8)} - f^8_\eta C^{(0)})\langle 0|\frac{\alpha_e}{2\pi} F\tilde{F}|\gamma\gamma\rangle. \tag{17}$$

This yields the value of the form factor N (9),

$$N(0,0,0) = \frac{N_c}{2\pi^2 f^8_\eta}(f^0_\eta C^{(8)} - f^8_\eta C^{(0)}). \tag{18}$$

4. Two-photon decays of η and η' and analysis of the ASR

In order to draw the conclusions for the processes of two-photon decays of the η and η', let us saturate the l.h.s. of the ASR (14) with resonances according to the quark-hadron duality. The first (lowest) contributions are given by the η and η', while the rest (higher) states are represented by the integral with a lower limit s_0,

$$f^0_\eta A_\eta + f^0_{\eta'} A_{\eta'} = \frac{1}{2\pi^2} N_c C_0 + B_0 + B_1, \tag{19}$$

where, for the sake of brevity, we introduced notations

$$B_0 \equiv N(0,0,0),\; B_1 \equiv -\frac{1}{\pi}\int_0^\infty ImR(s)ds - \frac{1}{\pi}\int_{s_0}^\infty ImF_3 ds. \tag{20}$$

The decay amplitudes of the two-photon decays of the η and η' mesons A_M ($M = \eta, \eta'$) can be expressed in terms of their decay widths

$$A_M = \sqrt{\frac{64\pi\Gamma_{M\to 2\gamma}}{e^4 m^3_M}}. \tag{21}$$

Let us have a closer look at the obtained ASR (19). The term B_0 stands for the subtraction constant in the dispersion representation of the gluon anomaly and we evaluated it from the the low energy theorem (18).

The term B_1 consists of the integral representing the spectral part of the gluon anomaly and the term covering higher resonances. The value of the lower limit s_0 ("continuum threshold") in

Table 1. Strong anomaly term contributions for different sets of meson decay constants

	$\begin{pmatrix} f_\eta^8 & f_{\eta'}^8 \\ f_\eta^0 & f_{\eta'}^0 \end{pmatrix} \frac{1}{f_\pi}$	B_0	B_1	$(B_0 + B_1)$
[11], free analysis	$\begin{pmatrix} 1.11 & -0.42 \\ 0.16 & 1.04 \end{pmatrix}$	-5.55$\times 10^{-2}$	4.91$\times 10^{-2}$	-0.64$\times 10^{-2}$
[11], OS mix. sch.	$\begin{pmatrix} 0.85 & -0.22 \\ 0.20 & 0.81 \end{pmatrix}$	-5.36$\times 10^{-2}$	3.84$\times 10^{-2}$	-1.53$\times 10^{-2}$
[11], QF mix. sch.	$\begin{pmatrix} 1.38 & -0.63 \\ 0.18 & 1.35 \end{pmatrix}$	-5.58$\times 10^{-2}$	6.39$\times 10^{-2}$	0.81$\times 10^{-2}$
[18], free analysis	$\begin{pmatrix} 1.39 & -0.59 \\ 0.054 & 1.29 \end{pmatrix}$	-5.77$\times 10^{-2}$	5.86$\times 10^{-2}$	0.095$\times 10^{-2}$
[19], QF mix. sch.	$\begin{pmatrix} 1.17 & -0.46 \\ 0.19 & 1.15 \end{pmatrix}$	-5.51$\times 10^{-2}$	5.47$\times 10^{-2}$	-0.047$\times 10^{-2}$

the last integral of B_1 should range between the masses squared of the last taken into account resonance (η') and the first resonance included into the integral term, $s_0 \gtrsim 1$ GeV^2. This results in a α_s^2 suppression of this integral comparing to the first integral term in B_1 as the form factor F_3 is described by a triangle graph (no α_s corrections) plus diagrams with additional boxes ($\propto \alpha_s^2$ for the first box term).

By making use of the ASR for the 8th component of the axial current [11]

$$f_\eta^8 A_\eta + f_{\eta'}^8 A_{\eta'} = \frac{1}{2\pi^2} N_c C^{(8)}, \tag{22}$$

we can express the two-photon decay amplitudes,

$$A_\eta = \frac{1}{\Delta}\left(\frac{N_c}{2\pi^2}(C^{(8)} f_{\eta'}^0 - C^{(0)} f_{\eta'}^8) - (B_0 + B_1) f_{\eta'}^8\right), \tag{23}$$

$$A_{\eta'} = \frac{1}{\Delta}\left(\frac{N_c}{2\pi^2}(C^{(0)} f_\eta^8 - C^{(8)} f_\eta^0) + (B_0 + B_1) f_\eta^8\right), \tag{24}$$

where $\Delta = f_\eta^8 f_{\eta'}^0 - f_{\eta'}^8 f_\eta^0$.

In order to do a numerical analysis of the ASR (19), we use the experimental values of the two-photon decay widths of the η and η' mesons and the decay constants values $f_M^{(a)}$. We employ the values of decay constants estimated in different analyses basing on the octet-singlet (OS) mixing scheme [11], quark-flavor (QF) mixing scheme [11, 19] or mixing-scheme-free analyses [11, 18].

The $B_0 + B_1$ term can be evaluated directly from Eq. (19). The low energy theorem additionally gives estimation for B_0 (18), so we can separately evaluate B_0 and B_1. The results are shown in the Table 1. These results demonstrate that B_0 and B_1 enter the ASR with opposite signs and almost cancel each other, giving only a small total contribution to the two-photon decay widths of the η and η'. The contribution of the gluon anomaly and higher order resonances (expressed by the $B_0 + B_1$ term) to the two-photon decay amplitudes appears to be rather small numerically in comparison with the contribution of the electromagnetic anomaly.

5. Conclusions

Using the dispersive approach to axial anomaly in the singlet current, we have obtained the sum rule with the electromagnetic and strong anomaly contributions. The strong anomaly contribution consists of the spectral part (originated from p^2-dependent term) and the subtraction constant (independendent of p^2).

The gluon matrix element $\langle 0|G\tilde{G}|\gamma\gamma\rangle$ is related to the electromagnetic amplitude $\langle 0|F\tilde{F}|\gamma\gamma\rangle$ in the chiral limit at $p^\mu = 0$ by means of the low energy theorem, which we generalized for the case of mixing states. It gives us an estimation for the subtraction constant of the gluon anomaly contribution in the dispersive form of the axial anomaly.

Combining the low-energy theorem and the anomaly sum rule in the singlet channel we determined separately the subtraction and the spectral parts of the gluon anomaly. The spectral part of gluon anomaly is found to be significant: it is of the order of electromagnetic anomaly contribution. However, it is almost canceled out by the subtraction term, resulting in overall small contribution of gluon anomaly to the $\eta(\eta') \to \gamma\gamma$ decays.

The contributions of gluon and photon anomalies in the $\eta(\eta') \to \gamma\gamma$ decays have been evaluated using the anomaly sum rule for the singlet axial current. We found a relatively small contribution of the gluon anomaly part.

As an outlook, let us note that ASR also allows one to fix the sign of pseudoscalar meson pole contributions to the vector meson electroproduction amplitudes reflected in spin asymmetry, discussed at this conference [20].

Aknowledgments. We thank A. Kataev, M. Polyakov and N. Stefanis or useful discussions and comments. The work was supported in part by Heisenberg-Landau Program HL-2017.

References

[1] S. L. Adler, *Phys. Rev.* **177**, 2426 (1969).
[2] J. S. Bell and R. Jackiw, *Nuovo Cim.* A **60** (1969) 47.
[3] A. M. Bernstein and B. R. Holstein, *Rev. Mod. Phys.* **85**, 49 (2013)
[4] B. L. Ioffe and A. G. Oganesian, *Phys. Lett.* B **647**, 389 (2007)
[5] A. D. Dolgov and V. I. Zakharov, *Nucl. Phys.* B **27**, 525 (1971).
[6] J. Horejsi, *Phys. Rev.* D **32**, 1029 (1985).
[7] J. Horejsi and O. Teryaev, *Z. Phys.* C **65**, 691 (1995).
[8] Y. N. Klopot, A. G. Oganesian and O. V. Teryaev, *Phys. Lett.* B **695**, 130 (2011)
[9] Y. N. Klopot, A. G. Oganesian and O. V. Teryaev, *Phys. Rev.* D **84**, 051901 (2011)
Y. Klopot, A. Oganesian and O. Teryaev, *JETP Lett.* **94**, 729 (2011)
[10] D. Melikhov and B. Stech, *Phys. Lett.* B **718**, 488 (2012)
[11] Y. Klopot, A. Oganesian and O. Teryaev, *Phys. Rev.* D **87**, no. 3, 036013 (2013) Erratum: [*Phys. Rev.* D **88**, no. 5, 059902 (2013)]
[12] Y. Klopot, A. Oganesian and O. Teryaev, *JETP Lett.* **99**, 679 (2014)
[13] R. Escribano, S. Gonzlez-Sols, P. Masjuan and P. Sanchez-Puertas, *Phys. Rev.* D **94** (2016) no.5, 054033
R. Escribano, P. Masjuan and P. Sanchez-Puertas, *Eur. Phys.* J. C **75**, no. 9, 414 (2015)
A. G. Oganesian, A. V. Pimikov, N. G. Stefanis and O. V. Teryaev, *Phys. Rev.* D **93**, no. 5, 054040 (2016)
N. G. Stefanis, A. P. Bakulev, S. V. Mikhailov and A. V. Pimikov, *Phys. Rev.* D **87**, no. 9, 094025 (2013)
S. N. Nedelko and V. E. Voronin, *Phys. Rev.* D **95**, no. 7, 074038 (2017)
D. Gomez Dumm, S. Noguera and N. N. Scoccola, *Phys. Rev.* D **95**, no. 5, 054006 (2017)
H. M. Choi, H. Y. Ryu and C. R. Ji, *Phys. Rev.* D **96**, no. 5, 056008 (2017)
A. E. Dorokhov and E. A. Kuraev, *Phys. Rev.* D **88**, no. 1, 014038 (2013)
H. Czy, P. Kisza and S. Tracz, *Preprint* arXiv:1711.00820 [hep-ph].
[14] L. Rosenberg, *Phys. Rev.* **129**, 2786 (1963).
[15] V. L. Eletsky, B. L. Ioffe and Y. I. Kogan, *Phys. Lett.* **122B**, 423 (1983).
[16] A. V. Radyushkin, *Phys. Rev.* D **80**, 094009 (2009)
[17] M. A. Shifman, *Phys. Rept.* **209**, 341 (1991).
[18] R. Escribano and J. M. Frere, *JHEP* **0506**, 029 (2005)
[19] T. Feldmann, P. Kroll and B. Stech, *Phys. Rev.* D **58**, 114006 (1998)
[20] A. Sandacz, “COMPASS results on DVCS and exclusive π^0 production”, these proceedings.

Bread and butter physics: NLO electroweak corrections to polarized top quark decays

S Groote[1] **and J G Körner**[2]
[1] Füüsika Instituut, Tartu Ülikool, W Ostwaldi 1, EE-50411 Tartu, Estonia
[2] PRISMA Cluster of Excellence, Institut für Physik, Johannes-Gutenberg-Universität, D-55099 Mainz, Germany

E-mail: jukoerne@uni-mainz.de

Abstract. We report on the status of an ongoing calculation of the NLO electroweak corrections to sequential polarized top quark decays $t(\uparrow) \to X_b + W^+(\to \ell^+ + \nu_\ell)$.

1. Singly produced top quarks: Rates, polarization and integrated luminosities

The t-channel production of single top quarks is mediated by parity-violating weak interactions which is a necessary condition for a non-vanishing polarization of the produced top quarks. And, in fact, the polarization of singly produced top quarks is calculated to be $P_t \sim 90\,\%$ where the polarization is primarily along the spectator quark of the production process. The Standard Model (SM) rates for single top production are given by

$$\sigma^{8\,\mathrm{TeV}} = 55 \cdot 10^3\ fb \qquad \sigma^{13\,\mathrm{TeV}} = 136 \cdot 10^3\, fb \tag{1}$$

Typical numbers on integrated luminosities quoted by the CMS and ATLAS Collaborations in their analysis of singly produced top quarks are 20 $fb^{-}1$ @ 8 TeV and 3 $fb^{-}1$ @ 13 TeV which corresponds to $1.1 \cdot 10^6$ and $0.41 \cdot 10^6$ single top quark events, respectively. The present experimental situation is nicely reviewed in Ref. [1]. The experimental analysis of top quark polarization by the ATLAS Coll. is based on $20.2\,fb^{-1}$ @ 8 TeV ($\sim 1.1 \cdot 10^6$ events) [2]. ATLAS quotes a value of $P_t = 0.97 \pm 0.12$ for the polarization of a singly produced top quark [2].

2. The hunt is on

The $t \to b$ transition matrix element of an effective current $J_\mu^{\rm eff}(t \to b)$ is given by (see e.g. Ref. [3])

$$M_\mu^{\rm eff} = -\frac{g_w}{\sqrt{2}}\bar{u}_b\Big\{\gamma_\mu(f_L P_L + f_R P_R) + \frac{i\sigma^{\mu\nu}q_\nu}{m_W}(g_L P_L + g_R P_R)\Big\}u_t \tag{2}$$

In the SM one has $f_R, g_L, g_R = 0$ and $f_L = V_{tb}^* \sim 1$. Imaginary parts of the coupling factors can be generated by final state interactions (SM; CP-conserving) or by introducing non-SM CP-violating imaginary couplings by hand. Real contributions of f_L, f_R, g_L, g_R to the spin density elements of the W^+ compete with higher order perturbative corrections.

Large samples of polarized top quarks are presently being produced at the LHC. There is a strong ongoing experimental program (ATLAS and CMS) to unravel the structure underlying polarized top quark decay. In particular, the hunt is on for signals of non-SM physics in polarized top quark decays.

3. Scope of the problem

There are 8 T-even and 2 T-odd helicity structure function describing sequential polarized top quark decays $t(\uparrow) \rightarrow X_b + W^+(\rightarrow \ell^+ + \nu_\ell)$. Three of these describe unpolarized top quark decay. As concerns NLO electroweak contributions the following calculations exist in the literature:

i) Total unpolarized rate [4, 5]

ii) Three unpolarized helicity fractions [6] preceded by a soft photon calculation [7]

iii) Electroweak one-loop corrections to the real and imaginary parts of the left and right tensorial coupling factors $g_{R,L}$ [8, 9, 10, 11].

The NLO electroweak radiative corrections to 5 T-even and to 2 T-odd polarized helicity structure functions are still missing and need to be calculated.

The electroweak radiative corrections to the unpolarized structure functions amount to 1.55% in the rate and to 1.3% − 2.1% for the longitudinal and transverse structure functions. The NLO electroweak radiative corrections to the polarized structure functions are not likely to be much larger then those for the unpolarized case. One will therefore not win big merits and recognition for such a calculation. We have nevertheless decided to embark on such a “bread and butter physics” enterprise.

4. Size of radiative corrections to unpolarized structure functions

Let us list the radiative corrections to the longitudinal rate Γ_L which we normalize to the Born term rate. One has

$$\hat{\Gamma}_L = 0.703 \Big(1 - \underbrace{9.51\%}_{\text{NLO QCD [12, 13]}} - \underbrace{3.21\%}_{\text{NNLO QCD [14]}} + \underbrace{1.32\%}_{\text{NLO EW [6]}}\Big) \tag{3}$$

Contrast this with what is known/unknown for the corresponding polarized longitudinal rate given by

$$\hat{\Gamma}_L^P = 0.703 \Big(1 - \underbrace{9.62\%}_{\text{NLO QCD [12, 13]}} - \underbrace{?\,\%}_{\text{NNLO QCD}} + \underbrace{?\,\%}_{\text{NLO EW}}\Big) \tag{4}$$

J.G.K. had applied for a grant to the Deutsche Forschungsgemeinschaft (DFG) proposing, among others, to calculate the NNLO QCD and NLO electroweak radiative corrections to polarized top quark decays to fill in the missing pieces such as in Eq. (4). The money asked for was a mere pittance. One of the referees shot down the grant application with the following argument: “The polarization of singly produced top quarks has been measured by the ATLAS Collaboration with an error of 12 % [2]. Why calculate the above radiative corrections which are likely to be smaller than 12 %?” My answer: The error on P_t is based on pre-2016 data. In the meantime the data base has grown by a factor of 7 and will continue to grow in the next few years. The HL-LHC to come in the next years is expected to collect 250 fb^-1 per year and detector which corresponds to 34 $\times$ 10^6 polarized top quarks. The error on the top quark's polarization is likely to be reduced considerably during the next few years which makes the calculation of radiative corrections to polarized top quark decays indispensible.

5. Counting the number of structure functions

The spin density matrix elements of the W^+ (also called helicity structure functions) are hermitian. They satisfy

$$\left(H_{\lambda_W\,\lambda'_W}^{\lambda_t\,\lambda'_t}\right)^\dagger = \left(H_{\lambda'_W\,\lambda_W}^{\lambda'_t\,\lambda_t}\right) \tag{5}$$

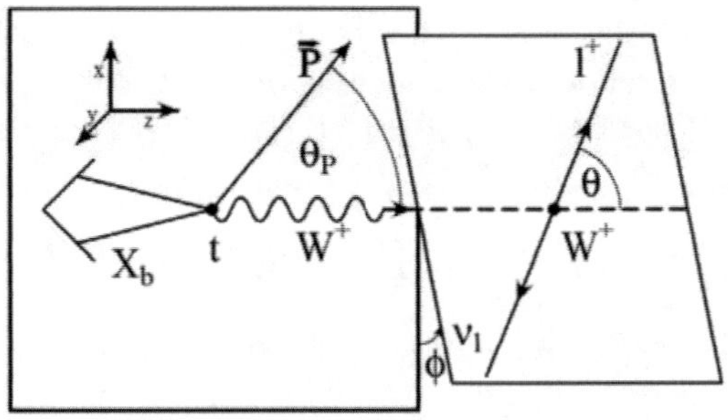

Figure 1. Definition of the polar angles θ and θ_P and the azimuthal angle ϕ in the sequential decay $t(\uparrow) \to X_b + W^+(\to \ell^+ + \nu_\ell)$.

One has to observe the angular momentum constraint $\lambda_W - \lambda'_W = \lambda_t - \lambda'_t$ which implies $|\lambda_W - \lambda'_W| \leq 1$. With the above constraint one counts ten independent double spin density matrix elements:

$$H^{++}_{++},\, H^{--}_{++},\, H^{++}_{--},\, H^{--}_{--},\, H^{++}_{00},\, H^{--}_{00},\, H^{+-}_{+0},\, H^{-+}_{0+},\, H^{-+}_{-0},\, H^{+-}_{0-}. \tag{6}$$

We mention that the counting by covariants is not straightforward because of a nontrivial Schouten identity among the covariants. We sort the $H^{\lambda_t\,\lambda'_t}_{\lambda_W\,\lambda'_W}$ into a set of three unpolarized structure functions and a set of seven polarized structure functions by writing

$$\begin{array}{lll} H_{++} = H^{++}_{++} + H^{--}_{++} & H_{00} = H^{++}_{00} + H^{--}_{00} & H_{--} = H^{++}_{--} + H^{--}_{--} \\ H^P_{++} = H^{++}_{++} - H^{--}_{++} & H^P_{00} = H^{++}_{00} - H^{--}_{00} & H^P_{--} = H^{++}_{--} - H^{--}_{--} \\ H^P_{+0} = H^{+-}_{+0} \quad H^P_{0+} = H^{-+}_{0+} & H^P_{-0} = H^{-+}_{-0} & H^P_{0-} = H^{+-}_{0-} \end{array} \tag{7}$$

It is convenient to consider particular linear superpositions of the spin density elements which feature in the angular decay distribution of the process. These are

$$\begin{array}{lll} H_U = H_{++} + H_{--} & H_L = H_{00} & H_F = H_{++} - H_{--} \\ H^P_U = H^P_{++} + H^P_{--} & H^P_L = H^P_{00} & H^P_F = H^P_{++} - H^P_{--} \\ H^P_I = \frac{1}{2}\left(\mathcal{R}eH^P_{+0} - \mathcal{R}eH^P_{-0}\right) & & H^P_A = \frac{1}{2}\left(\mathcal{R}eH^P_{+0} + \mathcal{R}eH^P_{-0}\right) \\ H^P_{\mathcal{I}I} = \frac{1}{2}\left(\mathcal{I}mH^P_{+0} - \mathcal{I}mH^P_{-0}\right) & & H^P_{\mathcal{I}A} = \frac{1}{2}\left(\mathcal{I}mH^P_{+0} + \mathcal{I}mH^P_{-0}\right) \end{array} \tag{8}$$

6. Angular decay distribution

We introduce the polar angles θ_P and θ which are defined in the top quark and W^+ rest frames, respectively, as depicted in Fig. 1. The azimuthal angle ϕ measures the azimuthal angle between the two planes as shown in Fig. 1.

The angular decay distribution in the sequential decay $t(\uparrow) \to X_b + W^+(\to \ell^+ + \nu_\ell)$ can then be obtained from the master formula [13]

$$W(\theta_P, \theta, \phi) \propto \sum_{\lambda_W - \lambda'_W = \lambda_t - \lambda'_t} e^{i(\lambda_W - \lambda'_W)\phi}\, d^1_{\lambda_W 1}(\theta)\, d^1_{\lambda'_W 1}(\theta)\, H^{\lambda_t\,\lambda'_t}_{\lambda_W\,\lambda'_W}\, \rho_{\lambda_t\,\lambda'_t}(\theta_P), \tag{9}$$

where the spin density matrix of the top quark is given by

$$\rho_{\lambda_t\,\lambda'_t}(\theta_P) = \frac{1}{2}\begin{pmatrix} 1 + P_t\cos\theta_P & P_t\sin\theta_P \\ P_t\sin\theta_P & 1 - P_t\cos\theta_P \end{pmatrix}. \tag{10}$$

IOP Conf. Series: Journal of Physics: Conf. Series **938** (2017) 012053 doi:10.1088/1742-6596/938/1/012053

P_t is the magnitude of the polarization of the top quark. The second lower index in the small Wigner $d(\theta)$-function $d^1_{\lambda_W 1}$ is fixed at $m = 1$ for zero mass leptons because the total m-quantum number of the lepton pair along the l^+ direction is $m = 1$. One then obtains the normalized angular decay distribution [13, 15, 16, 17, 18]

$$\begin{aligned} W(\cos\theta,\,\cos\theta_P,\,\phi) = \frac{1}{4\pi}\frac{3}{8}\Big\{ & \Big(\hat{H}_U + \hat{H}^P_U\, P_t \cos\theta_P\Big)(1+\cos^2\theta) \\ & + \Big(\hat{H}_L + \hat{H}^P_L\, P_t \cos\theta_P\Big) 2\sin^2\theta + \Big(\hat{H}_F + \hat{H}^P_F P_t \cos\theta_P\Big) 2\cos\theta \\ & + \hat{H}^P_I P_t \sin\theta_P\, 2\sqrt{2}\sin 2\theta\cos\phi + \hat{H}^P_A P_t \sin\theta_P\, 4\sqrt{2}\sin\theta\cos\phi \\ & + \hat{H}^P_{II} P_t \sin\theta_P\, 2\sqrt{2}\sin 2\theta\sin\phi + \hat{H}^P_{IA} P_t \sin\theta_P\, 4\sqrt{2}\sin\theta\sin\phi \Big\} \end{aligned}$$

where we define hatted normalized structure functions by

$$\hat{H}^{(P)}_i = \frac{H^{(P)}_i}{H_{\rm total}} = H^{(P)}_i/(H_U + H_L)$$

such that $\int d\cos\theta\, d\cos\theta_P\, d\phi\, W(\cos\theta,\,\cos\theta_P,\,\phi) = 1$.

7. T-odd correlations

The last two terms in the angular decay distribution correspond to T–odd angular correlations. This becomes evident by rewriting the angular factors in terms of the normalized three-vectors of the process (see Fig. 1):

$$\hat{P}_t = (\sin\theta_P, 0, \cos\theta_P) \qquad \hat{p}_\ell = (\sin\theta\cos\phi, \sin\theta\sin\phi, \cos\theta) \qquad \hat{q} = (0,0,1) \tag{11}$$

One finds

$$\begin{aligned} \sin\theta_P\, \sin\theta\, \sin\phi &= \hat{q}\cdot(\hat{P}_t \times \hat{p}_\ell) \\ \sin\theta_P\, \sin 2\theta\, \sin\phi &= 2\,(\hat{p}_\ell\cdot\hat{q})\,\hat{q}\cdot(\hat{P}_t\times\hat{p}_\ell) \end{aligned} \tag{12}$$

Under time reversal $t \to -t$ one has $(\hat{p}, \hat{P}_t) \to (-\hat{p}, -\hat{P}_t)$. One therefore calls the above two angular correlations T–odd correlations. As mentioned before there are two possible sources of T–odd correlations:

i) SM source: Imaginary parts from absorptive contributions,
ii) Non-SM source: CP-violating imaginary couplings.

8. NLO tree-level Feynman diagrams

There are four NLO electroweak tree level Feynman diagrams that contribute to $t \to b + W^+ + \gamma$ since the photon can couple to the top quark, the bottom quark and the W^+-boson (three diagrams) plus the diagram where the internal W^+ is replaced by the Goldstone boson ξ^+. The procedure to calculate the tree-level diagrams is standard. One splits the tree-level matrix element into a hard and a soft photon part according to

$$M^\mu M^{\nu *}(\text{hard}+\text{soft}) = \underbrace{\Big(M^\mu M^{\nu *}(\text{hard}+\text{soft}) - |M|^2(\text{soft})\Big)}_{\text{IR and M safe}} + \underbrace{M^\mu M^{\nu *}(\text{soft})}_{\text{universal}} \tag{13}$$

The first piece is infrared and collinear finite and can be safely integrated in $D = 4$ dimension. The second piece is universal in the sense that it is proportional to the Born term. The finite piece has to be projected onto the relevant eight T-even structure functions and then has to be integrated over the photon phase space. This has been done. For the projection onto the structure functions it is convenient to peruse the eight covariant projectors listed in Ref. [13]. The second universal piece has been known since long.

IOP Conf. Series: Journal of Physics: Conf. Series **938** (2017) 012053 doi:10.1088/1742-6596/938/1/012053

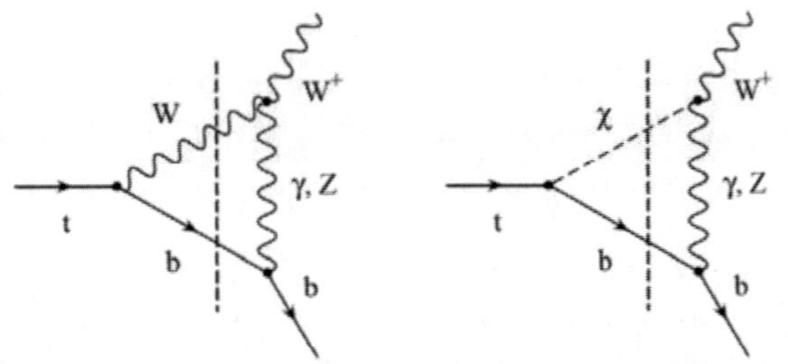

Figure 2. Absorptive parts of the four Feynman diagrams that contribute to the T-odd correlations in polarized top decays.

9. NLO electroweak one-loop vertex graphs

In the Feynman 't Hooft gauge there are altogether 18 NLO electroweak one-loop Feynman vertex diagrams that contribute to $t \to b + W^+$. The corresponding one-loop integrals are not simple in that they have five mass scales: m_t, m_b, m_W, m_Z, m_H.

In the following we shall set $m_b = 0$. The number of vertex diagrams is reduced to 13 since $g_{Hbb} = 0$ and $g_{\chi_3 bb} = 0$. The complexity of the one-loop integrals is also somewhat reduced because now there are only four mass scales. The calculation of the 13 one-loop diagrams plus the counter-term diagrams needed for renormalization has been done.

In the limit $m_b = 0$ one has the simplifying feature that the longitudinal and transverse projection of the matrix element M_μ are proportional to the corresponding projections of the Born matrix element, i.e. one has ($x = m_W/m_t$) [7]

$$\begin{aligned} \varepsilon^{*\mu}(0)M_\mu &= \left(f_L - x\, g_R\right)\varepsilon^{*\mu}(0)M_\mu(Born) \\ \varepsilon^{*\mu}(-)M_\mu &= \left(f_L - \frac{1}{x}\, g_R\right)\varepsilon^{*\mu}(-)M_\mu(Born) \end{aligned} \tag{14}$$

10. Electroweak one-loop vertex graphs that admit of absorptive cuts

There are four ($m_B = 0$) NLO electroweak one-loop Feynman diagrams for $t \to b + W^+$ that admit of absorptive cuts (often also referred to as final state interactions/rescattering corrections) which we show in Fig. 2. We have carefully extracted the imaginary parts of the four diagram in analytical form. Numerically one obtains

$$\mathcal{I}m\, g_R = -0.217\,\% \tag{15}$$

The two T-odd contributions can be projected out by taking the moment of the decay distribution w.r.t. $\sin\phi$. The T-odd contribution $\hat{H}^P_{\mathcal{I}A}$ can be isolated by limiting the ϕ-integration to the interval $[0, \pi]$. One obtains

$$\int_{-1}^{1} d\cos\theta \int_{-1}^{1} d\cos\theta_P \int_0^\pi d\phi\, W(\cos\theta, \cos\theta_P, \phi) = \frac{1}{2}\left(1 + \frac{3\pi}{4\sqrt{2}}\hat{H}^P_{\mathcal{I}A} P_t\right) \tag{16}$$

where we now use the Born term total structure function B_{U+L} as normalizing denominator. For the ratio of structure functions one obtains

$$\hat{H}^P_{\mathcal{I}A} = \frac{H^P_{\mathcal{I}A}}{B_{U+L}} = -\frac{1}{\sqrt{2}}\frac{(1-x^2)}{(1+2x^2)}\,\mathrm{Im} g_R = -0.1965\,\mathrm{Im} g_R = 0.0427\% \tag{17}$$

The azimuthal analyzing power is thus given by a tiny

$$\hat{H}^P_{\mathcal{I}A}\frac{3\pi}{4\sqrt{2}} = 0.0711\% \tag{18}$$

11. Final remarks

Let us discuss some of the features of the NLO electroweak corrections to sequential polarized top quark decays treated in this talk. The soft photon contributions are proportional to the Born terms with a universal proportionality factor $H_i^{(P)}(\text{soft}) = aB_i^{(P)}$ while the hard photon contributions are *not* proportional to the Born terms. In the limit $m_b = 0$ we have found several simplifications for the contributions of the one-loop vertex graphs. First, the number of contributing NLO one-loop vertex graphs is reduced. Second, the structure functions H_{--}, H_{--}^P, H_{-0}^P and H_{0-}^P vanish, and third, the contributions of the one-loop vertex graphs to the structure functions are proportional to the corresponding Born terms with non-universal proportionality factors $H_i^{(P)}(\text{one-loop}) = b_i\, B_i^{(P)}$. The non-universal factors b_i are not difficult to obtain.

The happy news is that we have assembled all ingredients necessary to obtain the NLO electroweak corrections to sequential polarized top quark decays. What remains to be done is to put the various pieces together and to evaluate them numerically. A first result was reported on that the azimuthal analyzing power of the SM absorptive T-odd contribution is a tiny 0.07 %. If experimentalists discover large T-odd effects in polarized top quark decays they would be most certainly due to non-SM CP-violating imaginary coupling terms with only a tiny contamination from SM absorptive contributions.

Acknowledgments

This work was supported by the Estonian Science Foundation under grant No. IUT2-27. S.G. acknowledges the hospitality and support of the theory group THEP at the Institute of Physics at the University of Mainz. J.G.K. would like to thank the Heisenberg-Landau Foundation for financial support and M.A. Ivanov for logistic support in the broadest sense and his fine hospitality while J.G.K. was visiting Dubna.

References

[1] Faltermann N 2017 Single top t-channel in ATLAS and CMS *Preprint* arXiv:1709.00841
[2] Aaboud M et al. 2017 Probing the Wtb vertex structure in t-channel single-top-quark production and decay in pp collisions at $\sqrt{s} = 8$ TeV with the ATLAS detector *JHEP* **1704** 124
[3] Bernreuther W, Gonzalez P and Wiebusch M 2009 *Eur. Phys. J.* C **60** 197
[4] Denner A and Sack T 1991 *Nucl. Phys.* B **358** 46
[5] Eilam G, Mendel R R, Migneron R and Soni A 1991 *Phys. Rev. Lett.* **66** 3105
[6] Do H S, Groote S, Körner J G and Mauser M C 2003 *Phys. Rev.* D **67** 091501
[7] Kuruma T 1993 *Z. Phys.* C **57** 551
[8] González-Sprinberg G A, Martinez R and Vidal J 2011 *JHEP* **1107** 094
[9] González-Sprinberg G A and Vidal J 2015 *Eur. Phys. J.* C **75** 615
[10] Arhrib A and Jueid A 2016 *JHEP* **1608** 082
[11] Fischer M, Groote S and Körner J G 2017 T-odd correlations in polarized top quark decay $t(\uparrow) \to X_b + W^+(\to \ell^+ + \nu_\ell)$ (to be published)
[12] Fischer M, Groote S, Körner J G, Mauser M C and Lampe B 1999 *Phys. Lett.* B **451** 406
[13] Fischer M, Groote S, Körner J G and Mauser M C 2002 *Phys. Rev.* D **65** 054036
[14] Czarnecki A, Körner J G and Piclum J H 2010 *Phys. Rev.* D **81** 111503
[15] Bialas P, Körner J G, Krämer M and Zalewski K 1993 *Z. Phys.* C **57** 115
[16] Boudreau J, Escobar C, Mueller J, Sapp K and Su J 2013 *Preprint* arXiv:1304.5639
[17] Aguilar-Saavedra J A and Bernabéu J 2016 *Phys. Rev.* D **93** 011301
[18] Aguilar-Saavedra J A, Boudreau J, Escobar C and Mueller J 2017 *Eur. Phys. J.* C **77** 200

IOP Conf. Series: Journal of Physics: Conf. Series **938** (2017) 012054 doi:10.1088/1742-6596/938/1/012054

Signals of the double intrinsic heavy quark at the current experiments

Sergey Koshkarev, Stefan Groote
Institute of Physics, University of Tartu, 10139, Estonia
E-mail: sergey.koshkarev@ut.ee

Abstract. In this paper we review existing data and analyze experimental opportunities for modern experiments to investigate the signals of intrinsic heavy quarks.

1. Introduction

Even though the existence of intrinsic heavy quarks was proposed more than 35 years ago, the mechanism is still under construction [1].

The existence of heavy quarks in the proton's light-front (LF) wavefunction at a large LF momentum fraction x is in fact predicted by QCD if one analyzes the higher Fock states $|uudc\bar{c}\rangle$ and $|uudc\bar{c}c\bar{c}\rangle$ in the hadronic eigenstate, i.e., Fock states where the heavy quark pairs are multi-connected to the valence quarks. LF wavefunctions, the eigensolutions of the QCD LF Hamiltonian, are defined at fixed LF time $\tau = t + z/c$ and are thus off-shell in the invariant mass. In QED for example, positronium has an analogous $|e^+e^-\mu^+\mu^-\rangle$ Fock state due to the insertion of light-by-light scattering in the positronium self-energy amplitude.

In such an "intrinsic charm" Fock state $|uudc\bar{c}\rangle$, the maximum kinematic configuration occurs at minimum invariant mass where all quarks are at rest in the hadron's rest frame, i.e., at equal rapidity in the moving hadron. Equal rapidity implies $x_i \propto (m^2+\vec{k}_\perp^2)^{1/2}$ for each quark, so that the heavy quarks in the Fock state carry most of the hadron's LF momentum. The operator product expansion predicts that the probability of intrinsic heavy-quark Fock states $|uudQ\bar{Q}\rangle$ scales as $1/m_Q^2$ due to the non-Abelian couplings of QCD [2, 3].

2. Review of existing data

Using the CERN pion beam at 150 and 280 GeV/c to produce charm particles with incident on hydrogen and platinum targets, the NA3 experiment provided data on the production of the double J/ψ with very similar features: a high value for the ratio $\sigma(\psi\psi)/\sigma(\psi) = (3\pm1)\times10^{-4}$, values $x_{\psi\psi} > 0.6$ at 150 GeV/c and $x_{\psi\psi} > 0.4$ at 280 GeV/c for the Feynman-x, and a transverse momentum of $p_{T,\psi\psi} = 0.9\pm0.1$ GeV/c [4, 5]. Note that perturbative QCD neither can explain the NA3 cross section nor the x_F distribution [6] (see also Fig. 1).

Utilizing the Fermilab negative and positive charged beams at 600 GeV/c to produce charmed particles in a thin foil of copper or in a diamond, the SELEX collaboration claimed

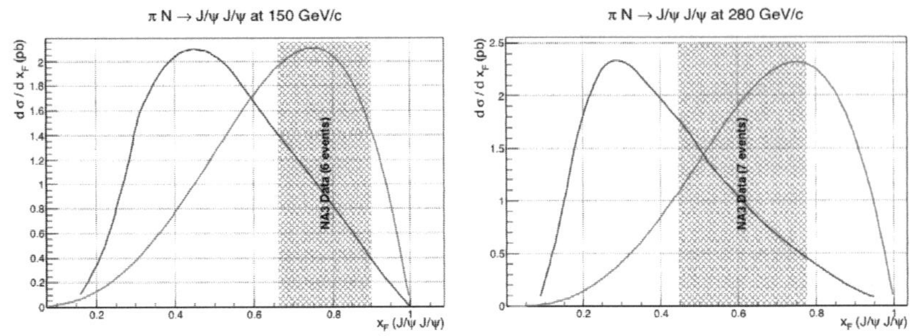

Figure 1. NA3 events (shaded area), pQCD prediction [7] (blue left curve) and prediction of the heavy quark mechanism (red right curve) at the 150 GeV/c (left panel) and at the 280 GeV/c π beam (right panel)

to observe two decay channels for the $\Xi^+_{cc}(ccd)$ state at mass around 3520 MeV/c^2. The experiment was operated in the kinematic region $x_F > 0.1$. The negative beam composition was about 50% Σ^- and 50% π^- while the positive beam was composed of 50% protons. The experimental data record used both positive and negative beams. 67% of the events were induced by Σ^-, 13% by π^-, and 18% by protons. In the first observation using the sample of $\Lambda_c^+ \to pK^-\pi^+$ [8, 9], a signal of 15.9 events over 6.1 ± 0.1 background events of $\Xi^+_{cc} \to \Lambda_c^+ K^-\pi^+$ was found [10]. To complement this result, SELEX published an observation of 5.62 signal events over 1.38 ± 0.13 background events for the decay mode $\Xi^+_{cc} \to pDK^-$ from a sample of $D^+ \to K^-\pi^+\pi^+$ decays [11].

It is interesting to compare the production properties of the Ξ^+_{cc} at the SELEX experiment with the production properties of the double J/ψ production at the NA3 experiment. Unfortunately, it is not possible to compare the results directly. However, we are still able to compare the following ratios $R = \sigma(c\bar{c}c\bar{c})/\sigma(c\bar{c})$:

$$R^{\mathrm{SELEX}} = R_{\Lambda_c^+} \times \frac{f(c \to \Lambda_c^+)}{f_{\Xi_{cc}}} \sim (1-4) \times 10^{-3}$$

and

$$R^{\mathrm{NA3}} = \frac{\sigma(\psi\psi)}{\sigma(\psi)} \times \frac{f_{J/\psi}}{f^2_{\psi/\pi}} \sim 2 \times 10^{-2},$$

where $f_{\psi/\pi} \approx 0.03$ [6] and $f_{J/\psi} \approx 0.06$ [12]. $f_{\Xi_{cc}} \approx 0.25$ [13] represents the fraction of cc pairs producing the sum of single charged baryons Ξ^+_{cc} and double charged baryons Ξ^{++}_{cc}, but this fraction cannot be less than the fraction to produce J/ψ. Therefore, R^{SELEX} should not be larger than 10^{-2}. It is clear that the SELEX production ration is complemented by the very trustable measurement of the double J/ψ production at the NA3 experiment. It is interesting to note that the intrinsic charm mechanism predicts $\langle x_F(\Xi_{cc})\rangle = 0.33$, as shown in Ref. [13]. This is in excellent agreement with the value $\langle x_F(\Xi^+_{cc})\rangle \sim 0.33$ measured by the SELEX experiment.

An analysis of the masses of the mass discrepancy between the Ξ^+_{cc}(3520 MeV) state measured by SELEX and the Ξ^{++}_{cc}(3621 MeV) state measured by LHCb [14] also complement the SELEX data [15].

IOP Conf. Series: Journal of Physics: Conf. Series **938** (2017) 012054 doi:10.1088/1742-6596/938/1/012054

3. Double intrinsic charm at modern experiments

3.1. Double J/ψ production at the COMPASS experiment

COMPASS is a fixed-target experiment using the 190 GeV/c π^- beam at the CERN SPS. Based on the already collected and planed statistics and compared to the NA3 statics we can expect up to 100 double J/ψ events [16] (see also Fig. 2).

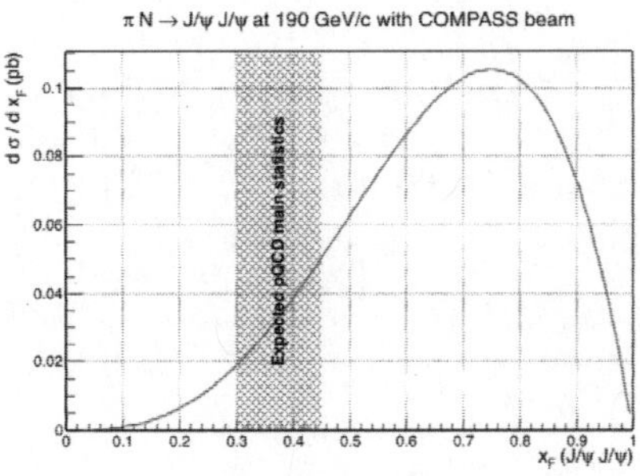

Figure 2. Prediction for the double J/ψ production at the COMPASS experiment

3.2. Production of the doubly charmed baryons at STAR

The STAR fixed-target program is a fixed-target experiment using the proton beam of the Relativistic Heavy Ion Collider (RHIC) with up to 250 GeV/c and the Au beam with up to 100 GeV/c colliding with a wired target. For $p_{\rm beam} = 200\,{\rm GeV}/c$ we may expect the production cross section of the Ξ_{cc} to be $\sigma(\Xi_{cc}) \approx (0.2 - 0.3) \times 75\,{\rm nb}$ [17].

The kinematic limits on the energy and the momentum of the doubly charmed baryon formed by the intrinsic charm from the target are given by

$$E_{lab} = \frac{1}{2m_{tar}}(m_{cc}^2 + m_{tar}^2), \quad p_{lab} = \frac{1}{2m_{tar}}(m_{cc}^2 - m_{tar}^2). \tag{1}$$

These last expressions depend solely on the two masses m_{cc} and m_{tar} and no longer on the beam energy. The momentum distribution and the distribution of the rapidity difference $\Delta y = y - y_{tar}$ in the laboratory frame are shown in Fig. 3. It is obvious that experiments at the STAR detector, typical at rapidities $|y| < 1$ for track selection, have the potential to observe doubly charmed baryons.

It is interesting as well to present the momentum distribution and the distribution of the rapidity difference for the J/ψ coming from the Fock state $|uudc\bar{c}\rangle$ (see Fig. 4).

4. Conclusion

Current theoretical and experimental knowledge suggests the evidence for the existence of intrinsic heavy quarks. Signals of the double intrinsic charm are widely accessible at modern experiments. The COMPASS experiment is already working on the analysis of the double J/ψ production [18]. The STAR experiment is able to confirm the beautiful prediction of the intrinsic heavy quark mechanism, producing a heavy quark state nearly at rest.

Acknowledgements

SK would like to thank Alexey Guskov for the invitation at the conference and much useful advice. This work was supported by the Estonian Research Council under Grant No. IUT2-27.

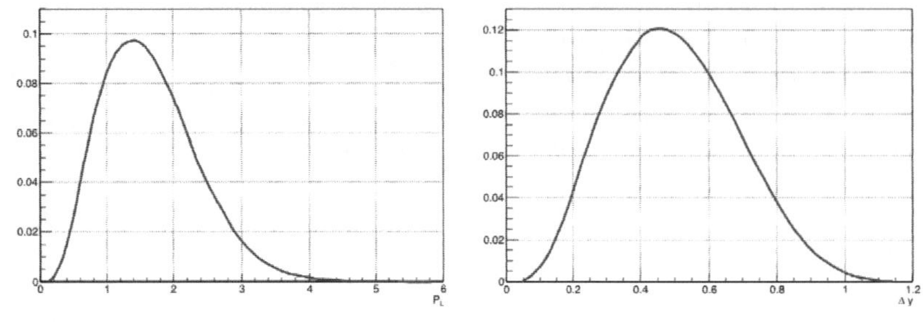

Figure 3. Prediction for the Ξ_{cc} momentum distribution (left panel) and for the distribution of the rapidity difference (right panel) in the laboratory frame

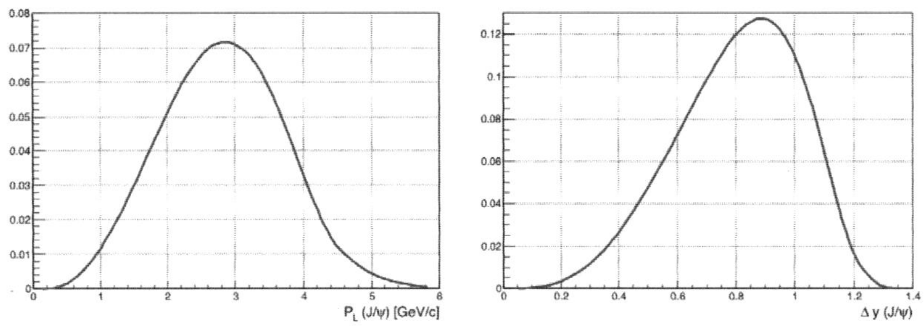

Figure 4. Prediction for the J/ψ momentum distribution (left panel) and for the distribution of the rapidity difference (right panel) in the laboratory frame

References

[1] S.J. Brodsky, P. Hoyer, C. Peterson and N. Sakai, Phys. Lett. **B93** (1980) 451
[2] S.J. Brodsky, J.C. Collins, S.D. Ellis, J.F. Gunion and A.H. Mueller, Snowmass Summer Study 1984:0227
[3] M. Franz, M.V. Polyakov and K. Goeke, Phys. Rev. **D62**, 074024 (2000)
[4] J. Badier *et al.* [NA3 Collaboration], Phys. Lett. **B114**, 457 (1982)
[5] J. Badier *et al.* [NA3 Collaboration], Phys. Lett. **B158**, 85 (1985)
[6] R. Vogt and S.J. Brodsky, Phys. Lett. **B349** (1995) 569
[7] R.E. Ecclestone and D.M. Scott, Phys. Lett. **B120** (1983) 237
[8] A. Kushnirenko *et al.* [SELEX Collaboration], Phys. Rev. Lett. **86**, 5243 (2001)
[9] A. Kushnirenko, Ph.D. thesis, Carnegie Mellon University, 2000
[10] M. Mattson *et al.* [SELEX Collaboration], Phys. Rev. Lett. **89**, 112001 (2002)
[11] A. Ocherashvili *et al.* [SELEX Collaboration], Phys. Lett. **B628**, 18 (2005)
[12] M. Mangano, H. Satz and U. Wiedermann (Eds.), CERN Yellow report, CERN-2004-009, 2004
[13] S. Koshkarev and V. Anikeev, Phys. Lett. **B765**, 171 (2017)
[14] R. Aaij *et al.* [LHCb Collaboration], Phys. Rev. Lett. **119**, 112001 (2017)
[15] S.J. Brodsky, S. Groote, S. Koshkarev, H.G. Dosch and G.F. de Téramond, arXiv:1709.09903 [hep-ph]
[16] M. Aghasyan *et al.* [COMPASS Collaboration], Phys. Rev. Lett. **119** (2017) no.11, 112002
[17] S. Groote and S. Koshkarev, Eur. Phys. J. **C77**, 509 (2017)
[18] A. Guskov [COMPASS Collaboration], Private communication

IOP Conf. Series: Journal of Physics: Conf. Series **938** (2017) 012055 doi:10.1088/1742-6596/938/1/012055

Bjorken sum rule in QCD with analytic coupling

C Ayala[1], G Cvetič[1], A V Kotikov[2] and B G Shaikhatdenov[2]

[1] Department of Physics, Universidad Técnica Federico Santa María Valparaíso, Chile
[2] Joint Institute for Nuclear Research, 141980 Dubna, Russia

E-mail: kotikov@theor.jinr.ru

Abstract. We present details of study of the Bjorken polarized sum rule carried out recently in [1] within the range of energies where the data were collected by JLAB collaboration, $0.05\ \mathrm{GeV}^2 < Q^2 < 3\ \mathrm{GeV}^2$. Three approaches to QCD with analytic (holomorphic) coupling are considered: Analytic Perturbation Theory (APT), Two-delta analytic QCD (2δanQCD), and Three-delta lattice-motivated analytic QCD in the three-loop and four-loop MiniMOM schemes (3l3δanQCD, 4l3δanQCD). The new frameworks (2δ and 3δ) with respective couplings give results which agree well with the experimental data for $0.5\ \mathrm{GeV}^2 < Q^2 < 3\ \mathrm{GeV}^2$ already when only one higher-twist term is taken into account.

1. Introduction

The well-known Bjorken polarized sum rule (BSR) $\Gamma_1^{p-n}(Q^2)$ [2] is chosen for testing the nonperturbative behavior in QCD. The experimental data on this quantity published by Jefferson Lab (JLAB) cover the low-Q^2 the range $0.05\ \mathrm{GeV}^2 < Q^2 < 3\ \mathrm{GeV}^2$ [3, 4, 5], and SLAC data [6] have $1\ \mathrm{GeV}^2 < Q^2$. The theoretical perturbation expansion of the leading-twist (LT) contribution to BSR is known to N^{3}LO ($\sim \alpha_s^4$) [7]. We present results of the research [1] into the applicability of perturbative QCD (pQCD) with inclusion of one higher-twist (HT) contribution $\sim 1/Q^2$ dictated by OPE. At low momenta $Q^2 < Q_0^2$ (≈ 0.4-$0.6\ \mathrm{GeV}^2$), we use a χPT-motivated ansatz [4][1] whose first term ($\sim Q^2$) is fixed by the Gerasimov–Drell–Hearn sum rule [9].

For the evaluation of the LT contribution we consider, in addition to pQCD, analytic frameworks of QCD (anQCD) which are a useful tool to evaluate physical quantities at low-momentum transfer. In these anQCD frameworks the running coupling has no spurious (Landau) singularities, unlike pQCD in the usual $\overline{\mathrm{MS}}$-like scheme. PQCD coupling $a(Q^2)_{\mathrm{pt}} \equiv \alpha_s(Q^2)/\pi$ possesses Landau singularities at small momenta $|Q^2| \lesssim 1\ \mathrm{GeV}^2$, while the general principles of quantum field theories (QFT) require that the spacelike QCD observables $\mathcal{D}(Q^2)$, such as current correlators and structure functions, be holomorphic (analytic) functions of Q^2 throughout an entire generalized spacelike region $Q^2 \in \mathbb{C}\backslash(-\infty, 0]$. The usual pQCD couplings $a_{\mathrm{pt}}(Q^2)$ do not reflect these properties, hence the LT part of $\mathcal{D}(Q^2)$ evaluated in terms of $a_{\mathrm{pt}}(Q^2)$ (as a truncated perturbation series) does not have these properties dictated by QFT. However, in anQCD we have $a_{\mathrm{pt}}(Q^2) \mapsto \mathcal{A}(Q^2)$, where $\mathcal{A}(Q^2)$ is the anQCD coupling holomorphic in $Q^2 \in \mathbb{C}\backslash(-\infty, 0]$. Therefore, the evaluation of the LT contribution $\mathcal{D}_{\mathrm{eval.}}(Q^2) \mapsto \mathcal{F}(\mathcal{A}(Q^2))$ has the correct holomorphic properties (the same is true for the HT contribution).

[1] Another low Q^2 expression based on the light-front holographic (LFH) coupling [8] was also successfully used in [1]; however, discussion of the results when this particular form was used is beyond the scope of this short note.

IOP Conf. Series: Journal of Physics: Conf. Series **938** (2017) 012055 doi:10.1088/1742-6596/938/1/012055

We consider three different anQCD frameworks. One is the Analytic Perturbation Theory (APT) of Shirkov *et al.* [10], in which the discontinuity function $\rho_1^{(\rm pt)}(\sigma) \equiv {\rm Im}\, a_{\rm pt}(Q^2 = -\sigma - i\epsilon)$ of the underlying QCD is kept unchanged. The coupling $\mathcal{A}^{(\rm APT)}(Q^2)$ [the APT analog of $a_{\rm pt}(Q^2)$] for $Q^2 \in \mathbb{C}\backslash(-\infty, 0]$ is then obtained by using a dispersion relation involving ${\rm Im}\, a_{\rm pt}(Q^2 = -\sigma - i\epsilon)$ in the entire interval $0 \leq \sigma < \infty$. The APT-analogs of the powers $a_{\rm pt}(Q^2)^\nu$, $\mathcal{A}_\nu^{(\rm APT)}(Q^2)$, were obtained in the works [11, 12]; this is known as Fractional APT (FAPT).

Other considered analytic frameworks are the Two-delta analytic QCD (2δanQCD [13]) and the (lattice-motivated) Three-delta analytic QCD (3δanQCD [14, 15]). They are less closely (than APT) based on the underlying pQCD coupling $a_{\rm pt}(Q^2)$: the equality ${\rm Im}\,\mathcal{A}(-\sigma - i\epsilon) = \rho_1^{(\rm pt)}(\sigma)$ is enforced only for sufficiently large $\sigma \geq M_0^2$ (where $M_0 \sim 1$ GeV is a "pQCD-onset" scale). For $0 < \sigma < M_0^2$, the otherwise unknown discontinuity function $\rho_1(\sigma) \equiv {\rm Im}\,\mathcal{A}(Q^2 = -\sigma - i\epsilon)$ is parametrized by two or three delta functions, respectively. Such a parametrization is partly motivated by the Padé approximant approach to the coupling $\mathcal{A}(Q^2)$. The renormalization schemes of the underlying pQCD coupling in 2δanQCD are constrained by the requirement that acceptable values of $M_0 \sim 1$ GeV and $\mathcal{A}(0) \sim 1$ are obtained [13, 16, 17]. In 3δanQCD, the lattice calculations provide two conditions for the coupling at very low $Q^2 < 1\ {\rm GeV}^2$, which give two additional constraints on the delta functions. In both 2δanQCD and 3δanQCD the coupling $\mathcal{A}(Q^2)$ practically agrees with the underlying pQCD coupling $a_{\rm pt}(Q^2)$ at $|Q^2| \gg \Lambda^2_{\rm QCD}$, namely: $[\mathcal{A}(Q^2) - a_{\rm pt}(Q^2)] \sim (\Lambda^2_{\rm QCD}/Q^2)^5$. This is not the case in (F)APT where $[\mathcal{A}(Q^2) - a_{\rm pt}(Q^2)] \sim (\Lambda^2_{\rm QCD}/Q^2)^1$. The construction of the analytic analogs $\mathcal{A}_n(Q^2)$ of powers $a_{\rm pt}(Q^2)^n$, for general anQCD, was formulated for integer n in [18], and for general (noninteger) n in [19].

At $Q^2 = 0$, $\mathcal{A}(Q^2)$ is finite and positive in (F)APT and in 2δanQCD, $\mathcal{A}(0) \approx 0.44$ and 0.66, respectively ($N_f = 3$). Lattice results [20] suggest the condition $\mathcal{A}(0) = 0$ which is enforced in 3δanQCD. Interestingly, the holomorphic coupling of Refs. [21] also has $\mathcal{A}(0) = 0$.

2. Bjorken sum rule

BSR Γ_1^{p-n} is the nonsinglet combination given by the difference between the proton and neutron polarized structure functions and integrated over the entire Bjorken-x interval

$$\Gamma_1^{p-n}(Q^2) = \int_0^1 dx\, \left[g_1^p(x, Q^2) - g_1^n(x, Q^2)\right] . \tag{1}$$

This is the first moment of the nonsinglet contribution to the polarized structure functions.

BSR can be written in terms of a sum of two terms, one coming from pQCD as an expansion of the running coupling $a_{\rm pt}(Q^2) = \alpha_s(Q^2)/\pi$ and the other is the twist-4 contribution dictated by the OPE [2]

$$\Gamma_1^{p-n}(Q^2) = \frac{g_A}{6} E_{\rm NS}(Q^2) + \frac{\mu_4^{p-n}(Q^2)}{Q^2} . \tag{2}$$

In the limit $Q^2 \to \infty$ we have $\Gamma_1^{p-n}(\infty) = g_A/6$, where g_A is the nucleon axial charge, $g_A = 1.2723 \pm 0.0023$ [22]. Higher-twist contributions are neglected here.

If we use the OPE formalism, the elastic contribution (at $x = 1$) to BSR (1) should in principle be included. It is convenient to exclude the elastic contribution, since the Q^2-dependence of the nonsinglet inelastic BSR at low Q^2 is constrained by the Gerasimov–Drell–Hearn (GDH) sum rule [9], as was highlighted in [23]. Further, at high Q^2 the elastic contribution is not noticable [23]. Hence we investigate the behavior of the pure inelastic contribution as a continuation to low-energy regime [24].

The twist-2 contribution $E_{\rm NS}(Q^2)$ in (2) is known up to N^{3}LO [7]:

$$E_{\rm NS}(Q^2) = 1 + e_1^{\rm NS} a_{\rm pt}(Q^2) + e_2^{\rm NS} a_{\rm pt}(Q^2)^2 + e_3^{\rm NS} a_{\rm pt}(Q^2)^3 + e_4^{\rm NS} a_{\rm pt}(Q^2)^4 . \tag{3}$$

Table 1. The HT coefficient $\mu_4^{p-n}(Q^2_{\rm in})$ ($Q^2_{\rm in} = 1$ GeV^2) and $Q^2_{\rm min}$, both in GeV^2, for pQCD extracted from data at $Q^2 \leq 3$ GeV^2, in various perturbative QCD orders (up to N^{3}LO).

pQCD	LO	NLO	N^{2}LO	N^{3}LO
$\mu^{p-n}_{4,pQCD}(1.)$	-0.059±0.002	-0.037±0.002	-0.031±0.002	-0.008±0.002
$Q^2_{\rm min}$	0.660	0.660	0.844	0.844

We will evaluate this in analytic QCD frameworks, where the perturbation series (3) must be expressed as a nonpower series via the transformation $a(Q^2)^n \mapsto \mathcal{A}_n^{(j)}(Q^2)$, where index j indicates the analytic QCD framework (j =APT, 2δanQCD, and 3δanQCD). For the numerical evaluation of $\mathcal{A}_n^{(j)}(Q^2)$'s, various programs [13, 14, 15, 17] written in Mathematica are used.[2]

We will use $N_f = 3$ and various renormalization scales (RScl) $\mu^2 \neq Q^2$ in the evaluation of the quantity $E_{{\rm NS},j}(Q^2)$. In such a case, the dependence on the RScl parameter $C \equiv \ln(\mu^2/Q^2)$ enters the coefficients $e_j^{\rm NS} \mapsto e_j^{\rm NS}(C)$ and the couplings $\mathcal{A}_n^{(j)}(Q^2 \exp(C))$.

The last term in BSR (2) has known evolution [25] in pQCD and, consequently, in general analytic versions

$$\mu_4^{p-n}(Q^2) = \mu_4^{p-n}(Q^2_{\rm in}) \left(\frac{a_{\rm pt}(Q^2)}{a_{\rm pt}(Q^2_{\rm in})} \right)^{\gamma_0/8\beta_0} \quad , \quad \mu_{4,j}^{p-n}(Q^2) = \mu_{4,j}^{p-n}(Q^2_{\rm in}) \frac{\mathcal{A}^{(j)}_{\gamma_0/8\beta_0}(Q^2)}{\mathcal{A}^{(j)}_{\gamma_0/8\beta_0}(Q^2_{\rm in})} \ . \tag{4}$$

These contributions are important in the low-energy regime $Q^2 \sim 1$ GeV^2. At very low $Q^2 < 1$ GeV^2, they grow quickly and the OPE series diverges. This is a general problem in OPE. However, we can address this problem by replacing the OPE expression (2) at very low $Q^2 < Q_0^2$ (≈ 0.5 GeV^2) with a χPT-motivated expression [4][3]:

$$\Gamma_1^{p-n}(Q^2) = \frac{(\kappa_n^2 - \kappa_p^2)}{8M_p^2} Q^2 + A\,(Q^2)^2 + B\,(Q^2)^3 \quad (Q^2 \lesssim 0.5 \ {\rm GeV}^2). \tag{5}$$

Here, κ_X is the anomalous moment of the nucleon X ($\kappa_p = 1.793$, $\kappa_n = 1.916$), A and B are fit parameters. The first term ($\sim Q^2$) originates from the Gerasimov–Drell–Hearn sum rule [9]. At higher Q^2 ($Q^2 \gtrsim 0.5$ GeV^2), we use OPE (2).

3. Numerical Results

Here we first test only the OPE approach (2) with pQCD $\overline{\rm MS}$ LT term. We search at each order for a minimum scale $Q^2 = Q^2_{\rm min}$ where χ^2 is minimal, while the other fit parameter is μ_4^{p-n}. In order to fix the $\overline{\rm MS}$ QCD scale $\overline{\Lambda}$, we perform the standard extraction, i.e., $\overline{\Lambda}_{({\rm N_f}=3)}$ value is obtained from a reference value, $a(Q^2 = M_Z^2) = 0.1185/\pi$ [27]. The RGE evolution of $a_{\rm pt}(Q^2)$ down from $Q^2 = M_Z^2$ ($N_f = 5$) to low Q^2 (where $N_f = 3$) is carried out with four-loop $\overline{\rm MS}$ beta function and with the corresponding three-loop quark threshold conditions [28] (at thresholds $\kappa \overline{m}_q(\overline{m}_q)$ with $\kappa = 1$).

In Table 1 we present the obtained values of $\mu^{p-n}_{4,pQCD}$ (only the statistical errors were considered), and $Q^2_{\rm min}$, to various orders in the perturbation expansion (3). As noted in previous works [23, 29, 30], a duality between HT contribution and the order of a perturbation series appears: when we use higher order in pQCD, the HT contribution becomes smaller in its absolute

[2] The programs can be downloaded from the web page http://gcvetic.usm.cl

[3] There are also other methods to address the low-Q^2 regime, e.g., an extension from the GDH sum rule made via a QCD-improved model [24], or with a resummation of perturbative series in [26], or with the LFH QCD-motivated ansatz [8].

IOP Conf. Series: Journal of Physics: Conf. Series **938** (2017) 012055 doi:10.1088/1742-6596/938/1/012055

Table 2. The fit parameter values obtained with various approaches to BSR.

Approach (j)	$\mu_{4,j}^{p-n}(1.)$	C	Q_0^2	A	B	χ^2
$\overline{\rm MS}$ pQCD	-0.0344	1.801	0.646	0.658	-0.840	24.44
(F)APT	-0.0498	1.019	0.633	0.658	-0.840	13.53
2δanQCD	-0.0238	-0.859	0.500	0.831	-1.269	5.49
(31)3δanQCD	-0.0105	0.795	0.467	0.752	-1.065	4.97
(41)3δanQCD	-0.0187	1.017	0.431	0.842	-1.342	4.95

value. But this apparent property is unstable, because the $\mu_{4,pQCD}^{p-n}$ coefficient is very sensitive to $\Lambda^{(\rm pQCD)}$ parameter at N^{3}LO [23]. The extracted values are consistent with those obtained at LO in [31], and at NLO in [32].

In Fig. 1 we show the pQCD fit of BSR function Γ_1^{p-n} at NLO, N^{2}LO and N^{3}LO. Increasing the perturbation order, the range of applicability of pQCD becomes smaller, covering fewer points of data in the low-Q^2 region.

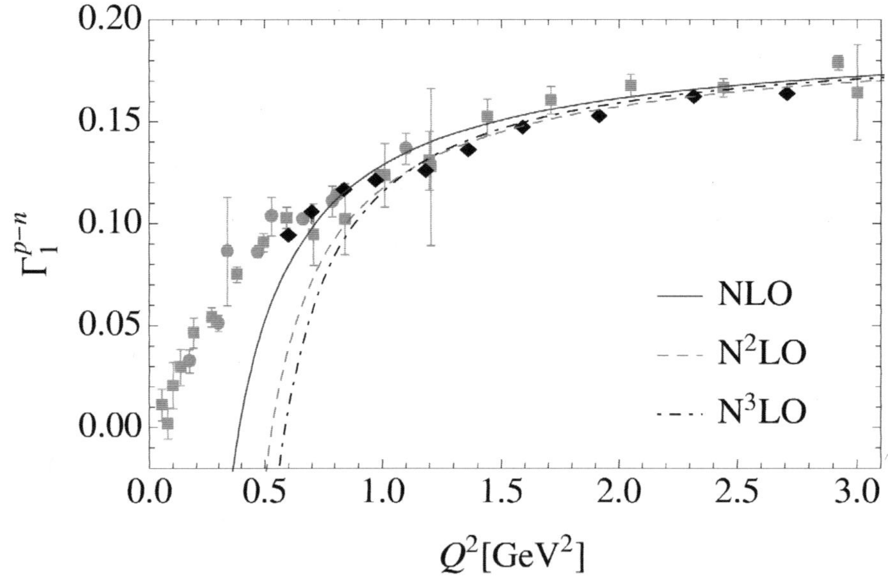

Figure 1. The described fits of data [3, 4, 5, 6] on BSR Γ_1^{p-n} as a function of Q^2, to various orders of perturbation series (2). The newer data [5] have small statistical errors and are in black, and the older data [3, 4, 6] are in light grey (orange online).

Now we employ analytic (holomorphic) QCD approaches in the fits. In Table 2 we show, for five different evaluations of the LT contribution $E_{\rm NS}(Q^2)$, the resulting values of the fit parameters: HT coefficient $\mu_4^{p-n}(Q_{\rm in}^2)$; RScl parameter $C \equiv \ln(\mu^2/Q^2)$ of the LT contribution; matching point scale Q_0^2; parameter A in the χPT-motivated expression (5). The values of the B parameter of the χPT-motivated expression were extracted by the matching condition at $Q^2 = Q_0^2$. The last column gives the values of χ^2 for the obtained curves. The resulting values of A are approximately consistent with the value $A = 0.74$ obtained in χPT calculations in [34] but not with the value $A = 2.4$ obtained in [35].

The obtained curves are shown in Fig. 2. As mentioned, these curves consist of two curves "glued together" at a matching point Q_0^2: the OPE curve (2) for $Q^2 \geq Q_0^2$ and the χPT-motivated curve (5) for $Q^2 \leq Q_0^2$.

IOP Conf. Series: Journal of Physics: Conf. Series **938** (2017) 012055 doi:10.1088/1742-6596/938/1/012055

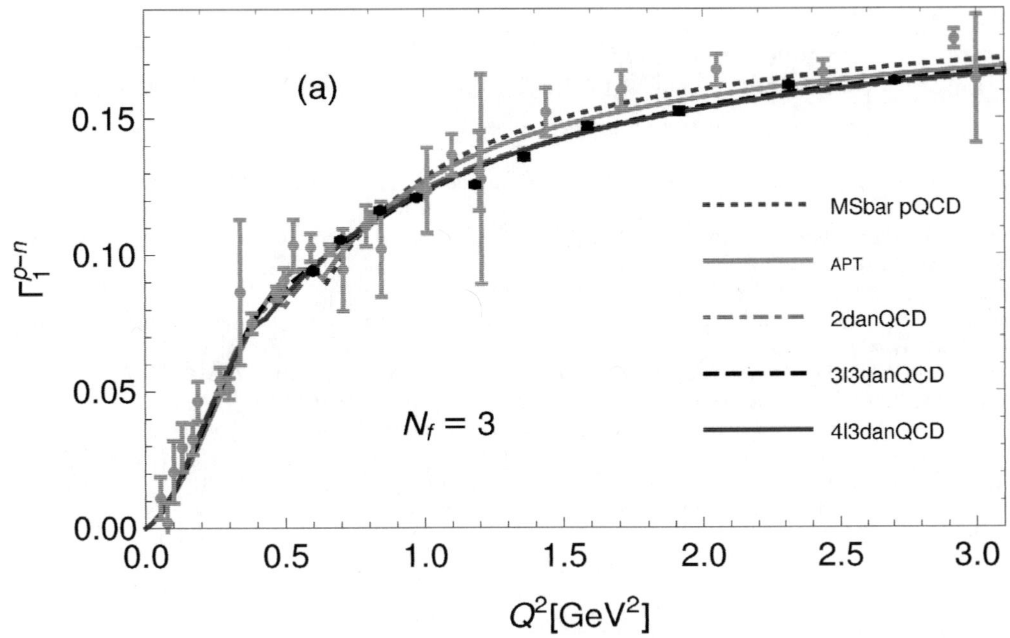

Figure 2. Fits of data [3, 4, 5, 6] on BSR $\Gamma_1^{p-n}(Q^2)$ as a function of Q^2, using at $Q^2 \geq Q_0^2$ (four-loop) $\overline{\text{MS}}$ pQCD and analytic QCD frameworks.

These curves indicate that the pQCD $\overline{\text{MS}}$ approach and, to a lesser extent, the (F)APT approach, give a visible slope discontinuity (kink) at the matching point $Q^2 = Q_0^2$ between the OPE and the χPT-motivated expression, i.e., they cannot bridge well the gap between the high and low-Q^2 regimes. However, 2δanQCD and 3δanQCD are able to bridge this gap without a visible kink, cf. Fig. 2.

4. Conclusions

In [1] we investigated the Bjorken polarized sum rule (BSR) $\Gamma_1^{p-n}(Q^2)$ (with the elastic contribution excluded) as a function of squared momentum transfer Q^2, at $Q^2 \leq 3$ GeV^2 in various QCD approaches, and compared it with the available experimental data. For $Q^2 \geq Q_0^2$ (≈ 0.5 GeV^2) we used the theoretical expressions for the leading-twist (LT) contribution plus one higher-twist (HT) term μ_4^{p-n}/Q^2, Eqs. (3) and (2). For $Q^2 \leq Q_0^2$, the χPT-motivated expression (5) was used. The fit parameters were the HT coefficient $\mu_4^{p-n}(Q_{\text{in}}^2)$ (at $Q_{\text{in}}^2 = 1$ GeV^2), the renormalization scale (RScl) parameter $C \equiv \ln(\mu^2/Q^2)$ in the LT contribution, the transition scale Q_0^2, and the free parameter A in the χPT-motivated expression (5). For the evaluation of the LT contribution at $Q^2 \geq Q_0^2$ we used the usual $\overline{\text{MS}}$ pQCD, and various QCD versions with holomorphic infrared-safe coupling $\mathcal{A}(Q^2)$: (F)APT [10]; 2δanQCD [13, 16, 17]; and a lattice-motivated 3δanQCD coupling in the three- and four-loop lattice MiniMOM scheme: 3l3δanQCD [14] and 4l3δanQCD [15]. It turned out that the latter three holomorphic (analytic) QCD versions give the best fit results and the lowest values of $Q_0^2 \approx$ 0.4-0.5 GeV^2 and $\chi^2 \approx$ 5.0-5.5. (F)APT requires a higher value $Q_0^2 \approx 0.63$ GeV^2 and gives $\chi^2 \approx 13.5$. The $\overline{\text{MS}}$ pQCD gives the worst results, $\chi^2 \approx 24.$; the principal reason for this lies in the fact that the $\overline{\text{MS}}$ pQCD coupling $a_{\text{pt}}(Q^2)$ has Landau singularities at positive $Q^2 \leq 0.37$ GeV^2, and this makes the evaluation of low-Q^2 BSR unreliable.

The presented evaluation of low-Q^2 BSR shows that it is imperative to use QCD frameworks whose couplings have no Landau singularities in this region.

Acknowledgments

The work of C.A. was supported by FONDECYT Postdoctoral Grant No. 3170116. The work of A.V.K. and B.G.S. was supported in part by the RFBR Grant No. 16-02-00790-a.

References

[1] C. Ayala, G. Cvetič, A. V. Kotikov and B. G. Shaikhatdenov, arXiv:1708.06284 [hep-ph].
[2] J. D. Bjorken, Phys. Rev. **148**, 1467 (1966); Phys. Rev. D **1**, 1376 (1970).
[3] A. Deur *et al.*, Phys. Rev. Lett. **93**, 212001 (2004).
[4] A. Deur *et al.*, Phys. Rev. D **78**, 032001 (2008).
[5] A. Deur *et al.*, Phys. Rev. D **90**, 012009 (2014).
[6] K. Abe *et al.* [E154 Collaboration], Phys. Rev. Lett. **79**, 26 (1997); P. L. Anthony *et al.* [E155 Collaboration], Phys. Lett. B **493**, 19 (2000).
[7] P. A. Baikov, K. G. Chetyrkin and J. H. Kühn, Phys. Rev. Lett. **104**, 132004 (2010).
[8] S. J. Brodsky, G. F. de Teramond and A. Deur, Phys. Rev. D **81**, 096010 (2010)
[9] S. B. Gerasimov, Sov.J.Nucl. Phys. **2**, 430 (1966); S. D. Drell and A. C. Hearn, Phys.Rev. Lett. **16**, 908 (1966).
[10] D. V. Shirkov, I. L. Solovtsov, Phys. Rev. Lett. 79 (1997) 1209–1212; K. A. Milton, I. L. Solovtsov, Phys. Rev. D55 (1997) 5295–5298; I. L. Solovtsov, D. V. Shirkov, Phys. Lett. B442 (1998) 344–348; D. V. Shirkov, Theor. Math. Phys. **127** (2001) 409; Eur. Phys. J. C **22** (2001) 331.
[11] A. P. Bakulev, S. V. Mikhailov, N. G. Stefanis, Phys. Rev. D **72** (2005) 074014; A. P. Bakulev, A. I. Karanikas and N. G. Stefanis, Phys. Rev. D **72** (2005) 074015.
[12] A. P. Bakulev, S. V. Mikhailov and N. G. Stefanis, Phys. Rev. D **75** (2007) 056005; JHEP **1006** (2010) 085.
[13] C. Ayala, C. Contreras and G. Cvetič, Phys. Rev. D **85** (2012) 114043.
[14] C. Ayala, G. Cvetič and R. Kögerler, J. Phys. G **44**, no. 7, 075001 (2017)
[15] C. Ayala, G. Cvetič, R. Kögerler and I. Kondrashuk, arXiv:1703.01321 [hep-ph].
[16] G. Cvetič and C. Villavicencio, Phys. Rev. D **86**, 116001 (2012).
[17] C. Ayala and G. Cvetič, Comput. Phys. Commun. **190**, 182 (2015).
[18] G. Cvetič and C. Valenzuela, Phys. Rev. D **74**, 114030 (2006); J. Phys. G **32**, L27 (2006).
[19] G. Cvetič and A. V. Kotikov, J. Phys. G **39**, 065005 (2012).
[20] I. L. Bogolubsky, E.-M. Ilgenfritz, M. Müller-Preussker and A. Sternbeck, Phys. Lett. B **676**, 69 (2009); E.-M. Ilgenfritz, M. Müller-Preussker, A. Sternbeck and A. Schiller, hep-lat/0601027; A. G. Duarte, O. Oliveira and P. J. Silva, Phys. Rev. D **94**, no. 1, 014502 (2016).
[21] M. Baldicchi, A. V. Nesterenko, G. M. Prosperi, D. V. Shirkov and C. Simolo, Phys. Rev. Lett. **99**, 242001 (2007); M. Baldicchi, A. V. Nesterenko, G. M. Prosperi and C. Simolo, Phys. Rev. D **77**, 034013 (2008).
[22] C. Patrignani *et al.* [Particle Data Group Collaboration], Chin. Phys. C **40**, no. 10, 100001 (2016).
[23] R. S. Pasechnik, D. V. Shirkov and O. V. Teryaev, Phys. Rev. D **78**, 071902 (2008); R. S. Pasechnik, D. V. Shirkov, O. V. Teryaev, O. P. Solovtsova and V. L. Khandramai, Phys. Rev. D **81**, 016010 (2010); Phys. Lett. B **706**, 340 (2012); V. L. Khandramai, O. P. Solovtsova and O. V. Teryaev, Nonlin. Phenom. Complex Syst. **16**, 93 (2013).
[24] J. Soffer and O. Teryaev, Phys. Rev. Lett. **70**, 3373 (1993); Phys. Rev. D **70**, 116004 (2004).
[25] E. V. Shuryak and A. I. Vainshtein, Nucl. Phys. B **201**, 141 (1982); H. Kawamura, T. Uematsu, J. Kodaira and Y. Yasui, Mod. Phys. Lett. A **12**, 135 (1997).
[26] A. V. Kotikov and B. G. Shaikhatdenov, Phys. Part. Nucl. **45**, 26 (2014).
[27] K. A. Olive *et al.* [Particle Data Group Collaboration], Chin. Phys. C **38**, 090001 (2014)
[28] K. G. Chetyrkin, B. A. Kniehl and M. Steinhauser, Phys. Rev. Lett. **79**, 2184 (1997); Nucl. Phys. B **510**, 61 (1998); K. G. Chetyrkin, J. H. Kühn and M. Steinhauser, Comput. Phys. Commun. **133**, 43 (2000).
[29] A. L. Kataev, A. V. Kotikov, G. Parente and A. V. Sidorov, Phys. Lett. B **417**, 374 (1998); A. L. Kataev, G. Parente and A. V. Sidorov, Nucl. Phys. B **573**, 405 (2000); Phys. Part. Nucl. **34**, 20 (2003); S. Narison and V. I. Zakharov, Phys. Lett. B **679**, 355 (2009).
[30] G. Parente, A. V. Kotikov and V. G. Krivokhizhin, Phys. Lett. B **333**, 190 (1994); B. G. Shaikhatdenov, A. V. Kotikov, V. G. Krivokhizhin and G. Parente, Phys. Rev. D **81**, 034008 (2010); A. V. Kotikov, V. G. Krivokhizhin and B. G. Shaikhatdenov, Phys. Atom. Nucl. **75**, 507 (2012); J. Phys. G **42**, no. 9, 095004 (2015).
[31] G. G. Ross and R. G. Roberts, Phys. Lett. B **322**, 425 (1994).
[32] A. V. Sidorov and C. Weiss, Phys. Rev. D **73**, 074016 (2006).
[33] J. Beringer *et al.* [Particle Data Group Collaboration], Phys. Rev. D **86**, 010001 (2012).
[34] X. D. Ji, C. W. Kao and J. Osborne, Phys. Lett. B **472**, 1 (2000).
[35] V. Bernard, T. R. Hemmert and U. G. Meissner, Phys. Rev. D **67**, 076008 (2003).

Possible effect of mixed phase and deconfinement upon spin correlations in the $\Lambda\bar{\Lambda}$ pairs generated in relativistic heavy-ion collisions

Valery V Lyuboshitz[1] and **Vladimir L Lyuboshitz**

Joint Institute for Nuclear Research, 141980 Dubna, Moscow Region, Russia

E-mail: `Valery.Lyuboshitz@jinr.ru`

Abstract. Spin correlations for the $\Lambda\Lambda$ and $\Lambda\bar{\Lambda}$ pairs, generated in relativistic heavy-ion collisions, and related angular correlations at the joint registration of space-parity nonconserving hadronic decays of two hyperons are theoretically analyzed. The correlation tensor components can be derived from the double angular distribution of products of two decays by the method of "moments". The properties of the "trace" of the correlation tensor (a sum of three diagonal components), determining the angular correlations as well as the relative fractions of the triplet and singlet states of respective pairs, are discussed. In the present report, spin correlations for two identical ($\Lambda\Lambda$) and two non-identical ($\Lambda\bar{\Lambda}$) particles are generally considered within the conventional model of one-particle sources, implying that correlations vanish at sufficiently large relative momenta. However, under these conditions (especially at ultrarelativistic energies), for two non-identical particles ($\Lambda\bar{\Lambda}$) the two-particle annihilation sources – quark-antiquark and two-gluon ones – start playing a noticeable role and lead to the difference of the correlation tensor from zero. In particular, such a situation may arise, when the system passes through the "mixed phase" and – due to the multiple production of free quarks and gluons in the process of deconfinement of hadronic matter – the number of two-particle sources strongly increases.

1. General structure of the spin density matrix of the pairs $\Lambda\Lambda$ and $\Lambda\bar{\Lambda}$

Spin correlations for $\Lambda\Lambda$ and $\Lambda\bar{\Lambda}$ pairs, generated in relativistic heavy-ion collisions, and respective angular correlations at joint registration of space-parity nonconserving hadronic decays of two hyperons give important information on the character of multiple processes .

The spin density matrix of the $\Lambda\Lambda$ and $\Lambda\bar{\Lambda}$ pairs, just as the spin density matrix of two spin-1/2 particles in general, can be presented in the following form [1,2,3]:

$$\hat{\rho}^{(1,2)} = \frac{1}{4}\left[\hat{I}^{(1)} \otimes \hat{I}^{(2)} + (\hat{\boldsymbol{\sigma}}^{(1)}\mathbf{P}_1) \otimes \hat{I}^{(2)} + \hat{I}^{(1)} \otimes (\hat{\boldsymbol{\sigma}}^{(2)}\mathbf{P}_2) + \sum_{i=1}^{3}\sum_{k=1}^{3} T_{ik}\hat{\sigma}_i^{(1)} \otimes \hat{\sigma}_k^{(2)}\right]; \qquad (1)$$

in doing so, $tr_{(1,2)}\hat{\rho}^{(1,2)} = 1$. Here $\hat{I}$ is the two-row unit matrix, $\hat{\boldsymbol{\sigma}} = (\ \hat{\sigma}_x,\ \hat{\sigma}_y,\ \hat{\sigma}_z\ )$ is the vector Pauli operator ($x, y, z \to 1, 2, 3$), $\mathbf{P}_1$ and $\mathbf{P}_2$ are the polarization vectors of first and second particle ($\mathbf{P}_1 = \langle\hat{\boldsymbol{\sigma}}^{(1)}\rangle, \mathbf{P}_2 = \langle\hat{\boldsymbol{\sigma}}^{(2)}\rangle$), $T_{ik} = \langle\hat{\sigma}_i^{(1)} \otimes \hat{\sigma}_k^{(2)}\rangle$ are the correlation tensor components. In the general case $T_{ik} \neq P_{1i}\, P_{2k}$. The tensor with components $C_{ik} = T_{ik} - P_{1i}P_{2k}$ describes the spin correlations of two particles .

[1] Speaker at DSPIN-17

2. Spin correlations and angular correlations at joint registration of decays of two Λ particles into the channel $\Lambda \to p + \pi^-$

It is essential that any decay of an unstable particle may serve as an analyzer of its spin state .

The normalized angular distribution at the decay $\Lambda \to p + \pi^-$ takes the form:

$$\frac{d\,w(\mathbf{n})}{d\,\Omega_{\mathbf{n}}} = \frac{1}{4\,\pi}(1 + \alpha_\Lambda \mathbf{P}_\Lambda\, \mathbf{n}). \tag{2}$$

Here $\mathbf{P}_\Lambda$ is the polarization vector of the Λ particle, $\mathbf{n}$ is the unit vector along the direction of proton momentum in the rest frame of the Λ particle, α_Λ is the coefficient of P-odd angular asymmetry ($\alpha_\Lambda = 0.642$). The decay $\Lambda \to p + \pi^-$ selects the projections of spin of the Λ particle onto the direction of proton momentum; the analyzing power equals $\boldsymbol{\xi} = \alpha_\Lambda \mathbf{n}$.

Now let us consider the double angular distribution of flight directions for protons formed in the decays of two Λ particles into the channel $\Lambda \to p + \pi^-$, normalized by unity (the analyzing powers are $\boldsymbol{\xi}_1 = \alpha_\Lambda \mathbf{n}_1$, $\boldsymbol{\xi}_2 = \alpha_\Lambda \mathbf{n}_2$). It is described by the following formula [2,3]:

$$\frac{d^2\,w(\mathbf{n}_1, \mathbf{n}_2)}{d\,\Omega_{\mathbf{n}_1}\,d\,\Omega_{\mathbf{n}_2}} = \frac{1}{16\,\pi^2}\Big[1 + \alpha_\Lambda \mathbf{P}_1\,\mathbf{n}_1 + \alpha_\Lambda \mathbf{P}_2\,\mathbf{n}_2 + \alpha_\Lambda^2 \sum_{i=1}^{3}\sum_{k=1}^{3} T_{ik} n_{1i} n_{2k}\Big], \tag{3}$$

where $\mathbf{P}_1$ and $\mathbf{P}_2$ are polarization vectors of the first and second Λ particle, T_{ik} are the correlation tensor components, $\mathbf{n}_1$ and $\mathbf{n}_2$ are unit vectors in the respective rest frames of the first and second Λ particle, defined in the common (unified) coordinate axes of the c.m. frame of the pair ($i, k = \{1, 2, 3\} = \{x, y, z\}$) .

The polarization parameters can be determined from the angular distribution of decay products by the method of moments [2,3] .

The angular correlation, integrated over all angles except the angle between the vectors $\mathbf{n}_1$ and $\mathbf{n}_2$ and described by the formula [2,3,4,5]

$$d\,w(\cos\theta) = \frac{1}{2}\,(\,1 + \frac{1}{3}\,\alpha_\Lambda^2 T\,\cos\theta)\sin\theta d\,\theta = \frac{1}{2}\,[\,1 - \alpha_\Lambda^2\,(W_s - \frac{W_t}{3})\cos\theta]\sin\theta d\,\theta, \tag{4}$$

is determined only by the “trace” of the correlation tensor $T = W_t - 3W_s$ (W_s and W_t are relative fractions of the singlet state and triplet states, respectively), and it does not depend on the polarization vectors (single-particle states may be unpolarized).

3. Correlations at the joint registration of the decays $\Lambda \to p + \pi^-$ and $\bar\Lambda \to \bar p + \pi^+$

Due to CP invariance, the coefficients of P-odd angular asymmetry for the decays $\Lambda \to p + \pi^-$ and $\bar\Lambda \to \bar p + \pi^+$ have equal absolute values and opposite signs: $\alpha_{\bar\Lambda} = -\alpha_\Lambda = -0.642$. The double angular distribution for this case is as follows [2,3]:

$$\frac{d^2\,w(\mathbf{n}_1, \mathbf{n}_2)}{d\,\Omega_{\mathbf{n}_1}\,d\,\Omega_{\mathbf{n}_2}} = \frac{1}{16\,\pi^2}\Big[1 + \alpha_\Lambda \mathbf{P}_\Lambda\,\mathbf{n}_1 - \alpha_\Lambda \mathbf{P}_{\bar\Lambda}\,\mathbf{n}_2 - \alpha_\Lambda^2 \sum_{i=1}^{3}\sum_{k=1}^{3} T_{ik} n_{1i} n_{2k}\Big] \tag{5}$$

(here $-\alpha_\Lambda = +\alpha_{\bar\Lambda}$ and $-\alpha_\Lambda^2 = +\alpha_\Lambda \alpha_{\bar\Lambda}$) .

Thus, the angular correlation between the proton and antiproton momenta in the rest frames of the Λ and $\bar\Lambda$ particles is described by the expression:

$$d\,w(\cos\theta) = \frac{1}{2}\,(\,1 - \frac{1}{3}\,\alpha_\Lambda^2 T\,\cos\theta)\sin\theta d\,\theta = \frac{1}{2}\,[\,1 + \alpha_\Lambda^2\,(W_s - \frac{W_t}{3})\cos\theta]\sin\theta d\,\theta, \tag{6}$$

where θ is the angle between the proton and antiproton momenta .

4. Spin correlations at the generation of $\Lambda\bar{\Lambda}$ pairs in multiple processes

Further we will use the model of one-particle sources [6], which is the most adequate one in the case of collisions of relativistic ions .

As far as two Λ particles (identical particles) are concerned, the spin correlations and angular correlations at their decays were considered previously, within the model of one-particle sources, in the works [2,7] – taking into account Fermi statistics and final-state interaction (see also the detailed analysis in the paper [8]).

In the present report, we will be interested in spin correlations at the decays of $\Lambda\bar{\Lambda}$ pairs . In the framework of the model of independent one-particle sources, spin correlations in the $\Lambda\bar{\Lambda}$ system arise only on account of the difference between the interaction in the final triplet state ($S = 1$) and the interaction in the final singlet state . At small relative momenta, the s-wave interaction plays the dominant role as before, but, contrary to the case of identical particles ($\Lambda\Lambda$), in the case of non-identical particles ($\Lambda\bar{\Lambda}$) the total spin may take both the values $S = 1$ and $S = 0$ at the orbital momentum $L = 0$. In doing so, the interference effect, connected with quantum statistics, is absent .

If the sources emit unpolarized particles, then, in the case under consideration, the correlation function describing momentum-energy correlations has the following structure (in the c.m. frame of the $\Lambda\bar{\Lambda}$ pair):

$$R(\mathbf{k},\mathbf{v}) = 1 + \frac{3}{4}\, B_t^{(\Lambda\bar{\Lambda})}(\mathbf{k},\mathbf{v}) + \frac{1}{4}\, B_s^{(\Lambda\bar{\Lambda})}(\mathbf{k},\mathbf{v}). \tag{7}$$

The spin density matrix of the $\Lambda\bar{\Lambda}$ pair is given by the formula :

$$\hat{\rho}^{(\Lambda\bar{\Lambda})} = \hat{I}^{(1)} \otimes \hat{I}^{(2)} + \frac{B_t^{(\Lambda\bar{\Lambda})}(\mathbf{k},\mathbf{v}) - B_s^{(\Lambda\bar{\Lambda})}(\mathbf{k},\mathbf{v})}{4\, R(\mathbf{k},\mathbf{v})}\, \hat{\boldsymbol{\sigma}}^{(1)} \otimes \hat{\boldsymbol{\sigma}}^{(2)}, \tag{8}$$

and the components of the correlation tensor are as follows:

$$T_{ik} = \frac{B_t^{(\Lambda\bar{\Lambda})}(\mathbf{k},\mathbf{v}) - B_s^{(\Lambda\bar{\Lambda})}(\mathbf{k},\mathbf{v})}{4 + 3\, B_t^{(\Lambda\bar{\Lambda})}(\mathbf{k},\mathbf{v}) + B_s^{(\Lambda\bar{\Lambda})}(\mathbf{k},\mathbf{v})}\, \delta_{ik}; \tag{9}$$

here the contributions of final-state triplet and singlet $\Lambda\bar{\Lambda}$ interaction are determined by the expression obtained in the works [2,7] .

At sufficiently large values of k, one should expect that [7]:

$$B_s^{(\Lambda\bar{\Lambda})}(\mathbf{k},\mathbf{v}) = 0, \qquad B_t^{(\Lambda\bar{\Lambda})}(\mathbf{k},\mathbf{v}) = 0.$$

In this case the angular correlations in the decays $\Lambda \to p + \pi^-$, $\bar{\Lambda} \to \bar{p} + \pi^+$, connected with the final-state interaction, are absent :

$$T_{ik} = 0, \quad T = 0.$$

5. Angular correlations in the decays $\Lambda \to p + \pi^-$ and $\bar{\Lambda} \to \bar{p} + \pi^+$ and the “mixed phase”

Thus, at sufficiently large relative momenta (for example, $k \gtrsim m_\pi$) one should expect that the angular correlations in the decays $\Lambda \to p + \pi^-$ and $\bar{\Lambda} \to \bar{p} + \pi^+$, connected with the interaction of the Λ and $\bar{\Lambda}$ hyperons in the final state (i.e. with one-particle sources) are absent . But, if at the considered energy the dynamical trajectory of the system passes through the so-called “mixed phase”, then the two-particle sources, consisting of the free quark and antiquark , start playing a noticeable role . For example, the process $s\,\bar{s} \to \Lambda\bar{\Lambda}$ may be discussed .

In this process, the charge parity of the pairs $s\bar{s}$ and $\Lambda\bar{\Lambda}$ is equal to $C = (-1)^{L+S}$, where L is the orbital momentum and S is the total spin of the fermion and antifermion. Meantime, the CP parity of the fermion-antifermion pair is $CP = (-1)^{S+1}$.

In the case of one-gluon exchange, $CP = 1$, and then $S = 1$, i.e. the $\Lambda\bar{\Lambda}$ pair is generated in the triplet state; in doing so, the "trace" of the correlation tensor $T = 1$.

Even if the frames of one-gluon exchange are overstepped, the quarks s and $\bar{s}$, being ultrarelativistic, interact in the triplet state ($S = 1$) . In so doing, the primary CP parity $CP = 1$, and, due to the CP parity conservation, the $\Lambda\bar{\Lambda}$ pair is also produced in the triplet state. Let us denote the contribution of two-quark sources by x . Then at large relative momenta

$$T = x > 0.$$

Apart from the two-quark sources, there are also two-gluon sources being able to play a comparable role. Analogously with the annihilation process $\gamma\gamma \to e^+e^-$ [9], in this case the "trace" of the correlation tensor is described by the formula (here the process $g\,g \to \Lambda\bar{\Lambda}$ is implied):

$$T = 1 - \frac{4\,(1-\beta^2)}{1 + 2\,\beta^2\,\sin^2\theta - \beta^4 - \beta^4\,\sin^4\theta}, \tag{10}$$

where β is the velocity of Λ (and $\bar{\Lambda}$) in the c.m. frame of the $\Lambda\bar{\Lambda}$ pair, θ is the angle between the momenta of one of the gluons and Λ in the c.m. frame (see [9]). At small β ($\beta \ll 1$) the $\Lambda\bar{\Lambda}$ pair is produced in the singlet state (total spin $S = 0$, $T = -3$), whereas at $\beta \approx 1$ – in the triplet state ($S = 1$, $T = 1$). Let us remark that at ultrarelativistic velocities β (i.e. at extremely large relative momenta of Λ and $\bar{\Lambda}$) both the two-quark and two-gluon mechanisms lead to the triplet state of the $\Lambda\bar{\Lambda}$ pair ($T = 1$).

6. Summary

So, it is surely advisable to investigate the spin correlations of $\Lambda\Lambda$ and $\Lambda\bar{\Lambda}$ pairs produced in relativistic heavy ion collisions (see also, e.g., [8,10]) .

The spin correlations, as well as the momentum–energy ones, make it possible to determine the space–time characteristics of the multiple particle generation region . In doing so, the best way of studying the spin correlations is the method of angular correlations – method of moments.

Finally, it should be emphasized that, in the general case, the appearance of angular correlations in the decays $\Lambda \to p + \pi^-$ and $\bar{\Lambda} \to \bar{p} + \pi^+$ with the nonzero values of the "trace" of the correlation tensor T at large relative momenta of the Λ and $\bar{\Lambda}$ particles may testify to the passage of the system through the "mixed phase" [8,10] .

References

[1] Lyuboshitz V L and Podgoretsky M I 1997 *Yad. Fiz.* **60** 45
[2] Lyuboshitz V L 2000 *Proc. of the XXXIV PNPI Winter School "Physics of Atomic Nuclei and Elementary Particles"* (Saint-Petersburg) p 402
[3] Lednicky R and Lyuboshitz V L 2001 *Phys. Lett.* B **508** 146
[4] Alexander G and Lipkin H J 1995 *Phys. Lett.* B **352** 162
[5] Lednicky R, Lyuboshitz V V and Lyuboshitz V L 2003 *Yad. Fiz.* **66** 1007
[6] Podgoretsky M I 1989 *Fiz. Elem. Chast. At. Yadra* **20** 628
[7] Lednicky R and Lyuboshitz V L 1982 *Yad. Fiz.* **35** 1316
[8] Lyuboshitz V L and Lyuboshitz V V 2010 *Yad. Fiz.* **73** 836 [*Phys. At. Nucl.* **73** 805]
[9] McMaster W H 1961 *Rev. Mod. Phys.* **33** 8
[10] Lyuboshitz V V and Lyuboshitz V L 2013 *Proc. of the First International Large Hadron Collider Physics Conference – LHCP 2013 (Barcelona, Spain, May 13 - 18, 2013), EPJ Web Conf.* **60** 20021 (3 pages)

IOP Conf. Series: Journal of Physics: Conf. Series **938** (2017) 012057 doi:10.1088/1742-6596/938/1/012057

Polarization effects in the reactions $p +^3 He \to \pi^+ +^4 He$, $\pi^+ + {}^4He \to p + {}^3He$ and quantum character of spin correlations in the final $(p,\ {}^3He)$ system

Valery V Lyuboshitz[1] and **Vladimir L Lyuboshitz**

Joint Institute for Nuclear Research, 141980 Dubna, Moscow Region, Russia

E-mail: Valery.Lyuboshitz@jinr.ru

Abstract. The general consequences of T invariance for the direct and inverse binary reactions $a + b \to c + d$, $c + d \to a + b$ with spin-1/2 particles a, b and unpolarized particles c, d are considered. Using the formalism of helicity amplitudes, the polarization effects are studied in the reaction $p + {}^3He \to \pi^+ + {}^4He$ and in the inverse process $\pi^+ + {}^4He \to p + {}^3He$. It is shown that in the reaction $\pi^+ + {}^4He \to p + {}^3He$ the spins of the final proton and ${}^3He$ nucleus are strongly correlated. A structural expression through helicity amplitudes, corresponding to arbitrary emission angles, is obtained for the correlation tensor. It is established that in the reaction $\pi^+ + {}^4He \to p + {}^3He$ one of the "classical" incoherence inequalities of the Bell type for diagonal components of the correlation tensor is necessarily violated and, thus, the spin correlations of the final particles have the strongly pronounced quantum character.

1. Consequences of T invariance for binary reactions

Let us consider the reaction $a+b \to c+d$, where a, b are the spin-1/2 particles and the particles c, d have arbitrary spins. The effective cross-section $\sigma_{a+b\to c+d}$ in the c.m. frame of the particles a and b, summed over spin projections of the final particles c and d, has the structure [1]:

$$\sigma_{a+b\to c+d}(\mathbf{k}_a, \mathbf{P}^{(a)}, \mathbf{P}^{(b)}; \mathbf{k}_c) = \sigma_0(E,\theta)\, L(\mathbf{k}_a, \mathbf{P}^{(a)}, \mathbf{P}^{(b)}; \mathbf{k}_c), \tag{1}$$

where $\sigma_0(E,\theta)$ is the respective cross-section for unpolarized particles a, b and L is the linear function of the polarization vectors $\mathbf{P}^{(a)}$ and $\mathbf{P}^{(b)}$:

$$L(\mathbf{k}_a, \mathbf{P}^{(a)}, \mathbf{P}^{(b)}; \mathbf{k}_c) = 1 + A(E,\theta)(\mathbf{P}^{(a)}\mathbf{n}) + B(E,\theta)(\mathbf{P}^{(b)}\mathbf{n}) + C(E,\theta)(\mathbf{P}^{(a)}\mathbf{P}^{(b)}) +$$
$$+D(E,\theta)(\mathbf{P}^{(a)}\mathbf{l})(\mathbf{P}^{(b)}\mathbf{l}) + F(E,\theta)(\mathbf{P}^{(a)}\mathbf{m})(\mathbf{P}^{(b)}\mathbf{m}) + G(E,\theta)(\mathbf{P}^{(a)}\mathbf{l})(\mathbf{P}^{(b)}\mathbf{m}) + H(E,\theta)(\mathbf{P}^{(a)}\mathbf{m})(\mathbf{P}^{(b)}\mathbf{l}). \tag{2}$$

Here $\mathbf{l}$, $\mathbf{m}$, $\mathbf{n}$ are the mutually orthogonal unit vectors, defined as:

$$\mathbf{l} = \mathbf{k}_a/k_a; \quad \mathbf{m} = \frac{\mathbf{l}' - \mathbf{l}(\mathbf{l}'\mathbf{l})}{\sin\theta}; \quad \mathbf{n} = \frac{\mathbf{l}\times\mathbf{l}'}{\sin\theta} \tag{3}$$

[1] Presenter at DSPIN-17

($\mathbf{l}' = \mathbf{k}_c/k_c$); $\mathbf{k}_a$, $\mathbf{k}_c$ are the respective momenta of the particles a and c, E is the total energy in the c.m. frame and $\theta = \arccos\,(\mathbf{l}\,\mathbf{l}')$ is the emission angle.

Meantime, for the inverse reaction $c + d \to a + b$ with the unpolarized particles c, d and the fixed polarization vectors of the final particles $\boldsymbol{\zeta}^{(a)}$, $\boldsymbol{\zeta}^{(b)}$ the effective cross-section takes, due to T invariance and the principle of detailed balance, the following form [1]:

$$\sigma_{c+d\to a+b}(\mathbf{k}_c; \mathbf{k}^{(a)}, \boldsymbol{\zeta}^{(a)}, \boldsymbol{\zeta}^{(b)}) = \frac{1}{4}\,\widetilde{\sigma}_0(E,\theta) L(-\mathbf{k}_a, -\boldsymbol{\zeta}^{(a)}, -\boldsymbol{\zeta}^{(b)}; -\mathbf{k}_c), \tag{4}$$

where

$$\widetilde{\sigma}_0(E,\theta) = \frac{4k_a^2}{k_c^2(2j_c+1)(2j_d+1)}\sigma_0(E,\theta) \tag{5}$$

is the cross-section of the inverse reaction, summed over the spin projections of the final particles a, b. The two-particle spin density matrix $\hat{\rho}^{(a,b)}$ for the final particles a, b can also be expressed through the same function L, replacing the polarization vectors by the vector Pauli operators:

$$\hat{\rho}^{(a,b)} = \frac{1}{4}\Big[\,\hat{I}^{(a)} \otimes \hat{I}^{(b)} + (\mathbf{P}^{(a)}(E,\theta)\,\hat{\boldsymbol{\sigma}}^{(a)}) \otimes \hat{I}^{(b)} + \hat{I}^{(a)} \otimes (\mathbf{P}^{(b)}(E,\theta)\,\hat{\boldsymbol{\sigma}}^{(b)}) +$$
$$+ \sum_{i=1}^{3}\sum_{k=1}^{3} T_{ik}(E,\theta)\,\hat{\sigma}_i^{(a)} \otimes \hat{\sigma}_k^{(b)}\Big] = \frac{1}{4}\,\hat{L}(-\mathbf{k}_a, -\hat{\boldsymbol{\sigma}}^{(a)}, -\hat{\boldsymbol{\sigma}}^{(b)}; -\mathbf{k}_c). \tag{6}$$

In Eq. (6) $\hat{I}^a$, $\hat{I}^{(b)}$ are two-row unit matrices, $\mathbf{P}^{(a)}(E,\theta) = -A(E,\theta)\mathbf{n}$, $\mathbf{P}^{(b)}(E,\theta) = -\,B(E,\theta)\mathbf{n}$ are the polarization vectors of the particles a and b,

$$T_{ik}(E,\theta) = C(E,\theta)\delta_{ik} + D(E,\theta)l_i l_k + F(E,\theta)m_i m_k + G(E,\theta)l_i m_k + H(E,\theta)m_i l_k \tag{7}$$

are components of the correlation tensor describing spin correlations in the final two-particle system (a, b). In doing so, all the functions $A, B, C, D, F, G, H, \sigma_0$ and the unit vectors $\mathbf{l}$, $\mathbf{m}$, $\mathbf{n}$ in Eqs. (1)−(7) are *the same* as for the direct reaction $a + b \to c + d$.

Thus, due to T invariance, the dependence of the effective cross-section of the direct reaction $a + b \to c + d$ upon the polarizations of initial particles *completely determines* the polarization vectors and spin correlations for the same particles a, b produced in the inverse reaction $c + d \to a + b$ with unpolarized primary particles.

2. Polarization effects in the reaction $p + {}^3He \to \pi^+ + {}^4He$

This reaction belongs to the type $1/2 + 1/2 \to 0 + 0$ (the proton and ${}^3He$ nucleus have spin 1/2, π^+ and ${}^4He$ have zero spin). Thus, on account of the negative internal parity of the π^+ meson, this reaction can proceed *only from triplet* states of the system $(p, {}^3He)$ [2,3] (as follows from the parity and angular momentum conservation).

Let us choose the axis of the total spin quantization z along the vector $\mathbf{l} = \mathbf{k}_p/k_p$. There exist three possible triplet states of the $(p, {}^3He)$ system with the spin projections $+1$, -1 and 0 onto z:

$$|+1,\mathbf{l}\rangle = |+1/2,\mathbf{l}\rangle^{(p)} \otimes |+1/2,\mathbf{l}\rangle^{(He)}, \quad |-1,\mathbf{l}\rangle = |-1/2,\mathbf{l}\rangle^{(p)} \otimes |-1/2,\mathbf{l}\rangle^{(He)},$$
$$|0,\mathbf{l}\rangle = \frac{1}{\sqrt{2}}\Big(|+1/2,\mathbf{l}\rangle^{(p)} \otimes |-1/2,\mathbf{l}\rangle^{(He)} + |-1/2,\mathbf{l}\rangle^{(p)} \otimes |+1/2,\mathbf{l}\rangle^{(He)}\Big). \tag{8}$$

The two-particle spin density matrix for the $(p, {}^3He)$ system is:

$$\hat{\rho}^{(p,He)} = \frac{1}{4}(\hat{I}^{(p)} + \mathbf{P}^{(p)}\boldsymbol{\sigma}^{(p)}) \otimes (\hat{I}^{(He)} + \mathbf{P}^{(He)}\boldsymbol{\sigma}^{(He)}) \tag{9}$$

($\mathbf{P}^{(p)}$ and $\mathbf{P}^{(He)}$ are the independent polarization vectors). Using the technique of helicity amplitudes (the helicity amplitude $R_\lambda(E,\theta)$, $\lambda = \pm 1, 0$, is the amplitude of the reaction $p + {}^3He \to \pi^+ + {}^4He$ proceeding from the state $|\lambda, \mathbf{l}\rangle$ (8)), we may write [1]:

$$\sigma_{p+{}^3He\to\pi^+ + {}^4He} = \langle\psi|\,\hat{\rho}^{(p,He)}\,|\psi\rangle = \sum_{\lambda}\sum_{\lambda'} R_\lambda(E,\theta)\langle\lambda,\mathbf{l}|\,\hat{\rho}^{(p,He)}\,|\,\lambda',\mathbf{l}\rangle R^*_{\lambda'}(E,\theta); \tag{10}$$

$$|\psi\rangle = \sum_{\lambda=\pm1,0} R^*_\lambda(E,\theta)|\lambda,\mathbf{l}\rangle =$$

$$= R^*_1(E,\theta)\left(|+1/2,z\rangle^{(p)}\otimes|+1/2,z\rangle^{(He)} - |-1/2,z\rangle^{(p)}\otimes|-1/2,z\rangle^{(He)}\right) +$$
$$+\frac{1}{\sqrt{2}}R^*_0(E,\theta)\left(|+1/2,z\rangle^{(p)}\otimes|-1/2,z\rangle^{(He)} + |-1/2,z\rangle^{(p)}\otimes|+1/2,z\rangle^{(He)}\right) \tag{11}$$

is the non-normalized initial two-particle spin state selected by the reaction (due to the parity conservation, $R_{+1}(E,\theta) = -R_{-1}(E,\theta) \equiv R_1(E,\theta)$).

Finally, using Eq. (11) and the formula (9) for the spin density matrix, we find that the cross-section $\sigma_{p+{}^3He\to\pi^+ + {}^4He}$ (10) is described by the general structural formula for $\sigma_{a+b\to c+d}$ (1)-(2), where the functions σ_0, A, B, C, D, F, G, H are bilinear combinations of the helicity amplitudes R_1, R_0 [1]:

$$\sigma_0(E,\theta) = \frac{1}{4}\langle\psi\mid\psi\rangle = \frac{1}{4}(|R_0(E,\theta)|^2 + 2|R_1(E,\theta)|^2), \tag{12}$$

$$A(E,\theta) = B(E,\theta) = \frac{1}{\sqrt{2}\sigma_0(E,\theta)}\mathrm{Im}(R_1(E,\theta)R^*_0(E,\theta)), \tag{13}$$

$$C(E,\theta) = 1, \quad D(E,\theta) = -\frac{|R_0(E,\theta)|^2}{2\sigma_0(E,\theta)}, \quad F(E,\theta) = -\frac{|R_1(E,\theta)|^2}{2\sigma_0(E,\theta)}, \tag{14}$$

$$G(E,\theta) = H(E,\theta) = \frac{1}{\sqrt{2}\sigma_0(E,\theta)}\mathrm{Re}(R_1(E,\theta)R^*_0(E,\theta)). \tag{15}$$

For the particular cases $\theta = 0$ and $\theta = \pi$, when $R_1(E,\theta) = 0$, the expression for cross-section $\sigma_{p+{}^3He\to\pi^+ + {}^4He}$ takes a considerably simpler form:

$$\sigma_{p+{}^3He\to\pi^+ + {}^4He} = \frac{1}{4}|R_0|^2\left(1 + \mathbf{P}^{(p)}\mathbf{P}^{(He)} - 2(\mathbf{P}^{(p)}\mathbf{l})(\mathbf{P}^{(He)}\mathbf{l})\right). \tag{16}$$

3. Spin effects in the inverse reaction $\pi^+ + {}^4He \to p + {}^3He$

In the reaction $\pi^+ + {}^4He \to p + {}^3He$ the $(p, {}^3He)$ system is produced *in the triplet state only.* This state, normalized to unity, is as follows [1]:

$$|\,\widetilde{\psi}\rangle = \frac{1}{(|R_0(E,\theta)|^2 + 2|R_1(E,\theta)|^2)^{1/2}}\times$$

$$\times\Big[R_1(E,\theta)\left(|+1/2,\mathbf{l}\rangle^{(p)}\otimes|+1/2,\mathbf{l}\rangle^{(He)} - |-1/2,\mathbf{l}\rangle^{(p)}\otimes|-1/2,\mathbf{l}\rangle^{(He)}\right) +$$
$$+\frac{1}{\sqrt{2}}R_0(E,\theta)\left(|+1/2,\mathbf{l}\rangle^{(p)}\otimes|-1/2,\mathbf{l}\rangle^{(He)} + |-1/2,\mathbf{l}\rangle^{(p)}\otimes|+1/2,\mathbf{l}\rangle^{(He)}\right)\Big], \tag{17}$$

and it is symmetric under the interchange of spin quantum numbers of the proton and ${}^3He$.

Basing on the T invariance, we obtain for the effective cross-section of the reaction $\pi^+ + {}^4He \to p + {}^3He$ in the c.m. frame, summed over the spin projections in the final state, for the polarization vectors of the proton and ${}^3He$ in the final system and for the correlation tensor of the $(p, {}^3He)$ system [1] (all notations in Eqs. (17)-(20) are the same as in Section 2):

$$\widetilde{\sigma}_0(E,\theta) = (k_p/k_\pi)^2(|R_0(E,\theta)|^2 + 2|R_1(E,\theta)|^2); \tag{18}$$

$$\mathbf{P}^{(p)}(E,\theta) = \langle\widetilde{\psi}|\,\hat{\boldsymbol{\sigma}}^{(p)}\,|\widetilde{\psi}\rangle = \mathbf{P}^{(He)}(E,\theta) = \langle\widetilde{\psi}|\,\hat{\boldsymbol{\sigma}}^{(He)}\,|\widetilde{\psi}\rangle =$$
$$= -A(E,\theta)\mathbf{n} = -2\sqrt{2}\,\frac{\mathrm{Im}(R_1(E,\theta)R_0^*(E,\theta))}{|R_0(E,\theta)|^2 + 2|R_1(E,\theta)|^2}\,\mathbf{n}; \tag{19}$$

$$T_{ik}(E,\theta) = \langle\widetilde{\psi}|\,\hat{\sigma}_i^{(p)}\hat{\sigma}_k^{(He)}\,|\widetilde{\psi}\rangle = \delta_{ik} - \frac{2}{|R_0(E,\theta)|^2 + 2|R_1(E,\theta)|^2}\times$$
$$\times\Big[|R_0(E,\theta)|^2 l_i l_k + 2|R_1(E,\theta)|^2 m_i m_k - \sqrt{2}\,\mathrm{Re}(R_1(E,\theta)R_0^*(E,\theta))(l_i m_k + m_i l_k)\Big]. \tag{20}$$

In accordance with Eqs. (19), (20), the spins of the proton and the ${}^3He$ nucleus in the reaction $\pi^+ + {}^4He \to p + {}^3He$ must be *tightly correlated* (see [1] for more details).

4. Violation of the incoherence inequalities for the correlation tensor

As it was established in the paper [4], in the case of incoherent mixtures of factorizable two-particle states of spin-1/2 fermions the following inequalities for the diagonal components of the correlation tensor should be satisfied:

$$|\,T_{11} + T_{22} + T_{33}\,| \le 1; \quad |\,T_{11} + T_{22}\,| \le 1; \quad |\,T_{11} + T_{33}\,| \le 1; \quad |\,T_{22} + T_{33}\,| \le 1. \tag{21}$$

However, for *non-factorizable* quantum-mechanical superpositions these inequalities may be violated. The triplet state $|\widetilde{\psi}\rangle$ (17) of the final system in the reaction $\pi^+ + {}^4He \to p + {}^3He$ represents a characteristic example of such non-factorizable spin states (it is well seen that the state $|\widetilde{\psi}\rangle$ cannot be reduced to the product of one-particle spin states).

Let us calculate the components of the correlation tensor T_{ik} (20) for the system $(p, {}^3He)$ in the coordinate frame with $z \parallel \mathbf{l}$, $x \parallel \mathbf{m}$, $y \parallel \mathbf{n}$. Finally, we obtain the following expressions :

$$T_{11} = \frac{|R_0(E,\theta)|^2 - 2|R_1(E,\theta)|^2}{|R_0(E,\theta)|^2 + 2|R_1(E,\theta)|^2}; \quad T_{22} = 1; \quad T_{33} = \frac{2|R_1(E,\theta)|^2 - |R_0(E,\theta)|^2}{|R_0(E,\theta)|^2 + 2|R_1(E,\theta)|^2} = -T_{11}; \tag{22}$$

$$T_{13} = T_{31} = \frac{2\sqrt{2}}{|R_0(E,\theta)|^2 + 2|R_1(E,\theta)|^2}\,\mathrm{Re}(R_1 R_0^*); \quad T_{12} = T_{21} = T_{23} = T_{32} = 0 \tag{23}$$

(indexes: $1 \to x$, $2 \to y$, $3 \to z$); in doing so, $tr(T) = 1$) .

Thus, as follows from Eq. (22), in the reaction $\pi^+ + {}^4He \to p + {}^3He$ *one* of the incoherence inequalities (21) for the diagonal components of the correlation tensor is *necessarily* violated, irrespective of the concrete mechanism of generation of the system $(p, {}^3He)$. Indeed, if $|R_0|^2 > 2|R_1|^2$, we obtain that $|\,T_{11} + T_{22}\,| > 1$; if, on the contrary, $|R_0|^2 < 2|R_1|^2$, then $|\,T_{22} + T_{33}\,| > 1$. Meantime, in both the cases the other three incoherence inequalities (21) are satisfied.

References

[1] Lyuboshitz V V and Lyuboshitz V L 2000 *Phys. At. Nucl.* **63** 767
[2] Bilenky S M and Ryndin R M 1963 *Phys. Lett* **6** 217
[3] Lyuboshitz V L 1970 *Sov. J. Nucl. Phys.* **12** 107
[4] Lednicky R and Lyuboshitz V L 2001 *Phys. Lett.* **B 508** 146

IOP Conf. Series: Journal of Physics: Conf. Series **938** (2017) 012058 doi:10.1088/1742-6596/938/1/012058

Spin dynamics of fermion particles in gravitational and electromagnetic fields

Yuri N. Obukhov[1], **Alexander J. Silenko**[23], **and Oleg V. Teryaev**[24]

[1] Theoretical Physics Laboratory, Nuclear Safety Institute, Russian Academy of Sciences, B.Tulskaya 52, 115191 Moscow, Russia

[2] Bogoliubov Laboratory of Theoretical Physics, Joint Institute for Nuclear Research, Dubna 141980, Russia

E-mail: obukhov@ibrae.ac.ru, alsilenko@mail.ru, teryaev@theor.jinr.ru

Abstract. We present the results of a comprehensive study of the classical and quantum dynamics of spin $\frac{1}{2}$ Dirac fermion particles with dipole moments under the simultaneous action of arbitrary external fields, including the gravitational, inertial and electromagnetic ones. We demonstrate the complete consistency of the quantum and classical dynamics. Physical applications are discussed.

1. Introduction

Understanding the motion of fermion particles with dipole moments (electrons, protons, neutrons, neutrinos) in arbitrary electromagnetic, gravitational and inertial fields is important for many physical and astrophysical problems [1]. We continue the study of the classical and quantum dynamics of spinning particles with dipole moments in the framework of the general-relativistically covariant Dirac theory [2, 3, 4, 5, 6]. In this approach, the spacetime can have either the Riemannian or the post-Riemannian geometry. The exact Foldy-Wouthuysen transformation for the most general case of a fermion moving in combined configurations of arbitrary strong gravitational (inertial) and electromagnetic fields is derived. These results are used to obtain the quantum and quasiclassical equations of motion of fermion particles. Applications to the precision experiments are considered. In particular, effects of terrestrial gravity and rotation in the high-energy physics are discussed, the bounds on the new physical coupling parameters are obtained, and gravitational wave action on spin is analysed.

Our basic notations and conventions are the same as in [6].

2. Fermion in external classical fields

The relativistic theory of a fermion particle with spin $\frac{1}{2}$ is based on the Dirac Lagrangian

$$L = \frac{i\hbar}{2}\left(\overline{\psi}\gamma^\alpha D_\alpha\psi - D_\alpha\overline{\psi}\gamma^\alpha\psi\right) - mc\overline{\psi}\psi + \frac{\mu'}{2c}F_{\alpha\beta}\overline{\psi}\sigma^{\alpha\beta}\psi + \frac{\delta'}{2}G_{\alpha\beta}\overline{\psi}\sigma^{\alpha\beta}\psi. \tag{1}$$

[3] Also at: Research Institute for Nuclear Problems, Belarusian State University, Minsk 220030, Belarus.

[4] Also at: National Research Nuclear University "MEPhI" (Moscow Engineering Physics Institute), Kashirskoe Shosse 31, 115409 Moscow, Russia.

IOP Conf. Series: Journal of Physics: Conf. Series **938** (2017) 012058 doi:10.1088/1742-6596/938/1/012058

The first term accounts for the minimal coupling of a fermion with the mass m and the electric charge q to the external gauge fields – the electromagnetic potential A_i and the gravitational coframe and the Lorentz connection $(e_i^\alpha, \Gamma_i{}^{\alpha\beta})$ – via the covariant spinor derivative

$$D_\alpha\psi = e_\alpha^i\left(\partial_i\psi - \frac{iq}{\hbar}A_i\psi + \frac{i}{4}\Gamma_i{}^{\beta\gamma}\sigma_{\beta\gamma}\psi\right), \tag{2}$$

whereas the two Pauli terms describe the non-minimal coupling of the electromagnetic field strength $F_{\alpha\beta}, G_{\alpha\beta} = \frac{1}{2}\eta_{\alpha\beta\mu\nu}F^{\mu\nu}$ to an anomalous magnetic moment (AMM) and an electric dipole moment (EDM). The corresponding coupling parameters have the dimension $[\mu'] = [q\hbar/2m]$ of the magnetic dipole (nuclear magneton), $[\delta'] = [q\,l]$ of the electric dipole (charge times length). The anomalous magnetic moment can be written in terms of the magneton $\mu_0 = \frac{q\hbar}{2m}$, and in a similar way one can introduce a convenient unit of an electric dipole moment. A reasonable definition is the electric charge times the electron Compton length: $\delta_0 = q\,\frac{\hbar}{mc}$. Then one can write for both types of the dipole moments:

$$\mu' = a\,\frac{q\hbar}{2m}, \qquad \delta' = b\,\frac{q\hbar}{2mc}, \tag{3}$$

The dimensionless constant parameters $a = (g-2)/2$ (where g is the gyromagnetic factor) and b characterize the magnitude of the anomalous magnetic and electric dipole moments, respectively.

3. Electrodynamics in curved spacetime

A nontrivial feature of the system under consideration is the absence of superposition. The motion of a spinning particle only in the electromagnetic field or only in the gravitational field was extensively studied in the past. However, when both external fields are acting on a fermion together, their influence on spin and particle's trajectory is not just a sum of two separate effects. Indeed, whereas electromagnetic field affects only electric charges and currents, gravity is universal and it couples to all types of matter, including the electromagnetic field. Accordingly, a spinning particle feels gravity both directly –via the coframe and connection– and indirectly –via the electromagnetic field which carries an imprint of the curved spacetime geometry.

Denoting $x^i = (t, x^a)$ the local coordinates, we write the spacetime interval as

$$ds^2 = V^2c^2dt^2 - \delta_{\widehat{a}\widehat{b}}W^{\widehat{a}}{}_cW^{\widehat{b}}{}_d\,(dx^c - K^c cdt)\,(dx^d - K^d cdt). \tag{4}$$

The functions $V = V(x^i)$ and $K^a = K^a(x^i)$, and the 3×3 matrix $W^{\widehat{a}}{}_b = W^{\widehat{a}}{}_b(x^i)$ depend arbitrarily on the local spacetime coordinates. The total number of independent components is $1+3+9 = 13$, but in view of the rotation freedom $W^{\widehat{a}}{}_b \longrightarrow L^{\widehat{a}}{}_{\widehat{c}}W^{\widehat{c}}{}_b$ with $L^{\widehat{a}}{}_{\widehat{c}}(x^i) \in SO(3)$, this number is reduced to $13 - 3 = 10$. In other words, the line element (4) describes an arbitrary 4-dimensional geometry. Explicitly, the components of the metric g_{ij} read

$$g_{00} = c^2(V^2 - \underline{g}_{ab}K^aK^b), \qquad g_{0a} = c\underline{g}_{ab}K^b, \qquad g_{ab} = -\,\underline{g}_{ab}, \tag{5}$$

$$g^{00} = \frac{1}{c^2V^2}, \qquad g^{0a} = \frac{K^a}{cV^2}, \qquad g^{ab} = -\,\underline{g}^{ab} + \frac{1}{V^2}\,K^aK^b. \tag{6}$$

The spatial 3-dimensional metric is given by $\underline{g}_{ab} = \delta_{\widehat{c}\widehat{d}}W^{\widehat{c}}{}_aW^{\widehat{d}}{}_b$, and $\underline{g}^{ab} = \delta^{\widehat{c}\widehat{d}}W^a{}_{\widehat{c}}W^b{}_{\widehat{d}}$. The 3×3 matrix $W^b{}_{\widehat{a}}$ is inverse to $W^{\widehat{a}}{}_b$.

In Maxwell's theory, the electromagnetic field is described in terms of the field strength 2-form $F = \frac{1}{2}F_{ij}dx^i \wedge dx^j$ and the electromagnetic excitation 2-form $H = \frac{1}{2}H_{ij}dx^i \wedge dx^j$. These fundamental variables satisfy the generally covariant Maxwell equations [7]:

$$dF = 0, \qquad dH = J, \qquad H = \lambda_0 \star F. \tag{7}$$

The constitutive relation is linear and local, with $\lambda_0 = \sqrt{\frac{\varepsilon_0}{\mu_0}}$ (where ε_0 and μ_0 are the electric and magnetic constants of the vacuum), and the star $\star$ denotes the Hodge duality operator of the spacetime metric. The current 3-form $J = \frac{1}{6}J_{ijk}dx^i \wedge dx^j \wedge dx^k$ describes the distribution of the electric charges and currents which are the sources of the electromagnetic field.

The equations (7) can be written in the equivalent vector form in terms of the components of the electric and magnetic fields, $\boldsymbol{E}_a = \{F_{10}, F_{20}, F_{30}\}$ and $\boldsymbol{B}^a = \{F_{23}, F_{31}, F_{12}\}$, and the components of the magnetic and electric excitations, $\boldsymbol{H}_a = \{H_{01}, H_{02}, H_{03}\}$ and $\boldsymbol{D}^a = \{H_{23}, H_{31}, H_{12}\}$. Identifying the components of the source 3-form J with the electric current density $\boldsymbol{J}^a = \{-J_{023}, -J_{031}, -J_{012}\}$ and the charge density $\rho = J_{123}$, we recast (7) into [8]

$$\boldsymbol{\nabla}\times\boldsymbol{E} + \dot{\boldsymbol{B}} = 0, \quad \boldsymbol{\nabla}\cdot\boldsymbol{B} = 0, \qquad \boldsymbol{\nabla}\times\boldsymbol{H} - \dot{\boldsymbol{D}} = \boldsymbol{J}, \quad \boldsymbol{\nabla}\cdot\boldsymbol{D} = \rho. \tag{8}$$

The influence of the inertia and gravity is encoded in the Maxwell-Lorentz constitutive relation between the electric and magnetic fields $\boldsymbol{E}, \boldsymbol{B}$ and the electric and magnetic excitations $\boldsymbol{D}, \boldsymbol{H}$:

$$D^a = \frac{\varepsilon_0 w}{V}\,\underline{g}^{ab}E_b - \lambda_0\frac{w}{V}\,\underline{g}^{ad}\epsilon_{bcd}K^c\,B^b, \tag{9}$$

$$H_a = \frac{1}{\mu_0 w V}\left\{(V^2 - K^2)\underline{g}_{ab} + K_aK_b\right\}B^b - \lambda_0\frac{w}{V}\,\epsilon_{adc}K^c\,\underline{g}^{db}E_b. \tag{10}$$

Here $K_a = \underline{g}_{ab}K^b$, $K^2 = K_aK^a = \underline{g}_{ab}K^aK^b$, and $w = \det W^{\widehat{c}}{}_d$.

For the description of Dirac fermions, one needs a coframe e_i^α, such that $g_{\alpha\beta}e_i^\alpha e_j^\beta = g_{ij}$, with $g_{\alpha\beta} = \mathrm{diag}(c^2, -1, -1, -1)$, and one should carefully distinguish the *holonomic* components $\boldsymbol{E}, \boldsymbol{B}$ of F_{ij}, from the *anholonomic* components $\mathfrak{E}, \mathfrak{B}$ of the Maxwell tensor $F_{\alpha\beta} = e^i_\alpha e^j_\beta F_{ij}$, defined as $\mathfrak{E}_a = \{F_{\widehat{1}\widehat{0}}, F_{\widehat{2}\widehat{0}}, F_{\widehat{3}\widehat{0}}\}$ and $\mathfrak{B}^a = \{F_{\widehat{2}\widehat{3}}, F_{\widehat{3}\widehat{1}}, F_{\widehat{1}\widehat{2}}\}$. In the Schwinger gauge $e_a^{\widehat{0}} = 0$ (also $e_{\widehat{a}}^0 = 0$), $a = 1, 2, 3$, the coframe reads explicitly:

$$e_i^{\widehat{0}} = V\,\delta_i^0, \qquad e_i^{\widehat{a}} = W^{\widehat{a}}{}_b\left(\delta_i^b - cK^b\,\delta_i^0\right), \qquad a = 1, 2, 3. \tag{11}$$

Accordingly, the holonomic and and anholonomic electromagnetic fields are related via

$$\mathfrak{E}_a = \frac{1}{V}\,W^b{}_{\widehat{a}}(\boldsymbol{E} + c\boldsymbol{K}\times\boldsymbol{B})_b, \qquad \mathfrak{B}^a = \frac{1}{w}\,W^{\widehat{a}}{}_b\,\boldsymbol{B}^b, \tag{12}$$

Hereafter the vector product is defined by $\{\boldsymbol{A}\times\boldsymbol{B}\}_a = \epsilon_{abc}A^bB^c$ for any 3-vectors A^b and B^c.

4. Schrödinger and Foldy-Wouthuysen pictures for Dirac particle

To evaluate the covariant spinor derivative (2), we need the local Lorentz connection. For the general spacetime metric (4) with the tetrad (11), the connection components read explicitly

$$\Gamma_{i\,\widehat{a}\widehat{0}} = \frac{c^2}{V}\,W^b{}_{\widehat{a}}\,\partial_b V\,e_i{}^{\widehat{0}} - \frac{c}{V}\,\mathcal{Q}_{(\widehat{a}\widehat{b})}\,e_i{}^{\widehat{b}}, \qquad \Gamma_{i\,\widehat{a}\widehat{b}} = \frac{c}{V}\,\mathcal{Q}_{[\widehat{a}\widehat{b}]}\,e_i{}^{\widehat{0}} + (\mathcal{C}_{\widehat{a}\widehat{b}\widehat{c}} + \mathcal{C}_{\widehat{a}\widehat{c}\widehat{b}} + \mathcal{C}_{\widehat{c}\widehat{b}\widehat{a}})\,e_i{}^{\widehat{c}}, \tag{13}$$

where we introduced

$$\mathcal{Q}_{\widehat{a}\widehat{b}} = g_{\widehat{a}\widehat{c}}W^d{}_{\widehat{b}}\left(\frac{1}{c}\dot{W}^{\widehat{c}}{}_d + K^e\partial_e W^{\widehat{c}}{}_d + W^{\widehat{c}}{}_e\partial_d K^e\right), \qquad \mathcal{C}_{\widehat{a}\widehat{b}}{}^{\widehat{c}} = W^d{}_{\widehat{a}}W^e{}_{\widehat{b}}\,\partial_{[d}W^{\widehat{c}}{}_{e]}, \tag{14}$$

and $\mathcal{C}_{\widehat{a}\widehat{b}\widehat{c}} = g_{\widehat{c}\widehat{d}}\mathcal{C}_{\widehat{a}\widehat{b}}{}^{\widehat{d}}$. The dot denotes the time derivative ($\dot{} = \partial_t$).

We derive the Dirac equation from the action principle with the Lagrangian (1), and after a rescaling of the wave function $\psi \longrightarrow (\sqrt{-g}e^{0}_{\widehat{0}})^{\frac{1}{2}}\psi$, we recast the resulting wave equation into a Schrödinger form $i\hbar\frac{\partial\psi}{\partial t} = \mathcal{H}\psi$ with the Hermitian (and self-adjoint) Hamiltonian

$$\begin{aligned}\mathcal{H} &= \beta mc^2 V + q\Phi + \frac{c}{2}\left(\pi_b \mathcal{F}^b{}_a\alpha^a + \alpha^a\mathcal{F}^b{}_a\pi_b\right) + \frac{c}{2}\left(\boldsymbol{K}\cdot\boldsymbol{\pi} + \boldsymbol{\pi}\cdot\boldsymbol{K}\right)\\ &\quad + \frac{\hbar c}{4}\left(\boldsymbol{\Xi}\cdot\boldsymbol{\Sigma} - \Upsilon\gamma_5\right) - \beta V\left(\boldsymbol{\Sigma}\cdot\boldsymbol{\mathcal{M}} + i\boldsymbol{\alpha}\cdot\boldsymbol{\mathcal{P}}\right). \qquad (15)\end{aligned}$$

Here: $\alpha^a = \beta\gamma^a$ $(a,b,c,\dots = 1,2,3)$ and the spin matrices $\Sigma^1 = i\gamma^{\hat{2}}\gamma^{\hat{3}}, \Sigma^2 = i\gamma^{\hat{3}}\gamma^{\hat{1}}, \Sigma^3 = i\gamma^{\hat{1}}\gamma^{\hat{2}}$ and $\gamma_5 = i\alpha^{\hat{1}}\alpha^{\hat{2}}\alpha^{\hat{3}}$. Boldface notation is used for 3-vectors $\boldsymbol{K} = \{K^a\}$, $\boldsymbol{\alpha} = \{\alpha^a\}$, $\boldsymbol{\Sigma} = \{\Sigma^a\}$, $\boldsymbol{\pi} = \{\pi_a\}$. The latter is the kinetic momentum operator, $\boldsymbol{\pi} = -i\hbar\boldsymbol{\nabla} - q\boldsymbol{A}$ (with $A_0 = -\Phi$, and $\boldsymbol{A} = \{A_1, A_2, A_3\}$). The minimal coupling gives rise to the terms in (15) with the objects

$$\mathcal{F}^b{}_a = VW^b{}_{\widehat{a}}, \qquad \Upsilon = V\epsilon^{\widehat{a}\widehat{b}\widehat{c}}\Gamma_{\widehat{a}\widehat{b}\widehat{c}} = -V\epsilon^{\widehat{a}\widehat{b}\widehat{c}}C_{\widehat{a}\widehat{b}\widehat{c}}, \qquad \Xi^a = \frac{V}{c}\epsilon^{\widehat{a}\widehat{b}\widehat{c}}\Gamma_{\widehat{0}\widehat{b}\widehat{c}} = \epsilon_{\widehat{a}\widehat{b}\widehat{c}}Q^{\widehat{b}\widehat{c}}, \qquad (16)$$

whereas the nonminimal coupling is encoded in

$$\mathcal{M}^a = \mu'\mathfrak{B}^a + \delta'\mathfrak{E}^a, \qquad \mathcal{P}_a = c\delta'\mathfrak{B}_a - \mu'\mathfrak{E}_a/c. \qquad (17)$$

The physical contents of the theory is revealed when one transitions from the Schrödinger picture (15) to the Foldy-Wouthuysen (FW) representation. We apply the general method developed in Refs. [9, 10, 11] to construct the FW transformation for the Dirac Hamiltonian (15) and to derive the FW Hamiltonian which is exact in all terms of the zero and the first orders in the Planck constant $\hbar$ without making assumptions about the weakness of the external fields.

Omitting the technical details (see [2, 3, 4, 5]) we find for the FW Hamiltonian [12]:

$$\mathcal{H}_{FW} = \beta\epsilon' + q\Phi + \frac{c}{2}\left(\boldsymbol{K}\cdot\boldsymbol{\pi} + \boldsymbol{\pi}\cdot\boldsymbol{K}\right) + \frac{\hbar}{2}\boldsymbol{\Pi}\cdot\boldsymbol{\Omega}_{(1)} + \frac{\hbar}{2}\boldsymbol{\Sigma}\cdot\boldsymbol{\Omega}_{(2)}, \qquad (18)$$

where we introduced the 3-vector operators

$$\begin{aligned}\Omega^a_{(1)} &= \frac{mc^4}{2}\left\{\frac{1}{\mathcal{T}}, \{\pi_e, \epsilon^{abc}\mathcal{F}^e{}_b\mathcal{F}^d{}_c\partial_d V\}\right\}\\ &\quad + \frac{c^2}{8}\left\{\frac{1}{\epsilon'}, \{\pi_e, (2\epsilon^{abc}\mathcal{F}^d{}_b\partial_d\mathcal{F}^e{}_c + \delta^{ab}\mathcal{F}^e{}_b\Upsilon)\}\right\}\\ &\quad + \frac{qc^2}{4}\epsilon^{abc}\left\{\frac{1}{\mathcal{T}}, \left(\{\mathcal{F}^d{}_b, \pi_d\}V^2\mathfrak{E}_c - V^2\mathfrak{E}_b\{\mathcal{F}^d{}_c, \pi_d\}\right)\right\}\\ &\quad - \frac{c}{4\hbar}\epsilon^{abc}\left\{\frac{1}{\epsilon'}, \left(\{\mathcal{F}^d{}_b, \pi_d\}V\mathcal{P}_c - V\mathcal{P}_b\{\mathcal{F}^d{}_c, \pi_d\}\right)\right\}, \qquad (19)\end{aligned}$$

$$\begin{aligned}\Omega^a_{(2)} &= \frac{c^2}{8}\left\{\frac{1}{\mathcal{T}}, \left\{\{\pi_e, \mathcal{F}^e{}_b\}, \left\{\pi_f, \left[\epsilon^{abc}\left(\frac{1}{c}\dot{\mathcal{F}}^f{}_c - \mathcal{F}^d{}_c\partial_d K^f + K^d\partial_d\mathcal{F}^f{}_c\right)\right.\right.\right.\right.\\ &\quad \left.\left.\left.\left. - \frac{1}{2}\mathcal{F}^f{}_d\left(\delta^{db}\Xi^a - \delta^{da}\Xi^b\right)\right]\right\}\right\}\right\} + \frac{c}{2}\Xi^a - \frac{qc^2}{2}\left\{\frac{1}{\epsilon'}, V^2\mathfrak{B}^a\right\}\\ &\quad - \frac{2V}{\hbar}\mathcal{M}^a + \frac{c^2}{2\hbar}\left\{\frac{1}{\mathcal{T}}, \{\{\delta^{ab}\mathcal{F}^d{}_b\mathcal{F}^e{}_c V\mathcal{M}^c, \pi_d\}, \pi_e\}\right\}, \qquad (20)\end{aligned}$$

with the curly brackets $\{\ ,\ \}$ denoting anticommutators, and

$$\epsilon' = \sqrt{m^2c^4V^2 + \frac{c^2}{4}\delta^{ac}\{\pi_b, \mathcal{F}^b{}_a\}\{\pi_d, \mathcal{F}^d{}_c\}}, \qquad \mathcal{T} = 2\epsilon'^2 + \{\epsilon', mc^2V\}. \qquad (21)$$

5. Quantum and classical spin dynamics

To find the dynamics of the spin, we evaluate the commutator of the polarization operator $\boldsymbol{\Pi} = \beta\boldsymbol{\Sigma}$ with the FW Hamiltonian (18):

$$\frac{d\boldsymbol{\Pi}}{dt} = \frac{i}{\hbar}[\mathcal{H}_{FW}, \boldsymbol{\Pi}] = \boldsymbol{\Omega}_{(1)} \times \boldsymbol{\Sigma} + \boldsymbol{\Omega}_{(2)} \times \boldsymbol{\Pi}. \tag{22}$$

The result describes the precession of the spin in the external gravitational and electromagnetic fields, and hence $\boldsymbol{\Omega}_{(1)}$ and $\boldsymbol{\Omega}_{(2)}$ have a clear meaning of the angular velocity operators.

It is straightforward to derive the corresponding semiclassical expressions by evaluating all anticommutators and neglecting the powers of $\hbar$ higher than 1 (the classical limit of the relativistic quantum mechanics is discussed in [13]). The equation (22) then gives rise to the semiclassical equations describing the precession of the average spin $\boldsymbol{s}$ vector:

$$\frac{d\boldsymbol{s}}{dt} = \boldsymbol{\Omega} \times \boldsymbol{s} = (\boldsymbol{\Omega}_{(1)} + \boldsymbol{\Omega}_{(2)}) \times \boldsymbol{s}, \tag{23}$$

$$\Omega^a_{(1)} = \frac{c^2}{\epsilon'}\mathcal{F}^d{}_c\pi_d\Big[\frac{1}{2}\Upsilon\delta^{ac} - \epsilon^{abe}V C_{be}{}^c + \frac{\epsilon'}{\epsilon' + mc^2V}\epsilon^{abc}W^e{}_{\widehat{b}}\partial_e V + \frac{qV^2}{\epsilon' + mc^2V}\epsilon^{acb}\mathfrak{E}_b - \frac{2V}{c\hbar}\epsilon^{acb}\mathcal{P}_b\Big], \tag{24}$$

$$\Omega^a_{(2)} = \frac{c}{2}\Xi^a - \frac{c^3}{\epsilon'(\epsilon' + mc^2V)}\epsilon^{abc}\mathcal{Q}_{(bd)}\delta^{dn}\mathcal{F}^k{}_n\pi_k\mathcal{F}^l{}_c\pi_l - \frac{qc^2V^2}{\epsilon'}\mathfrak{B}^a + \frac{2V}{\hbar}\Big[-\mathcal{M}^a + \frac{c^2}{\epsilon'(\epsilon' + mc^2V)}\delta^{an}\mathcal{F}^c{}_n\pi_c\mathcal{F}^d{}_b\pi_d\mathcal{M}^b\Big]. \tag{25}$$

Here, in the semiclassical limit, we have $\epsilon' = \sqrt{m^2c^4V^2 + c^2\delta^{cd}\mathcal{F}^a{}_c\mathcal{F}^b{}_d\pi_a\pi_b}$.

Furthermore, from (18) we obtain the velocity operator in the semiclassical approximation:

$$\frac{dx^a}{dt} = \frac{i}{\hbar}[\mathcal{H}_{FW}, x^a] = \beta\frac{\partial\epsilon'}{\partial\pi_a} + cK^a = \beta\frac{c^2}{\epsilon'}\mathcal{F}^a{}_b\delta^{bc}\mathcal{F}^d{}_c\pi_d + cK^a, \tag{26}$$

which allows to identify the *anholonomic velocity* operator in the Schwinger frame (11) with

$$\beta\frac{c^2}{\epsilon'}\mathcal{F}^b{}_a\pi_b = \widehat{v}_a. \tag{27}$$

As a result, $\delta^{cd}\mathcal{F}^a{}_c\mathcal{F}^b{}_d\pi_a\pi_b = (\epsilon')^2\widehat{v}^2/c^2$, hence $(\epsilon')^2 = m^2c^4V^2 + (\epsilon')^2\widehat{v}^2/c^2$ and $\epsilon' = \gamma mc^2 V$, thus, recalling that $\gamma = (1 - \widehat{v}^2/c^2)^{-1/2}$ is the Lorentz factor ($\widehat{v}^2 = \delta_{ab}\widehat{v}^a\widehat{v}^b$), we find:

$$\frac{\epsilon'}{\epsilon' + mc^2V} = \frac{\gamma}{1+\gamma}, \qquad \frac{c^3}{\epsilon'(\epsilon' + mc^2V)}\mathcal{F}^b{}_a\pi_b\mathcal{F}^d{}_c\pi_d = \frac{\gamma}{1+\gamma}\frac{\widehat{v}_a\widehat{v}_c}{c}. \tag{28}$$

Making use of (27) and (28) in (24) and (25), we then establish a fundamental agreement of the quantum and classical spin dynamics. Indeed, after the replacement of quantum operators by their classical counterparts, we recover (23) as the equation motion of the classical spin with the precession angular velocity $\boldsymbol{\Omega} = \boldsymbol{\Omega}_{(1)} + \boldsymbol{\Omega}_{(2)}$, obtained in the rest frame for the relativistic particle with mass m, electric charge q and dipole moments μ', δ' in the gravitational and electromagnetic fields, described by the dynamical system [6]

$$\frac{DU^\alpha}{d\tau} = \frac{dU^\alpha}{d\tau} + U^i\Gamma_{i\beta}{}^\alpha U^\beta = -\frac{q}{m}g^{\alpha\beta}F_{\beta\gamma}U^\gamma, \tag{29}$$

$$\frac{DS^\alpha}{d\tau} = \frac{dS^\alpha}{d\tau} + U^i\Gamma_{i\beta}{}^\alpha S^\beta = -\frac{q}{m}g^{\alpha\beta}F_{\beta\gamma}S^\gamma - \frac{2}{\hbar}\Big[M^\alpha{}_\beta + \frac{1}{c^a}\left(M_{\beta\gamma}U^\alpha U^\gamma - M^{\alpha\gamma}U_\beta U_\gamma\right)\Big]S^\beta. \tag{30}$$

Here U^α is the 4-velocity and S^α is the spin 4-vector of the particle, and the polarization tensor is introduced by

$$M_{\alpha\beta} = \mu' F_{\alpha\beta} + c\delta' G_{\alpha\beta}. \tag{31}$$

Its components, $M_{\hat{0}\hat{a}} = c\mathcal{P}_a$ and $M_{\hat{a}\hat{b}} = \epsilon_{abc}\mathcal{M}^c$, are obviously identified with the 3-vectors $\boldsymbol{\mathcal{M}}$ and $\boldsymbol{\mathcal{P}}$ which were defined in (17).

Quite remarkably, we can recast the precession angular velocity into a compact form

$$\boldsymbol{\Omega} = \frac{q}{m}\left[-\mathfrak{B}_{\text{eff}} + \frac{\gamma}{\gamma+1}\frac{\widehat{\boldsymbol{v}}\times\mathfrak{E}_{\text{eff}}}{c^2}\right] \tag{32}$$

by introducing the effective magnetic and electric fields

$$\mathfrak{B}_{\text{eff}} = \mathfrak{B} + \frac{2m}{q\hbar}\left[\boldsymbol{\mathcal{M}} - \frac{\gamma^2}{c}\widehat{\boldsymbol{v}}\times\left(\boldsymbol{\mathcal{P}} - \frac{\widehat{\boldsymbol{v}}\times\boldsymbol{\mathcal{M}}}{c}\right)\right] + \frac{m}{q}\boldsymbol{\mathcal{B}}, \tag{33}$$

$$\mathfrak{E}_{\text{eff}} = \mathfrak{E} - \frac{2m\gamma^2}{q\hbar}\widehat{\boldsymbol{v}}\times\left(\boldsymbol{\mathcal{M}} + \frac{\widehat{\boldsymbol{v}}\times\boldsymbol{\mathcal{P}}}{c}\right) + \frac{m}{q}\boldsymbol{\mathcal{E}}. \tag{34}$$

These expressions encompass the true magnetic and electric fields (the first terms on the right-hand sides), the nonminimal contributions due to AMM and EDM (the second terms), and the gravitomagnetic and gravitoelectric fields constructed from the metric of the spacetime:

$$\mathcal{B}^a = \frac{\gamma}{V}\left(-\frac{c}{2}\Xi^a - \frac{1}{2}\Upsilon\widehat{v}^a + \epsilon^{abc}V\mathcal{C}_{bc}{}^d\widehat{v}_d\right), \qquad \mathcal{E}^a = \frac{\gamma}{V}\delta^{ac}\left(c\mathcal{Q}_{(\widehat{c}\widehat{b})}\widehat{v}^b - c^2 W^b{}_{\widehat{c}}\partial_b V\right). \tag{35}$$

6. Physical effects

The resulting general framework can be applied to the analysis of the possible physical effects of the spin dynamics. Let us briefly overview the three important applications.

6.1. Manifestations of terrestrial gravity and inertia in high-energy physics

The spacetime geometry [6], which correctly describes the terrestrial inertia and (weak) gravity of a rotating source, corresponds to the choice $W^{\hat{a}}{}_c = W\delta^{\hat{a}}_c$, and

$$V = 1 - \frac{GM}{c^2 r}, \qquad W = 1 + \frac{GM}{c^2 r}, \qquad \boldsymbol{K} = -\frac{\boldsymbol{\omega}\times\boldsymbol{r}}{c}. \tag{36}$$

Here $M = M_\oplus$, and $\boldsymbol{\omega} = (0, 0, \omega_\oplus)$, with $\omega_\oplus = \frac{2\pi}{T_\oplus} = 7.29\times 10^{-5}\text{s}^{-1}$. The inertial (due to rotation) and the gravitational fields of the Earth affect the motion of an elementary particle and its spin dynamics. This influence is not negligible and should be taken into account in high-energy physics experiments.

Earth's rotation manifests itself as the Coriolis and the centrifugal forces and produces the additional rotation of the spin. The corresponding corrections to particle's trajectory and the spin motion are rather small. Bigger oscillatory corrections average to zero, but the inhomogeneity of the inertial field gives rise to small non-oscillatory corrections which should be taken into consideration only for the storage ring EDM experiments.

Earth's gravity produces an additional force that acts on particle's momentum as well as an additional torque affecting the spin, although no direct electromagnetic effects caused by Earth's gravity were observed so far. The additional forces include the Newton-like force and the reaction force provided by a focusing system. The additional torques are caused by the focusing field and by the geodetic effect. In the storage ring EDM experiments, these forces and torques lead to the additional spin rotation about the radial axis. Theoretical analysis shows that terrestrial gravity can produce the same effect as deuteron's EDM of $\delta' = 1.5\times 10^{-29}$ $e\cdot$cm in the planned dEDM experiment with magnetic focusing [14].

6.2. Limits on the non-Riemannian spacetime structure

Spin is a natural tool to probe the spacetime structure beyond Einstein's general relativity. Importantly, the torsion can be measured only by the matter with intrinsic spin [15, 16, 17]; orbital angular momentum (e.g., mechanically rotating gyroscopes) does not feel the torsion.

The spin of the Dirac particle is totally antisymmetric, and hence it can couple only to the axial torsion vector $\check{T}^\alpha = -\frac{1}{2}\eta^{\alpha\mu\nu\lambda}T_{\mu\nu\lambda} = \{\check{T}^{\hat{0}}, \check{T}^{\hat{a}}\}$. As a result, the hypothetical spin-torsion coupling modifies the Dirac Hamiltonian (15) via the additional terms [5]

$$-\frac{\hbar cV}{4}\left(\boldsymbol{\Sigma}\cdot\check{\boldsymbol{T}} + c\gamma_5\check{T}^{\hat{0}}\right). \tag{37}$$

We can straightforwardly take the spin-torsion coupling into account in (15) by the shift

$$\Upsilon \longrightarrow \Upsilon + Vc\check{T}^{\hat{0}}, \qquad \Xi^{\hat{a}} \longrightarrow \Xi^{\hat{a}} - V\check{T}^{\hat{a}}, \tag{38}$$

Making use of the theoretical framework established here, we can find observational bounds on spin-torsion coupling from the dynamics of freely precessing nuclear spins in a uniform magnetic field $\boldsymbol{B}$. In particular, using the data reported [18] for the experiment with ^{199}Hg and ^{201}Hg atoms devoted to the search of a hypothetical scalar-pseudoscalar interactions, we derive the restriction on the absolute value of the spacetime torsion:

$$\frac{\hbar c}{4}|\check{\boldsymbol{T}}|\cdot|\cos\Theta| < 2.2\times 10^{-21}\,\mathrm{eV}, \qquad |\check{\boldsymbol{T}}|\cdot|\cos\Theta| < 4.3\times 10^{-14}\,\mathrm{m}^{-1}. \tag{39}$$

Here Θ is the angle between $\boldsymbol{B}$ and the torsion $\check{\boldsymbol{T}}$. These bounds can be further improved by approximately one order on the basis of the more recent experimental data [19].

6.3. Spin in a gravitational wave

Fermion in the field of a gravitational wave was discussed earlier in [20, 21, 22].

In the local coordinates (t, x, y, z), the spacetime geometry of a weak gravitational wave is given by the metric (4) with

$$V = 1, \qquad \boldsymbol{K} = 0, \qquad W^{\hat{a}}{}_b = \begin{pmatrix} 1 + w_\oplus & w_\otimes & 0 \\ w_\otimes & 1 - w_\oplus & 0 \\ 0 & 0 & 1 \end{pmatrix}. \tag{40}$$

The functions $w_\otimes = w_\otimes(\varphi)$ and $w_\oplus = w_\oplus(\varphi)$ of the phase $\varphi = \omega(t - z/c)$, describe the two independent polarizations of a plane gravitational wave with the frequency ω propagating along the z direction [23].

One can probe the gravitational wave by observing the spin dynamics in the magnetic resonance setup [12]. Assuming that the wave has only one polarization, $w_\oplus = 0$, whereas $w_\otimes = g_0\cos\varphi$ (with amplitude g_0), we can arrange the constant homogeneous magnetic field in the plane of the wave front: $\boldsymbol{B} = (B_0, 0, 0)$, with $B_0 = $const. Then (12) yields $\mathfrak{B} = (B_0, B_0 w_\otimes, 0)$, and hence we find the field configuration that reproduces the magnetic resonance conditions, when the spin is affected by the constant homogeneous magnetic field along x and an additional alternating field in the perpendicular plane (y, z). Then [24], we find

$$P_{-\frac{1}{2}} = \frac{\sin^2\{\omega_0 g_0(t - t_0)\Lambda/4\}}{\Lambda^2} \tag{41}$$

the probability of a flip, at time t, for the spin as compared to its initial opposite orientation at t_0. Here $\omega_0 = \frac{2(\mu_0+\mu')B_0}{\hbar}$ is the Larmor frequency, $\Lambda^2 = 1 + \frac{4(1-\xi)^2}{g_0^2}$ and $\xi = \frac{\omega}{\omega_0}\left(1 - \frac{g_0^2}{16\xi^2}\right)$.

We thus confirm the qualitative conclusions [20, 22] on the prospects of the gravitational wave searches using the magnetic resonance type experiments. The effect (41) is quadratic in the small quantity g_0, however, polarization effects which are linear in g_0 can possibly show up in the analysis of the spin components orthogonal to initial spin polarization [25, 26].

7. Discussion

Continuing the study of the dynamics of the quantum and classical Dirac fermions with spin $\frac{1}{2}$ and dipole moments in the gravitational and electromagnetic fields, we extended here our earlier findings in [27, 28, 29, 2, 3, 4] to the case of an arbitrary spacetime metric plus an arbitrary electromagnetic field. This is a nontrivial problem, because there is no direct superposition of the electromagnetic effects on spin with the gravity-spin effects, and the action of gravity on spin is twofold: via the coframe and connection (16) and via the electromagnetic field which gets modified (12) in the curved spacetime [12].

We use the general framework, in which we derived the exact FW Hamiltonian (18) and the quantum equations of motion, to analyse the effects of terrestrial gravity and rotation in the high-energy physics, to find the new strong bounds on the spacetime torsion, and to develop a magnetic resonance setup for spin in the gravitational wave.

Acknowledgments

The work was supported in part by the Belarusian Republican Foundation for Fundamental Research (Grant No. Φ16D-004), by the Heisenberg-Landau program of the German Ministry for Science and Technology (Bundesministerium für Bildung und Forschung), and by the Russian Foundation for Basic Research (Grants No. 14-01-00647 and No. 16-02-00844-A). YNO is grateful to the organizers of the DSPIN17 conference for the warm hospitality and support.

References

[1] Ni W-T 2010 *Rep. Prog. Phys.* **73** 056901
[2] Obukhov Y N, Silenko A J, and Teryaev O V 2009 *Phys. Rev. D* **80** 064044
[3] Obukhov Y N, Silenko A J, and Teryaev O V 2011 *Phys. Rev. D* **84** 024025
[4] Obukhov Y N, Silenko A J, and Teryaev O V 2013 *Phys. Rev. D* **88** 084014
[5] Obukhov Y N, Silenko A J, and Teryaev O V 2014 *Phys. Rev. D* **90** 124068
[6] Obukhov Y N, Silenko A J, and Teryaev O V 2016 *Phys. Rev. D* **94** 044019
[7] Hehl F W and Obukhov Y N 2003 *Foundations of Classical Electrodynamics: Charge, Flux, and Metric* (Birkhäuser: Boston)
[8] Hehl F W, Itin Y, and Obukhov Y N 2016 *Int. J. Mod. Phys. D* **25** 1640016
[9] Silenko A J 2008 *Phys. Rev. A* **77** 012116
[10] Silenko A J 2015 *Phys. Rev. A* **91** 012111
[11] Silenko A J 2015 *Phys. Rev. A* **91** 022103
[12] Obukhov Y N, Silenko A J, and Teryaev O V 2017 arXiv:1708.05601
[13] Silenko A J 2013 *Particles and Nuclei Letters* **10** 144
[14] Anastassopoulos D et al. (EDM Collaboration) 2008 *AGS Proposal: Search for a permanent electric dipole moment of the deuteron nucleus at the* 10^{-29} *e· cm level*, `https://www.bnl.gov/edm/files/pdf/deuteron_proposal_080423_final.pdf`
[15] Yasskin P B and Stoeger W R 1980 *Phys. Rev. D* **21** 2081
[16] Puetzfeld D and Obukhov Y N 2007 *Phys. Rev. D* **76** 084025
[17] Hehl F W, Obukhov Y N, and Puetzfeld D 2013 *Phys. Lett. A* **377** 1775
[18] Venema B J, Majumder P K, Lamoreaux S K, Heckel B R, and Fortson E N 1992 *Phys. Rev. Lett.* **68** 135
[19] Gemmel C et al. 2010 *Eur. Phys. J. D* **57** 303
[20] Gonçalves B, Obukhov Y N, and Shapiro I S 2007 *Phys. Rev. D* **75** 124023
[21] Quach J Q 2015 *Phys. Rev. D* **92** 084047
[22] Quach J Q 2016 *Phys. Rev. D* **93** 104048
[23] Stephani H, Kramer D, MacCallum M, Hoenselaers C, and Herlt R 2003 *Exact solutions of Einstein's field equations*, 2nd ed. (Cambridge University Press: Cambridge) Secs. 24 and 31
[24] Bloch F and Siegert A 1940 *Phys. Rev.* **57** 522
[25] Silenko A J 2017 *Eur. Phys. J. C* **77** 341
[26] Silenko A J 2017 *Europhys. Lett.* **118** 61003
[27] Silenko A J and Teryaev O V 2005 *Phys. Rev. D* **71** 064016
[28] Silenko A J and Teryaev O V 2007 *Phys. Rev. D* **76** 061101(R)
[29] Silenko A J 2008 *Acta Phys. Polon. B Proc. Suppl.* **1** 87

IOP Conf. Series: Journal of Physics: Conf. Series **938** (2017) 012059 doi:10.1088/1742-6596/938/1/012059

Model identification of new heavy Z' bosons at ILC with polarized beams

A.A. Pankov[1,2,3], **A.V. Tsytrinov**[1]

[1] Abdus Salam ICTP Affiliated Centre at Pavel Sukhoi Gomel State Technical University, Gomel, 246746, Belarus

[2] Institute for Nuclear Problems, Belarusian State University, Minsk, 220030, Belarus

[3] Gelepov Laboratory of Nuclear Problems, Joint Institute for Nuclear Research, Dubna, 141980, Russia

E-mail: pankov@ictp.it

Abstract. Extra neutral gauge bosons, Z's, are predicted by many theoretical scenarios of physics beyond the Standard Model, and intensive searches for their signatures will be performed at present and future high energy colliders. It is quite possible that Z's are heavy enough to lie beyond the discovery reach expected at the CERN Large Hadron Collider LHC, in which case only indirect signatures of Z' exchanges may occur at future colliders, through deviations of the measured cross sections from the Standard Model predictions. We here discuss in this context the expected sensitivity to Z' parameters of fermion-pair production cross sections at the planned International Linear Collider (ILC), especially as regards the potential of distinguishing different Z' models once such deviations are observed. Specifically, we evaluate the discovery and identification reaches on Z' gauge bosons pertinent to the E_6, LR, ALR and SSM classes of models at the ILC.

1. Introduction

Electroweak theories beyond the Standard Model (SM) based on spontaneously broken extended gauge symmetries naturally envisage the existence of heavy, neutral, vector bosons Z'. The variety of the proposed Z' models is somewhat broad, and for definiteness in the sequel we shall focus on the so-called $Z'_{\rm SSM}$, Z'_{E_6}, $Z'_{\rm LR}$ and $Z'_{\rm ALR}$ models [1, 2]. Particular attention has recently been devoted to the phenomenological properties and the search reaches on such scenarios, and in some sense we may consider these Z' models as representative of this New Physics (NP) sector.

A typical manifestation of the production of such states is represented by (narrow) peaks observed in the cross sections for processes among SM particles at high energy accelerators, for example, in the dilepton invariant mass distributions for Drell-Yan (DY) process $pp \to Z' \to \ell^+\ell^- + X$ with $\ell = e, \mu$ at the CERN LHC hadronic colliders. Current experimental search limits on $M_{Z'}$ at 95% C.L., at the LHC with $\sqrt{s} = 13$ TeV using ≈ 36 fb^{-1} in DY, generally range in the interval 3.8–4.5 TeV, depending on the particular Z' model being tested [3, 4].

Clearly, the eventual discovery of a peak should be supplemented by the verification of the spin-1 of the assumed underlying Z', vs. the alternative spin-2 and spin-0 hypotheses corresponding, e.g., to exchanges of a Randall-Sundrum graviton resonance or a sneutrino. This kind of analysis relies on appropriate angular differential distributions and/or angular asymmetries. Finally, once the spin-1 has been established, the particular Z' scenario pertinent to the observed signal should be identified, see, e.g., Refs. [5, 6]. From studies of Drell-Yan processes at the LHC with a time-integrated luminosity of 100 fb^{-1}, it turns out that one can expect, at the 5-σ level, discovery limits on $M_{Z'}$ of the order of 4–4.5 TeV, spin-1 identification up to $M_{Z'} \simeq 2.5$–3.0 TeV and potential of distinction among the individual Z' models up to $M_{Z'} \simeq 2.1$ TeV (95% C.L.).

An alternative resource for the observation of virtual heavy gauge boson exchanges should be represented by the next generation e^+e^- ILC, with center of mass energy $\sqrt{s} = 0.5$–1 TeV and typical time-integrated luminosities $\mathcal{L}_{\rm int} \sim 0.5$–1 ab^{-1}, and the really high precision measurements that will be possible there. Indeed, the baseline configuration envisages a very high electron beam polarization (larger than 80%) and positron beam polarization of order 60% [7].

We will here focus on the fermion-antifermion production reactions at the polarized ILC [8]:

$$e^+ + e^- \to f + \bar{f}, \qquad f = e, \mu, \tau, c, b. \tag{1}$$

Particular emphasis will be given to the comparison between the cases of unpolarized and polarized initial beams, as regards the expected potential of ILC in identifying the Z' models of interest here, for $M_{Z'}$ values beyond the limits accessible at the LHC. Concerning the Z' mass, we will follow the scenario where the Z' mass range is above the LHC discovery limit and, here, with $M_{Z'}$ unknown, both discovery and identification reaches should be assessed for the ILC.

2. Observables and Z' models

The polarized differential cross section for the Bhabha process $e^+ + e^- \to e^+ + e^-$, where γ and Z can be exchanged also in the t-channel, can be written at leading order as [8]:

$$\begin{aligned}\frac{d\sigma(P^-,P^+)}{dz} &= \frac{(1+P^-)(1-P^+)}{4}\frac{d\sigma_{\rm R}}{dz} + \frac{(1-P^-)(1+P^+)}{4}\frac{d\sigma_{\rm L}}{dz} \\ &+ \frac{(1+P^-)(1+P^+)}{4}\frac{d\sigma_{{\rm RL},t}}{dz} + \frac{(1-P^-)(1-P^+)}{4}\frac{d\sigma_{{\rm LR},t}}{dz},\end{aligned} \tag{2}$$

with the decomposition

$$\frac{d\sigma_{\rm L}}{dz} = \frac{d\sigma_{\rm LL}}{dz} + \frac{d\sigma_{{\rm LR},s}}{dz}, \qquad \frac{d\sigma_{\rm R}}{dz} = \frac{d\sigma_{\rm RR}}{dz} + \frac{d\sigma_{{\rm RL},s}}{dz}. \tag{3}$$

In Eqs. (2) and (3), the subscripts t and s denote helicity cross sections with SM γ and Z exchanges in the corresponding channels, $z = \cos\theta$ and the subscripts L, R denote the respective helicities, P^- and P^+ denote the degrees of longitudinal polarization of the e^- and e^+ beams, respectively.[1] In terms of helicity amplitudes:

$$\begin{aligned}\frac{d\sigma_{\rm LL}}{dz} &= \frac{2\pi\alpha^2_{\rm e.m.}}{s}|G^{ee}_{{\rm LL},s} + G^{ee}_{{\rm LL},t}|^2, & \frac{d\sigma_{\rm RR}}{dz} &= \frac{2\pi\alpha^2_{\rm e.m.}}{s}|G^{ee}_{{\rm RR},s} + G^{ee}_{{\rm RR},t}|^2, \\ \frac{d\sigma_{{\rm LR},t}}{dz} &= \frac{d\sigma_{{\rm RL},t}}{dz} = \frac{2\pi\alpha^2_{\rm e.m.}}{s}|G^{ee}_{{\rm LR},t}|^2, & \frac{d\sigma_{{\rm LR},s}}{dz} &= \frac{d\sigma_{{\rm RL},s}}{dz} = \frac{2\pi\alpha^2_{\rm e.m.}}{s}|G^{ee}_{{\rm LR},s}|^2.\end{aligned} \tag{4}$$

According to the previous considerations the amplitudes $G^{ee}_{\alpha\beta,i}$, with $\alpha, \beta = {\rm L}, {\rm R}$ and $i = s, t$, are given by the sum of the SM γ, Z exchanges plus deviations representing the effect induced

[1] In the review [7], the opposite sign convention for positron polarization was adopted.

by a Z' boson:

$$\begin{aligned}
G^{ee}_{\rm LL,s} &= u\left(\frac{1}{s}+\frac{(g^e_{\rm L})^2}{s-M^2_Z}+\frac{(g^{e\prime}_{\rm L})^2}{s-M^2_{Z'}}\right), &\quad G^{ee}_{\rm LL,t} &= u\left(\frac{1}{t}+\frac{(g^e_{\rm L})^2}{t-M^2_Z}+\frac{(g^{e\prime}_{\rm L})^2}{t-M^2_{Z'}}\right),\\
G^{ee}_{\rm RR,s} &= u\left(\frac{1}{s}+\frac{(g^e_{\rm R})^2}{s-M^2_Z}+\frac{(g^{e\prime}_{\rm R})^2}{s-M^2_{Z'}}\right), &\quad G^{ee}_{\rm RR,t} &= u\left(\frac{1}{t}+\frac{(g^e_{\rm R})^2}{t-M^2_Z}+\frac{(g^{e\prime}_{\rm R})^2}{t-M^2_{Z'}}\right),\\
G^{ee}_{\rm LR,s} &= t\left(\frac{1}{s}+\frac{g^e_{\rm R}\,g^e_{\rm L}}{s-M^2_Z}+\frac{g^{e\prime}_{\rm R}\,g^{e\prime}_{\rm L}}{s-M^2_{Z'}}\right), &\quad G^{ee}_{\rm LR,t} &= s\left(\frac{1}{t}+\frac{g^e_{\rm R}\,g^e_{\rm L}}{t-M^2_Z}+\frac{g^{e\prime}_{\rm R}\,g^{e\prime}_{\rm L}}{t-M^2_{Z'}}\right).
\end{aligned} \tag{5}$$

Here, $u,t=-s(1\pm z)/2$ (we are neglecting fermion masses), $g_{\rm L}=-\cot 2\theta_W$ and $g_{\rm R}=\tan\theta_W$ with θ_W the electroweak mixing angle, whereas $g'_{\rm L}$ and $g'_{\rm R}$ are characteristic of the particular Z' model. In the annihilation channel, below the Z' mass, the Z' interference with the SM will be destructive in the LL and RR cross sections, whereas it can be of either sign in the LR and RL cross sections.

The polarized differential cross section for the leptonic channels $e^+e^-\to l^+l^-$ with $l=\mu,\tau$ can be obtained directly from Eq. (2), basically by dropping the t-channel contributions. The same is true, after some obvious substitutions, for the annihilations into $c\bar{c}$ and $b\bar{b}$ final states, in which case also the color (N_C) and QCD correction factors, $C_s\simeq N_C\,[1+\alpha_s/\pi+1.4\,(\alpha_s/\pi)^2]$, must be taken into account. The s-channel helicity amplitudes for the process (1) with $f\neq e,t$ can be written as:

$$G^{ef}_{\alpha\alpha,s}=u\left(\frac{Q_e\,Q_f}{s}+\frac{g^e_\alpha g^f_\alpha}{s-M^2_Z}+\frac{g'^e_\alpha g'^f_\alpha}{s-M^2_{Z'}}\right),\qquad G^{ef}_{\alpha\beta,s}=t\left(\frac{Q_e\,Q_f}{s}+\frac{g^e_\alpha g^f_\beta}{s-M^2_Z}+\frac{g'^e_\alpha g'^f_\beta}{s-M^2_{Z'}}\right), \tag{6}$$

where in the latter expression $\alpha\neq\beta$.

We evaluate the discovery and identification reaches on Z' gauge bosons pertinent to the E_6, LR, ALR and SSM classes of models at the ILC.

3. Discovery reach on the Z' mass

In the absence of available data, the assessment of the expected 'discovery reaches' on the various Z's needs the definition of a 'distance' between the NP model predictions and those of the SM for the basic observables that will be measured. The former predictions parametrically depend on the Z' mass and its corresponding coupling constants, while the latter ones are calculated using the parameters known from the SM fits. Such a comparison can be performed by a standard χ^2-like procedure. We divide the full angular range into bins and identify the basic observables with the polarized differential angular distributions for processes (1), $\mathcal{O}={\rm d}\sigma(P^-,P^+)/{\rm d}z$, in each bin. Correspondingly, the relevant χ^2 can symbolically be defined as:

$$\chi^2(\mathcal{O})=\sum_f\ \sum_{\{P^-,\,P^+\}}\ \sum_{\rm bins}\frac{[\mathcal{O}({\rm SM}+Z')-\mathcal{O}({\rm SM})]^2_{\rm bin}}{(\delta\mathcal{O}_{\rm bin})^2}. \tag{7}$$

To derive the expected 'discovery' limits on Z' models at the ILC, for the 'annihilation' channels in Eq. (1), with $f\neq e,t$, we restrict ourselves to combining in Eq. (7) the $(P^-,P^+)=(|P^-|,-|P^+|)$ and $(-|P^-|,|P^+|)$ beam polarization configurations, that are the predominant ones. For the Bhabha process, $f=e$, we combine in (7) the cross sections with all four possible polarization configurations, i.e., $(P^-,P^+)=(|P^-|,-|P^+|)$, $(-|P^-|,|P^+|)$, $(|P^-|,|P^+|)$, $(-|P^-|,-|P^+|)$. Numerically, we take for the electron beam $|P^-|=0.8$ and for the positron beam $|P^+|=0.6$.

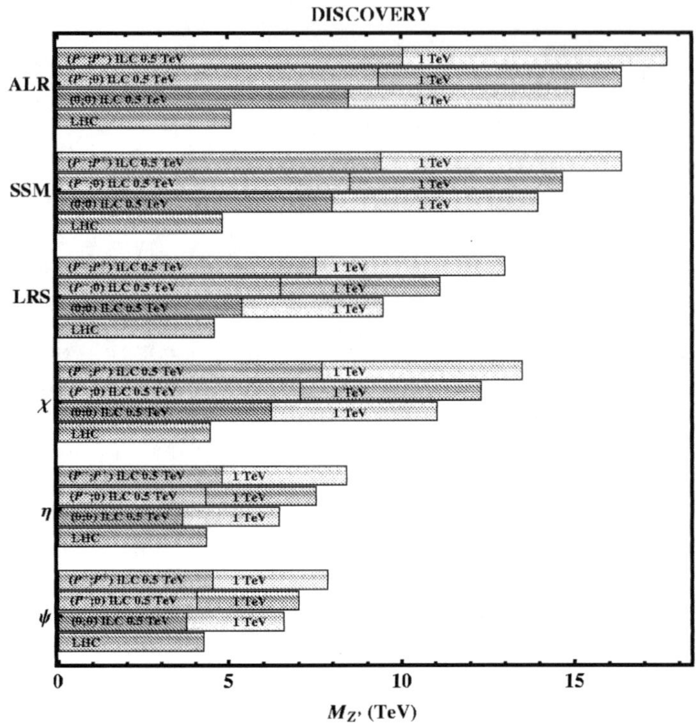

Figure 1. Discovery reaches on Z' models obtained from combined analysis of the unpolarized and polarized processes (1) (95% C.L.) at the ILC with $\sqrt{s} = 0.5$ TeV (1 TeV) and $\mathcal{L}_{\text{int}} = 500$ fb^{-1} (1000 fb^{-1}), compared to the results expected from Drell-Yan processes at the LHC at the 5-σ level [6]. Three options of polarization are considered at the ILC: unpolarized beams, $P^- = P^+ = 0$; polarized electron beam, $|P^-| = 0.8$; both beams polarized, $|P^-| = 0.8$ and $|P^+| = 0.6$.

Regarding the ILC energy and the time-integrated luminosity (which, for simplicity, we assume to be equally distributed among the different polarization configurations defined above), we will give explicit numerical results for c.m. energy $\sqrt{s} = 0.5$ TeV with time-integrated luminosity $\mathcal{L}_{\text{int}} = 500$ fb^{-1}, and for the 'ultimate' upgrade values $\sqrt{s} = 1.0$ TeV with $\mathcal{L}_{\text{int}} = 1000$ fb^{-1}. The assumed final state identification efficiencies governing, together with the luminosity, the expected statistical uncertainties, are: 100% for e^+e^- pairs; 95% for l^+l^- events ($l = \mu, \tau$); 35% and 60% for $c\bar{c}$ and $b\bar{b}$.

As for the major systematic uncertainties, they originate from errors on beam polarizations, on the time-integrated luminosity, and the final-state reconstruction and energy efficiencies. For the longitudinal polarizations, we adopt the values $\delta P^-/P^- = \delta P^+/P^+ = 0.25\%$, rather ambitious, especially as far as P^+ is concerned, but strictly needed for conducting the planned measurements at the permille level. As regards the other systematic uncertainties mentioned above, we assume for the combination the (perhaps conservative) lumpsum value of 0.5%. The systematic uncertainties are included using the covariance matrix approach.

The Fig. 1 includes a comparison with the discovery potential of the LHC with luminosity 100 fb^{-1}, from the Drell-Yan processes $pp \to l^+l^- + X$ ($l = e, \mu$) (at the 5-σ level). These values provide a representative overview of the sensitivities of the reach in $M_{Z'}$ on the planned energy and luminosity, as well as on beam polarization.

4. Distinction of Z' models

Basically, in the previous subsection we have assessed the extent to which Z' models can give values of e^+e^- differential cross sections that can *exclude* the SM hypothesis to a prescribed C.L. Such 'discovery reaches' are represented by upper limits on Z' masses, for which the observable deviations between the corresponding Z' models and SM predictions are sufficiently large compared to the foreseeable experimental uncertainties on the cross sections at the ILC).

However, since different models can give rise to similar deviations, we would like to determine the ILC potential of identifying, among the various competing possibilities, the source of a deviation, should it be effectively observed. These ID-limits should obviously be expected to lie below the corresponding ILC discovery reaches and, for an approximate but relatively simple assessment, we adapt the naive χ^2-like procedure applied in the previous subsection.

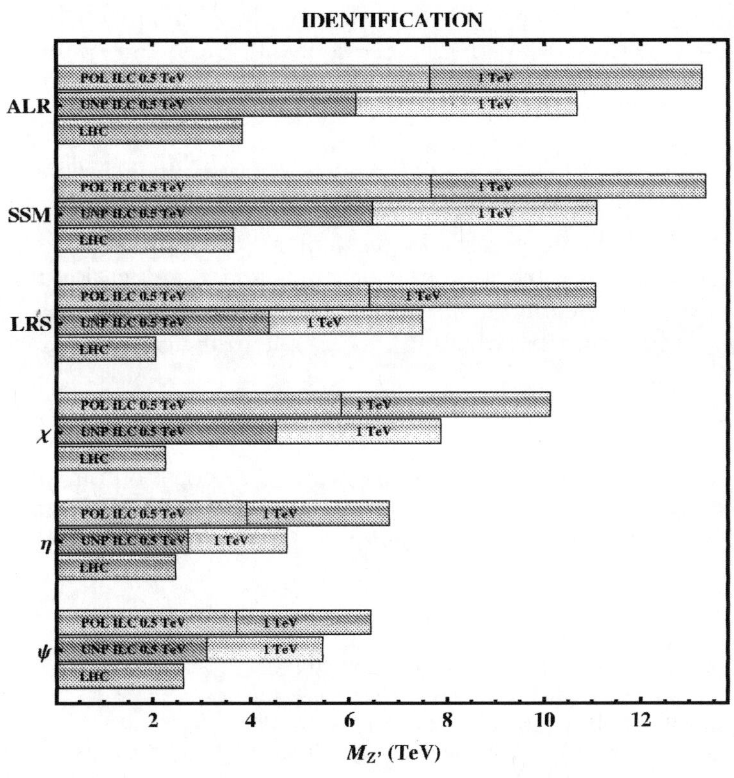

Figure 2. Comparison of the Z'-model distinction bounds on $M_{Z'}$ obtained from combined analysis of the unpolarized and polarized processes (1) at the ILC with $\sqrt{s} = 0.5$ TeV (1 TeV) and $\mathcal{L}_{\text{int}} = 500$ fb^{-1} (1000 fb^{-1}), compared to the results expected from Drell-Yan processes at the LHC at 95% C.L. [6] Two options of polarization are considered: unpolarized beams $P^- = P^+ = 0$ and both beams are polarized, $|P^-| = 0.8$ and $|P^+| = 0.6$.

To this purpose, we start by defining a 'distance' between pairs of Z' models, i and j with i, j denoting any of the SSM, SM, ALR, LRS, ψ, η, χ, but $i \neq j$. We assume for example model i to be the 'true' model, namely, we consider 'data' sets obtained from the dynamics i, with corresponding 'experimental' uncertainties, compatible with the expected 'true' experimental data. The assessment of its distinguishability from a j model, that we call 'tested' model, can be performed by a χ^2 comparison analogous to (7), with the χ^2 defined as:

$$\chi^2(\mathcal{O})_{i,j} = \sum_f \sum_{\{P^-,\ P^+\}} \sum_{\text{bins}} \frac{\left[\mathcal{O}(Z'_i) - \mathcal{O}(Z'_j)\right]^2_{\text{bin}}}{(\delta_i \mathcal{O}_{\text{bin}})^2}. \tag{8}$$

Here, we study a scenario where the Z' mass cannot be known *a priori*, e.g., the Z' is too heavy to be discovered at the LHC (say, $M_{Z'}$ >4–5 TeV), but deviations from the SM predictions can still be observed at the ILC. Actually, models with different Z' masses and coupling constants can in principle be the source of a deviation from the SM predictions observed at the ILC. With the coupling constants held fixed numerically at the theoretical values pertinent to the Z'_i and Z'_j models under consideration, the χ^2_{ij} of Eq. (8) becomes a function of the two masses, $M_{Z'_i}$ and $M_{Z'_j}$, both assumed to lie in the respective ILC discovery ranges.

Finally, Fig. 2 shows the comparison of identification reaches or distinction bounds on the Z'-models considered in Fig. 1, together with the corresponding bounds on $M_{Z'}$ obtained from the process $pp \to l^+l^- + X$ at the LHC with c.m. energy 14 TeV and time-integrated luminosity 100 fb^{-1}. We assume, for the ILC, the same c.m. energy, luminosty and beam polarization as in Fig. 1. The figure speaks for itself, and in particular clearly exhibits the roles of the ILC parameters.

In summary, one might be able to distinguish among the considered Z' models at 95% C.L. up to $M_{Z'} \simeq 3.1$ TeV (4.0 TeV) for unpolarized (polarized) beams at the ILC (0.5 TeV) and 5.3 TeV (7.0 TeV) at the ILC (1 TeV), respectively. In particular, the figure explicitly manifests the substantial role of electron beam polarization in sharpening the identification reaches. Positron polarization can also give a considerable enhancement in this regard (if measurable with the same high accuracy as for electron polarization), although to a more limited extent in some cases.

Acknowledgments
Authors would like to thank Per Osland for the enjoyable collaboration on the subject matter covered here. This research has been partially supported by the Abdus Salam ICTP (TRIL Programme) and the Belarusian Republican Foundation for Fundamental Research.

References

[1] Langacker P. 2009 *Rev. Mod. Phys.* **81** 1199
[2] Erler J., Langacker P., Munir S. and E.Rojas E. 2009 *JHEP* **0908** 017
[3] Aaboud M. *et al.* [ATLAS Collaboration], arXiv:1707.02424 [hep-ex]
[4] The CMS Collaboration [CMS Collaboration], CMS-PAS-EXO-15-005
[5] Osland P., Pankov A. A., Paver N. and Tsytrinov A. V. 2008 *Phys. Rev.* D **78** 035008
[6] Osland P., Pankov A. A., Tsytrinov A. V. and Paver N. 2009 *Phys. Rev.* D **79** 115021
[7] Moortgat-Pick G. A. *et al.* 2008 *Phys. Rept.* **460** 131
[8] For details of the analysis and original references, see Osland P., Pankov A. A. and Tsytrinov A. V. 2010 *Eur. Phys. J.* C **67** 191; 2017 *Phys. Rev.* D **96** 055040

IOP Conf. Series: Journal of Physics: Conf. Series **938** (2017) 012060 doi:10.1088/1742-6596/938/1/012060

Momentum transfer dependence of the hadron GPDs and Compton form factors

O.V. Selyugin
BLTPh, JINR, Dubna, Russia
E-mail: selugin@theor.jinr.ru

Abstract.
The generalized parton distributions (GPDs) of the meson and nucleon at small and large values of the momentum transfer were determined on the basis of the comparative analysis of different sets of experimental data of electromagnetic form factors of the proton and neutron, using the different sets of the parton distribution functions (PDFs). As a result, different form factors of the nucleons and meson were calculated. The t-dependence of these form factors was checked up in the description of the nucleon-nucleon and meson-nucleon elastic scattering and the differential cross sections of the real Compton scattering.

1. Introduction

The recent results from the LHC gave plenty of new information about the elastic and dip-inelastic processes which leads to new questions in the study of the structure of hadrons. At first, it is needed to note that the new data of the TOTEM and ATLAS Collaborations on the elastic proton-proton scattering indeed show that none of the old model predictions give the true descriptions of the elastic cross sections at the LHC [1]. The different reactions can be related with the different form factors. The strong hadron-hadron scattering can be proportional to the matter distribution of the hadrons. The Compton scattering is described by the Compton form factors. Now we have more general parton distributions which depend on different variables - $GTMDs(x, \vec{k}, \xi, \vec{\Delta})$ - Generalized transverse momentum dependent parton distributions. [2, 3, 4]. They parameterize the unintegrated off-diagonal quark-quark correlator, depending on the three-momentum $\vec{k}$ of the quark and on the four-momentum Δ, which is transferred by the probe to the hadron. Taking $\Delta = 0$, we can obtain the $TMD(x, \vec{k})$ - transverse momentum-dependent parton distribution. In other way, after integration over $\vec{k}$ we obtain $GPDs(x, \xi, \Delta)$ - Generalized parton distributions. The remarkable property of the $GPDs$ is that the integration of the different momentum of GPDs over x gives us the different hadron form factors [5, 6, 7].

The x dependence of $GPDs$ in most part is determined by the standard PDFs, which are obtained by the different Collaborations from the analysis of the dip-inelastic processes. Many different forms of the t-dependence of GPDs were proposed. In the quark diquark model [8] the form of GPDs consists of three parts - PDFs, function distribution and Regge-like. In other works (see e.g. [9]), the description of the t-dependence of GPDs was developed in a more complicated picture using the polynomial forms with respect to x.

2. Momentum transfer dependence of GPDs

Let us modify the original Gaussian ansatz and choose the t-dependence of GPDs in a simple form $\mathcal{H}^q(x,t) = q(x)\ exp[a_+\ f(x)\ t]$, with $f(x) = (1-x)^2/x^\beta$ [10]. The isotopic invariance can be used to relate the proton and neutron GPDs.

The complex analysis of the corresponding description of the electromagnetic form factors of the proton and neutron by the different PDF sets (24 cases) was carried out in [11]. These PDFs include the leading order (LO), next leading order (NLO) and next-next leading order (NNLO) determination of the parton distribution functions. They used the different forms of the x dependence of PDFs. We slightly complicated the form of GPDs in comparison with the equation used in [10], but it is the simplest one as compared to other works (for example [12]).

$$\mathcal{H}^u(x,t) = q(x)^u\ e^{2a_H f(x)_u\ t};\quad \mathcal{H}_{\lceil}{}^d(x,t) = q(x)^d\ e^{2a_H f_d(x)\ t}; \tag{1}$$

$$\mathcal{E}^u(x,t) = q(x)^u(1-x)^{\gamma_u}\ e^{2a_E\ f(x)_u\ t};\quad \mathcal{E}_{\lceil}{}^d(x,t) = q(x)^d(1-x)^{\gamma_d}\ e^{2a_E f_d(x)\ t}, \tag{2}$$

where $f_u(x) = \frac{(1-x)^{2+\epsilon_u}}{(x_0+x)^m}$ and $f_d(x) = (1+\epsilon_0)(\frac{(1-x)^{1+\epsilon_d}}{(x_0+x)^m})$.

The hadron form factors will be obtained by integration over x in the whole range $0-1$. Hence, the obtained form factors will be dependent on the x-dependence of the forms of PDF at the ends of the integration region. The different Collaborations determined the PDF sets from the inelastic processes only in some region of x, which is only approximated to $x=0$ and $x=1$. Some PDFs have the polynomial form of x with different power. Some other have the exponential dependence of x. As a result, the behavior of PDFs, when $x \to 0$ or $x \to 1$, can impact the form of the calculated form factors.

On the basis of our GPDs with ABM12 [13] PDFs we calculated the hadron form factors by the numerical integration and then by fitting these integral results by the standard dipole form with some additional parameters $F_1(t) = (4m_p - \mu t)/(4m_p - t)\ \tilde{G}_d(t)$ with $\tilde{G}_d(t) = 1/(1+q/a_1+q^2/a_2^2+q^3/a_3^3)^2$ which is slightly different from the standard dipole form on two additional terms with small sizes of coefficients. The matter form factor

$$A(t) = \int_0^1 x\ dx[q_u(x)e^{2\alpha_H f(x)_u/t} + q_d(x)e^{2\alpha_H f_d(x)/t}] \tag{3}$$

is fitted by the simple dipole form $A(t) = \Lambda^4/(\Lambda^2-t)^2$. These form factors will be used in our model of the proton-proton and proton-antiproton elastic scattering.

3. Magnetic transition form factor $G^*_{M(\gamma^* N\Delta)}$

To check the momentum dependence of the spin-dependent part of GPDs $E_{u,d}(x,\xi=0,t)$, we can calculate the magnetic transition form factor which is determined by the difference of $E_u(x,\xi=0,t)$ and $E_d(x,\xi=0,t)$. For the magnetic $N \to \Delta$ transition form factor $G^*_M(t)$, in the large N_c limit, the relevant $GPD_{N\Delta}$ can be expressed in terms of the isovector GPD yielding the sum rule [14]

$$G^*_M(t) = \frac{G^*_M(t=0)}{k_v}\int_{-1}^1 dx(E_u(x,\xi,t) - E_d(x,\xi,t)) \tag{4}$$

where $k_v = k_p - k_n = 3.70$

$$E_u(x,\xi=0,t) = d(x)e^{[2\alpha_1(\frac{(1-x)^{p_1}}{(x_0+x)^{p_2}})]};\quad E_d(x,\xi=0,t) = d(x)e^{[2\alpha_1(\frac{(1-x)^{p_1}k_d}{(x_0+x)^{p_2}}+dx(1-x)t)]}. \tag{5}$$

There are two different conventions - Jones-Scadron and Ash convention. They are related as $G^*_{M,J-S}(Q^2) = G^*_{M,Ash}\sqrt{1+\frac{Q^2}{(M+m)^2}}$. We will be present the data in the Ash convention.

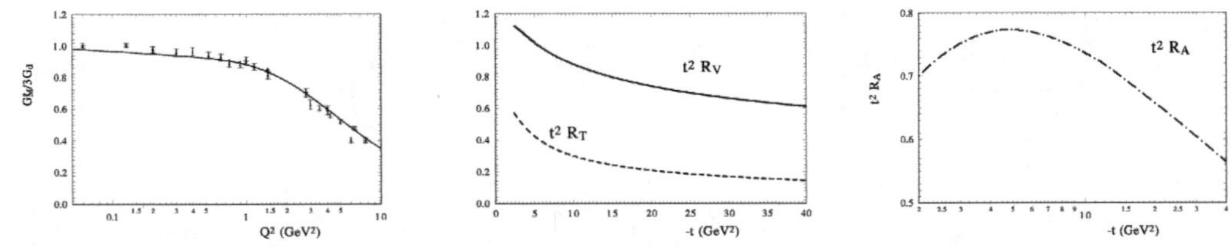

Figure 1. a) [left] The transition magnetic form factors $G^*_{M,Ash}$ (line- our calculations, points the experimental data [21]; b) [middle] the Compton form factors $t^2R_V(t)$ and $t^2R_T(t)$, c) [right] $t^2R_A(t)$

The results of our calculations are presented in Fig.1 (a). The experimental data exist up to $-t = 8$ GeV^2 and our results show a sufficiently good coincidence with experimental data. It is confirmed that the form of the momentum transfer dependence of $E(x, \xi, t)$ determined in our model is right.

4. The Compton cross sections

Now let us calculate the moments of the GPDs with inverse power of x. It gives us the Compton form factors. Using the obtained form factors the reaction of the real Compton scattering can be calculating. The differential cross section for that reaction can be written as [12]

$$\frac{d\sigma}{dt} = \frac{\pi\alpha_{em}^2}{s^2}\frac{(s-u)^2}{-us}[R_V^2(t) \ - \ \frac{t}{4m^2}R_T^2(t) + \frac{t^2}{(s-u)^2}R_A^2(t)], \tag{6}$$

where $R_V((t)$, $R_T(t)$, $R_A(t)$ are the form factors given by the $1/x$ moments of the corresponding GPDs $H^q(x,t)$, $E^q(x,t)$, $\tilde{H}^q(x,t)$. The last is related with the axial form factors. As noted in [12], this factorization, which bears some similarity to the handbag factorization of DVCS, is formulated in a symmetric frame where the skewness $\xi = 0$. For $H^q(x,t)$, $E^q(x,t)$ we used the PDFs obtained from the works [15] with the parameters obtained in our fitting procedure of the description of the proton and neutron electromagnetic form factors in [11].

$$R_i(t) = \sum_q e_q^2 \int_0^1 \frac{dx}{x} \mathcal{F}j_q(x, \xi = 0, t), \tag{7}$$

where $\mathcal{F}j_q$ are equal H_q, E_q and $\tilde{H}_q$ and give the the form factors $R_V(t)$, $R_T(t)$, $R_A(t)$, respectively.

In the present work for $\tilde{H}^q(x,t)$ we take Δq^e in the form [16] for NNLO $Q_0 = 2$ GeV^2

$$x\Delta_q(x, Q_0) = N_q \eta_q x^{a_q}(1-x)^{b_q}(1 + c_q x); \tag{8}$$

Assuming $SU(3)$ flavor symmetry of $\Delta\bar{q}$ the coefficient N_q is determined as $\frac{1}{N_q} = (1 + c_q\frac{a_q}{1+a_q+b_q})\ B(a_q, b_q+1)$, where $B(a_q, b_q+1)$ is determined by $B(a,b) = \Gamma(a)\Gamma(b)/\Gamma(a+b) = \int_0^1 t^{a-1}(1-t)^{b-1}dt$. The results of our calculations of the Compton form factors are shown in Fig. 1(b,c). Obviously, $R_V(t)$ and $R_T(t)$ have a similar momentum transfer dependence, but essentially differ in size. On the contrary, the axial form factor R_A has an essentially different t dependence. The results for the cross sections are presented in Fig.2 (a). Except the very large angles at low energy the coincidence with the experimental data is sufficiently good. The calculations of R_i on the whole, correspond to the calculations [12].

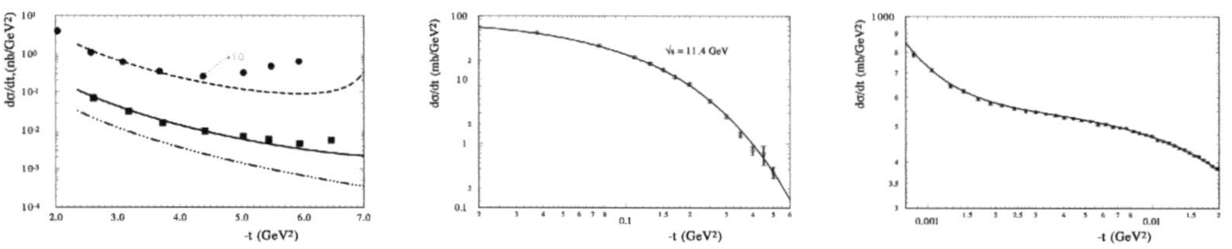

Figure 2. a) [left] Differential Compton cross sections (the lines are our calculations at $s = 8.9$ GeV^2, $s = 10.92$ GeV^2 and $s = 20$ GeV^2 the data points are for $s = 8.9$ GeV^2 (circles); $s = 10.92$ GeV^2 (squares) [22]. The differential cross sections of the elastic pp scattering b) [middle] at $\sqrt{s} = 11.4$ GeV c) [right] at $\sqrt{s} = 13$ TeV; (lines are the calculation in the framework of the HEGS model; points - the experimental data [29]; and data of the TOTEM Collaboration [30]

5. Hadron form factors and elastic nucleon-nucleon scattering

Both hadron electromagnetic and gravitomagnetic form factors were used in the framework of the high energy generalized structure model of the elastic nucleon-nucleon scattering. This allows us to build the model with minimum fitting parameters [17, 18, 19].

The Born term of the elastic hadron amplitude can now be written as

$$\begin{aligned} F_h^{Born}(s,t) = \quad & h_1\ G^2(t)\ F_a(s,t)\ (1 + r_1/\hat{s}^{0.5}) + h_2\ A^2(t)\ F_b(s,t) \\ & \pm h_{odd}\ A^2(t) F_b(s,t)\ (1 + r_2/\hat{s}^{0.5}), \end{aligned} \tag{9}$$

where $F_a(s,t)$ and $F_b(s,t)$ have the standard Regge form: $F_a(s,t) \;=\; \hat{s}^{\epsilon_1}\ e^{B(\hat{s})\ t}$; $F_b(s,t) \;=\; \hat{s}^{\epsilon_1}\ e^{B(\hat{s})/4\ t}$, $\hat{s} \;=\; s\ e^{-i\pi/2}/s_0$; $s_0 \;=\; 4m_p^2$ GeV^2, and $h_{odd} \;=\; ih_3 t/(1 - r_0^2 t)$. The intercept $1 + \epsilon_1 = 1.11$ was chosen from the data of the different reactions and was fixed by the same size for all terms of the Born scattering amplitude. The slope of the scattering amplitude has the standard logarithmic dependence on the energy $B(s) \;=\; \alpha'\ ln(\hat{s})$ with $\alpha' \;=\; 0.24$ GeV^{-2} and with some small additional term [18], which reflects the small non-linear behavior of α' at small momentum transfer [20]. The final elastic hadron scattering amplitude is obtained after unitarization of the Born term. So, at first, we have to calculate the eikonal phase $\chi(s,b) \;=\; -\frac{1}{2\pi}\int d^2q\ e^{i\vec{b}\cdot\vec{q}}\ F_h^{\rm Born}(s,q^2)$ and then obtain the final hadron scattering amplitude

$$F_h(s,t) = is \int b\ J_0(bq)\ \Gamma(s,b)\ db; \qquad \text{with} \quad \Gamma(s,b) = 1 - \exp[\chi(s,b)]. \tag{10}$$

At large t our model calculations are extended up to $-t \;=\; 15$ GeV^2. We added a small contribution of the energy independent part of the spin flip amplitude in the form similar to the proposed in [23] and analyzed in [24]. $F_{sf}(s,t) \;=\; h_{sf} q^3 F_1^2(t) e^{-B_{sf} q^2}$. The model is very simple from the viewpoint of the number of fitting parameters and functions. There are no any artificial functions or any cuts which bound the separate parts of the amplitude by some region of momentum transfer. In the framework of the model the description of the experimental data was obtained simultaniously at the large momentum transfer and in the Coulomb-hadron region in the energy region from $\sqrt{s} = 9$ GeV up to LHC energies. Figure 2(b) represents the description at $\sqrt{s} \;=\; 11.4$ GeV and Figure 2(c) represents the model predictions for $\sqrt{s} \;=\; 13$ TeV, which coincide well with the preliminary data presented at the conference BLOIS-17 (2017) [30].

6. Nucleon structure and radii

Let us now consider, as an example, the neutron structure in the impact parameter [25, 26] representation. We are particularly motivated by recent discussion of the definition of charge density of the neutron at small impact parameters corresponding to the "center" of the neutron [27, 28]. In [28], the charge density of the neutron is related to $F_1^n(t)$ and calculated using the phenomenological representation of $G_E^n(t)$ and $G_M^n(t)$. It differs essentially from the definition of the neutron charge distribution in the Breit frame.

$$\begin{aligned} \rho_{G_E}(b) &= \int d^2q[F_1(q^2) + \tau F_2(q^2)]\ e^{i\vec{q}\vec{b}} \\ &= \sum_q e_q \int_0^1 dx \int d^2q[H_q(x,\xi=0,q^2) + \tau E_q(x,\xi=0,q^2)]e^{i\vec{q}\vec{b}} \end{aligned} \tag{11}$$

with , as usual, $\tau = (q/m_p)^2$. Using our model of t-dependence of GPDs, we may calculate both forms of the neutron charge distribution in the impact parameter representation and, moreover, determine separate contributions of u and d quarks. The respective separate contributions of u and d-quarks are shown in Fig. 3(a). We can see that u-quarks have the large negative charge density in the centre of the neutron.

Let us compare the distribution of the electric charge and matter (that is, gravitational charge) in the nucleon with the matter density

$$\rho_0^{Gr}(b) = \frac{1}{2\pi} \int_\infty^0 dq\ q\ J_0(qb) A(q^2). \tag{12}$$

The difference of the matter density and the charge density for proton is shown in Fig.3(b). The radius of the hadron can be determined by the slope of its form factor $\langle r_h^2 \rangle = -6/f_h(0)d/dt[f_h(t)]_{t=0}$. For the electromagnetic form factor with our form of $\tilde{G}(t)$ we have

$$-\frac{6}{\tilde{G}(0)}\frac{d}{dt}[\tilde{G}(0)] = 2a_2t - (2(1/L^2 + a_1/(2\sqrt{t})))/(1 + a_1\sqrt{-t} - t/L^2)^3 \tag{13}$$

As the coefficients a_i and t are very small eq.(13) gives us the simple relations $\langle r_h^2 \rangle \approx 1/L^2$. The electromagnetic form factors and the gravitomagnetic form factors are describe almost the same form - the dipole form factor but with different sizes of the L_i^2. Hence we have for the electromagnetic radius the ratio of the radiuses of the electromagnetic and the gravitomagnetic form factors - $r_E/r_A = (1/L_E^2)/(1/L_A^2) \approx 1.5$. It is shown that the matter distribution is concentrated in the center of the hadron and at large distances the charge distribution has a longer tail than the matter distribution.

7. Conclusion

The obtained new momentum transfer dependence of GPDs, based on the analysis of practically all existing experimental data on the electromagnetic form factors of the proton and neutron, shows a good description of the t dependence of a wide circle of different reactions. As a result, the description of various reactions is based on the same representation of the hadron structure. Especially it concerns the real Compton scattering and high energy elastic hadron scattering. The new high energy generelized structure (HEGS) model, based on the electromagnetic and gravitomagnetic form factors, gives a good quantitative description of the existing experimental data of the proton-proton and proton-antiproton elastic scattering in a wide region of the energy scattering and momentum transfer, including the Coulomb-hadron interference region, with the minimum number of the fitting parameters. Their predictions coincides well with the new experimental data obtained at the LHC at $\sqrt{s} = 7$ and $\sqrt{s} = 8$ TeV The predictions of the

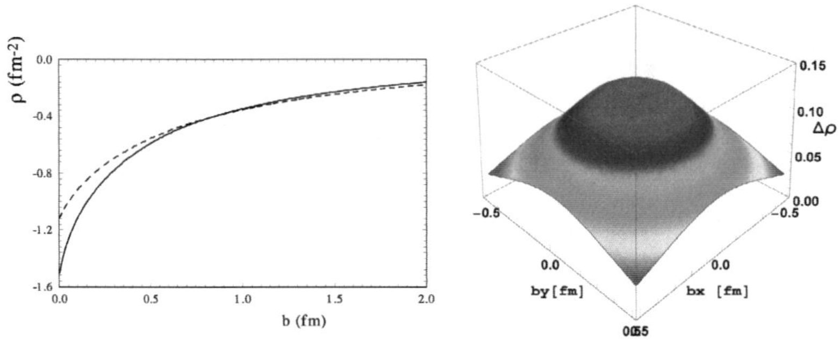

Figure 3. a) [left] u-quark (hard line) and $-d$-quark (dashed line) density of the neutron.

model based on the obtained electromagnetic and gravitomagnetic form factors coincides well with the recent preliminary data at $\sqrt{s} = 13$ TeV [30]. The investigation of the nucleon structure shows that the density of the matter is more concentrated than the charge density in the nucleon.

Acknowledgments *The authors would like to thank J.-R. Cudell and O.V. Teryaev for fruitful discussion of some questions considered in the paper.*

References

[1] R. Fiore, L. Jenkovszky, R. Orava, E. Predazzi, A. Prokudin, O. Selyugin, Mod.Phys., A24 (2009) 2551.
[2] S. Meissner, A. Metz, M. Schlegel, and K. Goeke, JHEP, **0908** (2009) 056.
[3] C. Lorce and B. Pasquini, JHEP, **1309** (2013) 138.
[4] M. Burkardt and B. Pasquini, EPJA Special Issue on "3D Structure of the Nucleon"; EPJ,
[5] D. Muller *et al.*, Fortsch. Phys. **42**, (1994) 101;
[6] X.D. Ji, Phys. Lett. **78** , (1997) 610; Phys. Rev D **55** (1997) 7114;
[7] Radyushkin, A.V., Phys. Rev. D **56**, 5524 (1997).
[8] G.R. Goldstein, J.O. Hernandez, S. Liuti, Phys.Rev. **D84** 034007 (2011).
[9] M.Diehl *et al.*, Eur.Phys. J. C **39** (2005) 1.
[10] O. Selyugin, O. Teryaev, Phys. Rev. **D 79** 033003 (2009);
[11] O.V. Selyugin, Phys. Rev. **D 89** 093007 (2014) .
[12] M. Diehl and P. Kroll, Eur.Phys.J. **C73** 2397 (2013).
[13] S. Alekhin, J. Blu"mlein, and S. Moch, Phys.Rev. D86 , 054009 (2012).
[14] M. Guidal, M. V. Polyakov, A. V. Radyushkin, M. Vanderhaeghen, PhysRevD. **D72** (2004) 054013.
[15] H. Khanpour *et al.*, arXiv:1205.5194
[16] F.Taghavi-Shahri, H. Khanpour .. 1603.03157
[17] O.V. Selyugin, Eur.Phys.J. **C72**, 2073 (2012).
[18] O.V. Selyugin, Phys. Rev. **D 91** 113003 (2015) .
[19] O. V. Selyugin, Nucl.Phys. A **903** 54 (2013).
[20] O.V. Selyugin, "Diffraction (2016)
[21] F. Hagelstein, arxiv: 1710.00874.
[22] A. Danagoulian, et. al. (Jefferson Lab Hall A Collaboration), Phys.Rev.Lett., **98** 152001 (2007).
[23] M.V. Galynskii, E.A. Kuraev, Phys.Rev. D, **89** (2014) 054005.
[24] O.V. Selyugin, Particel Nucleii Letters
[25] M. Burkhardt, [hep-ph]/0509316.
[26] H. Dahiya, A. Mukherjee, S. Ray, [hep-ph]/0705.3580.
[27] M. Burkardt, [hep-ph]/0709.2966v2(October, 2007).
[28] G. A. Miller, Phys.Rev.Lett., **99** (2007) 112001.
[29] http://durpdg.dur.ac.uk/hepdata/reac.html.
[30] M. Deile (TOTEM Collaboration), talk in the Workshop on Diffraction in High-Energy Physics, Praha (Czech. Resp.), June 26-30 (2017).

IOP Conf. Series: Journal of Physics: Conf. Series **938** (2017) 012061   doi:10.1088/1742-6596/938/1/012061

Manifestations of geometric phases in a proton electric-dipole-moment experiment in an all-electric storage ring

Alexander J. Silenko

Research Institute for Nuclear Problems, Belarusian State University, Minsk 220030, Belarus
Bogoliubov Laboratory of Theoretical Physics, Joint Institute for Nuclear Research, Dubna 141980, Russia

E-mail: alsilenko@mail.ru

Abstract. We consider a proton electric-dipole-moment experiment in an all-electric storage ring when the spin is frozen and local longitudinal and vertical electric fields alternate. In this experiment, the geometric (Berry) phases are very important. Due to the these phases, the spin rotates about the radial axis. The corresponding systematic error is rather important while it can be canceled with clockwise and counterclockwise beams. The geometric phases also lead to the spin rotation about the radial axis. This effect can be canceled with clockwise and counterclockwise beams as well. The sign of the azimuthal component of the angular velocity of the spin precession depends on the starting point where the spin orientation is perfect. The radial component of this quantity keeps its value and sign for each starting point. When the longitudinal and vertical electric fields are joined in the same sections without any alternation, the systematic error due to the geometric phases does not appear but another systematic effect of the spin rotation about the azimuthal axis takes place. It has opposite signs for clockwise and counterclockwise beams.

1. Introduction

The geometric (Berry) phase [1, 2] is an important effect of classical and quantum mechanics. It is a phase difference acquired over the course of a cycle, when a system is subjected to cyclic adiabatic processes, which results from the geometrical properties of the parameter space of the Hamiltonian. The assumption that the action of the perturbing fields is much weaker than that of the main field means the adiabatic approximation, wherein the time scale over which a time-dependent Hamiltonian varies is long compared to typical quantum-mechanical oscillation periods. In this case, a quantum-mechanical system remains in the same eigenstate but develops a dynamical phase factor being the geometric (Berry) phase.

The geometric phases are rather important in spin physics. In particular, they manifest themselves in the muon g–2 experiment [3]. In this experiment, the motional magnetic field in the muon rest frame caused by electric focusing leads to a change of the g–2 frequency (so-called pitch correction [4, 5, 6, 7]). An importance of these phases for electric-dipole-moment (EDM) experiments is caused by their potential to mimic the EDM. They condition the most important systematic error in the neutron EDM experiment [8] and play an important role in storage-ring EDM experiments. The Berry phases resulting in the false EDM effect can originate from the noncommutativity of spin rotations about different axes [9]. In three dimensions, when the spin rotates first about one axis and then about a second one, and then the rotations repeat with opposite sign, the total spin rotation is equal to the product of the two rotation angles with respect to the third axis [9, 10]. For an EDM experiment in an all-electric storage ring, this situation takes place for the false EDM effect conditioned by a vertical and a longitudinal component of a magnetic field. However, the corresponding systematic error [9, 10, 11, 12] is small due to magnetic shielding.

We consider another effect conditioned by the geometric phase. It takes place in the proton EDM experiment in an all-electric storage ring and appears owing to the noncommutativity of impacts of the vertical and longitudinal electric fields.

2. Spin behavior in an all-electric storage ring

A spin behavior in an all-electric storage ring is defined by the Thomas-Bargmann-Michel-Telegdi equation [13] generalized to the case of a particle with the EDM [14, 15, 16, 17, 18]. The main distinguishing feature of the all-electric storage ring for the proton EDM experiment is the use of radial electric field electrodes for bending, quadrupoles for focusing, and sextupoles for a chromaticity and a long spin coherence time [9]. When the magnetic field is absent, the spin motion equation in the Frenet-Serret (FS) coordinate system (see Ref. [19] for details) reduces to

$$\frac{d\boldsymbol{\zeta}}{dt} = \boldsymbol{\Omega}\times\boldsymbol{\zeta}, \qquad \boldsymbol{\Omega} = \frac{e}{mc}\left\{\left(G - \frac{1}{\gamma^2-1}\right)(\boldsymbol{\beta}\times\boldsymbol{E}) - \frac{\eta}{2}\left[\boldsymbol{E} - \frac{\gamma}{\gamma+1}\boldsymbol{\beta}(\boldsymbol{\beta}\cdot\boldsymbol{E})\right]\right\}, \tag{1}$$

where $G = (g-2)/2$, $\boldsymbol{\beta} = \boldsymbol{v}/c$, $\eta = 4dm/e$, d is the EDM. When the FS coordinate system is used, one describes the momentum evolution in the cylindrical coordinate system and the spin motion relative to the momentum direction. Except for field misalignments and imperfections, the electric field is orthogonal to $\boldsymbol{\beta}$. Thus, we can neglect the term proportional to $\boldsymbol{\beta}\cdot\boldsymbol{E}$. When the particle momentum in the all-electric EDM storage ring satisfies the condition

$$p = mc\sqrt{\gamma^2-1} = \frac{mc}{\sqrt{G}}, \tag{2}$$

the spin in frozen. In this case, the spin follows the momentum even if the beam is deviated from the ideal orbit. When the condition (2) is exactly satisfied, systematic errors are vanished. We should mention that the cylindrical coordinates are usually more convenient for a description of the spin dynamics than the FS ones [19]. However, the latter coordinates define the spin motion relative to the momentum direction. Since this motion is the only reason for an appearance of systematic errors in the considered case, the use of the FS coordinates is preferable. In the cylindrical coordinate system, the beam with the initial azimuthal polarization can acquire a vertical polarization component even if Eq. (2) is exact. As a result, the description of a spin evolution in this system may lead to a consideration of some fictitious sources of systematic errors.

While the effect of the geometric phase owing to the vertical and longitudinal magnetic fields is rather small, the residual *radial* magnetic field can lead to a significant false EDM signal. This field produces a vertical spin rotation signal which cannot be canceled with clockwise (CW) and counterclockwise (CCW) beams. In addition, the radial magnetic field causes a vertical separation of two counter-rotating beams [10].

It has been mentioned in Ref. [10] that systematic errors in the proton EDM experiment can be conditioned by the vertical electric field and deviations of the beam momentum from the value (2). In the present work, we show that these errors can be rather significant and evaluate them. We also determine main properties of spin dynamics caused by these systematic errors. We consider the spin rotation about the radial and longitudinal axes. Unlike precedent investigations, we give a detailed analysis for some special conditions.

In the all-electric storage ring, the longitudinal electric field does not rotate the spin and the main radial electric field rotating the spin about the vertical axis cannot cause a false EDM signal. Therefore, the electric fields cannot create the Berry phase from the noncommutativity of spin rotations. The false EDM signal appears owing to an alternation of the vertical and longitudinal electric fields. When the condition (2) is violated, the first field causes the spin rotation about the radial axis imitating the EDM signal. The second field changing the beam momentum brings a violation of this condition. The resulting effect is caused by misalignments and imperfections of the both fields and these fields do not have a strong impact on the spin. It can be easily checked that the effect considered meets conditions of an appearance of the Berry phase.

If the *averaged* beam momentum does not satisfy Eq. (2), the spin rotation about the vertical axis appears. Since this rotation can be easily recognized, we suppose that the beam momentum satisfies Eq. (2) on the average. To simplify the analysis, we specify starting points of the beam trajectory where the beam momentum complies with Eq. (2) ($\Delta\gamma = 0$) and neglect very small deviations of Ω_z from zero. When Eq. (2) is satisfied even locally, the spin rotates about the radial axis only due to the EDM.

Any spin effect conditioned by the magnetic dipole moment (MDM) and mimicking the EDM signal may take place only when the condition (2) is violated. This situation takes place when the particle momentum is changed by a longitudinal electric field. We consider second-order systematic errors (bilinear in small effects caused by the MDM). A violation of the above-mentioned condition by a vertical electric field and by imperfections of the main radial electric field can be neglected because p_r^2 and p_z^2 are small quantities of the second order. Certainly, the average longitudinal electric field is equal to zero because *perpetuum mobile* cannot exist. However, this field can be nonzero in some sections of the ring.

3. Spin rotation mimicking the EDM signal

It can be easily shown that the angular velocity of the spin motion in the all-electric storage ring with small perturbations of the azimuthal particle momentum is given by

$$\boldsymbol{\Omega} = \frac{2e}{mc}G^2\gamma\left(\boldsymbol{\beta}\times\boldsymbol{E}\right)\Delta\gamma - \frac{e\eta}{2mc}\boldsymbol{E}, \tag{3}$$

where $\Delta\gamma$ is the deviation from (2). The main systematic error consists in local spin rotations about the radial axis and is caused by the vertical electric field $\boldsymbol{E}^{(v)} = \pm\left|\boldsymbol{E}^{(v)}\right|\boldsymbol{e}_z \equiv \pm E^{(v)}\boldsymbol{e}_z$.

Let us consider the simplest case leading to this systematic error. This case is shown in Figs. 1, 2. The main electric field is always antiparallel to $\boldsymbol{e}_r$. We denote $\beta = |\boldsymbol{\beta}|$. Regions with nonzero $\Delta\gamma$ and $\boldsymbol{E}^{(v)}$ bounded by the points A, B, C, and D alternate. We will analyze four cases when both the momentum and the spin are collinear to the azimuthal direction, $\boldsymbol{e}_\phi$, in these starting points.

3.1. Clockwise beam motion (Fig. 1)

a) The starting point is A. At the points B and C, $\Delta\gamma > 0$. The angular velocity of the spin motion on the path BC (without the EDM effect) is

$$\boldsymbol{\Omega}_{BC} = -\frac{2e}{mc}G^2\beta\gamma E^{(v)}|\Delta\gamma|\boldsymbol{e}_z. \tag{4}$$

At the points D and A, $\Delta\gamma = 0$. On the path CA, $\boldsymbol{\Omega}_{CA} = 0$. As a result, the average angular velocity is given by [20]

$$\boldsymbol{\Omega} = \frac{1}{4}\boldsymbol{\Omega}_{BC} = -\alpha\boldsymbol{e}_z, \quad \alpha = \frac{e}{2mc}G^2\beta\gamma E^{(v)}|\Delta\gamma|. \tag{5}$$

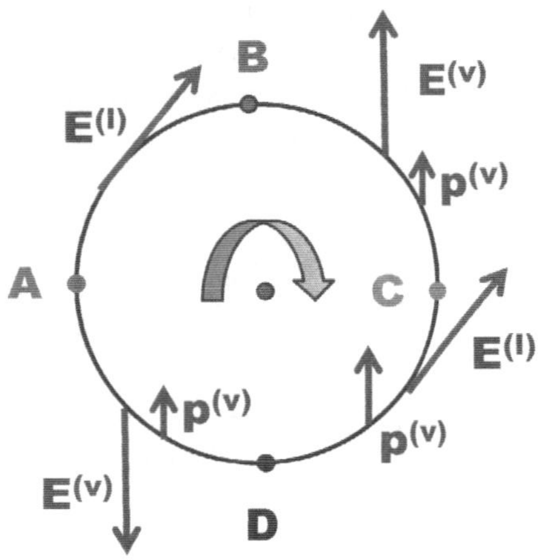

Figure 1. Clockwise beam motion.

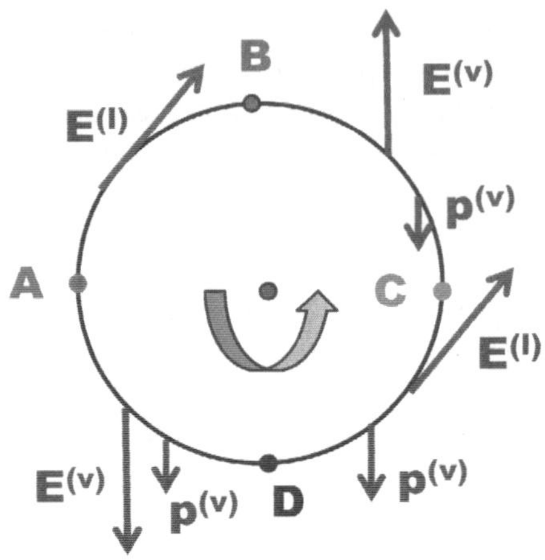

Figure 2. Counterclockwise beam motion.

This is a systematic error imitating the EDM effect.

We should note that the longitudinal (azimuthal) electric field brings also the radial velocity component, $\boldsymbol{\beta}^{(r)}$. However, the vector product $\boldsymbol{\beta}^{(r)} \times \boldsymbol{E}$ with allowance for the factor $\Delta\gamma$ leads to a third-order correction to $\boldsymbol{\Omega}$. This correction can be neglected.

b) The starting point is B. At the points C and D, $\Delta\gamma = 0$ and $\Delta\gamma < 0$, respectively. The false effect is only on the path DA: $\boldsymbol{\Omega}_{DA} = -4\alpha\boldsymbol{e}_z$ and $\boldsymbol{\Omega}_{AB} = 0$. As a result, $\boldsymbol{\Omega} = -\alpha\boldsymbol{e}_z$.

The analysis is similar for the starting points C and D. The false effect does not depend on the starting point. This effect can be calculated [20] and is given by

$$|\boldsymbol{\Omega}| = \alpha = \tfrac{e}{2mc}G^2\beta\gamma E^{(v)}\Delta\gamma_{max}. \tag{6}$$

3.2. Counterclockwise beam motion (Fig. 2)

a) The starting point is A. At the points D and C, $\Delta\gamma = 0$ and $\Delta\gamma < 0$, respectively. The false effect is only on the path CB: $\boldsymbol{\Omega}_{CB} = 4\alpha\boldsymbol{e}_z$. On the path BA, $\boldsymbol{\Omega}_{BA} = 0$. As a result, $\boldsymbol{\Omega} = \alpha\boldsymbol{e}_z$.

b) The starting point is B. At the points A and D, $\Delta\gamma < 0$. The false effect is only on the path AD: $\boldsymbol{\Omega}_{AD} = 4\alpha\boldsymbol{e}_z$ and $\boldsymbol{\Omega}_{DB} = 0$. As a result, $\boldsymbol{\Omega} = \alpha\boldsymbol{e}_z$.

For the starting points C and D, the analysis is similar [20]. The systematic error due to the geometric phases has different signs for the two directions of the beam rotation. The EDM effect remains the same in this case. Thus, the considered systematic error can be canceled with CW and CCW beams.

4. Spin rotation about the longitudinal direction

There exists also an effect which does not influence spin dynamics of longitudinally polarized beams in all-electric storage rings. This is the spin rotation about the azimuthal axis which is also conditioned by the longitudinal and vertical electric fields. We suppose that the spin precession about the vertical axis is vanished.

Let us return to Figs. 1, 2. The vertical electric field in the sections BC and DA changes the vertical component of the particle momentum. The maximum value of this component is equal to

$$p^{(v)}_{max} = \frac{eR}{c\beta}\int_{\phi_1}^{\phi_2} E^{(v)}d\phi. \tag{7}$$

IOP Conf. Series: Journal of Physics: Conf. Series **938** (2017) 012061 doi:10.1088/1742-6596/938/1/012061

In the considered case,

$$p^{(v)}_{max} = \frac{\pi e R}{2c\beta} E^{(v)}. \tag{8}$$

The vertical component of the particle velocity is given by

$$\beta^{(v)}_{max} = \frac{\pi e R}{2mc^2\beta\gamma} E^{(v)}. \tag{9}$$

Evidently,

$$E = \frac{mc^2\beta^2\gamma}{eR}, \qquad \beta^{(v)}_{max} E = \frac{\pi}{2}\beta E^{(v)}, \tag{10}$$

where E is the main electric field.

We analyze the same cases which have been considered in the precedent section. In the current section, we calculate only the angular velocity of the spin motion acting on the radial spin component.

4.1. Clockwise beam motion (Fig. 1)

The momentum distribution shown at this figure is valid when the starting points are A and B.

a) The starting point is A. At the points B and C, $\Delta\gamma > 0$. The average value of β_z on the path BC is equal to $\beta^{(v)}_{max}/2$. The angular velocity of the spin motion on this path reads

$$\boldsymbol{\Omega}_{BC} = -\frac{e}{mc} G^2 \beta^{(v)}_{max} \gamma E \Delta\gamma_{max} \boldsymbol{e}_\phi = -\pi\alpha \boldsymbol{e}_\phi. \tag{11}$$

At the points D and A, $\Delta\gamma = 0$. On the path CD, $\beta_z = \beta^{(v)}_{max} = const$ and the average value of $\Delta\gamma$ is equal to $\Delta\gamma_{max}/2$. As a result, the angular velocity of the spin motion on this path is given by $\boldsymbol{\Omega}_{CD} = -\pi\alpha\boldsymbol{e}_\phi$. The average angular velocity of the spin motion takes the form

$$\boldsymbol{\Omega} = -\frac{\pi}{2}\alpha\boldsymbol{e}_\phi. \tag{12}$$

b) The starting point is B. At the points B and C, $\Delta\gamma = 0$. On the path BC, $\boldsymbol{\Omega}_{BC} = 0$. The average value of $\Delta\gamma$ on the path CD is equal to $-\Delta\gamma_{max}/2$. The angular velocity of the spin motion on this path reads $\boldsymbol{\Omega}_{CD} = \pi\alpha\boldsymbol{e}_\phi$. On the path DA, the average value of β_z is equal to $\beta^{(v)}_{max}/2$ and $\boldsymbol{\Omega}_{DA} = \pi\alpha\boldsymbol{e}_\phi$. As a result, $\boldsymbol{\Omega} = \frac{\pi}{2}\alpha\boldsymbol{e}_\phi$.

For the starting points C and D, the analysis is similar [20].

4.2. Counterclockwise beam motion (Fig. 2)

The momentum distribution shown at Fig. 2 is valid when the starting points are A and B.

a) The starting point is A. At the points C and B, $\Delta\gamma = \Delta\gamma_{max}$. The average value of $\Delta\gamma$ on the path DC is equal to $\Delta\gamma_{max}/2$. The angular velocity of the spin motion on this path reads $\boldsymbol{\Omega}_{DC} = \pi\alpha\boldsymbol{e}_\phi$. On the path CB, $\Delta\gamma = \Delta\gamma_{max} = const$ and the average value of β_z is equal to $-\beta^{(v)}_{max}/2$. As a result, the angular velocity of the spin motion on this path is given by $\boldsymbol{\Omega}_{CB} = \pi\alpha\boldsymbol{e}_\phi$. The average angular velocity of the spin motion takes the form $\boldsymbol{\Omega} = \frac{\pi}{2}\alpha\boldsymbol{e}_\phi$.

b) The starting point is B. At the points A and D, $\Delta\gamma = -\Delta\gamma_{max}$. On the path AD, the average value of β_z is equal to $-\beta^{(v)}_{max}/2$ and $\boldsymbol{\Omega}_{AD} = -\pi\alpha\boldsymbol{e}_\phi$. On the path DC, the average value of $\Delta\gamma$ is equal to $-\Delta\gamma_{max}/2$ and $\boldsymbol{\Omega}_{DC} = -\pi\alpha\boldsymbol{e}_\phi$. As a result, $\boldsymbol{\Omega} = -\frac{\pi}{2}\alpha\boldsymbol{e}_\phi$.

The analysis is similar for the starting points C and D [20]. The longitudinal component of the angular velocity of the spin precession has different signs for the CW and CCW directions of the beam rotation. Therefore, this systematic error can be canceled with CW and CCW beams. Even if the spin precession about the vertical axis is not vanished, this property allows one to eliminate the systematic errors caused by the Berry phases. The azimuthal component of $\boldsymbol{\Omega}$, contrary to the radial one, depends on the starting point where the spin orientation is perfect.

5. Spin evolution in joined longitudinal and vertical electric fields

In this section, we consider the spin evolution on condition that the longitudinal and vertical electric fields are joined in the same sections without any alternation. This consideration shows that only an alternation of these fields leads to a systematic error mimicking the EDM signal.

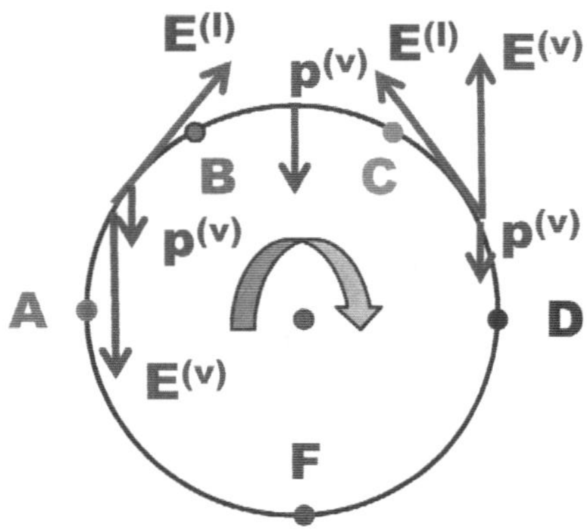

Figure 3. Spin motion in joined longitudinal and vertical electric fields.

Let us consider the simple case shown in Fig. 3. In this case, the longitudinal and vertical electric fields are constant into the paths AB and CD and the fields averaged over the ring are equal to zero. The starting point is F. Since $\Delta\gamma = 0$ at the points A and D and the beam rotates CW, the average value of $\Delta\gamma$ on the path AB is given by Eq. (6) and is positive. The average value of $\Delta\gamma$ on the path CD is the same. However, the values of $\boldsymbol{E}^{(v)}$ are opposite. As a result, the total spin rotation is equal to zero. The same situation takes place when the beam rotates CCW. Thus, the systematic error due to the Berry phases is absent when the longitudinal and vertical electric fields are joined in the same sections without any alternation.

However, another systematic effect of the spin rotation about the azimuthal axis takes place. Figure 3 shows that the signs of the vertical components of the momentum and velocity are the same in the sections AB and CD. Specifically, we obtain the relations

$$\beta_z = -\beta^{(v)}_{max}x, \qquad \Delta\gamma = \Delta\gamma_{max}x, \qquad x = \frac{\phi(X)-\phi(A)}{\phi(B)-\phi(A)}, \tag{13}$$

for any point $X \in [A, B]$. As a result, the average angular velocity of the spin motion is given by

$$\boldsymbol{\Omega}_{AB} = \frac{2e}{3mc}G^2\beta^{(v)}_{max}\gamma E\Delta\gamma_{max}\boldsymbol{e}_\phi = \frac{2}{3}\pi\alpha\boldsymbol{e}_\phi.$$

The average angular velocity of the spin motion on the path CD is the same: $\boldsymbol{\Omega}_{CD} = \boldsymbol{\Omega}_{AB}$. As a result [20],

$$\boldsymbol{\Omega} = \frac{2[\phi(B)-\phi(A)]}{3}\alpha\boldsymbol{e}_\phi. \tag{14}$$

When the beam rotates CCW, γ remains the same but the vertical components of the momentum and velocity change the signs. In this case,

$$\boldsymbol{\Omega} = -\frac{2[\phi(B)-\phi(A)]}{3}\alpha\boldsymbol{e}_\phi.$$

Thus, the angular velocity of the spin rotation has opposite directions for the CW and CCW beams.

6. Discussion and conclusions

The present study shows that local longitudinal and vertical electric fields in the all-electric storage ring may lead to the systematic effects caused by the Berry phases. This takes place when sections with the longitudinal and vertical electric fields alternate. If these fields are perfectly joined, such systematic effects do not appear. Estimates show that the systematic error caused by the Berry phases is not small. Nevertheless, it can be canceled with CW and CCW beams. This possibility is conditioned by different signs of the radial component of the angular velocity of the spin precession $\mathbf{\Omega}$ for the CW and CCW directions of the beam rotation. The azimuthal component of $\mathbf{\Omega}$ also possesses this property. However, the sign of this component depends on the starting point (where the spin orientation is perfect) while the radial component of $\mathbf{\Omega}$ keeps its value for each starting point. When the longitudinal and vertical electric fields are joined in the same sections without any alternation, the systematic error due to the Berry phases does not appear. However, another systematic effect of the spin rotation about the azimuthal axis takes place. The angular velocity of the spin rotation has opposite signs for CW and CCW beams.

It is important that both components of $\mathbf{\Omega}$ are caused by the same local longitudinal and vertical electric fields. Therefore, the azimuthal component of $\mathbf{\Omega}$ can be even helpful because it allows one to evaluate an order of magnitude of the radial component even without the use of the beam with the opposite direction of rotation.

Another important systematic error in the proton EDM experiment in the all-electric storage ring is the residual radial magnetic field. However, this field can be measured by high-precision magnetometers and the map of magnetic field over the ring can be obtained. In this case, the systematic error due to the radial magnetic field can be transformed into the systematic correction.

Acknowledgments

The work was supported in part by the Belarusian Republican Foundation for Fundamental Research (Grant No. Φ16D-004) and by the Heisenberg-Landau program of the German Ministry for Science and Technology (Bundesministerium für Bildung und Forschung).

References

[1] Pancharatnam S 1956 Generalized theory of interference, and its applications. Part I. Coherent pencils *Proc. Indian Acad. Sci.* **44** 247

[2] Berry M 1984 Quantal Phase Factors Accompanying Adiabatic Changes *Proc. Roy. Soc. A* **392** 45

[3] Bennett G W *et al.* (Muon (g–2) Collaboration) 2006 Final report of the E821 muon anomalous magnetic moment measurement at BNL *Phys. Rev. D* **73** 072003

[4] Granger S and Ford G W Electron 1972 Spin Motion in a Magnetic Mirror Trap *Phys. Rev. Lett.* **28** 1479

[5] Farley F J M 1972 Pitch correction in (g-2) experiments *Phys. Lett. B* **42** 66; Farley F J M and Picasso E 1990 *The muon* g*-2 experiments* (Quantum Electrodynamics) ed T Kinoshita (Singapore: World Scientific) pp 479-559

[6] Field J H and Fiorentini G 1974 Corrections to the g − 2 frequency in weak focusing storage devices due to betatron oscillations *Nuovo Cim. A* **21** 297

[7] Silenko A J 2006 Equation of spin motion in storage rings in the cylindrical coordinate system *Phys. Rev. ST Accel. Beams* **9** 034003

[8] Steyerl A, Kaufman C, Müller G, Malik S S, Desai A M, Golub R 2014 Calculation of geometric phases in electric dipole searches with trapped spin-1/2 particles based on direct solution of the Schrödinger equation *Phys. Rev. A* **89** 052129; Golub R, Kaufman C, Müller G, and Steyerl A 2015 Geometric phases in electric dipole searches with trapped spin-1/2 particles in general fields and measurement cells of arbitrary shape with smooth or rough walls *Phys. Rev. A* **92** 062123

[9] Anastassopoulos D *et al* (Storage Ring EDM Collaboration) 2008 AGS Proposal: Search for a permanent electric dipole moment of the deuteron nucleus at the $10^{-29}e\cdot$ cm level, https://www.bnl.gov/edm/files/pdf/deuteron_proposal_080423_final.pdf

[10] Anastassopoulos D *et al* (Storage Ring EDM Collaboration) 2011 A proposal to measure the proton electric dipole moment with 10^{-29} e·cm sensitivity by the storage ring EDM collaboration, https://www.bnl.gov/edm/files/pdf/proton_EDM_proposal_20111027_final.pdf

[11] Semertzidis Y K 2014 Geometrical phase from B-field in the proton EDM ring. Internal Storage Ring EDM Collaboration note, unpublished.

[12] Haciomeroglu S 2017 Spin tracking studies for pEDM experiment. EDM Kickoff meeting @ CERN, https://indico.cern.ch/event/609422/contributions/2463327/attachments/1426326/2188381/spin_tracking_selcuk.pdf

[13] Thomas L H 1926 The motion of the spinning electron *Nature (London)* **117** 514; 1927 The kinematics of an electron with an axis *Philos. Mag.* **3** 1; Bargmann V, Michel L, and Telegdi V L 1959 Precession of the polarization of particles moving in a homogeneous electromagnetic field. *Phys. Rev. Lett.* **2**, 435 (1959). This equation was also derived by Frenkel J 1926 Die Elektrodynamik des rotierenden Elektrons *Z. Phys.* **37** 243

[14] Nelson D F, Schupp A A, Pidd R W and Crane H R 1959 Search for an electric dipole moment of the electron *Phys. Rev. Lett.* **2** 492; Khriplovich I B 1998 Feasibility of search for nuclear electric dipole moments at ion storage rings *Phys. Lett. B* **444** 98

[15] Silenko A J 2005 Quantum-mechanical description of the electromagnetic interaction of relativistic particles with electric and magnetic dipole moments *Russ. Phys. J.* **48** 748

[16] Silenko A J 2013 Quantum-mechanical description of spin-1 particles with electric dipole moments *Phys. Rev. D* **87** 073015

[17] Fukuyama T and Silenko A J 2013 Derivation of generalized Thomas-Bargmann-Michel-Telegdi equation for a particle with electric dipole moment *Int. J. Mod. Phys. A* **28** 1350147

[18] Silenko A J 2015 Spin precession of a particle with an electric dipole moment: contributions from classical electrodynamics and from the Thomas effect *Phys. Scripta* **90** 065303

[19] Silenko A J 2015 Comparison of spin dynamics in the cylindrical and Frenet-Serret coordinate systems *Phys. Part. Nucl. Lett.* **12** 8

[20] Silenko A J 2017 Berry phases in an electric-dipole-moment experiment in an all-electric storage ring *Preprint* 1710.01609 [nucl-ex]

Nucleon structure functions in the truncated moments approach

D Strózik-Kotlorz[1,2], S V Mikhailov[2] O V Teryaev[2] and A Kotlorz[3]

1 Opole University of Technology, Department of Physics, Prószkowska 76, 45-758 Opole, Poland

2 Bogoliubov Laboratory of Theoretical Physics, JINR, 141980 Dubna, Russia

3 Opole University of Technology, Department of Mathematics and IT Applications, Prószkowska 76, 45-758 Opole, Poland

E-mail: dorota@theor.jinr.ru

Abstract. We demonstrate advantages of the truncated Mellin moments (TMM) approach in the analysis of DIS data. We present a novel method for determination of the Bjorken sum rule (BSR) from restricted in x variable experimental data. We show how to incorporate different uncertainties for each kinematic bin. We apply our analysis to recent COMPASS data.

1. Introduction

Our understanding of the matter structure and fundamental particle interactions at high energies is mostly provided by deep inelastic scattering (DIS) of leptons on hadrons and hadron-hadron collisions. In virtue of QCD factorization in these processes hadron properties can be described in terms of the parton distribution functions (PDFs) $f_p(x, \mu^2)$. They are universal process-independent densities explaining how the whole hadron momentum P is partitioned in $x \cdot P$ between partons of type p. Here hard momentum transfer q: $-q^2 = Q^2 \gg P^2 = m_h^2$, and the Bjorken variable x satisfies $0 < x = Q^2/(2Pq) < 1$. These distributions $f_p(x, \mu^2)$ are formed by nonperturbative strong interaction at hadronic scale m_h^2, while the dependence on the normalization scale μ^2 is governed within perturbative QCD by the well-known Dokshitzer-Gribov-Lipatov-Altarelli-Parisi (DGLAP) evolution equations [1]. Alternatively, one can study how to evolve with this scale μ^2 the Mellin moments of the parton densities $f(n, \mu^2)$, which are integrals of PDFs weighted with x^n over the whole range (0,1) of x. These moments provide a natural framework of QCD analysis as they originate from the basic formalism of operator product expansion and are essential in testing sum rules. However, these standard moments, in principle, cannot be extracted from any experiment due to kinematic constraints inevitably appearing in real DIS of lepton-hadron and hadron-hadron collisions. Therefore, it is demanded to invent new "real observables" with a goal to overcome these kinematic constraints. Generalized moments of the parton distribution $f(x, \mu^2)$, $f(z; n, \mu^2) = \int_z^1 f(x, \mu^2) x^{n-1} dx$, contains the unavoidable lower limit of integration $z \equiv x_{\min} = Q^2_{\min}/(2(Pq)_{\max}) > 0$, and in this way the kinematic constraint can be taken into account. The idea of "truncated" Mellin moments of the parton densities in QCD analysis was introduced and developed in the late 1990s [2]. Later on, diagonal integro-differential DGLAP-type evolution equations for the single and double truncated moments of the parton densities were derived in [3] and [4,5]. The main finding of the truncated Mellin moments (TMM) approach is that the nth moment of the parton density also obeys the DGLAP equation, but with a rescaled evolution kernel $P'(z) = z^n P(z)$ [3]. The TMM approach has already been successfully applied, e.g., in spin physics to derive a generalization of the Wandzura-Wilczek relation in terms of the truncated moments and to obtain the evolution equation for the structure function g_2 [5]. The evolution equations for double cut moments and their application to study the quark-hadron duality were also discussed in [6]. The advantages of the TMM approach to QCD factorization for DIS structure functions were presented in [7]. A valuable generalization of the TMM approach incorporating multiple integrations as well as multiple differentiations of the original parton distribution has been obtained in [8]. Recently, based on the TMM, we constructed a device which allows one to improve an experimental determination of the Bjorken sum rule (BSR) [9]. Here, we present the main results of our recent approach [10] - the generalized BSR that allows an effective determination of the BSR value from the experimental data in a restricted kinematic range of x. We also show how to incorporate different uncertainties for each kinematic bin. We apply our analysis to recent COMPASS data on the spin structure function g_1.

2. Generalized Bjorken sum rule

The generalized truncated moment $f(z, n, \omega)$, obtained as a Mellin convolution of the PDF f with any normalized function $\omega(x)$,

$$f(x, n; \omega) = (\omega * x^n f) \equiv \int_x^1 \omega(x/z)\; f(z, \mu^2)\, z^n \frac{dz}{z}, \tag{1a}$$

$$\int_0^1 \omega(t) dt = 1\,, \tag{1b}$$

obeys the DGLAP evolution equation with the rescaled kernel [9]:

$$\mathcal{P}(y) = P(y) \cdot y^n. \tag{2}$$

In a case of the non-singlet polarized structure function g_1 and for $n = 0$, one obtains the generalized structure function $g_{1;\,\omega}$,

$$g_{1;\,\omega}(x) \equiv (\omega * g_1)(x) \equiv \int_x^1 \omega(x/z)\; g_1(z, Q^2)\, \frac{dz}{z} \tag{3}$$

with the same DGLAP evolution kernel as g_1, [10], namely $P(y)$. In this way, we define the truncated BSR, $\Gamma_1(x_0)$, and simultaneously, the generalized cut Bjorken sum rules (gBSR), $\Gamma_{1;\,\omega}(x_0)$,

$$\Gamma_1(x_0) = \int_{x_0}^1 g_1(x)\, dx\,, \tag{4}$$

$$\Gamma_{1;\,\omega}(x_0) = \int_{x_0}^1 g_{1;\,\omega}(x)\, dx, \tag{5}$$

which are equal to the ordinary BSR as $x_0 \to 0$:

$$\Gamma_{1;\,\omega}(0) = \int_0^1 g_{1;\,\omega}(x)\, dx = \int_0^1 g_1(x)\, dx \equiv \Gamma_1(0). \tag{6}$$

Now, one can estimate the value of $\Gamma_1(0)$ from the smooth extrapolation of the truncated moments $\Gamma_{1;\omega}(x_0)$ in x_0. To this aim, one constructs *a bunch* of different $\Gamma_{1;\omega}(x_0)$ based on the simple sign-changing normalized function $\omega(x)$ depending on three parameters z_1, z_2, A,

$$\omega(z) = -A\,\delta(z - z_1) + (1 + A)\,\delta(z - z_2). \tag{7}$$

Here the ω-model parameters are $z_2 > z_1 > x_0 > 0$ and $A > 0$ for the sign change. This leads to a "shuffle" of the initial function g_1 in x variable and the generalized truncated BSR $\Gamma_{1;\omega}(x_0)$,

$$\Gamma_{1;\omega}(x_0) = \int_{x_0/z_2}^{1} g_1(x)\,dx + A \int_{x_0/z_2}^{x_0/z_1} g_1(x)\,dx\,, \tag{8}$$

approaches the limit $\Gamma_1(0)$ more quickly than the ordinary $\Gamma_1(x_0)$ in Eq. (4). In this way, the total BSR limit $\Gamma_1(0)$ can be determined very effectively with use of a few first orders of Taylor expansion:

$$\Gamma_1(0) = \Gamma_{1;\omega}(x_0 - x_0) = \Gamma_{1;\omega}(x_0) - x_0\,\Gamma'_{1;\omega}(x_0) + x_0^2\,\frac{1}{2}\,\Gamma''_{1;\omega}(x_0) + \cdots\,. \tag{9}$$

Figs. 1 and 2 illustrate the features of $\Gamma_{1;\omega}$, where we plot the bunch $\Gamma_{1;\omega}(x_0)$, Eq. (8) for different values of A, including: "constant behavior" value $A = A_{00}$, "quasi-linear behavior" value $A = A_{02}(\bar{x})$ and the standard truncated BSR $\Gamma_1(x_0)$, Eq. (4) (thick black curve). We use the popular form of parametrization of g_1 at $Q_0^2 = 1\,\mathrm{GeV}^2$,

$$g_1(x, Q_0^2) = N \cdot x^a (1 - x)^b (1 + \gamma x), \tag{10}$$

where $a = 0$ in Fig. 1 and $a = -0.4$ in Fig. 2, respectively, $b = 3$, $\gamma = 5$ and the coefficient N is the norm. In our tests, in order to obtain a smooth approach of the bunch in the experimentally available x region, we fixed $z_1 = 0.7$ and $z_2 = 0.9$. One can see significantly rapid saturation

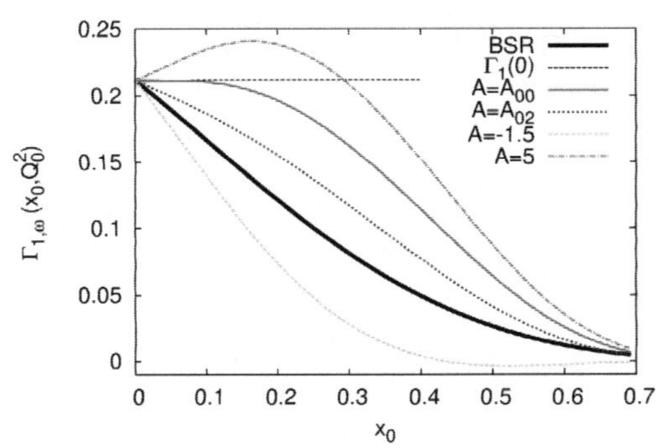

Figure 1. $\Gamma_{1;\omega}(x_0)$, Eq. (8), for different values of A and the truncated BSR $\Gamma_1(x_0)$, Eq. (4) (thick black curve) as a function of x_0. Input $g_1(x, Q_0^2)$, Eq. (10), with $a = 0$.

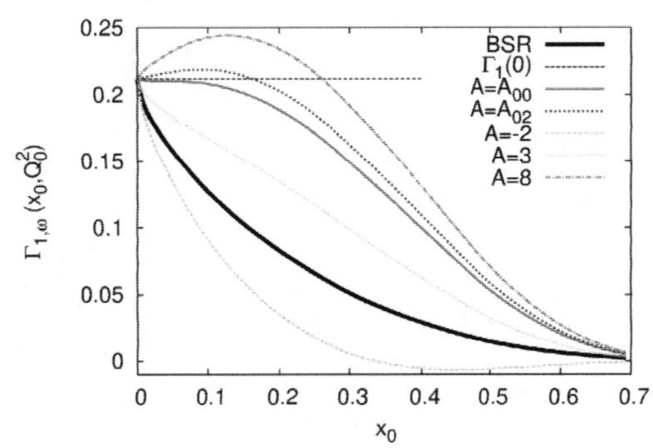

Figure 2. $\Gamma_{1;\omega}(x_0)$, Eq. (8), for different values of A and the truncated BSR $\Gamma_1(x_0)$, Eq. (4) (thick black curve) as a function of x_0. Input $g_1(x, Q_0^2)$, Eq. (10), with $a = -0.4$.

of $\Gamma_{1;\omega}(x_0)$ to $\Gamma_1(0)$ in comparison with $\Gamma_1(x_0)$. The quasi-linear regime of gBSR near 0 visibly starts at rather large values of $x_0 \gtrsim 0.1$ for the different parametrization in Eq. (10). This should ensue the applicability of the first order approximation (IAPX),

$$\Gamma_1(0) \approx \Gamma_1^{\mathrm{IAPX}}(x_0) = \Gamma_{1;\omega}(x_0) - x_0\,\Gamma'_{1;\omega}(x_0), \tag{11}$$

even for JLab experimental conditions, where the admissible x bunches are rather far from 0. In practice, one can use fit to the data instead of the ready input parametrization. It is worthy to notice that the analysis based on the bunch behavior allows one to shift the available region of x to smaller values, $x_0 = x \cdot z_2$. In this manner, using data from large x and choosing suitable values of z_1 and z_2, one is able to get an answer in a much smaller x region.

Presented above idea of the generalized BSR can be used to analysis of the experimental data incorporating different uncertainties for each measurement. Usually, the results for g_1 are extracted from the data for kinematic bins $\{Q_i^2, x_i\}$ with uncertainties $\Delta(g(x_i))$. In order to compute moments of g_1 and verify the BSR, bins must be evolved to a common scale Q^2. Then, the experimental bins $\{x_i\}_1^n$ can be used to construct multi-point weight function ω in Eq. (1),

$$\omega(z) = -\sum_{i=1}^{n-1} A_i\, \delta(z - z_i) + (1 + A)\, \delta(z - z_n)\,, \tag{12}$$

where

$$A = \sum_{i=1}^{n-1} A_i > 0\,. \tag{13}$$

This multi-point ($\{z_i,\ A_i\}_1^n$) ansatz is a generalization of the previous two-point ($\{z_1,\ z_2,\ A\}$) one, Eq. (7), and leads to a new presentation for $\Gamma_{1;\omega}$,

$$\Gamma_{1;\omega}(x_0) \;=\; \underbrace{\int_{x_0/z_n}^{1} g_1(x)\, dx + A \int_{x_0/z_n}^{x_0/z_{n-1}} g_1(x)\, dx}_{\text{two-point result } \Gamma_1} + A \sum_{i=1}^{n-2} w_i \int_{x_0/z_{i+1}}^{x_0/z_i} g_1(x)\, dx. \tag{14}$$

In order to use the generalized BSR to the determination of $\Gamma_1(0)$, one only needs to construct the set of $\{A_i\}$. Choosing the most appropriate weights w_i and hence A_i,

$$w_i = \frac{1}{A}\sum_{k=1}^{i} A_k\,; \qquad A_i = A \cdot (w_i - w_{i-1})\,, \tag{15}$$

one is able to tune the analysis to the real experimental constraints. The most natural way of implementing the experimental uncertainties of g_1 into the ansatz of $\Gamma_{1;\omega}$ is to use w_i inversely proportional to the relative uncertainties $\Delta(g_1)/g_1$ at each x_i:

$$w_i \sim \frac{g_1(x_i)}{\Delta(g_1(x_i))}\,, \tag{16}$$

to increase the weights of the data with smaller uncertainties.

3. Analysis of data

In this section we present practical estimation of $\Gamma_1(0)$ from the recent COMPASS data [11] using two-point and multi-point versions of the gBSR approach described in the previous section. We apply the simplified equations of our approach, written in terms of experimental parameters.

3.1. Two-point estimation of $\Gamma_1(0)$

For practical purposes, we write here the essential formulas for the generalized BSR in terms of experimental data and demonstrate the effective method for the estimation of $\Gamma_1(0)$. Thus, the gBSR, Eq. (8), where the lower limit of integrations has to be strictly related to the minimal x accessible experimentally, x_{min}, takes the form

$$\Gamma_{1;\,\omega}(x_{min}, r) = \int_{x_{min}}^{1} g_1(x)\, dx + A \int_{x_{min}}^{x_{min}/r} g_1(x)\, dx. \tag{17}$$

The experimental lower value x_{min} in the above equation is related to x_0 from Eq. (8) via $x_0 = x_{min}{\cdot}z_2$. The ratio parameter $r \equiv z_1/z_2$ and $x_0 < x_{min} < r < 1$. We have tested all methods of estimation of $\Gamma_1(0)$, described in [10] and have found that a very effective method, universal for the different small-x behavior of g_1 and for $x_{min} \lesssim 0.1$, is the first order approximation, Eq. (11). With use of the parameters x_{min} and r, it reads

$$\Gamma_1(0) \approx \Gamma_1^{\rm IAPX}(x_{min}, r) = \Gamma_{1;\,\omega}(x_{min}, r) + (A+1)\, x_{min}\, g_1(x_{min}) - A\frac{x_{min}}{r} g_1(x_{min}/r) \tag{18}$$

with A obtained requiring the second derivative of $\Gamma_{1;\,\omega}(x_0)$ to vanish (first order approximation),

$$A = \left[r^2 \frac{g_1'(x_{min}/r)}{g_1'(x_{min})} - 1 \right]^{-1}. \tag{19}$$

Below we present our results on determination of the BSR based on the recent COMPASS data, where $x_{min} = 0.0036$. We follow the method described above using Eqs. (17)–(19). Our fit to the data at $Q^2 = 3$ GeV^2 is $g_1(x) \sim x^{-0.36}(1-x)^3(1+3.9\,x)$. One can see from Fig. 3 that $\Gamma_{1;\,\omega}(x)$ approaches $\Gamma_1(0)$ more quickly than the original BSR $\Gamma_1(x)$.

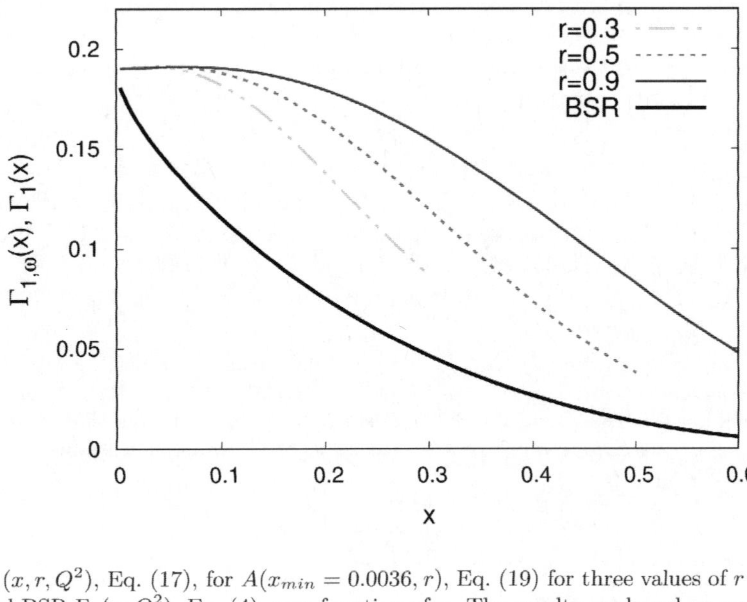

Figure 3. $\Gamma_{1,\omega}(x, r, Q^2)$, Eq. (17), for $A(x_{min} = 0.0036, r)$, Eq. (19) for three values of r : 0.9, 0.5, 0.3, together with the truncated BSR $\Gamma_1(x, Q^2)$, Eq. (4), as a function of x. The results are based on our fit to the COMPASS data.

We find the estimated total BSR $\Gamma_1(0)$ from Eqs. (17)–(19) for $x_{min} = 0.0036$ and different r (0.9, 0.5, 0.3, 0.1) $\Gamma_1^{\rm IAPX}(x_{min}, r) = 0.190$. This should be compared to the final estimation provided by COMPASS collaboration, $\Gamma_1(0) = 0.192 \pm 0.007_{\rm stat} \pm 0.015_{\rm syst}$.

3.2. *Multi-point estimation of* $\Gamma_1(0)$

Here we develope the idea of the generalized BSR to analysis of the COMPASS data on $g_1(x)$ incorporating uncertainties for each measurement. We consider the results for n experimental values of x: $x_{min} \equiv x_1 < x_2 < x_3 < \cdots < x_n \equiv x_{max} < 1$, where $x_i = x_0/z_{n-i+1}$ and corresponding $g_1(x_i)$ together with their relative statistical uncertainties $U_i = \Delta g_1(x_i)/g_1(x_i)$. In terms of the experimental parameters the multi-point gBSR, Eq. (14), has the form

$$\Gamma_{1;\omega}(\{x_i\}_1^n) = \int_{x_1}^{1} g_1(x)\,dx + A\sum_{i=1}^{n-1}\tilde{w}_i \int_{x_i}^{x_{i+1}} g_1(x)\,dx\,, \tag{20}$$

where we choose the weights $\tilde{w}_i = U_1/U_i$ reflecting the increase of the experimental uncertainties of g_1 with decreasing x. The first order approximation for $\Gamma_1(0)$ in the multi-point case is a generalization of Eqs. (18) and (19):

$$\Gamma_1(0) \approx \Gamma_1^{\mathrm{IAPX}}(\{x_i\}_1^n) = \Gamma_{1;\omega}(\{x_i\}_1^n) + x_1\,g_1(x_1) + A\sum_{i=1}^{n-1}\tilde{w}_i\left[x_i\,g_1(x_i) - x_{i+1}\,g_1(x_{i+1})\right]\,, \tag{21}$$

$$A = \frac{x_1^2\,g_1'(x_1)}{\sum\limits_{i=1}^{n-1}\tilde{w}_i\left[x_{i+1}^2\,g_1'(x_{i+1}) - x_i^2\,g_1'(x_i)\right]}\,. \tag{22}$$

In our analysis we use fit function to evolve $g_1(x_1, Q_i^2)$ to the common value of $Q^2 = 3$ GeV^2 and also to find $g_1'(x_i)$. We obtain from Eqs. (20)–(22) $\Gamma_1(0) \approx \Gamma_1^{\mathrm{IAPX}}(\{x_i\}_1^n) = 0.190$, the same value as in the previous case of 2-point estimation. This result is in a good agreement with the value $0.192 \pm 0.007_{\mathrm{stat}} \pm 0.015_{\mathrm{syst}}$ provided by COMPASS [11] from reanalyzed data.

4. Conclusions

The generalized BSR, based on the truncated Mellin moment approach provides a powerful tool to test QCD at experimental constraints.

Acknowledgments

We thank Eva-Maria Kabuss and Yann Bedfer for their help with COMPASS data. This work is supported by the Bogoliubov-Infeld Program, Grant No. 01-3-1113-2014/2018. S.V.M. acknowledges support from the BelRFFR-JINR, Grant No. F16D-004.

References

[1] Gribov V N and Lipatov L N 1972 *Sov. J. Nucl. Phys.* **15** 438, *ibid.* 675; Dokshitzer Yu L 1977 *Sov. Phys. JETP* **46** 641; Altarelli G and Parisi G 1977 *Nucl. Phys.* B **126** 298
[2] Forte S and Magnea L 1999 *Phys. Lett.* B **448** 295; Forte S, Magnea L, Piccione A and Ridolfi G 2001 *Nucl. Phys.* B **594** 46; Piccione A 2001 *Phys. Lett.* B **518** 207; Forte S, Latorre J I, Magnea L and Piccione A 2001 *Nucl. Phys.* B **643** 477
[3] Kotlorz D and Kotlorz A 2007 *Phys. Lett.* B **644** 284
[4] Kotlorz D and Kotlorz A 2009 *Acta Phys. Pol.* B **40** 1661
[5] Kotlorz D and Kotlorz A 2011 *Acta Phys. Pol.* B **42** 1231
[6] Psaker A, Melnitchouk W, Christy M E and Keppel C 2008 *Phys. Rev.* C **78** 025206
[7] Kotlorz D and Kotlorz A 2016 *Int. J. Mod. Phys.* A **31** 1650181
[8] Kotlorz D and Mikhailov S V 2014 *J. High Energy Phys.* JHEP06(2014)065
[9] Bjorken J D 1966 *Phys. Rev.* **148** 1467, *ibid.* 1970 D **1** 1376
[10] Kotlorz D, Mikhailov S V, Teryaev O V and Kotlorz A 2017 *Phys. Rev.* D **96** 016015
[11] COMPASS Collaboration, Alekseev M G *et al.* 2010 *Phys. Lett.* B **690** 466; COMPASS Collaboration, Adolph C *et al.* 2016 *Phys. Lett.* B **753** 18; COMPASS Collaboration, Adolph C *et al.* 2017 *Phys. Lett.* B **769** 34

IOP Conf. Series: Journal of Physics: Conf. Series **938** (2017) 012063 doi:10.1088/1742-6596/938/1/012063

Polarization in heavy-ion collisions: magnetic field and vorticity

M. Baznat[1], K. Gudima[1], G. Prokhorov[2], A. Sorin[2,3], O. Teryaev[2,3] and V. Zakharov[4]

1 Institute of Applied Physics, Academy of Sciences of Moldova, MD-2028 Kishinev, Moldova
2 Joint Institute for Nuclear Research, 141980 Dubna, Russia
3 National Research Nuclear University (MEPhI), 115409 Moscow, Russia
4 Alikhanov Institute for Theoretical and Experimental Physics (ITEP) B. Cheremushkinskaya ul. 25, RU-117218 Moscow, Russia

E-mail: teryaev@theor.jinr.ru

Abstract. The polarization of hyperons due to axial chiral vortical effect is discussed. The effect is proportional to (strange) chemical potential and is pronounced at lower energies, contrary to that of magnetic field. The polarization of antihyperons has the same sign and larger magnitude. The emergence of vortical structures is observed in kinetic QGSM models. The hydrodynamical helicity separation receives the contribution of longitudinal velocity and vorticity implying the quadrupole structure of the latter. The transition from the quark vortical effects to baryons in confined phase may be achieved by exploring the axial charge. At the hadronic level the polarization corresponds to the cores of quantized vortices in pionic superfluid. The chiral vortical effects may be also studied in the frmework of Wigner function establishing the relation to the thermodynamical approach to polarization.

1. Introduction

The experimental evidences for polarization of hyperons in heavy-ion collisions found by STAR collaboration [1] attracted recently much attention [2–6].

The studies of polarization are often performed [7] in the framework of approach exploring local equilibrium thermodynamics [8] and hydrodynamical calculations of vorticity [9–11].

There is another approach to polarization first proposed in [12] and independently in [17]. The so-called axial vortical effect (see e.g. [13] and references therein) being the macroscopic manifestation of axial anomaly [14]leads to induced axial current of strange quarks which may be converted to polarization of Λ-hyperons [12, 17].

The effect is proportional to vorticity and helicity of the strong interacting medium which was calculated [15] in the kinetic Quark-Gluon-String Model(QGSM) [18–20]. In this way, in fact, some particular case of fundamental relation between kinetic and hydrodynamic description was considered.

This calculation were later made more detailed [21], including the spatial and temporal dependence of strange chemical potential which is also the ingredient of anomalous approach to polarization.

2. Helicity separation and quadrupole structure

The effect of helicity separation was found in [15]. It was also found [15] that it receives contributions from both transverse and longitudinal velocity and vorticity. The latter, as the longitudinal velocity changes sign in each half-space, should lead to the quadrupole structure of longitudinal vorticity, in accordance to recent results [16]. This conclusion, following from observations in [15] is now confirmed by explicit calculation (Fig.1).

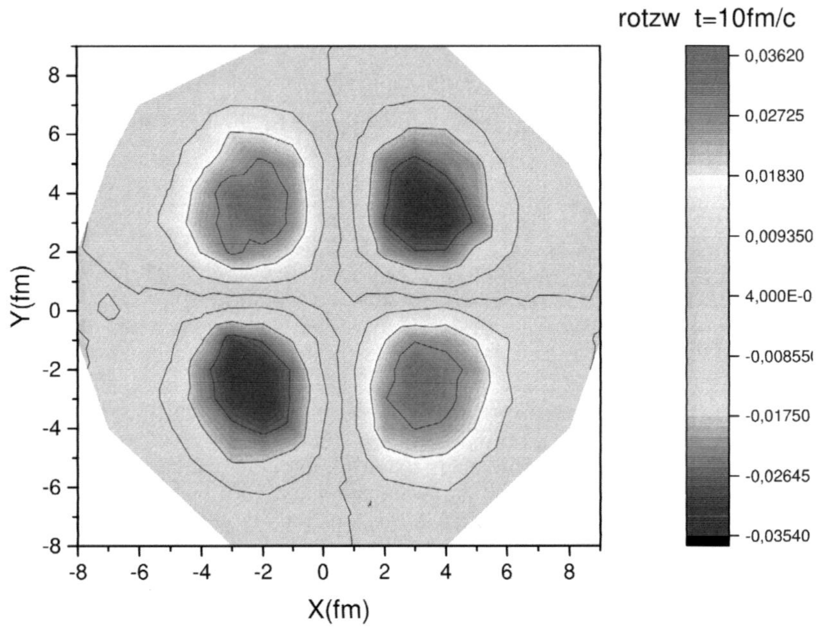

Figure 1. Quadrupole structure of longitudinal vorticity.

3. Anomalous mechanism of hyperon polarization

Anomalous mechanism of polarization makes this effect qualitatively similar to the local violation [22] of discrete symmetries in strongly interacting QCD matter. The most well known here is the Chiral Magnetic Effect (CME) which uses the (C)P-violating (electro)magnetic field emerging in heavy ion collisions in order to probe the (C)P-odd effects in QCD matter. At the same time, magnetic field is typically growing with energy, in the constrast to vorticity and chemical potential effects which we will discuss below.

The polarization, in turn, is similar to Chiral Vortical Effect (CVE) [23] due to coupling to P-odd medium vorticity leading to the induced electromagnetic and all conserved-charge currents [12], in particular the baryonic one.

Recently the pioneering preliminary results on global polarization of Λ and $\bar{\Lambda}$ hyperons in $Au-Au$ collisions in Beam Energy Scan at RHIC were released [1] and the qualitative tendency of polarization decrease with energy in agreement with the prediction [12] was revealed. The recent theoretical analysis [24] suggested that decrease of polarization with energy is related to the suppression of Axial Magnetic effect contribution in strongly correlated QCD matter found in lattice simulations.

IOP Conf. Series: Journal of Physics: Conf. Series **938** (2017) 012063 doi:10.1088/1742-6596/938/1/012063

Indeed, the chiral vorticity coefficient describing the axial vortical effect

$$c_V = \frac{\mu_s^2 + \mu_A^2}{2\pi^2} + \frac{T^2}{6}, \tag{1}$$

contains temperature dependent term related to gravitational anomaly [25], and naively it can be quite substantial and increase with energy. However, lattice simulations [26] lead to the zero result in the confined phase and to the suppression by one order of magnitude at high temperatures. Neglecting axial chemical potential, the coefficient c_V takes the form

$$c_V = \frac{\mu_s^2}{2\pi^2} + k\frac{T^2}{6}, \tag{2}$$

As soon as for free fermion gas the $T^2/6$ term is recovered [27] for large lattice volume at fixed temperature, the above-mentioned suppression should be attributed to the correlation effects. It was suggested [24], that the accurate measurements of polarization energy dependence may serve a sensitive probe of strongly correlated QCD matter.

4. Numerical simulations of axial anomaly contributions to (anti)hyperon polarization

We performed the numerical simulations in QGSM model [18–20]. We decomposed the space-time to the cells, allowing to define velocity and vorticity in the kinetic model, as described in detail in [15]. To define the strange chemical potential (assuming that Λ polarization is carried by strange quark) we used the matching procedure [21] of distribution functions to its (local) equilibrium values. In this paper, we also determine in this way the values of temperature. In general, let us stress that we realized in our particular case the relation between kinetics, hydrodynamics and thermodynamics.

We first neglect the gravitational anomaly contribution and start by considering the energy dependence of polarization (described in detail in [24]) for three values of impact parameter. The results are presented at Fig.1. The curves correspond to $b = 8.0fm, 6.4fm, 4.8fm$.

We continue by the inclusion of contribution related to gravitational anomaly, which is the central issue of this paper. The results are presented at Fig.2. We consider as a starting point the original value of anomaly coefficient [25] $T^2/6$ which is reproduced for large lattice volume at fixed temperature [27]. We present the curves following from the coefficients suppressed by factor k (2) resulting from the lattice calculations [26]. We compare values of $k = 1$ with $k = 0, 1/15, 1/10$. As one can see, the lattice-supported value $1/15$ is most close to the behavior of preliminary data which may be considered as a signal of strongly correlated matter formation. The closeness of $k = 0$ curve to the experimental points may be related to the contribution of confinement phase, where lattice calculations [26] lead to zero temperature-dependent effect. At the same time, already $k = 1/10$ leads to the curve growing with energy, while $k = 1$ leads to extremely strong growth.

The $\bar{\Lambda}$ polarization is emerging due to the polarization of $\bar{s}-$quarks, which has the same sign, as the axial current and charge are C-even operators. The magnitude of the $\bar{\Lambda}$ is larger as the same axial charge is distributed between the polarizations of the smaller number of particles [24]. It is mandatory to take into account the contribution of K^* mesons. In the case of Λ the $K^{*-}, \bar{K}^0$ mesons contain two sea(anti)quarks and does not change the polarization significantly. At the same time, for $\bar{\Lambda}$ the relevant K^{*+}, K^0 mesons are more numerous and make the excess of $\bar{\Lambda}$ polarization less pronounced.

Note that this excess is larger for smaller energies, where suppression of $\bar{\Lambda}$ is larger. This differs from the (C-odd) effect of magnetic field, which is increasing with energy, although more detailed studies taking into account the finite time of magnetic field existence are required.

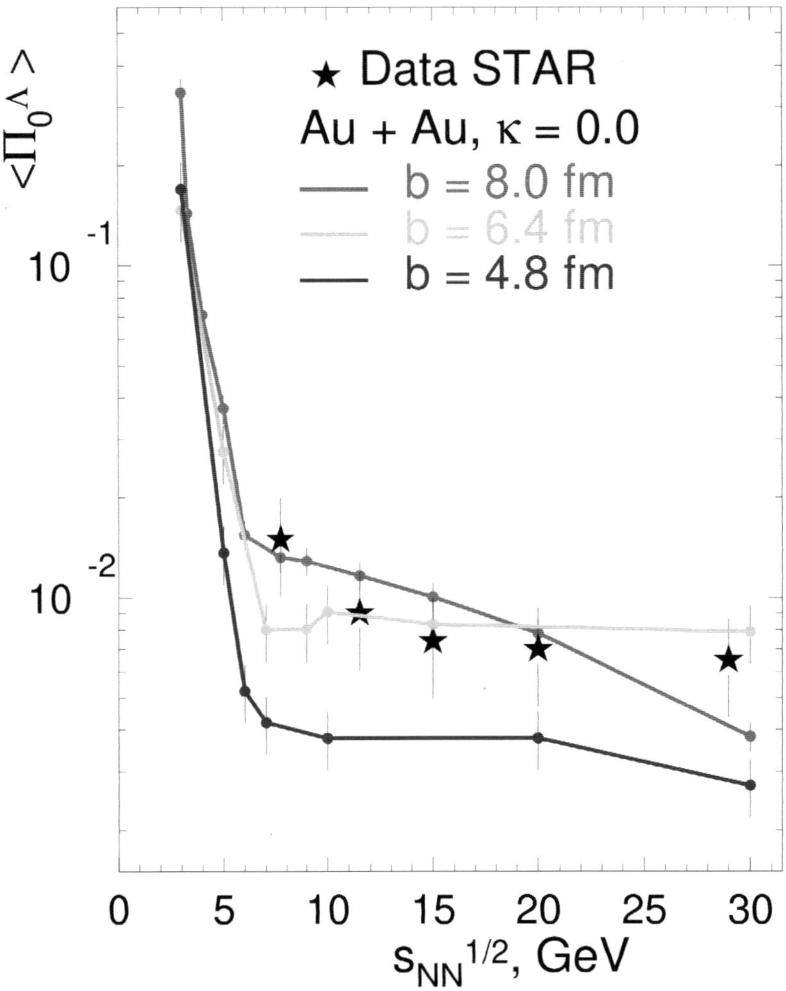

Figure 2. Energy dependence of polarization for different values of impact parameter.

The quantitative analysis of these effects, taking into account the gravitational anomaly contribution, is presented at Figure 3. The result is in reasonable agreement with STAR data, although further analysis is required.

5. Quantized vortices in pionic superfluid

The description of polarization at the ahdronic level may be achieved by consideration of quantized vortices in mesonic superfluid [28]. The baryonic polarization emerges as a radiative correction at their cores leading to the current

$$(j^5_B)_\alpha = \frac{N_c}{36\pi^2 f_\pi^2}\epsilon_{\alpha\beta\gamma\delta}(\partial^\beta\pi^0)(\partial^\gamma\partial^\delta\pi^0) \ . \quad (3)$$

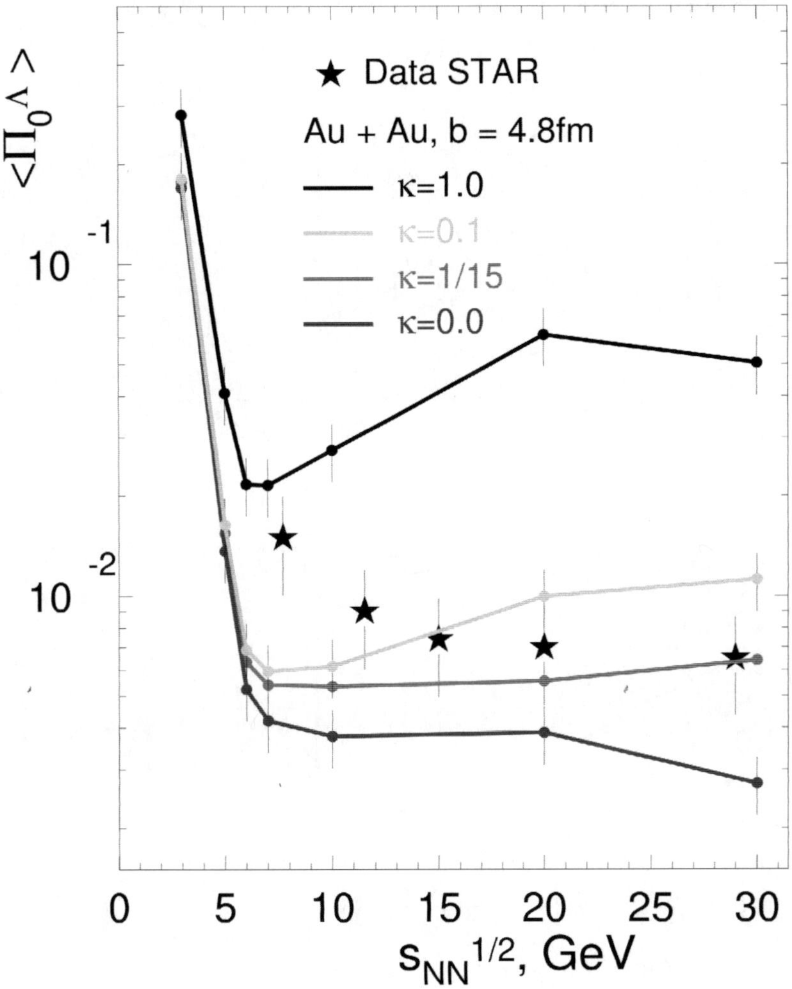

Figure 3. Energy dependence of polarization for different values of gravitational anomaly contribution.

One may also explain that in such a way: as the quantized circulation should lead to the growing velocity for small distances, the pointlike spin contribution at the vortex core is required in order not to exceed the velocity of light. In the liquid helium, to the contrary, the velocity of sound is not exceeded because the quantized circulation is not manifested at the distances smaller than interatomic ones.

6. Chiral vortical effect from Wigner function

The approach based on chiral vortical effects and making therefore the polarization a kind of anomalous transport seems to be rather different from the thermodynamical approach [2]. However, the recent study [29] shows that the covariant Wigner function should lead to the

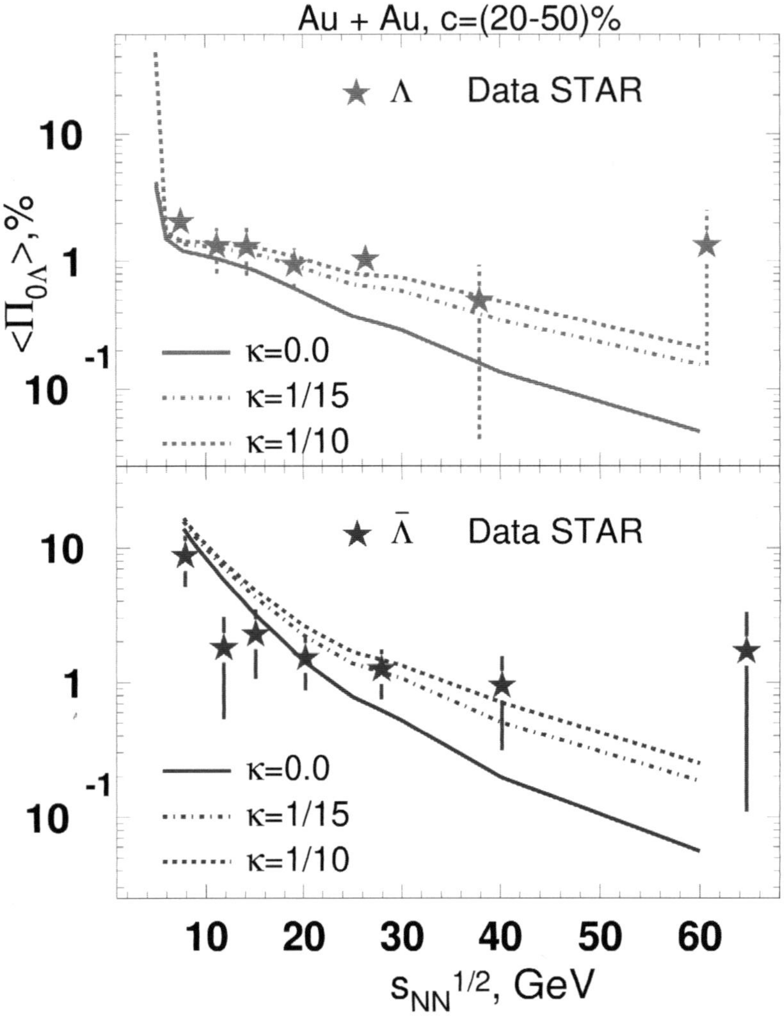

Figure 4. Comparison of Λ and $\bar{\Lambda}$ polarizations for different values of gravitational anomaly contribution.

expression analogous to induced anomalouys current. This may be related to the fact that the in-medium current correlators should include triangle diagrams when the expansion over the external parameters is performed. This however does not imply the anomaloud non-conservation of axial current. The appearance of extra contributions to axial current [29] may be due to the ambiguity in Wigner function definition [30]. One can even guess that their absence may be used as a constraint for the proper definition of Wigner function.

7. Conclusions and Outlook

We found that the anomalous mechanism may naturally explain decrease of polarization with energy.

The polarization of antihyperons of the same sign and larger magnitude as hyperons with the difference decreasing with energy, contrary to the effect of magnetic field.

The polarization at the hadronic level may be attributed to the cores of quantized vortices in pionic superfluid.

The anomalus contribution may be obtained in the framework of Wigner function approach.

The further experimental theoretical and experimental studies should be performed. In particular, this is related to the polarization os other (Σ, Ξ) hyperons, which should have the same (up to the mass rescaling) polarization in thermodynamical approach, but may differ significantly due to the different chemical potentials in the anomalous approach.

8. Acknowledgments

We are indebted to F. Becattini, M. Dvornikov, A. Braguta, A. Sadofyev, M. Ulybyshev and S. Voloshin for discussion and correspondence. The work is supported by RFBR grants 17-02-01108 and 18-02-01107.

References

[1] L. Adamczyk *et al.* [STAR Collaboration], "Global Λ hyperon polarization in nuclear collisions: evidence for the most vortical fluid," Nature **548**, 62 (2017) doi:10.1038/nature23004 [arXiv:1701.06657 [nucl-ex]].

[2] F. Becattini, I. Karpenko, M. Lisa, I. Upsal and S. Voloshin, "Global hyperon polarization at local thermodynamic equilibrium with vorticity, magnetic field and feed-down," Phys. Rev. C **95**, no. 5, 054902 (2017) doi:10.1103/PhysRevC.95.054902 [arXiv:1610.02506 [nucl-th]].

[3] I. Karpenko and F. Becattini, Eur. Phys. J. C **77**, no. 4, 213 (2017) doi:10.1140/epjc/s10052-017-4765-1 [arXiv:1610.04717 [nucl-th]].

[4] I. Karpenko and F. Becattini, "Vorticity in the QGP liquid and hyperon polarization at the RHIC BES energies," J. Phys. Conf. Ser. **779**, no. 1, 012068 (2017) doi:10.1088/1742-6596/779/1/012068 [arXiv:1611.08136 [nucl-th]].

[5] Y. Xie, R. C. Glastad and L. P. Csernai, "Λ polarization in an exact rotating and expanding fluid dynamical model for peripheral heavy ion reactions," Phys. Rev. C **92** (2015) no.6, 064901 doi:10.1103/PhysRevC.92.064901 [arXiv:1505.07221 [nucl-th]].

[6] R. h. Fang, J. y. Pang, Q. Wang and X. n. Wang, "Pseudoscalar condensation induced by chiral anomaly and vorticity for massive fermions," arXiv:1611.04670 [nucl-th].

[7] F. Becattini, L. Csernai and D. J. Wang, "Λ polarization in peripheral heavy ion collisions," Phys. Rev. C **88**, no. 3, 034905 (2013) [arXiv:1304.4427 [nucl-th]].

[8] F. Becattini, L. Bucciantini, E. Grossi and L. Tinti, "Local thermodynamical equilibrium and the beta frame for a quantum relativistic fluid," Eur. Phys. J. C **75**, no. 5, 191 (2015) doi:10.1140/epjc/s10052-015-3384-y [arXiv:1403.6265 [hep-th]].

[9] B. Betz, M. Gyulassy and G. Torrieri, "Polarization probes of vorticity in heavy ion collisions," Phys. Rev. C **76** (2007) 044901 doi:10.1103/PhysRevC.76.044901 [arXiv:0708.0035 [nucl-th]].

[10] L. P. Csernai, V. K. Magas and D. J. Wang, "Flow Vorticity in Peripheral High Energy Heavy Ion Collisions," Phys. Rev. C **87** (2013) no.3, 034906 doi:10.1103/PhysRevC.87.034906 [arXiv:1302.5310 [nucl-th]].

[11] L. P. Csernai, D. J. Wang and T. Csorgo, "Rotation in an exact hydrodynamical model," Phys. Rev. C **90** (2014) no.2, 024901 doi:10.1103/PhysRevC.90.024901 [arXiv:1406.1017 [hep-ph]].

[12] O. Rogachevsky, A. Sorin and O. Teryaev, "Chiral vortaic effect and neutron asymmetries in heavy-ion collisions," Phys. Rev. C **82**, 054910 (2010) [arXiv:1006.1331 [hep-ph]].

[13] T. Kalaydzhyan, "Temperature dependence of the chiral vortical effects," Phys. Rev. D **89**, no. 10, 105012 (2014) [arXiv:1403.1256 [hep-th]].

[14] D.T. Son and P. Surowka, "Hydrodynamics with Triangle Anomalies," Phys. Rev. Lett. **103**, 191601 (2009) [arXiv:0906.5044 [hep-th]].

[15] M. Baznat, K. Gudima, A. Sorin and O. Teryaev, "Helicity separation in Heavy-Ion Collisions," Phys. Rev. C **88**, 061901 (2013) [arXiv:1301.7003 [nucl-th]].

[16] F. Becattini and I. Karpenko, "Global longitudinal polarization in relativistic heavy-ion collisions at very high energy," arXiv:1707.07984 [nucl-th].

[17] J. -H. Gao, Z. -T. Liang, S. Pu, Q. Wang and X. -N. Wang, "Chiral Anomaly and Local Polarization Effect from Quantum Kinetic Approach," Phys. Rev. Lett. **109**, 232301 (2012) [arXiv:1203.0725 [hep-ph]].

[18] V.D. Toneev, K.K. Gudima, Nucl. Phys. A **400**, 173c (1983).

[19] V.D. Toneev, N.S. Amelin, K.K. Gudima, S.Yu. Sivoklokov, Nucl. Phys. A **519**, 463c (1990).

[20] N.S. Amelin, E.F. Staubo, L.S. Csernai et al., Phys.Rev. C **44**, 1541 (1991).

[21] M. I. Baznat, K. K. Gudima, A. S. Sorin and O. V. Teryaev, "Femto-vortex sheets and hyperon polarization in heavy-ion collisions," Phys. Rev. C **93**, no. 3, 031902 (2016) doi:10.1103/PhysRevC.93.031902 [arXiv:1507.04652 [nucl-th]].

[22] K. Fukushima, D.E. Kharzeev and H.J. Warringa, "The Chiral Magnetic Effect," Phys. Rev. D **78**, 074033 (2008) [arXiv:0808.3382 [hep-ph]].

[23] D. Kharzeev and A. Zhitnitsky, "Charge separation induced by P-odd bubbles in QCD matter," Nucl. Phys. A **797**, 67 (2007) [arXiv:0706.1026 [hep-ph]].

[24] A. Sorin and O. Teryaev, "Axial anomaly and energy dependence of hyperon polarization in Heavy-Ion Collisions," arXiv:1606.08398 [nucl-th]; Phys. Rev. C, in print.

[25] K. Landsteiner, E. Megias, L. Melgar and F. Pena-Benitez, "Holographic Gravitational Anomaly and Chiral Vortical Effect," JHEP **1109** (2011) 121

[26] V. Braguta, M.N. Chernodub, K. Landsteiner, M.I. Polikarpov, M.V. Ulybyshev Phys.Rev. D88 (2013) 071501 DOI: 10.1103/PhysRevD.88.071501 e-Print: arXiv:1303.6266 [hep-lat]; V. Braguta, M.N. Chernodub, V.A. Goy, K. Landsteiner, A.V. Molochkov, M.I. Polikarpov, Phys.Rev. D89 (2014) no.7, 074510 DOI: 10.1103/PhysRevD.89.074510 e-Print: arXiv:1401.8095 [hep-lat]; V. Braguta, M. N. Chernodub, V. A. Goy, K. Landsteiner, A. V. Molochkov and M. Ulybyshev, "Study of axial magnetic effect," AIP Conf. Proc. **1701**, 030002 (2016); doi:10.1063/1.4938608; V. Goy, "Investigation of SU(2) gluodynamics in the framework of the latice approach", PhD Thesis (in Russian), Vladivostok, 2015.

[27] P. V. Buividovich, "Axial Magnetic Effect and Chiral Vortical Effect with free lattice chiral fermions," J. Phys. Conf. Ser. **607**, no. 1, 012018 (2015) doi:10.1088/1742-6596/607/1/012018 [arXiv:1309.4966 [hep-lat]].

[28] O. V. Teryaev and V. I. Zakharov, "From the chiral vortical effect to polarization of baryons: A model," Phys. Rev. D **96**, no. 9, 096023 (2017). doi:10.1103/PhysRevD.96.096023; "Chiral vortical effect in pionic superfluid vs spin alignment of baryons," arXiv:1705.01650 [hep-th].

[29] G. Prokhorov and O. Teryaev, "Anomalous current from covariant Wigner functions," arXiv:1707.02491 [hep-th].

[30] F. Becattini, Private communication.

IOP Conf. Series: Journal of Physics: Conf. Series **938** (2017) 012064 doi:10.1088/1742-6596/938/1/012064

Spin and model determination of Z′ - boson in lepton pair production at CERN LHC

A.V. Tsytrinov[1], **A.A. Pankov**[1,2,3], **I.A. Serenkova**[1], **V.A. Bednyakov**[3]

[1] Abdus Salam ICTP Affiliated Centre at Pavel Sukhoi Gomel State Technical University, Gomel, 246746, Belarus

[2] Institute for Nuclear Problems, Belarusian State University, Minsk, 220030, Belarus

[3] Gelepov Laboratory of Nuclear Problems, Joint Institute for Nuclear Research, Dubna, 141980, Russia

E-mail: tsytrin@gstu.by

Abstract. Many new physics models predict production of heavy resonances in Drell-Yan channel and can be observed at the CERN LHC. If a new resonance is discovered as a peak in the dilepton invariant mass distribution at the LHC, the identification of its spin and couplings can be done by measuring production rates and angular distributions of the decay products. Here we discuss the spin-1 identification of Z'-boson for a set of representative models (SSM, E_6, LR, and ALR) against the spin-2 RS graviton resonance and a spin-0 sneutrino resonance with the same mass and producing the same number of events under the resonance peak. We use the center-edge asymmetry for spin identification, as well as the total dilepton production cross section for the distinguishing the considered Z'-boson models from one another.

1. Introduction

New heavy resonances are predicted by numerous New Physics (NP) scenarios, candidate solutions of conceptual problems of the standard model (SM). In particular, this is the case of models of gravity with extra spatial dimensions, grand-unified theories (GUT), and supersymmetric (SUSY) theories with R-parity breaking ($\not{R}_p$). These new heavy resonances, with mass $M \gg M_Z$, may be either produced or exchanged in reactions among SM particles at the high energy collider LHC. A particularly interesting process to be studied in this regard at the LHC is the Drell-Yan (DY) dilepton production ($l = e, \mu$)

$$p + p \to l^+l^- + X, \tag{1}$$

where exchanges of the new particles can occur and manifest themselves as peaks in the (l^+l^-) invariant mass M. Once the heavy resonance is discovered at some $M = M_R$, further analysis is needed to identify the theoretical framework for NP to which it belongs. Correspondingly, for any NP model, one defines as *identification* reach the upper limit for the resonance mass range where it can be identified as the source of the resonance, against the other, potentially competitor scenarios, that can give a peak with the same mass and same number of events under the peak. This should be compared to the *discovery* reach, which specifies the (naturally more extended) mass range where the peak in the cross section pertaining to the model can just be observed experimentally. Clearly, the determination of the spin of the resonance with center-edge asymmetry $A_{\rm CE}$ represents an important aspect of the selection among different classes of non-standard interactions giving rise to the observed peak. The potential advantages of $A_{\rm CE}$ to discriminate the spin-2 graviton resonance against the spin-1 and spin-0 hypotheses were discussed in Refs. [1, 2, 3, 4].

Here, we apply $A_{\rm CE}$ to the spin-1 identification of a peak observed in the dilepton mass distribution of process (1) at the LHC, against the spin-2 and spin-0 alternative hypotheses.

The existence of heavy neutral Z' vector bosons are a feature of many extensions of the SM. They arise in extended gauge theories including grand unified theories, superstring theories, and Left-Right symmetric models and in other models such as the BESS model and models of composite gauge bosons. For explicit NP realizations, for the spin-1 Z' models we refer to Refs. [5]; for the alternative spin-2 and spin-0 hypotheses we refer for the Randall-Sundrum [6] graviton excitation (RS) and for the SUSY $\not R_p$ sneutrino exchange to [7], respectively.

The search reach at a collider for new gauge bosons is somewhat model dependent due to the rather large variations in their couplings to the SM fermions which are present in extended gauge theories currently on the market. This implies that any overview of the subject is necessarily incomplete. Hence, we will be forced to limit ourselves to a few representative models. To be specific we consider the so-called $Z'_{\rm SSM}$, Z'_{E_6}, $Z'_{\rm LR}$, $Z'_{\rm ALR}$ models. In this note we study the discovery potential of the experiments that will be performed at the LHC. In addition to the discovery reach we also examine the diagnostic power of the LHC for heavy gauge boson physics.

2. Observables and considered NP models

The parton model cross section for inclusive production of a dilepton with invariant mass M can be written as

$$\frac{{\rm d}\sigma(R_{ll})}{{\rm d}M\,{\rm d}y\,{\rm d}z}=K\frac{2M}{s}\sum_{ij}f_i(\xi_1,M)f_j(\xi_2,M)\frac{{\rm d}\hat\sigma}{{\rm d}z}(i+j\to l^++l^-). \tag{2}$$

Here, s is the proton-proton center-of-mass energy squared; $z=\cos\theta_{\rm c.m.}$ with $\theta_{\rm c.m.}$ the lepton-quark angle in the dilepton center-of-mass frame; y is the dilepton rapidity; $f_{i,j}(\xi_{1,2},M)$ are parton distribution functions in the protons P_1 and P_2, respectively, with $\xi_{1,2}=(M/\sqrt{s})\exp(\pm y)$ the parton fractional momenta; finally, ${\rm d}\hat\sigma_{ij}$ are the partonic differential cross sections. In (2), the factor K accounts for next-to-leading order QCD contributions. For simplicity, and to make our procedure more transparent, we will use as an approximation a global flat value $K=1.3$.

Since we are interested in a (narrow) peak production and subsequent decay into the DY pair, $pp\to R\to l^+l^-$, we consider the lepton differential angular distribution, integrated over an interval of M around M_R:

$$\frac{{\rm d}\sigma(R_{ll})}{{\rm d}z}=\int_{M_R-\Delta M/2}^{M_R+\Delta M/2}{\rm d}M\int_{-Y}^{Y}\frac{{\rm d}\sigma}{{\rm d}M\,{\rm d}y\,{\rm d}z}\,{\rm d}y. \tag{3}$$

The number of events under the peak, that determines the statistics, is therefore given by:

$$\sigma(R_{ll})\equiv\sigma(pp\to R)\cdot{\rm BR(R\to l^+l^-)}=\int_{-{\rm z_{cut}}}^{\rm z_{cut}}{\rm dz}\int_{\rm M_R-\Delta M/2}^{\rm M_R+\Delta M/2}{\rm dM}\int_{-\rm Y}^{\rm Y}{\rm dy}\frac{{\rm d}\sigma}{\rm dM\,dy\,dz}. \tag{4}$$

For the full final phase space, $z_{\rm cut}=1$ and $Y=\log(\sqrt{s}/M)$. Concerning the size of the bin ΔM, it should include a number (at least one) of peak widths to enhance the probability to pick up the resonance. In our analysis, we adopt the parametrization of ΔM vs. M exploited in Ref. [10] and, denoting by N_B and N_S the number of 'background' and 'signal' events in the bin, the criterion $N_S=5\sqrt{N_B}$ or 10 events, whichever is larger, as the minimum signal for the peak discovery.

2.1. Z′ models

The list of Z' models that will be considered in our analysis is the following:

(i) The three possible $U(1)$ Z' scenarios originating from the exceptional group E_6 spontaneous breaking. They are defined in terms of a mixing angle β. The specific values $\beta = 0$, $\beta = \pi/2$ and $\beta = \arctan - \sqrt{5/3}$, correspond to different E_6 breaking patterns and define the popular scenarios Z'_χ, Z'_ψ and Z'_η, respectively.

(ii) The left-right models, originating from the breaking of an $SO(10)$ grand-unification symmetry, and where the corresponding $Z'_{\rm LR}$ couples to a combination of right-handed and $B - L$ neutral currents (B and L denote lepton and baryon currents), specified by a real parameter $\alpha_{\rm LR}$ bounded by $\sqrt{2/3} \lesssim \alpha_{\rm LR} \lesssim \sqrt{2}$. We fix $\alpha_{\rm LR} = \sqrt{2}$, which corresponds to a pure L-R symmetric model.

(iii) The $Z'_{\rm ALR}$ predicted by the 'alternative' left-right scenario.

(iv) The so-called sequential $Z'_{\rm SSM}$, where the couplings to fermions are the same as those of the SM Z.

Current Z' mass limits from the LHC are 4.5 TeV [8] (ATLAS) and 4 TeV [9] (CMS).

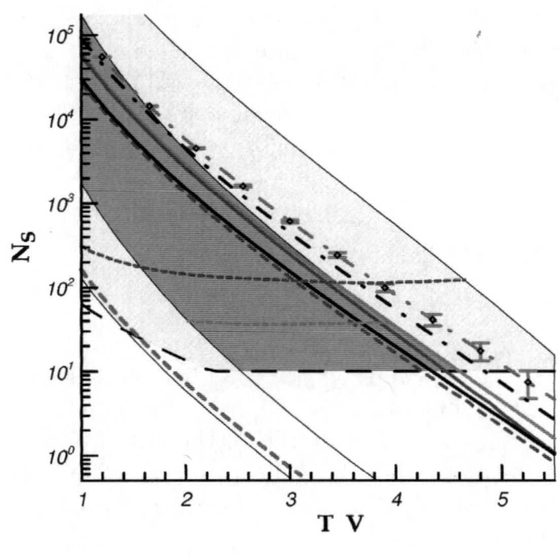

Figure 1. Expected number of resonance (signal) events N_S vs. M_R ($R = Z', G, \tilde{\nu}_\tau$) at the 14 TeV LHC with $\mathcal{L}_{\rm int} = 100\ fb^{-1}$ for the process $pp \to R \to l^+l^- + X$ ($l = e, \mu$) [10]. Event rates for various Z' models are shown. Green area corresponds to graviton signature space for $0.01 < c < 0.1$ while the yellow area is the sneutrino signature space for $10^{-5} < X < 10^{-1}$. Minimum number of signal events needed to detect the resonance (5-σ level) above the background and the minimum number of events to exclude the spin-2 and spin-0 hypotheses at 95% C.L. are shown. Error bars correspond to the statistical uncertainties for the ALR model.

2.2. RS graviton excitation

We consider the simplest scenario in the class of models based on one compactified warped extra dimension and two branes, proposed in the context of the SM gauge-hierarchy problem in [6]. The model predicts a tower of narrow Kaluza–Klein (KK), spin-2, graviton excitations $G^{(n)}$ ($n \geq 1$) with the peculiar mass spectrum $M^{(n)} = M^{(1)} x_n / x_1$ (x_i are the zeros of the Bessel function, $J_1(x_i) = 0$). Their masses and couplings to the SM particles are proportional to Λ_π and $1/\Lambda_\pi$, respectively, with Λ_π the gravity effective mass scale on the SM brane. For Λ_π of the TeV order, such RS graviton resonances can be exchanged in the process (1) and mimic Z' exchange. The independent parameters of the model can be chosen as the dimensionless ratio $c = k/\overline{M}_{\text{Pl}}$ (with k the 5-dimensional curvature and $\overline{M}_{\text{Pl}} = 1/\sqrt{8\pi G_{\text{N}}}$ the reduced Planck mass), and the mass M_G of the lowest KK resonance $G^{(1)}$. Accordingly, $\Lambda_\pi = M_G/cx_1$.

There are two partonic subprocesses, $q\bar{q} \to \gamma, Z, G \to l^+l^-$ and $gg \to G \to l^+l^-$, needed to describe hadronic production of lepton pair within KK models. The theoretically 'natural' ranges for the RS model parameters are $0.01 \leq c \leq 0.1$ and $\Lambda_\pi < 10\,\text{TeV}$. Current lower bounds at 95% C.L. from the LHC are: $M_G > 4.1$ TeV for $c = 0.1$ (ATLAS) [11].

2.3. Sneutrino exchange

Sneutrino ($\tilde{\nu}$) exchange can occur in SUSY with R-parity breaking, and represents a possible, spin-0, interpretation of a peak in the dilepton invariant mass distribution of the process (1). The cross section for the relevant partonic process, $q\bar{q} \to \tilde{\nu} \to l^+l^-$, is flat in z and expressed in terms of two Yukawa couplings, λ and λ', are the R-parity-violating sneutrino couplings to l^+l^- and $d\bar{d}$, respectively. Actually, in the narrow-width approximation, the partonic cross section turns out to depend on the product $X = (\lambda')^2 B_l$, with B_l the sneutrino leptonic branching ratio. Current limits on X are rather loose, and we may consider for this parameter the range $10^{-5} \leq X \leq 10^{-1}$. For $10^{-4} \leq X \leq 10^{-2}$, the range is $M_{\tilde{\nu}} \gtrsim 280 - 800$ GeV [12].

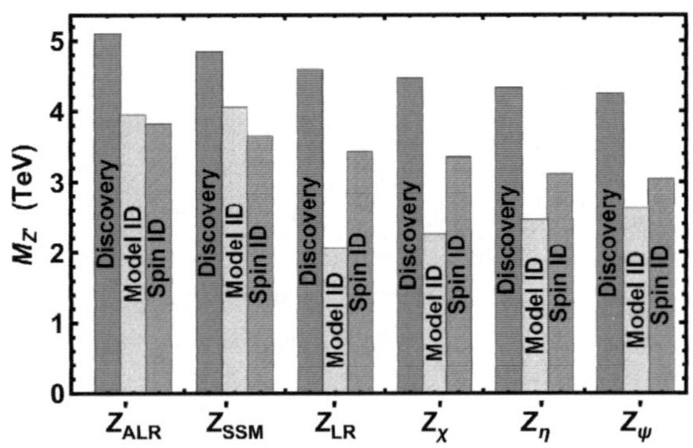

Figure 2. Discovery limits on $M_{Z'}$ (5-σ level) and Z' spin identification reaches (95% C.L.) for neutral gauge bosons of representative models, using the lepton-pair production cross section $\sigma \cdot B_l$ ($l = e, \mu$) and center-edge asymmetry A_{CE}, respectively, at the 14 TeV LHC with integrated luminosity of 100 fb^{-1}. Also, Z'-model distinction reaches (95% C.L.) are obtained from the analysis of the leptonic event rates.

3. Model signature spaces

In Fig. 1, we show the predicted number of resonance (signal) events N_S in the Drell-Yan process (1) at LHC, *vs.* M_R, where $R = Z', G, \tilde{\nu}$ denotes the three alternative possibilities outlined in the previous subsections. The assumed integrated luminosity is $\mathcal{L}_{\text{int}} = 100 fb^{-1}$, the cuts in phase space relevant to the foreseen detector acceptance specified above have been imposed, and the channels $l = e, \mu$ have been combined. Also, the minimum signal for resonance discovery above the 'background' at 5σ is represented by the long-dashed line.

For any model, one can define a corresponding *signature space* as the region, in the (M_R, N_S) plot of Fig. 1, that can be 'populated' by the model by varying its parameters in the domains mentioned above. Clearly, in regions where the signature spaces overlap, the values of M_R are such that it is not possible to distinguish a model as the source of the peak against the others, because the number of signal events under the peak can be the same. Further analyses are needed in these cases to perform the identification of the peak source. For example, the 'blue' area in Fig. 1 corresponds to the graviton signature space for $0.01 \leq c \leq 0.1$, while the yellow area (which has substantial overlap with the blue one—indicated as green) is that for the sneutrino signature space corresponding to $10^{-5} \leq X \leq 10^{-1}$.

As regards the discovery and identification of Z' we are interested in, the signature spaces in Fig. 1 reduce to the lines labelled by the different models, because the event rates are fixed,

IOP Conf. Series: Journal of Physics: Conf. Series **938** (2017) 012064 doi:10.1088/1742-6596/938/1/012064

Table 1. Discovery and model identification (in TeV) for Z'-models at 14 TeV LHC in high luminosity modes.

Z' model	Discovery 300 fb^{-1}	Model ID 300 fb^{-1}	Discovery 3000 fb^{-1}	Model ID 3000 fb^{-1}
Z'_{ALR}	5.7	4.4	6.9	4.9
Z'_{SSM}	5.4	4.4	6.5	4.9
Z'_{LR}	5.1	2.2	6.3	2.4
Z'_{χ}	5.0	2.5	6.0	2.8
Z'_{η}	4.9	2.7	6.1	2.9
Z'_{ψ}	4.8	2.9	5.9	3.3

Table 2. Spin identification (in TeV) of Z'-models under considertion at 14 TeV LHC.

	100 fb^{-1}	300 fb^{-1}	3000 fb^{-1}
Z' spin ID	3.0	3.8	5.0

once M'_Z is given, through the couplings in Table 1 of Ref. [10]. Fig. 1 shows that, with the assumed luminosity of 100 fb^{-1}, Z' gauge boson masses up to 4–5 TeV are in principle within the 5-σ reach of the LHC, consistent with earlier studies. We here assume that the Z' can only decay to pairs of SM fermions in order to obtain the leptonic branching ratio B_l. It is important to note that in many models, where Z' can also decay to exotic fermions and/or SUSY particles this overestimates B_l and, thus, the search reach.

On the other hand, Fig. 1 demonstrates that, as far as the production rate of DY pairs is concerned, there is a substantial overlap between the Z' and the $\tilde{\nu}$ signature spaces, which determines a domain in $(M_{\tilde{\nu}}, X)$ where spin-0 $\tilde{\nu}$ exchange and Z' exchanges are not distinguishable because they lead to the same event rate under the peak. The same is true for the spin-2, RS model. However, as shown by Fig. 1, in this case it is interesting to note that, if one literally takes the suggested range $c \leq 0.1$ as the 'naturally' preferred one, the ALR and SSM scenarios can be discriminated against the RS (spin-2) resonance *already* at the level of event rates in a wide range of mass values accessible to the LHC, with no need for further analyses. Conversely, only the E_6 and LR Z' models possess a 'confusion region' with the RS resonance G, concentrated near the upper border of the graviton signature domain.

In fact, the spin-0 exclusion is more restrictive than that for spin-2, as discussed above. Fig. 2 gives the 'translation' of the discovery reach on Z' models as well as identification reach on Z' spin presented in Fig. 1, in the form of the histograms. As one can see from Fig. 2, the spin-1 identification (or, actually, the spin-0 and spin-2 exclusion) can be obtained up to Z' mass of the order of 2.5–3.5 TeV, depending on the specific model. The model dependence of the spin identification reach is due to the difference in statistics, stemming from the different cross sections associated with these models.

4. Concluding remarks

Table 1 shows that Z' boson can be observed at 14 TeV LHC in high luminosity modes up to $M'_Z \approx 6.9$ TeV. The center-edge asymmetry, A_{CE}, will allow to determine the spin-1 (i.e., exclude spin-0 and spin-2) of heavy Z' gauge boson up to $M'_Z \approx 5.0$ TeV, at 95% C.L. (Table 2). With 3000 fb^{-1} of integrated luminosity, one can distinguish among different Z'-models under study up to $M'_Z \approx 2.4$ TeV (95% C.L.).

IOP Conf. Series: Journal of Physics: Conf. Series **938** (2017) 012064 doi:10.1088/1742-6596/938/1/012064

Acknowledgments
Authors would like to thank Prof. P. Osland and Prof. N. Paver for the enjoyable collaboration on the subject matter covered here. This research has been partially supported by the Abdus Salam ICTP (TRIL Programme) and the Belarusian Republican Foundation for Fundamental Research. One of the authors [I.S.] acknowledges the receipt of the grant from the Abdus Salam International Centre for Theoretical Physics, Trieste, Italy. A.T. acknowledges the support from Gelepov Laboratory of Nuclear Problems, JINR.

References

[1] Osland P., Pankov A. A. and Paver N. 2003 *Phys. Rev.* D **68**, 015007
[2] Dvergsnes E. W., Osland P., Pankov A. A. and Paver N. 2004 *Phys. Rev.* D **69**, 115001
[3] Osland P., Pankov A. A., Paver N. and Tsytrinov A. V. 2008 *Phys. Rev.* D **78**, 035008
[4] Osland P., Pankov A. A. and Tsytrinov A. V. 2010 *Eur. Phys. J.* C **67** 191
[5] For reviews and original references see, e.g., Langacker P., Rizzo T.G., Leike A. 1999 *Phys. Rept.* **317**, 143; Hewett J. L. and Rizzo T. G. 1989 *Phys. Rept.* **183**, 193
[6] Randall L. and Sundrum R. 1999 *Phys. Rev. Lett.* **83** 3370; Randall L. and Sundrum R. 1999 *Phys. Rev. Lett.* **83** (1999) 4690
[7] Kalinowski J., Ruckl R., Spiesberger H. and Zerwas P. M. 1997 *Phys. Lett.* B **406** 314; Rizzo T. G. 1999 *Phys. Rev.* D **59** 113004 For a review see, *e.g.*: Barbier R. *et al.* 2005 *Phys. Rept.* **420**, 1
[8] Aaboud M.[ATLAS Collaboration], ATLAS-CONF-2017-027
[9] Sirunyan A.M. [CMS Collaboration], CMS-PAS-EXO-15-005.
[10] Osland P., Pankov A. A., Tsytrinov A. V. and Paver N. 2009 *Phys. Rev.* D **79**, 115021
[11] Aaboud M.*et al.* [ATLAS Collaboration] 2017 CERN-EP-2017-132
[12] Aaltonen T.*et al.* [CDF Collaboration] 2007 *Phys. Rev. Lett.* **99**, 171802; 2009 *Phys. Rev. Lett.* **102**, 091805 Hooper R. J. [D0 Collaboration] 2005 *Int. J. Mod. Phys.* A **20**, 3277

IOP Conf. Series: Journal of Physics: Conf. Series **938** (2017) 012065 doi:10.1088/1742-6596/938/1/012065

Angular distributions in pion-nucleon Drell-Yan process

I V Anikin[1], L Szymanowski[2], O V Teryaev[1] and N Volchanskiy[1,3]

[1] Bogoliubov Laboratory of Theoretical Physics, JINR, 141980 Dubna, Russia
[2] National Centre for Nuclear Research (NCBJ), 00-999 Warsaw, Poland
[3] Research Institute of Physics, Southern Federal University, 344090 Rostov-on-Don, Russia

E-mail: nikolay.volchanskiy@gmail.com

Abstract. We study the Drell-Yan process in the polarized pion-nucleon scattering. We calculate the $U(1)_{\rm em}$ gauge invariant hadron tensors. To this end, we have to take into account two diagrams—the standard diagram and an additional one—both diagrams are required to get the gauge invariant hadron tensor. We obtain a new single transverse spin asymmetry that probes gluon poles and chiral-odd distribution functions. The process of pion-nucleon scattering in the limit of large Feynman x is described by a convolution of two non-collinear exclusive-channel distribution amplitudes, which demonstrates the duality between different factorization regimes.

1. Introduction

For decades the Drell-Yan (DY) process has been remaining one of the most convenient objects in the studies of hadron structure, both from the theoretical and experimental sides. Several facilities are currently carrying out or developing programs dedicated to measuring DY processes—SPS (COMPASS) [1, 2], RHIC [3], NICA [4, 5]. An important family of observables in these experiments is single transverse spin asymmetries (SS$_\perp$A), which are related to the three-dimensional hadron structure owing to the inextricable link between the transverse spin and the parton transverse momentum dependence [6, 7, 8, 9, 10, 11].

In this report based mostly on Ref. [12], we apply ideas developed in [13, 14, 15] (see also talk given by I. Anikin in this workshop) to DY process induced by the polarized pion-nucleon scattering at large Feynman x. It turns out that the inclusion of gluon poles in the spirit of [13, 14, 15] predicts a new SS$_\perp$A non-trivially dependent on the azimuthal angle of nucleon spin vector (in the Collins–Soper frame). Moreover, as we will see in the following, the non-trivial angular dependence of SS$_\perp$A can be considered as a signal of the parton transverse momentum dependence.

On the other hand, the limit of large Feynman x allows us to describe pions by wave functions and distribution amplitudes rather than parton distributions. In a sense, this reveals a duality between semi-inclusive and exclusive processes or, put in other words, transverse momentum and generalized parton distributions. Besides, it gives a possibility to work with the gluon poles in the most explicit form—represented by a hard gluon propagator rather than inclusive function $B^V(x_1, x_2)$.

IOP Conf. Series: Journal of Physics: Conf. Series **938** (2017) 012065   doi:10.1088/1742-6596/938/1/012065

2. Kinematics. Sudakov decompositions

We study DY process in pion-nucleon scattering (Fig. 1) with nucleon N being transversely polarized, $\pi(P_1) + N(P_2, S) \to \gamma^*(q) + \bar{q}(K) + \text{spectators} \to \ell(l_1) + \bar{\ell}(l_2) + \bar{q}(K) + \text{spectators}$. The virtual photon produces a pair of unpolarized leptons with an invariant mass squared $q^2 = Q^2$ assumed to be large enough to make use of the factorization theorem for the process.

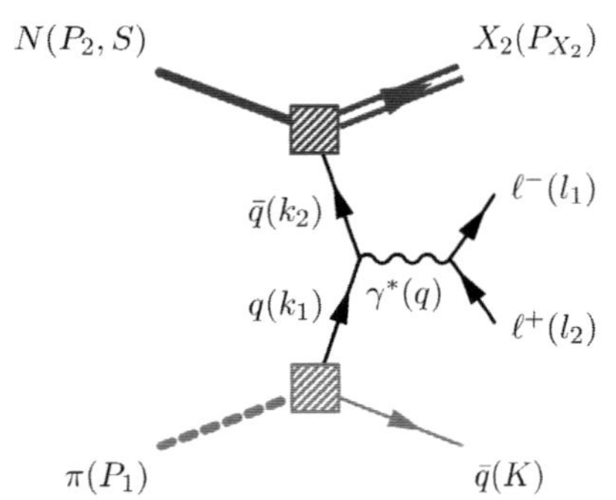

Figure 1. Pion-nucleon Drell-Yan process. In the case we consider, the nucleon is transversely polarized, leptons are unpolarized.

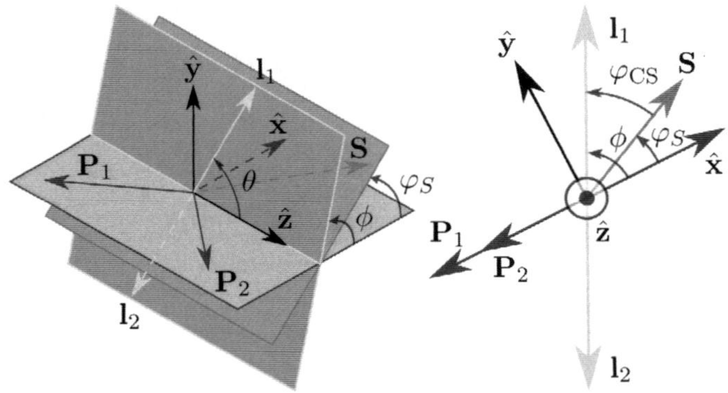

Figure 2. Collins–Soper frame. Cyan, olive and red planes are scattering, lepton and nucleon-spin planes, respectively. The colour of a vector corresponds to the colour of a plane the vector lies in.

The Sudakov decompositions of momenta and nucleon spin 4-vector read

$$P_1 \approx \frac{Q}{x_B\sqrt{2}}\, n^* + P_{1\perp}, \quad P_2 \approx \frac{Q}{y_B\sqrt{2}}\, n + P_{2\perp}, \quad S \approx \frac{\lambda}{M_N}\, P_2 + S_\perp \tag{1}$$

$$q = \frac{Q}{\sqrt{2}}\,(n^* + n) + q_\perp, \quad q_\perp^2 \ll Q^2, \tag{2}$$

where λ and M_N are nucleon helicity and mass, x_B and y_B are usual dimensionless Bjorken variables, n^* and n are plus and minus light-cone vectors,

$$n^{*\mu} = (1/\sqrt{2}, 0, 0, 1/\sqrt{2}), \quad n^\mu = (1/\sqrt{2}, 0, 0, -1/\sqrt{2}). \tag{3}$$

In what follows, all Lorentz non-invariant quantities are defined in the Collins–Soper frame [11, 16], which is a rest frame of the dilepton pair with unit vectors of the spatial axes

$$\hat{\mathbf{x}} \sim -\hat{\mathbf{P}}_1 - \hat{\mathbf{P}}_2, \qquad \hat{\mathbf{z}} \sim -\hat{\mathbf{P}}_1 + \hat{\mathbf{P}}_2, \qquad \hat{\mathbf{y}} = \hat{\mathbf{z}} \times \hat{\mathbf{x}}. \tag{4}$$

Here, a hatted vector is a vector of the same direction and unit magnitude, $\hat{\mathbf{A}} = \mathbf{A}/|\mathbf{A}|$. All relevant angles are depicted in Fig. 2—the polar θ and azimuthal ϕ angles of the lepton $\ell^-(l_1)$, the angle φ_S between scattering plane and nucleon-spin plane (azimuthal angle of the nucleon spin vector).

3. Hadron tensor

The cross section of the DY process can be written [11, 16] as

$$\frac{\mathrm{d}\sigma}{\mathrm{d}^4 q\, \mathrm{d}\Omega^{\mathrm{CS}}_{\mathbf{k}_1}} = \frac{\alpha_{\mathrm{em}}^2}{2Q^4 s} L_{\mu\nu} W^{\mu\nu}, \tag{5}$$

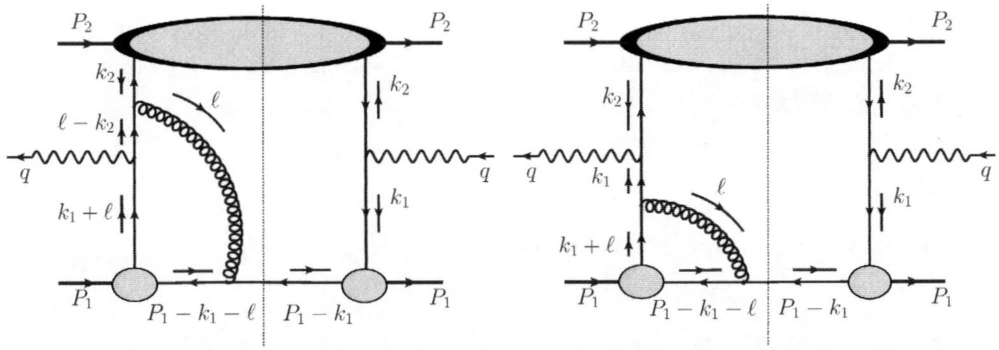

Figure 3. The Feynman diagrams contributing to the polarized DY hadron tensor: the standard diagram (left) and non-standard one (right).

where $s = (P_1 + P_2)^2$; $\Omega^{\rm CS}_{\mathbf{k}_1}$ is the solid angle of lepton in the Collins–Soper frame; $L_{\mu\nu}$ and $W_{\mu\nu}$ stand for the lepton and hadron tensor, respectively. Since the leptons are unpolarized in the case we consider, the lepton tensor is real. As a consequence, the hadron tensor has to be real as well.

Before the collinear factorization [13, 14, 15], the standard (Fig. 3a) and non-standard (Fig. 3b) diagrams give the following contributions to the hadron tensor:

$$W^{(\text{stand.})}_{\mu\nu} \sim \int d^4k_1\, d^4k_2\, \delta^{(4)}(k_1+k_2-q)$$
$$\times \int d^4\ell\, D_{\alpha\beta}(\ell)\, \text{Tr}\left[\gamma_\nu \Gamma \gamma_\alpha S(\ell-k_2)\gamma_\mu \Gamma_1 \gamma_\beta \Gamma_2\right] \bar{\Phi}^{[\Gamma]}(k_2)\, \Phi^{[\Gamma_1]}_{(2)}(k_1;\ell)\, \Phi^{[\Gamma_2]}_{(1)}(k_1)\, \delta\big((P_1-k_1)^2\big)\,, \quad (6)$$

$$W^{(\text{nonstand.})}_{\mu\nu} \sim \int d^4k_1\, d^4k_2\, \delta^{(4)}(k_1+k_2-q)$$
$$\times \int d^4\ell\, D_{\alpha\beta}(\ell)\, \text{Tr}\left[\gamma_\nu \Gamma \gamma_\mu S(k_1)\gamma_\alpha \Gamma_1 \gamma_\beta \Gamma_2\right] \bar{\Phi}^{[\Gamma]}(k_2)\, \Phi^{[\Gamma_1]}_{(2)}(k_1;\ell)\, \Phi^{[\Gamma_2]}_{(1)}(k_1)\, \delta\big((P_1-k_1)^2\big), \quad (7)$$

where

$$\bar{\Phi}^{[\Gamma]}(k_2) = \int \frac{d^4\eta_2}{(2\pi)^4} e^{-ik_2\eta_2} \langle P_2, S|\bar{\psi}(\eta_2)\Gamma\psi(0)]\,|S, P_2\rangle\,, \quad (8)$$

$$\Phi^{[\Gamma_1]}_{(2)}(k_1;\ell) = \int \frac{d^4\eta_1}{(2\pi)^4} e^{i(P_1-\ell-k_1)\eta_1} \langle 0|\bar{\psi}(\eta_1)\,\Gamma_1\psi(0)|P_1\rangle\,, \quad (9)$$

$$\Phi^{[\Gamma_2]}_{(1)}(k_1) = \int \frac{d^4\xi}{(2\pi)^4} e^{-ik_1\xi} \langle P_1|\bar{\psi}(\xi)\,\Gamma_2\psi(0)|0\rangle\,. \quad (10)$$

In Eqns. (6) and (7), the δ-functions $\delta\big((P_1-k_1)^2\big)$ indicate that the quark with the momentum $P_1 - k_1$ is on its mass shell.

Let us consider first the standard contribution, which exists even for real B^V-function. We constrain ourselves to the case with one of the pion distribution amplitudes being projected onto the matrix related to the chiral-odd distribution. Then, the pion chiral-odd distribution amplitudes "automatically" isolates the chiral-odd structure in the nucleon matrix element. So, the matrices Γ, Γ_1, Γ_2 and related distribution amplitudes in Eqn. (6) are as follows:

$$\Gamma\bar{\Phi}^{[\Gamma]} = \sigma^{+\perp}\bar{\Phi}^{[\sigma^{-\perp}]}, \qquad \Gamma_1\Phi^{[\Gamma_1]}_{(2)} = \gamma^-(\gamma_5)\Phi^{[\gamma^+(\gamma_5)]}_{(2)}, \qquad \Gamma_2\Phi^{[\Gamma_2]}_{(1)} = \sigma^{+-}(\gamma_5)\Phi^{[\sigma^{-+}(\gamma_5)]}_{(1)}. \quad (11)$$

These are the twist-2 nucleon parton distribution and the pion distribution amplitudes of twist two and three. Following the factorization procedure (see, for example, [17, 18, 19, 20]), we have for the upper blob in Fig. 3a

$$\bar{\Phi}^{[\sigma^{-\perp}]}(y) \stackrel{\text{def.}}{=} \int (\mathrm{d}^4 k_2)\delta(y - k_2^-/P_2^-)\bar{\Phi}^{[\sigma^{-\perp}]}(k_2) = \varepsilon^{-\perp S^\perp P_2}\bar{h}_1(y). \tag{12}$$

For the lower blobs in Fig. 3a, we obtain a following convolution

$$\mathbb{B}^{[\Gamma_2,\Gamma_1]}_{\alpha\beta}(x_1,x_2) = \frac{g^\perp_{\alpha\beta}}{2P_1^+(x_2-x_1)}\Phi^{[\Gamma_2]}_{(1)}(x_1) \overset{\mathbf{k}_1^\perp}{\circledast} \Phi^{[\Gamma_1]}_{(2)}(\bar{x}_2)\,, \tag{13}$$

where the integration over $\mathrm{d}k_1^-$ has been done with the help of $\delta((P_1-k_1)^2)$ and the convolution over the transverse momentum is defined as

$$F \overset{\mathbf{k}_1^\perp}{\circledast} G = \int (\mathrm{d}^2\mathbf{k}_1^\perp)\, F(\mathbf{k}_1^\perp) \int (\mathrm{d}\ell^- \mathrm{d}^2\boldsymbol{\ell}_\perp)\,\Delta(\ell^-,\boldsymbol{\ell}_\perp)\, G(\mathbf{k}_1^\perp;\ell^-\boldsymbol{\ell}_\perp)\,, \tag{14}$$

$$\Delta(\ell^-,\boldsymbol{\ell}_\perp) = \frac{1}{\ell^- - \boldsymbol{\ell}_\perp^2/(2(x_2-x_1)P_1^+) + i\,\mathrm{sign}(x_2-x_1)\,0}. \tag{15}$$

The convolution of the two non-collinear pion distribution amplitudes, the function $\mathbb{B}^{[\Gamma_2,\Gamma_1]}_{\alpha\beta}(x_1,x_2)$ of Eqn. (13), is an analogue of the function $B^V(x_1,x_2)$ that is introduced for the description of the inclusive channel [13]. The representation through two non-collinear functions is physically sound, if the invariant mass of pion-induced spectators is not large. In the limit of $|\boldsymbol{\ell}_\perp| \gg |\mathbf{k}_1^\perp|$, we can write a similar representation through two collinear distribution amplitudes. The resemblance between the functions $B^V(x_1,x_2)$ and $\mathbb{B}^{[\Gamma_2,\Gamma_1]}_{\alpha\beta}(x_1,x_2)$ can be seen as a manifestation of duality between exclusive and inclusive channels.

In Eqn. (13), the gluon pole at $x_1 = x_2$ can be conveniently dealt with in the contour gauge (see Refs. [13, 14, 15] and the talk given by I. Anikin in this workshop). This leads to a modification

$$\frac{1}{x_2-x_1} \stackrel{\text{c. g.}}{\Longrightarrow} \frac{1}{x_2-x_1-i\epsilon}, \tag{16}$$

which follows from the integral representation of the unit step function [14].

The non-standard term in the hadron tensor (Fig. 3b) can be considered along the same lines as the standard one [13]. Completing all proper decompositions and simplifications, we finally get the following expressions for the hadron tensor:

$$W^{(\text{stand.})}_{\mu\nu} \sim i\int \mathrm{d}x_1\,\mathrm{d}y\,\delta^{(4)}(x_1P + yP_2 - q)\overline{\Phi}^{[\Gamma]}(y)$$
$$\times \int \mathrm{d}x_2\ \mathrm{Tr}\left[\gamma_\nu\Gamma\gamma_\alpha\frac{(x_2-x_1)\not{P}_1^+}{-(x_2-x_1)ys+i0}\gamma_\mu\gamma^-\gamma_\beta\Gamma_2\right]\frac{1}{2}\frac{g^\perp_{\alpha\beta}\tilde{n}^{[-}P_1^{+]}}{x_2-x_1-i\epsilon}\Phi^{\text{tw-3}}_{(1)}(x_1) \overset{\mathbf{k}_1^\perp}{\circledast} \Phi^{\text{tw-2}}_{(2)}(x_2), \tag{17}$$

$$W^{(\text{nonstand.})}_{\mu\nu} \sim i\int \mathrm{d}x_1\,\mathrm{d}y\,\delta^{(4)}(x_1P_1 + yP_2 - q)\overline{\Phi}^{[\Gamma]}(y)$$
$$\times\ \mathrm{Tr}\left[\gamma_\nu\Gamma\gamma_\mu\frac{\gamma^+}{2x_1P_1^+ + i0}\gamma_\alpha\gamma^-\gamma_\beta\Gamma_2\right]\frac{1}{2}\int \mathrm{d}x_2\frac{g^\perp_{\alpha\beta}\,\tilde{n}^{[-}P_1^{+]}}{x_2-x_1-i\epsilon}\Phi^{\text{tw-3}}_{(1)}(x_1) \overset{\mathbf{k}_1^\perp}{\circledast} \Phi^{\text{tw-2}}_{(2)}(x_2). \tag{18}$$

It can be readily seen that both terms (17) and (18) are not gauge invariant in QED sense, while the sum thereof is invariant. The sum reads

$$\overline{W}^{\mu\nu} = \int \mathrm{d}^2\mathbf{q}_\perp W^{\mu\nu}$$
$$\sim i \int \mathrm{d}x_1\, \mathrm{d}y\, \delta(x_1 P_1^+ - q^+)\delta(yP_2^- - q^-)\bar{h}_1(y) \int \mathrm{d}x_2 \widetilde{B}(x_1, x_2) \frac{\varepsilon_\perp^{\nu S_\perp P_2 P_1}}{(P_1 P_2)} \left[\frac{P_1^\mu}{y} - \frac{P_2^\mu}{x_1}\right], \quad (19)$$

where

$$\widetilde{B}(x_1, x_2) = \frac{1}{2} \frac{\Phi_{(1)}^{\text{tw-3}}(x_1) \overset{\mathbf{k}_1^\perp}{\circledast} \Phi_{(2)}^{\text{tw-2}}(x_2)}{x_2 - x_1 - i\epsilon}. \quad (20)$$

The above expression coincides formally with the result obtained in [13] for the inclusive Drell-Yan process.

4. Single transverse spin asymmetry

$\mathrm{SS}_\perp\mathrm{A}$ is written as

$$\mathcal{A} = \frac{\mathrm{d}\sigma^{(\uparrow)} - \mathrm{d}\sigma^{(\downarrow)}}{\mathrm{d}\sigma^{(\uparrow)} + \mathrm{d}\sigma^{(\downarrow)}}, \quad (21)$$

where $\mathrm{d}\sigma^{(\uparrow)}$ and $\mathrm{d}\sigma^{(\downarrow)}$ are cross sections (5) with opposite directions of the nucleon transverse spin. For the process under consideration, we can obtain the $\mathrm{SS}_\perp\mathrm{A}$ from Eqns. (19) and (20). This asymmetry for the chiral-odd contribution can be put in terms of [2] as follows:

$$\mathcal{A} = \frac{|\mathbf{S}_\perp|}{Q} \frac{D_{[\sin 2\theta]} \sin\phi_S B_{UT}^{\sin\phi_S}}{\bar{f}_1(y_B)\, H_1(x_B)}, \qquad D_{[\sin 2\theta]} = \frac{\sin 2\theta}{1 + \cos^2\theta}, \quad (22)$$

where $B_{UT}^{\sin\phi_S} = 2\bar{h}_1(y_B)\, \Phi_{(1)}^{\text{tw-3}}(x_B) \overset{\mathbf{k}_1^\perp}{\circledast} \Phi_{(2)}^{\text{tw-2}}(x_B)$; $\bar{f}_1(y_B)$ and $H_1(x_B)$ parametrize the following matrix elements contributing to the unpolarized cross section

$$\langle P_2|\bar{\psi}\gamma^-\psi|P_2\rangle \overset{\mathcal{F}}{\sim} P_2^- \bar{f}_1(y), \qquad \langle P_1|\bar{\psi}_+|q(K)\rangle\langle q(K)|\psi_+|P_1\rangle \overset{\mathcal{F}}{\sim} P_1^+ H_1(x), \quad (23)$$

$$H_1(x) = \frac{1}{2\bar{x}_1 P_1^+} \int (d^2\mathbf{k}_1^\perp)\Phi_{(2)}^{[\gamma^+(\gamma_5)]}(\bar{x}_1, \mathbf{k}_1^\perp)\Phi_{(1)}^{[\mathbb{I}(\gamma_5)]}(x_1, \mathbf{k}_1^\perp). \quad (24)$$

It should be stressed that the asymmetry (22) is different from both the leading twist Sivers asymmetry $A_{UT}^{\sin\phi_S}$, which is accompanied by the depolarization factor $D_{[1+\cos^2\theta]}$, and the higher twist asymmetries $A_{UT}^{\sin(\phi_S \pm \phi)}$, which are related to a different tensor structure, although appearing with the same factor $D_{[\sin 2\theta]}$ as the asymmetry (22).

5. Conclusion

We derived the hadron tensor for pion-nucleon DY process with large Feynman x, transversely polarized nucleon and "primordial" transverse momenta. We took into account two diagrams (Fig. 3), standard diagram and non-standard [13] one—both diagrams are necessary to get manifestly $U(1)_{\text{em}}$ gauge invariant hadron tensor. The pion-related soft part of the diagrams is described by a convolution of two non-collinear distribution functions—a sort of decomposition of the inclusive function $B^V(x_1, x_2)$. This approximation implies that the invariant masses of undetected spectators in the pion sector are relatively small. We studied the case of the nucleon distribution and one of the pion amplitudes being chiral-odd. In this case, we obtained unexpected angular dependence of $SS_\perp A$ that can be measured experimentally. The asymmetry can be seen as a signal for the transverse "primordial" momentum dependence probing both gluon poles and chiral-odd functions and related to the duality of different factorization regimes in exclusive and inclusive channels (see, [21]).

Acknowledgments

We thank A.V. Efremov, L. Motyka and A. P. Nagaytsev for useful discussions. The work by I.V.A. was partially supported by the Bogoliubov–Infeld Program and Heisenberg–Landau Program HL-2017. L.Sz. is partly supported by grant No 2015/17/B/ST2/01838 from the National Science Center in Poland.

References

[1] Baum G *et al* [COMPASS Collaboration] 1996 COMPASS: A proposal for a common muon and proton apparatus for structure and spectroscopy CERN-SPSLC-96-14 CERN-SPSLC-P-297
[2] Aghasyan M *et al* [COMPASS Collaboration] 2017 First measurement of transverse-spin-dependent azimuthal asymmetries in the Drell-Yan process *Preprint* arXiv:1704.00488 [hep-ex].
[3] Bai M *et al* 2013 Status and plans for the polarized hadron collider at RHIC *Proc. 4th Int. Particle Accelerator Conference (IPAC 2013): Shanghai, China, May 12-17, 2013* TUYB103
[4] Kouznetsov O and Savin I 2017 *Nucl. Part. Phys. Proc.* **282** 20
[5] Savin I *et al* 2016 *Eur. Phys. J.* A **52** 215
[6] Angeles-Martinez R *et al* 2015 *Acta Phys. Polon.* B **46** 2501
[7] Boer D *et al* 2011 Gluons and the quark sea at high energies: Distributions, polarization, tomography *Preprint* arXiv:1108.1713 [nucl-th].
[8] Boer D, Mulders P J and Pijlman F 2003 *Nucl. Phys.* B **667** 201
[9] Kang Z B, Qiu J W, Vogelsang W and Yuan F 2011 *Phys. Rev.* D **83** 094001
[10] Boer D 2011 *Phys. Lett.* B **702** 242
[11] Arnold S, Metz A and Schlegel M 2009 *Phys. Rev.* D **79** 034005
[12] Anikin I V, Szymanowski L and Teryaev O V and Volchanskiy N 2017 *Phys. Rev.* D **95** 111501
[13] Anikin I V and Teryaev O V 2010 *Phys. Lett.* B **690** 519
[14] Anikin I V and Teryaev O V 2015 *Eur. Phys. J.* C **75** 184
[15] Anikin I V and Teryaev O V 2015 *Phys. Lett.* B **751** 495
[16] Barone V, Drago A and Ratcliffe P G 2002 *Phys. Rept.* **359** 1
[17] Anikin I V, Ivanov D Y, Pire P, Szymanowski L and Wallon S 2010 *Nucl. Phys.* B **828** 1
[18] Belitsky A V and Radyushkin A V 2005 *Phys. Rept.* **418** 1
[19] Diehl M 2003 *Phys. Rept.* **388** 41
[20] Braun V M and Manashov A N 2012 *J. High Energy Phys.* JHEP01(2012)085
[21] Anikin I V, Cherednikov I O, Stefanis N G and Teryaev O V 2009 *Eur. Phys. J.* C **61** 357

IOP Conf. Series: Journal of Physics: Conf. Series **938** (2017) 012066 doi:10.1088/1742-6596/938/1/012066

New Method for Taking into Account Radiative Events in the Inclusive Experiment MOLLER

Vladimir Zykunov

Veksler and Baldin Laboratory of High Energy Physics, 6, Joliot-Curie Str., 141980, Dubna, Russia

E-mail: zykunov@cern.ch

Abstract. A new method for taking into account of radiative events in inclusive polarized experiments that, in principle, makes no use of an unphysical separation into soft- and hard-photon regions is demonstrated. A procedure for taking into account bremsstrahlung in the MOLLER forthcoming ultraprecise experiment at the Thomas Jefferson Laboratory is developed on the basis of this method.

1. Introduction

Inclusive experiments play a role of paramount importance in high-energy physics. In a number of cases, this is because it is impossible in principle to detect all particles produced in the reaction being considered; in other cases (including deep-inelastic lepton scattering on nucleons, the Drell–Yan reaction, and identical-particle scattering), this is not needed for deducing information about the reaction mechanism. In almost any of the present-day experiments, reliable information can be obtained from experimental data only upon taking into account electroweak radiative corrections, which are higher order effects in the Standard Model. An infrared divergence is a well-known problem encountered in taking into account contributions of extra virtual particles to observables. A path toward solving this problem goes through incorporating the contribution of real bremsstrahlung photons into respective calculations. If bremsstrahlung photons are not detected in the reaction being considered, the problem of taking them into account should be addressed in dealing with inclusive reactions.

A new method (N-method) for precisely taking into account the emission of one bremsstrahlung photon as a component of one-loop radiative corrections is formulated in the present study by considering the example of observables in the inclusive Møller scattering reaction

$$e^-(k_1) + e^-(p_1) \to e^-(k_2) + e^-(p_2) + (\gamma(k)) \tag{1}$$

involving polarized electrons [1] (here in brackets the 4-momenta of particles are presented).

The nonradiative and radiative process of polarized Møller scattering was fully described in [2] and [3], correspondingly, and the notation used there is reproduced in the present study where, we place emphasis on the method used to calculate the reaction phase space, which originally has the form

$$d\Phi = \delta^4(k_1 + p_1 - k_2 - p_2 - k)\frac{d^3\mathbf{p_2}}{2p_{20}}\frac{d^3\mathbf{k_2}}{2k_{20}}\frac{d^3\mathbf{k}}{2k_0}. \tag{2}$$

2. Phase space

In [2], the phase space (2) was reduced to the form presented in [4], which involves a dependence on four (radiative) invariant quantities; that is,

$$d\Phi = \frac{\pi}{16s\sqrt{-\Delta_4}} dt dv dz dv_1, \tag{3}$$

where $\Delta_4 = \Delta_4(k_1, p_1, k_2, p_2)$ is Gram determinant. Although the expression for the phase space looks quite simple and admits a clear interpretation in the respective computer code, the phase-space in (3) presents a difficult challenge in integration. This served as a motivation of searches for new ways to improve the accuracy.

To improve the convergence we transform the phase space (2) in the reference frame comoving with the center-of-mass of primary particles, where $\mathbf{k_1} + \mathbf{p_1} = 0$, $k_{10} = p_{10} = \sqrt{s}/2$. First of all, we remove the three-dimensional integral with respect to $d^3\mathbf{p_2}$, whereupon we have

$$d\Phi = \delta(\sqrt{s} - k_{20} - p_{20} - k_0) \frac{1}{2p_{20}} \frac{d^3\mathbf{k_2}}{2k_{20}} \frac{d^3\mathbf{k}}{2k_0}. \tag{4}$$

We then go over to the system of spherical coordinates,

$$d^3\mathbf{k_2} = |\mathbf{k_2}|^2 d|\mathbf{k_2}| d\cos\theta_{k_2} d\varphi_{k_2}$$

and apply the relation $|\mathbf{k_2}|d|\mathbf{k_2}| = k_{20}dk_{20}$.

It is now necessary to remove the integral with respect to k_{20} with the aid of the remaining delta function by using the property $\delta[f(x)] = \delta(x - x_0)/|f'(x_0)|$, where x_0 is root of $f(x_0) = 0$. In the case being considered, the function $f(x)$ is the lefthand side of the energy-conservation law specified in the implicit form

$$f(x) = k_{20} + p_{20} + k_0 - \sqrt{s}, \tag{5}$$

where $\sqrt{s}$ is the total reaction energy and $x \equiv k_{20}$. One can readily evaluate the derivative

$$f'(x) = 1 + \frac{x(1 - |\mathbf{k}|A(x^2 - m^2)^{-1/2})}{\sqrt{x^2 - 2|\mathbf{k}|A\sqrt{x^2 - m^2} + |\mathbf{k}|^2}}. \tag{6}$$

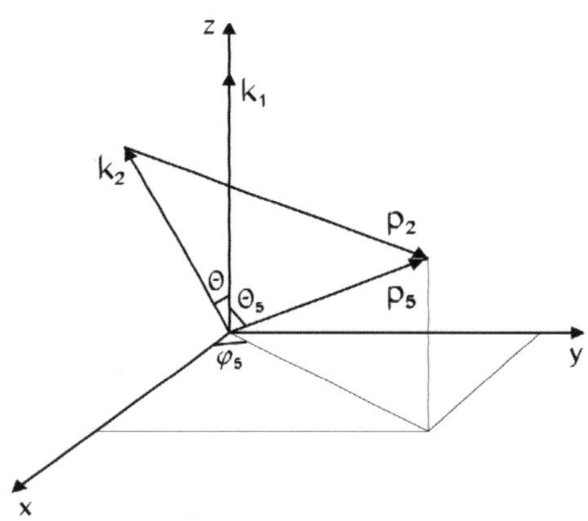

Figure 1. Configuration of the 3-momenta of the final-state particles in the c.m. frame.

In Fig. 1, we depict the 3-momentum vectors of the final-state particles in the c.m. frame by using the auxiliary vector $\mathbf{p_5} = -\mathbf{k}$. The possible energies of the final-state electron can be found by resolving the energy-conservation law $f(x_0^{\pm}) = 0$. The result has the form

$$x_0^{\pm} = \frac{BC \pm \sqrt{C^2 + m^2(1 - B^2)}}{1 - B^2}, \tag{7}$$

where the coefficients

$$B = \frac{\sqrt{s} - k_0}{A|\mathbf{k}|}, \quad C = \frac{|\mathbf{k}|^2 - (\sqrt{s} - k_0)^2}{2A|\mathbf{k}|} \tag{8}$$

are expressed in terms of the factor

$$A = \cos(\widehat{\mathbf{p_5}, \mathbf{k_2}}) = \sin\theta \sin\theta_5 \cos\varphi_5 + \cos\theta \cos\theta_5. \tag{9}$$

This factor specifies that sign on the right-side of Eq. (7) which corresponds to the the energy: at a positive value of A (in this situation the 3-momentum vector of the emitted photon forms an obtuse angle with respect to the vector $\mathbf{k_2}$) we choose x_0^-, and vice versa. With allowance for the foregoing, we ultimately obtain the phase space in the form

$$d\Phi = \frac{|\mathbf{k_2}|}{4p_{20}|f'(x_0)|} d\cos\theta_{k_2} d\varphi_{k_2} \frac{d^3\mathbf{k}}{2k_0}, \tag{10}$$

where $d\cos\theta_{k_2} \equiv dc$. Integration with respect to the azimuthal angle φ_{k_2} yields 2π because of symmetry under rotations of the system about the beam axis. Further, we proceed to perform integration with respect to $\mathbf{k}$. We have

$$d^3\mathbf{k} = |\mathbf{k}|^2 d|\mathbf{k}| d\cos\theta_k d\varphi_k, \;\; \theta_k = \pi - \theta_5, \;\; \varphi_k = \pi + \varphi_5.$$

Upon employing $\mathbf{p_5}$ the 3-momentum vectors of final-state particles in the c.m. frame assume a rather simple form; that is,

$$\begin{aligned} \mathbf{k_2} &= (|\mathbf{k_2}|\sin\theta, 0, |\mathbf{k_2}|\cos\theta), \\ \mathbf{p_5} &= (|\mathbf{k}|\sin\theta_5\cos\varphi_5, |\mathbf{k}|\sin\theta_5\sin\varphi_5, |\mathbf{k}|\cos\theta_5), \\ \mathbf{p_2} &= \mathbf{p_5} - \mathbf{k_2}. \end{aligned} \tag{11}$$

It is clear that $p_{20} = \sqrt{m^2 + |\mathbf{p_2}|^2}$ and is readily calculable on the basis of (11).

We will now express all radiative invariants in terms of the proton energy and azimuthal and polar angles. We have [2]:

$$z_1 = 2k_0k_{10} + 2|\mathbf{k}||\mathbf{k_1}|\cos\theta_5, \tag{12}$$
$$v_1 = 2k_0p_{10} - 2|\mathbf{k}||\mathbf{p_1}|\cos\theta_5, \tag{13}$$
$$z = 2k_0k_{20} + 2|\mathbf{k}||\mathbf{k_2}|A, \tag{14}$$
$$v = 2k_0(\sqrt{s} - k_{20}) - 2|\mathbf{k}||\mathbf{k_2}|A. \tag{15}$$

It is noteworthy that, in fact, k_0 and $|\mathbf{k}|$ are different in all of the equations used. In other words, the photon mass λ is retained in these expressions. In the following, this mass will be used as an infinitesimal parameter to regularize infrared divergences.

3. Integration over the total phase space

Everything is now ready for performing integration over the total bremsstrahlung-photon phase space. By employing the expressions obtained for the phase space, we remove the integral with respect to the detected-electron azimuthal angle φ_{k_2}, whereupon the R-contribution to the cross section assumes the form

$$\sigma^R \equiv \frac{d\sigma^R}{dc} = -\frac{\alpha^3}{4\pi s} \int\limits_0^{|\mathbf{k}|_{\max}} \frac{|\mathbf{k}|^2}{k_0} d|\mathbf{k}| \int\limits_{-1}^{1} d\cos\theta_k \int\limits_0^{2\pi} d\varphi_k \frac{|\mathbf{k_2}|}{p_{20}|f'(x_0)|} \sum_{i,j=\gamma,Z} MM^+. \tag{16}$$

Even in this form, the relevant numerical integration (we will employ the VEGAS Monte Carlo integrator [5]) involves no problems if one pays due attention to the condensation of points in the region of small $|\mathbf{k}|$. However, it is possible to represent the result in a more elegant form. First, we proceed to consider the integral with respect to the photon energy by employing the equality $|\mathbf{k}|d|\mathbf{k}| = k_0 dk_0$. We then have

$$\int\limits_0^{|\mathbf{k}|_{\max}} \frac{|\mathbf{k}|^2}{k_0} d|\mathbf{k}| ... = \int\limits_\lambda^\Omega |\mathbf{k}| dk_0 ...,$$

where Ω is the maximum (chosen) bremsstrahlung photon energy. By applying the change of variable (a similar change of variable was used, for example, in [6] to isolate the dependence of the cross section on the minimal photon energy)

$$k_0 = \lambda^{1-x}\Omega^x, \quad \frac{dk_0}{k_0} = \ln\frac{\Omega}{\lambda} dx, \tag{17}$$

we obtain the cross section in the form

$$\sigma^R = -\frac{\alpha^3}{4\pi s} \ln\frac{\Omega}{\lambda} \int\limits_0^1 dx |\mathbf{k}| k_0 \int\limits_{-1}^1 d\cos\theta_k \int\limits_0^{2\pi} d\varphi_k \frac{|\mathbf{k_2}|}{p_{20}|f'(x_0)|} \sum_{i,j=\gamma,Z} MM^+. \tag{18}$$

4. Conclusion

In order to obtain reliable data in the MOLLER experiment with a declared relative error below 0.01, not only is it necessary to take precisely into account one-loop and at least leading two-loop radiative effects, but also an exact interpretation of events involving bremsstrahlung-photon emission with allowance for the detector potential is mandatory. The new N-method supplemented with a respective FORTRAN code developed on the basis of this method solves successfully the problem in question. It is the opinion of the author that the N-method may become a standard tool for calculations of the above type, since, by construction, it does not involve unphysical parameters, with the exception of a fictitious photon mass, which cancels the infrared divergence in a natural way. From the methodological point of view, the new method in question is advantageous in the sense that it makes no use of terms such as a "soft photon", a "hard photon", and a "parameter that separates the soft and hard regions", so that a cumbersome and essentially immaterial test for the absence of a dependence on this parameter is not required. One should also bear in mind that the calculation for the region of soft photons within a traditional approach is approximate, inevitably leading to a poorly controllable computational error, but that a calculation on the basis of the N-method involves no such error.

5. Acknowledgments

I am grateful to A.G. Aleksejevs, A.B. Arbuzov, S.G. Barkanova, and Yu.M. Bystritsky for discussions. I thank the organizers of XVII WORKSHOP ON HIGH ENERGY SPIN PHYSICS DSPIN-2017. Work was supported by Belarus scientific program (grant BRFFI-F16-121).

References

[1] Benesch J *et al.*, The MOLLER Collab., at `http://hallaweb.jlab.org/12GeV/Moller/downloads/DOE_Proposal/DOE_Moller.pdf` (2011).
[2] Zykunov V 2015 *Phys. Atom Nucl.* **78** 453
[3] Zykunov V 2017 *Phys. Atom Nucl.* **80** 699
[4] Byckling E and Kajanti K 1975 *Kinematics of elementary particles*, Mir, Moscow
[5] Lepage P 1978 *J. Comput. Phys.* **27** 192
[6] Ishikawa T *et al.* [MINAMI-TATEYA Group Collab.] KEK-92-19

IOP Conf. Series: Journal of Physics: Conf. Series **938** (2017) 012067 doi:10.1088/1742-6596/938/1/012067

Monte-Carlo study of drift mobility in ultrathin organic layers: Miller-Abrahams and Marcus models

Ya. V. Burdakov and V.R. Nikitenko
National Research Nuclear University MEPhI (Moscow Engineering Physics Institute), Moscow, 115409, Russia

E-mail: vladronik@yandex.ru

Abstract. It is shown by the means of Monte-Carlo modeling that the drift mobility decreases considerably along with the increasing of the layer thickness from 10 to 100 nm, if the hopping rates are calculated in the framework of the Marcus model, according to approximately the same low, as in the case of the Miller-Abrahams model. The diffusion current can contribute approximately the same contribution, as an energetic relaxation of the generated non-equilibrium charge carriers, for the case of moderate disorder.

1. Introduction

The investigation of non-stationary transport of charge carriers in very thin (less than 100 nm) layers of organic materials is important both from the fundamental viewpoint (the investigation of peculiarities of energy distribution of hopping centers and morphology of the thin layers) and for applications, because just the layers of this thickness are applicable in light-emitting diodes, photovoltaic sells and other basic elements of organic electronics. In the several works it was shown, that the drift mobility decreases along with increasing of the film thickness, if the thickness is rather small, even for the case of quasi-equilibrium initial generation of charge carriers [3-5]. Calculations were performed in the framework of Miller-Abrahams (MA) model. However, the Marcus model [6], which accounts for the polaronic effect, is often used for the analyses of hopping transport in organic materials, in parallel with the MA model. The temperature and field dependence of both the "true" and drift mobility is well investigated in the framework of the MA model, both experimentally [7] and theoretically – by the means of numerical [7] and analytic [8,9] methods, while analytic description of mobility in Marcus model is less developed [10]. In this work, the drift mobility in ultrathin organic layers is investigated by the means of Monte Carlo simulation for the different dimensionality, generation conditions and boundary conditions. The comparison of the results from both the models is carried out. The simulations yield considerable decreasing of the drift mobility along with decreasing of the thickness L from 10 to 100 nanometers from both the models. The method for the determination of the quasi-equilibrium mobility from the analyses of the obtained thickness dependence of the drift mobility is proposed, and the analysis of the reasons of the decreasing of the drift mobility is carried out.

2. The method of simulations.

The model of MC simulations, that is similar to the work [4], is used here. The model is based on the well-known Gaussian disorder model [7]. The localized states (hopping centers) form a simple cubic lattice with the lattice constant a. Energies of the centers are distributed as a Gaussian function with the variance σ. At variance with the works [3,4], both MA and Marcus models were used for the description of the hopping rates from a state i to the state j, ν_{ij}. In the MA model,

$$\nu_{ij} = \nu_0 exp\left(-2\gamma r - \left(\frac{\Delta\varepsilon_{ij}+|\Delta\varepsilon_{ij}|}{2kT}\right)\right), \quad (1)$$

while in Marcus model

$$\nu_{ij} = \nu_0 exp\left(-2\gamma r - \left(\frac{(\varepsilon_j-\varepsilon_i+E_r)^2}{4E_r kT}\right)\right), \quad (2)$$

where r is the distance between the states i and j, γ is the inverse localization radius of the wave function, E_r is the reorganization energy, ν_0 is the frequency factor, T is the absolute temperature and k is the Boltzmann's constant. In these simulations the value $2\gamma a = 10$ is assumed.

Both three- and two-dimensional (molecular monolayer) structures were investigated (3D and 2D systems, respectively). Various boundary conditions were also considered. In the first variant, the layer is sandwiched between two electrodes, an electron generates next to the left electrode and it cannot penetrate across the interface. When the electron achieve the right electrode (in the direction of the electric force), the procedure is finished. This variant refers below as (0:0). In the second variant, both the boundaries of the layer are considered as virtual ones, i.e. the layer is a part of a prolonged media, hence a carrier can perform the same random walk outside the layer. The simulation finished in this case, then the electron crosses the right boundary at the last time. The longitude of the zone outside the layer, in which the random walk is possible, taken as $30a$ (it is enough to provide free walk of a carrier at all values of the field strength under consideration. This variant refers below as (30:30). A carrier is provided to move much longer in the transversal directions, than the longitudinal thickness of the layer.

In parallel, we consider different variants of an initial energy distribution of charge carriers. The initial energy distribution of generated carriers is the same as the Gaussian distribution of hopping centers, (referred as Notquas). The initial occupation probability is a Boltzmann's function of an energy of a center (referred below as Quas), hence the initial energy distribution of carriers is shifted down by the value of σ^2/kT. One can investigate an influence of the energetic relaxation and conventional diffusion on the drift mobility, comparing results of the models Notquas, Quas and (0:0), (30:30), respectively.

The mean inverse of the transit time, $\langle 1/t_{tr}\rangle$, defines from the series of simulations (the typical number is 3000 at a given set of parameters). Then the drift mobility calculates as the mean drift velocity, divided by the electric field strength, F: $\mu_{dr} = (L/F)\langle 1/t_{tr}\rangle$, where $L=Na$.

3. Results and discussion

The figure 1a shows, that the drift mobility varieties smaller with changes of the dimensionality and reorganization energy at smaller values of the layer thickness, than at larger ones.

a) b)

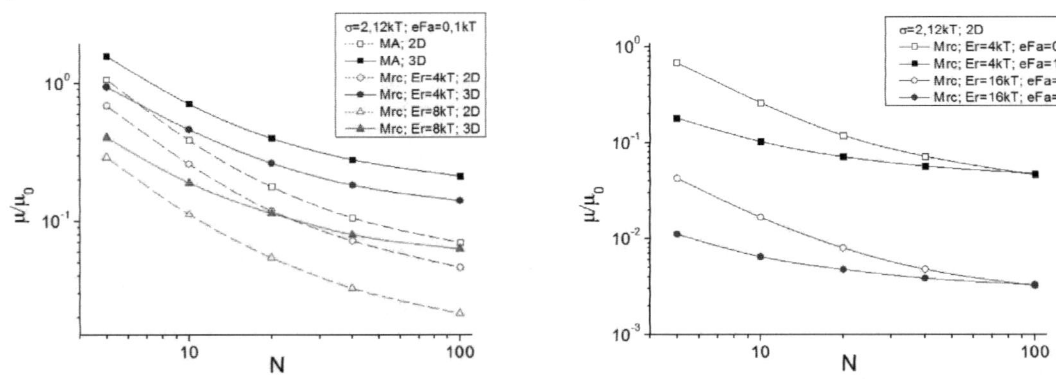

Figure 1 (a) The comparison of the thickness dependences of the drift mobility at various values of dimensionality (the solid curves 3D, the dashed curves – 2D) from the MA and Marcus models for the case of weak field, eFa/kT=0.1. (b) The same for the case of 2D conductivity by the use of Marcus model at two values of the field strength F, eFa/kT=0.1 и eFa/kT=1. $\mu_0 = (e/kT)\nu_0 \exp(-2\gamma a)$. The calculations performed in the variant Quas (0:0).

The drift mobility decreases and approaches to the infinite-medium value faster in 3D case, than in the 2D case, probably due to preferable possibility to avoid deep trapping. Figure 1b shows that the thickness dependence of the drift mobility is more pronounced, if the field is weaker.

Figure 2a shows, that the mobility from the Marcus model approaches to the respective results of the MA model along with decreasing of reorganization energy, E_r (providing that Er/kT>1). The calculations performed under this condition. However, the drift mobility, being normalized by the value $\mu_0 exp(-Er/4kT)$, tends to saturate at high values of E_r , see the figure 2b.

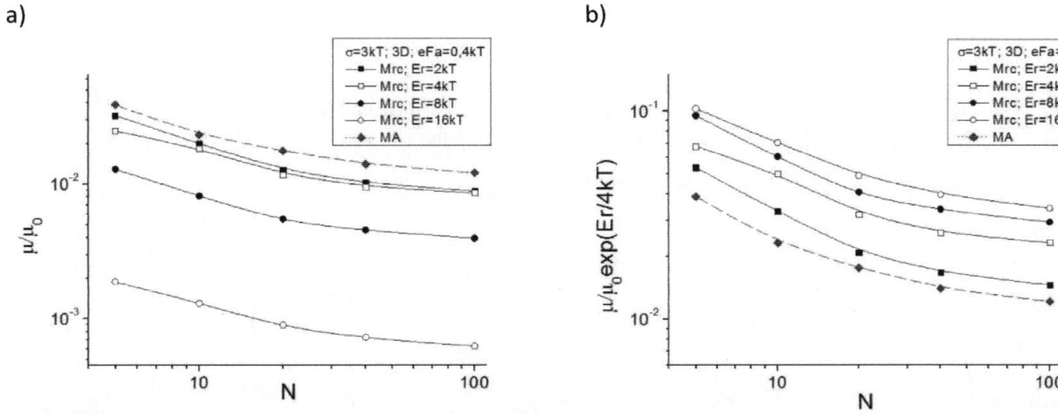

Figure 2 (a) The comparison of the thickness dependences of drift mobility from MA model and from Marcus model, parametric in reorganization energy, E_r, ($E_r/kT=2;4;8;16$), eFa/kT=0,4. (b) The same, but the mobility from Marcus model is normalized by the value $\mu_0 exp(-Er/4kT)$. Calculations performed in the variant Quas (0:0).

Figure 3a demonstrates that the mobility is smaller and more proximal to the quasi-equilibrium under the condition (30:30), than (0:0). It is not surprising, because the number of hops is larger in the letter case, than in the former. The transition from non-equilibrium to the quasi-equilibrium generation makes the thickness dependence weaker, but not eliminates it totally, see the fig. 3b.

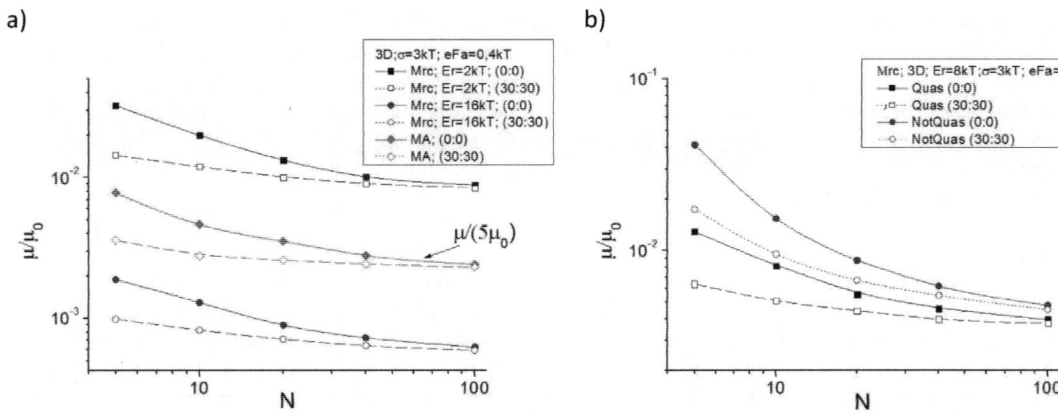

Figure 3 (a) The comparison of the drift mobility from MA and Marcus models, parametric in reorganization energy (Er/kT=2 and 16), for the case of quasi-equilibrium initial generation, at various boundary conditions, eFa/kT=0.4 (a); (b) the drift mobility from the Marcus model, E_r/kT=8, at various conditions of initial generation and boundary conditions.

Calculations of this work confirms for the case of Marcus model the decreasing of drift mobility along with increasing thickness, if values of the thicknesses are rather small,

$$\mu = \mu_\infty \left[1 + \left(L/L_0 \right)^{-\beta} \right], \qquad (3)$$

as it was founded earlier for the case of MA model [3,4].
The power low exponent β has the values close to unity. Accounting for the precision of calculations, we cannot find a systematic dependence of the value of β on the reorganization energy. Rather, it depends on the field strength and on the disorder parameter, σ/kT, in the same manner, as it was reported previously [3,4]. For example, $\beta = 0.9 \pm 0.15$ at $\sigma/kT = 3$ and $eFa/kT = 0.4$ from the model (0:0) Quas. The MA model yields $\beta = 1.16$. Apparently, the exponent from the Marcus model is smaller to some extent, than for the case of MA model. The reduction of the exponent β is more considerable under the change of the initial and boundary conditions: 1.7 ± 0.05 from (0:0) Notquas, 1.1 ± 0.1 from the (30:30) Notquas, 0.9 ± 0.15 from the (0:0) Quas, and 0.75 ± 0.1 from the (30:30) Quas model, i.e. along

with approaching to the quasi-equilibrium limit. It is important, that one can define the infinite-thickness limit (i.e. the quasi-equilibrium, "true" value of mobility, μ_∞, by the use of eq. (3). This procedure yields to the results, which are in a good agreement with the known values [7,8], for the case of the MA model [11], and one has a large economy of computing resource in comparison with the direct calculations in a rather thick layer.

a) b)

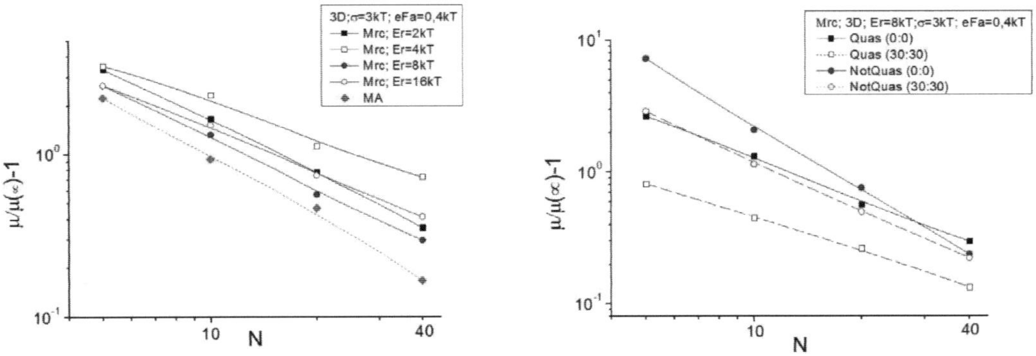

Figure 4 The same as in the figure 3, but the mobility is normalized in order to illustrate the eq. (3)

It is instructive to analyze the difference of results from the models with different initial and boundary conditions, see the figure 3. Obviously, the drift mobility is more proximal to the quasi-equilibrium limit for the case of the initial quasi-equilibrium generation, than for the case of non-equilibrium generation, but the thickness dependence remains considerable in the latter case. The contribution of diffusion current in the thin (about 10 nm) films seems to be the most important reason, rather than the lack of trapping to the deep states, due to the small value of disorder. This contribution is less pronounced, if carriers have a possibility to walk outside the region $0 < x < L$.

4. Conclusions

This paper shows that the Marcus model results in approximately the same power law of decreasing of the drift mobility along with the increasing of the thickness of the thin films, as the MA model. Meanwhile, the drift mobility has a weaker dependence on the reorganization energy, than the quasi-equilibrium (the "true") mobility. The diffusion current should be included in the analytic modeling for the case of ultrathin (about 10 nm), in order to have a quantitative description of the transient current.

Acknowledgement. This work was supported by the "Improving of the competitiveness" program of the National Research Nuclear University MEPhI (Moscow Engineering Physics Institute).

References

[1] Friend R H, Gymer R W, Holmes A B, Burroughes J H, Marks R N, Taliani C, Bradley D D C, DosSantos D A, Bredas J L, Loglund M and Salaneck W R 1999 Nature (London) **397** p 121
[2] Mozer A J and Sariciftci N S 2006 *C.R.Chimie* pp 568-577
[3] Korolev N A, Nikitenko V R, and Ivanov D V 2011 Semiconductors **45** 230
[4] Nikitenko V R, Sukharev V M, Sannikova N A 2014 J. of Physics: Conference Series **541** 012053
[5] S Raj Mohan, Manoranjan P Singh, M P Joshi 2010 *Organic Electronics* **11** 1642
[6] Marcus R A 1993 *Rev. Mod. Phys.* **65** 599
[7] Bässler H 1993 *Phys. Status Solidi B* **175** 15
[8] Nikitenko V R, Strikhanov M N 2014 *J. Appl. Phys.* **115** 073704
[9] Baranovskii S D, Faber T, Hensel F and Thomas P 1997 *J. Phys.: Condens. Matter* **9** 2699
[10] Ling Li, Nianduan Lu, Ming Liu and Heinz Bässler 2014 *Phys. Rev. B* **90** 214107
[11] Khan V D, Nikitenko V R, Burdakov Ya V J. 2017 J. of Physics: Conference Series **938** 1069

Adsorption of chemical species on nitrogen-doped [60]fullerene: tight-binding study

Ilya Y. Dolinskiy and Nikolai V. Novikov

National Research Nuclear University "MEPhI"

E-mail: dolinsky94@yandex.ru

Abstract. We present a theoretical study of the adsorption of common species COOH, H and OH on nitrogen-doped C_{60} cage. We find that these radicals tend to bind with the carbon atoms, nearest to the nitrogen defect. Structural and energetical characteristics of functionalized cages are studied in detail. The reactivity of the fullerene is found to be some more than that of graphene.

1. Introduction

In 1973, Russian scientists D.A. Bochvar and Ye.N. Halpern published an article with the results of quantum chemical calculations [1]. It was suggested that there should be a stable form of carbon containing 60 carbon atoms in the molecule. Also in this article was shown the form of such a hypothetical molecule [1]. However, this work was unnoticed. Only in 1985 Kroto and his co-workers found in the mass spectrum of graphite decomposition products under the action of a laser beam an intense peak with a mass of 720 a.m.u. [2]. It was belonged to the C_{60} molecule. Another, less intense peak, corresponding to a mass of 840 a.m.u., was belonged to the C_{70} molecule. A new allotropic modification of carbon was called as "fullerenes".

The most effective method of obtaining fullerenes is based on the thermal decomposition of graphite. In this process, high-frequency and resistive heating of graphite electrode [3,4], combustion of hydrocarbons [5,6], laser irradiation of the graphite surface [7,8], and evaporation of graphite by a focused solar ray [9] are used. Most often, an arc discharge with graphite electrodes in a helium atmosphere is used to obtain fullerenes [10,11].

The spherical fullerene molecule has unique physical, chemical and biological properties. Therefore, fullerene and its derivatives are widely used in various fields.

The biological activity of fullerenes and their derivatives is actively studied [12–15]. It is shown that representatives of this class are able to inhibit various enzymes [16], cause specific cleavage of DNA molecules [17], actively participate in various oxidation-reduction processes in the body [13]. The study of the metabolism of fullerenes has begun, special attention is paid to antiviral properties [13,18]. It is shown, in particular, that certain derivatives of fullerenes are capable of inhibiting the protease of the AIDS virus [13]. The idea of creating anti-cancer medicines based on fullerenes is widely discussed [19–22].

Also considered the possibility of obtaining super-hard materials based on fullerenes [23–25]. It was shown the possibility of obtaining large diamonds. After the addition of fullerene to graphite, the graphite-diamond phase transition begins at a lower temperature and pressure [26].

Fullerenes can be used as precursors for the growth of diamond films. It is shown that the C_{60} fragmentation on C_2 occurs in the microwave discharge. C_{60} is a suitable material for the growth of diamond crystals. It is predicted that such a high growth rate will significantly reduce the cost of CVD-grown diamonds [27].

Due to the ability to polymerize under the action of a laser or electron beam and form an insoluble phase in organic solvents, fullerene can be used as a resist for submicron lithography. Fullerene films can withstand considerable heating, they do not contaminate the substrate and they allow a dry manifestation [28].

Fullerene-containing materials are promising for use as materials for nonlinear optics [29]. Fullerites can be used to create a field-effect transistor [30], photovoltaic devices [31], solar cells [32–34].

To use fullerenes effectively, one need to learn how to tune its characteristics. This can be achieved by introducing structural defects in them [35,36], and also by doping them [37]. Another common method of altering the properties of fullerene is substitutional doping, in which some carbon atoms are replaced by atoms of other elements.

Adsorption of adenine to the surface of a Cr or Ni doped fullerene is being studied [38]. These studies can later be used to develop technologies for the delivery of genes, drugs and other biomolecules. The adsorption of hydrogen into fullerene doped with Na and Li [39–41] is investigated, since the development of solid-state fuel cells for storing hydrogen with high gravimetric and bulk density is of great importance in the development of hydrogen energy. Nitrogen-doped fullerene can be used as a cathode catalyst for hydrogen fuel cells [42]. Nitrogen-doped fullerene can also be used as a catalyst that is active for the oxygen reduction reaction and the oxygen evolution reaction and is therefore applicable to energy conversion and storage devices such as fuel cells and metal-to-air battery systems [43]. Fullerene, doped with rhodamine B, is considered as a material for improving the extraction of electrons when creating organic solar cells [44].

In this study, we investigate the influence of nitrogen doping on reactivity of C_{60} cage. Nitrogen-doped fullerene can be synthesized by vaporization of graphite under the atmosphere of nitrogen and helium mixture [45]. The present work is devoted to the investigation of the adsorption of COOH, H and OH radicals onto nitrogen-doped fullerene. The listed radicals are widespread adsorbates: H is interesting from the point of view of hydrogen storage [39–42,46,47]; $C_{60}(OH)$ can be useful in biomedicine [48–50]; $C_{60}(COOH)$ provides its biocompatibility [51]. In this article, computer simulation is used to study the effect of the nitrogen atom, embedded in the fullerene carbon framework, on the binding energy of fullerene with adsorbents.

2. Computational details

In our study, a computer simulation of the adsorption of the radicals H, OH, COOH on nitrogen doped fullerene. To optimize the molecule and calculate the binding energy, we used a non-orthogonal tight binding model (NTBM) [52,53] with the parameterization [53]. This semiempirical model describes well the interatomic interactions of $H_kC_lN_mO_n$ compounds. A more accurate method of the density functional theory (DFT) requires too much computer processing power, and the non-orthogonal tight binding model is accurate enough and is often used for fullerenes [54–56], and for systems with a carbon-nitrogen framework [57,58]. Optimization of the geometry of the initial and functionalized nitrogen-doped fullerene was carried out by the method of gradient descent: the atoms were displaced in the direction of the forces acting on them until these forces became less than 0.001 eV / Å.

We define the adsorption energy E_{ads} as $E_{ads} = E(C_{59}N+R) - E(C_{59}N) - E(R)$, where $E(C_{59}N+R)$, $E(C_{59}N)$ and $E(R)$ are the total energies of functionalized C_{60} cage, pristine C_{60} and isolated radical R (R= COOH, H or OH), respectively.

3. Results and discussion

We calculated the adsorption energies for each of the radicals COOH, H, OH attached to one of ten different atoms (A0-A9) of the nitrogen-doped fullerene, see figure 1a. These atoms were chosen so

IOP Conf. Series: Journal of Physics: Conf. Series **938** (2017) 012068 doi:10.1088/1742-6596/938/1/012068

that they were removed from the nitrogen atom (A0) built into the fullerene framework to the diametrically opposite fullerene atom (A9). As a result, we obtain oscillation behavior of the binding energy as a function of the distance between nitrogen atom and absorbent. The figure 1b shows that the effect of nitrogen defect on the adsorption energy is weaker, when the radical attached far from nitrogen atom.

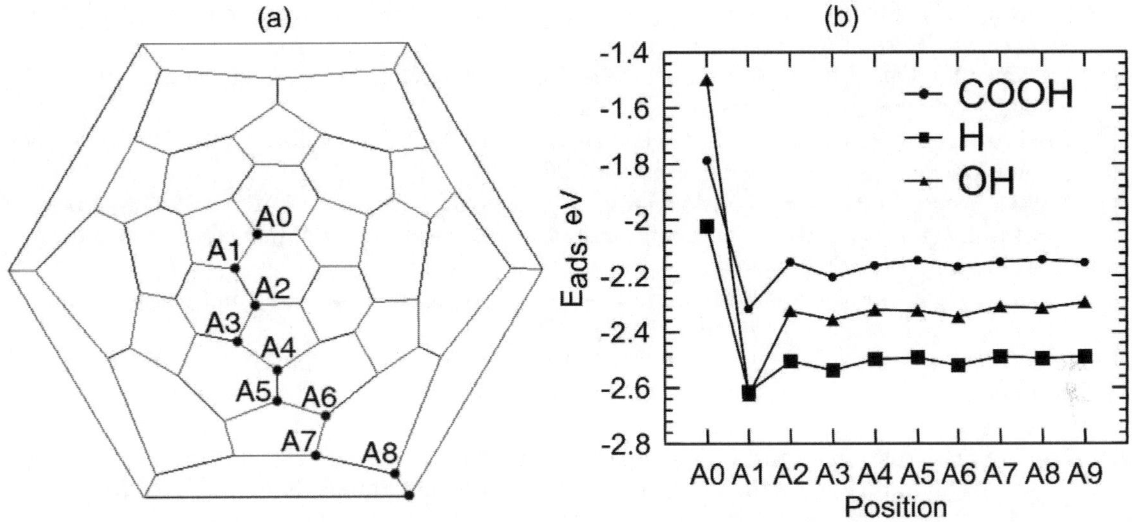

Figure 1. (a) Schlegel diagram of N-doped C_{60} cage. A0 position corresponds to the nitrogen atom. A0-A9 positions are regarded as the adsorption centres. (b) Adsorption energies for different adsorption positions.

Calculations show that all the selected radicals are capable of adsorbed on nitrogen atom, but this variant is not of interest to us, since on it the adsorption energy is the lowest of all. The most favorable position for adsorption was A1: the carbon atom closest to the nitrogen atom. A similar result was obtained by studying the adsorption of radicals on nitrogen-doped graphene according to our calculations in the frame of the same tight-binding model. The energy values for adsorption on graphene on the atom closest to nitrogen were -1.50 eV, -1.91 eV, -2.00 eV for COOH, H, OH, respectively, and for fullerene -2.32 eV, -2.62 eV, -2.63 eV for the same radicals, respectively. Thus, one can see a similarity of the adsorption properties of these carbon allotropes. The binding energy on this atom is higher than on the others. However, the adsorption energy for nitrogen-doped fullerene was greater than that of nitrogen-doped graphene. Most likely this is due to the fact that the fullerene can be considered as a convex curved graphene. It was previously shown that the adsorption of atomic hydrogen onto convex portions of graphene takes place with more energy than on flat or concave ones [59].

4. Conclusions

In the course of this work, we investigated the adsorption of the radicals COOH, H, OH on nitrogen-doped fullerene. As a result, we have obtained that the lowest adsorption energy corresponds to adsorption on the carbon atom, which is adjacent to the nitrogen atom, regardless of the structure and chemical composition of the adsorbed radical. Therefore, the nitrogen defect enhances the reactivity of the fullerene. Adsorption of both H and OH species results in the higher energy gain, than that for COOH. Calculations have shown that the adsorption properties of fullerene are similar to those of curved graphene on convex portions: the adsorption on fullerene is much more feasible than that on flat graphene. Thus, we believe that controlling the tunable local curvature of nitrogen-doped graphene will make it possible to control its adsorption properties.

References

[1] Bochvar D A and Halpern E G 1973 Electronic structure of molecules C_{20} and C_{60} *Dokl. Akad. Nauk SSSR* **209** 610–2

[2] Kroto H W, Heath J R, O'Brien S C, Curl R F and Smalley R E 1985 C_{60}: Buckminsterfullerene *Nature* **318** 162–3

[3] Churilov G N 2008 Synthesis of fullerenes and other nanomaterials in arc discharge *Fullerenes Nanotub. Carbon Nanostructures* **16** 395–403

[4] Kyesmen P I, Onoja A and Amah A N 2016 Fullerenes synthesis by combined resistive heating and arc discharge techniques *Springerplus* **5** 1–7

[5] Goel A, Hebgen P, Vander Sande J B and Howard J B 2002 Combustion synthesis of fullerenes and fullerenic nanostructures *Carbon* **40** 177–82

[6] Mitra S, Pope C J, Gleason K K, Makarovsky Y, Lafleur A L and Howard J B 1992 Synthesis of fullerenes (C_{60} and C_{70}) by combustion of hydrocarbons in a flat flame burner. *Mater. Res. Soc. Symp. Proc.* **270** 149–54

[7] Kozlov B N, Kirillov S N and Mamyrin B A 1997 Formation of fullerenes in laser ablation of graphite in a vacuum *SPIE* **3093** 233–8

[8] Afanas'ev D V, Baranov G A, Belyaev A A, Dyuzhev G A and Zinchenko A K 2001 Fullerenes obtained during graphite evaporation with stationary CO_2 laser radiation *Tech. Phys. Lett.* **27** 408–10

[9] Laplaze D, Bernier P, Journet C, Sauvajol J L, Bormann D, Flamant G and Lebrun M 1997 The Use of Solar Energy for the Production of Fullerenes and Porous Silicon *J. Phys. III* **7** 463–72

[10] Saïdane K, Razafinimanana M, Lange H, Huczko A, Baltas M, Gleizes A and Meunier J-L 2004 Fullerene synthesis in the graphite electrode arc process: local plasma characteristics and correlation with yield *J. Phys. D. Appl. Phys.* **37** 232–9

[11] Kyesmen P I, Onoja A D and Amah A N Fullerenes Synthesis Using Fabricated Arc Discharge System with Relatively Large Chamber Size *IOSR J. Appl. Phys.* **7** 77–83

[12] Yamakoshi Y, Sueyoshi S and Miyata N 1999 Biological activity of photoexcited fullerene *Bull. Natl. Inst. Heal. Sci.* **117** 50–60

[13] Bakry R, M Vallant R, Najam-ul-Haq M, Rainer M, Szabo Z, W Huck C and K Bonn G 2007 Medicinal applications of fullerenes *Int. J. Nanomedicine* **2** 639–49

[14] Nakamura S and Mashino T 2009 Biological activities of water-soluble fullerene derivatives *J. Phys. Conf. Ser.* **159** 12003

[15] Ros T Da, Spalluto G and Prato M 2001 Biological Applications of Fullerene Derivatives: A Brief Overview *Croat. J. Educ.* **74** 743–55

[16] Innocenti A, Durdagi S, Doostdar N, Amanda Strom T, Barron A R and Supuran C T 2010 Nanoscale enzyme inhibitors: Fullerenes inhibit carbonic anhydrase by occluding the active site entrance *Bioorganic Med. Chem.* **18** 2822–8

[17] Kumar A, Rao M V. and Menon S K 2009 Photoinduced DNA cleavage by fullerene-lysine conjugate *Tetrahedron Lett.* **50** 6526–30

[18] Shoji M, Takahashi E, Hatakeyama D, Iwai Y, Morita Y, Shirayama R, Echigo N, Kido H, Nakamura S, Mashino T, Okutani T and Kuzuhara T 2013 Anti-Influenza Activity of C_{60} Fullerene Derivatives *PLoS One* **8** e66337

[19] Liao F, Saitoh Y and Miwa N 2011 Anticancer Effects of Fullerene [C-60] Included in Polyethylene Glycol Combined With Visible Light Irradiation Through ROS Generation and DNA Fragmentation on Fibrosarcoma Cells With Scarce Cytotoxicity to Normal Fibroblasts *Oncol. Res.* **19** 203–16

[20] Prylutska S, Grynyuk I, Matyshevska O, Prylutskyy Y, Evstigneev M, Scharff P and Ritter U 2014 C_{60} Fullerene as Synergistic Agent in Tumor-Inhibitory Doxorubicin Treatment *Drugs R D* **14** 333–40

[21] Prylutska S, Panchuk R, Gołuński G, Skivka L, Prylutskyy Y, Hurmach V, Skorohyd N,

Borowik A, Woziwodzka A, Piosik J, Kyzyma O, Garamus V, Bulavin L, Evstigneev M, Buchelnikov A, Stoika R, Berger W, Ritter U and Scharff P 2017 C_{60} fullerene enhances cisplatin anticancer activity and overcomes tumor cell drug resistance *Nano Res.* **10** 652–71

[22] Fan J, Fang G, Zeng F, Wang X and Wu S 2013 Water-dispersible fullerene aggregates as a targeted anticancer prodrug with both chemo- and photodynamic therapeutic actions *Small* **9** 613–21

[23] Popov M, Kulnitskiy B and Blank V 2014 Superhard Materials Based on Fullerenes and Nanotubes *Compr. Hard Mater.* **3** 515–38

[24] Ivanovskaya V V. and Ivanovskii A L 2010 Simulation of novel superhard carbon materials based on fullerenes and nanotubes *J. Superhard Mater.* **32** 67–87

[25] Blank V D, Buga S G, Dubitsky G A, Serebryanaya N R, Prokhorov V M, Mavrin B N, Denisov V N, Chernozatonskii L A, Berezina S and Levin V M 2006 Synthesis of superhard and ultrahard materials by 3D-polymerization of C_{60}, C_{70} fullerenes under high pressure (15 GPa) and temperatures up to 1820 K *Zeitschrift fur Naturforsch. - Sect. B J. Chem. Sci.* **61** 1547–54

[26] Vul' A Y, Davidenko V M, Kidalov S V, Ordan'yan S S and Yashin V A 2001 Fullerenes catalyze the graphite-diamond phase transition *Tech. Phys. Lett.* **27** 384–6

[27] Gruen D M, Liu S, Krauss A R, Luo J and Pan X 1994 Fullerenes as precursors for diamond film growth without hydrogen or oxygen additions *Appl. Phys. Lett.* **64** 1502–4

[28] Jiao Y, Pello J, Mejia A M, Shen L, Smalbrugge B, Geluk E J, Smit M and van der Tol J 2014 Fullerene-assisted electron-beam lithography for pattern improvement and loss reduction in InP membrane waveguide devices *Opt. Lett.* **39** 1645

[29] Jiang J, Dong J M, Wan X G and Xing D Y 1998 A new kind of nonlinear optical material: the fullerene tube *J. Phys. B-Atomic Mol. Opt. Phys.* **31** 3079–86

[30] Jeong Y J, Yun D-J, Jang J, Park S, An T K, Kim L H, Kim S H and Park C E 2015 Solution-processed n-type fullerene field-effect transistors prepared using CVD-grown graphene electrodes: improving performance with thermal annealing *Phys. Chem. Chem. Phys.* **17** 6635–43

[31] Koehler M, Roman L S, Inganäs O and Da Luz M G E 2004 Modeling bilayer polymer/fullerene photovoltaic devices *J. Appl. Phys.* **96** 40–3

[32] Zhang F, Inganäs O, Zhou Y and Vandewal K 2016 Development of polymer-fullerene solar cells *Natl. Sci. Rev.* **3** 222–39

[33] Thompson B C and Fréchet J M J 2008 Polymer-fullerene composite solar cells *Angew. Chemie - Int. Ed.* **47** 58–77

[34] Mohajeri A and Omidvar A 2015 Fullerene-based materials for solar cell applications: Design of novel acceptors for efficient polymer solar cells-a DFT study *Phys. Chem. Chem. Phys.* **17** 22367–76

[35] Dinadayalane T C and Leszczynski J 2012 *Handbook of Computational Chemistry* (Jackson: Jackson State University) pp 793–868

[36] Pérez-Garrido A, Alhama F and Catalá J D 2002 Electronic structure of fullerenes with defects *Chem. Phys.* **278** 71–8

[37] Wang C and Gao Y 2014 Electronic structure evolution in doping of fullerene (C_{60}) by molybdenum trioxide *Appl. Phys. Lett.* **105** 111601

[38] Rad A S, Aghaei S M, Aali E and Peyravi M 2017 Study on the electronic structure of Cr- and Ni-doped fullerenes upon adsorption of adenine: A comprehensive DFT calculation *Diam. Relat. Mater.* **77** 116–21

[39] Ren H J, Cui C X, Li X J and Liu Y J 2017 A DFT study of the hydrogen storage potentials and properties of Na- and Li-doped fullerenes *Int. J. Hydrogen Energy* **42** 312–21

[40] Cheng Y H, Zhang C Y, Ren J and Tong K Y 2016 Hydrogen storage in Li-doped fullerene-intercalated hexagonal boron nitrogen layers *Front. Phys.* **11** 113101

[41] Guo J-H, Zhang H and Miyamoto Y 2013 New Li-doped fullerene-intercalated phthalocyanine

covalent organic frameworks designed for hydrogen storage *Phys. Chem. Chem. Phys.* **15** 8199

[42] Gao F, Zhao G L, Yang S and Spivey J J 2013 Nitrogen-doped fullerene as a potential catalyst for hydrogen fuel cells *J. Am. Chem. Soc.* **135** 3315–8

[43] Noh S, Kwon C, Hwang J, Ohsaka T, Kim B, Kim T Y, Yoon Y, Chen Z, Seo M H, Han B, Hyo Noh S, Kwon C, Hwang J, Ohsaka T, Kim B-J, Kim T-Y, Yoon Y-G, Chen Z, Ho Seo M and Han B 2017 Self-assembled nitrogen doped fullerenes and their catalysis for fuel cell and rechargeable metal-air battery applications *Nanoscale* **9** 7373-9

[44] Schneider T, Gärtner S, Ebenhoch B, Behrends J and Colsmann A 2016 Enhanced electron extraction from solution processed organic solar cells employing rhodamine B doped fullerene layers *Synth. Met.* **221** 201–5

[45] Jia Y F, Zhu Q and Thomas K M 2002 Synthesis of nitrogen-doped fullerenes by vaporization of graphite and characteristization using NEXAFS and temperature programmed combustion *Chem. Phys. Lett.* **364** 171–5

[46] Talyzin A V., Tsybin Y O, Purcell J M, Schaub T M, Shulga Y M, Noréus D, Sato T, Dzwilewski A, Sundqvist B and Marshall A G 2006 Reaction of hydrogen gas with C_{60} at elevated pressure and temperature: Hydrogenation and cage fragmentation *J. Phys. Chem. A* **110** 8528–34

[47] Luzan S M, Tsybin Y O and Talyzin A V. 2011 Reaction of C_{60} with hydrogen gas: In situ monitoring and pathways *J. Phys. Chem. C* **115** 11484–92

[48] Zhu J, Ji Z, Wang J, Sun R, Zhang X, Gao Y, Sun H, Liu Y, Wang Z, Li A, Ma J, Wang T, Jia G and Gu Y 2008 Tumor-inhibitory effect and immunomodulatory activity of fullerol $C_{60}(OH)_x$ *Small* **4** 1168–75

[49] Cai X, Hao J, Zhang X, Yu B, Ren J, Luo C, Li Q, Huang Q, Shi X, Li W and Liu J 2010 The polyhydroxylated fullerene derivative $C_{60}(OH)_{24}$ protects mice from ionizing-radiation-induced immune and mitochondrial dysfunction *Toxicol. Appl. Pharmacol.* **243** 27–34

[50] Rokitskaya T I and Antonenko Y N 2016 Fullerenol $C_{60}(OH)_{24}$ increases ion permeability of lipid membranes in a pH-dependent manner *Biochim. Biophys. Acta - Biomembr.* **1858** 1165–74

[51] Ye S, Zhou T, Cheng K, Chen M, Wang Y, Jiang Y and Yang P 2015 Carboxylic Acid Fullerene (C_{60}) Derivatives Attenuated Neuroinflammatory Responses by Modulating Mitochondrial Dynamics *Nanoscale Res. Lett.* **10** 246

[52] Katin K P and Maslov M M 2011 Thermal stability of nitro derivatives of hydrocarbon cubane *Russ. J. Phys. Chem. B* **5** 770–9

[53] Maslov M M, Podlivaev A I and Katin K P 2016 Nonorthogonal tight-binding model with H-C-N-O parameterisation *Mol. Simul.* **42** 305–11

[54] Katin K P and Podlivaev A I 2010 Dynamic Characteristics of the Low-Temperature Decomposition of the C_{20} Fullerene *Phys. Solid State* **52** 436-8

[55] Podlivaev A I, Katin K P, Lobanov D A and Openov L A 2011 Specific features of the stone-wales transformation in the C_{20} and C_{36} fullerenes *Phys. Solid State* **53** 215–20

[56] Openov L A, Podlivaev A I and Maslov M M 2012 Comparative kinetic stability of classical and non-classical fullerenes C_{46} *Phys. Lett. A* **376** 3146–9

[57] Katin K P and Maslov M M 2017 Toward CL-20 crystalline covalent solids: On the dependence of energy and electronic properties on the effective size of CL-20 chains *J. Phys. Chem. Solids* **108** 82–7

[58] Degtyarenko N N, Katin K P and Maslov M M 2014 Simulation of metastable CL-20 cluster structures *Phys. Solid State* **56** 1467–71

[59] Goler S, Coletti C, Tozzini V, Piazza V, Mashoff T, Beltram F, Pellegrini V and Heun S 2013 Influence of Graphene Curvature on Hydrogen Adsorption: Toward Hydrogen Storage Devices *J. Phys. Chem. C* **117** 11506–13

An analytic modelling of charge transport in thin layers of disordered organic materials

M D Khan, V R Nikitenko and Ya V Burdakov
National Research Nuclear University MEPhI (Moscow Engineering Physics Institute), Moscow, Russian Federation

E-mail: vladronik@yandex.ru

Abstract A simple analytic model for the description of transient current of charge carriers in thin layers of disordered organic materials with Gaussian energy distribution of hopping centers is proposed. Validity of this approximation is supported by the comparison with the data of Monte-Carlo modelling of the drift mobility dependence on the film thickness.

1. Introduction

Disordered organic semiconductors are the subject of intensive investigations due to applications in light-emitting diodes [1], photovoltaic devices [2] and other basic elements of electronics. Charge carrier transport, being characterized by mobility in general, is one of the basic processes, determining parameters of electronic devices. Time of flight method, which is the well-known experimental technique of mobility measurements [3,4], meets great difficulties, being applied to the films of thicknesses less than 100 nm (the typical values for the active layers in organic electronic devices), because the penetration length of the light exceeds this value. However, the method of charge carrier's generation by the use of a very thin (about 10 nm) dye generation layer has been proposed [4]. Meanwhile, the drift mobility in these thin films, can differs considerably from the "true" mobility, which measures in layers thicker than 1 mcm, due to variety of reasons [5].
Results of various non-stationary experimental methods show the decrease of the drift mobility along with the increase of the film thickness. The numerous reasons can be, for example, the following: 1) the morphology of the film can be thickness-dependent due to special alignment of organic molecules in the vicinity of electrode; 2) energetic relaxation of charge carriers, in the case if initial energetic distribution is far from equilibrium, so that equilibration occurs in course of transport, and the equilibration time exceeds the time of flight. At last, a carrier has not enough time to be captured in deep states in the course of the transit, if the layer is rather thin and the disorder is rather large [5]. This circumstance can also result in the decreasing mobility along with the increasing thickness, and the drift mobility exceeds the value, predicted by the theory of quasi-equilibrium transport, even if the initial energy distribution of charge carriers is quasi-equilibrium. In this work we perform calculations of transient current and drift mobility in time-of-flight conditions, by the use of various analytic approximations, derived previously. Proceeding from the notion that the carrier may not meet sufficiently deep traps during the time of flight, a new simplified analytic approximation is proposed.

2. Theoretical model

Charge carriers assumed to be generated as a thin layer next to one of electrodes. Their initial energy distribution is far from equilibrium, in accord with typical conditions of time-of-flight experiment. The main idea of the analytic model is the following: the states of rather deep energy, $E < E_*$, can be occupied only with small probability during the transit of a carrier through the layer. The energy E_* depends only on the film thickness, L, in the simplest approach, which was realized in the work [5]. In this work, the "cut-off energy", $E_*(t)$, is determined at a given time, t, from the condition, that the capture by the states with energies less, than E_*, is probable for the time not earlier than t. According to the formalism of the multiple-trapping model [6], which applies for the description of hopping transport on the base of transport level concept [7-9], the energy E_* can be defined by the condition

$$\tau\left(E_*(t)\right)=t, \tag{1}$$

$$\tau(E)=\tau_0\left[\int_{-\infty}^{E} dE\, g(E,t)/N_t\right]^{-1}, \tag{2}$$

$\tau(E)$ is the time before the carrier's capture on the states of the energy deeper than E, τ_0 is the characteristic capture time, N_t is the concentration of localized hopping states. Energy distribution of these states is described by the Gaussian function with the variance σ :

$$g(E)=\frac{N_t}{\sqrt{2\pi\sigma^2}}\exp\left(-\frac{E^2}{2\sigma^2}\right), \tag{3}$$

energies of the neighbor states are not correlated, according with the Gaussian disorder model [3,10]. By the use of eqs. (1)-(3), we obtain the equation for the energy $E_*(t)$ for the case of Gaussian energy distribution:

$$\frac{2\tau_0}{t}=Erfc\left(\frac{|E_*|}{\sqrt{2}\sigma}\right). \tag{4}$$

Approximate solution of eq. (4) is the following:

$$E_*(t)=-\sqrt{1,8\sigma ln\left(\frac{kT}{1,2\sigma}\frac{t}{\tau_0}\right)}, \tag{5}$$

providing values of σ / kT from 3 to 4. The numerical coefficient before σ in (5) differs slightly from 1.8 outside this interval.

One can found the concentration of charge carriers from the following differential equation:

$$\frac{\partial p(x,t)}{\partial t}+\mu(t)F_0\frac{\partial p(x,t)}{\partial x}-D(t)\frac{\partial^2 p(x,t)}{\partial x^2}=-\lambda(t)\left(p(x,t)-p(x,0)\right), \tag{6}$$

which describes non-equilibrium transport [9]. The demarcation energy $E_d(t)=E_c-kT\ln(\nu_0 t)$, where E_c is the effective transport level, ν_0 is a frequency factor, separates “currently deep” and “currently shallow” states [6,9]. Occupation probability of “currently deep” states is far from equilibrium up to the time t, as well as initially, while quasi-equilibrium occupation of “currently shallow” states has been established. However, the “deep” states are practically not populated, if $E_*(t)>E_d(t)$, while the populated states with energies $E>E_*(t)$ are in quasi-equilibrium, hence one can neglect the right-hand side of eq. (6), which describes the delay of carriers in “deep” states, and replace the bottom limit of integration from $E_d(t)$ to $E_*(t)$, in all the functions [9], presented below:

$$\mu(t)=\mu_0\,\theta_*(t)/\left[1+\theta_*(t)\right], \tag{7}$$

$$\theta_*(t)=N_0\tau_0\nu_0\left[\int_{E_*(t)}^{E_c} dEg(E)\exp\left(\frac{E_c-E}{kT}\right)\right]^{-1}, \tag{8}$$

$$D(t)=D_T(t)+D_F(t)=\mu(t)\frac{kT}{e}+\mu^2(t)F_0^2\left\langle t_{rel}^*(t)\right\rangle, \tag{9}$$

$$\left\langle t_{rel}^{*}(t)\right\rangle = \frac{\int\limits_{E_*(t)}^{E_c} dE g(E)\exp\left(-\frac{E}{kT}\right)\exp\left(\frac{E_c - E}{kT}\right)}{\nu_0 \int\limits_{E_*(t)}^{E_c} dE' g(E')\exp\left(-\frac{E'}{kT}\right)}. \tag{10}$$

Typically, $\theta_*(t) \ll 1$ for $t \gg \tau_0$ in eq. (7). The first term in eq. (9) is a conventional diffusion, while the second is the so-called field-stimulated diffusion [9]. The solution of eq. (6) is a Gaussian function, if one can neglect the influence of electrodes,

$$p_*(x,t) = \frac{\sigma_0}{\sqrt{4\pi\tilde{D}(t)}}\exp\left(-\frac{\left(x - F_0\mathrm{M}(t)\right)^2}{4\tilde{D}(t)}\right), \tag{11}$$

where $\tilde{D}(t) = \int\limits_0^t dt' D(t')$, $\mathrm{M}(t) = \int\limits_0^t dt' \mu(t')$.

Transient current density in the layer of the thickness L is the following:

$$j_*(t) = \frac{e\mu(t)F_0}{L}\int\limits_0^L dx p_*(x,t). \tag{12}$$

By the use of eqs. (11) ,(12), one obtains the following expression for the current for the case of the time-of flight conditions:

$$j_*(t) = \frac{e\mu(t)F_0\sigma_0}{2L}\left(1 + Erf\left[\frac{L - F_0M(t)}{2\sqrt{\tilde{D}(t)}}\right]\right). \tag{13}$$

3. Results and discussion

Figure 1 shows the transient currents, calculated under the approximation, described above (the solid line), see the eq. (13), and under various analytic approximations, described previously [9]: the quasi-equilibrium regime (dot-dashed lines), the extremely non-equilibrium, or dispersive transport (short dashes), non-equilibrium transport regime (dashed line). The latter approximation if more general and accurate, than the dispersive regime, which is applicable only in the limiting case of short times and large disorder [6]. One has to note, that the condition $E_*(t) > E_d(t)$ is fulfilled for the case $\sigma = 3kT$, see the fig. 1a, while this condition is violated for the case $\sigma = 4kT$. Exactly, the new approximation (13) is in a good agreement with the results of the non-equilibrium transport approximation in a broad interval around the time of flight, while the approximation of dispersive transport yields the greatly underestimated time of flight. Vice versa, the dispersive transport approximation agrees much better with the non-equilibrium transport, while the new approximation, for the case $\sigma = 4kT$, see the figure 1b. The major difference is the overestimation of the field-stimulation diffusion, because $E_*(t) < E_d(t)$, see the eq. (9).

Figure 2 shows the dependence of the normalized drift mobility on the field thickness for the different approximations and comparison with results of Monte-Carlo (MC) modeling [11], see the open symbols in the fig. 2. In the course of MC calculations, a carrier had a possibility for random walks in the both sides from the boundaries of the layer (the virtual layer), that is in the better agreement with eqs. (12), (13), because the diffusion current is not included explicitly in these equations. The drift mobility is calculated as follows:

$$\mu = L/(t_{tr}F_0). \tag{14}$$

Times of flight are calculated from the equality of the mean position of carriers to the film thickness, [6],

$$\langle x\rangle(t_{tr}) = \int\limits_0^\infty dx x p_*(x,t_{tr}) = L. \tag{15}$$

IOP Conf. Series: Journal of Physics: Conf. Series **938** (2017) 012069 doi:10.1088/1742-6596/938/1/012069

The figure 2 confirms, that the results of the model (13) are in good agreement with the results of the non-equilibrium transport model.

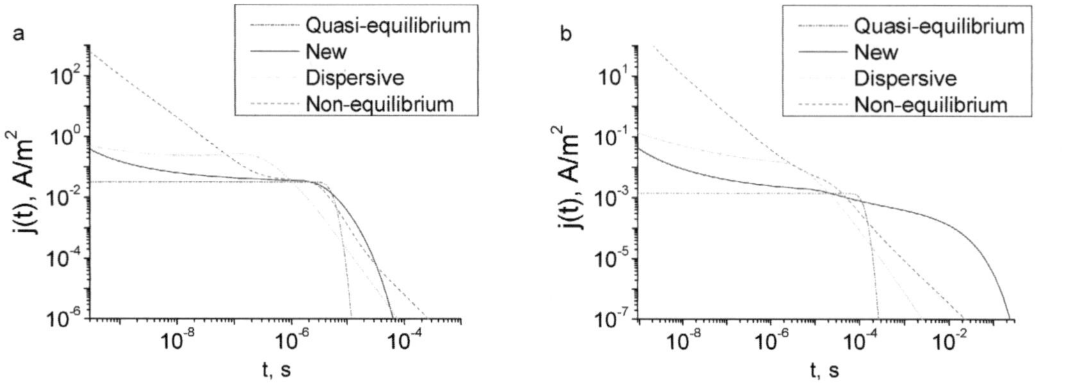

Figure 1. Transient current in the various regimes in the double logarithmic scale: a) $\sigma = 3kT$, b) $\sigma = 4kT$ .Other parameters: $\tau_0 = 1.806\times10^{-11}$ s , $\nu_0 = 10^{14}$ s^{-1}, $L = 100$ nm , $F_0 = 10^7$ V/m , $\sigma_0 = 10^{12}$ m^{-2}, $T = 300$ K, $\mu_0 = (e/kT)\nu_0 \exp(-2\gamma a_0) a_0^2$, $2\gamma a_0 = 10$, $a_0 = 1$ nm is the mean intermolecular distance.

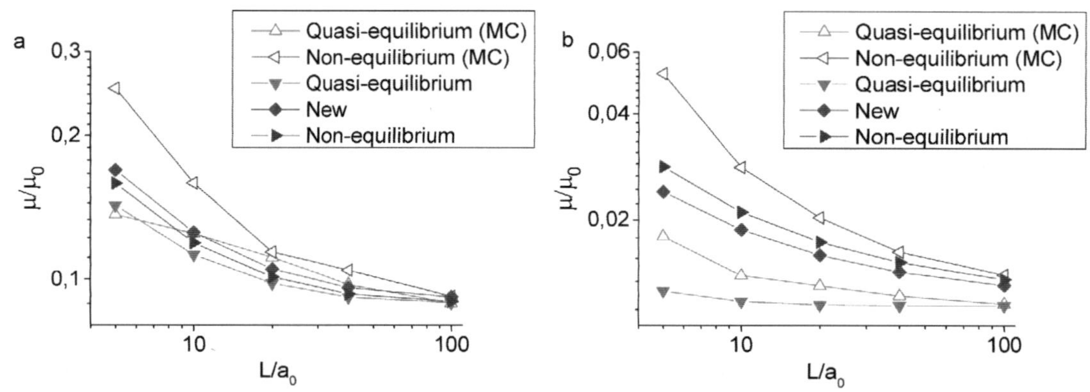

Figure 2. The normalized drift mobility, dependent on the film thickness, a) $\sigma = 2.12kT$, $F_0 = 2\times10^6$ V/m; б) $\sigma = 3kT$, $F_0 = 10^7$ V/m . Values of the other parameters are the same, as in the figure 1.

The condition $E_*(t) > E_d(t)$ is fulfilled in the both cases, shown in the figure 2. Obviously, the drift mobility decreases considerably along with the increase of thickness, because the function $\mu(t)$ in the eq. (6) decreases with time due to involvement of more and more deep states to the transport. The figure 2 include the MC- data for the 2 cases: non-equilibrium and quasi-equilibrium (in energy) initial energy distribution of carriers (the upper and the down curve, respectively). For the first case, the decrease occurs more sharply. But the decrease is considerable even in the latter case, although it seems that the drift mobility should be energy independent, because transport is quasi-equilibrium. Due to weakness of disorder, the most probable reason of the decreasing mobility (and the reason of difference of MC- and analytic results) in this case seems to be the contribution of the diffusion current, which was not accounted for explicitly in eq. (12), as in the work [9]. The contribution of diffusion current, which is typically neglected in the time of flight experiments, should be considerable in this case of so small thickness and relatively small field strength. In addition, the discreteness of the media can manifest itself, if the thickness is comparable to 10 nm. In the opposite case of large thickness, the drift mobility approaches to the quasi-equilibrium value, which was obtained previously in the multiple trapping approach [9, 12], $\mu = \xi \mu_0 \omega_0 \tau_0 \exp\left(-E_c/kT - 1/2(\sigma/kT)^2\right)$, $\omega_0 = \nu_0 \exp(-2\gamma a_0)$, $\tau_0 \omega_0 = kT/\sigma$ [12], $\xi \approx 1/4$. These results are in a good agreement also with the well-known expression, $\mu = \mu_0 \exp\left(-C(\sigma/kT)^2\right)$, $C \approx 0.44$ [10].

4. Conclusion

An analytic solution for the transient current is obtained, which simplifies calculations sufficiently. This circumstance is important for the more complicated tasks, than the modelling of time-of-flight experiment. The quantitative accuracy can be improved for the case of extremely thin layers (about 10 nm), incorporating rigorously the diffusion current together with the boundary conditions (remember, that the simple solution (11) refers to the case of an infinite medium.

Acknowledgement. This work was supported by the "Improving of the competitiveness" program of the National Research Nuclear University MEPhI (Moscow Engineering Physics Institute).

References

[1] Friend R H, Gymer R W, Holmes A B, Burroughes J H, Marks R N, Taliani C, Bradley D D C, DosSantos D A, Bredas J L, Loglund M and Salaneck W R 1999 *Nature (London)* **397** p 121
[2] Mozer A J and Sariciftci N S 2006 *C.R.Chimie* pp 568-577
[3] Borsenberger P M and Weiss D S Organic photoreceptors for xerography 1998 *Boca Raton: CRC Press* 799 pp
[4] Lupton J M, Samuel I D W, Beavington R, Frampton M J, Burn P L and Bässler H 2001 *Phys. Rev. B* **63** 155206
[5] Nikitenko V R, Sannikova N A, Sukharev V M and Strikhanov M N 2015 *Physics Procedia* **72** pp 444–449
[6] Arkhipov V I and Rudenko A I 1982 *Phil. Mag. B* **45** 2 pp 189-207
[7] Baranovskii S D, Faber T, Hensel F and Thomas P 1997 *J. Phys.: Condens. Matter* **9**, 2699
[8] Arkhipov V I, Emelianova E V and Adriaenssens G I 2001 *Phys. Rev. B* **64** 125125
[9] Nikitenko V R and von Seggern H 2007 *J. Appl. Phys.* **102** 103708 9 pp
[10] Bässler H 1993 *Phys. Status Solidi B* **175** 15
[11] Burdakov Ya V and Nikitenko V R 2017 J. of Physics: Conference Series **938** 1067
[12] Nikitenko V R and Strikhanov M N 2014 *J. Appl. Phys.* **115** 7 073704 9 pp

IOP Conf. Series: Journal of Physics: Conf. Series **938** (2017) 012070 doi:10.1088/1742-6596/938/1/012070

Thermal Stability of Heptaprismane $C_{14}H_{14}$

Nikolai V. Novikov and Ilya Y. Dolinskiy
National Research Nuclear University "MEPhI"

E-mail: NickNovikovvv@mail.ru

Abstract. Heptaprismane has been numerically simulated by the method of the nonorthogonal tight-binding model in this paper. Thermokinetic parameters of its decomposition (activation energy and the frequency factor) have been determined. The temperature dependence of the heptaprismane lifetime in the temperature range $T = 300-1000\text{K}$ has been investigated. Obtained data have been compared well with the earlier reported results concerning the "smaller" polyprismanes.

1. Introduction

Quantum chemical calculations are the powerful tools for design and investigation of new hypothetical molecules and nanostructures. So, new class of carbon molecules with a half-flat configuration of carbon atoms was recently designed [1]. These molecular systems are refer to as the carbon [n, m]prismanes or polyprismanes, where *m* is the number of vertices of a closed carbon ring, and *n* is the number of layers of dehydrogenated cycloalkane molecules [2]. There are problems with synthesis of the rings because of the technical difficulties arising at construction of them [2]. Therefore only some types of "small" prismanes, such as [2, 3]prismane, [2, 4]prismane, cubane [3], and [2, 5]prismane [4] have been obtained. Electronic properties and structural characteristics of the [n, m]prismanes have been actively studied in the density functional theory for the last time [5,6,7]. Their electronic characteristics were studied in detail for possible using of polyprismanes in nanoelectronics as a nanoconductors [8]. The mechanical properties of polyprismanes were actively studied as well. For example there was investigated the influence of edge doping with different atoms and functional groups on mechanical characteristics of the samples [9,10]. Existence of negative Poisson's ratio in these molecular systems has been also established [11]. Nevertheless, despite the great interest to polyprismanes, the problem of obtaining even "small" prismanes remains unsolved. A detailed study of their thermal stability is able to solve this problem. To date, the the following structures were already investigated: cubane C_8H_8 [12], hexaprismane $C_{12}H_{12}$ [13] and octaprismane $C_{16}H_{16}$ [13]. However, it is necessary to investigate structures consisting of an odd number of carbon atoms for a more detailed study of this problem. In that context, one of the most interesting system for consideration is the polyprismane $C_{14}H_{14}$ (Fig. 1) that occupies an intermediate position between $C_{12}H_{12}$ and $C_{16}H_{16}$. In this paper, we simulate the dynamics of heptaprismane $C_{14}H_{14}$ thermal decomposition in a wide temperature range. From obtained data we can determinate the thermal stability of a structure and to estimate the lifetime of the sample at different temperatures. Activation energy E_a and frequency factor A are also derived.

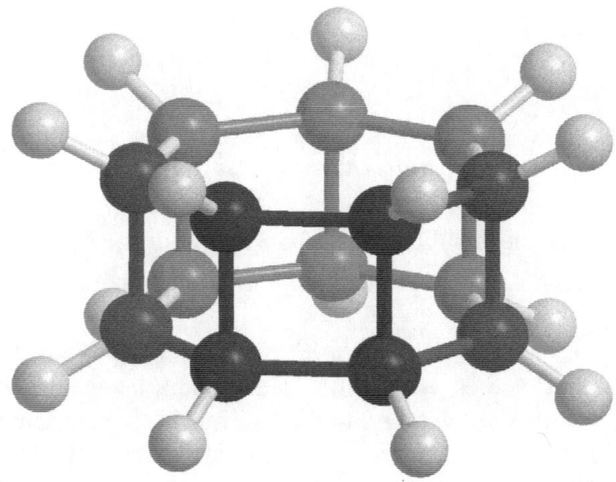

Fig. 1. Heptaprismane $C_{14}H_{14}$. Dark balls are carbon atoms, and light balls are hydrogen atoms.

2. Computational details

All quantum-mechanical calculations were carried out by GAMESS software complex [14] in terms of the DFT with the B3LYP hybrid exchange-correlation potential [15,16] and basis 6-311G as well as the use nonorthogonal tight-binding model [17] with appropriate parameters set [18]. As was pointed in recent studies [12], the model given admits the partition of total potential energy on two parts: electronic orbital and ion repulsive energy. This method matches well with structural and energetic characteristics every one of small hydrocarbon molecules and macroscopic systems. The advantage over an *ab initio* methods is higher computer performance that allows an exploration the evolution of the system with ~100 atoms within 1 ns to 1 μs – it's quite enough to obtain a statistical data. Earlier the same model was successfully applied to molecular dynamic studies of similar high-strained hydrocarbons: hexa- and octaprismane [13], cubane [12], its derivatives [19,20] and tetrahedrane [21]. The equilibrium structures and decomposition products of heptaprismane were obtained by structural relaxation where an initial configuration of the molecule under the influence of internal forces passed to states corresponding to the global or local minimum of the energy. The molecular dynamics simulation was carried out to trace back the temporal evolution of the molecule at high temperatures in the frame of the tight-binding potential [18]. The values of velocity and displacement for all atoms of heptaprismane at the initial time were randomly specified so that total momentum and total angular momentum of the system were zero. Forces acting between atoms were calculated with the Hellmann–Feynman theorem. After this step, classical Newton equations were solved using velocity Verlet method (fast integration method) with a time step of 0.3 fs. Total energy of the cluster consists of potential energy of the system and the sum of kinetic energies of its atoms. During the cluster thermal evolution, total energy of the system remains constant. As an analogue of the usual temperature, a microcanonical temperature T was introduced, which is the measure of the kinetic energy of the relative atomic motion. The value of T was calculated by formula:

$$\langle E_{kin} \rangle = (1/2) k_B T (3N - 6), \tag{1}$$

where $\langle E_{kin} \rangle$ is the time-averaged kinetic energy of the cluster, k_B is the Boltzmann's constant and N = 28 is number of atoms in the heptaprismane.

The atomic coordinates of the molecule were written to analyze the intermediate configurations arising during the evolution of heptaprismane. After the calculation, the computer visualization of the process was carried out.

IOP Conf. Series: Journal of Physics: Conf. Series **938** (2017) 012070 doi:10.1088/1742-6596/938/1/012070

3. Results and discussion

The binding energies of the heptaprismane molecule, its isomers, and the decomposition products were calculated by the equation:

$$E_b = \left(nE(\mathrm{C}) + mE(\mathrm{H}) - E(\mathrm{C}_{14}\mathrm{H}_{14})\right) / (n+m), \tag{2}$$

where $n=14$ and $m=14$ are the numbers of carbon and hydrogen atoms, respectively, $E(\mathrm{C})$ is the energy of isolated carbon atoms, $E(\mathrm{H})$ is the energy of isolated hydrogen atoms and $E(C_{14}H_{14})$ is the total energy of the heptaprismane.

The binding energies of heptaprismane in terms of the nonorthogonal tight-binding model [18] is 4.517 eV/atom. This result agrees well with data found using the density functional method (B3LYP/6311G [22]): 4.63 eV/atom. Thermal stability of heptaprismane is determined by the activation energy. Due to the fact that process of decay is random, it's correct to speak about average lifetime τ at some temperature rather than about critical temperature of decay. To determine the temperature dependence of the lifetime of the molecule, τ was calculated as a function of the initial temperature of the system in the temperature range from 300 to 1000 K. Each value of T corresponded to different sets of initial atomic velocities and displacements. The obtained dependence of natural logarithm of lifetime $\ln\tau$ on inverse temperature T^{-1} was approximated by a straight line (Fig. 2). Frequency factor A and activation energy E_a of decay process were derived from the line (see Fig. 2) using the Arrhenius equation.

$$\tau^{-1}(T) = A\exp\left(-\frac{E_a}{k_b T}\right). \tag{3}$$

For the average values and standard deviations, we obtained $E_a = 0.42 \pm 0.02$ eV and $A = 10^{14.07\pm0.15}\,s^{-1}$. It is worth noting that for the similar calculations for cubane $E_a = 1.9 \pm 0.1$ eV, $A = 10^{16.03\pm0.36}\,s^{-1}$ [12], for hexaprismane $E_a = 0.59 \pm 0.07$ eV, $A = 10^{13.8\pm0.5}\,s^{-1}$ [13] and for octaprismane $E_a = 0.24 \pm 0.04$ eV, $A = 10^{13.1\pm0.6}\,s^{-1}$ [13]. We can see from the plot of the activation energy against the number of vertices of "small" prismanes (Fig. 3) that with increasing of m, the activation energy of the hydrocarbon systems decreases monotonically. So, one can conclude that the kinetic stability of [2, m]prismanes decreases with increasing number of vertices of a closed carbon ring.

4. Conclusions

In the course of this work, we investigated the thermal stability of heptaprismane. Numerical simulation of this dependence was carried out by the nonorthogonal tight-binding model. Based on the obtained results, the temperature dependence of the lifetime of the investigated system was determined. This allowed us to find both the activation energy and the frequency factor of the cluster. Comparison with the results of earlier studies of the cubane, hexaprismane and octaprismane allowed us to make conclusions about the decrease in the thermal stability of "small" prismanes with an increase in the number of vertices of closed carbon rings. Results of this work may be used to provide some insight for new methods of synthesis of carbon heptaprismanes. We expect that obtained data will contribute to further researches of the physical and chemical properties of polyprismanes.

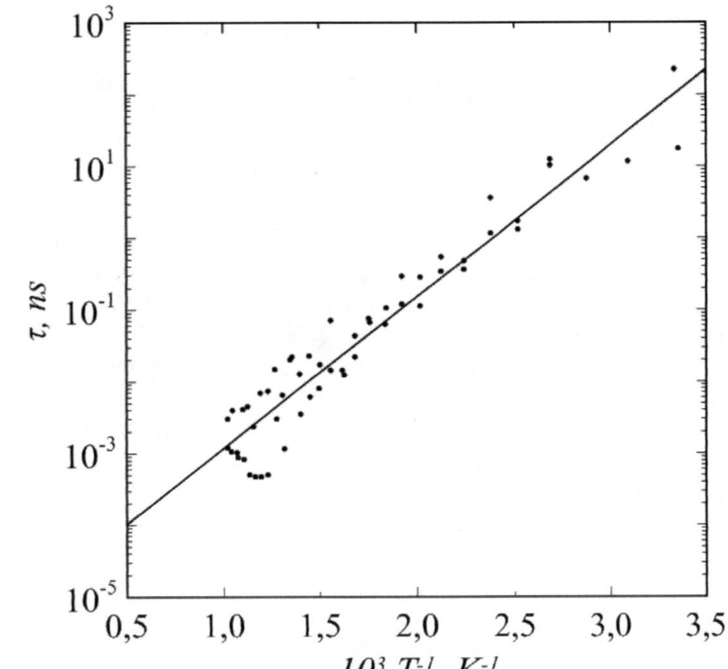

Fig. 2. Dependence of the lifetime τ of heptaprismane on the inverse initial temperature T^{-1}. Circles are calculation results, the solid line is their linear approximation.

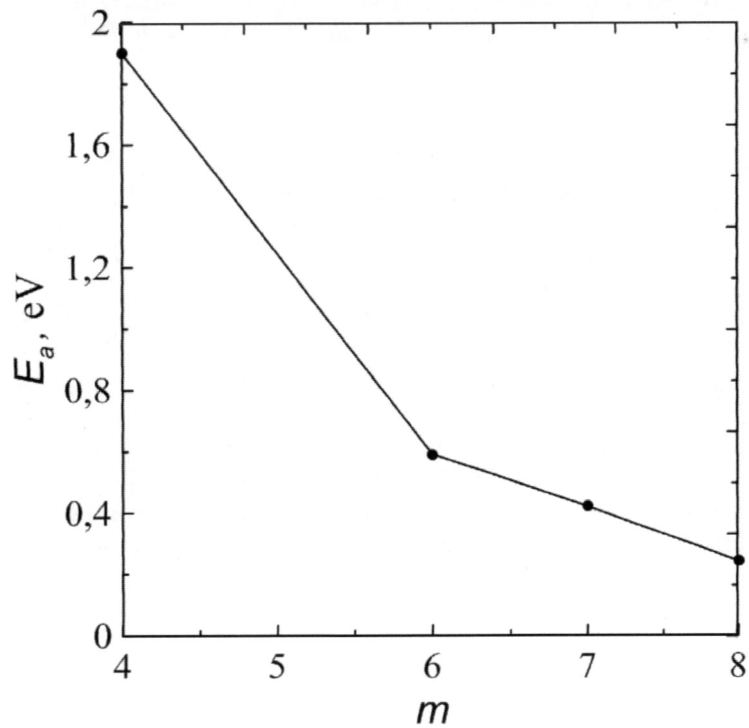

Fig.3. Dependences of the activation energy E_a on the number of vertices m of "small" prismanes. Circles represent calculated data for C_8H_8, $C_{12}H_{12}$, $C_{14}H_{14}$ and $C_{16}H_{16}$ prismanes, respectively.

References

[1] Minyaev R M, Minkin V I, Gribanova T N, Starikov A G and Hoffmann R 2003 Poly[n]prismanes: A Family of Stable Cage Structures with Half-Planar Carbon Centers *J. Org. Chem.* **68** 8588–94

[2] Lewars E G 2008 Modeling Marvels: Computational Anticipation of Novel Molecules (Dordrecht: Springer Science+Business Media) p 185

[3] Eaton P E and Cole T W 1964 The Cubane System *J. Am. Chem. Soc.* **86** 962–4

[4] Eaton P E, Or Y S and Branca S J 1981 Pentaprismane *J. Am. Chem. Soc.* **103** 2134–6

[5] Kuzmin S and Duley W W 2010 *Ab initio* calculations of a new type of tubular carbon molecule based on multi-layered cyclic C_6 structures *Phys. Lett. A* **374** 1374–8

[6] Kuzmin S and Duley W W 2012 *Ab initio* calculations of some electronic and vibrational properties of molecules based on multi-layered stacks of cyclic C_6 *Fullerenes Nanotub. Carbon Nanostructures* **20** 730–6

[7] Katin K P, Shostachenko S A, Avkhadieva A I and Maslov M M 2015 Geometry, energy, and some electronic properties of carbon polyprismanes: Ab Initio and tight-binding study *Adv. Phys. Chem.* **2015** Article ID 506894

[8] Kuzmin S L and Duley W W 2013 *Ab initio* calculation of the electronic and vibrational properties of metal-organic molecules based on cyclic C_6 intercalated with some group VIII transition metals *Ann. Phys.* **525** 297–308

[9] Pour N, Altus E, Basch H and Hoz S 2009 The Origin of the Auxetic Effect in Prismanes: Bowtie Structure and the Mechanical Properties of Biprismanes *J. Phys. Chem. C* **113** 3467–70

[10] Pour N, Altus E, Basch H and Hoz S 2010 Silicon vs Carbon in Prismanes: Reversal of a Mechanical Property by Fluorine Substitution *J. Phys. Chem. C* **114** 10386–9

[11] Pour N, Itzhaki L, Hoz B, Altus E, Basch H and Hoz S 2006 Auxetics at the Molecular Level: A Negative Poisson's Ratio in Molecular Rods *Angew. Chemie* **118** 6127–9

[12] Maslov M M, Lobanov D A, Podlivaev A I and Openov L A 2009 Thermal stability of cubane C_8H_8 *Phys. Solid State* **51** 645–8

[13] Shostachenko S A, Maslov M M, Prudkovskii V S and Katin K P 2015 Thermal stability of hexaprismane $C_{12}H_{12}$ and octaprismane $C_{16}H_{16}$ *Phys. Solid State* **57** 1023–7

[14] Schmidt M W, Baldridge K K, Boatz J A, Elbert S T, Gordon M S, Jensen J H, Koseki S, Matsunaga N, Nguyen K A, Su S, Windus T L, Dupuis M and Montgomery J A 1993 General atomic and molecular electronic structure system *J. Comput. Chem.* **14** 1347–63

[15] Lee C, Yang W and Parr R G 1988 Development of the Colle-Salvetti correlation-energy formula into a functional of the electron density *Phys. Rev. B* **37** 785–9

[16] Becke A D 1993 A new mixing of Hartree–Fock and local density-functional theories *J. Chem. Phys.* **98** 1372–7

[17] Katin K P and Maslov M M 2011 Thermal stability of nitro derivatives of hydrocarbon cubane *Russ. J. Phys. Chem. B* **5** 770–9

[18] Maslov M M, Podlivaev A I and Katin K P 2016 Nonorthogonal tight-binding model with H-C-N-O parametrization *Molecular Simulation* **42** 305–11

[19] Katin K P, Prudkovskiy V S and Maslov M M 2016 Influence of methyl functional groups on the stability of cubane carbon cage *Physica E* **81** 1–6

[20] Maslov M M, Katin K P, Avkhadieva A I and Podlivaev A I 2014 Dynamic modeling of cyclotetracubyl thermal decomposition *Russ. J. Phys. Chem. B* **8** 152–7

[21] Maslov M M and Katin K P 2011 On the thermal stability of tetrahedrane: Tight-binding molecular dynamics study *Chem. Phys.* **387** 66–8

[22] Krishnan R, Binkley J S, Seeger R and Pople J A 1980 Self-consistent molecular orbital methods. XX. A basis set for correlated wave functions *J. Chem. Phys.* **72** 650–4

IOP Conf. Series: Journal of Physics: Conf. Series **938** (2017) 012071 doi:10.1088/1742-6596/938/1/012071

Mechanical properties of fullerite of various composition

L. Kh. Rysaeva

Institute for Metals Superplasticity Problems of RAS, 450001, St. Khalturina 39, Ufa. Russia

E-mail: lesya813rys@gmail.com

Abstract. Molecular dynamics simulation is used to study the structures of fullerite of various composition as well as their mechanical properties. Fullerites based on fullerene C_{60} with simple cubic and face-centered packing, fullerene-like molecule C_{48} and fullerene C_{240} with simple cubic packing are studied. Compliance and stiffness coefficients are calculated for fullerites C_{60} and C_{48}. For fullerite C_{240}, C_{60}, and C_{48}, deformation behavior under the effect of hydrostatic compression is also investigated. It is shown that the fullerenes in the fullerite remain almost spherical up to high values of compressive strain, as a result of which the fullerite is an elastic medium up to densities of 2.5 g/cm^3. The increasing stiffness and strength under an applied compression is found for all the considered fullerites.

1. Introduction

In recent years, more and more attention has been attracted by bulk carbon nanomaterials, which demonstrate new interesting properties. For example, bulk carbon nanostructures created on the basis of graphene flakes, carbon nanotubes and fullerenes [1-4], diamond-like phases [5-16], to name a few are of considerable interest. Such properties of these materials as superconductivity, field emission, non-wetting of the surface, ultrahigh specific surface area combined with electrical conductivity seems very promising for various applications: in the fields of catalysis, separation, gas storage, electronics, electrode materials, supercapacitors, etc. [17-22]. Economical and environmentally friendly methods of obtaining such materials are being developed during last decades.

One of the interesting structures, which were studied to the present moment both experimentally and theoretically, is fullerite, a material formed by fullerenes deposited in an ordered lattice. The fullerite crystal C_{60} has a cubic structure with a face-centered cubic (FCC) lattice, which has a lattice constant of 14.2 Å and a density of 1.65±0.03 g/cm^3 can be easily produced experimentally. At room temperature, hexagonal packing of molecules is observed, although FCC stacking is preferable. At room temperature, the molecular centers form a regular FCC crystal lattice, but the molecules themselves freely rotate around their centers. In [23], where the influence of deformation on the physical properties of C_{240} fullerenes was studied, it is shown that even a small deformation leads to a shift in the electronic spectra. The bulk modulus of fullerite C_{60} increases rapidly with increasing pressure. To date, studies have been made on the elastic properties of polycrystalline fullerite samples, and measurements of Young's modulus on single-crystal samples of small dimensions [24, 25]. Earlier measurements of the elastic constants of some diamond-like structures and fullerites can be found in [26-29]. The elastic moduli $c_{11,}$ c_{12} and c_{44} of single-crystal C_{60} were determined by measurement of ultrasound velocities and found to be equal to ~15 GPa, ~9 GPa and ~6 GPa correspondingly [26]. The relaxation contribution [27] and temperature behavior [28, 29] of the elastic modulus is also discussed.

One of the interesting and important areas of research is the search for materials with unusual mechanical properties. Analysis of the elastic properties of various carbon nanostructures showed that they can have a negative Poisson's ratio [30-33] which is quite rare property [12, 14, 34-44]. To date, auxeticity was found mostly for carbon nanotubes [34-37] or graphene which is already known to have some negative parameters, for example, negative thermal expansion, potential auxeticity at various conditions, such as very high temperatures [38] or resonance under fully blocked boundary conditions [39]. Effect of different factors on auxeticity was studied for graphene [31, 40, 41]. For example, corrugation can be effectively used to tailor graphene properties [31, 42]. The lowering of Poisson's ratio was found due to an introduction of low concentrations of vacancies. In recent years, graphene kirigami has been investigated due to its specific properties under deformation [43] and unusual auxetic behavior [44]. Thus, carbon nanostructures can be considered as very promising in case of search for such auxetic structures.

The aim of this work is to consider the fullerite crystals of various compositions and study their mechanical properties by molecular dynamics simulation. Three types of fullerite with different crystal packing are considered.

2. Simulation details

The simulation cell of fullerite is created on the basis of fullerenes C_{60} and C_{240} and the fullerene-like molecule C_{48}. Initial structures are prepared with the help of home-made programs written by authors, which allow to combine different structural elements in three-dimensional structures, such as fullerite. The single structural element is shown in figure 1a. Structure based on C_{60} have SC (simple cubic) and FCC (face-centered) packing, fullerites based on the fullerene-like molecule C_{48} and fullerene C_{240} have SC packing. Examples of the structures with different crystal packing are shown in figure 1b, c.

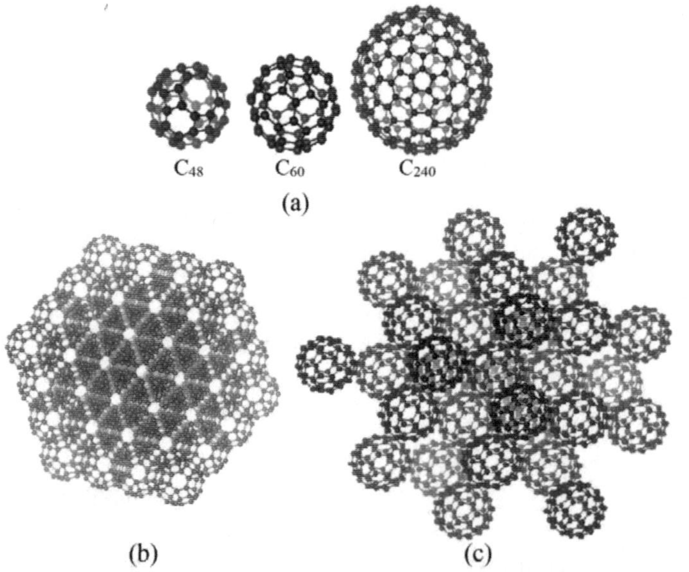

Figure 1. (a) Fullerenes C_{48}, C_{60} and C_{240} correspondingly. Fullerite based on (b) fullerene-like molecule C_{48} with SC stacking (c) fullerene C_{60} molecule with FCC stacking.

The simulation cell includes 4 structural units respectively along the *x*, *y* and *z* directions. Structural units of the material were randomly oriented in space. In view of the high symmetry of the fullerene molecules, it can be considered that this material is a crystalline body and has translational symmetry. Periodic boundary conditions along the *x*, *y* and *z* directions are applied.

The simulation is carried out using molecular dynamics simulation using the LAMMPS (Large-scale Atomic/Molecular Massively Parallel Simulator) software package [45], where the interaction of carbon atoms is described by the AIREBO interatomic interaction potential [46]. The Lennard-Jones potential, which is an integral part of the AIREBO potential, has a minimum potential energy of 0.024 eV at a distance $r = 3.4$ Å. This part of the AIREBO potential is used to describe the pairwise Van der Waals interatomic forces between the structural elements (fullerenes). This potential is based on the well-known Tertsoff potential, and have been actively used to study carbon structures and widely used to study various carbon structures and their properties [47-53]. Equations of motion of atoms were integrated according to the Verle method of the fourth order of accuracy with an integration step of 0.2 fs. All numerical experiments are carried out at 300 K. The Nose-Hoover thermostat is used for thermalisation of the system. At the initial moment, the structure relaxed until a local or global minimum of potential energy is reached.

To calculate the compliance moduli s_{ij}, a linearly increasing stress with one non-zero component is applied to the simulation cell and the corresponding strain components are calculated. It turned out that for strains less than 2%, stress-strain curves are practically linear and from the slopes of these curves the required constants s_{ij} can be calculated from Hooke's law. This technique is described in details in [12, 14].

For a cubic crystal, it is sufficient to calculate only three modulus of compliance from expression

$$\begin{pmatrix} \varepsilon_1 \\ \varepsilon_2 \\ \varepsilon_3 \\ \varepsilon_4 \\ \varepsilon_5 \\ \varepsilon_6 \end{pmatrix} = \begin{pmatrix} s_{11} & s_{12} & s_{12} & 0 & 0 & 0 \\ s_{12} & s_{11} & s_{12} & 0 & 0 & 0 \\ s_{12} & s_{12} & s_{11} & 0 & 0 & 0 \\ 0 & 0 & 0 & s_{44} & 0 & 0 \\ 0 & 0 & 0 & 0 & s_{44} & 0 \\ 0 & 0 & 0 & 0 & 0 & s_{44} \end{pmatrix} \begin{pmatrix} \sigma_1 \\ \sigma_2 \\ \sigma_3 \\ \sigma_4 \\ \sigma_5 \\ \sigma_6 \end{pmatrix}.$$

To study the deformation behavior, the simulation cell is subjected to deformation-controlled hydrostatic compression $\varepsilon_{xx} = \varepsilon_{yy} = \varepsilon_{zz} = -\varepsilon$, where $\varepsilon \geq 0$ it is a parameter that increases monotonically with the velocity $\dot{\varepsilon} = 0.02$ ps^{-1}. Test calculations with lower strain rates (four and eight times lower) did not reveal noticeable changes in the results. In the process of deformation, the hydrostatic pressure $p = (\sigma_{xx} + \sigma_{yy} + \sigma_{zz})/3$ and material density ρ are calculated. The bulk modulus is calculated as $B = \rho(\Delta p / \Delta \rho)$.

3. Results and discussion

3.1. *Fullerite C_{48} and C_{60}*

The values of the compliance and stiffness coefficients of fullerite C_{48} and C_{60}, obtained by molecular dynamics, are presented in tables 1 and 2. The results show a qualitative agreement with the previously known moduli of rigidity of fullerite C_{60}, determined experimentally in [26-29]. However, the calculated values are several times larger. This can be explained by the fact that the experimentally obtained single crystals of fullerite contain many defects and, in addition, experimental measurements are carried out at finite temperatures, while the calculations presented in this work were carried out for an ideal crystal at zero temperature.

Table 1. Compliance coefficients for three structures.

	s_{11}, TPa^{-1}	s_{44}, TPa^{-1}	s_{12}, TPa^{-1}
C_{48}	10.5	6.61	-4.68
C_{60} SC	43.3	218	-1.87
C_{60} FCC	0.052	0.0348	-0.0224

Table 2. Stiffness coefficients for three structures.

	c_{11}	c_{44}	c_{12}
C_{48}	331 MPa	151 MPa	265 MPa
C_{60} SC	23.2 MPa	4.59 MPa	1.05 MPa
C_{60} FCC	55.3 GPa	28.7 GPa	41.8 GPa

In figure 2 hydrostatic pressure as the function of density is shown for three types of fullerite. It is observed that for all the structures the pressure is zero when $\rho < 1$ g/cm^3 and increases monotonically with ρ afterwards. This is because when $\rho < 1$ g/cm^3 fullerenes are far away from each other, and only start to interact when $\rho < 1$ g/cm^3. Accordingly, the zero strain level ($\varepsilon = 0$) is set at the point when $\rho = 1$ g/cm^3 and this convention is used hereafter. As it can be seen, all fullerene crystals can be compressed until very high densities, even close to the diamond density. Since AIREBO potential cannot adequately describe bonds destruction and transition to the amorphous state, compression is stopped at the density 2.8 g/cm^3 while crystalline order still can be identified.

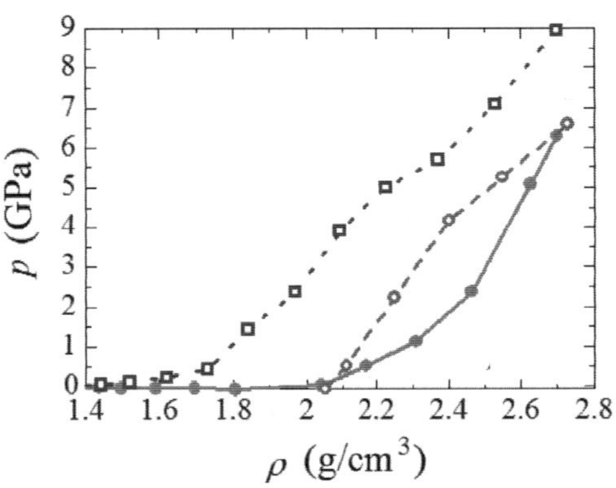

Figure 2. Hydrostatic pressure as the function of the material density at 300 K for fullerite C_{60} FCC (red solid line), fullerite C_{60} SC (blue dotted line) and fullerite C_{48} SC(green dashed line).

Two structures with SC packing behave differently from the fullerite C_{60} with FCC packing. For SC structures, pressure is much higher than for FCC structure because fullerenes in SC packing cannot reorganize in the simulation cell, while for FCC packing they have some free space for further compression. Fullerenes in FCC structures have the same round shape until very high densities, while fullerenes in SC structures slightly crumple under compression.

Elastic constants for three fullerite crystal are presented in table 3, where G_1 and G_2 are the maximum and minimum values of shear modulus. Bulk modulus is calculated as $B = 1/3 \cdot (c_{11}+2c_{12})$. Results are in good agreement with experimental work [26].

Table 3. Elastic coefficients for three structures.

	B	$G_1 = s_{44}^{-1}$	$G_2 = (2(s_{11} - s_{12}))^{-1}$
C_{48}	287 MPa	151 MPa	0.033 MPa
C_{60} SC	8.43 MPa	4.58 MPa	0.011 MPa
C_{60} FCC	46.3 GPa	28.7 GPa	6.72 GPa

3.2. Fullerite C_{240}

In figure 3a the dependence of the hydrostatic pressure on the density for $T = 300$ K is shown. From the inset in figure 3a it can be seen that there is a linear relationship between p and ε up to $\varepsilon = 0.03$. The stress-strain curve allows to estimate the bulk modulus, which is $B = 0.35$ GPa. Pressure-density curve

Loading of fullerite was followed by subsequent unloading. The corresponding curves are shown in figure 3b, where the loading curve is shown by a solid line, and the unloading curves from different compression strain are shown by dashed lines. Value ρ* corresponds to the value of the density, after which irreversible changes take place in the structure and its return to the original state is impossible. Unloading from a density higher than critical, results in a deviation of the unloading curve from the initial one. As can be seen from the figure 3b, fullerite remains almost an elastic medium till the high densities, and only at density values greater 2.5 g/sm^3 the discharge curve does not coincide with the loading curve, although the course of the curves remains the same.

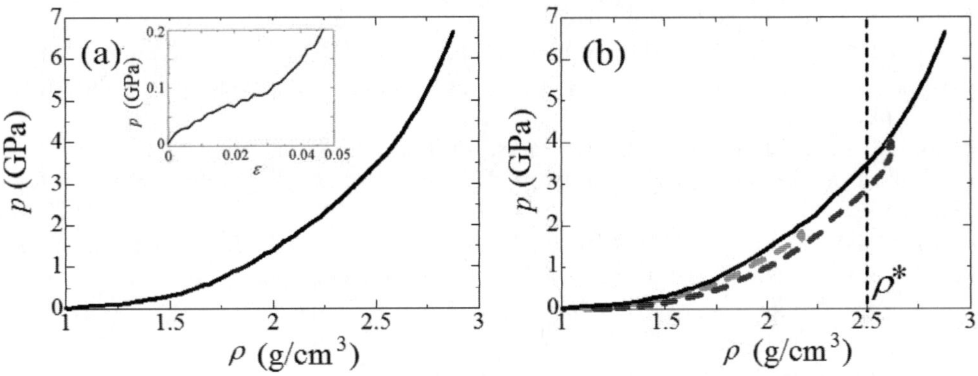

Figure 3. (a) Dependence of the hydrostatic pressure on the material density at $T = 300$ K. The inset shows the curve of the dependence of hydrostatic pressure on deformation for a small strain range. (b) Loading (solid lines) and unloading (dashed lines) curves from different strain levels.

4. Conclusion

The mechanical and structural properties of fullerite of various composition have been studied by molecular dynamics. The increasing stiffness and strength under an applied compression is found for all the considered fullerites. The estimation of the volume modulus of the structure considered gives $B = 0.35$ GPa, which is much lower than the values ~9 GPa, obtained at room temperature for fullerite C_{60} in [26] and present work. Such a difference in the values is easy to explain, taking into account the ratio of the sizes of fullerenes C_{60} and C_{240} and presence of various defects in experimental samples.

The high strength of fullerite is explained by the fact that fullerenes retain a convex shape close to spherical up to high values of strain and their collapse occurs instantaneously when the critical value of compression is reached. This also explains the fact that fullerite C_{240} is an elastic medium until the density lower than 2.5 g/cm^3 and unloading leads to an almost complete restoration of the initial structure. Fullerenes in fullerites C_{60} and C_{48} also remain stable and spherical with the well-defined crystalline order. However, FCC C_{60} sustain lower pressure in comparison with SC structures.

This study, showing a significant difference in the bulk modulus of fullerites of different dimensions and packing, poses the question of the effect of the size of a single building element of the cluster and its crystal packing. Obviously, the density of clusters consisting of different structural units will play an important role in assessing the mechanical properties and the effect of hydrostatic deformation on the behavior of the material, which, however, requires further research. Moreover, it is also interesting to study non-symmetric fullerenes, such as C_{70} and giant fullerenes as C_{960} because those fullerite crystals can also be produced experimentally.

Acknowledgments
Author acknowledges the support of the Russian Science Foundation project no14-13-00982.

References

[1] Baimova J A, Liu B, Dmitriev S V, Srikanth N, Zhou K. 2014 *Phys. Chem. Chem. Phys.* **16.** 19505
[2] Baimova J, Liu B, Dmitriev S V, Zhou K 2014 *Journal of Physics D: Applied Physics* **48** 095302
[3] Baimova J A, Liu B, Dmitriev S V, Zhou K 2014 *Phys. Stat. Solidi (rrl)* **8** 336

[4] Baimova Yu A, Murzaev R T, Dmitriev S V 2014 *Phys. Solid State* **56** 1946
[5] Katin K P, Maslov M M 2015 *Advances in condensed matter physics*. **2015** 754873
[6] Katin K P, Maslov M M 2011 *Russian journal of physical chemistry b* **5** 770
[7] Greshnyakov V A, Belenkov E A 2017 *Letters on Materials* **7** 318
[8] Belenkov E A, Greshnyakov V A 2016 *Letters on Materials* **6** 159
[9] Belenkov E A, Greshnyakov V A 2016 *Phys. Solid State* **58** 2145
[10] Belenkov E A, Greshnyakov V A 2015 *Phys. Solid State* **55** 1754
[11] Katin K P, Prudkovskiy V S, Maslov M M 2016 *Physica E* **81** 1
[12] Lisovenko D S, Baimova Yu A, Rysaeva L Kh, Gorodtsov V A, Rudskoy A I, Dmitriev S V 2016 *Phys. Stat. Solidi (b)* **253** 1295
[13] Belenkov E A, Greshnyakov V A 2016 *Phys. Solid State* **58** 2069
[14] Lisovenko D S, Baimova Yu A, Rysaeva L Kh, Gorodtsov V A, Dmitriev S V 2017 *Phys. Solid State* **59** 820
[15] Plotnikov V A, Dem'yanov B F, Yartsev V I, Solomatin C V 2017 *Letters on Materials* **7** 234
[16] Maslov M M, Katin K P 2016 *Chemical Physics Letters*. **644** 280
[17] Yang S, Kim T, Jung H, Park C, 2013 *Carbon* **53** 73
[18] Liu F, Song S, Xue D, Zhang H 2012 *Adv. Mater.* **24** 1089
[19] Meng L-Y, Park S-J 2012 *J. Colloid Interf. Sci.* **386** 285
[20] Miller J, Outlaw R, Holloway B 2010 *Science* **329** 1637
[21] Chen Y, Guo F, Jachak A, Kim S, et al. 2012 *Nano Lett.* **12** 1996
[22] Guo F, Silverberg G, Bowers S, Kim S-P, et al. 2012 *Environ. Sci. Technol.* **46** 7717
[23] Pudlak M, Pincak R, Osipov V A 2008 *Journal of Physics: Conference Series* **129** 012009.
[24] Hoen S, Chopra N G, Xiang X-D. et al 1992 *Phys. Rev. B*. **46** 737
[25] Shi X D, Kortan A R, Williams J M et al. 1992 *Phys. Rev. Lett.* **68** 827
[26] Kobelev N P, Nikolaev R K, Soifer Y M et al. 1998 *Phys. Solid State* **40** 154
[27] Kobelev N P 2002 *Phys. Solid State* **44** 195
[28] Kobelev N P, Nikolaev R K, Sidorov N S et al. 2002 *Phys. Solid State* **44** 429
[29] Kobelev N P, Nikolaev R K, Sidorov N S et al. 2001 *Phys. Solid State* **43** 2344
[30] Tretiakov K V, Wojciechowski K W 2013 *Phys. Status Solidi B* **250** 2020
[31] Grima J N, Winczewski S, Mizzi L, Grech M C, Cauchi R, Gatt R, Attard D, Wojciechowski K W, Rybicki J 2015 *Adv. Mater.* **27** 1455
[32] Goldstein R V, Gorodtsov V A, Lisovenko D S 2013 *Phys. Status Solidi B* **250** 2038
[33] Goldstein R V, Gorodtsov V A, Lisovenko D S, Volkov M A 2015 *Letters on Materials* **5** 409
[34] Coluci V R, Hall L J, Kozlov M E, Zhang M, Dantas S O, Galvao D S, Baughman R H 2008 *Phys. Rev. B* **78** 115408
[35] Hall L J, Coluci V R, Galvao D S, Kozlov M E, Zhang M, Dantas S O, Baughman R H 2008 *Science* **320** 504
[36] Jindal P, Jindal V K 2006 *J. Comput. Theor. Nanosci.* **3** 148
[37] Yao Y T, Alderson A, Alderson K L 2008 *Phys. Stat. Solidi B* **245** 2373
[38] Braghin F L, Hasselmann N 2010 *Phys. Rev. B* **82** 035407
[39] Scarpa F, Adhikari S, Phani A S 2010 *Int. J. Novel Mater.* **1** 39
[40] Dettori R, Cadelano E, Colombo L, 2012 *J. Phys.: Condens.* Matter. **24** 104020
[41] Tapia A, Peon-Escalante R, Villanueva C, Aviles F 2012*Comp. Mater. Sci.* **55** 255
[42] Qin H, Sun Y, Zhe Liu J, Lia M, Liu Y 2017 *Nanoscale* **9** 4135
[43] Blees M K, Barnard A W, Rose P A et al. 2015 *Nature* **524** 204
[44] Cai K, Luo J, Ling Y, Wan J, Qin Q 2016 *Sci. Rep.* **6** 35157
[45] http://lammps.sandia.gov/.
[46] Stuart S, Tutein A, Harrison J 2000 *J. Chem. Phys.* **11** 26472.
[47] Murzaev R T, Bachurin D V, Korznikova E A, Dmitriev S V 2017 *Phys. Lett. A* **381** 1003
[48] Baimova J A 2016 *Letters on Materials* **6** 31
[49] Dmitriev S V 2016 *Letters on Materials* **6** 86

[50] Jafary-Zadeh M, Zhang Y-W 2015 *Sci. Rep.* **5** 12848
[51] Gatchell M, Delaunay R, D'Angelo G, et al. 2017 *Phys. Chem. Chem. Phys.* **19** 19665
[52] Montazeri A, Ebrahimi S, Rafii-Tabar H 2015 *Mol. Simul.* **41** 1212
[53] Baimova J A, Murzaev R T, Rudskoy A I 2017 *Phys. Lett. A* **381** 3049

IOP Conf. Series: Journal of Physics: Conf. Series **938** (2017) 012072 doi:10.1088/1742-6596/938/1/012072

Temperature optimization of the Ti/Al/Ni/Au ohmic contact formation to the AlGaN/GaN heterostructure

S A Shostachenko*[1], Y A Porokhonko[1], R V Zakharchenko[1], M S Burdykin[1], R V Ryzhuk[1], N I Kargin[1], B V Kalinin[2], A A Belov[2] and A N Vasiliev[2]

[1]National Research Nuclear University MEPhI, Moscow, Russian Federation

[2]PLANETA-IRMIS, Saint Petersburg, Russian Federation

E-mail: Sashostachenko@mephi.ru

Abstract. This article is devoted to the experimental study of the ohmic contacts based on Ti/Al/Ni/Au metallization to the n+-doped region of the AlGaN / GaN heterostructure. Formed phases are studied at different temperatures. Based on the thermodynamic analysis of possible intermetallic reactions in the contact region, the optimum temperature range was found for the formation of ohmic contacts with a low contact resistance. The optimality of the modeled range is experimentally confirmed.

1. Introduction

Transistor devices based on III-nitride heterostructures are widely used both in the defense and civilian industries. For example, as the main component of microwave monolithic integrated circuits (microwave MIS) millimeter and submillimeter bands for communication systems, radiolocation, radio astronomy, radiometry, etc. [1], these materials, especially GaN, have also been used in the optoelectronic industry as a base for LED crystals.

In such devices ohmic contacts with a low contact resistance play an important role. To form such contacts to heterostructures based on GaN n-type conductivity, multilayer contacts based on Ti / Al.

The ohmic contact formation mechanisms on the basis of the Ti/Al metal system during thermal annealing are based on the appearance of a Ti_xN compound with a metallic type of conductivity. Ti_xN begins to form even when the titanium is deposited on the n-GaN surface and propagates in the bulk of the titanium metallization layer during thermal annealing. The formation of the Ti_xN layer makes nitrogen vacancies concentration in gallium nitride to increase near the TiN-GaN interface. These vacancies are donor levels that dramatically increase the electrical conductivity of an n-type semiconductor, which is one of the reasons for the low contact resistance [2]. In addition, Ti/Al-based metallization reduces the contact resistance by forming intermetallic compounds with a low work function on the GaN surface [3-7]. Needed intermetallic compounds are formed in the contact system during "rapid" annealing at high temperatures because of chemical reactions of the elements in the contact system.

Nitrides, intermetallides and free titanium and aluminum are present among the products of interaction. It sufficiently complicates the stability analysis of the Al/Ti/GaN contacts, which requires patterns detection of solid-phase reactions between contacting materials and changes in the phase composition of these structures in the temperature range of contact formation and their operation.

IOP Conf. Series: Journal of Physics: Conf. Series **938** (2017) 012072		doi:10.1088/1742-6596/938/1/012072

It is important to determine the optimum annealing temperature to form ohmic contacts with a low contact resistance. In this paper, this is done with help of the thermodynamic analysis. The results are confirmed by X-ray diffraction analysis of the obtained samples and experimentally obtained temperature dependence of the contact resistance.

2. Experimental method

In this work we used a substrate with an AlGaN/GaN heterostructure. Electron concentration in the upper AlGaN layer was 3×10^{18} cm^{-3}. The schematic profile of the heterostructure is shown in Fig. 1.

The surface was chemically treated in a mixture of NH4OH: H2O (DI) (1:10) before the metallization process to reduce the value of the contact resistance of the contacts [8].

Four-layer metallization of Ti/Al/Ni/Au (35/135/50/100 nm) was applied to the surface of the heterostructure by the resistive sputtering method in the Kurt J. Lesker PVD 75 [8]. The parameters of the ohmic contacts were measured by the long line method (TLM) [9]. The test structures (shown in Fig. 2) were manufactured to increase the accuracy of measuring the resistance of ohmic contacts [10]. The desired contact metallization pattern was created by inverse photolithography. Then the samples underwent rapid thermal annealing at the Modular RTP-600S unit at a given temperature for 30 seconds. Specific contact resistance was measured on an analyzer of Agilent B 1500 A semiconductor devices.

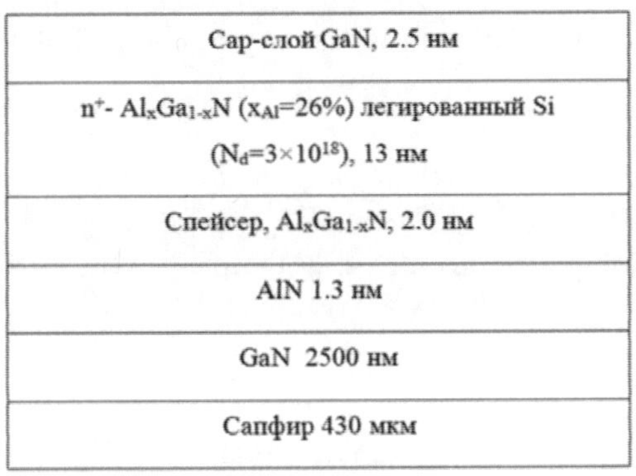

Figure 1. AlGaN/GaN heterostructure

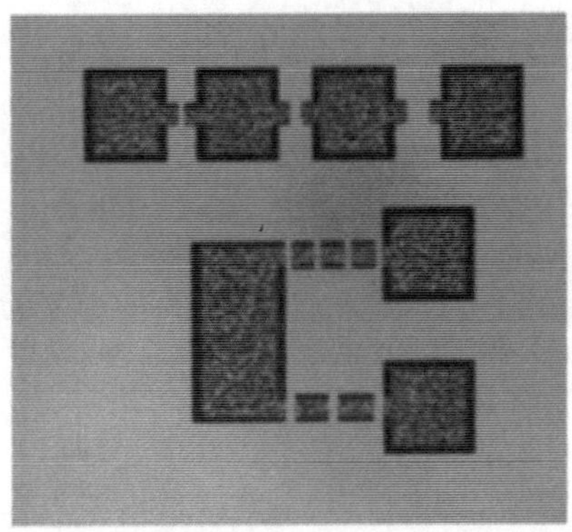

Figure 2. Test module for ohmic contacts resistance determination

An X-ray diffractometer Ultima IV was used to study the changes in the phase composition of the contact region, due to the appearance of intermetallic compounds of titanium, nitrogen and aluminum at different annealing temperatures.

3. Thermodynamic analysis of the Al-Ti-N system

Thermal treatment of Al/Ti/GaN structures is accompanied by the diffusion of nitrogen atoms, which interact with metal atoms in the temperature range 800-1000°C. Atoms form solid interstitial solutions and chemical compounds of nitride (AlN, TiN, Ti_2N), and in addition, during the interaction of titanium with aluminum compounds are formed by Al-Ti (AlTi, $AlTi_3$, Al_3Ti, Al_2Ti, Al_3Ti_2). The phase diagrams of possible binary systems from the initial three elements: aluminum, titanium and nitrogen, were studied in details in work [11].

To analyze the probability of occurrence and stability of these phases, it is necessary to take into account all possible chemical interactions of the initial three elements and for each reaction to calculate the temperature dependence of the change in the free Gibbs energy of the solid-phase reactions of formation of each compound and its temperature dependence. Phases are formed and are thermodynamically stable if the change in the free Gibbs energy (isobaric-isothermal potential) $\Delta G < 0$ in the considered temperature range for all reactions. To calculate this value, use the formula 1[11-12]:

$$\Delta G(T) = \Delta H^{O}_{298} - T\Delta S^{O}_{298} + \int_{298}^{T} \Delta c_p(T)dT - T\int_{298}^{T}\left(\frac{\Delta c_p(T)}{T}\right)dT \quad (1)$$

where $\Delta G(T)$ – change in Gibbs energy of the reaction; ΔH^{0}_{298} – change in enthalpy of reaction under standard conditions; ΔS^{0}_{298} – change in the entropy of the reaction under standard conditions; $\Delta c_p(T)$ – change in the heat capacity of the reaction, T – temperature in Kelvin.

In carrying out further calculations, reference data were used on the thermodynamic properties of elements, their nitrides and intermetallic compounds [13-15]. However, the properties of many compounds are not reflected in these sources. In this case, calculations were made using the thermodynamic properties of the compounds elements. The accuracy and limits of applicability of various calculation methods have been studied in sufficient detail in works [11-12, 16-17]. An analysis of various methods for calculating the thermodynamic properties of substances made it possible to conclude that the Neumann-Kopp method, which uses the rule of additivity of the thermodynamic properties of elements in the formation of a condensed binary compound, is the optimal method to ensure an acceptable accuracy of calculations [15]. As a result, the heat capacity of compound A_qB_r is determined by the formula:

$$C_{p,A_qB_r} = K_{NK}(qC_{p,A} + rC_{p,B}) \quad (2)$$

where K_{NK} – coefficient reflecting the deviation from the Neumann-Kopp rule.

Precisely the value of the K_{NK} coefficient can be determined using the Kelly rule, according to which, the heat capacity of the compound is determined by the expression 3 at the melting point or the first polymorphic transformation [18]:

$$C_p(T_{пл}) = K_K m \quad (3)$$

where m – number of atoms in the molecule of the compound; $K_K = 29{,}3$.

Then the expression for determining the coefficient K_{NK} can be written in the form:

$$K_{NK} = \frac{K_K(q+r)}{(qC_{p,A}(T_{пл}) + rC_{p,B}(T_{пл}))} \quad (4)$$

Using expressions (2) and (4), we can determine the temperature dependence of the specific heat of the compounds, using the reference data for the specific heat of the components and the values of their melting points and polymorphic transformations [15, 20].

It should be noted that at temperatures below 882°C, titanium has α-Ti modification with a hexagonal close-packed lattice, at temperatures above 882°C, β-Ti with a cubic body-centered lattice. [11] Consequently, the thermodynamic functions of titanium before and after 882°C are different. As a result, for some compounds it is necessary to separately consider the temperature ranges before and after 882°C.

As a result of calculations, it was revealed that the general form of the free Gibbs energy dependence on temperature is of the following character:

$$\Delta G = a + b\cdot T + \frac{c}{T} + d\cdot T^{\frac{3}{2}} + e\cdot T^2 + f\cdot T^3 + g\cdot T\cdot \ln T, \quad (5)$$

where T – temperature in Kelvins; a, b, c, d, e, f, g – approximating coefficients.

Table 1 presents the values of the approximating coefficients for various compounds at temperatures below 882°C, and in Table 2 at temperatures above 882°C.

IOP Conf. Series: Journal of Physics: Conf. Series **938** (2017) 012072 doi:10.1088/1742-6596/938/1/012072

Table 1. Values of coefficients at temperature up to 882°C

	a	b	c	d	e	f	g
TiN	$-3.36\cdot10^5$	92.9	0	0	0	0	0
AlTi	$7.22\cdot10^5$	$-7.52\cdot10^2$	$-3.66\cdot10^4$	27.1	$44.8\cdot10^{-5}$	0	7.62
Al_3Ti	$-4.12\cdot10^4$	$-1.09\cdot10^3$	$-5.32\cdot10^4$	39.4	$62.9\cdot10^{-5}$	0	11.1
$AlTi_3$	0	$9.07\cdot10^2$	$-4.25\cdot10^4$	-47.3	$-10.1\cdot10^{-4}$	0	-15.8
Ti_2Al_3	$-1.33\cdot10^9$	$3.02\cdot10^6$	$-1.12\cdot10^5$	177	$-46.4\cdot10^{-5}$	0	$-4.52\cdot10^5$
Al_2Ti	$2.82\cdot10^3$	$-4.57\cdot10^2$	$-3.12\cdot10^4$	15.8	$7.66\cdot10^{-5}$	0	7.17
AlN	$-3.22\cdot10^5$	93.3	0	0	0	0	0

Table 2. The values of the coefficients at the temperature after 882°C

	a	b	c	d	e	f	g
TiN	$-3.38\cdot10^6$	95.3	0	0	0	0	0
AlTi	$-7.10\cdot10^4$	-90.3	$-2.99\cdot10^4$	0	$57\cdot10^{-5}$	0	13.4
Al_3Ti	$-1.73\cdot10^5$	551	$2.48\cdot10^4$	0	$3.87\cdot10^{-3}$	0	-82
$AlTi_3$	$-7.81\cdot10^4$	-129	$-1.99\cdot10^4$	0	$38\cdot10^{-5}$	0	15.8
Ti_2Al_3	$-9.78\cdot10^9$	-116	$-3.58\cdot10^4$	0	$6.2\cdot10^{-5}$	0	13.9
Al_2Ti	0	52.2	$-1.99\cdot10^4$	0	$27.7\cdot10^{-5}$	0	-7.17
AlN	$-3.22\cdot10^5$	93.3	0	0	0	0	0

The results of calculating the change in the free Gibbs energy for the temperatures and compounds of interest to us are presented in Table 3.

Table 3. The calculated free Gibbs energy of the reactions of formation of double compounds at different temperatures.

Reaction	ΔG, kJ/mol					
	800°C	850°C	880°C	900°C	950°C	1000°C
1. Ti+N=TiN	-238	-231	-229	-226	-222	-217
2. Al+Ti=AlTi	1080	1150	1190	-64.3	-63.2	-62.1

3. $3Al+Ti=Al_3Ti$	254	302	331	-128	-126	-124
4. $3Ti+Al=AlTi_3$	-810	-889	-398	-98.0	-97.9	-97.8
5. $2Ti+3Al=Al_3Ti_2$	1650	1860	1910	-117	-117	-116
6. Al+N=AlN	-206	-200	-197	-194	-188	-183
7. $Ti+2Al=Al_2Ti$	121	141	152	-9.61	-0.761	-1.37

The reaction of formation of a given compound is thermodynamically probable for ΔG<0. Moreover, the reaction is thermodynamically more probable, in which ΔG is smaller. As can be seen from Table 3, the free energy of formation of all possible titanium aluminides, except $AlTi_3$ aluminide, is less than zero only at 900°C and above, hence, as indicated above, they appear and remain thermodynamically stable at these temperatures. Aluminide $AlTi_3$ and nitrides AlN and TiN nitrides appear and are stable even at 800°C.

4. Results and discussion

An x-ray diffraction analysis of the experimental samples was carried out to confirm the results of the above thermodynamic analysis. The results are shown in Fig. 6-8. After annealing at a temperature of 800°C, the presence of the following main phases was found: TiN, $AlTi_3$ (Fig. 3), which coincides with the calculated results (Table 3). Also a small peak is found on the roentgenogram, which is the titanium oxide formed by the interaction of the deposited titanium with the residuals of gallium oxide.

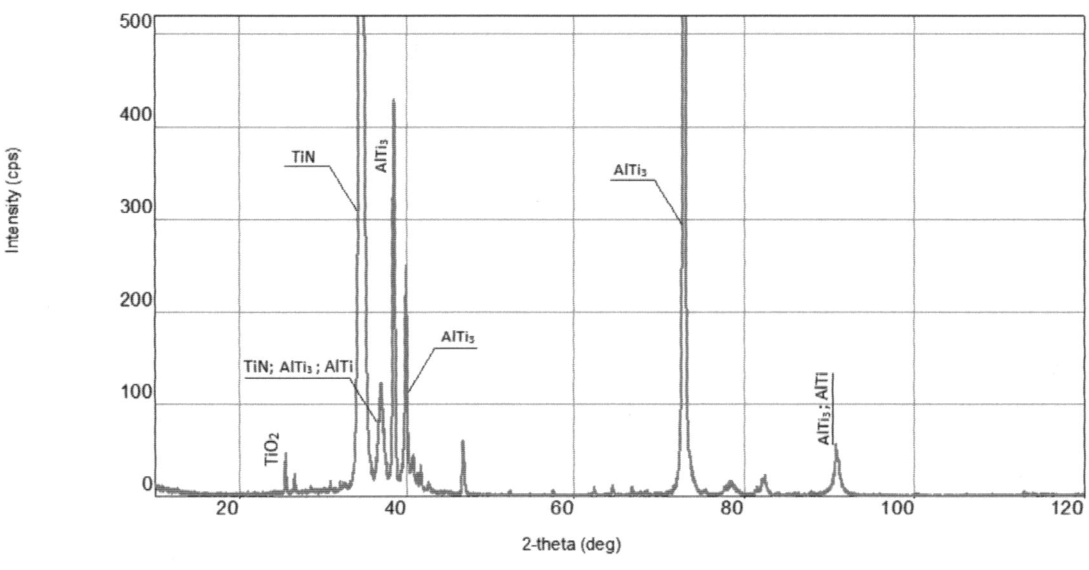

Figure 3. The X-ray diffraction pattern of the sample after annealing at 800°C

After annealing at 850°C (Fig. 4), the same phases were detected as in the case of 800 °C, but the height of the peaks indicates increasing in the content of TiN. Also, a low peak was found which corresponds to the residue of oxides Ga_2O_3.

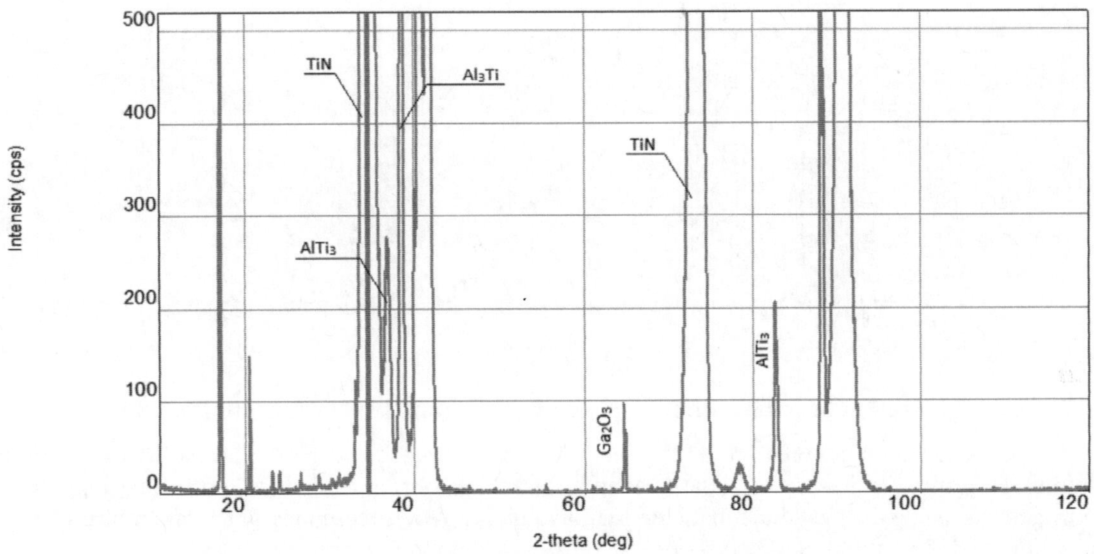

Figure 4. The X-ray diffraction pattern of the sample after annealing at 850°C

After annealing at 900°C, a greater number of phases are formed: AlN, TiN, $AlTi_3$, AlTi, Al_2Ti, Al_3Ti (fig.5), i.e. aluminides (Al_2Ti, AlTi и AlN) are formed additionally, which is also coincides with the thermodynamic calculations. The appearance of aluminum nitride, which is a dielectric, can increase the contact resistance. The peak corresponding to the titanium oxide was found again.

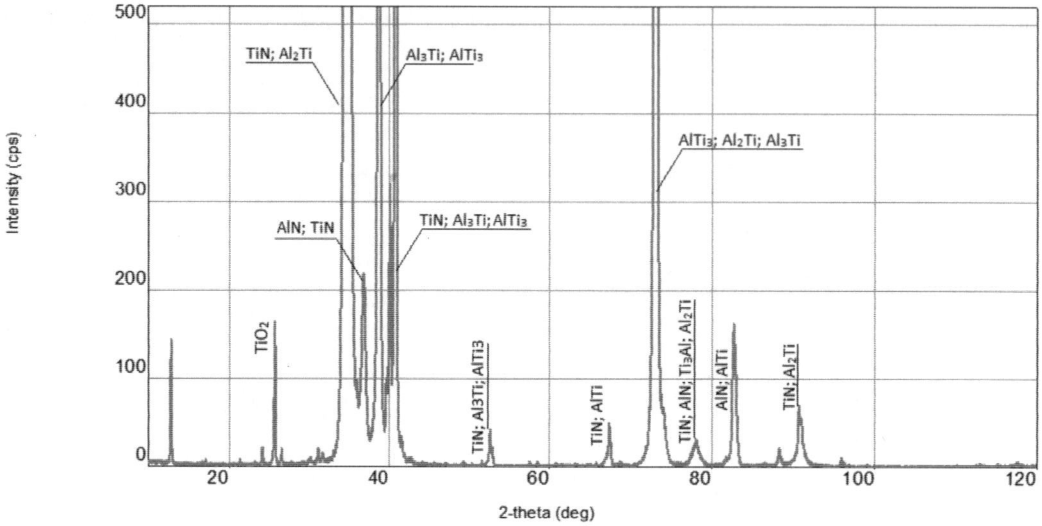

Figure 5. The X-ray diffraction pattern of the sample after annealing at 900°C.

Thus, at temperatures of 850°C and lower, as expected, less aluminum and titanium aluminide is formed, and as a result, the formation of a TiN compound and the corresponding amount of nitrogen vacancies in the semiconductor is more likely. When the annealing temperature changes from 800°C to 850°C, the peaks corresponding to the TiN compound increase.

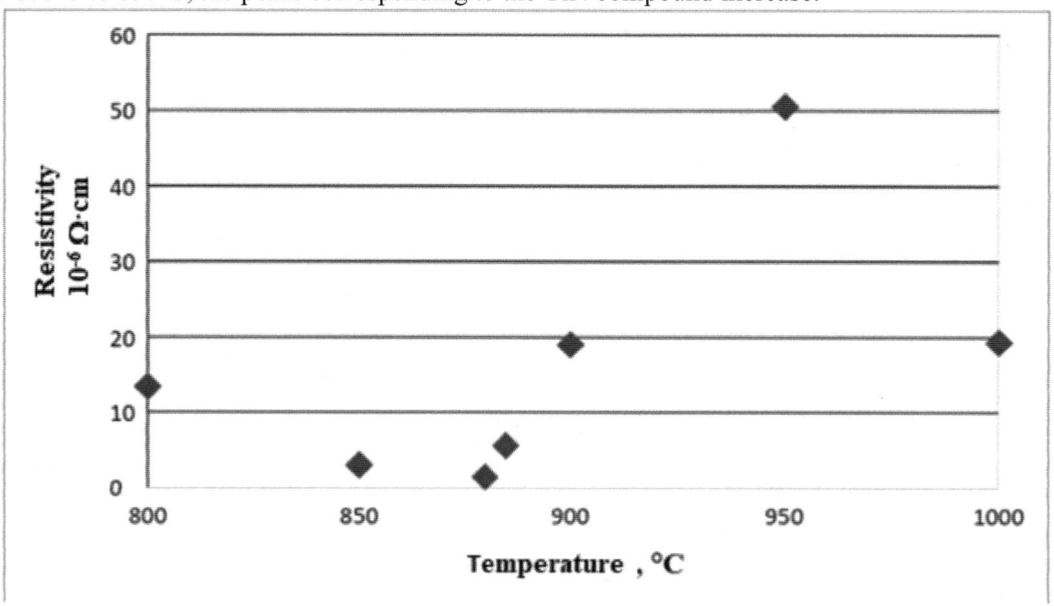

Figure 6. Specific contact resistance after annealing

Thus, we can expect that the optimal temperature conditions for the formation of the ohmic contacts are temperatures from 850 to 900°C, since free titanium is used to form fewer different aluminides, which follows from the data presented in Table 3. This increases the probability of the formation of titanium nitride, which is accompanied by the formation of nitrogen vacancies in the contact region of GaN. These vacancies create donor levels, due to which there is an increase in the n-type conductivity in the contact area and a decrease in the contact resistance.

Data on the phase composition of the contact region (Fig. 3-5) were compared with data on the contact resistance (Fig. 6). As expected, the minimum resistance is observed at an annealing temperature of 850 to 900 °C. That is, experimentally confirmed the assumption made on the basis of thermodynamic calculations for possible intermetallic reactions in the contact region. In this temperature range, the probability of TiN compound formation is higher, and titanium aluminides are lower than at higher temperatures.

5. Conclusion

In this paper we analyzed the effect of solid-state interaction processes in Al/Ti/GaN contacts during heat treatment on their final structure and electrical parameters. The analysis showed that after annealing the contact area is transformed into a structure that consists of phases that are the products of solid-phase interaction of three basic elements: titanium, nitrogen and aluminum.

As a result of the analysis of the thermodynamics of the reactions interaction of the three basic elements, it has been established that the optimal temperature range for the formation of good ohmic contacts is 850-900 °C. In this temperature range, the probability of formation of TiN compound is higher, and the aluminides of titanium and aluminum nitride are lower than at higher temperatures. This was confirmed by the results of direct measurements: the value of the specific contact resistance of the contact was $1.44 \cdot 10^{-6}\ \Omega \cdot cm^2$, which is lower than the values obtained in similar studies [18-21].

The presented approach is also valid for samples whose upper layer consists of an AlGaN ternary semiconductor with a Ti/Al-based metallization system applied to it, since the fast annealing contacts contain the same elements as when annealing contacts to a heterostructure with an upper GaN layer.

Acknowledgements

References

[1] Luther B P and Mohney S E 1997 Investigation of the mechanism for Ohmic contact formation in Al and Ti/Al contacts to n-type GaN *Appl. Phys. Lett.* V 70 P 57
[2] Schubert E F 2003 Light-emitting diodes *Cambridge University Press*
[3] Kong X, Wei K, Liu G and Liu X 2012 Role of Ti/Al relative thickness in the formation mechanism of Ti/Al/Ni/Au Ohmic contacts to AlGaN/GaN heterostructures *J. Phys. D: Appl. Phys.* V 45 P 265101
[4] Durbha, Pearton S J, Abernaty C R, Lee J W and Holloway P H 1996 Microstructural stability of ohmic contacts to $In_xGa_{1-x}N$ *J. Vac. Sci. Tech B* V 14 P 2582
[5] Morkoc H, Strike S, Gao G B, Lin M E, Sverdlov B and Burns M 1994 Large- band- gap SiC, III- V nitride, and II- VI ZnSe- based semiconductor device technologies *J. Appl. Phys.* V 76 P 1363
[6] Vanyukhin et al 2013 Investigation of the structure and morphology of the surface of a two-layer Ti / Al contact metallization *Proceedings of universities. Materials of electronic engineering(in Russian)* №3 p 60-65
[7] Vanyukhin K D et al 2016 Study of structure and surface morphology of two-layer contact Ti/Al metallization *Modern Electronic Materials (MOEM)* T 2 №2 p 54–59
[8] Lester F,Brown J M, Raner J C and Zhang L 1996 Nonalloyed Ti/Al ohmic contacts to n-type GaN using high-temperature premetallization anneal *Appl. Phys. Lett.* V 69 №18 pp 2737-2739
[9] Reeves G., Harrison H. B. Obtaining the specific contact resistance from transmission line model measurements // IEEE Electr. Device Lett. 1982 V. 3. P. 111.

[10] Shostachenko S 2015 Schottky-to-Ohmic Behavior in Annealed Ti/Si/Ti/Al/Ni/Au on AlGaN/GaN *Physics Procedia* V 72 P 419

[11] Tretyakov Y D 1978 Solid-phase reactions (in Russian) *Moscow: Chemistry* p 360

[12] Kireyev V A 1975 Methods of Practical Calculations in Thermodynamics of Chemical Reactions
(in Russian) *Moscow: Chemistry* p 536

[13] Samsonov G V and Vinitskiy I M 1976 Refractory compounds (in Russian) *Moscow: Metallurgy* p 560

[14] Chase M W Jr, Davies C A, Downey J R et al 1985 JANAF Thermochemical Tables. Third edition *J. Phys. Chem. Ref. Data* V 14 Suppl 1

[15] Kubashevsky O and Alkokk K B 1982 Metallurgical thermochemistry *Moscow: Metallurgy* p 392

[16] Morachevsky A G and Sladkov I B 1993 Thermodynamic calculations in metallurgy *Moscow: Metallurgy* p 304

[17] Moiseev G K and Sestak J 1995 Some calculations methods for estimation of thermodynamical and thermochemical properties of inorganic compounds *Prog. Crystal Grown and Charact* vol 30 pp 23

[18] Kong X, Wei K, Liu G and Liu X 2012 Role of Ti/Al relative thickness in the formation mechanism of Ti/Al/Ni/Au Ohmic contacts to AlGaN/GaN heterostructures *J Phys. D: Appl. Phys.*

[19] Xin H P et al 2010 *CS Mantech Conference* V45

[20] Butteri et al 2002 Systematic characterization of Cl2 reactive ion etching for improved ohmics in AlGaN/GaN HEMTs *IEEE Electron Device Letters* V 23 P 76.

[21] Jacobs B, Kramer M C J C M Geluk E J and Karouta F 2002 Optimisation of the Ti/Al/Ni/Au ohmic contact on AlGaN/GaN FET structures *Journal of Crystal Growth* V 241 Issues 1–2 P 15

Study of Ti/Si/Ti/Al/Ni/Au ohmic contact for AlGaN/GaN HEMT

S A Shostachenko*[1], Y A Porokhonko[1], R V Zakharchenko[1], M S Burdykin[1], R V Ryzhuk[1], N I Kargin[1], B V Kalinin[2], A A Belov[2] and A N Vasiliev[2]

[1]National Research Nuclear University MEPhI, Moscow, Russian Federation

[2]PLANETA-IRMIS, Saint Petersburg, Russian Federation

E-mail: Sashostachenko@mephi.ru

Abstract. This paper is dedicated to the experimental investigation of Ohmic contacts to the n^+-doped region of AlGaN/GaN transistor heterostructure based on Ti/Si/Ti/Al/Ni/Au metallization. Effect of annealing temperature on the specific resistance of Ohmic contact was studied. Ohmic contact with the resistance of $3.4 \cdot 10^{-6}$ $\Omega \cdot cm^2$ was formed by optimization of the annealing temperature and introduction of the additional doping silicon layer.

1. Introduction

Transistor devices on GaN-based heterostructures are widely used, both in the defense and civil industry. For example, as the main component of the microwave monolithic integrated circuits (RF MIC) millimeter and submillimeter bands for communication systems, radar, radio astronomy, radiometry, etc. [1]. Along with wide bandgap semiconductors based on GaN as a basis for the transistors, these materials have been applied in the optoelectronic industry as a base for the LED crystals.

Currently, in multilayer metallization contact microelectronics technologies are mostly used because of their best parameters (conductivity, adhesion to the surface morphology). For forming an ohmic contact to the n-type semiconductor is most often used by the multilayer contacts Ti / Al, the heat treatment in forming compounds with a low work function [2-3]. Standard metallization is a gallium nitride-based Ti / Al / Ni / Au.

The hypotheses on the mechanisms of formation of ohmic contact to the n-GaN-based Ti / Al metal system based on the appearance Ti_xN compound, which is a guide, begin to form during the deposition of titanium on the n-GaN surface and in the bulk of the titanium layer metallization during thermal annealing. Ti_xN formation temperature is in the range 200 ÷ 1000 °C, because of this titanium particles sprayed by various methods, can form a thin layer on the surface TixN.

In [4] indicated that the work function of Ti_xN is only 3.74 eV, which is on the theory corresponds to the formation of the Schottky ohmic contact to the n-GaN. The low specific contact resistance ohmic contact metal-GaN (to 10^{-6} - 10^{-7} Ohm·cm^2 at high concentrations of carriers in a semiconductor) is associated with the formation of nitrogen vacancies due to the interaction with the semiconductor contact metallization. These nitrogen vacancies form a layer under the contact, playing the role of high-alloyed layer [5-6]. In literature, there are several options for plating systems, basic presented in Table 1.

IOP Conf. Series: Journal of Physics: Conf. Series **938** (2017) 012073 doi:10.1088/1742-6596/938/1/012073

Table 1. Ohmic contact to n-GaN

Contact system	Layer thickness, nm	Annealing temperature, °C	Contact resistance, Ohm·mm	Source
Ti/Al/Ni/Au	30/200/40/10	800	0,50	[7]
Ti/Al/Ni/Au	20/220/55/45	870	0,30	[8]
Ti/Al/Mo/Au	15/60/35/50	850	0,32	[9]
Ti/Al/Ti/Au	-	750	0,54	[10]
Ta/Si/Ti/Al/Ni/Ta	5/5/20/120/40/30	850	0,27	[11]
Ta/Ti/Al/Mo/Au	10/30/90/40/25	850	0,41	[12]
Si/Ti/Al/Ni/Au	3 nm Si	800	0,23	[13]
	6 nm Si		0,26	

The thickness of metal system Ti / Al / Ni / Au is selected on the basis of experimental data. According to [14], on the one hand, should be sufficient for the titanium TixN formation layer, and the other titanium layer should not interfere with the formation of an aluminum contact. Aluminium thickness also has a certain range of variation [15]. Aluminium should establish contact and prevent the diffusion of nickel and gold, as well as contribute to the occurrence of phase $Al_{3+x}Ti$ ($x < 1$). In general, it should be emphasized that during the annealing is a series of complex physical and chemical reactions and the products of which are still not fully understood.

2. Experimental method

We used AlGaN/GaN heterostructure. Ti/Al/Ni/Au (35/135/50/100 nm) multilayer metallization was deposited on it by a resistive evaporation method with the use of Kurt J. Lesker PVD 75 film deposition system. Annealing was performed using rapid thermal annealing system Modular RTP-600S at a constant temperature for 30 s. Ohmic contact measurements were carried out on test structure for transmission line method (TLM) [15]. To improve the accuracy of contact resistance measurements we also have used test structure [4]. Specific contact resistance was measured by the use of the Agilent B 1500 A semiconductor devices analyzer.

3. Results and discussion

To investigate substrate heterostructure AlGaN/GaN metallization system Ti/Al/Ni/Au. (35/135/50/100 nm) was deposited and studied in details in work [4]. Now we propose Si doping contact system Ti/Si/Ti/Al/Ni/Au.

To change the height of the Schottky barrier it is necessary to change the material, drawing a laborious study of intermetallic phases formed during heat treatment. Another approach may be an increase in carrier density, which helps to give forming an ohmic contact at lower annealing temperatures and reduce the width of the barrier, which also affects the specific contact resistance in the direction of its decrease.

When growing a semiconductor heterostructure contact layer dopant is added (in this case silicon with a concentration of $3{\cdot}10^{18}$ cm^{-3} was used). To improve the art known concentration must be used, but time-consuming method of ion implantation. But there is another way to solve the problem - to introduce a system of dopant layers of metallization (between the heterostructure and metallization). This method makes sense since after annealing the mixing layers and silicon will still be required in the contact area and only. Also, this method is good because it is done in the same process area as the application of the metallization layers, that is, there is no need to make an extra process step.

The new system has the form metallization layers Ti/Si/Ti/Al/Ni/Au (5/6/30/135/50/100 nm) sublayer 5 nm titanium is required to provide greater adhesion to the semiconductor surface since layering occurred on the substrate without heating. The sample was treated at 800-1000 ° C (Figure 2). Introduction silicon is possible to reduce specific contact resistance to $3.76{\cdot}10^{-7}$ Om·cm^2 , which is lower than resistance obtained in previous work without Si [16].

Table 2. Specific contact resistance after annealing.

Sample	**Annealing temperature, °C**	**Concentration, cm^{-3}**	**Specific contact resistance, 10^{-6} Ohm·cm^2**
1	800	$7{,}50{\cdot}10^{19}$	19.9
2	850	$1{,}00{\cdot}10^{20}$	6,18
3	880	$1{,}20{\cdot}10^{20}$	1,6
4	885	$1{,}75{\cdot}10^{20}$	3,2
5	900	$4{,}00{\cdot}10^{20}$	0.38
6	950	$8{,}50{\cdot}10^{19}$	18.5
7	1000	$7{,}50{\cdot}10^{19}$	35

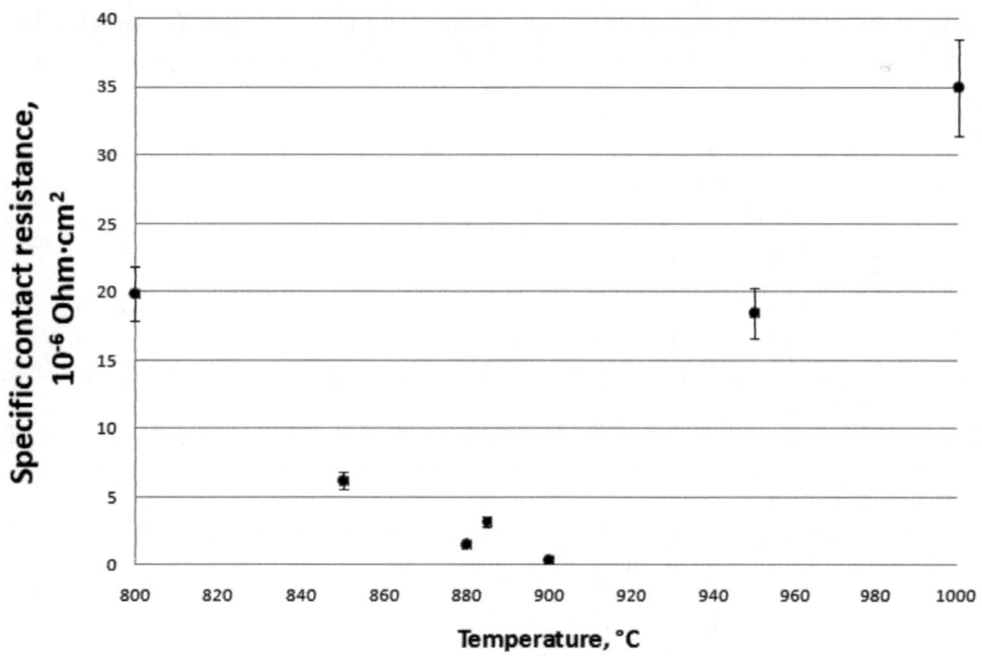

Figure 1. Contact resistance for Ti/Si/Ti/Al/Ni/Au

As seen from Table 2 and Figure 1, the minimum resistivity for samples with metallization Ti/Si/Ti/Al/Ni/Au is $3.76{\cdot}10^{7}$ Ohm·cm^2 and achieved at a temperature of 900 ° C. Therefore, to solve this problem, a study was conducted of processes of forming an ohmic contact with the subsequent optimization of the contact system.

The sequential annealing can determine the stages of forming an ohmic contact, but you can't explain the mechanism of formation of the contact and charge transport.

4. Conclusion

This study is about formation of an ohmic contact to the n-region AlGaN/GaN heterostructure during annealing the metallization system Ti/Si/Ti/Al/Ni/Au. To reduce contact resistance (resistivity and specific transition) we proposed to introduce dopant providing in the metallization during annealing increase the carrier concentration in the contact region. Optimum annealing time and temperature are respectively 900°C and 30 seconds. The specific transitional resistance is 3.4×10^{-6} Ohm $\times$ cm^2. The obtained resistivity value is lower than those obtained by other authors [13] because of the introduction of dopant in the metallization system, which is caused by the difference process and the difference of layer thicknesses, forming a contact. Also presumably selected metallization system may be more stable due to the presence of the underlayer Ti, performing the role of an adhesive layer.

Acknowledgements

This work was supported by the Ministry of Education and Science of the Russian Federation, project no. RFMEFI58117X0026 and performed on the equipment of the collective use center "Heterostructural microwave electronics and physics of wide-gap semiconductors of MEPhI".

References

[1] Streit D, Oki A, Lai R, Gutierrez-Aitken A 2000 How Cost-Efficient Phosphide Changes the Telecommunication Industry's Competitive Landscape *Compound Semiconductor Magazine* V 6(3) PP 234 – 235

[2] Blank T V and Gol'dberg Yu A 2007 Mechanisms of Current Flow in Metal–Semiconductor Ohmic Contacts *SEMICONDUCTORS* V 41(11) PP 1263-1292

[3] Padovani F and Stratton R 1966 Field and Thermionic Field Emission in Schottky Barriers *Solid-State Electronics* V 9(7) PP 695-707

[4] Shostachenko S 2015 Schottky-to-Ohmic Behavior in Annealed Ti/Si/Ti/Al/Ni/Au on AlGaN/GaN *Physics Procedia* V 72 PP 419-424

[5] Roccaforte F 2006 Nanoscale carrier transport in Ti/Al/Ni/Au Ohmic contacts on AlGaN epilayers grown on Si(111) *Appl. Phys. Lett.* V 89 022103

[6] Jacobs B 2002 Optimization of the Ti/Al/Ni/Au ohmic contact on AlGaN/GaN FET structures *J. Cryst. Growth.* V 241 PP 15—18

[7] Burm, Chu K, Schaff W J, Eastman L F, Khan M A, Chen Q, Yang J W and. Shur M S 1997 0.12 mkm Gate III-V Nitride HFET's with High Contact Resistances *IEEE Electron Device Letters* V 18(4) PP 141-143

[8] Kong X 2012 Role of Ti/Al relative thickness in the formation mechanism of Ti/Al/Ni/Au Ohmic contacts to AlGaN/GaN heterostructures *J. Phys. D: Appl. Phys.* V 45 265101

[9] Deen D 2010 Dependence of ohmic contact resistance on barrier thickness of AlN/GaN HEMT structures *Solid-State Electronics* V 54 PP 613–615

[10] Buttari D 2002 Systematic characterization of Cl reactive ion etching for Improved Ohmics in AlGaN/GaN HEMTs *IEEE Electron device letters V* 23(2) 782002

[11] Selvanathan D 2002 Low Resistance Ti/Al/Mo/Au Ohmic Contacts *J. Appl. Phys.* V 92 PP 1712-1714

[12] Yanxu Z 2014 "ffects of rapid thermal annealing on ohmic contact of AlGaN/GaN HEMTs *J. Semicond* V 35(2) 026004

IOP Conf. Series: Journal of Physics: Conf. Series **938** (2017) 012073 doi:10.1088/1742-6596/938/1/012073

[13] Li Y 2013 Low-Contact-Resistance Non-Gold Ta/Si/Ti/Al/Ni/Ta Ohmic Contacts on Undoped AlGaN/GaN High-Electron-Mobility Transistors Grown on Silicon *Appl. Phys. Express* V 6 116501

[14] Desmaris V, Eriksson J and Zirath H 2004 Low-Resistance Si/Ti/Al/Ni/Au Multilayer Ohmic Contact to Undoped AlGaN/GaN Heterostructures *Electrochemical and Solid-State Letters* V 7(4) PP 72-74

[15] Reeves Gand Harrison H B 1982 Obtaining the specific contact resistance from transmission line model measurements *IEEE Electr. Device Lett.* V 3 PP 111-113

[16] Shostachenko S 2016 Low resistance Ti/Si/Ti/Al/Ni/Au ohmic contact for AlGaN/GaN HEMT *Proceedings of SPIE* V 10224 1022403

Institute of Physics
Dirac House, Temple Back
Bristol BS1 6BE UK

ISSN: 1742-6588
ISBN 978-1-5108-5518-2

9 781510 855182

CURRAN ASSOCIATES INC.
proceedings.com